MICROBES

MICROBES
Concepts and Applications

Prakash S. Bisen
Emeritus Scientist
Defence Research Development Establishment
Defence Research Development Organization
Ministry of Defence, Government of India
Gwalior, India

Mousumi Debnath
Associate Professor
Department of Biotechnology
Central University of Rajasthan
City Road, Kishangargh
Ajmer, India

Godavarthi B. K. S. Prasad
Professor
School of Studies in Biochemistry and Biotechnology
Jiwaji University
Gwalior, India

Ⓦ **WILEY-BLACKWELL**
A JOHN WILEY & SONS, INC., PUBLICATION

Published by John Wiley & Sons, Inc., Hoboken, New Jersey
Published simultaneously in Canada

Wiley-Blackwell is an imprint of John Wiley & Sons, formed by the merger of Wiley's global Scientific, Technical, and Medical business with Blackwell Publishing.

For general information on our other products and services or for technical support, please contact our Customer Care Department within the United States at (800) 762-2974, outside the United States at (317) 572-3993 or fax (317) 572-4002.

Wiley also publishes its books in a variety of electronic formats. Some content that appears in print may not be available in electronic formats. For more information about Wiley products, visit our web site at www.wiley.com.

Library of Congress Cataloging-in-Publication Data:

Bisen, Prakash S.
  Microbes : concepts and applications/Prakash S. Bisen, Mousumi Debnath, Godavarthi B.K.S. Prasad.
     p. cm.
   Includes bibliographical references and index.
  ISBN 978-0-470-90594-4
 1. Microbiology. 2. Microbial diversity. 3. Microbial ecology. 4. Microbial biotechnology.
5. Microorganisms. I. Debnath, Mousumi. II. Prasad, Godavarthi B. K. S. III. Title.
  QR41.2.B576 2012
  616.9′041–dc23

                                                                            2011044258

Printed in the United States of America

10 9 8 7 6 5 4 3 2 1

CONTENTS

PREFACE

Microbiology embraces functional disciplines at all levels, from the behavior of population of atoms to the behavior of population of microorganisms. Microorganisms undergo profound changes during their transition from planktonic organisms to cells that are part of a complex, surface-attached community. These changes are reflected in the phenotypic characteristics developed in response to a variety of environmental conditions. This book gives an insight into the microbial world. It features a comprehensive account of various topics including systematics; diversity of microbes in the environment; microbes in agriculture, food, and industry; microbial interactions; microbial metabolism; microbial diseases; and the diagnostic approaches.

In the coming few decades, it is the Archaea and the extremophilic microorganisms found in harsh and unusual environment that will occupy the center stage of microbiology in view of their tremendous importance in industrial applications. However, the excitement of modern microbiology extends far beyond Archaea and extremophiles. Microbiologists are debating the novel, philosophical, and controversial but fascinating concept of bacteria as global superorganisms, and focus on bacterial biofilms is likely to have a profound influence on environmental and medical microbiology in the twenty-first century. We have attempted to cover these and all other similar aspects of modern microbiology that are not commonly seen in existing books. We have covered the older traditional microbiology rather briefly and concisely but the major focus has been on recent development.

The authors have designed a chapter on "Gene Technology," covering latest technologies relevant to the understanding of molecular aspects of microbes. This chapter offers a straightforward approach to learning the core principles associated with "Omics" technologies and their applications to the world of microbes. Omics technology if studied in isolation provides us certain specific snapshots of complex biological systems. Genomics, transcriptomics, proteomics, and metabolomics have enhanced our understanding of biological system by several folds and helped in providing insights into the biomedical research. Each of these omics technologies is limited in scope, as it can explain only one aspect of biological system. It is desirable to integrate snapshots provided by different omics technologies to get a holistic picture of complex biological system. Such challenging task of integration of omics technology is undertaken by the new discipline "Systems Biology."

Microbes are of major economic, environmental, and social importance and are being exploited for production of a wide range of products of commercial significance. The chapter on industrial applications of microbes deals with their role in production of enzymes, foods, probiotics, chemical feedstock, fuels and pharmaceuticals, biofertilizers, traditional fermentation-based beverages, and development of clean technologies used in waste treatment and pollution control. Developments in bacterial genomics of lactobacilli and bifidobacteria would dramatically alter the scope and impact of the probiotic field, offering tremendous new opportunities with accompanying challenges for research and industrial application. The genetic and molecular approaches used to study bacterial and fungal biofilms, host immune evasion, and drug resistance have been discussed. Current evidence indicates that several microbes, especially bacteria and viruses, encode miRNAs, and over 200 viral miRNAs involved in manipulation of both cellular and viral gene expression have now been identified. Small noncoding RNAs have been found in all organisms, primarily as regulators of translation bioterrorism. This book discusses our current knowledge of these regulatory genetic elements and their relationship to infection.

Each chapter starts with a prologue to the concept and finally ends with a vision for future approach and challenges. The authors have tried to foresight some of the innovative areas of microbes and their involvement in the near future. Some of the topics of interest are metagenomics, pharmacogenomics, biochip technology, aptamers, microorganisms on space, functional genomics for improvement of plants, and next generation diagnostic industry.

This book responds to the requirement of students of undergraduates and graduates in microbiology and medicine. Each topic has been approached as a separate topic, and large illustrations in the chapters can help students and researchers to understand better. An effort has been done to reach at the individual level. Students can master key concepts and develop insights into the latest technologies of microbial relevance. This book is self-explanatory and written in a lucid language. A large compilation of references have been added at the end of each chapter. This book can be a source reference to all innovative people interested in microbes. We welcome suggestions and comments for improvement of this book.

Prakash S. Bisen
Gwalior, India

Mousumi Debnath
Kishangarh, Ajmer, India

Godavarthi B. K. S. Prasad
Gwalior, India

ACKNOWLEDGMENTS

We wish to thank Dr. R. Vijayaraghavan, Outstanding Scientist and Director, Defence Research Development Establishment, Defence Research Development Organization, Ministry of Defence, Government of India, Gwalior, India, and Professor G. P. Agarwal, former Dean and Senior Microbiologist, Faculty of Life Sciences, R. D. University, Jabalpur, India, for their valuable guidance and encouragement and for extending all necessary facilities to complete the task smoothly.

Mr. Rakesh Singh Rathore, CEO, Vikrant Group of Institutions, Gwalior and Dr. Gurudev S. Davar and Mr. Puneet Davar, Tropilite Foods Pvt. Limited, Gwalior, are gratefully acknowledged for extending all secretarial assistance and for their help in various ways for completing this work without any hindrances.

We thank Mr. Devendra Singh and Mr. Avinash Dubey for all the computational help in preparation of figures and tables and our research students Ruchika Singh Raghuvanshi, Bhagwan S. Sanodiya, Gulab S. Thakur, Rakesh Baghel, and Rohit Sharma for their valuable assistance in the preparation of this book. Thanks are also due to Karen E. Chambers and Anna Ehler, Wiley-Blackwell Publishing, John Wiley & Sons, for their full support in publishing this book in time with patience and interest. Finally, we thank our families for their constant support, cooperation, and understanding. We are thankful to the Council of Scientific and Industrial Research (CSIR), New Delhi, for the award of Emeritus Scientist to Professor Prakash S. Bisen.

PRAKASH S. BISEN
Gwalior, India

MOUSUMI DEBNATH
Kishangarh, Ajmer, India

GODAVARTHI B. K. S. PRASAD
Gwalior, India

1

HUMAN AND MICROBIAL WORLD

1.1. PROLOGUE

The microbial world is vast, diverse, and dynamic. The Earth hosts over 10^{30} microorganisms, representing the largest component of the planet's biomass. Microbes include bacteria, archaea, mollicutes, fungi, microalgae, viruses, and protozoa, and many more organisms with a wide range of morphologies and lifestyles. All other life-forms depend on microbial metabolic activity. Microorganisms have colonized virtually every environment on earth ranging from deep sea thermal vents, polar sea ice, desert rocks, guts of termites, roots of plants, to the human body. Much as we might like to ignore them, microbes are present everywhere in our bodies, living in our mouth, skin, lungs, and gut. Indeed, the human body has 10 times as many microbial cells as human cells. They are a vital part of our health, breaking down otherwise indigestible foods, making essential vitamins, and even shaping our immune system. While microbes are often feared for the diseases they may cause, other microorganisms mediate the essential biogeochemical cycles of key elements that make our planet habitable. Ancient lineages of microorganisms may hold the key to understanding the earliest history of life on earth.

1.2. INNOVATIONS IN MICROBIOLOGY FOR HUMAN WELFARE

The human society is overburdened with infectious diseases. Despite worldwide efforts toward prevention and cure of these deadly infections, they remain major causes of

Microbes: Concepts and Applications, First Edition. Prakash S. Bisen, Mousumi Debnath, Godavarthi B. K. S. Prasad
© 2012 Wiley-Blackwell. Published 2012 by John Wiley & Sons, Inc.

human morbidity and mortality. Microbes play a role in diseases such as ulcers, heart disease, and obesity. Over the past century, microbiologists have searched for more rapid and efficient means of microbial identification. The identification and differentiation of microorganisms has principally relied on microbial morphology and growth variables. Advances in molecular biology over the past few years have opened new avenues for microbial identification, characterization, and molecular approaches for studying various aspects of infectious diseases. Perhaps, the most important development has been the concerted efforts to determine the genome sequence of important human pathogens. The genome sequence of the pathogen provides us with the complete list of genes, and, through functional genomics, a potential list of novel drug targets and vaccine candidates can be identified (Hasnain, 2001). The genetic variation inherent in the human population can modulate success of any vaccine or chemotherapeutic agent. This is why, the sequencing of the human genome has attracted not only the interest of all those working on human genetic disorders but also the interest of scientists working in the field of infectious diseases.

1.2.1. Impact of Microbes on the Human Genome Project

Technology and resources generated by the Human Genome Project and other genomic research are already having a major impact on research across life sciences. The elucidation of the human genome sequence will have a tremendous impact on our understanding of the prevention and cure of infectious diseases. The human genome sequence will further advance our understanding of microbial pathogens and commensals and vice versa (Relman and Falkow, 2001). This will be possible through efforts in areas such as structural genomics, pharmacogenomics, comparative genomics, proteomics, and, most importantly, functional genomics. Functional genomics includes not only understanding the function of genes and other parts of the genome but also the organization and control of genetic pathway(s). There is an urgent need to apply high throughput methodologies such as microarrays, proteomics (the complete protein profile of a cell as a function of time and space), and study of single nucleotide polymorphisms, transgenes and gene knockouts. Microarrays have a tremendous potential in

1. Determining new gene loci in diseases
2. Understanding global cellular response to a particular mode of therapy
3. Elucidating changes in global gene expression profiles during disease conditions; and so on.

Increasingly detailed genome maps have aided researchers seeking genes associated with dozens of genetic conditions, including myotonic dystrophy, fragile X syndrome, neurofibromatosis types 1 and 2, inherited colon cancer, Alzheimer's disease, and familial breast cancer.

The Human Microbiome Project (HMP) has published an analysis of 178 genomes from microbes that live in or on the human body. The core human microbiome is the set of genes present in a given habitat in all or the vast majority of humans (Fig. 1.1). The variable human microbiome is the set of genes present in a given habitat in a smaller subset of humans. This variation could result from a combination of factors such as host genotype, host physiological status (including the properties of the innate and

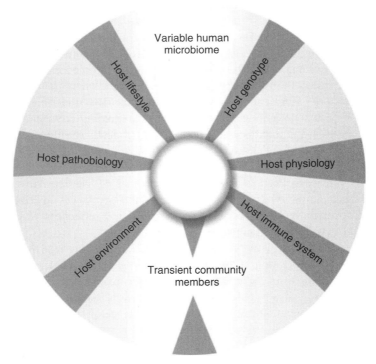

Figure 1.1. The concept of human microbiome.

adaptive immune systems), host pathobiology (disease status), host lifestyle (including diet), host environment (at home and/or work), and the presence of transient populations of microorganisms that cannot persistently colonize a habitat. The gradation in color of the core indicates the possibility that, during human microevolution, new genes might be included in the core microbiome, whereas other genes might be excluded (Turnbaugh et al., 2007). The researchers discovered novel genes and proteins that serve functions in human health and disease, adding a new level of understanding to what is known about the complexity and diversity of these organisms (http://www.nih.gov). Currently, only some of the bacteria, fungi, and viruses can grow in a laboratory setting. However, new genomic techniques can identify minute amounts of microbial DNA in an individual and determine its identity by comparing the genetic signature with known sequences in the project's database. Launched in 2008 as part of the NIH Common Fund's Roadmap for Medical Research, the HMP is a $157 million, five-year effort that will implement a series of increasingly complicated studies that reveal the interactive role of the microbiome in human health (http://www.eurekalert.org). The generated data will then be used to characterize the microbial communities found in samples taken from healthy human volunteers and, later, those with specific illnesses.

Studies were also conducted to evaluate the microbial diversity present in the HMP reference collection. For example, they found 29,693 previously undiscovered, unique proteins in the reference collection; more proteins than there are estimated genes in the human genome. The results were compared to the same number of previously sequenced microbial genomes randomly selected from public databases and reported 14,064 novel proteins.

These data suggest that the HMP reference collection has nearly twice the amount of microbial diversity than is represented by microbial genomes already in public databases (http://www.ncbi.nlm.nih.gov/genomeprj). One of the primary goals of the HMP reference collection is to expand researchers' ability to interpret data from metagenomic studies. Metagenomics is the study of a collection of genetic material (genomes) from a mixed community of organisms. Comparing metagenomic sequence data with genomes in the reference collection can help determine the novel or already existing sequences (Hsiao and Fraser-Liggett, 2009). A total of 16.8 million microbial sequences found in public databases have been compared to the genome sequences in the HMP reference collection and it was found that 62 genomes in the reference collection showed similarity with 11.3 million microbial sequences in public databases and 6.9 million of these (41%) correspond with genome sequences in the reference collection (http://www.ncbi.nlm.nih.gov/genomeprj).

On the horizon is a new era of molecular medicine characterized less by treating symptoms and more by looking at the most fundamental causes of diseases. Rapid and more specific diagnostic tests will make possible earlier treatment of countless maladies. Medical researchers will also be able to devise novel therapeutic regimens on the basis of new classes of drugs, immunotherapy techniques, avoidance of environmental conditions that may trigger disease, and possible augmentation or even replacement of defective genes through gene therapy.

Despite our reliance on the inhabitants of the microbial world, we know little of their number or their nature. Less than 0.01% of the estimated all microbes have been cultivated and characterized. Microbial genome sequencing will help to lay the foundation for knowledge that will ultimately benefit human health and the environment. The economy will benefit from further industrial applications of microbial capabilities.

Information gleaned from the characterization of complete microbial genomes will lead to insights into the development of such new energy-related biotechnologies as photosynthetic systems and microbial systems that function in extreme environments and organisms that can metabolize readily available renewable resources and waste material with equal facility. Expected benefits also include development of diverse new products, processes, and test methods that will open the door to a cleaner environment. Biomanufacturing will use nontoxic chemicals and enzymes to reduce the cost and improve the efficiency of industrial processes. Microbial enzymes have been used to bleach paper pulp, stone wash denim, remove lipstick from glassware, break down starch in brewing, and coagulate milk protein for cheese production. In the health arena, microbial sequences may help researchers to find new human genes and shed light on the disease-producing properties of pathogens.

Microbial genomics will also help pharmaceutical researchers to gain a better understanding of how pathogenic microbes cause disease. Sequencing these microbes will help to reveal vulnerabilities and identify new drug targets. Gaining a deeper understanding of the microbial world will also provide insights into the strategies and limits of life on this planet, and the human genome sequence will further strengthen our understanding of microbial pathogens and commensals, and vice versa.

Data generated in HMP have helped scientists to identify the minimum number of genes necessary for life and confirm the existence of a third major kingdom of life. Additionally, the new genetic techniques now allow us to establish, more precisely, the diversity of microorganisms and identify those critical to maintaining or restoring the function and integrity of large and small ecosystems; this knowledge can also be useful in monitoring and predicting environmental changes. Finally, studies on microbial communities provide models for understanding biological interactions and evolutionary history.

1.2.2. Microbial Biosensors

Biosensors are defined as analytical devices combining biospecific recognition systems with physical or electrochemical signaling. They have been used for many years to provide process control data in the pharmaceutical, fermentation, and food-processing industries. The generic system comprises three components: the biospecific interaction, the signal emitted when the target is bound, and the platform that transduces the binding reaction into a machine-readable output signal. Significant progress has been made in developing platforms that exploit recent technological advances in microfabrication, optoelectronics, and electromechanical nanotechnology. Dramatic improvements in device designs facilitated by new tools and instrumentation (Fig. 1.2) have increased biosensor sensitivities by several magnitudes (Tepper and Shlomi, 2010). The designed biosensors facilitate high throughput detection and quantification of chemicals of interest, enabling combinatorial metabolic engineering experiments aiming to overproduce

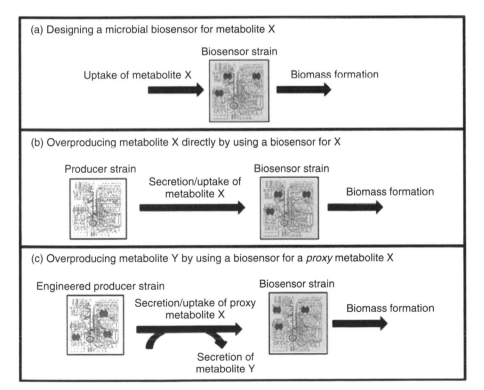

Figure 1.2. Concept of microbial biosensor (Tepper and Shlomi, 2010). (*See insert for color representation of the figure*).

them (http://www.cs.technion.ac.il). New platforms can be arrayed in panels to reduce costs and simplified methods can be employed to detect and validate against numerous hazards.

Majority of the described piezoelectric (PZ) devices are based on immunosensors. Targeting intact bacteria as antibody is relatively simple to immobilize, and entrapped bacterial cells can accumulate a significantly large, detectable mass. Many PZ devices are used in clinical and food pathogen identifications, for example, quartz crystal microbalance (QCM) devices coated with antibody, protein A, or other specific receptor molecules have been applied for a wide range of antibodies, pathogenic organisms (including *Vibrio, Salmonella, Campylobacter, Escherichia coli, Shigella, Yersinia*, viruses, and protozoa) and PCR amplicons (Hall, 2002). The simplicity, flexibility, and utility of the PZ systems in a range of foodborne and clinical applications make it well accepted for QCM format, although the sensitivity of these sensors is suitable only for very dense bacterial cultures. The QCM device coated with specific antibody and integrated into the culture enrichment tube resolves several problems in food safety testing, albeit at the expense of rapidity and cost. There are numerous biological components that can be coupled to mass aggregation and deposition chemistries, largely because the signal amplification required in most ELISA, western hybridization, and dot blot detection systems entails accumulation of precipitated mass in the process of generating a detectable signal. Among the ELISA systems that have been modified for mass sensors is Ag/Abs binding and DNA–DNA hybridization, and indirect, amplified systems such as double anti-angiogenic protein (DAAP) and avidin/streptavidin enzyme-conjugated secondary antibodies.

Microbial biosensors for environmental applications range in their development stages from proof of concept to full commercial availability, and the target detection specificity may fall in one of the following two groups (Bogue, 2003; Rodriguez-Mozaz et al., 2004):

- Biosensors that measure general biological effects/parameters.
- Biosensors for specific detection of target compounds.

The first group of biosensors is aimed to measure integral toxicity, genotoxicity, estrogenicity, or other general parameters of the sample, which affect living organisms. They essentially include whole microorganisms as biorecognition elements. The most-often reported cell-based biosensors include genetically modified bacteria with artificially constructed fusions of particular regulatory system (native promoter) with reporter genes. The presence of effectors (nonspecific stressor such as DNA damaging agents, heat shock, oxidative stress, toxic metals, and organic environmental pollutants) results in transcription and translation of fused target genes, generating recombinant proteins that produce some measurable response. Frequently used reporter genes are *lux* (coding for luciferase) and *gfp* (coding for green fluorescence protein), expression of which correlates with luminescence- or fluorescence-based light emission (Kohlmeier et al., 2006). Colorimetric determination of target gene expression is possible by fusing it to reporter genes coding for β-galalactosidase (*lacZ*) or alkaline phosphatase (*phoA*).

E. coli biosensor capable of detecting both genotoxic and oxidative damage has been developed. This was achieved by introducing two plasmids: fusion of *katG* (gene encoding for an important antioxidative enzyme) promoter to the *lux* reporter genes and *recA* (gene encoding a crucial enzyme for DNA repair) promoter with the *gfp* reporter gene (Mitchell and Gu, 2004). Besides genetically modified microorganisms (also named bioreporters), some other types of cellular biosensors have also been constructed. An example is the algal biosensor, which functions based on amperometric monitoring of photosynthetic O_2 evolution—the process affected by toxic compounds—was developed by coupling Clark electrode to the cyanobacterium *Spirulina subsalsa* (Campanella et al., 2000). Biosensors for specific determination of chemical compounds frequently contain molecules such as enzymes, receptors, and metal-binding proteins as recognition elements. A number of enzymes have been shown to be inhibited by toxic metals, pesticides, and some other important contaminants such as endocrine disrupting compounds. Limitations for the potential applications of many enzyme biosensors include limited sensitivity and selectivity, as well as interference by environmental matrices Marinšek Logar and Vodovnik, 2007. One recently introduced strategy to overcome the first two of these limitations uses inhibition ratio of two enzymes for the detection of specific compounds. Acetylcholinesterase and urease, coentrapped in the sol–gel matrix with the sensing probe (FITC-dextran), have successfully been used for the detection of Cu, Cd, and Hg (Tsai and Doong, 2005). Besides molecular biosensors, bioreporter cells may also be used for the detection of specific target compounds. A biosensor for nitrate monitoring has been constructed by transformation of plasmid containing nitrate reductase operon fused to *gfp* reporter gene to *E. coli* cells (Taylor et al., 2004).

1.2.3. Molecular Diagnostics

Traditionally, the clinical medical microbiology laboratory has functioned to identify the etiologic agents of infectious diseases through the direct examination and culture of clinical specimens. Direct examination is limited by the number of organisms present and by the ability of the laboratories to successfully recognize the pathogen. Similarly, the culture of the etiologic agent depends on the ability of the microbe to propagate on artificial media and the choice of appropriate media for the culture. When a sample of limited volume is submitted, it is often not possible to culture for all pathogens. In such instances, close clinical correlation is essential for the judicious use of the specimen available. Commercial kits for the molecular detection and identification of infectious pathogens have provided a degree of standardization and ease of use that has facilitated the introduction of molecular diagnostics into the clinical microbiology laboratory. The use of nucleic acid probes for identifying cultured organisms and for direct detection of organisms in clinical material was the first exposure that most laboratories had to explore commercially available molecular tests. Although these probe tests are still widely used, amplification-based methods are increasingly employed for diagnosis, identification, quantization of pathogens, and characterization of antimicrobial-drug-resistant genes.

The tools of molecular biology have proven readily adaptable for use in the clinical diagnostic laboratory and promise to be extremely useful in diagnosis, therapy, and epidemiologic investigations and infection control (Cormican and Pfaller, 1996; Pfaller, 2000, 2001). Although technical issues such as ease of performance, reproducibility, sensitivity, and specificity of molecular tests are important, cost and potential contribution to patient care are also of concern (Kant, 1995). Molecular methods may be an improvement over conventional microbiologic testing in many ways. Currently, their most practical and useful application is in detecting and identifying infectious agents for which routine growth-based culture and microscopy methods may not be adequate

(Fredricks and Relman, 1996; Fredricks and Relman, 1999; Tang and Persing, 1999; Woods, 2001).

Nucleic-acid-based tests used in diagnosing infectious diseases use standard methods for isolating nucleic acids from organisms and clinical material, and restriction endonuclease enzymes, gel electrophoresis, and nucleic acid hybridization techniques to analyze DNA or RNA (Tang and Persing, 1999). Because the target DNA or RNA may be present in very small amounts in clinical specimens, various signal amplification and target amplification techniques have been used to detect infectious agents in clinical diagnostic laboratories (Fredricks and Relman, 1999; Tang and Persing, 1999). Nucleic acid sequence analysis coupled with target amplification is clinically useful to detect and identify previously uncultivable organisms and characterize antimicrobial resistant gene mutations, thus aiding both diagnosis and treatment of infectious diseases (Fredricks and Relman, 1999). Automation and high-density oligonucleotide probe arrays (DNA chips) also hold great promise for characterizing microbial pathogens (Tang and Persing, 1999).

Although most clinicians and microbiologists enthusiastically welcome the new molecular tests for diagnosing infectious diseases, the high cost of these tests is of concern. Molecular methods will be increasingly used for pathogen identification, microbial quantification, and resistance testing.

The use of these detection methods in microbiology laboratories has resolved many problems and has initiated a revolution in the diagnosis and monitoring of infectious diseases. Some microorganisms are uncultivable at present, extremely fastidious, or hazardous to laboratory personnel. In these instances, the diagnosis often depends on the serologic detection of a humoral response or culture in an expensive biosafety level II–IV facility. In community medical microbiology laboratories, these facilities may not be available, or it may not be economically feasible to maintain the special media required for the culture of all of the rarely encountered pathogens. Thus, cultures are often sent to referral laboratories. During transit, fragile microbes may lose viability or become overgrown by contaminating organisms or competing normal flora.

Although direct detection of organisms in clinical specimens by nucleic acid probes is rapid and simple, it suffers from lack of sensitivity. Most direct probe detection assays require at least 10^4 copies of nucleic acid per microliter for reliable detection, a requirement rarely met in clinical samples without some form of amplification. Amplification of the detection signal after probe hybridization improves sensitivity to as low as 500 gene copies per microliter and provides quantitative capabilities. This approach has been extensively used for quantitative assays of viral load HIV, hepatitis B virus (HBV), and hepatitis C virus (HCV) but does not match the analytical sensitivity of target-amplification-based methods such as polymerase chain reaction (PCR) for detecting organisms.

Probe hybridization is useful in identifying organisms that grow slow after isolation from culture using either liquid or solid media. Identification of mycobacteria and other slow-growing organisms such as the dimorphic fungi (*Histoplasma capsulatum*, *Coccidioides immitis*, and *Blastomyces dermatitidis*), has certainly been facilitated by commercially available probes. All commercial probes for the identification of organisms are produced by Gen-Probe, which uses acridinium-ester-labeled probes directed at species-specific rRNA sequences. Gen-Probe products are available for the culture identification of *Mycobacterium tuberculosis, Mycobacterium avium-intracellulare* complex, *Mycobacterium gordonae, Mycobacterium kansasii, Cryptococcus neoformans*, the dimorphic fungi (listed above), *Neisseria gonorrhoeae, Staphylococcus aureus, Strepto coccus pneumoniae, E. coli, Haemophilus influenzae, Enterococcus* spp., *Streptococcus agalactiae*, and *Listeria monocytogenes*. The sensitivity and specificity of these probes are excellent, and they provide species identification within one working day. Because most of the bacteria listed and *C. neoformans* can be easily and efficiently identified by conventional methods within 1–2 days, many of these probes are not widely used. The mycobacterial probes, on the other hand, are accepted as mainstays for the identification of *M. tuberculosis* and related species.

Nucleic acid techniques such as plasmid profiling, various methods for generating restriction fragment length polymorphisms, and the PCR are making increasing inroads into clinical laboratories. PCR-based systems to detect the etiologic agents of diseases directly from clinical samples, without the need for culture, have been useful in rapid detection of uncultivable or fastidious microorganisms. Additionally, sequence analysis of amplified microbial DNA allows for the identification and better characterization of the pathogen. Subspecies variation, identified by various techniques, has been shown to be important in the prognosis of certain diseases. New advances in real-time PCR promise results that will come fast enough to revolutionize the practice of medicine.

Commercial-amplification-based molecular diagnostic systems for infectious diseases have focused largely on systems for detecting *N. gonorrhoeae, Chlamydia trachomatis, M. tuberculosis*, and specific viral infections [HBV, HCV, HIV, CMV (cytomegalovirus), and enterovirus]. Given the adaptability of PCR, numerous additional infectious pathogens have been detected by investigator-developed PCR assays. This novel, fully integrated device, coupled with appropriate databases, will insure better management of patients, should reduce health costs, and could have an impact on the spread of antibiotic resistance (Boissinot and Bergeron, 2002). Another exciting technology that has demonstrated clinical diagnostic utility is DNA microarray science. DNA microarray enables simultaneous analyses of global patterns of gene expression in microorganisms or host cells. In addition, genotyping and sequencing by microarray-based hybridization have been successfully used for organism identification and molecular resistance testing.

Microbial phenotypic characteristics, such as protein, bacteriophage, and chromatographic profiles, as well as biotyping and susceptibility testing, are used in most routine laboratories for identification and differentiation. Other important advances include the determination of viral load and the direct detection of genes or gene mutations responsible for drug resistance. Increased use of automation and user-friendly software makes these technologies more widely available. In all, the detection of infectious agents at the nucleic acid level represents a true synthesis of clinical chemistry and clinical microbiology techniques.

Ou et al. (2007) have beautifully illustrated an example of how to use whole genome sequence analyses of *Salmonella enterica* paratyphi A and existing comparative genomic hybridization data to design a highly discriminatory multiplex PCR assay that can be developed in any molecular diagnostic laboratory. In a time of overwhelmingly rapid expansion of genomic information, various navigation tools and a recipe for mining the genomic databases to design species, serovar, or pathotype specific PCR assays for accurate identification have been developed (Wenyong et al., 2007).

Molecular methods can rapidly detect antimicrobial drug resistance in clinical settings and have substantially contributed to our understanding of the spread and genetics of

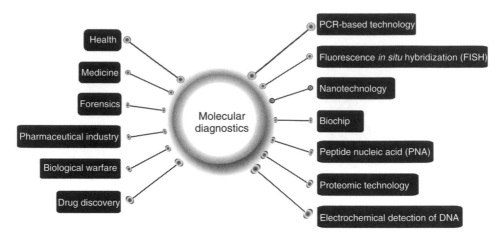

Figure 1.3. Applications and areas of interest related to molecular diagnostics.

resistance. Conventional broth and agar-based antimicrobial susceptibility testing methods provide a phenotypic profile of the response of a given microbe to an array of agents. Conventional methods are slow and fraught with problems, although useful for selecting potential therapeutic agents. The rapid evolution in microbial genomics will transform the process of accurate identification of novel, difficult to culture, or phenotypically indistinguishable pathogens and research and development in diagnostics, or "diagnomics."

Molecular typing methods have allowed investigators to study the relationship between colonizing and infecting isolates in individual patients. Most available DNA-based typing methods may be used in studying infections when applied in the context of a careful epidemiologic investigation. Molecular testing for infectious diseases includes testing for the host's predisposition to diseases, screening for infected or colonized persons, diagnosis of clinically important infections, and monitoring the course of infection or the spread of a specific pathogen in a given population.

There are many areas of molecular diagnostics and many tools related to this field (Fig. 1.3). Not all molecular diagnostic tests are extremely expensive. Direct costs vary widely, depending on the complexity and sophistication of the test performed. Inexpensive molecular tests are generally kit based and use methods that require little instrumentation or technologist experience. DNA probe methods that detect *C. trachomatis* or *N. gonorrhoeae* are examples of low-cost molecular tests. The more complex molecular tests such as resistance genotyping often have high labor costs because they require experienced, well-trained technologists. Although the more sophisticated tests may require expensive equipment (e.g., DNA sequencer) and reagents, advances in automation and the production of less expensive reagents promise to decrease these costs as well as technician time.

In general, molecular tests for infectious diseases have been more readily accepted for reimbursement; however, reimbursement is often on a case-by-case basis and may be slow and cumbersome. FDA approval of a test improves the likelihood that it will be reimbursed but does not ensure that the amount reimbursed will equal the cost of performing the test.

Molecular screening programs for infectious diseases are developed to detect symptomatic and asymptomatic diseases in individuals and groups. Persons at high risk, such as immuno-compromised patients or those attending family planning or obstetrical clinics, are screened for CMV and *Chlamydia*, respectively. Likewise, all blood donors are screened for blood borne pathogens. The financial outcome of such testing is unknown. The cost must be balanced against the benefits of earlier diagnosis and treatment and societal issues such as disease epidemiology and population management.

One of the most highly touted benefits of molecular testing for infectious diseases is the promise of earlier detection of certain pathogens. The rapid detection of *M. tuberculosis* directly in clinical specimens by PCR or other amplification-based methods is quite likely to be cost-effective in the management of tuberculosis. Other examples of infectious diseases that are amenable to molecular diagnosis and for which management can be improved by this technology include HSV (herpes simplex virus) encephalitis, *Helicobacter pylori* infection, and neuroborreliosis caused by *Borrelia burgdorferi*. For HSV encephalitis, detection of HSV in cerebrospinal fluid (CSF) can direct specific therapy and eliminate other tests including brain biopsy. Likewise, detection of *H. pylori* in gastric fluid can direct therapy and obviate the need for endoscopy and biopsy. PCR detection of *B. burgdorferi* in CSF is helpful in differentiating neuroborreliosis from other chronic neurologic conditions and chronic fatigue syndrome.

Molecular tests may be used to predict disease response to specific antimicrobial therapy (Fig. 1.4, Chinnery et al., 1999). Detection of specific resistance genes (*mec A, van A*) or point mutations resulting in resistance has been proved to be efficacious in managing diseases. Molecular based viral load testing has become a standard practice for patients with chronic hepatitis and AIDS. Viral load testing and genotyping of HCV are useful in determining the use of expensive therapy such as interferon therapy, and can be used to justify decisions on the extent and duration of therapy. With AIDS, viral load determinations and resistance genotyping have been used to select among the various protease inhibitor drugs available for treatment, improving patient's response and decreasing incidence of opportunistic infections.

Pharmacogenomics is the use of molecular based tests to predict the response to specific therapies and to monitor the response of the disease to the agents administered (Pfaller, 2001). The best examples of pharmacogenomics in infectious diseases are the use of viral load and resistance genotyping to select and monitor antiviral therapy of AIDS and chronic hepatitis. This application improves disease outcome, shortens the period of stay at hospital, reduces adverse events and toxicity, and facilitates cost-effective therapy by avoiding unnecessary expensive drugs, optimizing doses and timing, and eliminating ineffective drugs.

Molecular strain typing of microorganisms is now well recognized as an essential component of a comprehensive infection control program that also involves the infection control department, the infectious disease division, and pharmacy. The sequences of 16S ribosomal RNA sequences can be used to study the evolutionary relationship between bacteria. This region is highly conserved. The sequence in these variable regions is species specific. In this approach PCR primers, complementary to flanking conserved sequences are used to amplify the variable regions. The product is then sequenced and sequence compared against the database of the 16S sequence to identify the bacteria it is derived from. This approach to identification of the bacteria has particular advantage with organisms that cannot be easily cultured in the laboratory, as the DNA is amplified by PCR rather than the organisms being amplified by growing in a culture.

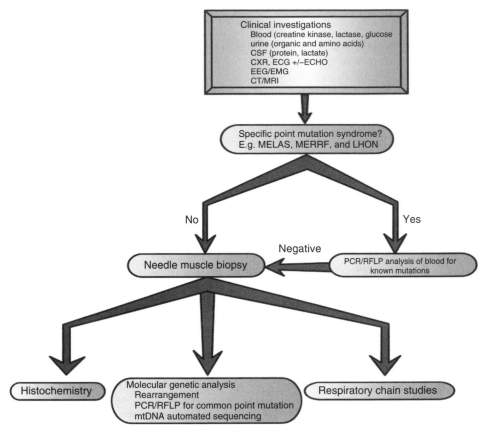

Figure 1.4. Molecular tests may be used to predict disease response to specific antimicrobial therapy.

Amplification of 16S genes by PCR can be very effective when combined with oligonucleotide hybridization probes or molecular beacon technology to identify bacteria in the mixture. A PCR reaction with a single set of primers complementary to conserved sequences will amplify species-specific sequence from a range of different bacteria in a mixture. These can then be probed with molecular beacons complementary to the individual species-specific sequences and labeled with different fluorophores. This makes it possible to identify more than one type of pathogenic bacterium in a mixture. By measuring the fluorescence levels in real time, it is also possible to make quantitative measurements and to detect the presence of a rare pathogen in the more abundant one.

One topical application of this technology is in devising tests to detect and identify bioterrorism agents. In the case of suspected bioterrorist attack, there is an urgent need for a robust and rapid assay for the selection of possible bioterrorist agents. A real-time PCR assay has been devised to simultaneously detect four bacteria having the potential to be used as bioterrorism agents by using a single set of PCR primers and four species-specific molecular beacons. In the case of bioterrorism accident, it would be vital to be able to identify the bacteria concerned. Technology has obvious applications in routine clinical laboratories where patient care could be improved by reducing the time taken.

1.2.4. Nanomedicine

The early genesis of the concept of nanomedicine sprang from the visionary idea that tiny nanorobots and related machines could be designed, manufactured, and introduced into the human body to perform cellular repairs at the molecular level. Today, it has branched out in hundreds of different directions, each of them embodying the key insight that the ability to structure materials and devices at the molecular scale can bring enormous immediate benefits in the research and practice of medicine. Nanomedicine is defined as the application of nanotechnologies including nanobiotechnologies in medicine. Evidently, dimensional parameters alone are insufficient to refer someone or other work to the field of nanotechnology (e.g., nanomedicine). Fundamental novelty of nanomedicine as a branch of knowledge and technology is exemplified by the developments in pharmacology and design of medicinal products that brought about new drugs (nanomedical/nanopharmaceutical). These products are multicomponent supramolecular compounds designed for a specific purpose whose intricate structure is intended not so much to impart new properties as to properly deliver the active ingredient to the biological target. Accordingly, nanomedicine should be regarded as the use of supramolecular complexes with a well-differentiated surface, manufactured by purposeful assembly of selected components for diagnostic and/or therapeutic application (Piotrovsky, 2010). Nanopharmaceuticals are defined as a big part of what nanomedicine is today. Nanopharmaceuticals can be developed either as drug delivery systems or as biologically active drug products (Table 1.1). Various types of nanomedicine are already in clinical use these days (Fig. 1.5).

DermaVir vaccine is a novel "pathogen-like" nanomedicine containing a plasmid DNA complexed with a polyethylenimine (pDNA/PEIm) that is mannobiosylated to target

TABLE 1.1. Various Types of Nanomedicine[a]

Type of Nanomedicine	Name of Nanomedicine	Use	Mode of Administration
Polymer as therapeutics	Copaxone	Multiple sclerosis	Parenteral injection
	Renagel	End-stage renal failure	Oral
	Emmelle	HIV/AIDS prevention	Topical
	Ampligen	Chronic fatigue syndrome	Parenteral injection
	Vivagel (Dendrimer)	HIV/AIDS prevention	Topical
	Macugen (PEG–aptamer)	Age-related macular degeneration	Topical
Polymer–protein conjugates	Adagen	SCID	Parenteral injection
	Zinostatin Stimaler (SMANCS)	Cancer	Local infusion
	Oncaspar	Cancer	Parenteral injection
	PEG–INTRON	Hepatitis C	Parenteral injection
	PEGASYS	Hepatitis C	Parenteral injection
	PEGvisomant	Acromegaly	Parenteral injection
	Neulasta	Cancer	Parenteral injection
	Cd870 (PEG–anti TNF–fab)	Crohn's disease Rheumatoid arthritis	Parenteral injection

[a] Source: From Ruth Ducan on nanopharmaceuticals.

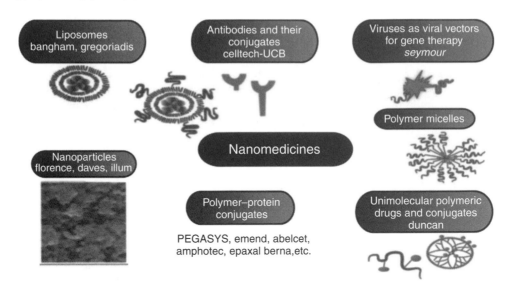

Figure 1.5. Various types of nanomedicines. (*See insert for color representation of the figure*).

antigen presenting cells and to induce immune responses. A commercially viable vaccine product was developed and the variability of raw materials and their relationship with the product's biological activity was investigated and found that the cGMP quality requirements are not sufficient to formulate the nanomedicine with optimal biological activity. The high cationic concentration of the pDNA favored the biological activity, but did not support the stability of the nanomedicine (Toke et al., 2010).

Nanomedicine also offers the prospect of powerful new tools for the treatment of human diseases and the augmentation of human biological systems (Table 1.2). They have been used as liposomes (3–100 nm) and nanoparticles (iron oxide, 5–50 nm) in clinical laboratories. To cure cancer, targeted drug delivery system is used using

TABLE 1.2. Application of Nanomedicine for Health

Application of Nanomedicine	Nanomaterial Name and Type	Pharmacological Function	Diseases
Nanomedicines in the clinic	Liposome (30–100 nm)	Targeted drug delivery	Cancer
	Nanoparticle (iron oxide, 5–50 nm)	Contrast agent for magnetic resonance imaging	Hepatic (liver)
Nanomedicines under development	Dendrimer (5–50 nm)	Contrast agent for magnetic resonance imaging	Cardiovascular phase III clinical trial
	Fullerenes (carbon buckyball 2–20 nm)	Antioxidant	Neurodegenerative, cardiovascular
	Nanoshells (gold-coated silica 60 nm)	Hyperthermia	Cancer preclinical

liposomes (Park, 2002). For hepatic diseases, nanoparticles are used as the contrast agent for generating resonance imaging (Thorek et al., 2006). Nanomedicine using dendrimers (Bharali et al., 2009), fullerenes (Partha and Conyers, 2009), and gold-coated nanoshells (Hirsch et al., 2003) is also under the developmental stage. In the cardiovascular phase III clinical trials, dendrimers are used as the contrast agent for magnetic resonance (Saha, 2009). Diamodoid-based medical nanorobotics may offer substantial improvements in capabilities over natural biological systems, exceeding even the improvements possible via tissue engineering and biotechnology. For example, the respirocytes, the artificial red blood cells comprise microscopic diamondoid pressure tanks that are operated at high atmospheric pressure and could carry >200 times respiratory gases than an equal volume of natural red blood cells.

Nanomedicine has been used to cure many diseases (Table 1.2). The clottocytes are artificial platelets that can stop human bleeding within ~ 1 s of physical injury, but using only 0.01% the bloodstream concentration of natural platelets in other words, nanorobotic clottocytes would be $\sim 10,000$ times more effective as clotting agents than an equal volume of natural platelets. In neurodegenerative diseases, carbon buckyballs (2–20 nm) are used as antioxidants. In preclinical cancer therapy during hyperthermia, gold-coated silica (60 nm) nanoshells are used.

Microbiovores constitute a potentially large class of medical nonorobots intended to be deployed in human patients for a wide variety of antimicrobial therapeutic purposes. They can also be useful in treating infections of the meninges or the CSF and respiratory diseases involving the presence of bacteria in the lungs or sputum, and can also digest bacterial biofilms. These handy nanorobots can quickly rid the blood of nonbacterial pathogens such as viruses (viremia), fungus cells (fungemia), or parasites (parasitemia).

A nanorobotic device that can safely provide quick and complete eradication of blood borne pathogens using relatively low doses of devices would be a welcome addition to the physician's therapeutic armamentarium Freitas, 2005. The ultimate tool of nanomedicine is the medical nanorobot—a robot to the size of a bacterium, composed of thousands of molecule-size mechanical parts perhaps resembling macroscale gears, bearings, and ratchets, possibly composed of a strong diamond-like material.

A nanorobot will need motors to make things move, and manipulator arms or mechanical legs for dexterity and mobility. It will have a power supply for energy, sensors to guide its actions, and an onboard computer to control its behavior. But, unlike a regular robot, a nanorobot will be very small. A nanorobot that would travel through the bloodstream must be smaller than the red cells—tiny enough to squeeze through even the narrowest capillaries in the human body.

Medical nanorobots could also be used to perform surgery on individual cells. In one proposed procedure, a cell repair nanorobot called a "chromallocyte," controlled by a physician, would extract all existing chromosomes from a diseased cell and insert fresh new ones in their place. This process is called *chromosome replacement therapy*. The replaced chromosomes are manufactured outside the patient's body using a desktop nanofactory optimized for organic molecules.

The patient's own individual genome serves as the blueprint to fabricate the new genetic material. Each chromallocyte is loaded with a single copy of a digitally corrected chromosome set. After injection, each device travels to its target tissue cell, enters the nucleus, replaces old worn out genes with new chromosome copies, then exits the cell and is removed from the body. If the patient chooses, inherited defective genes could be replaced with nondefective base pair sequences, permanently curing any genetic disease and even permitting cancerous cells to be reprogrammed to a healthy state. Perhaps, most importantly, chromosome replacement therapy could correct the accumulating genetic damage and mutations that lead to aging in every one of our cells. At present, medical nanorobots are just theory. Nanorobots will have several advantages. Firstly, they can physically enter cells and scan the chemicals present inside. Secondly, they can have onboard computers that allow them to do calculations not available to immune cells. Thirdly, nanorobots can be programmed and deployed after a cancer is diagnosed, whereas the immune system is always guessing about whether a cancer exists. Given such molecular tools, a small device can be designed to identify and kill cancer cells (Saha, 2009).

The potential impact of medical nanorobotics is enormous. Rather than using drugs that act statistically and have unwanted side effects, we can deploy therapeutic nanomachines that act with digital precision, have no side effects, and can report exactly what they did back to the physician. Continuous medical monitoring by embedded nanorobotic systems will provide automatic collection of long-baseline physiologic data permitting detection of slowly developing chronic conditions that may take years or decades to develop, such as obesity, diabetes, calcium loss, or Alzheimer's. Nanorobot life cycle costs can be very low because nanorobots, unlike drugs and other consumable pharmaceutical agents are intended to be removed intact from the body after every use, then refurbished and recycled many times, possibly indefinitely. Even if the delivery of nanomedicine does not reduce total health-care expenditures—which it should—it will likely free up billions of dollars that are now spent on premiums for private and public health-insurance programs.

1.2.5. Personalized Medicine

The war against infectious agents has produced a powerful arsenal of therapeutics, but treatment with drugs can sometimes exacerbate the problem. The drug-resistant strains and the infectious agents that are least susceptible to drugs survive to infect again. They become the dominant variety in the microbe population, a present day example of natural selection in action. This leads to an ever-present concern that drugs can be rendered useless when the microbial world employs the survival of the fittest strategy of evolution. And, frequently used drugs contribute to their own demise by strengthening the resistance of many enemies.

Some engineering challenges are to develop better systems to rapidly assess a patient's genetic profile; another is collecting and managing massive amounts of data on individual patients; and yet another is the need to create inexpensive and rapid diagnostic devices such as gene chips and sensors able to detect minute amounts of chemicals in the blood. In addition, improved systems are necessary to find effective and safe drugs that can exploit the new knowledge of differences in individuals. The current "gold standard" for testing a drug's worth and safety is the randomized controlled clinical trial—a study that randomly assigns people to a new drug or to nothing at all, a placebo, to assess how the drug performs, but this approach essentially decides a drug's usefulness based on average results for the group of patients as a whole, not for the individual (Bottinger, 2007).

New methods are also needed for delivering personalized drugs quickly and efficiently to the site in the body where the disease is localized (Kalow, 2006). For instance,

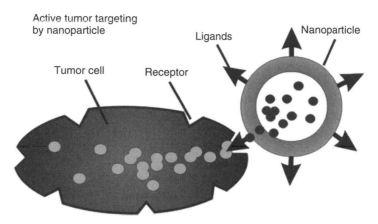

Figure 1.6. Drug delivery by nanoparticles.

researchers are exploring ways to engineer nanoparticles that are capable of delivering a drug to its target in the body while evading the body's natural immune response (Fig. 1.6). For example, when the drug-packed liposome is injected into the bloodstream, the amino acids on the nanoparticles attache to the proteins. The heat is pushed to the surface of the tumor and more of the drug is delivered to the tumor.

Such nanoparticles could be designed to be sensitive to the body's internal conditions, and, therefore, could, for example, release insulin only when the blood's glucose concentration is high. In a new field called *synthetic biology*, novel biomaterials are being engineered to replace or aid in the repair of damaged body tissues. Some are scaffolds that contain biological signals that attract stem cells and guide their growth into specific tissue types. Mastery of synthetic tissue engineering could make it possible to regenerate tissues and organs (Lutolf and Hubbell, 2005).

Ultimately, the personalization of medicine should have enormous benefits. It ought to make disease (and even the risk of disease) evident much earlier, when it can be treated more successfully or prevented altogether. It could reduce medical costs by identifying cases where expensive treatments are unnecessary or futile. It will reduce trial-and-error treatments and ensure that optimum doses of medicine are applied sooner.

More optimistically, personalized medicine could provide the path for curing cancer, by showing why some people contract cancer and others do not, or how some cancer patients survive when others do not. Thus, personalized medicine involves the use of laboratory-based molecular diagnostics and medical imaging to select suitable candidates for treatment with a particular drug(s), to rule out those patients who would suffer unacceptable side effects from the proposed drug treatment, and to monitor the health status of the patient postinitiation of therapy to assess therapeutic drug levels and the continuing efficacy of the agent in suppressing or curing the disease. Hence, personalized medicine is a new trend in drug development based on tailoring drugs to patients based on their individual genetic profiles.

Personalized-medicine-based pharmaceuticals avoid these issues by marketing to patients with specific genetic profiles that maximize both the safety and efficacy of the drug on each patient. Promising personalized-medicine-based pharmaceuticals in the pipeline have shown close to 100% efficacy in patients. Major pharmaceutical firms have responded to the growing emphasis on individualized therapy to improve drug efficacy and safety with large investments in pharmacogenomics research (Mancinelli et al., 2000); Table 1.3). Examples of drugs in the market that have used genetic markers to achieve improved safety and efficacy in patients include Gleevec [Novartis AG (NVS)] and Herceptin [Genentech (DNA)] (http://www.wikinvest.com). Of course, a transition to personalized medicine is not without its social and ethical problems.

Even if the technical challenges can be met, there are issues of privacy when unveiling a person's unique biological profile, and it is likely that there will still be masses of people throughout the world unable to access its benefits deep into the century.

The drug resistance problem is not limited to bacteria and antibiotics. Antiviral drugs for fighting diseases such as AIDS and influenza face similar problems from emerging strains of resistant viruses (Wright and Sutherland, 2007). In fact, understanding the development of resistance in viruses is especially critical for designing strategies to prevent pandemics. The use of any antimicrobial drug must be weighed against its contribution to speeding up the appearance of resistant strains.

The engineering challenges for enabling drug discovery mirror those needed to enable personalized medicine development: more effective tools and techniques. This helps in rapid analysis and diagnosis so that a variety of drugs can be quickly screened and proper treatments can be promptly applied (West et al., 2006). Current drugs are often prescribed incorrectly or unnecessarily, promoting the development of resistance without real medical benefit. Quicker, more precise diagnosis may lead to more targeted and effective therapies. Antibiotics that attack a wide range of bacteria have typically been sought, because doctors could not always be sure of the precise bacterium causing an infection. Instruments that can determine the real culprit right away could lead to the use of more narrowly targeted drugs, reducing the risk of promoting resistance. Developing organism-specific antibiotics could become one of the century's most important biomedical engineering challenges. Personalized medicine will reshape pharmaceutical research and development and the calculation of cost-effectiveness by health services. The previous business model was based on so-called blockbuster drugs, intended for general use in the population and generating annual global profits in excess of $1 billion. Profits from blockbuster drugs offset the expenses of regulatory approval and investment in research and development (http://www.parliament.uk/documents/post/postpn329.pdf).

This could be especially challenging in the case of biological agents specifically designed to be weapons. A system must be in place to rapidly analyze their methods of attacking the body and quickly produce an appropriate medicine. In the case of a virus, small molecules might be engineered to turn off the microbe's reproductive machinery. Instructions for making proteins are stored by genes in DNA. Another biochemical molecule, called *messenger RNA*, copies those instructions and carries them to the cell's protein factories (Dietel and Sers, 2006). Sometimes other small RNA molecules can attach to the messenger RNA and deactivate it, thereby preventing protein production by blocking the messenger, a process known as *RNA interference*. Viruses can be blocked by small RNAs in the same manner, if the proper small RNAs can be produced to attach to and deactivate the molecules that reproduce the virus. The key is to decipher rapidly the sequence of chemicals comprising the virus so that effective small RNA molecules can be designed and deployed.

TABLE 1.3. Selected Pharmaceutical Companies Focusing on Genomics
and Pharmacogenomics

Company	Web Address	Focus
ACLARA BioSciences, Inc.	http://www.aclara.com	Lab card microfluidic technology
Aeiveos Sciences Group, LLC	www.aeiveos.com	Aging-related genes and gene responses
Affymetrix, Inc.	http://www.affymetrix.com	GeneChip microarray technology
Aurora Bioscience Corp	http://www.aurorabio.com	Genomic and drug screening technology
Axys Pharmaceuticals Inc./PPGx	http://www.axyspharm.com	Pharmacogenomics (with PDD Inc.)
Caliper Technologies Corp	http://www.clipertech.com	Microfluidic Lab Chip, SNP scanning (with Agilent)
Celera Genomics	http://www.celera.com	Human Genome sequencing and SNP scanning
Cellomics, Inc.	http://www.cellomics.com/	Pharmacocellomics, cellular bioinformatics
Curagen Corp	http://www.curagen.com/	SNP scanning; gene expression and drug response
Epidauros	http://www.epidauros.com/	Pharmacogenomics in drug discovery and therapy
Exelixis, Inc.	http://www.exelixis.com	Model systems for drug discovery
Eurona Medical, AB	http://www.eurona.com/	Drug responses and genetic profiling
Gemini Research, Ltd	http://www.gemini-research.co.uk/	Gene discovery; dizygotic twin studies
Genaissance Pharmaceuticals, Inc.	http://www.genaissance.com	Genetic polymorphism in cancer, vascular lesions
Gene Logic, Inc.	http://www.genelogic.com	Gene expression databases
Genome Therapeutics Corp	http://www.crik.com/	Human high-resolution polymorphism database
Genometrix, Inc.	http://www.genometrix.com	DNA microarrays
Genomic Solutions, Inc.	http://www.genomesolutions.com/	Genomics
Genset, SA	http://www.genset.fr	High-density biallelic maps; SNP identification
Hexagen Pic	http://www.hexagen.co.uk/	Single-strand conformational assay of polymorphisms
Hyseq, Inc.	http://www.hyseq.com	Genomic methods for therapeutic discovery
Incyte Pharmaceuticals, Inc.	http://www.incyte.com	Bioinformatics, SNP scanning, functional genomics
Kiva Genetics	http://www.kivagen.com	Pharmacogenetic testing services

(continued)

TABLE 1.3. (*Continued*)

Company	Web Address	Focus
Lion Bioscience, AG	http://www.lion-ag.de/	Bioinformatics, drug targets from gene expression
Lynx Therapeutics	http://www.lynxgen.com	Microbead-based DNA/SNP scanning
Microcide Pharmaceuticals	http://www.microcide.com/	Microbial genomics and antibiotics
Mitokor, Inc.	http://www.mitokor.com/	Mitochondrial genome analysis
Nova Molecular, Inc.	http://www.cns-hts.com/	CNS disease profiling
Millennium Predictive Medicine	http://www.mlnm.com/subsid/ mpmx.html	Pharmacogenomics, predicting disease and therapy
Orchid Biocomputer, Inc.	http://www.orchidbio.com	Microfluidic devices and pharmacogenetic testing
PE Biosystems, Inc.	http://www.pebio.com	Genomics, drug discovery
PPGx	http://www.ppgx.com	Pharmacogenetic testing services
Protogene Laboratories	http://www.protogene.com	DNA microarray development
Rigel, Inc.	http://www.rigelinc.com/	Identification of genetic drug targets
Rosetta Inpharmatics	http://www.rii.com/	Oligonucleotide array studies
Third Wave Technologies, Inc.	http://www.twt.com/	SNP scanning, pharmacogenomics
Transgenomic, Inc.	http://www.transgenomic.com/	Discovery of genetic variations
Variagenics, Inc.	http://www.variagenics.com/	Cancer therapeutics based on loss of heterozygosity

Traditional vaccines have demonstrated the ability to prevent diseases, and even eradicate some such as smallpox. It may be possible to design vaccines to treat diseases as well. Personalized vaccines might be envisioned for either use. But, more effective and reliable manufacturing methods are needed for vaccines, especially when responding to a need for mass immunization in the face of a pandemic (Heymann, 2006). A healthy future for the world's population will depend on engineering new strategies to overcome multiple drug resistances (Gerard and Sutherland, 2007). One major challenge in this endeavor will be to understand more fully how drug resistance comes about, how it evolves, and how it spreads. Furthermore, the system for finding and developing new drugs must itself evolve, and entirely novel approaches to fighting pathogens may also be needed (Kalow, 2006).

Drug resistance is nothing new. The traditional approach to this problem is still potentially useful in expanding the search for new antibiotics. Historically, many drugs to fight disease-producing microbes have been found as naturally occurring chemicals in soil bacteria, which is still a source of promising candidate (Lesko, 2007). Even more drug candidates, though, may be available from microbes in more specialized ecological niches or from plants or from bacteria living in remote or harsh environments, namely, deep lakes and oceans.

1.2.6. Biowarfare

Biological warfare agents are a group of pathogens and toxins of biological origin that can be potentially misused for military or criminal purposes (Pohanka and Kuca, 2010). In early December 2002, the National Security Council learned of a smallpox outbreak in Oklahoma. Twenty cases were confirmed by the Centres for Disease Control and Prevention (CDC), with 14 more suspected. There were 16 more reported cases in Georgia and Pennsylvania. Federal and State authorities quickly informed the public and implemented a vaccine distribution policy to those people most at risk of being exposed to the smallpox virus (http://learn.genetics.utah.edu). Three days before the Christmas holiday and 13 days after the initial outbreak, a total of 16,000 smallpox cases were reported in 25 states, and 1000 people were dead. Ten other countries reported cases of smallpox, likely to be caused by visitors from the United States. Canada and Mexico closed their borders to the United States. Vaccine supplies were depleted, and health officials predicted that by February, there will be three million cases of smallpox, leading to as many as one million deaths. The above scenario was, in fact, a game.

In their natural state, bacteria, viruses, and fungi can make pretty good biological weapons. If genetic engineering is used, more harmful agents can emerge. During the Cold War, several offensive biowarfare programs were run to develop the so-called Super Bugs. One such program, Project Bonfire, worked to create bacteria that were resistant to about 10 varieties of antibiotics. This was done by identifying and cutting out genes that conferred antibiotic resistance in many different strains of bacteria. By pasting these genes into the DNA of the anthrax bacterium, the Project Bonfire created a strain of anthrax that resisted any existing cure, making it impossible to treat (Fig. 1.7). The Hunter Program was another biological warfare research program that focused on combining whole genomes of different viruses to produce completely new hybrid viruses (Fig. 1.8). These artificial viruses could cause unpredictable symptoms that have no known treatment. In an innovative twist, the Hunter Program also created bacterial strains that carried pathogenic viruses inside them. These strains would be double trouble: a person who contracted the bacterial disease would likely be treated with an antibiotic, which would stop the infection by disrupting the bacterial cells. This would release the virus, resulting in an outbreak of viral disease. Such a scenario would confuse medical personnel, making treatment very difficult.

It is not known whether the biological agents were ever actually used to infect people. At the same time, there was mounting fear over the offensive biological warfare agenda, focusing on the difficult task of delivering biological weapons such as anthrax to a population. In experiments conducted at sea or at the desert facility, large populations of animals such as guinea pigs, monkeys, and sheep were exposed to these deadly agents. Even more unsettling was the research to test the dispersal of bacterial spores on human populations; harmless bacterial spores that mimicked anthrax in size and shape were covertly dispersed in the ventilation system, over the bay and in the subways. Results from these experiments revealed not only the difficulty of delivering anthrax to a populace, but also the deadly consequences of a successful distribution (http://learn.genetics.utah.edu).

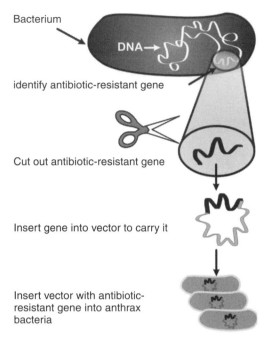

Bacterium

DNA→

identify antibiotic-resistant gene

Cut out antibiotic-resistant gene

Insert gene into vector to carry it

Insert vector with antibiotic-
resistant gene into anthrax
bacteria

Figure 1.7. Project Bonfire: creation of antibiotic-resistant bacteria.

The existence of these experiments was unknown to the public until the 1970s, and the actual results were not revealed until 1999, when some of the data was declassified by the Department of Defence. Advances in genetics may soon make possible the development of ethnic bioweapons that target specific ethnic or racial groups based on genetic markers (Appel, 2009). The tools for specific defense against bioweapons consist of vaccines against both viruses and bacteria, and of antibiotics and drugs against bacteria.

Vaccines and antimicrobials are of limited usefulness because of the large number of possible microbes that can be used for weapons, because of antimicrobial resistance to drugs and antibiotics, and because of limitations in technical feasibility for developing vaccines and antibacterials against certain agents. Induction of nonspecific innate immunity by immunostimulatory vaccines (at one time licensed) needs to be explored for possible immunoprophylactic–therapeutic activity when administered immediately following exposure to bioweapon pathogens.

Research into the offensive use of biological weapons has been carried out all over the world (Alibek, 2004; Guillemin, 2005; Karwa et al., 2005; Guillemin, 2006). The agents of biowarfare under category A are the high priority agents that include organisms that pose a high risk to national security because they can be disseminated/transmitted from person to person. The agents for biowarfare are botulism, small pox, marburg, plaque, and turaremia. The investigation on the use of *Bacillus anthracis* (anthrax), botulinum toxin (botulism), *Yersinia pestis* (plague), *Francisella tularensis* (tularemia), *Coxiella burnetii* (Q fever), Venezuelan equine encephalitis virus, *Brucella suis* (brucellosis), and Staphylococcal enterotoxin B has already been made.

This treaty prohibits the stockpiling of biological agents for offensive military purposes, and also forbids research into offensive use of biological agents. From 1975

Genomes of different viruses were combined
to produce new hybrid viruses

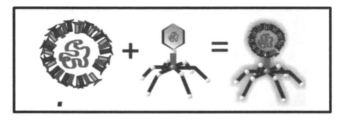

Viruses are inserted into bacteria

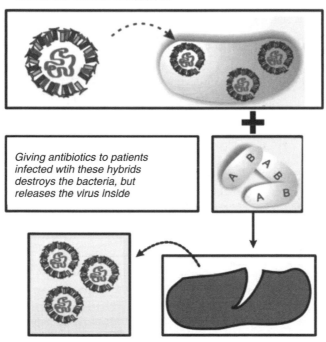

Giving antibiotics to patients
infected wtih these hybrids
destroys the bacteria, but
releases the virus inside

Figure 1.8. Hunter Program: creation of hybrid viruses.

to 1983, Soviet-backed forces in Laos, Cambodia, and Afghanistan allegedly used tricothecene mycotoxins (T-2 toxins) in what was called "Yellow Rain" (Seeley et al., 1985). After being exposed, people and animals became disoriented and ill, and a small percentage of those stricken died. The use of T-2 toxins has been denied and the presence of the yellow spots was reported as being caused by defecating bees. Various molecular methods of detection of the agents, such as PCR, hybridization, and strain typing, were performed on the basis of bacterial total cell protein profiles, random amplified polymorphic DNA (RAPD), and ribotyping, as well as of plasmid and DNA microrestriction analyses for the identification of agents of bioterrorism and biowarfare.

Various commercial tests utilizing biochemical, immunological, nucleic acid, and bioluminescence procedures are currently available to identify biological threat agents.

Newer tests have also been developed to identify such agents using aptamers, biochips, evanescent wave biosensors, cantilevers, living cells, and other innovative technologies (Lim et al., 2005; Carter and Cary, 2007; Garnier et al., 2009).

The pathogenic *Burkholderia mallei* and *Burkholderia pseudomallei* are the causative agents of glanders and melioidosis, respectively, in humans and animals, and are regarded as potential agents of bioterrorism. The existing bacteriological and immunological methods of identification of *B. mallei* and *B. pseudomallei* are not efficient enough for the rapid diagnosis and typing of strains (Antonov and Iliukhin, 2005). Schmoock et al. (2009) reported a DNA-microarray-based detection and identification of *Burkholderia mallei, Burkholderia pseudomallei*, and *Burkholderia* spp.

1.3. THE MICROBIAL WORLD

For most of its history, life on earth consisted solely of microscopic life-forms, and microbial life still dominates earth in many aspects (Table 1.4; (di Castri and Younes, 1990; Corliss, 1991; Hawksworth, 1991). The estimated 5×10^{30} prokaryotic cells inhabiting our planet sequester some 350–550 Pg (1 Pg = 1015 g) of carbon, 85–130 Pg of nitrogen, and 9–14 Pg of phosphorus making them the largest reservoir of those nutrients on earth (Whitman et al., 1998). Bacteria and archaea live in all environments capable of sustaining other life and in many cases are the sole inhabitants of extreme environments: from deep sea vents to rocks found in boreholes 6 km beneath the earth's surface. Bacteria, archaea, and microeukaryotes dominate earth's habitats, compound recycling, nutrient sequestration, and, according to some estimates, biomass. Microbes are not only ubiquitous, they are essential to all life, as they are the primary source for nutrients and the primary recyclers of dead matter back to available organic form. Along with all other animals and plants, the human condition is profoundly affected by microbes, from the scourges of human, farm animal, and crop pandemics, to the benefits in agriculture, food industry, and medicine, to name a few (Wooley et al., 2010).

Microbes are believed to be the common ancestors of all organisms. They not only grow virtually everywhere but also are present in abundance. In contrast to the relatively small number of humans (6×10^9), populations of terrestrial and marine bacteria are immense, 5×10^{30} and 1.2×10^{29}, respectively (Whitman et al., 1998). In fact, the human body contains 10 times more bacterial cells than human cells. Microbes carry

TABLE 1.4. Conservatively Estimated Total Microorganism Species in the World

Group	Known Species	Estimated Total Species	Percentage of Known Species
Viruses	5,000	1,30,000	4
Bacteria	4,760	40,000	12
Algae	40,000	60,000	67
Fungi and lichen	69,000	15,00,000	5
Protozoa	40,000	1,00,000	31
Total	1,57,000	18,20,000	9

out innumerable transformations of matter that are essential to life and thus have an enormous effect on climate and the geosphere.

We humans have more bacterial cells (10^{14}) inhabiting our body than our own cells (10^{13}) (Salvage, 1977; Berg, 1996). Scientists realized in the early 1990s that only a small fraction (1%) of microbes in natural communities was known and these communities became the focus of many studies. The sequencing of genomes of our own microbes is necessary (Wooley et al., 2010). Several years later, however, it is still shocking to realize the depth of our ignorance, which is well illustrated by the story of the SAR11 clade. When the technique of ribotyping (cloning and sequencing 16S rRNA genes) was first used to survey natural ecosystems, SAR11 was one of the first groups of novel microbes to be discovered (Giovannoni et al., 1990). We now know that the highest percentage of bacterial 16S ribosomal genes present in all oceanic and coastal waters is from members of the SAR11 clade, making this group "one of the most successful clades of organisms on the planet" (Giovannoni et al., 2005).

The basis for classification by ribotyping is the sequence of the 16S ribosomal RNA gene in prokaryotes and the 18S gene in eukaryotes. The 16S and 18S rRNA genes were selected for classification and identification of microbes because these genes are universal and essential; all living organisms must synthesize proteins to survive (Woese and Fox, 1977). These genes are also well suited for this purpose because they contain both conserved and variable regions, as is evident in the nucleotide sequence of the 16S gene.

Microbes are essential to human survival, performing tasks that the human body on its own cannot. This means the human body is more of an amalgam of different life-forms rather than being alone organism. It is believed that microbes coevolved with humans, with each supporting the other's survival. Interestingly, other research suggests people are not born with microbes, but pick them up from their environment. We know that microbial flora plays a significant role in human health and disease. There is also a shift in thinking about the role of microbes. While it was thought that individual microbes performed specific tasks on their own, there is now thought that some of them may even work together. Evidence suggests that signals are transmitted between some groups of microbes (http://www.accessexcellence.org). In addition, microbial flora populations can vary between individuals (Fig. 1.9).

1.3.1. Classification System

Whittaker based his kingdom groupings on the three main modes of nutrition in natural communities: absorption, ingestion, and autotrophy. He also credited the evolutionary sequence of unicellular to multicellular with central importance to his classification scheme. Utilizing these criteria as the basis for classification, Whittaker returned the bacteria to kingdom Protista (also based on their unicellular nature) and placed all algae (green, brown, and red) into kingdom Plantae. The protozoa were reassigned by Whittaker (1959) to his kingdom Protista. Whittaker's primary phyletic interest over-all was, however, in establishing a separate kingdom to contain macroscopic fungi. In particular, Whittaker observed the absorptive role of the fungi in the natural environ-ment. He rejected the common belief that the superficial resemblance of fungi to plants, with their nonmotile habit and cell walls, made them true plants. Whittaker did not believe that the fungi were derived from algae, but rather he thought they evolved from "colorless, flagellated protist ancestors." In 1969, Whittaker published a revision of his four-kingdom system to expand it to five kingdoms, now including a separate bacterial

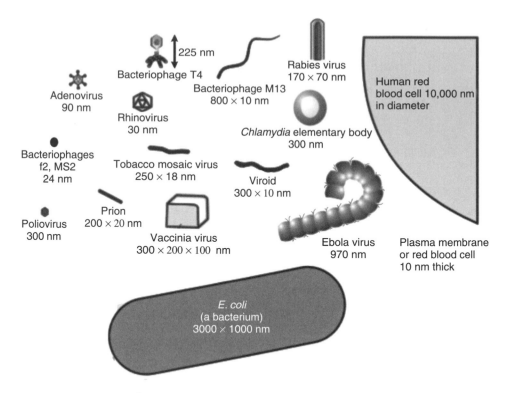

Figure 1.9. Morphological diversity in microbes.

kingdom named *Monera* in recognition of the fundamental division of life as "prokary-otic" versus "eukaryotic." Whittaker noted that this concept was now more evident. Otherwise, Whittaker's reasoning remained the same for retaining the other four king-doms in this five-kingdom system (Fig. 1.10). He reasserted his ecological model as well as the belief that inclusion of multicellular organisms into kingdom Protista would make the five-kingdom system an evolutionarily unnatural, heterogeneous grouping.

1.3.2. Viruses, Viroids, and Prions

In order to be alive, an organism must be composed of one or more cells. Viroids and prions are not living things and so are termed as agents. Many infectious agents consist of only a few of the molecules typically found in cells. Viruses consist of a piece of nucleic acid surrounded by a protective protein coat. They come in a variety of shapes, conferred by the shape of the coat. Viruses share with all organisms and agents the need to reproduce copies of themselves; otherwise they would not exist in nature. Viruses can only multiply inside living host cells, whose machinery and nutrients they must borrow for reproduction. Otherwise inside the host, they are inactive. One way of classifying viruses is to group them broadly into animal viruses, plant viruses.

A more fundamental classification separates them into two categories: DNA and RNA viruses (Dimijian, 2000; Table 1.5). Thus, viruses may be considered obligate intracellular parasites. All forms of life including members of the bacteria, archae, and eucarya can be infected by viruses. Although viruses frequently kill the cells in which they multiply,

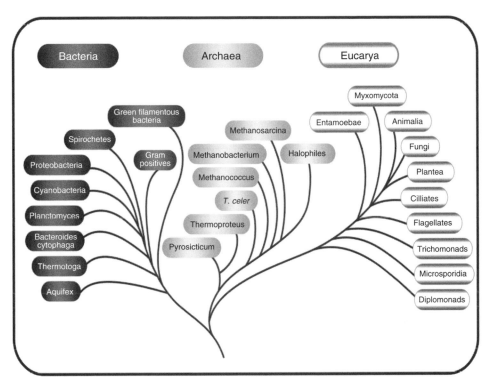

Figure 1.10. Phylogenetic tree of life.

TABLE 1.5. Classification of Viruses

Type	
DNA viruses	Herpes viruses
	Adenoviruses
	Hepatitis viruses
	Bacteriophages
RNA viruses	Retro viruses
	Influenza viruses
	Hepatitis virus A, C, and D
	Hemorrhagic fever viruses (dengue viruses,
	hantaviruses, Ebola virus)

some viruses exist harmoniously within the host cell without causing obvious ill effects. Some even cause clinical and special effects to human beings (Table 1.6).

Viriods define a group of pathogens that are also much smaller and simpler than viruses, consisting of a single short piece of single-stranded RNA that varies from 246 to 375 nucleotides without a protective coat. Hence, they are resistant to proteases. A single viriod RNA molecule is capable of infecting a cell. The viriod RNA is circular and is resistant to digestion by nucleases. They are smaller than viruses and like viruses; they can reproduce only inside cells. Viriods replicate autonomously within susceptible cells. No other virions or viriods are required for their replication. Viroids

TABLE 1.6. Families of Viruses that Affect Humans

Characteristics/ Dimensions	Viral Family	Important Genera	Clinical or Special Features
Single-stranded DNA nonenveloped 18–25 nm	Parvoviridae	Human parvovirus B19	Fifth disease; anemia in immuno-compromised patients
Double-stranded DNA nonenveloped 70–90 nm	Adenoviridae	Mastadenovirus	Medium-sized viruses that cause various respiratory infections in humans; some cause tumors in animals
40–57 nm	Papovariridae	Papillomavirus (human wort virus) Polyomavirus	Small viruses that induce tumors; the human wort virus (papilloma) and certain viruses that produce cancer in animals (polyoma and simian) belong to this family
Double-stranded DNA-enveloped 200–350 nm	Poxviridae	Orthopoxvirus (vaccinia and small pox viruses) Molluscipoxvirus	Very large, complex, brick-shaped viruses that cause diseases such as smallpox (variola), molluscum contagiosum (wortlike skin lesion), and cowpox
150–200 nm	Herpesviridae	Simplexvirus (HHV-1 & 2) Varicellovirus (HHV-3) Lymphocryptovirus (HHV-4) Cytomegalovirus (HHV-5) Roseolovirus (HHV-6) HHV-7 Kaposi's sarcoma (HHV-8)	Medium-sized viruses that cause various human diseases such as fever blisters, chickenpox, shingles, and infectious mononucleosis; implicated in a type of human cancer called Burkit's lymphoma
42 nm	Hepadnoviridae	Hepadnavirus (hepatitis B virus)	After protein synthesis, hepatitis B virus uses reverse transcriptase to produce its DNA from mRNA; causes hepatitis B and liver tumors

cause a number of plant disease, and some scientists speculate that they cause diseases in humans.

Prions are very unusual agents that are responsible for at least six neurodegenerative diseases in humans and animals, they are always fatal. They are proteinaceous infectious agents that apparently contain only protein and no nucleic acids. Although it is unlikely, it is possible that another agent that is very difficult to isolate might also be involved in causing the neurodegenerative disease. In all these infections, brain function degenerates as neurons die and brain tissue develop sponge-like holes. Thus, the term *transmissible spongiform encephalopathies* (*TSEs*) has been given to all of the diseases. Prions have many properties of viruses, but their evolutionary relationship to viruses is unclear. Both viruses and prions are obligate intracellular parasites, but prions are smaller than the smallest viruses, and unlike viruses or any other replicating agents contain no nucleic acids. Prions are not inactivated by UV light or nucleases, but are inactivated by chemicals that denature proteins as well as heat.

One of the most intriguing questions regarding prions is how they replicate if they do not contain any nucleic acids. The prion replicates by converting the normal host protein into a prion protein, thereby creating more prion protein molecules (Fig. 1.11). Thus, the prion protein is infectious because it catalyzes the conversion of normal protein into a prion protein by changing the folding properties of the protein.

In most cases, the prion disease is only transmitted to members of the same species, because the amino acid sequence of different prion protein in different species differs from one another. However, the barrier to prion transmission between species also depends on the strain of the prion. It is now clear that the strain of the prion that causes mad cow disease in England has killed more than 100 people by causing a disease very similar to Creutzfeldt–Jakob disease; presumably, these people ate beef of infected animals. Thus so far no deaths have been attributed to eating sheep infected with the scrapie agent or deer and eck infected with the prion causing chronic wasting disease. However, because

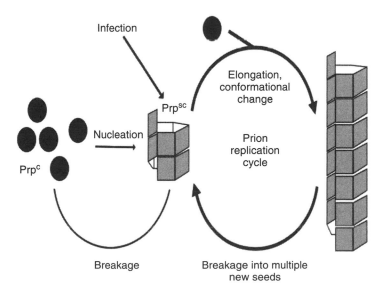

Figure 1.11. Replication of prions.

the incubation period extends over many years, the possibility that cross infection can occur in this situation also has not been ruled away completely.

1.3.2.1. Bacterial Viruses

A virus, injects its genome into a host bacterium, initiating production of new viruses and viral DNA. Bacterial viruses or bacteriophages are viruses that infect bacteria. It has a 48 kb double-stranded DNA genome packaged into an icosahedral head. Its genome is linear. An example is λ phage. The λ DNA when enters *E.coli* the ends of the linear genome join together to produce a circular genome. On infection, λ can either follow a lysogenic or lytic pathway. During the lysogenic pathway, the viral genome is integrated into the host genome and then replicated as part of the host genome; no expression of the viral structural proteins occurs in this case. If λ enters the lytic pathway, proteins required for the formation of the progeny virions are made and the viral genome is replicated and progeny virions are formed. The cell lyses and about 100 new virions are released. The virus contains a very compact genome coding for the 46 genes. λ has been exploited as a vector for DNA cloning predominantly with which it inserts its DNA into *E. coli*. They are perhaps the best understood viruses, yet at the same time, their structure can be extraordinarily complex (Fig. 1.12). They were originally distinguished by their smallness (hence, they were described as "filterable" because of their ability to pass through bacteria retaining filters) and their inability to replicate outside of a living host cell. Because these properties are shared by certain bacteria (rickettsiae, chlamydiae), viruses are further characterized by their simple organization and their unique mode of replication. A virus consists of genetic material, which may be either DNA or RNA, and is surrounded by a protein coat, and, in some viruses, by a membranous envelope.

Unlike cellular organisms, viruses do not contain all the biochemical mechanisms for their own replication; viruses replicate by using the biochemical mechanisms of a host cell to synthesize and assemble their separate components. When a complete virus

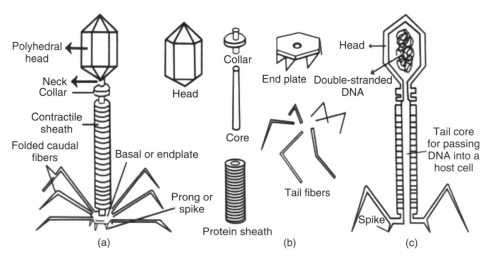

Figure 1.12. Bacteriophage: (a) T$_4$ phage, (b) isolated parts of the phage, (c) diagrammatic section showing the location of dsDNA in the head.

particle (virion) comes in contact with a host cell, the viral nucleic acid and, in some viruses, a few enzymes are introduced into the host cell.

Viruses vary in their stability, some such as poxviruses, parvoviruses, and rotaviruses are very stable and survive well outside the body while others, particularly those viruses such as herpes virus and influenza virus that are enveloped, do not survive well and therefore usually require close contact for transmission and are readily destroyed by disinfectants, particularly those with a detergent action. Some viruses produce acute disease while others, sometimes referred to as *slow viruses* such as retroviruses and lentiviruses and the scrapie agent, produce diseases that progress often to death over many years. Viruses in several families are transmitted by arthropod vectors. Bacteriophages continue to play a key role in bacterial genetics and molecular biology. Phage can confer key phenotypes on their host, for example, converting a nonpathogenic strain into a pathogen, and they play a key role in regulating bacterial populations in all sorts of environment (Fig. 1.13).

The use of bacteriophages was important in discovering that DNA in viruses can reproduce through two mechanisms: the lytic cycle and the lysogenic cycle. When viruses reproduce by the lytic cycle, they break open, or lyse, their host cells, resulting in the destruction of the host. In the lysogenic cycle, the phage's DNA (viral DNA) recombines with the bacterial chromosome. Once it has inserted itself, it is known as a prophage. A host cell that carries a prophage has the potential to lyse, thus it is called a lysogenic cell.

The phage−bacterium relationship varies enormously. From the simple predator−prey model to a complex, almost symbiotic relationship that promotes the survival and evolutionary success of both. While infection of bacteria used in the fermentation industry

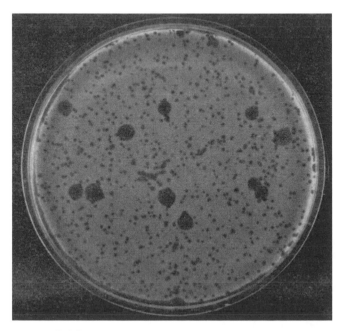

Figure 1.13. Plaques of different sizes caused by at least two bacteriophages on a carpet of actively multiplying bacteria.

can be very problematic and result in financial losses, in other scenarios phage infection of bacteria can be exploited for industrial and/or medical applications. In fact, interest in phage and phage gene products as potential therapeutic agents is increasing rapidly and is likely to have a profound impact on the pharmaceutical industry and biotechnology, in general, over the coming years. One potential application is the use of phage to combat the growing menace of antibiotic-resistant infections.

1.3.3. Bacteria

Bacteria are most widely distributed; they have simple morphology, are small in size, are the most difficult to classify and are the most hard to identify. It is even difficult to provide a descriptive definition of what a bacterial organism is because of considerable diversity in the group. About the only generalization that can be made for the entire group is that they are prokaryotic and some are photosynthetic. A few of them utilize a pigment that is chemically different from the chlorophyll a. It is called bacteriochlorophyll. Probably, the simplest definition that one can construct from these facts is bacteria are prokaryons without chlorophyll a.

Since they are prokaryons, bacteria share the kingdom Monera with the cyanobacteria. Although bacteria are generally smaller than cyanobacteria, some of the cyanobacteria range in size as that of bacteria. Most bacteria are only $0.5-2.0$ μm in diameter. The various shapes of bacteria encountered are mainly grouped into three types: rod, spherical, and helical or curved. Rod-shaped bacteria may vary considerably in length, may have square, round, or pointed ends, and may be motile or nonmotile.

The spherical- or coccus-shaped bacteria may occur as a single bacterium, or in pairs, in tetrads, in chains, and in irregular masses. The helical and curved bacteria exist as slender spirochetes, spirullum, and bent rods (vibrios) (Fig. 1.14).

1.3.3.1. Cyanobacteria

Cyanobacteria get their name from the bluish pigment phycocyanin, which they use to capture light for photosynthesis. The cyanobacteria, blue-green bacteria, constitute subkingdom I of the kingdom Monera. These microorganisms were formerly referred to as *algae*. Cyanobacteria can be found in almost every conceivable environment, from oceans to freshwater to bare rock to soil. They can occur as planktonic cells or form phototropic biofilms in freshwater and marine environments, they occur in damp soil, or even temporarily moistened rocks in deserts. The ability of cyanobacteria to perform oxygenic photosynthesis is thought to have converted the early reducing atmosphere into an oxidizing one, which dramatically changed the composition of life-forms on earth by stimulating biodiversity and leading to the near extinction of oxygen intolerant organisms. Their prokaryotic type nucleus definitely sets them apart from the eukaryotic algae. Although some bacteria are phototropic, the difference between phototropic bacteria and cyanobacteria is that the cyanobacteria have chlorophyll a and the phototropic bacteria do not have. Bacteriochlorophyll is the photosynthetic pigment in the phototropic bacteria. Over thousand species have been reported (Fig. 1.15). Aquatic cyanobacteria are probably best known for the extensive and highly visible blooms that can form in both freshwater and the marine environment and can have the appearance of blue-green paint or scum. The association of toxicity with such blooms has frequently led to the closure of recreational waters when blooms are observed.

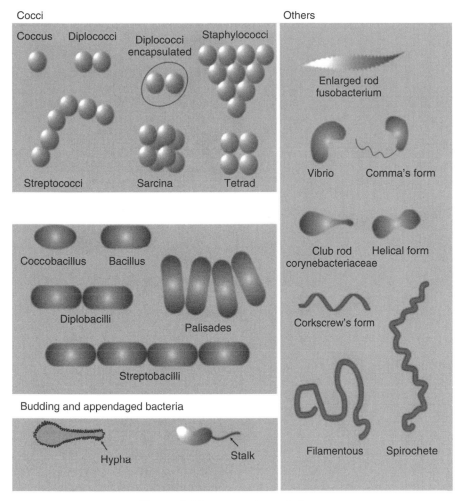

Figure 1.14. Bacterial morphology.

Cyanobacteria account for 20–30% of the earth's photosynthetic productivity and convert solar energy into biomass stored chemical energy at the rate of ~450 TW. Cyanobacteria utilize the energy of sunlight to drive photosynthesis, a process where the energy of light is used to split water molecules into oxygen, protons, and electrons. While most of the high energy electrons derived from water are utilized by the cyanobacterial cells for their own needs, a fraction of these electrons are donated to the external environment via electrogenic activity. Cyanobacterial electrogenic activity is an important microbiological conduit of solar energy into the biosphere.

Cyanobacteria are the only group of organism (Fig. 1.15) that are able to reduce nitrogen and carbon in aerobic conditions, a fact that may be responsible for their evolutionary and ecological success. The water-oxidizing photosynthesis is accomplished by coupling the activity of photosystem (PS) II and I (Z-scheme). In anaerobic conditions, they are also able to use only PS I—cyclic photophosphorylation with electron donors other than water (hydrogen sulfide, thiosulphate, or even molecular hydrogen)

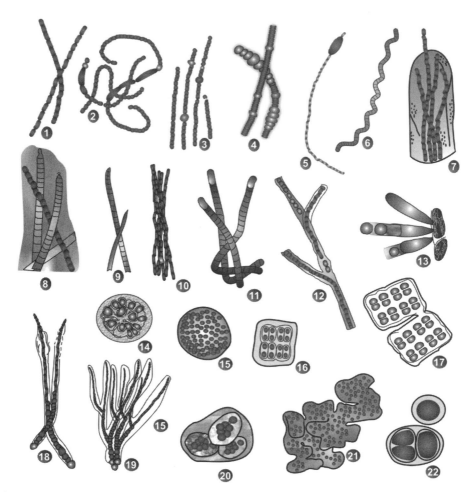

Figure 1.15. Schematic diagram of different forms of cyanobacteria (blue-green algae): (1) *Anabaena* (350×), (2) *Anabaena* (350×), (3) *Anabaena* (175×), (4) *Nodularia* (350×), (5) *Cylindrospermum* (175×), (6) *Arthrospira* (700×), (7) *Microcoleus* (350×), (8) *Phormidium* (350×), (9) *Oscillatoria* (175×), (10) *Aphanizomenon* (175×), (11) *Lyngbya* (700×), (12) *Tolpothrix* (350×), (13) *Entophysalis* (1000×), (14) *Gomphosphaeria* (1000×), (15) *Gomphosphaeria* (350×), (16) *Agmenellum* (700×), (17) *Agmenellum* (175×), (18) *Calothrix* (350×), (19) *Rivularia* (175×), (20) *Anacystis* (700×), (21) *Anacystis* (175×), (22) *Anacystis* (700×).

just like purple photosynthetic bacteria. Furthermore, they share an archaeal property, the ability to reduce elemental sulfur by anaerobic respiration in the dark. Their photosynthetic electron transport shares the same compartment as the components of respiratory electron transport. Their plasma membrane contains only components of the respiratory chain, while the thylakoid membrane hosts both respiratory and photosynthetic electron transport. Cyanobacteria have an elaborate and highly organized system of internal membranes that function in photosynthesis. They are present in almost all moist environments from the tropics to the poles, including both freshwater and marine. The different colors are due to the presence of the pigments like chlorophyll *a*, carotene, xanthophylls, *c*-phycocyanine that are produced in varying proportions. Phycocyanin and red *c*-phycoerythrin are unique to cyanobacteria and red algae. Cellular structure is considerably different from that of the eukaryotic algae. Nuclear membranes are absent in cyanobacteria. The nuclear material consists of DNA granules in a more or less colorless area in the center of the cell. Unlike the algae, the pigments of the cyanobacteria are not contained in the chloroplast; instead they are located in granules . Phycobilisomes are attached to membranes that permeate the cytoplasm. According to endosymbiotic theory, chloroplasts in plants and eukaryotic algae have evolved from cyanobacteria via endosymbiosis. Owing to their ability to fix nitrogen in aerobic conditions they are often found as symbionts with a number of other groups of organisms such as fungi (lichens), corals, pteridophytes (*Azolla*), and angiosperms (*Gunnera*), and provide energy to the host.

1.3.3.2. Archaea

Archaea are microbes: most live in extreme environments (called *extremophiles*; Delong, 1992) and others live in normal temperatures and salinities, some even live in our guts. Some extremophile species love heat; they like to live in boiling water, like the geysers of Yellowstone Park, and inside volcanoes (Wächtershäuser, 2006). They like the heat so much that it has earned the nickname "thermophile," which means "loving heat," and it would probably freeze to death at room temperature. Other extremophile Archaea love to live in very salty, called *hypersaline*, environments. They are able to survive in these extreme places where other organisms cannot. These salt loving archaea are called *halophiles*.

Archaea was originally thought to be just like bacteria, but archaea is a much different and simpler form of life. It may also be the oldest form of life on earth (Wächtershäuser, 2006). Archaea requires neither sunlight for photosynthesis as do plants, nor oxygen. Archaea absorbs CO_2, N_2, or H_2S and gives off methane gas as a waste product the same way humans breathe in oxygen and breathe out carbon dioxide. The relationship between archaea and eukaryotes remains problematic.

Complicating factors include claims that the relationship between eukaryotes and the archaeal phylum Euryarchaeota is closer compared to the relationship between the Euryarchaeota and the phylum Crenarchaeota and the presence of archaean-like genes in certain bacteria, such as *Thermotoga maritima*, from horizontal gene transfer. The leading hypothesis is that the ancestor of the eukaryotes diverged early from the archaea, and that eukaryotes arose through fusion of an archaeon and eubacterium, which became the nucleus and cytoplasm; this accounts for various genetic similarities but runs into difficulties explaining cell structure.

1.3.4. Eucarya

All members of the living world except the prokaryotes are in the domain Eukarya, and all members consist of eukaryotic cells. The microbial world is composed of single-celled members of the Eucarya as well as their close multicellular relatives. These members include algae, fungi, and protozoa. Algae and protozoa are also referred to as *protists*. The Eukarya are the structurally simple eukaryotes that lack mitochondria and some other key organelles. Most of these eukaryotes are metabolically deficient and are pathogenic parasites of humans and other animals. Like prokaryotes, there exists a diverse array of eukaryotic microorganisms.

Some of these such as algae are phototropic; they contain chlorophyll-rich organelles called *chloroplast* and live in an environment that contains only a few minerals, carbohydrates, and CO_2 and light.

The Eucarya are usually classified into four kingdoms: plants, animals, fungi, and protists. Kingdom under domain Archaea is Archaeabacteria and kingdom under domain Bacteria is Eubacteria. Eukarya is the one which has four kingdoms under, namely Plantae, Animalia, Protista, and Fungi. The first three of these correspond to phylogenetically coherent groups as well. However, the eukaryotic protists do not form a group, but rather comprise many phylogenetically disparate groups (including slime molds, multiple groups of algae, and many distinct groups of protozoa). Just as molecular analyses were required to see the natural relationships among prokaryotes, they are also allowing us to infer relationships among these nonplant, nonanimal, nonfungus eukaryotes (Figs. 1.16 and 1.17).

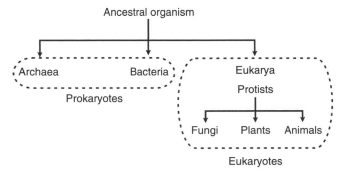

Figure 1.16. Interrelation between prokaryotes and eukaryotes.

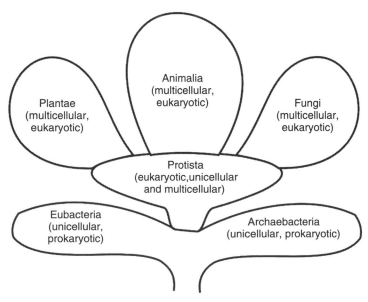

Figure 1.17. Phylogenetic origin of Eukarya.

Many groups of fungi are important. In addition to providing yeast for making bread and fermenting sugar to alcohol, the fungi include the causative agents responsible for conditions such as blight, ergot, rot, rust, smut, wilt, ringworm, and athlete's foot. Although it might seem that it is straightforward to identify an organism as a fungus, this is not always the case. The unambiguous identification of the opportunistic pathogen, *Pneumocystis* (which has not been grown in cultures), as a fungus required analysis of its ribosomal RNA. The "eukaryotic protists" also provide many organisms of clinical importance, including *Crithidia, Trypanosoma, Plasmodium, Amoeba*, and *Giardia*.

Eukarya differ fundamentally from the Bacteria and Archaea in terms of the organization of the genetic machinery with multiple chromosomes being housed within a nucleus bounded by a double membrane. This unique and complex organization must be explained by any theory on the origin of the Eukarya. The theory presented here delivers an explanation in addition to solving the lipid problem. It explains the eukaryal genetic machinery as an automatic, irreversible, almost deterministic consequence of the endosymbiosis, whereby a precell was converted into the nucleus. The following 11 features that are shared by all extant Eukarya may be explained by the historic origin of the nucleus in an endosymbiotic precell within a bacterial host:

1. The nonorganellar genome is polychromosomal and the chromosomes are housed inside the nucleus. This can be explained by the hypothesis that a polychromosomal precell turned into the nucleus, whereby the precell chromosomes became trapped and separated from the outer cell membrane by the nuclear double membrane. Therefore, their replication could not become linked with the outer cell membrane during cell division. This means that there was no selective advantage in their unification into one large circular chromosome as was the case with the Bacteria and Archaea.

2. The nuclear membrane is endowed with pores and connected with an endoplasmic reticulum (Moreira and Lopez-Garcia, 1998).

3. The nonorganellar chromosomes are linear as opposed to the circular bacterial and archaeal chromosomes. This can be explained by their possible origin in the linear intermediates of the hypothesized rolling circle replication of the precell chromosomes, perhaps with a multiplication in length.

4. Reproduction is based on a process of mitosis. The higher Eukarya, such as plants and animals, have an open mitosis, while all others have a closed mitosis. This finds a simple explanation in the hypothesis that the nucleus originated from a polychromosomal precell, which means that closed mitosis is original and open mitosis is a derived process.

5. The nuclear chromosomes contain genes of bacterial origin. Approximately 75% of these bacterial genes in the nuclear chromosomes are clearly of nonorganellar vintage (Esser et al., 2004). This can be explained with an evolutionary process, whereby the outer bacterial chromosomes were lost and their indispensable genes became relocated into the nuclear chromosomes.

6. The intermediary and energy metabolism is restricted to the outer cytoplasm. As a corollary, protein synthesis by translation is also restricted to the outer cytoplasm, that is, to exist in close quarters with the metabolism, where the enzymes are needed. This can be explained by the hypothesis that in the endosymbiosis, it would have been the bacterial host cell, which took charge of intermediary and

energy metabolism, with the advantage that this metabolism moved closest to the influx of nutrients. Incidentally, some translation appears to have remained to this day inside the nucleus (Brogna, 2001). This is simply explainable as a leftover from precell days.

7. Transcription occurs exclusively inside the nucleus and the mRNAs become short and capped. They are exported through the nuclear pores into the cytoplasm. This can be explained by the hypothesis that the expression of the genes became separated from their original location in the precell turned nucleus and that the requirement of their transport through the nuclear pores did not permit preservation of long operons as they are active in extant Bacteria and Archaea.

8. The nonorganellar ribosomes are synthesized in a peculiar manner. The rRNAs are synthesized inside the nucleus. The ribosomal proteins are synthesized outside the nucleus. The ribosomal proteins are synthesized outside the nucleus in the cytoplasm. Some of the ribosomal proteins are then imported into the nucleus for assembly with the ribosomal RNAs to form incomplete ribosomal subunits. The incomplete ribosomal subunits are then exported through the pores of the nuclear membrane into the cytoplasm, where they are completed by the addition of further ribosomal proteins. The completed ribosomes are then used in the cytoplasm for translation. This roundabout pattern can be explained by the hypothesis that the incipient precell endosymbiont inside the bacterial host would have maintained a precellular process of translation for some time. The incomplete ribosomal subunits would simply reflect their origin in primitive precellular ribosomes, thus give us insights into the most ancient ribosomes and their early evolution.

9. Most genes in the nuclear chromosomes are infested with introns of unknown origin. This infestation with introns is not lethal or even particularly deleterious due to a peculiar processing (maturation) of the primary mRNAs inside the nucleus. The introns are spliced out and the remaining mRNA segments are interconnected to form a mature, functional mRNA. This can be explained by the hypothesis that a physical isolation of the intranuclear, precell-derived space from the extranuclear bacteria-derived cytoplasmic space would have ensued with endosymbiosis. This means that the intranuclear space would have become a safe compartment for intron splicing so that immature intron-laden mRNAs would have been prevented from entering the cytoplasm, thereby the metabolism would have been protected from being messed up with products of a translation of immature mRNAs.

10. The genes for the intermediary metabolism are most closely related to the bacteria while the genetic machinery is most closely related to and actually more complex than the genetic machinery of the archaea. This can now be explained with reference to the hypothetical phylogeny. The Bacteria are shown as diverging first from relatively primitive precells. The Archaea are shown as diverging later from evolutionarily more advanced precells. The Eukarya are shown as diverging last (by endosymbiosis) from precells that would have been still farther evolved than the precells, from which the Archaea originated. This rendered the precell-derived genetic machinery of the Eukarya similar to, but more evolved than, the genetic machinery of the Archaea. The close relationship of the metabolic genes of the Eukarya with those of the Bacteria is simply due to their bacterial origin.

11. The genetic machinery inside the nucleus has somewhat "primitive" features with numerous small RNAs being involved in replication, transcription, and mRNA maturation. This can now be explained by the hypothesis that these "primitive" features would date back as leftovers to the workings of the "primitive" precells that were involved in endosymbiosis.

With these 11 explanatory points, the presentation of theory of the origin of the Eukarya by endosymbiosis of precells and wall-less bacteria is completed.

1.3.5. Algae

The subkingdom Algae includes all the photosynthetic eukaryotic organisms in the protista. Being true protista, they lack tissue differentiation. In Whittaker's five-kingdom system, some algae have been included with protozoan. The algae may be unicellular, colonial, or filamentous. The undifferentiated algal structure is often referred to as *thallus*. They lack the stem, root, leaf structure that result from tissue specialization. Algae are organisms that use light energy to convert CO_2 and H_2O to carbohydrates and other cellular products with the release of oxygen. Algae contain chlorophyll *a* which is necessary for photosynthesis. In addition, many other algae contain other pigments that extend the range of light waves that can be used by these organisms for photosynthesis. Algae include both microscopic and macroscopic multicellular organisms (Figs. 1.18–1.20).

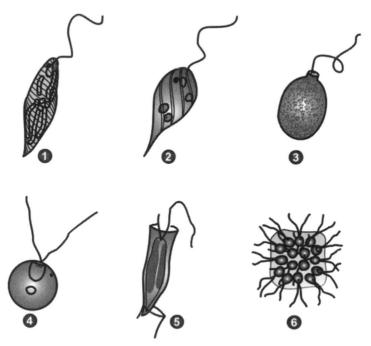

Figure 1.18. Microscopic view of flagellated algae: (1) *Euglena* (700×), (2) *Phacus* (1000×), (3) *Trachelomonas* (1000×), (4) *Chlamydomonas* (1000×), (5) *Dinobyron* (1000×), (6) *Gonium* (350×).

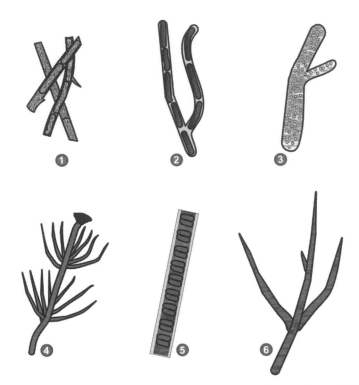

Figure 1.19. Microscopic view of filamentous algae: (1) *Rhizoclonium* (175×), (2) *Cladophora* (100×), (3) *Vaucheria* (100×), (4) *Chara* (3×), (5) *Zygnema* (175×), (6) *Stigeoclonium* (100×).

Algae, however, lack a well-organized vascular system. Algae do not directly infect humans but some produce toxins that cause paralytic shell fish poisoning. Some of these toxins do not cause illness in the shellfish that feed on the algae but accumulate in their tissues and when eaten by man cause nerve damage. As one of the primary producers of carbohydrate and other cellular products, the algae are essential in the food chains of the world. In addition, they produce a large proportion of the oxygen in the atmosphere.

These microorganisms are universally present where ample moisture, favorable temperature, and sufficient sunlight exist. Although a great majority of them live submerged in water, some grow on soil, others grow on the bark of trees or on the surface of the rocks. Algae have distinct, visible nuclei and chloroplast. Chloroplast is the organelle that contains chlorophyll *a* and other pigments. Photosynthesis takes place within these bodies. The size, shape, distribution, and number of chloroplasts vary considerably from species to species. In some instances a single chloroplast may occupy most of the cell space.

Although there are seven divisions of algae, only five will be listed here. Since the two groups cryptomonads and red algae are usually not encountered in freshwater ponds.

1.3.5.1. Euglenophycophyta (Euglenoids)

All members of this division are flagellated and appear to be intermediates between the algae and protozoa. Protozoan-like characteristics seen in euglenoids are the absence of

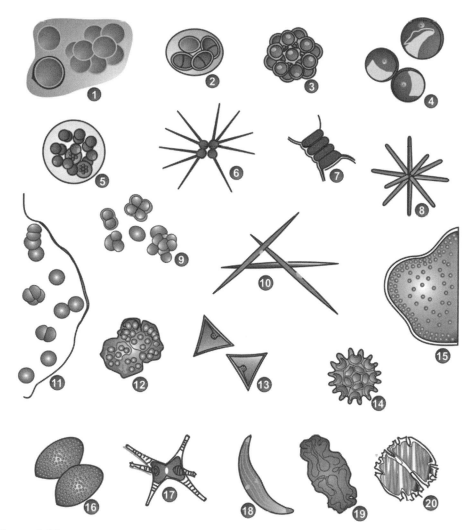

Figure 1.20. Microscopic view of nonfilamentous and nonflagellated algae: (1) *Chlorococcum* (700×), (2) *Oocystis* (700×), (3) *Coelastrum* (350×), (4) *Chlorellia* (350×), (5) *Sphaerocystis* (350×), (6) *Micractinium* (700×), (7) *Scendesmus* (700×), (8) *Actinastrum* (700×), (9) *Phytoconis* (700×), (10) *Ankistrodesmus* (700×), (11) *Pamella* (700×), (12) *Botryococcus* (700×), (13) *Tetraedron* (1000×), (14) *Petraspora* (100×), (15) *Tetraspora* (100×), (16) *Staurastrum* (700×), (17) *Staurastrum* (350×), (18) *Closterium* (175×), (19) *Euastrum* (350×), (20) *Micrasterias* (175×).

cell wall, presence of gullet, ability to assimilate organic substances, and absence of chloroplast in some species. The absence of cell wall makes these protists very flexible. Instead of cell wall, they possess a semirigid outer pellicle, which gives the organism a definite form. Photosynthesis types contain chlorophyll *a* and *b*, and they always have a red stigma or eyespot that is light sensitive. Their characteristic food storage is a lipopolysaccharide, paramylum. The photosynthetic euglenoids can be bleached experimentally by various means in the laboratory. The colorless forms that develop, however, cannot be induced to revert back to phototrophy.

1.3.5.2. Chlorophycophyta (Green Algae)

The green algae are present in ponds. They are grass green in color, resembling the euglenoids in having chlorophyll *a* and *b*. They differ from the euglenoids in that they synthesize starch instead of paramylum for food storage.

The diversity of this group is too great to explore; however, the small flagellated *Chlamydomonas* appears to be the archetype of the entire group. Many colonial forms such as *Pandorina, Eudorina, Gonium, Volvox* consist of organisms similar to *Chlamydomonas*. It is the consensus that all the filamentous algae that have evolved from this flagellated form are Chlorophycophyta. A unique group of green algae are the desmids. With the exceptions of a few species, the cells of desmids consist of two similar halves or semicells. The two halves usually are separated by a constriction, the isthmus.

1.3.5.3. Chrysophycophyta (Golden Brown Algae)

This large diversified division consists of over 6000 species (Fig. 1.21). They differ from the euglenoids and green algae in food storage in the form of oils and leucosin, and in pigments such as chlorophyll *a* and *c* and fucoxanthin, a brownish pigment. It is a combination of fucoxanthin, other yellow pigments, and chlorophylls that causes most of these algae to appear golden brown. Representatives of this division are *Chrysococcus, Synura*, and *Dinobryon*, which are typical flagellated chrysophycophytes. *Vaucheria* and *Tribonema* are the only filamentous chrysophycophytes. All these organisms fall under a special category of algae called *diatoms*. The diatoms are unique in that they have hard cell walls of pectin, cellulose, or silicon oxide that are constructed in two halves. The two halves fit like a lid and box. Skeletons of dead diatoms accumulate on the ocean

Figure 1.21. Microscopic view of diatoms: (1) *Gomphonema* (175×), (2) *Cymbella* (175×), (3) *Nitschia* (1500×), (4) *Tabellaria* (1000×), (5) *Navicula* (750×), (6) *Meridion* (750×).

bottom to form diatomite or diatomaceous earth that is commercially available as an excellent polishing compound. It is postulated that much of our petroleum compounds may have been formulated by the accumulation of oil from dead diatoms over millions of years.

1.3.5.4. Lichens (Symbionts)

Lichens are composite organisms consisting of a symbiotic association of a fungus (the mycobiont) with a photosynthetic partner (the photobiont or phycobiont), usually either a green alga (commonly *Trebouxia*) or a cyanobacterium (commonly *Nostoc*). The morphology, physiology, and biochemistry of lichens are very different from those of the isolated fungus and cyanobacteria in cultures (Fig. 1.22). Lichens occur in some of the most extreme environments on earth: arctic tundra, hot deserts, rocky coasts, and toxic slag heaps. However, they are also seen abundant as epiphytes on leaves and branches in rain forests and temperate woodland, on bare rock, including walls and gravestones, and on exposed soil surfaces (e.g., *Collema*) in otherwise mesic habitats. Lichens are widespread and maybe long-lived; however, many are also vulnerable to environmental disturbance, and may be useful to scientists in assessing the effects of air pollution, ozone depletion, and metal contamination. Lichens have also been used in making dyes and perfumes, as well as in traditional medicines. Thallus is the body of the lichen. The fungal hyphae (filaments) branch and then fuse together (anastomose) to form a mesh of hair-like threads. The top surface is normally a layer of tightly packed hyphae known as *cortex* covering below the algal layer where the photobiont lives. Below this is the medulla an area of loose hyphae in which nutrients are stored. Sometimes, a lower cortex exists; in others, the medulla rests on the surface. Filamentous lichens consist of chains of algal cells wrapped around with fungal hyphae. Cladonia are a successful group of lichens, which have a primary thallus and a secondary thallus. The primary thallus is small and clings closely to the substrate while the secondary thallus is a shrubby growth, for example, *fruticose lichens*. Once the lichen is established, the primary thallus

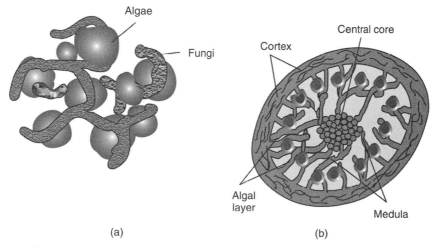

(a)   (b)

Figure 1.22.   (a) Symbiotic association in lichen, (b) TS of fruticose lichens.

often dies off. Lichens have been classified on the basis of their morphology, symbiotic relationship, and the type of structure:

1. *Crustose Lichens*. These are classified on the basis of forming a crust on the surface of the substrate on which they are growing. This crust can be quite thick and granular or actually embedded within the substrate. In this latter case, the fruiting bodies still rise above the surface. In many crustose lichens the surface of the thallus breaks up into a cellular, crazy-paving-like pattern. Crustose lichens tend to grow out from their edges and have their fruiting bodies in their center. Crustose lichens are very difficult to remove from their substrates.

2. *Squamulose Lichens*. These have a portion of their thallus lifted off the substrate to form "squamules." They are otherwise similar to crustose lichens in that they possess an upper cortex but no lower cortex (Fig. 1.23).

3. *Foliose Lichens*. These have an upper and lower cortex. They are generally raised to some extent above the substrate but connected to it by rhizines (specialized root-like hyphae). They are easier to remove from their substrate.

4. *Fruticose Lichens*. These are attached to their substrate by a single point and rise, or, more usually, dangle from this and are also known as shrubby lichens. Some foliose lichens can be stubby like fruticose lichens; however, close examination will reveal that the algal part exists only on one side of the flattish thallus, whereas in fruticose lichens it exists as a ring around the thallus, even when it is flattened as in *Ramalina* sp.

5. *Leprose Lichens*. These are an odd group of lichens that have never been observed to produce fruiting bodies and have not yet been identified properly or, at least, not yet given full scientific names. They not only lack an inner cortex but also lack an outer one, that is, no cortex, only an algal cell layer and sometimes a weakly defined medulla (Fig. 1.24).

1.3.6. Fungi

Fungi comprise of a large group of eukaryotic nonphotosynthetic organisms that include such diverse forms as slime molds, water molds, mushrooms, puff balls, bracket fungi, yeasts, and molds. They belong to the kingdom Mycetae and the study is called *mycology*. Myceteae consists of three divisions: Gymnomycota, the slime moulds; mastigomycota,

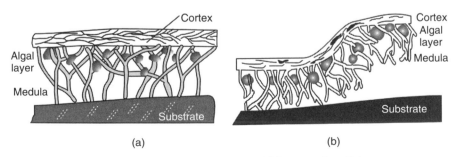

(a)   (b)

Figure 1.23. (a) Crustose lichen and (b) squamulose lichen.

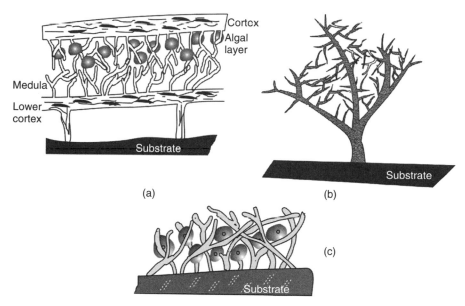

Figure 1.24. (a) Foliose lichen, (b) fructicose lichen, (c) leprose lichen.

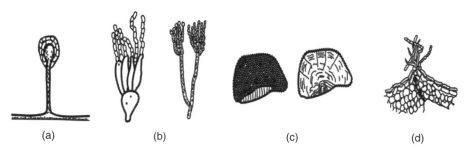

Figure 1.25. Schematic diagram of fungi: (a) *Rhizopus*, (b) *Penicillium*, (c) *Polyporus*, (d) *Cercospora*.

the water moulds; and amastigomycota, the yeast, the molds, bracket fungi, and so on. Most fungi are aerobic or facultative anaerobic. Only a few fungi are anaerobic. A large number of fungi cause diseases in plants. Fortunately, only a few species cause diseases in animals and humans. Along with bacteria, fungi are the principal decomposers of carbon compounds on the earth. The decomposition releases carbon dioxide into the atmosphere and nitrogen compounds into the soil, which are then taken up by plants and converted into organic compounds. Without this breakdown of organic material, the world would quickly be overrun with organic waste.

Fungi may be saprophytic or parasitic and unicellular or filamentous organism (Fig. 1.25). Some organisms such as slime molds are borderline between fungi and protozoa in that amoeboid characteristics are present and fungi-like spores are produced. The distinguishing characters of the group as a whole are that they are eukaryotic, nonphotosynthetic, lack tissue differentiation, have cell walls made up of chitin and other polysaccharides, and are propagated by spores both sexually and asexually.

Species within the Amastigomycota may have cottony or mold-like appearance, or moist or yeasty appearance that set them apart. Some species exist as molds under some conditions and some as yeast-like under other conditions. Such species are said to be dimorphic or diphasic.

1.3.6.1. Molds

The molds have microscopic filaments called *hyphae*. If the filaments have cross walls, it is referred to as *septate hyphae*. If no cross walls are present, it is called *nonseptate* or *aseptate*. Actually, most of the fungi that are classified as being septate are incompletely septate, have central openings that allow the streaming of the cytoplasm from one compartment to another. A mass of unmeshed microscopic hyphae is called *mycelium*. Two kinds of asexual spores are seen in molds: sporangia and conidia. Sporangiospores are spores that are formed within a sac called *sporangium*. The sporangia are attached to stalks called *sporangiospore*. Conidia are asexual spores that form on specialized hyphae called *conidiospore*. If the conidia are small they are called *microconidia*; large multicellular conidia are known as *macroconidia*. The four types of conidia are as follows:

Phialospores. Conidia of this type are produced by vase shaped cells called phialids. *Penicillium* and *Gliocadium* are produced by this type.

Blastoconidia. Conidia of this type are produced by budding from the cells of the preexisting conidia as in *Cladosporium*, which typically has lemon-shaped spores.

Arthospores. This type of conidia forms by separation from the preexisting hyphal cells, for example, *Oospora*.

Chlamydomonas. These spores are large, thick-walled, round or irregular structure formed within or at the ends of hyphae. Common to most fungi, they generally form on old cultures, for example, *Candida albicans*.

Sexual spores: Three kinds of sexual spores are seen in molds: zygospores, ascospores, and basidospores. Zygospores are formed by the union of nuclear material from the hyphae of the two different strains. Ascospores, on the other hand, are sexual spores produced in enclosures, which may be oval sacs or elongated tubes. Basidiospores are sexually produced on club-shaped bodies called *basidia*. A basidium is considered by some to be a modified type of ascus.

1.3.6.2. Yeast

Unlike molds, yeast does not have true hyphae. Instead, they form multicellular structures called *pseudohyphae*. The only asexual spores formed are called *blastospores* or *bud*. These spores form as an out pouching of the cell by the budding process, resulting in the formation of *pseudohyphae*.

Division Amastigomycota consists of four subdivisions: Zygomycotina, Ascomycotina, Basidiomycotina, and Deuteromycotina.

ZYGOMYCOTINA. These fungi have septate hyphae and produce zygospores. They also produce sporangiospores. *Rhizopus, Mucor*, and *Syncephalastrum* are representative genera of this subdivision.

ASCOMYCOTINA. Since all fungi in this subdivision produce ascospores, they are grouped into one class. They are commonly referred to as the *Ascomycetes* and are also called *sac fungi*. All of them have septate hyphae and most of them have chitinous walls (Fig. 1.26).

Fungi in this group produce a single ascus, *ascomycete yeasts*. Other ascomycetes produce numerous asci in complex flask-shaped fruiting bodies (*perithecia* or *pseudothecia*), in cup-shaped structures, or in hollow spherical bodies, as in powdery mildews, *Eupenicillium* or *Talaromyces*, the sexual stages for *Penicillum*.

BASIDIOMYCOTINA. All fungi in this subdivision belong to one class, the Basidiomycetes. Puffballs, mushrooms, smuts, rust, and shelf fungi on tree branches are also basidiomycetes. The sexual spores of this class are basidiospores.

DEUTEROMYCOTINA. This fourth division of the Amastigomycotina is an artificial group that was created to place any fungi that has not been shown to have some means of sexual reproduction. Often, species that are relegated to this division remain here for only a short period of time; as soon as the right condition has been provided for the sexual spores to form, they are reclassified into one of the subdivisions. Sometimes, however, the sexual and asexual stages of the fungus are discovered and named separately by different mycologists, with the result that a single species acquires two different names. Generally, there is a switch over to the sexual stage name. Members of this group are called *fungi imperfecti* or *deuteromycetes*.

1.3.7. Protozoa

The phylum Protozoa consists of about 15,000 species of protozoans. The phylum consists of microscopic organisms in which all the vital activities are performed by a single cell, commonly referred to as *single-celled organisms*. They are cosmopolitan in distribution and include aquatic (freshwater and marine) forms too. Phylum Protozoa includes free-living animals (Fig. 1.27) such as *Euglena, Paramecium, Amoeba, Noctiluca*, and *Elphidium* and parasitic animals such as *Plasmodium* species, *Monocystic, Entamoeba, Trypanosoma*, and *Giardia*. The organisms of this phylum greatly vary in shape, size, locomotory organelles, and method of reproduction.

The subkingdom Protozoa includes single-celled organisms; however, some of them do form colonial aggregates. Externally, the cells are covered with cell membranes or a pellicle, cell walls are absent, and distinct nuclei with nuclear membranes are present. Specialized organelles such as contractile vacuoles, cytostomes, mitochondria, ribosomes, flagella, and cilia may also be present.

All protozoa produce cysts, which are resistant dormant stages that enable them to survive drought, heat, and freezing. They produce asexually by cell division and exhibit various degrees of sexual reproduction. The subkingdom Protozoa is divided into three phyla: Sarcomastigophora, Ciliophora, and Apicomplexa. Type of locomotion plays an important role in classifying them. A brief description of each phylum follows.

1.3.7.1. Sarcomastigophora

Members of this phylum are divided into two subphyla: Sarcodina and Mastigophora. Members of the subphylum Sarcodina (*Amoeba*), move about by the formation of flowing

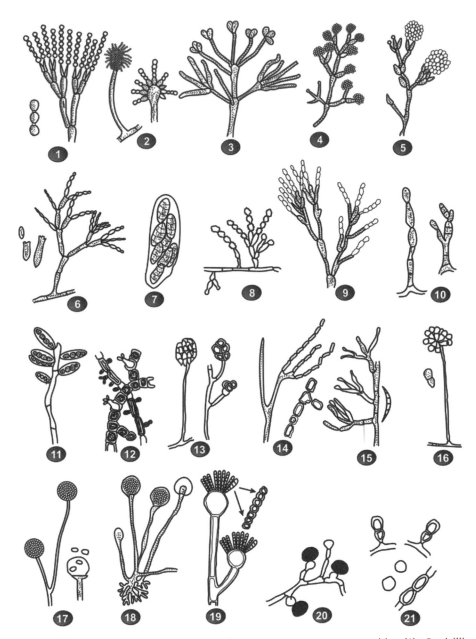

Figure 1.26. Microscopic appearance of some more common molds: (1) *Penicillium*, (2) *Aspergillus*, (3) *Verticillium*, (4) *Trichoderma*, (5) *Gliocadium*, (6) *Cladosporium*, (7) *Pleospora*, (8) *Scopulariopsis*, (9) *Paecilomyces*, (10) *Alternaria*, (11) *Bipolaris*, (12) *Pullularia*, (13) *Diplosporium*, (14) *Oospora*, (15) *Fusarium*, (16) *Trichothecium*, (17) *Mucor*, (18) *Rhizopus*, (19) *Syncephalastrum*, (20) *Nigrospora*, (21) *Montospora*.

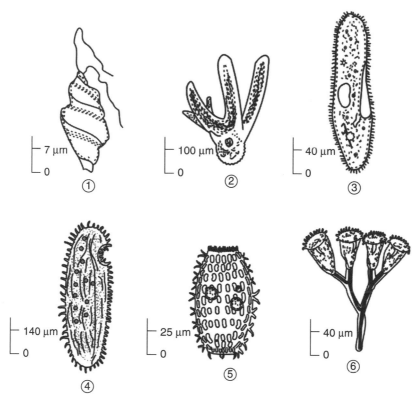

Figure 1.27. Microscopic view of different protozoans: (1) *Heteronema*, (2) *Amoeba*, (3) *Paramecium*, (4) *Loxodes*, (5) *Coleps*, (6) *Zoothamnium*.

protoplasmic projections called *pseudopodia*. The formation of pseudopodia is commonly referred to as *amoeboid movement*. The subphylum Mastigophora (*Zooflagellates*) possess whip-like structures called *flagella*. There is a considerable diversity among members of this group.

1.3.7.2. Ciliophora

Microorganisms of this phylum are undoubtedly the most advanced and structurally complex of all protozoans. Evidence seems to indicate that they have evolved from the zooflagellates. Movement and procuring food are accomplished with a short hair-like structure called *cilia*.

1.3.7.3. Apicomplexa

This phylum has only one class, the Sporozoa. Members of this phylum lack locomotor organelles, and are all internal parasites. Their life cycle includes spore forming stages. *Plasmodium*, the malaria parasite, is a significant pathogenic sporozoan of humans.

1.3.7.4. Genome of Protozoan

The completion of five new protozoan parasite genomes led to an improved understanding of their biology, and comparative genomics shed light on the diversity of protozoan genome and evolution. *Entamoeba histolytica* is a human gut parasite that causes amebiasis, a significant health problem in developing countries. The 23.7 Mb draft *E. histolytica* genome contains 9938 predicted genes and a striking abundance of tandemly repeated transferRNA-containing arrays.

The genomes of *E. histolytica* and the amitochondrial protist pathogens *Giardia lamblia* and *Trichomonas vaginalis* share several metabolic adaptations. These include reduced or eliminated mitochondrial metabolic pathways. Indeed, the genome data are consistent with the lack of a mitochondrial genome and enzymes of tricarboxylic acid cycle and mitochondrial electron transport chain. Secondary gene loss and lateral gene transfer, mainly from prokaryotes, seem to have shaped *E. histolytica* metabolism. Moreover, *Theileria parva* and *Theileria annulata* have a higher gene density (4035 and 3792 genes, respectively) and more spliced genes. Despite conserved gene sequences and synteny—3265 orthologous gene pairs between the two species—several species-specific genes were identified. Most genes without orthologues are members of unequally expanded gene families with only a small proportion is present as single copies (34 in *T. annulata* and 60 in *T. parva*). The chromosomes of *T. annulata* and *T. parva* have syntenic regions with only a few inversions of small blocks and no movement of blocks between chromosomes. Short breaks in synteny correspond to gene insertions or deletions and often involve members of large gene families such as the *Tpr* genes (*T. parva* repeat) and their counterparts in *T. annulata*, the *Tar* (*T. annulata* repeat) genes.

Like many parasitic protozoa, both *Theileria* species have tandem arrays of genus-specific, hypervariable gene families that are located in the subtelomeres and are predicted to encode secreted proteins. The overall arrangement of subtelomeric genes is conserved with one (or more) ABC-transporter gene(s) marking the boundaries between subtelomeric gene families and the house-keeping genes. Many *Theileria*-specific protein family members incorporate one or more copies of a polymorphic FAINT (frequently associated in *Theileria*) domain. Over 900 copies of FAINT domain are present in ~166 proteins of both genomes. Like the trypanosomatids, evidence of positive immune selection was found in the macroschizont- and merozoite-stage expressed genes. Also, several candidate genes for host cell transformation have been identified, by assuming that these genes are expressed in macroschizonts and that their products are released into the host cell cytoplasm or expressed on the parasite surface.

Compared with *Plasmodium falciparum*, the metabolism of *Theileria* spp. is streamlined. Several biosynthetic pathways are absent, including those for haem, type II fatty acids, polyamines, and shikimic acid. *Theileria* spp. have lost the ability to salvage purines and have limited ability to interconvert amino acids, but isoprenoid biosynthesis is, however, present. This reduced metabolism suggests substantial dependence on the host cell for many substrates (Hertz-Fowler et al., 2005).

Protozoa represent one of the oldest forms of animal life. They have become adapted to almost all types of environment, the availability of food and water being the most important factors governing their prevalence in different microenvironments. Few protozoa can synthesize their food from inorganic materials and therefore they depend on the available organic substances such as disintegrating plant or animal material or, relevant here, living microorganisms. Hence, soils that contain high levels of organic matter and bacteria will support an abundance of protozoa. The survival of protozoa under fluctuating environmental conditions is enhanced by the ability to form cysts. Amoebic cysts are resistant to many different environmental stresses such as disinfectants and desiccation. Free-living protozoa are known to occur concurrently with pathogenically important bacteria. These predatory protozoa graze on bacterial biofilms in the environment as a food source. Some bacteria, however, are resistant to killing by protozoa. This intra-amoebic survival of bacterial pathogens and the use of the amoebic cyst to avoid environmental stress and aid dissemination have highlighted the protozoa as the Trojan Horses of the bacterial world. Free-living protozoa have been shown to aid the survival and persistence of numerous pathogenic bacterial species.

The possibility that this might also apply to mycobacterial pathogens is receiving increased attention. Importantly, these relationships between protozoa and bacterial pathogens are highly exploitable and are being used to investigate virulence factors in a broad range of species, including *Legionella pneumophila*, *L. monocytogenes*, *Vibrio cholera*, *Chlamydia pneumoniae*, *Burkholderia*, *E. coli*, and methicillin-resistant *S. aureus*. As a laboratory host cell that can differentiate mycobacteria on the basis of virulence, the amoeba provides us with a biologically relevant and robust tool for the investigation of mycobacterial virulence genes, information that is essential to the generation of new treatments and vaccines (Rhodes et al., 2007).

1.3.8. Microscopic Invertebrates

While looking for protozoa, algae, and cyanobacteria in the pond, one invariably encounters large, transparent, complex microorganisms that, to the inexperienced, appear as protozoan. In most instances, these moving "monsters" are "rotifers"; in some cases, they are copepods, daphnis, or any one of the other forms.

All organisms are multicellular with an organ system. If an organ system is present, then these organisms cannot be protists, because the organ system represents tissue differentiation. Collectively, these microscopic forms are designated as *invertebrates* (Fig. 1.28). A few invertebrates such as *Dugesia* and *Hydra* are macroscopic in adult form but they are microscopic when immature. The following phyla are listed according to the degree of complexity, the simplest being the first.

1.3.8.1. Coelenterata

Members of this phylum/family are almost exclusively found in marine habitat. The only common freshwater form is *Hydra*. In addition, there are a few less-common freshwater genera similar to the marine hydroids.

The hydra is very common in ponds and attached to rocks, twigs, or other substrata. Around the mouth at the free end are five tentacles of various lengths, depending on the species. Smaller organisms such as *Daphnia* are grasped by the tentacles and passed on to the mouth. These animals have a digestive cavity that makes up the bulk of the interior. Since no anus is present, undigested remains of the food are expelled through the mouth.

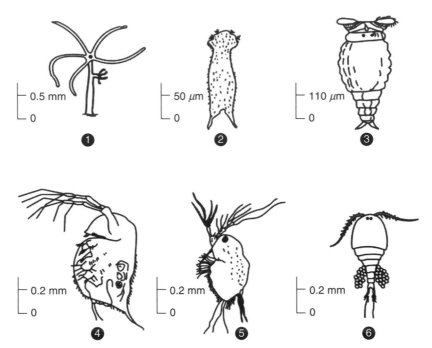

Figure 1.28. Microscopic invertebrates: (1) *Hydra*, (2) *Lepidermella*, (3) *Philodina*, (4) *Daphnia*, (5) *Latonopis*, (6) *Cyclops*.

1.3.8.2. Platyhelminthes

The invertebrates of this phylum are commonly referred to as *flatworms*. The phylum contains two parasitic classes and one free-living organism, the *Turbullaria*. It is the organism that is encountered in freshwater. The characteristics common to all these organisms are dorsoventral flatness, a ciliated epidermis, a ventral mouth, and eyespots on the dorsal surface near the anterior end. As in the coelenterates, undigested food must be ejected through the mouth since no anus is present. Reproduction may be asexual by fission or fragmentation; generally, however, reproduction is sexual, each organism having both male and female reproductive organs. Species identification of the turbellarians is exceedingly difficult and is based, to a greater extent, on the details of the reproductive system.

1.3.8.3. Nematoda

The members of this phylum are the roundworms. They are commonly known as *Nemas* or *nematodes*. They are characteristically round in cross section, have an external cuticle without cilia, lack eyes, and have a tubular digestive system complete with the mouth, intestine, and anus. The males are generally much smaller than the females and have a hooked posterior end. The number of the named species is only a fraction of the total nematodes in existence. Species identification of these invertebrates requires very detailed study of the many minute anatomical features, which requires complete knowledge of anatomy.

1.3.8.4. Aschelminthes

This phylum includes classes Gastrotricha and Rotifera. Most of the members of this phylum are microscopic. This proximity of the nematodes in classification is due to the type of the body cavity pseudocoel that is present in both phyla.

The gastrotrichs are very similar to the ciliated protozoans in size and in habits. The typical gastrotrich is elongate, flexible, forked at the posterior end and is covered with bristles. The digestive system consists of an anterior mouth surrounded by bristles, a pharynx, intestine, and posterior anus. Species identification is based partially on the shape of the head, tail structure and size, and distribution of the spines. Overall length is also an important characteristic of identification. They feed on unicellular algae.

The rotifers are most easily differentiated by the wheel-like arrangement of the cilia at the anterior end and by the presence of chewing pharynx within the body. They are considerably diverse in their food habits; some feed on algae and protozoa, others on juices of the plant cells, and some are parasitic. They play an important role in keeping the water clean. They also serve as food for small worms and crustaceans, being an important part of the link in the food chain of freshwaters.

1.3.8.5. Annelida

This phylum includes three classes: Oligochaeta, Polychaeta, and Hirudinea. Polychaetes are primarily marine and leeches are macroscopic and parasitic. Some oligochaetes are found in marine habitats, but most are found in freshwater and soil. These worms are characterized by body segmentation, bristles on each segment, and an anterior mouth. Although most of the oligochaetes breathe through the skin, some of the aquatic forms possess gills at the posterior end or along the sides of the segments. Most oligochaetes feed on vegetation; some feed on the mud at the bottoms of polluted waters, aiding in purifying such places.

1.3.8.6. Tardigrada

These invertebrates are of uncertain taxonomic position. They appear to be closely related to both Annelids and Arthropods. They are commonly referred to as *water bears*. They have a head, four trunk segments and four pairs of legs. The ends of the legs may have claws, fingers, or disc-like structures. The anterior ends have retractable snouts with teeth. Sexes are separate and females are oviparous. They are primarily herbivorous. Locomotion is by crawling, and not by swimming. During desiccation of their habitat, they contract to form barrel shaped *tuns* and are capable of surviving years of dryness, even in extremes of heat and cold. Widespread distribution is due to dispersal of the tuns by the wind.

1.3.8.7. Arthropoda

This phylum contains most of the known Animalia, almost a million species. The three representatives groups Cladocera, Ostracoda, and Copepoda of the class Crustacaea have in common jointed appendages, an exoskeleton, and gills. The cladocera are represented by *Daphnia* and latonopsis. They are commonly known as water fleas. All cladocera have a distinct head. The body is covered by a bivalve-like carapace. There is often a distinct cervical notch between the head and body. A compound eye may be present; when present it is movable. They have many appendages; antennules, antennae, mouth parts, and four to six pairs of legs. The ostracods are bivalved crustaceans that are distinguished from minute clams by the absence of lines of growth on the shell. Their bodies are not distinctly segmented. They have seven appendages. The ends terminate with a pair of caudal furca. The copepods such as *Cyclops* and *Canthocamptus* lack the shell-like covering of the ostracods and cladocera.

1.3.9. Microbial Interrelationships

Organisms do not exist alone in nature but in a matrix of other organisms of many species. Thus, interaction is the rule of nature. When one or more types of organisms reciprocate each others effect, it is called interaction. The evidence for such interaction is direct. Population of one species is different in the absence and in the presence of a second species. Interaction between two different biological population can be classified according to whether both population are unaffected by the interaction, one/both population benefit, or one/both population are adversely affected.

A microbial ecosystem represents a delicately balanced population of microorganisms each interacting with and influencing the other members of the population (Fig. 1.29). An understanding of the nature and effects of these interactions is essential to improve the performance of these ecologies, which are important, in such diverse processes as biological waste treatment procedures, water pollution abatement, industrial fermentations, soil microbial association, and human or animal digestive processes. The frequency of microbial interaction may be visualized if one stops for a moment to think of the enormous numbers of microbes. Therefore, while speaking of microbial interactions in a natural habitat such as soil or water, we must realize that the microbes live there in microenvironments as populations of millions or billions and those they intermingle on a scale that is impossible for higher plants and animals. Microbes usually interact in a positive or negative manner. Neutralism is not frequent in the case of this kind of relationship.

Among single population of microbes (microbes of the same species), microbes cooperate at low densities and compete at high cell densities. Beneficial interactions among microbes are facilitated by close physical proximity, for example, biofilms and flocs. Commonly, positive interactions dominate at low population densities and negative interactions dominate at high population densities. Hence, there is an optimal population density for maximal growth rate. The various positive interactions within microbial populations are as follows.

There are many symbiotic relationships among microbes and between microbes and higher organisms. Microorganisms have developed mechanisms to defeat an animal's defense against disease. There is deep symbiosis between some bacteria and their human hosts. Three approaches have been used to visualize the interactions of the world of microscopic organisms: *The first approach* is to isolate the numerous microbial sp., define their biochemical functions, and deduce their activity in a mixed culture of natural environment. *The second approach* seeks to overcome this dilemma by direct observation under microscope. Thus, a series of snapshots of microbial populations, their types, and interactions are taken directly from the soil to the microscope. *The third approach* uses laboratory models of microbial associations, whereby pure cultures are combined in the

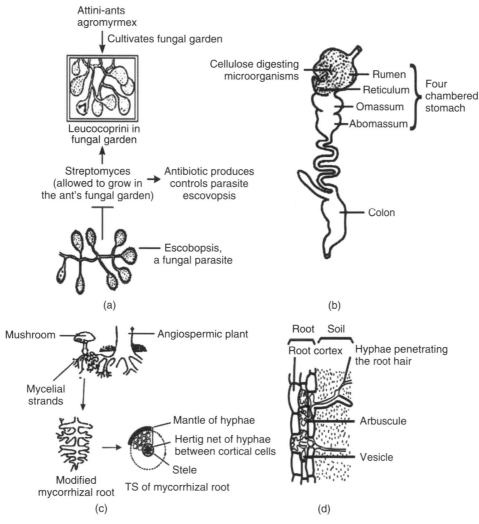

Figure 1.29. Microbial association: (a) schematic diagram showing the use of antibiotic-producing *Streptomyces* by attini ants to control the growth of fungal parasite in their fungal garden; (b) alimentary canal of ruminants; (c) ectotrophic mycorrhiza; (d) VA mycorrhiza.

laboratory under conditions that imitate natural habitat but can be controlled and analyzed. This approach is an excellent method to study interactions between two organisms under varying environment (Weindling and Ingols, 1956).

There are several types of microbial interactions, such as commensalism, inhibition, food competition, predation, parasitism, and synergism, which either singly or in combination may influence the functioning of the microbial ecology. To understand interactions, it is necessary to perform a detailed study of the physiology of the individual predominating microorganism to establish their requirements with respect to such environmental factors as nutrients, temperature, pH, oxidation–reduction potential, removal of waste products, or toxic materials, which may be involved in control processes, and to determine how these factors affect their capabilities. The sum total of this information will indicate the possible interactions between the microorganisms and will form the basis for conducting experiments either in the laboratory or with mathematical models. Such experiments will lead to an understanding of microbial activities and to the formulation of control measures, often using an alteration of the environmental factors for regulation of the microbial ecologies. Extensive research remains to be done on the microbial interactions to obtain the desired and precise control of these ecological processes (Gall, 1970).

1.3.10. Probiotic Microbes

Probiotics are live microorganisms that confer benefit to the host. They can ameliorate or prevent diseases including antibiotic-associated diarrhea, irritable bowel syndrome, and inflammatory bowel disease. Probiotics are likely to function through enhancement of barrier function, immunomodulation, and competitive adherence to the mucus and epithelium (Ohland and Macnaughton, 2010). Probiotics are bacteria that we eat and they are good for our health. They are found in a number of foods that are readily available in the supermarket, and they taste good. Since probiotic microbes do not cause diseases, there is no such thing as having too much of them. Probiotics are, in most cases, bacteria that are similar to beneficial microorganisms found in the human gut. They are also called *friendly bacteria* or *good bacteria*.

Probiotic bacteria are beneficial bacteria living in the human gut that are now widely used as food additives for their health-promoting effects (Anukam, 2007). These bacteria have coevolved with their human host over millions of years. Their contributions to health and to the development of the host's immune system depend on an intricate web of bacteria–bacteria and bacteria–host relationships that if thrown out of balance will most likely result in a disease.

Probiotics are available in foods and dietary supplements (for example, capsules, tablets, and powders) and in some other forms as well. Examples of foods containing probiotics are yogurt, fermented and unfermented milk, miso, tempeh, and some juices and soy beverages. In probiotic foods and supplements, the bacteria may have been present originally or added during preparation.

Most often, the bacteria come from two groups, *Lactobacillus* or *Bifidobacterium*. Within each group, there are different species (for example, *Lactobacillus acidophilus* and *Bifidobacterium bifidus*), and within each species, different strains (or varieties). A schematic diagram showing the various functions of *Bifidobacterium* are shown in Figure 1.30. A few common probiotics, such as *Saccharomyces boulardii*, are yeasts, which are different from bacteria.

Some probiotic foods date back to ancient times, such as fermented foods and cultured milk products. Interest in probiotics, in general, has been growing. There are several reasons that people are interested in probiotics for health purposes. Researchers are exploring whether probiotics could halt unfriendly microorganism in the first place and/or suppress their growth and activity in conditions such as infectious diarrhea, irritable bowel syndrome, inflammatory bowel disease (e.g., ulcerative colitis and Crohn's disease), infection with *H. pylori* (a bacterium that causes most ulcers and many types of chronic stomach inflammation), tooth decay and periodontal disease, vaginal infections, stomach and respiratory infections that children acquire in day care, and skin infections. Another part of interest in probiotics stems from the fact that there are cells in the digestive tract connected with the immune system (Fig. 1.32).

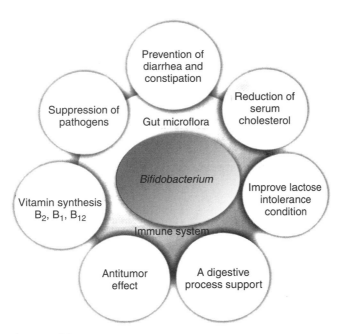

Figure 1.30. Various functions of *Bifidobacterium* as probiotics.

The scientific understanding of probiotics and their potential for preventing and treating diseased conditions are at an early stage, but moving ahead. In general, probiotic side effects, if they occur, tend to be mild and digestive (such as gas or bloating), but in some people more serious effects have also been seen. Probiotics might theoretically cause infections that need to be treated with antibiotics, especially in people with underlying health conditions. They could also cause unhealthy metabolic activities, too much stimulation of the immune system, or gene transfer.

The use of probiotics for prevention and/or treatment of gastrointestinal maladies are becoming increasingly more common within the health-care sector than ever before. The commercial and public interest in live microbes taken as food supplements is equally having a paradigm shift for urogenital health benefits (Kingsley, 2007).

1.4. FUTURE CHALLENGES: METAGENOMICS

Metagenomics enables the DNA from all microbes to be sequenced at once, without any culturing. Such an approach was impossible even a decade ago. As the use of metagenomics has become increasingly common, scientists have had to address the challenge of analyzing an enormous number of genomic sequences (Handelsman, 2004).

Comparing such a huge number of metagenomes is an enormous computational task. But, with metagenomics, job seems shortened. This automated technology revolutionizes the steps needed to acquire an accurately annotated genome. The database allows an overview of the microbial communities and the ability to focus on one metabolic area and detect differences in the proteins being used by the microbes in each environment (Fig. 1.32).

Probiotic benefits

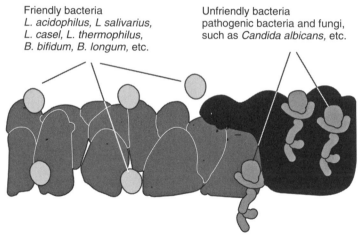

Acidophilus and other
probiotic bacteria
secrete:
Antiviral
antibacterial and
antifungal chemicals

Probiotics form a
physical barrier to
hinder invasion of
bacteria and yeasts

Probiotics like
acidophilus create
an acidic
microenvironment
which promotes
iron and other
mineral absorption

Friendly bacteria
L. acidophilus, L salivarius,
L. casel, L. thermophilus,
B. bifidum, B. longum, etc.

Unfriendly bacteria
pathogenic bacteria and fungi,
such as *Candida albicans,* etc.

Figure 1.31. Intestinal association of probiotics and their effects.

Comparative metagenomics is a technique that characterizes the DNA content of whole communities of organisms rather than individual species. Statistical analysis of the frequency distribution of 14,585,213 microbial and viral sequences explained the functional potential of nine biomes. In contrast to researchers expectation to find similar behaviors among the metagenomes in every environment, they have distinctive metabolic profiles.

This discovery could lead to innovations in curing viral or bacterial diseases, as well as help to develop new methods of environmental conservation. Metagenomics is a discipline that enables the genomic study of uncultured microorganisms. This evidence was derived from analyses of 16S rRNA gene sequences amplified directly from the environment, an approach that avoided the bias imposed by culturing and led to the discovery

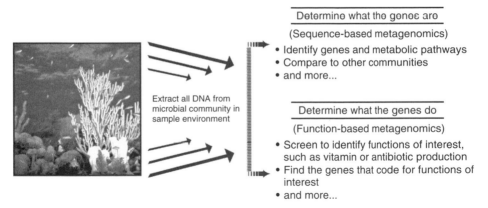

Determine what the genes are

(Sequence-based metagenomics)
• Identify genes and metabolic pathways
• Compare to other communities
• and more...

Extract all DNA from
microbial community in
sample environment

Determine what the genes do

(Function-based metagenomics)
• Screen to identify functions of interest,
  such as vitamin or antibiotic production
• Find the genes that code for functions of
  interest
• and more...

Figure 1.32. The metagenomic process.

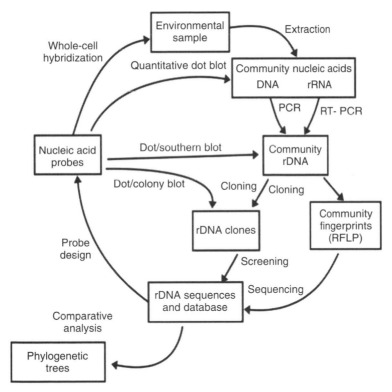

Figure 1.33. Strategies based on rRNA sequences for characterizing microbial communities without cultivation.

of vast new lineages of microbial life (Fig. 1.33). Although the portrait of the microbial world was revolutionized by the analysis of 16S rRNA genes, such studies yielded only a phylogenetic description of community membership, providing little insight into the genetics, physiology, and biochemistry of the members (Handelsman, 2004). Faster, cheaper sequencing technologies and the ability to sequence uncultured microbes sampled

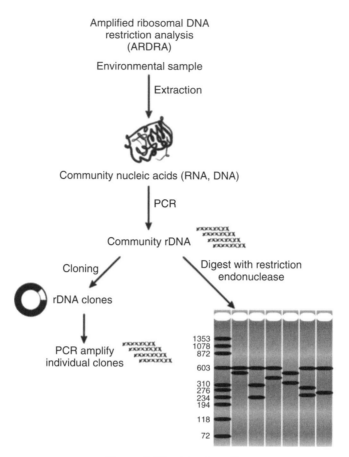

Figure 1.33. (*Continued*).

directly from their habitats are expanding and transforming our view of the microbial world. Distilling meaningful information from the millions of new genomic sequences presents a serious challenge to bioinformaticians. In cultured microbes, the genomic data come from a single clone, making sequence assembly and annotation tractable. In metagenomics, the data come from heterogeneous microbial communities, sometimes containing more than 10,000 species, with the sequence data being noisy and partial. From sampling, to assembly, to gene calling and function prediction, bioinformatics faces new demands in interpreting voluminous, noisy, and often partial sequence data. Although metagenomics is a relative newcomer to science, the past few years have seen an explosion in computational methods applied to metagenomic-based research.

REFERENCES

Alibek K. 2004. Smallpox: a disease and a weapon. *Int J Infect Dis* **2**, S3–S8.

Antonov VA, Iliukhin VI. 2005. Molecular-genetic approaches to diagnosis and intraspecific typing of causative agents of glanders and melioidosis. *Mol Gen Mikrobiol Virusol* **2**, 3–9.

Anukam KC. 2007. From gut to urogenital tract: probiotic-microbes descending and ascending. AgroFood Industry hi-tech. *Eur J Nut Func Foods* **18**, 10–13.

Appel JM. 2009. Is all fair in biological warfare? The controversy over genetically engineered biological weapons. *J Med Ethics* **35**, 429–432.

Berg R. 1996. The indigenous gastrointestinal microflora. *Trends Microbiol* **4**, 430–435.

Bharali JD, Khalil M, Gurbuz M, Simone TM, Mousa SA. 2009. Nanoparticles and cancer therapy: a concise review with emphasis on dendrimers. *Int J Nanomed* **4**, 1–7.

Bogue RW. 2003. Biosensors for monitoring the environment. *Sens Rev* **23**, 302–310.

Boissinot M, Bergeron MG. 2002. Toward rapid real-time molecular diagnostic to guide smart use of antimicrobials. *Curr Opin Microbiol* **5**, 478–482.

Bottinger EP. 2007. Foundations, promises, and uncertainties of personalized medicine. *Mt Sinai J Med* **74**, 15–21.

Brogna S. 2001. Pre-mRNA processing: insights from nonsense. *Curr Biol* **11**, R838–R841.

Campanella L, Cubadda P, Sammartino MP, Saoncella A. 2000. An algal biosensor for the monitoring of water toxicity in estuarine environments. *Water Res* **35**, 69–76.

Carter DJ, Cary RB. 2007. Lateral flow microarrays: a novel platform for rapid nucleic acid detection based on miniaturized lateral flow chromatography. *Nucleic Acids Res* **35**, e74.

di Castri F, Younes T. 1990. Ecosystem function of biological diversity. *Biol Int* **22** (Special Issue), 1–20.

Chinnery PF, Howell N, Andrews RM, Turnbull DM. 1999. Clinical mitochondrial genetics. *J Med Genet* **36**, 425–436.

Corliss JO. 1991. Introduction to the protozoa. In: Harrison FW, Corliss JJ, eds., *Microscopic Anatomy of Invertebrates*. Vol. 1, pp. 1–12. Wiley–Liss, New York.

Cormican MG, Pfaller MA. 1996. Molecular pathology of infectious diseases. In: Henry JB, ed., *Clinical Diagnosis and Management by Laboratory Methods*, 19th ed., pp. 1390–1399. W.B. Saunders Company, Philadelphia.

Delong EF. 1992. Archaea in coastal marine environments. *Proc Nat Acad Sci USA* **89**, 5685–5689.

Dietel M, Sers C. 2006. Personalized medicine and development of targeted therapies: the upcoming challenge for diagnostic molecular pathology. *A Rev Virchows Arch* **448**, 744–755.

Dimijian GG. 2000. Pathogens and parasites: strategies and challenges. *BUMC Proc* **13**, 19–29.

Esser C, Ahmadinejad N, Wiegand C, Rotte C, Sebastiani F, et al., 2004. A genome phylogeny for mitochondria among α-proteobacteria and a predominantly eubacterial ancestry of yeast nuclear genes. *Mol Biol Evol* **21**, 1643–1660.

Fredricks DN, Relman DA. 1996. Sequence-based identification of microbial pathogens: a reconsideration of Koch's postulates. *Clin Microbiol Rev* **9**, 18–33.

Fredricks DN, Relman DA. 1999. Application of polymerase chain reaction to the diagnosis of infectious disease. *Clin Infect Dis* **29**, 475–488.

Freitas RA Jr. 2005. Microbivores: artificial mechanical phagocytes using digest and discharge protocol. *J Evol Technol* **14**, 55–106..

Gall LS. 1970. Significance of microbial interactions in control of microbial ecosystems. *Biotechnol Bioeng* **12**, 333–340.

Garnier L, Gaudin JC, Bensadoun P, Rebillat I, Morel Y. 2009. Real-time PCR assay for detection of a new simulant for poxvirus biothreat agents. *Appl Environ Microbiol* **75**, 1614–1620.

Gerard DW, Sutherland AD. 2007. New strategies for combating multidrug-resistant bacteria. *Trends Mol Med* **13**, 260–267.

Giovannoni SJ, Britschgi TB, Moyer CL, Field KG. 1990. Genetic diversity in Sargasso Sea bacterioplankton. *Nature* **345**, 60–63.

Giovannoni SJ, Tripp HJ, Givan S, Podar M, Vergin KL, et al., 2005. Genome streamlining in a cosmopolitan oceanic bacterium. *Science* **309**, 1242–1245.

Guillemin J. 2005. Biological weapons and secrecy (WC 2300). *FASEB J* **19**, 1763–1765.

Guillemin J. 2006. Scientists and the history of biological weapons. A brief historical overview of the development of biological weapons in the twentieth century. *EMBO Rep* **7**, Spec No S45–49.

Hall RH. 2002. Biosensor technologies for detecting microbiological food borne hazards. *Microbes Infect* **4**, 425–432.

Handelsman J. 2004. Metagenomics: application of genomics to uncultured microorganisms. *Microbiol Mol Biol Rev* **68**, 669–685.

Hasnain SE. 2001. Impact of human genome sequencing on microbiology. *Ind J Med Microbiol* **19**, 114–115.

Hawksworth DI. 1991. The Fungal dimension of bio-diversity: magnitude, significance and conservation. *Mycol Res* **95**, 641–655.

Hertz-Fowler C, Berriman M, Pain A. 2005. Genome Watch: a feast of protozoan genomes. *Nat Rev Microbiol* **3**, 670–671.

Heymann DL. 2006. Resistance to anti-infective drugs and the threat to public health. *Cell* **124**, 671–675.

Hirsch LR, Jackson JB, Lee A, Halas NJ, West JL. 2003. A whole blood immunoassay using gold nanoshells. *Anal Chem* **75**, 2377–2381.

Hsiao WW, Fraser-Liggett CM. 2009. Human microbiome project–paving the way to a better understanding of ourselves and our microbes. *Drug Discov Today* **14**, 331–333.

Kalow W. 2006. Pharmacogenetics and pharmacogenomics: origin, status, and the hope for personalized medicine. *Pharm J* **6**, 162–165.

Kant JA. 1995. Molecular diagnostics: reimbursement and other selected financial issues. *Diagn Mol Pathol* **4**, 79–81.

Karwa M, Currie B, Kvetan V. 2005. Bioterrorism: preparing for the impossible or the improbable. *Crit Care Med* **33**, S75–S95.

Kingsley CA. 2007. The potential role of probiotics in reducing poverty-associated infections in developing countries. *J Infect Dev Ctries* **1**, 81–83.

Kohlmeier S, Mancuso M, Tecon R, Harms H, van den Meer JR, Wells M. 2006. Bioreporters: gfp versus lux revisited and single-cell response. *Biosens Bioelectron* **22**, 1578–1585.

Lesko LJ. 2007. Personalized medicine: elusive dream or imminent reality? *Clin Pharmacol Ther* **81**, 807–816.

Lim DV, Simpson JM, Kearns EA, Kramer MF. 2005. Current and developing technologies for monitoring agents of bioterrorism and biowarfare. *Clin Microbiol Rev* **18**, 583–607.

Lutolf MP, Hubbell JA. 2005. Synthetic biomaterials as instructive extracellular microenvironments for morphogenesis in tissue engineering. *Nat Biotechnol* **23**, 47–55.

Mancinelli L, Cronin M, Sadee W. 2000. Pharmacogenomics: the promise of personalized medicine. *AAPS PharmSci* **2**, E4 article 4. doi: 10.1208/ ps020104.

Marinšek Logar R, Vodovnik M. 2007. The applications of microbes in environmental monitoring In: Méndez-Vilas A, ed., *Current Research and Educational Topics and Trends in Applied Microbiology*, 577–585. Formatex Research Center, Badajoz, Spain.

Mitchell RJ, Gu MB. 2004. An Escherichia coli biosensor capable of detecting both genotoxic and oxidative damage. *Appl Microbiol Biotechnol* **64**, 46–52.

Moreira D, Lopez-Garcia P. 1998. Symbiosis between methanogenic Archaea and ∂-proteobacteria as the origin of eukaryotes: the synthrophy hypothesis. *J Mol Evol* **47**, 517–530.

Ohland CL, Macnaughton WK. 2010. Probiotic bacteria and intestinal epithelial barrier function. *Am J Physiol Gastrointest Liver Physiol* **298**, G807–G819.

Ou H, Ju C, Thong K, Ahmad N, Deng Z, Barer M, Rajakumar K. 2007. Translational genomics to develop a *Salmonella enterica* serovar Paratyphi A multiplex PCR assay: a rapid generic strategy for preexisting and emerging pathogens. *J Mol Diagn* **9**, 624–630.

Park JW. 2002. Liposome-based drug delivery in breast cancer treatment. *Breast Cancer Res* **4**, 95–99.

Partha R, Conyers LJ. 2009. Biomedical applications of functionalized fullerene-based nanomaterials. *Int J Nanomed* **4**, 261–275.

Pfaller MA. 2000. Diagnosis and management of infectious diseases: molecular methods for the new millennium. *Clin Lab News* **26**, 10–13.

Pfaller MA. 2001. Molecular approaches to diagnosing and managing infectious diseases: practicality and costs. *Emerg Infect Dis* **7**, 312–318.

Piotrovsky LB. 2010. Nanomedicine as a form of nanotechnologies. *Vestnik Rossiiskoi Akademii Meditsinskikh Nauk* **3**, 41–46.

Pohanka M, Kuca K. 2010. Biological warfare agents. *EXS* **100**, 559–578.

Relman DA, Falkow S. 2001. The meaning and impact of the human genome sequence for microbiology. *Trends Microbiol* **9**, 206–208.

Rhodes SG, De Leij FAAM, Dale JW. 2007. Protozoa as an environmental reservoir of bovine tuberculosis. *Trends Microbiol* **15**, 338–339.

Rodriguez-Mozaz S, Marco MP, Lopez de Alda MJ, Barceló D. 2004. Biosensors for environmental monitoring of endocrine disruptors: a review article. *Anal Bioanal Chem* **378**, 588–598.

Saha M. 2009. Nanomedicine: promising tiny machine for the healthcare in future-a review. *OMJ* **24**, 242–247.

Salvage DC. 1977. Microbial ecology of the gastrointestinal tract. *Annu Rev Microbiol* **31**, 107–133.

Schmoock G, Ehricht R, Melzer F, Rassbach A, Scholz CH, et al., 2009. DNA microarray-based detection and identification of *Burkholderia mallei, Burkholderia pseudomallei and Burkholderia spp*. *Mol Cell Probes* **23**, 178–187.

Seeley TD, Nowicke JW, Meselson M, Guillemin J, Akratanakul P. 1985. Yellow rain. *Sci Am* **253**, 122–131.

Tang YW, Persing DH. 1999. Molecular detection and identification of microorganisms. In: Murray PR, Baron EJ, Pfaller MA, Tenover FC, Yolken RH, eds., *Manual of Clinical Microbiology*, 7th ed., pp. 215–244. American Society for Microbiology, Washington.

Taylor CJ, Bain LA, Richardson DJ, Spiro S, Russell DA. 2004. Construction of a whole-cell gene reporter for the fluorescent bioassay of nitrate. *Anal Biochem* **328**, 60–66.

Tepper N, Shlomi T. 2010. Predicting metabolic engineering knockout strategies for chemical production: accounting for competing pathways. *Bioinformatics* **26**, 536–543.

Thorek DL, Chen AK, Czupryna J, Tsourkas A. 2006. Super paramagnetic iron oxide nanoparticle probes for molecular imaging. *Ann Biomed Eng* **34**, 23–38.

Toke ER, Lorincz O, Somogyi E, Lisziewicz J. 2010. Rational development of a stable liquid formulation for nanomedicine products. *Int J Pharm* **392**, 261–267.

Tsai H, Doong R. 2005. Simultaneous determination of pH, urea, acetylcholine and heavy metals using array-based enzymatic optical biosensor. *Biosens Bioelectron* **20**, 1796–1804.

Turnbaugh PJ, Ley RE, Hamady M, Fraser-Liggett CM, Rob Knight R, Gordon JI. 2007. The human microbiome project. *Nature* **449**, 804–810.

Wächtershäuser G. 2006. From volcanic origins of chemoautotrophic life to bacteria, archaea and eukarya. *Philos Trans R Soc Lond B Biol Sci* **361**, 1787–1808.

Weindling R, Ingols R. 1956. Microbial associations and antagonisms. *Ind Eng Chem* **48**, 1407–1410.

Wenyong M, Zhang W, James Versalovic J. 2007. Expanding the diagnostic capabilities of molecular microbiology by Genomic Methods. *J Mol Diagn* **9**, 572–573.

West M, Ginsburg GS, Huang AT, Nevins JR. 2006. Embracing the complexity of genomic data for personalized medicine. *Genome Res* **16**, 559–566.

Whitman WB, Coleman DC, Wiebe WJ. 1998. Prokaryotes: the unseen majority. *Proc Natl Acad Sci USA* **95**, 6578–6583.

Whittaker RH. 1959. On the broad classification of organisms. *Quart Rev Biol* **34**, 210–226.

Woese CR, Fox GE. 1977. Phylogenetic structure of the prokaryotic domain: the primary kingdoms. *Proc Natl Acad Sci USA* **74**, 5088–5090.

Woods GL. 2001. Molecular techniques in mycobacterial detection. *Arch Pathol Lab Med* **125**, 122–126.

Wooley JC, Godzik A, Friedberg I. 2010. A primer on metagenomics. *PLoS Comput Biol* **6**, e1000667. doi:10.1371/journal.pcbi.1000667.

Wright GD, Sutherland AD. 2007. New strategies for combating multidrug-resistant bacteria. *Trends Mol Med* **13** (6), 260–267.

2

GENE TECHNOLOGY: APPLICATIONS AND TECHNIQUES

2.1. PROLOGUE

Over the past decades, the development of new and powerful techniques for studying and manipulating DNA has revolutionized genetics. These techniques have allowed biologists to intervene directly in the genetic fate of organisms for the first time. In this chapter, we explore these technologies and consider how they are applied to address specific problems of great practical importance. Few areas of biology will have as great an impact on our future lives. Modern molecular medicine allows the use of many molecular biological techniques in the analysis of disease, disease genes, and disease gene function. The term *gene cloning* covers a range of techniques that make it possible to manipulate DNA in a test tube and also to return it to living organisms where it functions normally.

2.2. INTRODUCTION TO GENE TECHNOLOGY

The study of disease genes and their function in an unaffected individual has been made possible by the development of recombinant DNA (rDNA, Fig. 2.1) and cloning techniques. rDNA technology has now made it possible to produce proteins for pharmaceutical applications (Manning et al., 1989). This has been successfully accomplished in microbial cells such as bacteria and yeasts. In the early 1980s, the FDA (Food and Drug Administration) approved the clinical use of recombinant human insulin from

Microbes: Concepts and Applications, First Edition. Prakash S. Bisen, Mousumi Debnath, Godavarthi B. K. S. Prasad
© 2012 Wiley-Blackwell. Published 2012 by John Wiley & Sons, Inc.

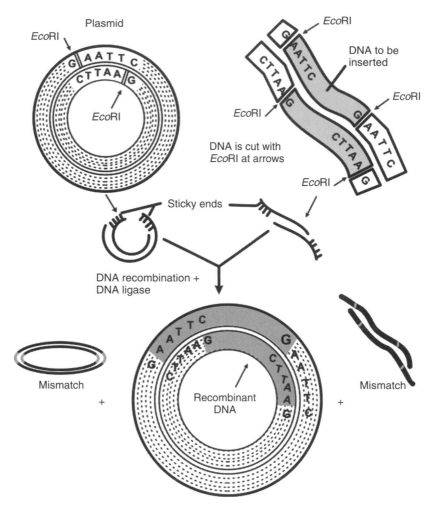

Figure 2.1. Recombinant DNA technology.

recombinant *Escherichia coli* (Humulin-US/Humuline-EU) for the treatment of diabetes, being the first recombinant pharmaceutical to enter the market. A total of 151 recombinant pharmaceuticals have so far been approved for human use by the FDA and/or by the European Medicines Agency (EMEA) (Ferrer-Miralles et al., 2009). The versatility and scaling up possibilities of the recombinant protein production opened up new commercial opportunities for pharmaceutical companies (Loumaye et al., 1995). Since the approval of recombinant insulin, other rDNA drugs have been marketed in parallel with the development and improvement of several heterologous protein production systems. This has generated specific strains of many microbial species adapted to protein production, and has allowed the progressive incorporation of yeasts and eukaryotic systems for this purpose. Among the approved protein-based recombinant pharmaceuticals licensed up to January 2009 by the FDA and EMEA, 45 (29.8%) are obtained from *E. coli*, 28 (18.5%) from *Saccharomyces cerevisiae*, 17 (11.2%) from hybridoma cells, 1 from transgenic goat milk, 1 from insect cells, and 59 (39%) from mammalian cells.

2.2.1. Genes and Bacteria

The ability of bacteria to mediate gene transfer has only recently been established and these observations have led to the use of various bacterial strains in gene therapy (Vassaux et al., 2006). Recent advances in gene therapy can be attributed to improvements in gene delivery vectors. New viral and nonviral transport vehicles that considerably increase the efficiency of transfection have been prepared. However, these vectors still have many disadvantages that are difficult to overcome; thus, a new approach is needed. The approach of bacterial delivery could in the future be important for gene therapy applications (Fig. 2.2; Pálffy et al., 2006). The types of bacteria used include attenuated strains of *Salmonella, Shigella, Listeria*, and *Yersinia*, as well as nonpathogenic *E. coli*. For some of these vectors, the mechanism of DNA transfer from the bacteria to the mammalian cell (Grillot-Courvalin et al., 1998) is not yet fully understood but their potential to deliver therapeutic molecules has been demonstrated *in vitro* and *in vivo* in experimental models.

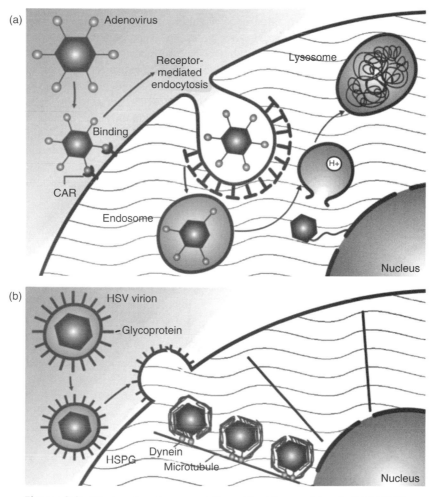

Figure 2.2. Viral vectors for gene delivery (Davidson and Breakefield, 2003).

2.2.2. Reporter Genes

Reporter genes have become an invaluable tool in the study of gene expression. They represent a gene whose phenotypic expression is easy to monitor and used to study promoter activity in different tissues and developmental stages. Most reporter genes are placed downstream to the promoter region, but close to the gene under study. This ensures that these genes are expressed together, and are not separated during cell division by crossover events.

rDNA constructs are made in which the reporter gene is attached to a promoter region of particular interest and the construct is transfected into a cell or an organism.

A reporter gene enables to track and study another gene in cell cultures, animals, and plants. Because most gene therapy techniques only work on a small number of individuals, there is a need to use a reporter gene to identify which cells have taken up the gene that is currently under study, and which have incorporated it into their chromosomes. They are widely used in biomedical and pharmaceutical research and also in molecular biology and biochemistry. The purpose of the reporter gene assay (Fig. 2.3) is to measure the regulatory potential of an unknown DNA sequence. This can be done by linking a promoter sequence to an easily detectable reporter gene such as that encoding for the firefly luciferase. Common reporter genes are β-galactosidase, β-glucuronidase, and luciferase. Various detection methods are used to measure expressed reporter gene protein. These include luminescence, absorbance, and fluorescence.

A commonly used reporter gene is the one that encodes green fluorescent proteins in jellyfish, which cause cells that have taken it up and expressed it to glow green under an ultraviolet lamp. Luciferase enzyme genes are also commonly used as reporter genes, causing the cell that expresses it to catalyze luciferins and produce light without external interference. Sometimes, reporter genes are simply placed in a vector independent of a chromosome, and other techniques are used to identify the gene under study. The availability of new bioluminescent proteins, obtained by cDNA cloning and mutagenesis

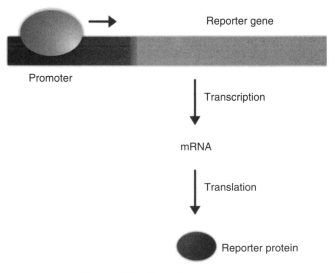

Figure 2.3. Reporter gene assay.

of wild type genes, expanded the applicability of these reporters from the perspective of using more proteins emitting at different wavelengths in the same cell-based assay. By spectrally resolving the light emitted by different reporter proteins, it is, in fact, possible to simultaneously monitor multiple targets. A new luciferase isolated from *Luciola italica* has been recently cloned, and thermostable red and green emitting mutants were obtained by random and site-directed mutagenesis. Different combinations of luciferases were used *in vitro* as purified proteins and expressed in bacterial and mammalian cells to test their suitability for multicolor assays. A mammalian triple color reporter model system was then developed using a green emitting wild-type *Photinus pyralis* luciferase, a red thermostable mutant of *L. italica* luciferase, and a secreted *Gaussia princeps* luciferase (GLuc) to monitor the two main pathways of bile acid biosynthesis. The two firefly luciferases were used to monitor cholesterol 7-α-hydroxylase and sterol 27-hydroxylase, while secreted constitutively expressed GLuc was used as an internal vitality control. By treating the cells with chenodeoxycholic acid, it was possible to obtain dose-dependent inhibitions of the two specific signals together with a constant production of GLuc, which allowed for a dynamic evaluation of the metabolic activity of the cells. This approach is suitable for high content analysis of gene transcription in living cells to shorten the time for screening assays, increasing throughput and cost-effectiveness. Reporter gene technology is widely used to monitor the cellular events associated with signal transduction and gene expression. On the basis of the splicing of transcriptional control elements to a variety of reporter genes (with easily measurable phenotypes), it "reports" the effects of a cascade of signaling events on gene expression inside cells. The principal advantage of these assays is their high sensitivity, reliability, convenience, and adaptability to large-scale measurements. With the advances in this technology and in detection methods, it is likely that luciferase and green fluorescent protein will become increasingly popular for the non-invasive monitoring of gene expression in living tissues and cells. Such techniques will be important in defining the molecular events associated with gene transcription, which has implications for our understanding of the molecular basis of disease and will influence our approach to gene therapy and drug development. The future appears to be promising for the continued expansion of the use of reporter genes in the many evolving biomedical related arenas (Min and Gambhir, 2008).

2.2.3. Recombinant DNA Pharmaceuticals

rDNA technology has now made it possible to produce proteins for pharmaceutical application (Manning et al., 1989). Consequently, proteins produced via biotechnology now comprise a significant portion of the drugs currently under development. Successful development of rDNA-derived pharmaceuticals, a new class of therapeutic agents, is determined by a variety of factors affecting the selection and positioning of the compound under development. High quality recombinant products covering monoclonal antibodies, vaccines thrombolytics, and so on, have been developed by these techniques and more are expected to join in (Table 2.1).

Techniques for gene transfer have been improved to introduce synthetic sequences into the cellular genomes. Candidate cells and culture systems have been screened for their capacity to synthesize the proper recombinant molecules. rDNA technologies now offer a very potent set of technical platforms for the controlled and scalable production of polypeptides of interest by relatively inexpensive procedures. When massively expressed

TABLE 2.1. Top Innovative Biotech Companies

Company	Innovation
Novartis	H1N1 vaccines based on cells—faster and more reliable than traditional method of growing viruses in chicken eggs
Synthetic Genomics	Teamed up with Exxon Mobil to develop microbes that produce biofuel. Exxon's $600 million stake signals that algae fuel is going mainstream
Cytori Therapeutics	Its Celution system harvests adult stem cells with maximum efficiency, and its regenerative therapies have shown promise in trials
Roche/Genentech	Roche acquired biotech giant Genentech for $46.8 billion in 2009. Genentech has three of the five best-selling biotech drugs and widely licensed patents on DNA sequencing technology
Human Genome Sciences	Human Genome Sciences announced that its lupus drug, Benlysta, had produced substantial improvements in clinical-trial participants, a conspicuous success since so many lupus treatments have flopped
Osiris Therapeutics	Like Cytori, Osiris is at the vanguard of the stem-cell revolution. According to Osiris, its marquee drug, Prochymal, a preparation of adult stem cells from connective tissue, may work for ailments as varied as diabetes and graft-versus-host disease.
Amyris Biotechnologies	Amyris is engineering yeast molecules to churn out hydrocarbon-based biofuels.
Biogen Idec	Biogen has 20 drugs in Phase II clinical trials or beyond
Novavax	Novavax develops vaccines using proprietary virus-like particles (VLPs) rather than live viruses. The particles trigger the same immune response as a virus but cannot cause infection. In December 2009, Novavax released the results of Phase II clinical trials of a VLP vaccine that proved effective against three kinds of flu, including H1N1
Regeneron Pharmaceuticals	Regeneron's pipeline of human antibody products, now in clinical trials, treats rheumatoid arthritis, gout, chronic pain, and cancer. Sanofi-Aventis has promised Regeneron an additional $1 billion over seven years to broaden its drug development efforts

Source: FastCompany, *Top 10 Most Innovative Companies by Industry*, Feb. 2010.

in a host such as bacteria, recombinant proteins often tend to misfold and accumulate as soluble and insoluble aggregates (de Marco et al., 2005).

After more than half a century of treating diabetics with animal insulin, rDNA technologies and advanced protein chemistry could succeed in human insulin preparations. Human insulin produced by rDNA technology is the first commercial health care product derived from this technology (Johnson, 1983). Large quantities of biosynthetic human proinsulin have become available through rDNA technology (Revers et al., 1984). A chimeric bovine growth hormone (GH, amino acids Met-Asp-Gln- >1–23) and human growth hormone (hGH) (amino acids 24–191) plasmids were constructed and expressed in *E. coli* (Binder et al., 1989). Human follicle stimulating hormone (FSH) (Loumaye et al., 1995) and tissue-type plasminogen activator (Tiefenbrunn et al., 1985) have now been produced *in vitro* by rDNA technology.

Approved therapeutic-protein-based products from *E. coli* include hormones (human insulin and insulin analogs, calcitonin, parathyroid hormone, human growth hormone, glucagons, somatropin, and insulin growth factor 1), interferons (α-1, α-2a, α-2b, and γ-1b), interleukins 11 and 2, light and heavy chains raised against vascular endothelial growth factor A, tumor necrosis factor α, cholera B subunit protein, B-type natriuretic peptide, granulocyte colony stimulating factor, and plasminogen activator. It is noteworthy that most of the recombinant pharmaceuticals produced in *E. coli* are addressed for the treatment of infectious diseases or endocrine, nutritional, and metabolic disorder disease groups (Werner and Pommer, 1990; Redwan et al., 2008). The approved protein products produced in yeast are obtained exclusively from *S. cerevisiae* and correspond to hormones (insulin, insulin analogs, nonglycosylated hGH somatotropin, glucagon), vaccines (hepatitis B virus surface antigen in the formulation of 15 out of the 28 yeast-derived products) and VLPs of the major capsid protein L1 of human papillomavirus (HPV) type 6, 11, 16, and 18, urate oxidase from *Aspergillus flavus*, granulocyte–macrophage colony stimulating factor, albumin, hirudin of *Hirudo medicinalis*, and human platelet-derived growth factor. As in the case of *E. coli*, most of the recombinant pharmaceuticals from yeast address either infectious diseases or endocrine, nutritional, and metabolic disorders, these therapeutic areas being the most covered by microbial products. Interestingly, several yeast species, other than *S. cerevisiae*, are explored as sources of biopharmaceuticals and other proteins of biomedical interest (Porro et al., 2005). In addition, current metabolic engineering approaches and optimization of process procedures (Graf et al., 2008; Mattanovich et al., 2004) are dramatically expanding the potential of yeast species for the improved production of recombinant proteins.

The top 10 innovations of the biopharmaceutical are given in Table 2.1. A large number of products under various categories are produced by these companies. There is only one approved biopharmaceutical product that contains recombinant proteins from infected insect cell line Hi Five, Cervarix, which consists of recombinant papillomavirus C-terminal truncated major capsid protein L1 types 16 and 18. Nonetheless, this expression system has been extensively used in structural studies, since correctly folded eukaryotic proteins can be obtained in a secreted form in serum-free media (SFM), which enormously simplifies protein capture in purification protocols. Genetic engineering has been applied to obtain humanized monoclonal antibodies using either recombinant mammalian cells producing chimeric antibodies or genetically modified mice to produce human-like antibodies. One such product, Remicade, which binds tumor necrosis factor-α, is a pharmaceutical blockbuster used in the treatment of Crohn's disease (Table 2.2). Most of the therapeutic proteins approved so far have been obtained using transgenic hamster cell lines, namely, 49 from Chinese hamster ovary (CHO) cells and 1 from baby hamster kidney (BHK) cells. The main advantage of this expression system is that cells can be adapted to grow in suspension in SFM, protein-free, and chemically defined media. This fact increases the biosafety of final products reducing the risk of introducing prions of bovine spongiform encephalopathy (BSE) from bovine serum albumin and of infectious variant Creutzfeldt Jakob disease (vCJD) from human serum albumin. In addition, recombinant products can be secreted into the chemically defined media, which simplifies both upstream and downstream purification process.

TABLE 2.2. Top Recombinant Proteins and Projected Sales by 2014
(http://www.newinnovationsguide.com/BiotechCompanies.html)

Rank	Product	Company	Therapeutic Subcategory	Technology	Worldwide (WW) Sales ($m)
1.	Avastin	Roche	Antineoplastic MAbs	MAb	9.232
2.	Humira	Abbott + Eisai	Other antirheumatics	MAb	9.134
3.	Rituxan	Roche	Antineoplastic MAbs	MAb	7.815
4.	Enbrel	Wyeth+Amgen+ Takeda	Other antirheumatics	Recombinant product	6583
5.	Kabtys	Sabifu-Aventis	Antineoplastic MAbs	MAb	5796
6.	Herceptin	Roche	Antineoplastic MAbs	MAb	5796
7.	Crestor	Astra Zeneca	Antihyperlipidaemics	Small-molecule chemistry	5739
8.	Spiriva	Boehringer lngelheim	Anticholinergics	Small-molecule chemistry	5552
9.	Remicade	SGP + J&J + Mitsubishi	Other antirheumatics	MAb	5520
10.	Gleevec/ Glivec	Novartis	Other cytostatics	Small-molecule chemistry	5520

Abbreviation: MAb, monoclonal antibody.

Recombinant structural proteins include hepatitis B virus vaccine and CD4 protein, and recombinant modifier proteins include tissue plasminogen activator and superoxide dismutase (agents that split or splice organic molecules). In the future, gene defects associated with genetic diseases will be unraveled, leading to the production of new therapeutic agents designed to counteract or actually reverse those defects. Recombinant protein drugs will be further tailored to enhance their activity and specificity. These advances are so novel and momentous that patent protection has been extended not only to recombinant drugs but also to the recombinant microorganisms in which they are manufactured. In cloning genes, investigators directly use the protein designs that occur naturally. To use rDNA technology to functionally analyze mutations introduced into cloned eukaryotic genes, a rapid procedure is necessary to assay the steps along the gene expression pathway (Lomedico, 1982). If one assays shortly after its introduction into mammalian cells, it can be shown that this recombinant plasmid programs the synthesis of correctly spliced and polyadenylylated insulin mRNA that functions in the synthesis and secretion of rat proinsulin. This system permits rapid analysis of cloned *in vitro* engineered mutations and the programming of eukaryotic cells to manufacture proteins that they normally do not synthesize. Basic research will soon lead to the engineering of novel proteins with specified functions (Drake and Holland, 1999).

Overcoming the biological and methodological obstacles posed by cell factories to the production of rDNA pharmaceuticals is the main challenge in the further development of protein-based molecular medicine. rDNA technologies might have exhausted conventional cell factories and new production systems need to be deeply explored and incorporated into the production pipeline. On the other hand, a more profound comprehension of host cell physiology and stress responses to protein production would necessarily offer improved tools (at genetic, metabolic, or system levels) to favor high yield and high quality protein production. Apart from the expected incorporation into unusual mammalian hosts such as transgenic animals or plants, microbial cells appear as extremely robust and convenient hosts, and gaining knowledge about the biological aspects of protein production would hopefully enhance the performance of such hosts beyond the current apparent limitations. In this regard, not only commonly used bacteria and yeasts but unconventional strains or species are observed as promising cell factories for forthcoming recombinant drugs. Their incorporation into productive processes for human pharmaceuticals would hopefully push the trend of marketed products and fulfill the increasing demands of the pharmacological industry.

Desirably, recent insights about system's biology of recombinant cells and hosts and, especially, about arising novel concepts on recombinant protein quality (Gonzalez-Montalban, 2007; de Marco, 2007; de Marco et al., 2005) and host stress responses would enlarge the possibilities for metabolic and process engineering aiming to the economically feasible production of new, more complex drugs. The fast advances in the field of molecular medicine are urgently demanding improved production systems and novel and cheaper drugs. Their incorporation into productive processes for human pharmaceuticals would hopefully push the trend of marketed products and fulfill the increasing demands of the pharmacological industry.

2.2.4. *In vitro* DNA Selection

Effective catalysis involves the specific recognition of a target molecule with high affinity. Although many catalytic functions have evolved naturally, mostly in proteins, the need for new catalytic functions is growing; for example, to convert industrial byproducts to useful products. To develop chemicals that recognize new targets such as industrial byproducts, combinatorial molecular biology is currently being used. Aptamers are such new products that form unique structures to meet the requirements for more selective binding and sensitive recognition modules for increasingly sensitive detection methods that are required in natural sciences and medicine for the detection and quantization of molecules.

Aptamers are short, single-stranded nucleic acid (DNA or RNA) oligomers that have been selected *in vitro* to bind with high affinity and specifically to a certain molecular target (Lou et al., 2009; Ellington and Szostak, 1990). They are 50–100 base pairs in length that bind to proteins with K_ds (equilibrium constant) in the range of 1 pM to 1 nM similar to monoclonal antibodies. The molecular target may be amino acids, drugs, proteins, nucleic acids, small organic compounds, and even entire organisms (Joshi et al., 2009; Mairal et al., 2008). The term *aptamer* is derived from Latin *aptus* (= fitting) and Greek *meros* (= part) and was chosen to illustrate the "lock and key model" between aptamers and their binding partners. Owing to the multitude of structures that can be formed by aptamers, they are able to recognize virtually all classes of substrates and bind them perfectly fitting in analogy to antigen antibody interactions. In such a way, aptamers were developed for atoms and small molecules, for amino acids, peptides, polysaccharides, and proteins, and for complex targets such as viruses and protozoa. The selected sequences have the ability to recognize specific ligands by forming binding pockets and can bind to nucleic acids (Le Tinevez et al., 1998; Pileur et al., 2003), proteins (Bock et al., 1992; Lupold et al., 2002) or small organic compounds (Berens et al., 2001; Burgstaller and Famulok, 1996). The aptamer–target interaction takes place via structural compatibilities of the two binding partners, via electrostatic interactions such as van der Waals interactions, ionic or dipole forces, hydrogen bonds, or via stacking interactions.

These nucleic acid ligands bind to nucleic acids and proteins. Aptamer recognition affinity and specificity is comparable to those of monoclonal antibodies. Aptamers can be selected to recognize and bind a wide range of targets, including toxic compounds and inherently nonimmunogenic molecules that antibodies cannot be raised against. Furthermore, as aptamers are 10–100 times smaller than antibodies, they are expected to achieve higher tumor penetration than their counterparts (Jayasena, 1999).

Aptamers have many potential uses in intracellular processes, medicine, and technology. In addition to the genetic information encoded by nucleic acids they also function as highly specific affinity ligands by molecular interaction based on the three-dimensional folding pattern. The three-dimensional complex shape of a single-stranded oligonucleotide is primarily due to the base-composition-led intramolecular hybridization that initiates folding to a particular molecular shape. This molecular shape assists in binding through shape specific recognition to its targets leading to considerable three-dimensional structure stability, and thus the high degree of affinity. Natural examples of molecular shape recognition interactions of nucleic acids with proteins are tRNA, ribozymes, DNA binding proteins, and DNAzymes (Meyer, 2008).

Aptamers offer the utility for biotechnological and therapeutic applications as they offer molecular recognition properties that reveal the commonly used biomolecule, antibodies. Aptamers are functional molecules, usually DNA or RNA oligonucleotides, with the appropriate sequence and structure to form a complex with a target molecule. They are able to bind tightly and selectively to disease markers. This can greatly benefit disease diagnosis and therapy. In addition to their discriminate recognition, aptamers offer advantages over antibodies as they can be engineered completely in a test tube. They are readily produced by chemical synthesis, possess desirable storage properties, and elicit little or no immunogenicity in therapeutic applications. When compared to antibodies, aptamers have advantages and disadvantages as therapeutic and biological reagents (Jayasena, 1999). Aptamers present faster tissue penetration and wider applicability and present the opportunity for simple base modifications to improve functionality by comparison. Furthermore, their small size (molecular weights between 3000 and 20,000 daltons) may reduce steric hindrance, increasing surface coverage during immobilization (Deng et al., 2001). The *in vitro* selection process (SELEX) can be more precisely monitored than organismal immunization, and the affinities and specificities of aptamers can thus be better tailored than can those of antibodies. Aptamers can be more readily engineered than antibodies for biological/medicinal use (Jayasena, 1999).

Their applications are very diverse (Fig. 2.4). Aptamers have been used as molecular recognition elements in analytical systems for detection, separation, and purification of target molecules. They also play an important role in medical therapy (Pan et al., 1995; Homann and Goringer, 2001). Aptamers have been selected for various purposes, making use of their high specificity, versatility, and affinity in target recognition. Examples include aptamers that recognize proteins, peptides, dyes, amino acids, nucleotides, and drugs among others, to act as biosensors (Potyrailo et al., 1998), probes (Pavski and Le, 2001), and anticlotting agents (Bock et al., 1992).

Although the application of antibodies for the detection and therapy of infectious agents is well established, aptamers have some potential advantages over antibodies that may fill the gaps that antibody-based applications possess and that expedite the use of aptamers in the virologist's laboratory. These advantages include their small size, eased cell penetration, rapid and cheap synthesis including a variety of chemical modifications, and the fact that aptamers are nonimmunogenic. Probably, their crucial benefit is the

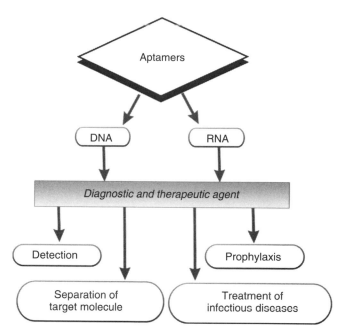

Figure 2.4. Aptamers have diverse applications in the field of diagnostics and therapeutics.

selection process performed *in vitro*. Thus, aptamers can be selected against targets that are either weakly immunogenic or toxic, such as toxin proteins. These targets are usually not applicable for *in vivo* production of monoclonal antibodies in mice; however, high affinity aptamers have been selected against some of these problematic targets. Moreover, the chemical and physical conditions for aptamer selection can be adapted to the real environment in which the aptamer will finally be applied. This includes cross-selection against similar targets that have to be excluded from an aptamer's detection pattern.

With the FDA approval of the first aptamer-based drug Macugen and the recent publication of an aptamer-based cocaine sensor, aptamers are starting to tap their full potential with the increasing need for additional specific detection tools. With the improving automation of the aptamer selection process and the growing knowledge of pathogen-specific targets, the number of aptamers used in diagnostics and therapy of infectious diseases will dramatically increase in the future.

2.3. NUCLEIC ACID HYBRIDIZATION

Nucleic acid hybridization techniques have contributed significantly to the understanding of gene organization, regulation, and expression. *In situ* hybridization (ISH) is a method for detecting specific nucleotide sequences by using a labeled complementary nucleic acid probe. The power of ISH can be greatly extended by the simultaneous use of multiple fluorescent colors. Multicolor fluorescence *in situ* hybridization, in its simplest form, can be used to identify as many labeled features as there are different fluorophores used in the hybridization. Key methodological advances have allowed facile preparation of low noise hybridization probes, and technological breakthroughs now permit multitarget visualization and quantitative analysis—both factors that have made FISH accessible to all and applicable to any investigation of nucleic acids. In future, this technique is likely to have more significant impact on live cell imaging and on medical diagnostics.

2.3.1. Colony Blot and Southern Blot

Both colony blot and Southern blot use probes to locate a specific nucleotide sequence. The probe is a single-stranded piece of a nucleic acid that has been tagged, or labeled with a detectable marker such as a radioactive isotope or a fluorescent dye, or biotin; the probe is complementary to either of the two strands being detected. The DNA of the specimen being examined is affixed to a solid surface such as a nylon membrane and treated with an alkaline solution to denature it. Once the probe has been given enough time to hybridize, unbound probe is washed off. The presence and location of bound probe can be determined by assaying for a detectable marker.

Colony blot is used to detect a given nucleotide sequence in a crude preparation of genomic DNA obtained from colonies grown on agar plate (Fig. 2.5). This procedure is done in such a way that one can readily determine which colonies on the plate contain the sequence of interest. The technique can be used to determine which cells in a collection of clones contain the DNA of interest. Southern blot is used to detect a given sequence in restriction fragments that have been separated according to size and transferred to a membrane filter. A technique called *gel electrophoresis* is used to separate the fragments. Ahuja et al. (2006) used colony filtration blot for the identification of soluble protein expression in *E. coli*. *E. coli* colonies expressing domains ATS, CIDR1α, and DBL1α could be induced and lysed on a membrane filter. Soluble proteins from each cell within the colony diffuse through the filter and are captured on a nitrocellulose membrane placed under the membrane filter. Detection with antibody reagents directed against the fusion tag shows spots indicating the presence of soluble proteins. Intensity of the spot corresponds to the yield of soluble proteins (Fig. 2.5).

The most obvious application of a Southern blot (Fig. 2.6) is to locate sequences that are homologous to the ones being studied, which is how many functionally related genes have been discovered and characterized. This information can also be used to simplify a cloning experiment.

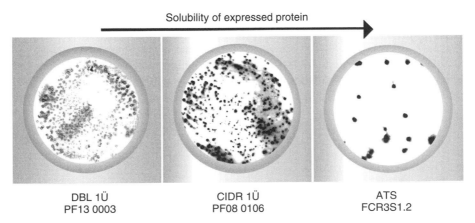

Figure 2.5. Colony filtration blot for the identification of soluble protein expression in *E. coli* (Ahuja et al., 2006).

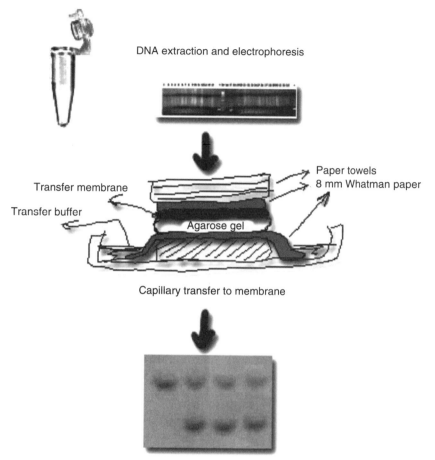

DNA extraction and electrophoresis

Transfer membrane

Transfer buffer

Paper towels
8 mm Whatman paper

Agarose gel

Capillary transfer to membrane

Hybridization membrane and autoradiography

Figure 2.6. Southern blot.

A less apparent but equally important use of the Southern blot is to distinguish different strains of a given species by detectable subtle variations in their nucleotide sequence. Certain mutations will create, others will modify the nucleotide sequence of the genome, and restriction enzyme recognizes these sequence variations. Thus, when genomic DNA of different strains is digested with the same restriction enzyme, each will give rise to different assortment of restriction fragment sizes. The differences are called *restriction fragment length polymorphism* (RFLP), and is the basis of the methods of DNA fingerprinting. Southern blot hybridization is used to selectively visualize restriction fragments that often vary in size. The probe used to selectively visualize restriction fragments, often varies in size. It hybridizes to fragments that demonstrate maximal differences among various strains.

2.3.2. Dot Blot and Zoo Blot

Dot blot is a simple method to detect biomolecules (Fig. 2.7). Often the mRNA has been characterized and the amount of mRNA for a particular gene is to be determined. In

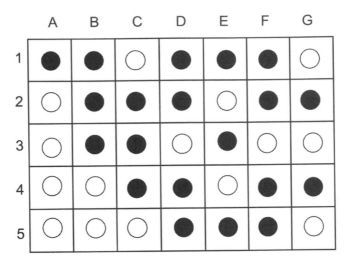

Figure 2.7. Scan of a dot blot result.

this case, a modified version of northern blot analysis, called dot blot, can be used. In a dot blot, the biomolecules are detected by directly applying as a dot. It is followed by detection by either nucleotide probes (Southern blot) or antibodies (Western blot). This technique helps in saving time. In dot blot, nucleic acids are blotted as a circular blot while, in slot blot, rectangular slots are made. But, these types of blots do not allow visualization of the size of the target biomolecule, and, secondly, when two molecules of different sizes are detected they will still appear as a single blot. Dot blot, therefore, confirms the presence of different biomolecules, especially RNA. Many RNA samples can be applied on the same membrane; these samples may be from different cell types, from cells growing under different conditions, or cells at different stages of development. The membrane with the RNA samples bound is placed in a sealed container with a labeled probe complementary to the gene of interest the probe will hybridize. Probe that has not been hybridized to its target mRNA is washed off from the membrane. The presence of hybridized probe is then detected by autoradiography or phospho imager analysis. A spot on your autoradiography or phosphor imager tells you that the transcript from the gene was present in the RNA sample and therefore the gene was being transcribed in the cell from which the RNA was purified. In the examples shown in Fig. 2.7, the filled circles represent female samples, and the open circles represent male samples. A1, A2, A3, A4, and A5 are Umbrella Cockatoos; B1, B2, B3, B4, and B5 are Alexandrian Parakeets; C1, C2, C3, C4, C5, D1, and D2 are Congo African Greys; D3, D4, and D5 are Chattering Lories; E1, E2, E3, E4, and E5 are Quaker Parakeets; F1, F2, F3, and F4 are Blue and Gold Macaws; F5 and G1 are Senegal Parrots; and G2, G3, G4, and G5 are Yellow-naped Amazons. The dot blot procedure is best suited for determining the sex of most Psittacines when fresh whole-blood samples are submitted.

A radioactive sample can be hybridized to detect the variations between the samples. To investigate the expression of entire genes or large set of genes, the northern blot procedure is reversed to follow the dot blot technique. In dot blot hybridization, the gene-specific nucleotide sequences or probes are applied and bound to membranes in specific patterns or arrays and hybridized to radioactive or fluorescent RNA or cDNA preparations. The amount of RNA or cDNA hybridized to each probe can be measured

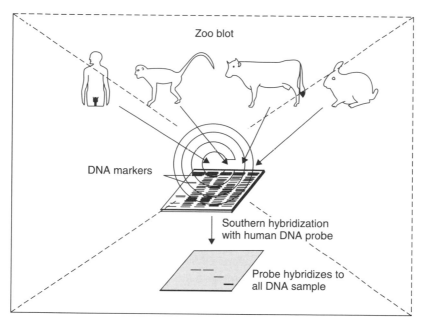

Figure 2.8. Zoo blot.

by scanning the blot with an imaging system that measures the amount of radioactivity or fluorescence, analyzes the results with the computer program, and compares the signal with known probes, an RNA or cDNA.

Zoo blot avoids the problem with poorly expressed and tissue-specific genes by searching not for RNA but for related sequences in the DNA of other organisms (Fig. 2.8). This approach is based on the observed fact that the homologous genes in related organisms have similar sequences, whereas the noncoding DNA sequences are quite different. If a DNA fragment from one species is used as a probe, a Southern blot for DNA from the related species and one more hybridization signals are obtained, which implies that the probe contains one or more genes. The objective is to determine if a fragment of human DNA hybridizes to DNA from the related species. Samples of human, chimps, cow, and rabbit DNA were prepared, restricted, and electrophored in an agarose gel. Southern hybridization was carried out using human DNA fragments as the probe. Positive hybridization signals were seen with each of the animal's DNAs suggesting that the human DNA fragments contain an expressed gene. The hybridizing restriction fragments from the cow and rabbit DNAs were smaller than the hybridization fragments in the chimp samples, which indicates that the restriction map around the expressed sequence is different in cows and rabbit but does not affect the conclusion that a homologous gene is present in all species.

2.3.3. *In situ* Hybridization

ISH is a method of detecting and studying DNA sequences in chromosomes and cells (Gall et al., 1969). ISH presents a unique set of problems as the sequence to be detected will be at a lower concentration, be masked because of associated proteins, or protected within a cell or cellular structure. Therefore, in order to probe the tissue or cells of interest, one has to increase the permeability of the cell and the visibility of the nucleotide sequence to the probe without destroying the structural integrity of the cell or tissue. ISH with radiolabeled probes is a long-established method in cytogenetics. This is distinct from immunohistochemistry, which localizes proteins in tissue sections. The technique of ISH is vital to molecular biologists and their understanding of the pathophysiology of cellular functions. The basic principles of ISH are the same, except that one is utilizing the probe to detect specific nucleotide sequences within cells and tissues (Hougaard et al., 1997). The sensitivity of the technique is such that threshold levels of detection are in the range of 10–20 copies of DNA/mRNA per cell.

Fluorescent ISH (FISH) is a molecular cytogenetic technique in which fluorescently labeled DNA probes are hybridized to metaphase spreads or interphase nuclei to detect or confirm gene or chromosome abnormalities that are generally beyond the resolution of routine cytogenetics. FISH have much higher rates of sensitivity and specificity. FISH also provides results more quickly because no cell culture is required. The first application of fluorescence *in situ* detection came in 1980, when RNA that was directly labeled at the 3′-end with fluorophore was used as the probe for specific DNA sequences (Bauman et al., 1980). Enzymatic incorporation of fluorophore-modified bases throughout the length of the probe has been widely used for the preparation of fluorescent probes; one color is synthesized at a time (Wiegant et al., 1991).

The use of amino-allyl modified bases, which could later be conjugated to any sort of hapten or fluorophore, was critical for the development of *in situ* technologies, because it allowed production of an array of low noise probes by simple chemistry (Wiegant et al., 1991). Methods of indirect detection allowed signal output to be increased artificially by the use of secondary reporters that bind to the hybridization probes. In the early 1980s, assays featuring nick translated, biotinylated probes, and secondary detection by fluorescent streptavidin conjugates were used for the detection of DNA (Manuelidis et al., 1982) and mRNA (Singer and Ward, 1982) targets. Approximately a decade later, improved labeling of synthetic, single-stranded DNA (ssDNA) probes allowed the chemical preparation of hybridization probes carrying enough fluorescent molecules to allow direct detection. Many variations on these themes of indirect and direct labeling have since been introduced, giving a wide spectrum of detection schemes from which to choose (Pachmann, 1987); the specifications, sensitivity, and resolution of these techniques are well described elsewhere (Raap, 1998).

The technique of ISH (Fig. 2.9) is used to locate the chromosomal location of a specific DNA (or RNA) probe (Femino et al., 1998). The theory is the same as for Southern hybridization, except that the DNA to which the probe will hybridize is the actual chromosome. The probe is labeled with a fluorescent dye. The sample DNA (metaphase chromosomes or interphase nuclei) is first *denatured*—a process that separates the complimentary strands within the DNA double helix structure (Fig. 2.10). The fluorescently labeled probe of interest is then added to the denatured sample mixture, which hybridizes with the sample DNA at the target site as it reanneals (or reforms itself) back into a double helix. The probe signal can then be seen through a fluorescent microscope and the sample DNA can be scored for the presence or absence of the signal. If the species has been characterized cytogenetically, the marker can be assigned to the appropriate chromosome. Because this technique uses a fluorescent probe, it is called *fluorescence in situ hybridization* or FISH. It is also called *chromosome painting*.

Chromosome analyses by FISH have led to a marked progress in cytogenetics research (Pinkel et al., 1986). A prime example of the power of hybridization approaches in

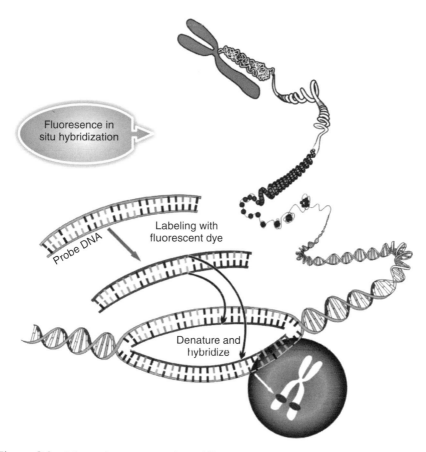

Figure 2.9. Schematic representation of fluorescence *in situ* hybridization technique.

genome investigation is comparative genomic hybridization (CGH), during which deletions and duplications of chromosomal regions are detected by differential fluorescence signals (Forozan et al., 1997). However, because the assay does not benefit from preservation of tissue structure or cellular architecture, its future applications are more likely to be *in silico* than *in situ* (Lichter et al., 1988). Initially, RNA assays could reliably detect only rather abundant messages, using clone-derived probes (Lawrence and Singer, 1986). Enhancements in detection and computer processing algorithms have subsequently allowed detection of rare targets (Tanke et al., 1995; Piper et al., 1990, 1994). Advances in microscope and detector hardware have allowed the low light level produced by FISH to be recorded and analyzed with increasing sensitivity (Tanke et al., 1999). Mathematical image-processing algorithms have been built on this progress to yield super-resolution technology to probe at submicroscopic resolution, using digital image stacks (Carrington et al., 1995).

New targets led to new applications of the FISH procedure. The new avenues of research opened by these applications required that more and more species be simultaneously detected (Fauth and Speicher, 2001). A major milestone in the detection of chromosome targets was the discrimination of all human chromosomes simultaneously, using computed interpretation of a five-color scheme (Schrock et al., 1996; Speicher

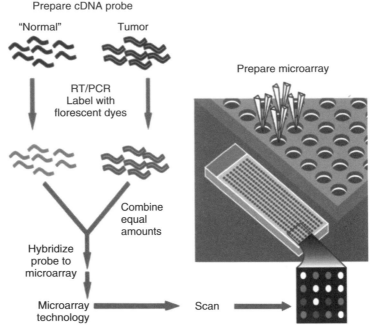

Figure 2.10. Concept of microarray.

et al., 1996). Although mRNAs can also be visualized in a multiplex manner (Levsky et al., 2003), FISH analysis of the entire transcriptome *in situ* is a daunting thought. One can only speculate that future technologies will feature increasingly higher order multiplexing, until the number of interesting nucleic acid targets is reached. The technical means for color coding such a large number of entities is already in place (Nederlof et al., 1989, 1990), although reduction to practice will be difficult and a means of deciphering spatially overlapping signals would be required to develop.

FISH was originally used with metaphase chromosomes. These chromosomes, prepared from nuclei that are undergoing division, are highly condensed with each chromosome in a set taking up a recognizable appearance characterized by the position of centromere and the banding pattern that emerges after the chromosome preparation is stained. With metaphase chromosomes, a fluorescent signal obtained by FISH is mapped by measuring its position relative to the end of the short arm of the chromosome.

A disadvantage is that the highly condensed nature of the metaphase chromosomes means that only low resolution mapping is possible, two markers needing to be at least 1 Mb apart to be resolved as separate hybridization signals. This degree of resolution is insufficient for the construction of useful chromosome maps, and the main application of metaphase FISH has been in determining which chromosome a new marker is located on, and providing a rough idea of its map position, as a preliminary to a finer scale mapping by other methods. For several years these "other methods" did not involve any form of FISH, but since 1995 a range of higher resolution FISH is not restricted to intact chromosomes. It also works well on interphase nuclei, where it is used to visualize the order of closely spaced (from 100 to 500 kb apart) probes on a chromosome and to measure their distances apart.

The development of *in situ* technologies has provided us with a wealth of information regarding the locations and expression patterns of genes in single cells. ISH is an invaluable tool for research and diagnostics, dramatically advancing the study of cell- and tissue-specific expression of many genes. Some of ISH applications include determination of chromosome structure, function, and evolution; chromosomal gene mapping; expression of genes; localization of viral DNA sequences; diagnosis of viral diseases (Ahtiluoto et al., 2000); localization of oncogenes (Suciu et al., 2008); sex determination; and chromosomal abnormalities (Siffroi et al., 2000; Brown et al., 2000). The uses and different approaches for ISH continue to increase, thus impacting many different research fields. New labeling techniques for probes, new detection systems, and advanced computer software increase the availability and efficiency of ISH (Leitch et al., 1994).

Lately, FISH technology has introduced the possibility of chromosome classification based on 24-color chromosome painting (M-FISH). Reciprocal chromosome painting and hybridizations with probes such as yeast artificial chromosomes, cosmids, and fiber FISH allow subchromosomal assignments of chromosome regions and can identify the breakpoints of rearranged chromosomes. Recent advances in FISH using probes specific for entire chromosomes or chromosome segments can quickly and economically provide a cytogenesis map of homologics. These probes cover large segments on chromosomes of various species and confirm previous hypotheses of extensive linkage conservation in mammals.

Cytopathological diagnosis of Merkel cell carcinoma (MCC) using interphase FISH analysis was performed for research purposes using centromeric probes of chromosomes 6 and 8 (Suciu et al., 2008). Trisomy of chromosome 6 was found in 85% of tumor cells in the first case of MCC, and case 2 exhibited trisomy 8 in 77% of tumor cells. In the absence of specific molecular markers, detection of trisomy 6 and/or trisomy 8 could help in identifying MCC. As FISH analysis is easily and quickly performed on interphase nuclei, it may be extended to the study of other relevant genetic abnormalities. Leversha et al. (2009) reported FISH analysis to analyze the gene copy number alterations of circulating tumor cells in metastasis prostrate cancer patients. FISH assay has been proved to be a promising tool for diagnosis, surveillance, and monitoring of carcinoma in the upper urinary tract (Luo et al., 2009).

2.3.4. Microarray Technology

The complete sequencing of several genomes, including that of humans, has signaled the beginning of a new era in which scientists are becoming increasingly interested in functional genomics; that is, uncovering both the functional roles of different genes and how these genes interact with, and/or influence, each other. Increasingly, this question is being addressed through the simultaneous analysis of hundreds to thousands of unique genetic elements with microarrays (Hinton et al., 2004). Microarrays and computational methods are playing a major role in attempts to reverse engineer gene networks from various observations (Kothapalli et al., 2002).

Microarrays exploit the preferential binding of complementary single-stranded nucleic acid sequences (Barrett and Kawasaki, 2003; Barrett et al., 2004). They are miniaturized biological devices consisting in molecules, for example, DNA or protein, named probes, which are orderly arranged at a microscopic scale onto a solid support such as a membrane or a glass microscopic slide (Fig. 2.10). A microarray is typically a glass (or some other material) slide, onto which DNA molecules are attached at fixed locations (spots). There may be tens of thousands of spots on an array, each containing a huge number of identical DNA molecules (or fragments of identical molecules), of lengths from twenty to hundreds of nucleotides. For gene expression studies, each of these molecules ideally should identify one gene or one exon in the genome; however, in practice, this is not always so simple and may not even be generally possible due to families of similar genes in a genome. Microarrays that contain all of the about 6000 genes of the yeast genome have been available since 1997. The spots are either printed on the microarrays by a robot, or synthesized by photolithography (similar to computer chip productions) or by ink jet printing (Hughes et al., 2001). The light generated oligonucleotide array, developed by Affymetrix, Inc. (Santa Clara, CA), involves synthesizing short 25-mer oligonucleotide probes directly onto a glass slide using photolithographic masks. Sample processing includes the production of labeled cRNA, hybridization to a microarray, and quantification of the obtained signal after laser scanning. Regardless of the array used, the output can be readily transferred to commercially available data analysis programs for the selection and clustering of significantly modified genes (Culhane et al., 2003).

DNA microarrays are widely used to analyze genome-wide gene expression patterns and to study genotypic variations. Before a nucleic acid (target) can be used for hybridization to the probes of a microarray, it needs to be extracted from the tissue and labeled. The total mRNA from the cells in two different conditions (Kuo et al., 2002) is extracted and labeled with two different fluorescent labels: for example, a green dye for cells at condition 1 and a red dye for cells at condition 2 (to be more accurate, the labeling is typically done by synthesizing ssDNAs that are complementary to the extracted mRNA by a enzyme called *reverse transcriptase* (RT). Frequently, it also needs to be amplified to increase detection sensitivity. Labeled gene products from the extracts hybridize to their complementary sequences in the spots because of the preferential binding—complementary single-stranded nucleic acid sequences tend to attract each other and the longer the complimentary parts, the stronger the attraction. To increase throughput and minimize costs without reducing gene expression, different fluorescent dyes, namely, Alexa 488, Alexa 594, Cyanine 3, and Cyanine 5, can also be used for hybridization.

During a hybridization process, labeled target molecules with sequences complementary to the probes are captured quantitatively. Subsequently, a reader measures the amount of label on each probe. To generate accurate and informative data, one of the most critical aspects of these experiments is the quality of both the isolated and the labeled nucleic acid samples (Lockhart et al., 1996). The dyes enable the amount of sample bound to a spot to be measured by the level of fluorescence emitted when it is excited by a laser. If the RNA from the sample in condition 1 is in abundance, the spot will be green and if the RNA from the sample in condition 2 is in abundance, it will be red. If both are equal, the spot will be yellow, while if neither is present it will not fluoresce and appear black. Thus, from the fluorescence intensities and colors for each spot, the relative expression levels of the genes in both samples can be estimated.

The raw data that are produced from microarray experiments are the hybridized microarray images. To obtain information about gene expression levels, these images should be analyzed—each spot on the array is identified, its intensity is measured, and compared to the background. This is called *image quantitation*.

Microarrays are miniature arrays of gene fragments attached to glass slides. Monitoring the expression of thousands of genes simultaneously is possible with DNA microarray analysis. These devices have been highly effective for the simultaneous detection of large numbers of analytes in a sample, and microarrays have quickly emerged as important analytical tools in many branches of the biological sciences. A microarray-based analytical strategy is quicker and more convenient than serial testing for each analyte, and it has been successfully applied to both immunoassays and DNA-based assays. The current scope of microarray applications includes sequencing by hybridization, resequencing, mutation detection, assessment of gene copy number, CGH, drug discovery, expression analysis, and immunoassay (protein microarrays). In addition, oligonucleotide microarrays have been used for a nonbiological application, for example, computing (Hacia, 1999).

At present, two technologies dominate the field of high density microarrays: cDNA arrays and oligonucleotide arrays (Kane et al., 2000). The cDNA array has a long history of development stemming from immunodiagnostic work in the 1980s; however, it has been most widely developed in recent years by Stanford University (California) researchers depositing cDNA tags onto glass slides, or chips, with precise robotic printers (Lipshutz et al., 1999). Labeled cDNA fragments are then hybridized to the tags on the chip, scanned, and differences in mRNA between samples identified and visualized using a variation in the red/green matrix. A tissue microarray using paraffin-embedded formalin-fixed tissues can be used to evaluate the immunoprofiles of a large set of uterine adenocarcinomas with an extended panel of antibodies, comparing the profile of primary cervical and andometrial adenocarcinomas.

Differentially expressed genes will be defined herein as gene data determined to be statistical outliers from some standard state, and which cannot be ascribed to chance or natural variability (Li and Wong, 2001). Various creative techniques have been proposed and implemented for the selection of differentially expressed genes; however, none have yet gained widespread acceptance for microarray analysis (Kane et al., 2000). Despite this, there remains a great impulse to develop new data analysis techniques, partly driven by the obvious need to move beyond setting simple fold change cutoffs, which are out of context with the rest of the experimental and biological data at hand. Selecting differentially regulated genes based only on a single fold change across the entire range of experimental data preferentially selects genes expressed at a low level. This commonly used approach does not accommodate for background noise, variability, nonspecific binding, or low copy numbers—characteristics typical of microarray data, which may not be homogeneously distributed. Other approaches entail the use of standard statistical measures such as Student's t-test or ANOVA for every individual gene (Johnson and Wichern, 2002). However, owing to the cost of repeating microarray experiments, the number of replicates usually remains low, leading to inaccurate estimates of variance (Baldi and Long, 2001). Furthermore, due to the low number of replicates, the power of these "gene-by-gene" statistical tests to differentiate between regulated and nonregulated genes also remains very low.

Protein microarrays are versatile tools for parallel, miniaturized screening of binding events involving large numbers of immobilized proteins in a time- and cost-effective manner (Fu et al., 2009). They are increasingly applied for high throughput protein analyses in many research areas, such as protein interactions, expression profiling, and target discovery (McGrath et al., 2007). While conventionally made by the spotting of purified proteins, advances in technology have made it possible to produce protein microarrays through *in situ* cell-free synthesis directly from corresponding DNA arrays (Stoevesandt et al., 2009).

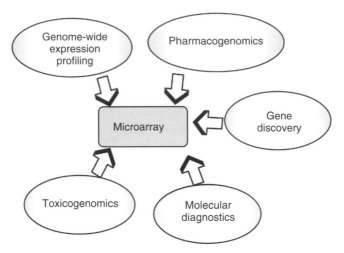

Figure 2.11. The wide applications of microarrays.

The DNA microarray technology is currently an area of great interest. This technology can rapidly provide a detailed view of the simultaneous expression of entire genomes and provide new insights into gene function, disease pathophysiology, disease classification, and drug development (Steven and Greenberg, 2001). The application of microarrays (Fig. 2.11) will have a significant role in clinical and basic science. Improved understanding of the gene changes occurring in human diseases will lead to the development of new diagnostic tools as well as more effective drugs (Hacia, 1999). In the basic sciences, microarray technology will permit researchers to elucidate the genetic pathways that are followed in the development, aging, and pathology of specific cells or tissues of the body.

With the help of microarray technology, however, they will be able to further classify different types of cancers on the basis of the patterns of gene activity in the tumor cells. DNA microarray analysis helped to identify several of those genes involved in pathological processes relating to arthritis, including apoptosis, inflammation, and cellular proliferation (Sasidharan et al., 2009a,b). One interesting gene, follistatin-like gene, is highly expressed along the margin of contact between inflammatory synovial pannus and eroding bone, suggesting a role in joint destruction. The identification of genes predisposing to human diseases is of paramount importance. Numerous genes may play a role in drug response and toxicity, introducing a daunting level of complexity into the search for candidate genes. Gene discovery in bladder cancer progression and gastrointestinal tract origin cancer cells have also been identified (Golub et al., 1999). This technology has provided insights into gene function, disease pathophysiology, disease classification and drug development. This technology is also called *genechip* technology as it incorporates molecular genetics and computer science on a massive scale (Steven and Greenberg, 2001).

Expression profiling by DNA microarray permits the identification of genes underlying clinical heterogeneity of many diseases which might contribute to disease progression, thereby improving assessment of treatment and prediction of patient outcome. The gene expression profiles provide a molecular fingerprint of the transcriptome.The last several years have witnessed remarkable development of new microarray-based automated techniques allowing parallel analysis of multiple DNA samples. Two major current applications of DNA microarrays are gene expression profiling (Lockhart et al., 1996;

Mecham et al., 2004) and gene mutation analysis (Hacia et al., 1996). Gene expression profiling allowed the molecular classification of leukemias (Golub et al., 1999) and other types of cancers (Brenton et al., 2001), and revelations about genetic network architecture (Tavazoie et al., 1999) using DNA microarrays, which were not otherwise achievable a few years ago.

However, studies typically result in the identification of hundreds of genes that may or may not be of relevance *in vivo*, particularly when large, genetically diverse study populations are used. The challenge is to design experimental systems and approaches that minimize variability in the data, increase the reproducibility amongst experiments, allow array data from multiple experiments to be assessed by a variety of statistical, supervised learning, and data clustering approaches, and provide a clear link between drug response and the expression of specific genes (Yuen et al., 2002). Now the possibility to simultaneously analyze the expression of thousands of genes using DNA microarrays has allowed exploring the relationships between gene expression and sensitivity to several anticancer drugs. A number of studies using microarrays for identifying genes governing tumor chemosensitivity focused on tumor cell lines. Some clinical studies have also been carried out to investigate whether tumor gene expression patterns could predict clinical response to chemotherapy. Results of these studies are encouraging; indicating that individualization of drug treatment based on multigenic response predictive markers is feasible. Faced with the skyrocketing expenses of modern health care, microarrays offer a solution with the prospects for more efficient, cost-effective, and personalized approaches to patient care.

2.3.5. Molecular Diagnostic Research

Molecular diagnostics is a rapidly advancing field in which insights into disease mechanisms are being elucidated by use of new gene-based biomarkers. Until recently, diagnostic and prognostic assessment of diseased tissues and tumors relied heavily on indirect indicators that permitted only general classifications into broad histologic or morphologic subtypes and did not take into account the alterations in individual gene expression. Global expression analysis using microarrays now allows for simultaneous interrogation of the expression of thousands of genes in a high throughput manner and offers unprecedented opportunities to obtain molecular signatures of the state of activity of diseased cells and patient samples. Microarray analysis may provide invaluable information on disease pathology, progression, resistance to treatment, and response to cellular microenvironments, and ultimately may lead to improved early diagnosis and innovative therapeutic approaches. Microarray technology has been widely used to investigate tumor classification, cancer progression, and chemotherapy resistance and sensitivity.

We are now in the "postgenomic era," during which the diagnostic, prognostic, and treatment response biomarker genes identified by microarray screening will be interrogated to provide personalized management of patients. Clinicians will be able to use microarrays during early clinical trials to confirm the mechanisms of action of drugs and to assess drug sensitivity and toxicity. Coupled with more conventional biochemical analysis such as IHC and ELISA, microarrays will be used for diagnostic and prognostic purposes.

Array-based CGH measures copy number variations at multiple loci simultaneously, providing an important tool for studying cancer and developmental disorders and for developing diagnostic and therapeutic targets (Barrett et al., 2004). The oligonucleotide arrays designed for CGH provide a robust and precise platform for detecting chromosomal alterations throughout a genome with high sensitivity even when using full complexity genomic samples. Using cDNA microarrays can acquire gene expression profiles that characterize anticancer drug sensitivity at the same time evaluating the genome-wide gene expression profiles of various cancer cell lines to identify the genes related to gastrointestinal tract cancer cell.

DNA microarrays are an integral part of the process for therapeutic discovery, optimization, and clinical validation. At an early stage, investigators use arrays to prioritize a few genes as potential therapeutic targets on the basis of various criteria. Subsequently, gene expression analysis assists in drug discovery and toxicology by eliminating poor compounds and optimizing the selection of promising leads. Integral to this process is the use of sophisticated statistics, mathematics, and bioinformatics to define statistically valid observations and to deduce complex patterns of phenotypes and biological pathways. In short, microarrays are redefining the drug discovery process by providing greater knowledge at each step and by illuminating the complex working of biological systems. With the development of microarray-based tests for diagnosing a wide range of diseases, the ability to gain more information from a limited clinical sample by using highly parallel expression techniques have emerged. Several research groups have focused on identifying subsets of genes that show differential expression (Favis and Barany, 2000; Favis et al., 2000) between healthy tissues or cell lines and their tumor counterparts to identify biomarkers for several solid tumors, including ovarian carcinomas, oral cancer, melanoma, colorectal cancer, and prostate cancer. SNP (single nucleotide polymorphism) analysis raises the possibility of individual tumor genome-wide allelotyping with potential prognostic and diagnostic applications (Hoque et al., 2003). The technology provides a basis for DNA-based cancer classification and helps to define the genes being modulated, improving the understanding of cancer genesis and potential therapeutic targets (Dutt and Beroukhim, 2007). Although the major limiting factors for routine use in a clinical setting, at present, are cost and access to the microarray technology, it is likely that costs will decrease in the near future and that the technology will become increasingly user-friendly and automated.

2.4. DNA SEQUENCING

DNA consists of four different subunits or nucleotides that constitute a four letter "language" coding for every other molecule found in living organisms. Nucleotides are linked together to form chains that are, in turn, united to form chromosomes. Each segment of a chromosome that codes for a specific molecule is called a *gene*. To determine the function of specific genes, scientists have learned to read or "decode" the sequence of nucleotides composing DNA in a process referred to as *DNA sequencing*. In 1975, Frederick Sanger and his colleagues and Maxam and Gilbert developed two different methods for determining the exact base sequence of a defined piece of DNA. These spectacular breakthroughs revolutionized molecular biology which has allowed us to determine the primary structure of a gene and, ultimately, made possible the sequencing of all 3 billion base pairs of the human genome.

Once a particular gene is selected for sequencing, many copies of this target DNA must be isolated. Although there are several strategies for obtaining large quantities of DNA, the most widely used is the polymerase chain reaction (PCR) method, a synthetic, cyclical process for amplifying DNA, which exponentially increases the number of copies of the target sequence.

PCR products are sequenced using a procedure whereby the amplification product acts as a template for synthesizing new strands of target DNA. However, because of the inclusion of specific chemical inhibitors in the reaction mixture, the newly synthesized DNA strands vary in length. Because they differ in size, these strands can be separated electrophoretically on polyacrylamide gels. In the past, these strands were labeled with radioisotopes and visualized by autoradiography (X-ray film). Modern sequencing procedures employ fluorescent chemical markers that are detected as the DNA fragments pass a stationary laser. The DNA sequence is recorded as a chromatogram where each "peak" corresponds to a single nucleotide. The chromatogram is recorded by a computer and automatically converted into data which can be compared with homologous sequences from other individuals, species, or genera.

The first ribosomal molecule to be used extensively in sequencing studies was the 5S rRNA, the smallest and least complex rRNA. Sequencing this molecule increased our understanding of the relationships between bacteria and resulted in the construction of a 5S rRNA sequence database. Major breakthroughs in phylogenetic systematics came with the sequencing of the larger and more complex small (16S or 18S) and large (26S or 28S) subunits. More recently, taxonomists have used sequences from the noncoding spacer regions between the small and large subunits to assess genetic variation among closely related species and between different populations of the same species.

For many years, DNA sequencing was an expensive and labor-intensive technique restricted to a few research laboratories, but now, the development of automated sequencing systems and computer programs for sequence data comparison has made this technology more cost efficient and more widely accessible. DNA sequencing is now considered the most powerful tool available to microbial systematists (Harris et al., 2008). Aside from its importance in the investigation of the evolutionary histories of microorganisms, sequence analysis is also a standard technique used in bacteria and yeast identification. Advances in sequencing technology have improved other methods for rapid microbial identification and typing, including *in situ* detection, by facilitating the design of nucleic acid probes used to detect bacteria and fungi in mixed microbial communities. This method has also greatly increased our understanding of the number and diversity of unculturable bacteria and fungi through studies that target species known only from their DNA sequences.

Bacterial and fungal DNA sequence data are collected by many research groups throughout the world and deposited in data banks accessible via the Internet. Although only a few organisms genetic complements are known in their entirety, portions of many thousands of species genomes are available for comparison and homology searches in these sequence databases. These databases are expected to expand, given the rapid advances in molecular technology, the decreasing cost of automated sequencing systems, and the recognition of the importance of DNA sequence data in organism classification and identification.

2.4.1. Dideoxychain Termination Method

The first method described by Sanger and Coulson for DNA sequencing was called *plus and minus* (Sanger and Coulson, 1975). This method used *E. coli* DNA polymerase I and DNA polymerase from bacteriophage T4 (Englund, 1971, 1972) with different limiting nucleoside triphosphates. The products generated by the polymerases were resolved by ionophoresis on acrylamide gels. Owing to the inefficacy of the "plus and minus" method, 2 years later, Sanger and his coworkers described a new breakthrough method for sequencing oligonucleotides via enzymic polymerization (Sanger et al., 1977a). This method, which would revolutionize the field of genomics in the years to come, was initially known as the chain termination method or the dideoxynucleotide method. The original method of sequencing a piece of DNA by Sanger method began with cloning the DNA into a vector, such as M13 phage or a phagemid, that would give us the cloned DNA, which should simply be heated to create ssDNA for sequencing. To the ssDNA, an oligonucleotide primer of about 20 bases long is hybridized; this primer is designed to hybridize right beside the multiple cloning region of the vector and is oriented with its 3′-end pointing toward the insert in the multiple cloning regions. If extended, the primer using the klenow fragments of DNA polymerase will produce DNA complementary to the insert.

The trick to Sanger's method is to carry out such DNA synthesis; it consisted of a catalyzed enzymic reaction that polymerizes the DNA fragments complementary to the template DNA of interest (unknown DNA). Briefly, a labeled primer (short oligonucleotide with a sequence complementary to the template DNA) was annealed to a specific known region on the template DNA, which provided a starting point for DNA synthesis. In the presence of DNA polymerases, catalytic polymerization of deoxynucleoside triphosphates (dNTPs) onto the DNA occurred. The polymerization was extended until the enzyme incorporated a modified nucleoside [called a *terminator* or *dideoxynucleoside triphosphate* (*ddNTP*)] into the growing chain. This method was performed in four different tubes, each containing the appropriate amount of one of the four terminators. All the generated fragments had the same 5′-end, whereas the residue at the 3′-end was determined by the dideoxynucleotide used in the reaction. After all four reactions were completed, the mixture of different-sized DNA fragments was resolved by electrophoresis on a denaturing polyacrylamide gel, in four parallel lanes (Fig. 2.12). The pattern of bands showed the distribution of termination in the synthesized strand of DNA, and the unknown sequence could be read by autoradiography. The enzymic method for DNA sequencing has been used for genomic research as the main tool to generate the fragments necessary for sequencing, regardless of the sequencing strategy.

2.4.2. Automated DNA Sequencing

Although the Sanger method was fast and convenient, it still suffered from the use of radioisotopic detection, which was slow and potentially risky. Additionally, it required four lanes to run one sample because the label was the same for all reactions. To overcome such problems, Smith et al. (1986) developed a set of four different fluorescent dyes (fluorescein, 4-chloro-7-nitrobenzo-2-1-diazole (NBD chloride), tetramethyl-rhodamine, and Texas Red) that allowed all four reactions to be separated in a single lane(Smith et al. 1985, 1986). Each of the four dyes was attached to the 5′-end of the primer and each labeled primer was associated with a particular ddNTP. For example, the fluorescein-labeled primer reaction was terminated with ddCTP (dideoxycytidine triphosphate), the tetramethyl-rhodamine labeled primer reaction with ddATP (dideoxyadenosine triphosphate), and so on. All four reactions were then combined and introduced onto a slab gel in a single lane. The bands were detected on excitation of the fluorescent moiety attached to the DNA with a laser beam at the end of the gel. The fluorescent light was separated by means of four different colored filters. After the 4-color data was generated, the sequence read out was straightforward, with the association of each color to only one base.

DNA sequencing in slab gels with fixed-point fluorescence detection then became automated DNA sequencing rather than "manual DNA sequencing," which required exposure of the whole slab gel to a photographic plate for a fixed-time and postanalysis detection (Griffin and Griffin, 1993; Adams et al., 1996). Automated DNA sequencing has been performed via two different labeling protocols. The first used a set of four fluorescent labels attached at the 5'-end of the primer, as described earlier. In the second method, the fluorescent moiety was linked to the ddNTP terminators, allowing the synthesis of all four ladders in a single vial. In the latter case, when the labeled ddNTP was incorporated, the enzyme terminated the extension at the same time as the ladder was labeled. Thus, the C-terminated ladder contained one fluorescent dye, and the G-, A-, and T-terminated ladders had their own respective labels. The protocols are known as dye-labeled primer chemistry and dye-labeled terminator chemistry, respectively, and both labeling arrangements are shown in Figure 2.13.

Alternative dyes were synthesized and linked to an M13 sequencing primer via a sulphydryl group and conjugated with tetramethyl-rhodamine iodoacetamide (Ansorge et al., 1986). This alternative dye used tetramethyl-rhodamine as the only fluorophore because

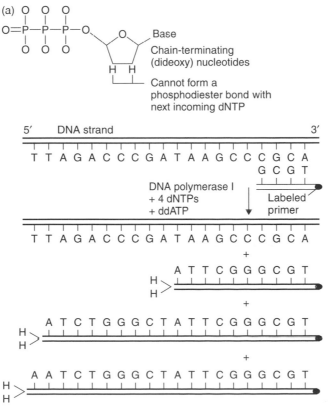

Figure 2.12. Sanger's method of sequencing.

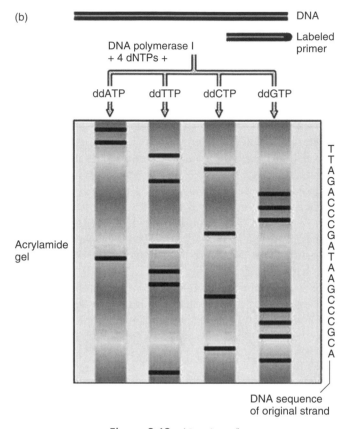

Figure 2.12. (*Continued*)

of its high extinction coefficient, high quantum yield, and long wavelength of absorption. One year later, the same group proposed a sulphydryl-containing M13 sequencing primer end labeled with fluorescein iodoacetamine (Ansorge et al., 1987). Other dyes commonly linked to the primers include carboxy fluorescein (FAM), carboxy-4′,5′-dichloro-2′,7′-dimetoxyfluorescein (JOE), carboxytetramethyl-rhodamine (TAMRA), and carboxy-X-rhodamine (ROX) (Swerdlow and Gesteland, 1990; Karger et al., 1991; Carson et al., 1993). These dyes have emission spectra with their maxima relatively well spaced, which facilitates color or base discrimination. One drawback of this group of dyes was the need for two wavelengths for excitation; one at 488 nm for FAM and JOE dyes, and the other at 543 nm for TAMRA and ROX dyes.

2.4.3. Maxam and Gilbert Method

A sequencing method based on chemical degradation was described by Maxam and Gilbert (1977). In this method, end-labeled DNA fragments are subjected to random cleavage at adenine, cytosine, guanine, or thymine positions using specific chemical agents. The chemical attack is based on three steps: base modification, removal of the modified base from its sugar, and DNA strand breaking at that sugar position (Maxam and Gilbert, 1977). The products of these four reactions are then separated

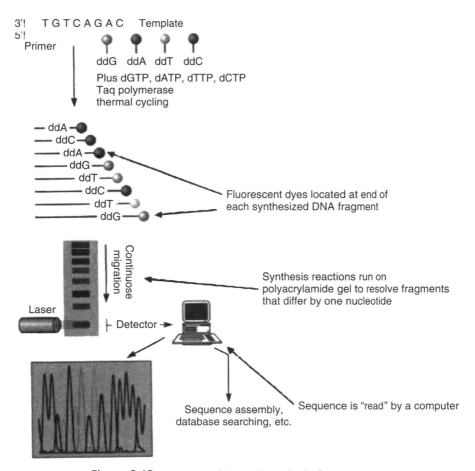

Figure 2.13. Automated Sanger's method of sequencing.

using polyacrylamide gel electrophoresis. The sequence can be easily read from the four parallel lanes in the sequencing gel (Fig. 2.14).

The template used in this sequencing method can be either double-stranded DNA (dsDNA) or ssDNA from chromosomal DNA. In general, the fragments are first digested with an appropriate restriction enzyme (Maxam and Gilbert, 1980), but they can also be prepared from an inserted or rearranged DNA region (Maxam, 1980). These DNA templates are then end labeled on one of the strands. Originally, this labeling was done with [ATP] phosphate or with a nucleotide linked to ADP and enzymically incorporated into the end fragment (Maxam and Gilbert, 1977). Alternatively, restriction fragments through dideoxyadenosine 5'-(α-thio) triphosphate (ddATPαS) and terminal deoxynucleotidyltransferase were used (Ornstein and Kashdan, 1985). These substitutions showed several advantages, including a longer lifetime, low emission energy, increase in the autoradiograph resolution and higher stability after labeling. Nevertheless, the use of radioactive labels is hazardous and a strategy based on a 21-mer fluorescein labeled M13 sequencing primer was therefore proposed. The fluorescent dye and its bound form to the oligonucleotide were shown to be stable during the chemical reactions used for the base-specific degradations (Ansorge et al., 1988). For instance, fluorescein attached via a mercaptopropyl or aminopropyl linker arm to the 5'-phosphate of an oligonucleotide

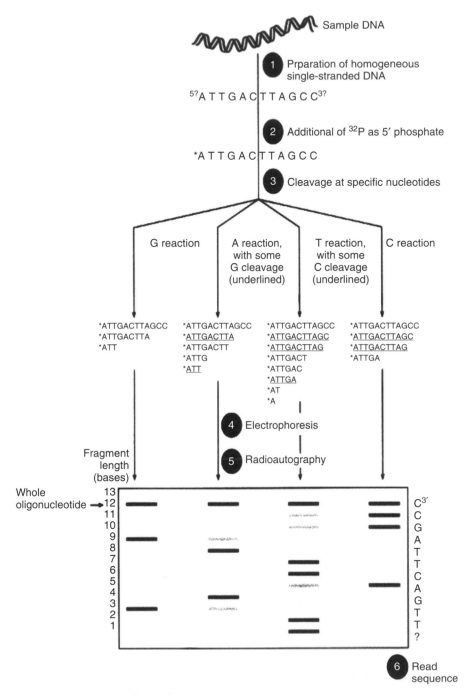

Figure 2.14. Maxam and Gilbert method.

was described and shown to be stable during the reactions used in the chemical cleavage procedures (Rosenthal et al., 1990).

Another nonradioactive labeling strategy that was stable during the chemical reactions uses a biotin marker molecule chemically or enzymatically attached to an oligonucleotide primer or enzymatically attached to an end labeling reaction of restriction enzyme sites (Richterich, 1989). After fragment separation by direct blotting electrophoresis, the membrane-bound sequence pattern can be visualized by a streptavidin-bridged enzymic color reaction.

An approach that made the automation of this labeling step possible was the use of PCR to amplify the products, where one of the primers was end labeled (Nakamaye et al., 1988; Stamm and Longo, 1990; Tahara et al., 1990). Among many dye and fluorophore labeling strategies, the chemiluminescent detection method showed competitive results. In this strategy, the chemically cleaved DNA fragment is transferred from a sequencing gel onto a nylon membrane. Specific sequences are then selected by hybridization to DNA oligonucleotides labeled with alkaline phosphatase or with biotin, leading directly or indirectly to the deposition of the enzyme. If a biotinylated probe is used, an incubation step with avidin alkaline phosphatase conjugate follows. The membrane is soaked in the chemiluminescent substrate "3-(2'-spiroadamantane)-4-methoxy-4-(3''-phosphoryloxy)phenyl- 1,2-dioxetane" (AMPPD) and exposed to a photographic film (Tizard et al., 1990).

Initially, all steps of these chemical sequencing methods were performed manually (Maxam and Gilbert, 1977, 1980). Years later, a system composed of a computer-controlled microchemical robot that carries out one of the four reactions (G, A + C, C + T, or C) in less than 2 h was described (Wada et al., 1983; Wada, 1984).

In conclusion, the main advantages of the Maxam–Gilbert and other chemical methods compared with Sanger's chain termination reaction method are as follows: (i) A fragment can be sequenced from the original DNA fragment, instead of from enzymic copies. (ii) No subcloning and no PCR reactions are required. Consequently, for the location of rare bases, the chemical cleavage analysis cannot be replaced by the dideoxynucleotide terminator method, as the latter analyzes the DNA of interest via its complementary sequence and, thus, gives sequence information only in terms of the four canonical bases. (iii) This method is less susceptible to mistakes with regard to sequencing of secondary structures or enzymic mistakes (Boland et al., 1994). (iv) Some of the chemical protocols are recognized by different authors as being simple and easy to control, and the chemical distinctions between the different bases are clear (Negri et al., 1991).

Therefore, the chemical degradation methods are used (i) for genomic sequencing, where information about DNA methylation and chromatin structure could be obtained (Church and Gilbert, 1984); (ii) to confirm the accuracy of synthesized oligonucleotides or to verify the sequence of DNA regions with hairpin loops (Ornstein and Kashdan, 1985); (iii) to locate rare bases, such as Hoogsteen base pairs (Sayers and Waring, 1993); (iv) to detect point mutations (Ferraboli et al., 1993); (v) to resolve ambiguities that arise during dideoxysequencing (Goszczynski and McGhee, 1991); (vi) to analyze DNA protein interactions (Isola et al., 1999); and (vii) to sequence short DNA fragments, in general. This method, when described in 1977, had a read-length of approximately 100 nucleotides (Maxam and Gilbert, 1977). In 1980, it achieved 250 bases per assay (Maxam, 1980; Maxam and Gilbert, 1980). At present, with general improvements during the last few years, read-lengths close to 500 bp and automatic processing of multiple samples have been achieved (Dolan et al., 1995).

Despite all advantages, most of the protocols have some drawbacks. First, the chemical reactions of most protocols are slow and the use of hazardous chemicals requires special handling care. The worst problem, however, is the occurrence of incomplete reactions that decrease the read-lengths. The explanation for this is that incomplete reactions introduce electrophoretic mobility polidispersion (caused by chemical and physical inhomogeneities among the DNA chains within a given band); which enlarges the bandwidths and this, in turn, reduces the interband resolution.

2.4.4. Primer Walking

The other approach for genomic sequencing is the direct sequencing of unknown DNA within sites whose sequence is known. For example, an unknown sequence of DNA is inserted into a vector and amplified. The first sequencing reaction is performed using the primers that hybridize to the vector sequence and polymerize the strand complementary to the template. A second priming site is then chosen inside the newly generated sequence, following the same direction as the first one (Fig. 2.15). This approach is known as primer walking (Studier, 1989; Martin-Gallardo et al., 1992), and its major advantage is the reduced redundancy (Voss et al., 1993) because a 6-mer can be employed due to the direct nature of the approach in many priming sites at different positions.

The use of such short oligonucleotides leads to the possibility of mispriming since uniqueness is reduced with the reduced size of the oligonucleotide. For example, the use of three small oligonucleotides could result in several sites where one or two of them could hybridize to the template and initiate mispriming. To avoid this situation, an ssDNA binding protein (SSB) (Kieleczawa et al., 1992) or the stacking effects of selected modular primers (Kotler et al., 1993) were used. At present, the appeal of a cost-effective and time-saving method that uses small oligonucleotide libraries has disappeared with improvements in primer synthesis technology (Lashkari et al., 1995). However, the demand for a sequencing method that was capable of providing long read-length (number of bases read per run), short analysis time, low cost, and high accuracy has led to several modifications of the original Sanger method. In addition to several improvements in the procedures and in the reagents used in the sequencing reaction, further development in DNA separation technology was of paramount importance for the completion of the Human Genome Project.

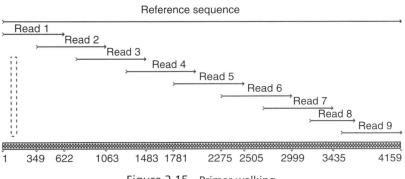

Figure 2.15. Primer walking.

2.4.5. Contig Sequencing

A *contig* is also sometimes defined as the DNA sequence reconstructed from a set of overlapping DNA segments. Fragmentation of a contig into 1–2 kb segments provides DNA segments of suitable size for sequencing. A computer program is used to assemble the segments back to one single uninterrupted piece (contig). Also a single gene or marker that identifies the comparable region in a target genome, but does not cluster with other genes or markers to form a segment, is called a *singleton*. Expressed sequence tag (EST) contains both contigs and singletons.

In order to make it easier to talk about data gained by the shotgun method of sequencing, the word "contig" was invented. A contig is a set of gel readings that are related to one another by overlap of their sequences. All gel readings belong to one and only one contig, and each contig contains at least one gel reading. The gel readings in a contig can be summed to form a contiguous consensus sequence and the length of this sequence is the length of the contig.

This defines a contig to be a set of overlapping segments of DNA. It naturally allows a set to contain a single element (ready for further comparison). It also defines the length of a contig to be the length of the consensus sequence derived from it. In the light of the current confusion, note that consensus sequences and contigs are entirely different classes of object. A "fingerprint clone contig" is assembled by using the computer program FPC to analyze the restriction enzyme digestion patterns of many large insert clones. Clones are then selected for sequencing to minimize overlap between adjacent clones. For a clone to be selected, all of its restriction enzyme fragments (except the two vector insert junction fragments) must be shared with at least one of its neighbors on each side in the contig. Once these overlapping clones have been sequenced, the set is a "sequenced clone contig." When all selected clones from a fingerprint clone contig have been sequenced, the sequenced clone contig will be the same as the fingerprint clone contig. Until then, a fingerprint clone contig may contain several sequenced clone contigs. After individual clones (for example, A and B) have been sequenced to draft coverage and the clones have been mapped, the data are analyzed by GigAssembler, producing merged sequence contigs from initial sequence contigs and linking these to form sequence contig scaffolds (Fig. 2.16). In the fingerprint comparison software, the groups of clones should be to contain single clones, otherwise they would miss overlaps. The length of the contigs could be estimated, but the Sanger Centre had to be created before it was possible to calculate their consensus sequences.

2.4.6. Shotgun Sequencing

Shotgun sequencing is a random process because there is no control of the region that is going to be sequenced (Lander et al., 2001). Genomic DNA is randomly fragmented (by sonication, nebulization, or other scission methods) into smaller pieces, normally ranging from 2 to 3 kb. The fragments, inserted into a vector, are replicated in a bacterial culture. Several positive amplifications are selected, and the DNA is extensively sequenced. Owing to the random nature of this process, the sequences generated overlap in many regions (Adams et al., 1996). The process of overlaying or alignment of the sequences is called sequence assembly (Fig. 2.17). Shotgun sequencing normally produces a high level of redundancy (the same base is sequenced 6 ± 10 times, in different reactions), which affects the total cost. A new variation of the method introduced by Venter et al. (1996)

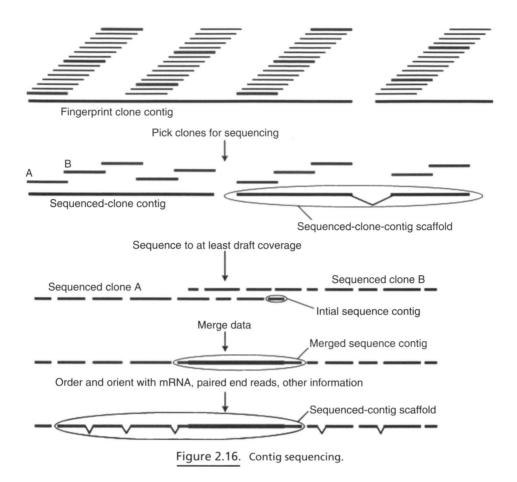

Figure 2.16. Contig sequencing.

involved shot gunning a whole genome at once. This strategy depended enormously on computational resources to align all generated sequences. However, the efforts were rewarded with the sequencing of the *Haemophilus infuenzae* genome in only 18 months (Fleischmann et al., 1995) and, more recently, the human genome (Venter et al., 2001).

Shotgun sequencing is well established, with ready availability of optimized cloning vectors, fluorescently labeled universal primers, and software for base calling and sequence assembly. The whole process has a high level of automation, from the cloning of the vectors and colony selection to the bases called. A simplified diagram of the shotgun process is summarized in Figure 2.17. Although the random approach is fully compatible with automation, it can produce gaps in the sequence that can only be completed by direct sequencing of the region. The DNA was cut into 150 Mb fragments and arranged into overlapping contiguous fragments. These contigs were cut into smaller pieces and sequenced completely.

Using pyrosequencing method, no cloning is required and hence it is quick; a draft sequence can be obtained in a few days. In addition to the speed, the method has another advantage over traditional sequencing protocol. Because there is no cloning there is no bias toward sequencing of particular regions of the genome. The genome of *Mycoplasma genitalium* was resequenced using this approach, and a slightly better assembly was obtained than that was obtained with the initial clone-based shotgun approach. It has

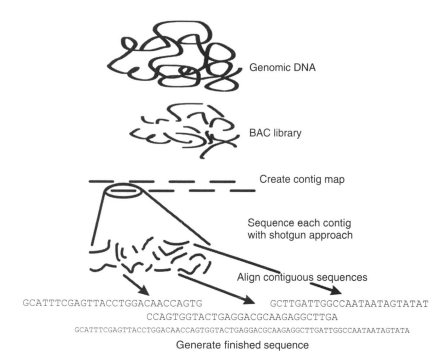

Genomic DNA

BAC library

Create contig map

Sequence each contig
with shotgun approach

Align contiguous sequences

GCATTTCGAGTTACCTGGACAACCAGTG   GCTTGATTGGCCAATAATAGTATAT
                    CCAGTGGTACTGAGGACGCAAGAGGCTTGA
        GCATTTCGAGTTACCTGGACAACCAGTGGTACTGAGGACGCAAGAGGCTTGATTGGCCAATAATAGTATA

Generate finished sequence

Figure 2.17. Shotgun sequencing.

been successfully used to identify point mutation within the genome of *Mycobacterium tuberculosis*, which is responsible for resistance of novel antibiotics.

2.4.7. Importance of Sequencing

The determination of the nucleotide sequence of a genome from different organisms is an important part in modern biology. It is now relatively easy to sequence a bacterial genome, with hundreds being completed to date. These includes the genomes of many human pathogens including *M. tuberculosis* (which causes TB), *Treponema palladium* (which causes syphilis), *Rickettsia prowazekii* (the cause of typhus), *Vibrio cholera* (the cause of cholera), and *Yersinia pestis* (the cause of plague). The genome analysis is a valuable tool for understanding and analyzing the biology of the organisms. It is almost certain that it plays a vital role in understanding the pathogenicity. Genome sequences can also be exploited to develop new drugs. Sequencing is also a valuable tool in our battle against emerging viruses as emphasized in the global response to the cornavirus that is believed to cause severe acute respiratory syndrome (SARS). The genome sequence was determined within a matter of months of it being identified and has helped a lot our understanding of SARS.

2.5. POLYMERASE CHAIN REACTION

PCR, like DNA sequencing, is based on the DNA polymerization reaction. PCR is used in genome sequencing, including the Human Genome Project. Using random primers,

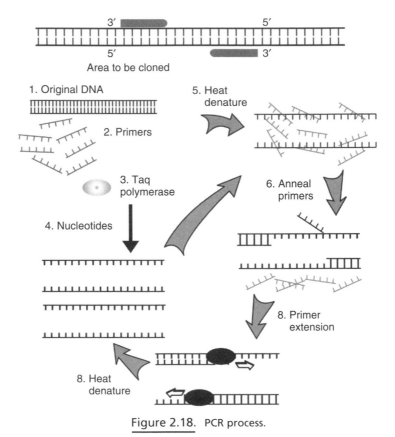

Figure 2.18. PCR process.

the entire sequence of a genome can be amplified in pieces. A primer and dNTPs are added along with a DNA template and the DNA polymerase (in this case, *Taq*). The main concern with PCR is that, in addition to using a primer that sits on the 5'-end of the gene and makes a new strand in that direction, a primer is made to the opposite strand to go in the other direction. The original template is melted (at 94 °C), the primers anneal (at 45–55 °C), and the polymerase makes two new strands (at 72 °C), doubling the amount of DNA present. This provides two new templates for the next cycle. The DNA is again melted, primers anneal, and the *Taq* makes four new strands. A wrong temperature during the annealing step can result in primers not binding to the template DNA at all, or binding at random. The elongation temperature depends on the DNA polymerase. The time for this step depends both on the DNA polymerase itself and on the length of the DNA fragment to be amplified (Fig. 2.18).

Before the use of thermostable DNA polymerases in PCR, researchers had to laboriously replenish the reaction with fresh enzyme (such as Klenow or T4 DNA polymerase) after each denaturation cycle. Thermostable DNA polymerases revolutionized and popularized PCR because of their ability to withstand high denaturation temperatures. The use of thermostable DNA polymerases also allowed higher annealing temperatures, which improved the stringency of primer annealing.

Thermostable DNA polymerases can also be used for either one-enzyme or two-enzyme real-time polymerase chain reaction (RT PCR) (Myers and Gelfand, 1991; Chiocchia and Smith, 1997). For example, *Tth* DNA polymerase can act as an RT in the presence of Mn^{2+} for one enzyme RT PCR (Myers and Gelfand, 1991). All of the DNA polymerases mentioned can be used for the amplification of the first-strand cDNA produced by an RT, such as AMV RT, in two-enzyme RT PCR. The thermostable DNA polymerases can be divided into two groups, those with a $3' \rightarrow 5'$ exonuclease (proofreading) activity, such as *Pfu* DNA polymerase, and those without the proofreading function, such as *Taq* DNA polymerase. These two groups have some important differences. Proofreading DNA polymerases are more accurate than nonproofreading polymerases due to the $3' \rightarrow 5'$ exonuclease activity, which can remove a misincorporated nucleotide from a growing chain of DNA. When the amplified product is to be cloned, expressed, or used in mutation analysis, *Pfu* DNA polymerase is a much better choice due to its high fidelity. However, for routine PCR, where simple detection of an amplification product is the goal, *Taq* DNA polymerase is the most commonly used enzyme because yields tend to be higher with a nonproofreading DNA polymerase (Lee et al., 2007).

PCR provides an extremely sensitive means of amplifying small quantities of DNA. Each step of the cycle should be optimized for each template and primer pair combination. Successful amplification depends on many factors, namely, DNA template quantity and quality and magnesium ions as cofactors of *Taq* polymerase.

PCR is widely used in different areas of science and biotechnology. The Human Genome Project is a worldwide endeavor to map the DNA base sequence of every gene in the human genome.

PCR has a major place in human diagnostics, and is gradually emerging as a tool in plant diagnostics, and is now a routine tool in epidemiological and ecological studies, and for genetic mapping—RAPD (random amplified polymorphic DNA) and AFLP (amplified fragment length polymorphisms) (Power, 1996). The molecular diagnostic laboratory is responsible for the development and performance of molecular diagnostic tests for nucleic acid targets found in a variety of settings in medicine. PCR can become an important tool for medical diagnosis. PCR can detect and identify bacteria and viruses that cause infections such as tuberculosis, chlamydia, viral meningitis, viral hepatitis, HIV, cytomegalovirus and many others. Primers are designed for DNA of a specific organism using PCR to detect the presence or absence of a pathogen in a patient's blood or tissue in a simple experiment. The three broad areas of testing in which PCR is extensively used are genetics, hematopathology, and infectious disease.

PCR is a powerful and reliable technique for rapid diagnosis of *M. tuberculosis* (Baek et al., 2000). The usefulness of PCR in the diagnosis of TB by using a variety of unselected clinical specimens is not clear as studies have differed in techniques including lysing method and target nucleic acid to detect products as well as the number and type of samples used, making the reported sensitivities and specificities difficult to compare (Pahwa et al., 2005). A recent study comparing four conventional techniques, FNA (fine needle aspiration) cytology, ZN (Ziehl–Neelsen) staining, culture and lymph node biopsies, and TB PCR, indicated 94.8% diagnosis but PCR was found to be highly sensitive (94.4%) though less specific (38.2%, Goel et al., 2001).

Two types of John Cunningham virus (JCV) are found in infected brain and kidney tissues. A highly reliable allele-specific PCR is used to detect point mutations in cellular genes. Specific amplification of two fragments, using four pairs of type-specific primers, is based on a single nucleotide difference at the $3'$-ends of the primers. A combination of three conditions in the PCR reaction was required for specificity: "hot start," a ramped ("touchdown") cycle profile, and a slightly lowered molar concentration of the specific primers and dNTPs (Ault et al., 1994).

Diagnosis of bacterial disease has improved in recent years because of the routine use of PCR in the clinical laboratory (Clarke, 2002). The multiplex PCR is a sensitive assay with the ability to detect a mean of one to two genome copy units per 100 µl clinical sample and allows nonculture-based diagnosis of the disease-causing organism within 3 h. A fluorescence-based multiplex PCR was automated for the simultaneous detection of *Neisseria meningitidis, Streptococcus pneumoniae*, and *Haemophilus influenzae* in clinical samples from patients with suspected meningitis (Smith et al., 2004). Pulsed-field gel electrophoresis and PCR were applied for the first time to the molecular characterization of *Clostridium tetani* (Plourde-Owobi et al., 2005).

Specific identification of *Bacillus anthracis* and differentiation from closely related *Bacillus cereus* and *Bacillus thuringiensis* strains is still a major diagnostic problem. Commercially available diagnostic kits that target plasmid markers cannot differentiate between *B. anthracis*, nonanthracis *Bacillus* species harboring anthrax-specific virulence plasmids, and plasmidless *B. anthracis* strains. A specific TaqMan PCR assay was successfully designed, which targets sequences of gene locus BA 5345 of the *B. anthracis* strain on the basis a chromosomal marker. In another study by Lee et al. (2009), a multiplex RT PCR assay that could sensitively detect *Salmonella* spp. and specifically differentiate *Salmonella typhimurium* from *Salmonella enteritidis* in meats was developed.

PCR-denaturing gradient gel electrophoresis (PCR-DGGE) method is used for the detection and identification of *Campylobacter, Helicobacter*, and *Arcobacter* species (Epsilobacteria) in clinical samples and for the evaluation of its efficacy on saliva samples from humans and domestic pets. The PCR-DGGE method should allow determination of the true prevalence and diversity of Epsilobacteria in clinical and other samples. Contact with the oral cavity of domestic pets may represent a route of transmission for epsilobacterial enteric diseases.

Using "hot start" PCR, the distribution pattern of HPV in vulvar lesions was analyzed. After amplification, *in situ* analysis showed that many of the cells that lacked halos contain HPV DNA and that the hybridization signal often localized to areas where there was a thickened granular layer. HPV DNA was not noted in the basal cells. The one copy of HPV 16 was detectable after PCR with a single primer pair by *in situ* analysis only if the hot start modification was employed (Nuovo et al., 1992). Gong et al. (2002) employed nested PCR for diagnosis of *Tuberculous lymphadenitis*, and PCR amplification with sequence-specific primers (PCR-SSCP) for identification of rifampicin resistance in fine needle aspirates.

RT PCR (Fig. 2.19) can be applied to traditional PCR applications as well as new applications that would have been less effective with traditional PCR. A fluorogenic probe hydrolysis (TaqMan) reverse transcriptase PCR has been used for the evaluation of diagnostic sensitivity of classical swine fever virus using clinical samples (Risatti et al., 2005). With the ability to collect data in the exponential growth phase, the power of PCR has been expanded into applications such as viral quantitation, pathogen detection including CMV detection, rapid diagnosis of meningococcal infection, penicillin susceptibility of *S. pneumoniae, M. tuberculosis* and its resistant strains, waterborne microbial pathogens in the environment, genotyping (allelic discrimination) by melting curve analysis or specific probes/beacons, trisomies, single-gene copy numbers, microdeletion genotypes, haplotyping, quantitative microsatellite analysis, prenatal diagnosis/sex determination using single cell isolated from maternal blood or fetal DNA in maternal circulation, and prenatal diagnosis of hemoglobinopathies intraoperative cancer diagnostics (Wittwer et al.,

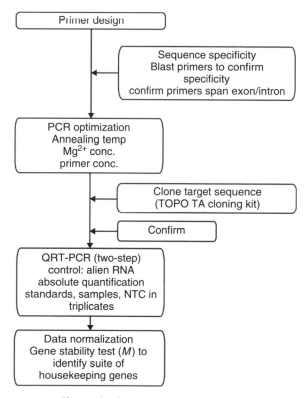

Figure 2.19. Real-time PCR protocol.

2001). Linear-after-the-exponential (LATE) PCR is a new method for real-time quantitative analysis of target numbers in small samples. It is adaptable to high throughput applications in clinical diagnostics, biodefense, forensics, and DNA sequencing.

Pulsed-field gel electrophoresis and colony PCR were applied for the first time by Plourde-Owobi et al. (2005) for the molecular characterization of *C. tetani*. Among five strains tested, one turned out to contain a mixture of two genetically different clones and two (D11 and G761) to contain bacteria differing by the presence or absence of the 74-kb plasmid harboring the tetX gene.

Genetic analysis of organisms (animals, plants, and bacteria) at the molecular level is an important and widely practiced area of genetic science. A number of techniques developed over more than a decade offers the opportunity to identify each individual or type of individual in a species uniquely and unambiguously. PCR has impacted this area of analysis because of its ease of use and simplicity over traditional variable number tandem repeats (VNTR) and RFLP-based methods (Jeffreys et al., 1985; Tsuchizaki et al., 2000; Sambrook and Russell, 2001).

2.6. OMICS TECHNOLOGY AND MICROBES

Biomedical research has undergone a fundamental change over the last decade with the advent of new techniques for drug discovery and basic research, which should be attributed to the impact of plethora of information available from Human Genome Project (Bilello, 2005).

Biological research is conducted in high throughput manner, which allows simultaneous study of multiple constituents such as genes, proteins, metabolites. This kind of study of high throughput data is fundamentally different from traditional studies focused on studying one gene/protein to understand the biological system. Analysis of human genome has widespread presence of sequence variations (Sachidanandam et al., 2001). Association of these sequence variations with human diseases susceptibility, adverse drug response, and drug resistance have been studied in detail for quite a few cases. The application of "omic-" technologies may lead to a change of paradigms and open new vistas in medicine.

Omics technologies are collections of methods to study various biomolecules such as RNA, proteins, and intermediary metabolites in a high throughput manner with an objective to characterize entire biomolecules in a single analysis. Omics technology has given us unparalleled opportunities for complete assessment of biochemical changes, and objectively compares these findings with other individuals and species. The terms *Ome* and *Omics* are derivations of the suffix–ome, which has been appended to a variety of previously existing biological terms to create names for fields of endeavor, such as genome, proteome, transcriptome, and metabolome, that are either speculative or have some tangible meaning in a particular context.

"Omic" technologies in a broader view include genomics, transcriptomics, proteomics, and metabolomics. Owing to the widespread use of various omic technologies, biology has transformed into a data-rich science from the earlier data-poor science; however, it comes with the challenge of deriving useful information from integrating information churned out from various omic technology. Systems biology provides the methods, computational capabilities, and interdisciplinary expertise to integrate data from various omic technologies to get a holistic picture of the underlying complex biological phenomenon.

Genomics, transcriptomics, proteomics, and metabolomics technologies are inherently high throughput. They increase substantially the number of proteins/genes that can be detected simultaneously and provide better insight into the physiological condition with availability of information about changes at complete genome and/or proteome level, rather than data from single gene or protein.

Complex biological processes are manifestation of several biochemical events happening at genomic, proteomic, and metabolic levels; therefore, it is of prime importance to study all these events to understand underlying complexities (Fig. 2.20). Omics technologies allow efficient monitoring of gene transcripts, proteins, and intermediary metabolites, making it possible to monitor a large number of key cellular pathways simultaneously and thus equipping us with data to simulate cellular processes and study concerned system in great detail.

2.6.1. Genomics: A Tool for Understanding Genes

The genome of an organism is its whole hereditary information encoded in the DNA (or, for some viruses, RNA). The genome of an organism is a complete DNA sequence of one set of chromosomes; for example, one of the two sets that a diploid individual carries in every somatic cell. The term *genome* can not only be applied specifically to mean the complete set of nuclear DNA (i.e., the "nuclear genome") but can also be applied to organelles that contain their own DNA, as with the mitochondrial genome or

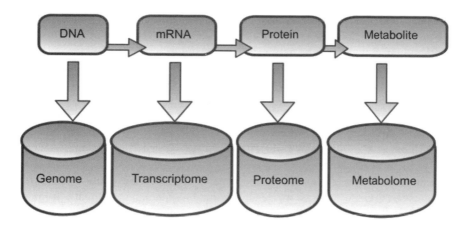

Figure 2.20. The central dogma and the interacting "ome" include the study of genome, proteome, transcriptome, and metabolome.

the chloroplast genome. The genome contains the coded instructions necessary for the organism to build and maintain itself. Genomics is the study of an organism's genome, or genetic material. Sequencing the genome of an organism was the pioneer step in the field of genomics. Development of tools and algorithm for sequence analysis facilitated several research areas such as phylogenetics, functional genomics, and target discovery.

The increasing amount of genomic and molecular information is the basis for understanding higher order biological systems, such as the cell and the organism, and their interactions with the environment, as well as for medical, industrial, and other practical applications (Kanehisa et al., 2006). Sequence data from Human Genome Project has created the foundation for the field of functional genomics, with an objective of assigning generic/spatial/temporal biological role of complete set of genes in an organism. Genomics-enabled technologies such as microarray allow us to study patterns of gene expression during various conditions.

Sequencing of all organism of clinicopathological importance is pivotal to our understanding of diseases or biological process at genomics level. In 1972, Walter Fiers and his team at the Laboratory of Molecular Biology of the University of Ghent (Ghent, Belgium) were the first to determine the sequence of a gene: the gene for bacteriophage MS2 coat protein. In 1976, the team determined the complete nucleotide sequence of bacteriophage MS2-RNA. The first DNA-based genome to be sequenced in its entirety was that of bacteriophage Φ-X174 (5368 bp), sequenced by Sanger (Sanger et al., 1977b). The first free-living organism's genome to be sequenced was that of *H. influenzae* (1.8 Mb) in 1995, and since then genomes are being sequenced at a rapid pace. A rough draft of the human genome was completed by the Human Genome Project in early 2001, creating much fanfare.

As of January 2005, the complete sequence of about 1,000 viruses, 220 bacterial species, and roughly 20 eukaryote organisms was known, of which half are fungi. Most bacteria whose genomes have been completely sequenced are disease-causing agents, such as *H. influenzae*. Of the other sequenced species, most were chosen because they were well-studied model organisms or promised to become good models. Yeast (*S. cerevisiae*) has long been an important model organism for the eukaryotic cell, while the fruit fly (*Drosophila melanogaster*) has been a very important tool (notably in early pre-molecular genetics). The worm *Caenorhabditis elegans* is an often-used simple model for multicellular organisms. The zebrafish (*Brachydanio rerio*) is used for many developmental studies on a molecular level and the flower *Arabidopsis thaliana* is a model organism for flowering plants. The Japanese pufferfish (*Takifugu rubripes*) and the spotted green pufferfish (*Tetraodon nigroviridis*) are interesting because of their small and compact genomes, containing very little noncoding DNA compared to most species. Among mammals, dog (*Canis familiaris*), brown rat (*Rattus norvegicus*), mouse (*Mus musculus*; Lein et al., 2007) and chimpanzee (*Pan troglodytes*) are all important model animals in medical research.

Clinical practice can be tailored on the basis of the outcomes from genomics-based experiments/tests. Gene expression, genetic variations among individuals, from different cohorts or study groups varies significantly, providing explanation for their varied response toward drug efficacy and toxicity. Moreover, drug administered in combination with other drugs, induces a gene expression profile different from their constituent's profile, which emphasizes the importance of study of gene–drug and drug–drug interaction network for better understanding. This kind of system-level study can open routes to "personalized medicine," which aims at providing safe and effective medication.

Genomics technologies have generated huge amount of data, which is developed by different platforms such as microarrays, SAGE (serial analysis of gene expression), and RT PCR. Nearly 1000 microbial genomes have been completely sequenced, and the number of sequences is still growing exponentially (Uchiyama et al., 2010). The growth of this number will be even further accelerated by the recent advancement of next-generation sequencing technologies. As a result of the numerous bacterial genome sequences currently available, the concept of a core genome, a set of orthologous genes commonly derived in bacterial genomic studies, is being increasingly used to explore genomic relationships among bacteria (Waters and Storz, 2009). For example, important insights into the origin of photosynthesis were recently obtained from the analysis of 892 core genes identified among 15 cyanobacteria genomes (Mulkidjanian et al., 2006). In another comparative genomic study of 4 magnetotactic bacteria, several unique genes from a core of 891 genes were identified as potentially important to the magnetic field sensing and taxis (movement) abilities of this group of prokaryotes (Richter et al., 2007). A general observation from these studies is that the number of genes that make up the core genome depends on the number and diversity of organisms being compared (Callister et al., 2008).

While the use of the core genome concept has led to important insights into the evolution of bacterial species and identification of potentially important novel genes, there has been little discussion regarding the actual expression of the core genome genes as proteins and the extent of this expression across the set of bacteria under study. The assumption that a gene will always produce a gene product, that is, protein, is debatable as evidence suggests that genes are silenced by evolutionary mechanisms and, as such, will not be expressed. Thus, the expression of a core gene in one organism, but not in another, can provide insight into the effects of both evolution and environmental pressures on the expressed phenotype. Yet, the expression of identified genes within core genomes is rarely verified by experimental observation because of the extensive resources and rigorous experimental design required to do so.

This information can help in genomic studies toward understanding microbial diversity. One of the promising approaches is the comparison of dozens of closely related or moderately related genomes, which is effective for analyzing critical differences among organisms and understanding the evolutionary process generating such diversity. Another important new advancement is the metagenomic approach, by which researchers can investigate the community structures of microbes and their gene contents in various environmental samples. To facilitate genomic diversity studies, however, effective utilization of the existing genomic data in terms of comparative genomics is crucial, although this becomes more difficult as the size of the genomic database increases.

Microbial Genome Database (MBGD) is a microbial database for large-scale comparative analysis of completely sequenced microbial genomes, the number of which is now growing rapidly (Uchiyama, 2003). It is based on comprehensive ortholog classification (Uchiyama, 2007) generated by a hierarchical clustering method, DomClust (Uchiyama, 2006). As compared to other comparative genomics resources covering complete microbial genomes, such as CMR (Peterson et al., 2001), Microbes Online (Alm et al., 2005), IMG (Markowitz et al., 2008), eggNOG (Jensen et al., 2008), and OMA (Schneider et al., 2007), a prominent feature of MBGD is that it allows users to create ortholog groups using a specified subgroup of organisms. This feature is useful for various types of comparative analysis, including comparisons among closely related as well as among distantly related organisms (Uchiyama, 2003), and also to facilitate comparative genomics from various points of view such as ortholog identification, paralog clustering, and motif analysis and gene order comparison.

Genomic technologies also offer a new alternative to the classic approach in the field of food production and processing. They allow quick identification of microorganisms present in the (raw) product and they could help to directly measure the total response of the target spoilage microorganisms to the applied preservation methods. By making use of this new approach, known as *applied microbial genomics*, it might be possible one day to reduce the number of experiments needed to measure all relevant responses. The tools used for this are small chips containing the information of thousands of genes of food spoilage microorganisms, which are attached to a solid surface (like a glass slide) in a grid-like array. These so-called microarrays could enable in the long run the outcome prediction of a preservation treatment and the definition of additional preservation steps if necessary. A number of campylobacter species, grouped under the name of thermophilic campylobacters, are "formidable" pathogens. Among them, *Campylobacter jejuni* is one of the world's most "successful" food poisoning bacteria and is probably responsible for more than twice as many cases of poisoning as *Salmonella*. *C. jejuni* is thought to have 1700 genes. Genomics is being used to explore the activity of individual genes and to look at the variety of proteins produced by the organism when it faces different environmental challenges. The adaptability of *C. jejuni* derives from a powerful set of regulatory genes, which enable it to change its metabolism rapidly depending on its environment, for example, whether it is in contaminated raw chicken or residing in the human gut. Applying microbial genomic research, a food sample can be easily tested for the presence of those genes that would indicate a contamination with *C. jejuni*. This test is significantly faster than conventional techniques, and might enable the use of sterilization methods with less damage to the food product.

Genomics of food microbes generate valuable knowledge that can be used for metabolic engineering, improving cell factories, and development of novel preservation methods. Furthermore, pre- and probiotic studies, characterization of stress responses, studies of microbial ecology, and, last but not least, development of novel risk assessment procedures will be facilitated. Genomics technology can even be applied as measures for traceability from farm to table (http://www.eufic.org/article/en/artid/microbial-genomics-food-quality-safety/).

2.6.2. Transcriptomics: A Tool for Deciphering Gene Function

Transcriptomics is the study of gene products or mRNA, which are transcribed from over 3 billion DNA base pairs. For any organism, gene expression is not a static phenomenon across cell type or different time-points. Gene expression profiles are very dynamic, which are context, temporal, tissue/cellular microenvironment dependent. Microarray technologies can be excellent for diagnostic purposes. One of the early applications of microarray was in mode of action (MOA) studies for isoniazid (INH), an antibiotic effective in the control of tuberculosis. Studies of *M. tuberculosis* exposed to INH indicated upregulation of a number of genes relevant to the MOA. In addition to possibly increasing the number of targets by identifying genes within pathways, the generation of "signature transcriptome analysis (TA) profiles" may be useful in predicting the MOA of new chemical entities. This approach has been shown to be quite useful in the search for tumor-specific markers and in distinguishing potential therapeutic targets in neoplastic cells. Gene expression profiling, helps in the identification of the set of genes that can be used as predictor for metastasis, recurrence, and drug resistance, and these predictive gene set can be used to segregate patients who require aggressive therapy (Lu et al., 2006; Xi et al., 2005).

Microarrays can provide almost a clear picture of changes on a transcriptome level for study system; however, there is no one-to-one correlation between mRNA expression and protein expression (Fig. 2.21; Selinger et al., 2000). Proteins are translated from coded information in respective mRNA; however, protein expression is modulated/regulated by various mechanisms such as posttranslation modification and miRNA regulation (Katayama et al., 2005). Although transcriptomics can help us immensely in understanding the disease mechanism, hypothesis generation, target identification, the results derived from transcriptomics, however, should be considered as a reflection of variations happening at a transcription level and other tools such as proteomics should be used as a supplement for proper validation or formulation of critical hypothesis.

With the dawn of next-generation (or deep) sequencing technologies in recent years (Ansorge, 2009; Metzker, 2010), their application to high depth sequencing of whole transcriptomes, a technique now referred to as *RNA-seq*, has been explored (Morozova et al., 2009; Wang et al., 2009; Wilhelm and Landry, 2009). RNA-seq requires the conversion of mRNA into cDNA by reverse transcription, followed by deep sequencing of this cDNA (Fig. 2.22).

RNA-seq was initially used only for analyzing eukaryotic mRNA, as prokaryotic mRNA is less stable and lacks the poly(A) tail that is used for enrichment and reverse transcription priming in eukaryotes. But these technological difficulties are being overcome, as various methods for enrichment of prokaryote mRNA and appropriate cDNA library construction protocols have been developed, some generating strand-specific libraries that provide valuable information about the orientation of transcripts.

In June 2008, the first report of RNA sequencing of whole microbial transcriptomes, that is, the yeasts *Saccharomyces cerevisae* (Nagalakshmi et al., 2008) and *Schizosaccharomyces pombe* (Wilhelm et al., 2008) appeared. Both studies demonstrated that most of the nonrepetitive sequence of the yeast genome is transcribed, and provided detailed information of novel genes, introns and their boundaries, 3′ and 5′ boundary mapping,

(a) Transcriptome analysis with a small genome

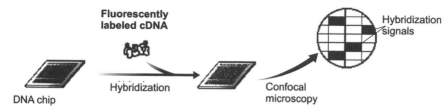

(b) Microarray analysis of a large genome

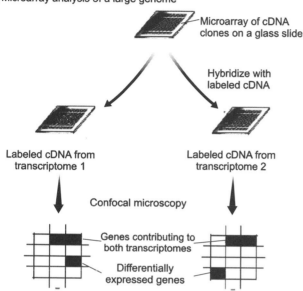

Figure 2.21. Microarray for transcriptomic analysis.

3′-end heterogeneity and overlapping genes, antisense RNA, and more. Since 2009, several examples have been reported of prokaryote whole-transcriptome analysis using tiling arrays and/or RNA-seq, and these are summarized in Table 2.3. The first reviews of prokaryote transcriptome sequencing have just appeared (Croucher et al., 2009; Ozsolak et al., 2009; van Vliet and Wren, 2009; Sorek and Cossart, 2010; van Vliet, 2010).

During microbial transcriptomic sequencing (Fig. 2.22), the starting material is a mix of RNA, followed by optional subtraction of tRNA and rRNA, generation of cDNA libraries, sequencing, bioinformatics, and interpretation of cDNA-sequencing-read histograms. Transcriptomic sequencing histograms can be done for monocistronic and polycistronic mRNAs, noncoding RNA (ncRNA), cis-acting RNAs, and antisense RNA.

Numerous new insights into genomic elements, gene expression, and complexity of regulation are emerging from these new high throughput and high resolution studies of microbial transcriptomes.

Whole-transcriptome mapping can identify contiguous expression extending into flanking regions of a protein encoding gene, indicative of 5′- or 3′-untranslated regions (UTRs). Long 5′-UTRs are often indicative of upstream regulatory elements, such as riboswitches (Toledo-Arana et al., 2009). Archaea have much shorter or no 5′-UTRs compared to

TABLE 2.3. Whole-Transcriptome Analysis of Microbes

Microorganisms	Techniques	Corrected Genes	New Genes	ncRNA	Antisense RNA	Reference
Bacteria						
Mycoplasma pneumonia	TA, RNA-seq, spotted arrays	5	4	108	89	Guell et al. (2009)
Salmonella enteric sv Typhi	ss RNA-seq			40		Perkins et al. (2009)
Chlamydia trachomatis L2b	RNA-seq	5		41	25	Albrecht et al. (2009)
Listeria monocytogenes EGD-e	TA		5	45	7	Toledo-Arana et al. (2009)
Listeria monocytogenes 10403S	RNA-seq			67		Oliver et al. (2009)
Burkholdderia cenocepacia	RNA-seq			13		Yoder-Himes et al. (2009)
Bacillus anthracis Sterne 34eF2	RNA-seq	11	57			Passalacqua et al. (2009)
Bacillus subtilis 168	TA		119	84	127	Rasmussen et al. (2009)
Vibrio cholera	RNA-seq			520	127	Liu et al. (2009)
Archaea						
Sulfolobus solfataricus P2	RNA-seq	162	80	310	185	Wurtzel et al. (2010)
Halobacterium salinarum	TA	61	10	61		Koide et al. (2009)
Eukaryotes						
Schizosaccharomyces pombe	TA, RNA-seq	75	26	427	37	Wilhelm et al. (2008)
Saccharomyces cerevisiae	RNA-seq	64		487		Nagalakshmi et al. (2008)

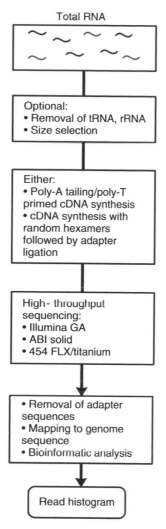

Figure 2.22. Steps involved in microbial transcriptome sequencing.

bacteria (Koide et al., 2009; Wurtzel et al., 2010), suggesting alternative modes of regulation. Long 3′-UTRs could affect expression of downstream genes or genes on the opposite strand, as found in archaea (Brenneis and Soppa, 2009).

Whole-transcriptome data allow operons to be better defined, and the first experimentally determined operon maps show that 60–70% of bacterial genes are transcribed as operons, but only 30–40% in archaea (Siezen et al., 2010). Staircase-like expression within operons appears to be common (Guell et al., 2009).

Whole-transcriptome analysis of *Mycoplasma pneumoniae* using a mixture of tiling arrays, deep sequencing, and 137 different growth conditions showed that there is context-dependent modulation of operon structure (Guell et al., 2009). This involves repression or activation of operon internal genes as well as genes located at the operon ends. This adds a whole new level of complexity to gene regulation. Similar "conditional operons" were found in *Halobacterium salinarum* (Koide et al., 2009).

Whole-transcriptome analysis of several prokaryotes has now identified large numbers of ncRNAs (Table 2.3), some of which are induced during niche switching, such as in *Burkholderia cenocepacia* (Yoder-Himes et al., 2009).

cis-Antisense RNA was previously thought to be extremely rare in prokaryotes, but whole-transcriptome analysis has recently detected hundreds of antisense transcripts in bacteria and archaea. Some of these have been experimentally shown to downregulate their sense counterparts (Toledo-Arana et al., 2009). This is an area in which much is still to be discovered, as *cis*-antisense may be a common form of regulation in prokaryotes.

The ultimate goal is to obtain a complete and bias-free view on microbial transcriptomes (Pan et al., 2005). The question remains in how far RNA-seq has the potential to provide such a view. Clearly, RNA-seq has a number of advantages above microarray technology, since RNA-seq offers both a single-base resolution and a high-mapping resolution. RNA-seq is especially suited to identify novel transcripts, alternative splice variants, and ncRNA (Marioni et al., 2008; Mortazavi et al., 2008; Nagalakshmi et al., 2008; Wilhelm et al., 2008).

2.6.3. Proteomics: A Tool to Unravel the Mysteries of Protein

A core genome could be supported by a set of conserved proteins or core proteome, where the proteome is defined as the collection of structural and functional proteins actually present in the cell and is thus a direct expression of cell phenotype (Callister et al., 2008). Proteome is a complete set of protein expressed by an organism at a given time and/or under the given condition. Proteins play a critical role in biochemical mechanism by directly taking part in signaling and biochemical pathways. Disease processes can be understood better by careful study of protein's function and behavior under different biological conditions. Proteomic aims to study various properties such as protein expression profiles, protein modifications, protein interactions, and protein networks of entire proteome in relation to cell function and biological processes (such as development, disease, and stress response). Proteomics-based experiments are high throughput in nature, with an objective of studying entire proteome of concerned system. Proteomics has become an invaluable tool in biomedical and drug development research.

Proteomics use various technologies such as (i) one- and two-dimensional gel electrophoresis to identify the relative mass of a protein and its isoelectric point; (ii) X-ray crystallography to resolve protein structure; (iii) nuclear magnetic resonance (NMR) to characterize the three-dimensional structure of peptides and proteins; (iv) low resolution techniques such as circular dichroism; (v) Fourier transform infrared spectroscopy and small-angle X-ray scattering (SAXS) to study the secondary structure of proteins; and (vi) mass spectrometry (no tandem), often MALDI TOF, is used to identify proteins by peptide mass fingerprinting (PMF). Protein profiling with MALDI TOF MS can be of high use in clinical diagnostics. Van de Velde et al. (2009) isolated and performed 2D DIGE (difference gel electrophoresis) proteomic analysis of intracellular and extracellular forms of *Listeria monocytogenes*. Comparative proteomics is used to reveal vaccine candidates and pathogenicity factors. Immunoproteomics identifies specific and nonspecific antigens. For the management of a huge amount of data, bioinformatics is a valuable instrument for the construction of complex protein databases.

Jungblut (2001) identified proteins of pathogenic microorganisms by using two-dimensional electrophoresis and mass spectrometry. This classical proteome analysis is now complemented by combination of capillary chromatography/mass spectrometry, miniaturization by chip technology, and protein interaction investigations.

Diversity of proteome is far greater than genome (20,000–25,000 genes vs about 1,000,000 proteins). Therefore, cellular processes cannot be fully understood by studying gene expression alone, and proteomics can provide more insights to cellular processes. The protein diversity is thought to be due to alternative splicing and posttranslational modification of proteins. Proteome diversity offers a great challenge to the research community for complete cataloging of all human proteins and their function and interactions, and such research efforts are coordinated by the Human Proteome Organization (HUPO).

Proteins do not work in isolation, there are complex network of protein modules which are responsible for particular cellular process, and within a network these proteins often interact with each other. Proteomics is used to identify such interactions, and these interaction data provide insights into cellular mechanism and also helps in assigning putative role to newly discovered proteins. A global approach was used to analyze protein synthesis and stability during the cell cycle of the bacterium *Caulobacter crescentus*. Approximately one-fourth (979) of the estimated *C. crescentus* gene products were detected by two-dimensional gel electrophoresis, 144 of which showed differential cell cycle expression patterns. Eighty one of these proteins were identified by mass spectrometry and were assigned to a wide variety of functional groups. Pattern analysis revealed that coexpression groups were functionally clustered. A total of 48 proteins were rapidly degraded in the course of one cell cycle. More than half of these unstable proteins were also found to be synthesized in a cell-cycle-dependent manner, establishing a strong correlation between rapid protein turnover and the periodicity of the bacterial cell cycle (Grünenfelder et al., 2001).

Several methods are available to probe protein–protein interactions. The traditional method is yeast two-hybrid analysis. New methods include protein microarrays, immunoaffinity chromatography followed by mass spectrometry, and combinations of experimental methods such as phage display and computational methods (Shiaw-Lin et al., 2006). Proteomics is also used to study microbial interactions. One aspect of competition between microbes can be simulated by treatment of one microbe with antibiotics produced by a competing microbe. However, an approach based on proteomics alone may not be sufficient to obtain a complete data set for describing microbial interactions. Therefore, further studies are necessary for proteins whose quantitative profile changes, for example, by generating knockout strains for phenotypic analysis. Despite some inherent limitations, proteomics is a useful method, and an important complement to other approaches for studies of microbial interactions (Melin, 2004).

Affinity chromatography, yeast two-hybrid techniques, fluorescence resonance energy transfer (FRET), and surface plasma resonance (SPR) are used to identify protein–protein and protein–DNA binding reactions. X-ray tomography is used to determine the location of labeled proteins or protein complexes in an intact cell. Software-enabled image analysis tools are used to automate the quantification and detection of spots within and among gel samples. These techniques are widely used in experiments on proteomics; however, there are still few technologies, experimental and analytical lacunae, which must be addressed for the evolution of well-defined protocols. Proteomics-based assays have potential to replace traditional toxicological experiments; toxic insult induces cellular response in the form of cell cycle arrest, apoptosis, and so on, which can be monitored by proteomics.

Bacterial genomes, with a simple gene structure, are a particularly attractive target for such methods. The identified peptides validate the predicted genes, correct erroneous gene annotations, and reveal some completely missed genes. While bacterial genome annotations have significantly improved in recent years, techniques for bacterial proteome annotation (including posttranslational chemical modifications, signal peptides, and proteolytic events) are still in their infancy. At the same time, the number of sequenced bacterial genomes is rising sharply, far outpacing our ability to validate the predicted genes, let alone annotate bacterial proteomes. Gupta et al. (2007) used tandem mass spectrometry (MS/MS) to annotate the proteome of *Shewanella oneidensis* MR-1, an important microbe for bioremediation. In particular, they provided the first comprehensive map of posttranslational modifications in a bacterial genome, including a large number of chemical modifications, signal peptide cleavages, and cleavage of N-terminal methionine residues. They also detect multiple genes that were missed or assigned incorrect start positions by gene prediction programs, and suggest corrections to improve the gene annotation.

M. tuberculosis gene annotation is an example, where the most-used data sets from two independent institutions (Sanger Institute and Institute of Genomic Research-TIGR) differ up to 12% in the number of annotated open reading frames, and 46% of the genes contained in both annotations have different start codons. Such differences emphasize the importance of the identification of the sequence of protein products to validate each gene annotation including its sequence coding area. With this objective, de Souza et al. (2008) submitted a culture filtrate sample from *M. tuberculosis* to a high accuracy LTQ-Orbitrap mass spectrometer analysis and applied refined N-terminal prediction to perform comparison of two gene annotations. From a total of 449 proteins identified from the MS data, they validated 35 tryptic peptides that were specific to one of the two data sets, representing 24 different proteins, and, from those, only 5 proteins were annotated in the Sanger database. In the remaining proteins, the observed differences were due to differences in annotation of transcriptional start sites. The results indicated that, even in a less complex sample likely to represent only 10% of the bacterial proteome, they were still able to detect major differences between different gene annotation approaches. This gives hope that high throughput proteomics techniques can be used to improve and validate gene annotations, and, in particular, for verification of high throughput, automatic gene annotations.

The Microbial Proteomic Resource (MPR) is a repository service that contains nonredundant protein databases of related bacterial strains, which were generated through in-house developed software called *Multi Strain Mass Spectrometry Prokaryotic Database Builder* (MSMSpdbb). MSMSpdbb merges and clusters protein sequences inferred from genomic sequences, and provides a protein list in FASTA format that covers for divergence in gene annotation, translational start site choice, and the presence of SNPs and other mutations. MSMSpdbb was developed in C++ using the Qt libraries (Nokia) and licensed under the GNU General Public License version 2. MSMSpdbb is freely available, and its installation files, instructions for use, and additional documentation can be found at the MPR web site http://org.uib.no/prokaryotedb/, and also found at Proteomecommons.org (de Souza et al., 2010).

Even diverse diseases share similar proteomic profile, as diseases induce many common pathways for inflammation, cell adhesion, and so on, which make disease characterization somewhat challenging. Scientific community is actively working on characterizing diseases based on biomarkers that are unique for a particular disease. Resolution of constituent proteins from a complex mixture is the most fundamental aim of proteomics technologies. Various methods are used for the isolation of more complex mixture of proteins: (i) low throughput sequencing through Edman degradation and (ii) MS-based high throughput PMF. PMF-based signatures can be used to build predictive models for diseased sample identification, which forms the platform for proteomics-based diagnostic assays (Petricoin et al., 2002; Tirumalai et al., 2003; Zhu et al., 2003; Yanagisawa et al., 2003; Pan et al., 2005). However, there are certain issues regarding reproducibility, robustness, and standardized detection protocol, which should be considered while designing a profiling-based diagnostic assay (Diamandis 2004, 2007; Baggerly et al., 2005; Ransohoff, 2005).

Gene expression profiling technologies in general and proteomic technologies in particular have proven extremely useful to study the physiological response of bacterial cells to various environmental stress conditions. Complex protein toolkits coordinated by sophisticated regulatory networks have evolved to accommodate bacterial survival under ever-present stress conditions such as varying temperatures, nutrient availability, or antibiotics produced by other microorganisms that compete for habitat. In the last decades, application of synthetic antibacterial agents resulted in additional bacterial exposure to antibiotic stress. While the targeted use of antibiotics has remarkably reduced human suffering from infectious diseases, the ever-increasing emergence of bacteria that are resistant to antibiotics has led to an urgent need for novel antibiotic strategies (Brötz-Oesterhelt et al., 2005).

Proteomics technology plays a very critical role in modern drug discovery process (Fig. 2.23) for the treatment of different diseases and tailoring the drugs for individualized patient therapy. Proteomics technology helps in understanding disease mechanism and evaluation of proteins based on their significance in relation to disease progression. Once a protein has been characterized as the target based on its significant role in concerned disease process, downstream analysis is focused to control the disease by modulating the protein to bring it back to its normal behavior. This modulation could be inhibition of an abnormally activated target protein or activation of a silenced target protein. Traditionally, bioassay experiments are conducted to detect candidate drug/molecule from a large pool of in-house compound repository. Similarly, computational methods make use of virtual high throughput screening (vHTS) protocols to detect potential compounds that can modulate the target protein. These potential compounds or candidates are tested further satisfactorily for their drugability and toxicity profile, and are taken forward for clinical trials.

Following the identification of a new drug candidate, knowledge on organ- and system-level responses help to prioritize the drug targets and design clinical trials based on their efficacy and safety. Toxicoproteomics is playing an important role in that respect (Chakravarti et al., 2010). Designing cancer drugs that act on specific targets offers the advantage that the mechanism of drug action can be understood and accurately monitored in clinical trials leading to the development of better drugs. Deciphering cancer drug targets requires the understanding of biochemical pathways that are affected in the cancer genome. Achieving successful new cancer drug development schemes will require a merger of research disciplines that include pharmacology, genomics, comparative genomics, functional genomics, proteomics, and bioinformatics (Onyango, 2004).

Proteomics has been very useful in our understanding of cellular mechanisms; however, its task will be considered accomplished only when it addresses unresolved challenges such as complete characterization of proteome with details of its spatial, temporal role under various conditions (stress, disease, etc.). With availability of such detailed

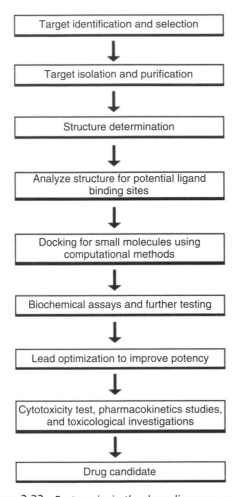

Figure 2.23. Proteomics in the drug discovery process.

data, it would be possible to simulate the system of interest, and thereby significantly expediting drug development cycle.

2.6.4. Metabolomics: A Tool to Ultimately Understand the Whole Metabolome

Metabolomics offers a comprehensive picture of metabolism by metabolome profiling techniques. This critical information about metabolism provides insight into biochemical mechanisms, which complements information deduced through genomics and proteomics technologies. Metabolome refers to the complete set of small-molecule metabolites with molecular mass less than 1000 Da (such as metabolic intermediates, hormones and other signaling molecules, and secondary metabolites) to be found within a biological sample, such as a single organism (Oliver et al., 1998).

Cell liberates hundreds of context-specific metabolites in response to energy changes, toxic insult, disease state, therapeutic treatment, which can be used to understand the biological system. These changes in metabolite were traditionally studied by metabolite fingerprinting, with an objective of identifying metabolites produced by an organism. Metabolomics studies differ from metabolite fingerprinting, as it is of high throughput in nature and conducted with an objective of relative comparison between samples rather than simply identifying metabolites and its broader aim is to quantify every single metabolite in its functional role.

Metabolomics is the comprehensive study of the metabolome present in cells, tissue, and body fluids. Cellular processes are manifestation of flux level changes in enzymatic reaction from biochemical pathways. Metabolomics is a technology to directly observe and study these flux level changes by quantitative and qualitative measurement of metabolites which are end product of enzymatic reaction (Fig. 2.24). The comprehensive metabolomic analysis of *E.coli*, including fluxome, transcriptome and proteome analyses, was reported by Ishii et al. (2007). They showed evidence of homeostasis in the level of metabolites using single-gene deletion mutant strains. Diversity in dynamic range of metabolites flux offers additional challenge for simultaneous estimation of metabolites with slow, extremely fast, transient fluxes. Flux analysis is a very important method for understanding metabolic pathways. Metabolic flux analysis is an analytical technique that quantifies intracellular metabolic fluxes and analyzes functional aspects of their network in greater detail. This analysis is based on mass balances of cellular metabolites normally at a steady state of growth (Mori and Begley, 2008; Sugimoto et al., 2010). The latest achievement is the comprehensive metabolic model of *E.coli*, composed of 2077 reactions and 1039 metabolites (Feist et al., 2007; Ravikirthi et al., 2011).

Transcriptomics and proteomics give details about high level regulation of biochemical pathways, whereas metabolomics gives direct estimates to study biochemical pathways. Metabolomics is a tool to complement other omics technology and has many applications

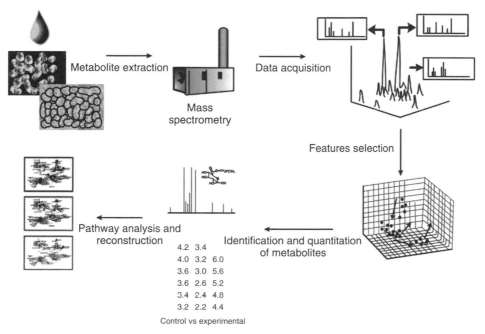

Figure 2.24. Conventional untargeted metabolomics workflow.

such as (i) assigning novel function to a protein, (ii) disease diagnosis, (iii) optimizing production by detailed understanding of the relationship between protein (enzyme) and metabolite production, and (iv) drug toxicity assay (Lindon et al., 2004). Demonstrating that metabolomic technique is useful in microbial studies as well, Bundy et al. (2005) were able to successfully discriminate between nonpathogenic strains of *B. cereus* and pathogenic strains isolated from meningitis patients.

Metabolomics influences many aspects of microbiology. A systematic optimization of metabolic quenching and a sample derivatization method for GC–MS metabolic fingerprint analysis was reported by Tian et al. (2009). Methanol, ethanol, acetone, and acetonitrile were selected to evaluate their metabolic quenching ability, and acetonitrile was regarded as the most efficient agent. The optimized derivatization conditions were determined by full factorial design considering temperature, solvent, and time as parameters. The best conditions were attained with N, O-bis(trimethylsiyl) trifluoroacetamide as derivatization agent and pyridine as solvent at 75 °C for 45 min. Method validation ascertained the optimized method to be robust. The above method was applied to metabolomic analysis of six different strains and it is proved that the metabolic trait of an engineered strain can be easily deduced by clustering analysis of metabolic fingerprints.

Environmental processes in ecosystems are dynamically altered by several metabolic responses in microorganisms, including intracellular sensing and pumping, battle for survival, and supply of or competition for nutrients. Notably, intestinal bacteria maintain homeostatic balance in mammals via multiple dynamic biochemical reactions to produce several metabolites from undigested food, and those metabolites exert various effects on mammalian cells in a time-dependent manner. Fukuda et al. (2009) established a method for the analysis of bacterial metabolic dynamics in real time and used it in combination with statistical NMR procedures and this novel method is called real-time metabolotyping (RT-MT), which performs sequential ^{1}H-NMR profiling and two-dimensional ^{1}H, ^{13}C-HSQC (heteronuclear single quantum coherence) profiling during bacterial growth in an NMR tube. The profiles were evaluated with such statistical methods as Z-score analysis, principal components analysis, and time series of statistical *to*tal *c*orrelation *s*pectroscopy (TOCSY). In addition, using 2D ^{1}H, ^{13}C-HSQC with the stable isotope labeling technique, the metabolic kinetics of specific biochemical reactions based on time-dependent 2D kinetic profiles was also observed. Using these methods, the pathway for linolenic acid hydrogenation by a gastrointestinal bacterium, *Butyrivibrio fibrisolvens*, was further clarified and identified *trans* 11 : *cis* 13 conjugated linoleic acid as the intermediate of linolenic acid hydrogenation by *B. fibrisolvens*, based on the results of ^{13}C-labeling RT-MT experiments. In addition, it was concluded that the biohydrogenation of polyunsaturated fatty acids serve as a defense mechanism against their toxic effects.

This method, RT-MT, is useful for the characterization of beneficial bacterium that shows potential for use as a probiotic by producing bioactive compounds.

Metabolic profiling through metabolomics comes with unprecedented challenges when compared with transcriptomics and proteomics. Genomics and transcriptomics technologies are developed for targeted analysis of sequences of four different nucleotides, and proteomics technology are based on the analysis of sequence of 22 amino acids. In metabolomics technological challenge is to develop and refine analytical methods for accurate estimation of metabolites having wide chemical diversity; metabolites belong to huge number of chemical compounds such as hydrophobic lipids, amino/nonamino organic acids, natural products, volatile alcohols, ketones, and hydrophilic carbohydrates.

It has been found that many major metabolic pathway genes are regulated by RpoS at the early stationary growing phase (Rahman and Shimizu, 2008). It might be the case that the regulation of global dynamic changes in cellular metabolism is primarily controlled by transcription factors, especially by so-called global regulators, such as RpoS, CRP, and SoxRS. Gene regulation in metabolic pathways is important (Ono et al., 2008). The expression of genes involved in lysine biosynthesis in *E. coli* is to obtain a quantitative understanding of the gene regulatory network. Promoter fusion is used with green fluorescent protein (GFP) of genes associated with lysine biosynthesis and monitored time-dependent changes in gene expression in response to changes in lysine concentration. On the basis of the quantitative data from flow cytometry, dynamic and theoretical models of gene expression were constructed to estimate the parameters of gene regulation (Mori and Begley, 2008).

Genome-scale modeling and simulation has been pointed out as one of the important directions in the systems approach. Using genome-scale modeling, the extreme halophile *H. salinarum* was found to consist of 557 metabolites, 600 reactions associated with 417 genes and 111 transport reactions with 73 genes mainly from the KEGG database (Gonzalez et al., 2009). The computational analysis of the aerobic growth of this organism was investigated using dynamic simulations in media with 15 available carbon and energy sources. This approach is useful for interpretation and for making hypotheses for next-step experiments and for showing some biologically interesting predictions for ribose and shikimate biosynthesis. The experiments to confirm their predictions are now underway and we look forward to seeing the evidence in the near future.

Soga et al. (2003) developed an efficient method to identify and quantify metabolites in a cell in a high throughput manner by combination of capillary electrophoresis with mass spectrometry by analyzing the alteration of metabolite concentration profiles after histidine starvation in *E. coli*. The extraction method of metabolites from *E. coli* cells was improved for this analysis and analyzed 375 charged, hydrophilic intermediates in a primary metabolism. The response of the *E. coli* cell to histidine starvation has been described; however, this method still remains to be improved for analyzing the oxidation of metabolite pools.

Biofilm formation is a critical step for infections by pathogens and has been implicated in the barrier for resistance of microbes to antibiotics and immune responses. Therefore, the efficient detection and quantification of key components are required to assist the design of clinical infection responses. To identify and quantify these biofilm components, which are secreted outside of the cell, is another important metabolome target, especially for environmental response and bacterial pathogenicity. Mark Howard and his colleagues describe here a new method using NMR and its application for the identification of carbohydrate polymer components in crude biofilm extracts from *Staphylococcus epidermidis* (Wagstaff et al., 2008).

Metabolome analysis is mainly carried out in a targeted or nontargeted (metabolite profiling) manner. It is designed for the analysis of selected class of metabolites such as flavonoids in targeted experiments, whereas, in nontargeted profiling, high throughput analysis is conducted with an objective to analyze all metabolites of the system of interest. NMR and MS are the most common detection techniques used in metabolomics. However, any spectroscopic method designed to detect low molecular weight compounds can be used for metabolomics. MS and NMR methods are preferred because of qualities such as detection capability, sensitivity, and ability to analyze a broad range of metabolites within one spectral acquisition. Small-molecule metabolites are extracted from the sample matrix. Metabolites are separated using chromatographic steps, ionized, and analyzed using MS. Features of interest are selected from raw data using univariate and multivariate statistical approaches. Features are then identified using database searches, comparisons to authentic standards, and MS/MS. Identified features can then be used for reaction monitoring, pathway analysis, and metabolic network reconstruction (Lee et al., 2010).

In metabolomics research, there are several steps between sampling of the biological condition under study and the biological interpretation of the results of the data analysis (German et al., 2002). Spectral data from MS and NMR studies often have noise, and hence must be preprocessed before final biological interpretation. First, the biological samples are extracted and prepared for analysis. Subsequently, different data preprocessing steps (Fiehn, 2002) are applied in order to generate "clean" data in the form of normalized peak areas that reflect the (intracellular) metabolite concentrations. Data is further transformed using appropriate methods such as log/relative transformation. Peaks detected in NMR spectra give an idea about the structure of a metabolite, and peaks detected by MS give an idea about the molecular weight of the metabolite. NMR and MS are often used in conjugation for study of novel metabolites.

Various statistical methods such as PCA (principal component analysis) and multivariate analysis are used to identify significant markers in spectral data. Metabolites can be identified by comparing their processed spectra with repository of spectra from known compounds. The use of metabolomic data to predict the disease susceptibility will require robust set of bioinformatics tools and quantitative reference databases. These databases containing metabolite profiles from the population must be built, stored, and indexed according to metabolic and health status. Building and annotating these databases with the knowledge to predict how a specific metabolic pattern from an individual can be adjusted with diet, drugs, and lifestyle to improve health represents a logical application of the biochemistry knowledge that life sciences have produced over the past 100 years (German et al., 2005). In January 2007, scientists at the University of Alberta and the University of Calgary completed the first draft of the human metabolome (Wishart et al., 2007). They have cataloged and characterized 2500 metabolites, 1200 drugs, and 3500 food components that can be found in the human body (Harrigan and Goodacre, 2003).

Metabolic signatures provide prognostic, diagnostic, and surrogate markers for a disease state; the ability to subclassify a disease; biomarkers for drug response phenotypes (pharmacometabolomics); and information about mechanisms of diseases (Ellis and Goodacre, 2006). Indeed, sophisticated metabolomic analytical platforms and informatic tools have been developed that make it possible to define initial metabolic signatures for several diseases (Kaddurah-Daouk, 2006). Therefore, metabolomics is a valuable platform for studies of complex diseases and the development of new therapies, both in nonclinical disease model characterization and clinical settings.

Applications in preclinical drug safety studies are illustrated by the Consortium for Metabolomic Toxicology, a collaboration involving several major pharmaceutical companies. The use of metabolomics in disease diagnosis and therapy monitoring, and the concept of pharmacometabonomics as a way of predicting an individual's response to treatment can be studied (Ellis and Goodacre, 2006). Combination drug therapies with individualized optimization are likely to become a major focus. Metabolomics incorporates the most advanced approaches to molecular phenotype system readout and provides the ideal diagnostic platform for the discovery of biomarker patterns associated with healthy and disease states, for use in personalized health monitoring programs and for the design of individualized interventions (German et al., 2003). Today, clinicians capture only a very small part of the information contained in the metabolome, as revealed by a defined set of blood chemistry analyses to define health and disease states. Examples include measuring glucose to monitor diabetes and measuring cholesterol for cardiovascular health. Such a narrow chemical analysis could potentially be replaced in the future with a metabolic signature that captures global biochemical changes in a disease and on treatment.

2.6.5. Integration of Omics

Genomics, transcriptomics, proteomics, and metabolomics have enhanced our understanding of biological system by several folds. Genomics has provided us explanation of differences in response among different species as well as among different individual. Transcriptomics has equipped us with a tool to study gene expression changes and their correlation with studied condition. Proteomics has given us a technology to study the expression and interactions of proteins. Metabolomics had helped us to directly study metabolites under different settings. Each of these omics technology is limited in scope, as it can explain only one aspect of biological system. Results of transcriptomics cannot be directly extrapolated to protein expression, as it has its limitation to address features such as posttranslation modification, phosphorylation, and alternative splicing. Protein expression is not perfectly correlated with its metabolic efficiency, that is, it cannot predict flux of metabolites. Similarly metabolomics cannot directly relate protein/gene controlling particular metabolite, as a single metabolite is known to be common across different metabolic pathways.

Omics technology if studied in isolation provides us certain specific snapshots of a complex biological system. Individual contributions of omics technology cannot be disputed; however, it is always desirable to integrate snapshots provided by different omics technologies to get a holistic picture of complex biological system (Fig. 2.25; Fischer, 2003; Loughlin, 2007). Such a challenging task of integration of omics technology is undertaken by a new discipline "systems biology."

During the last decade, we witnessed the huge impact of the comparative genomics for understanding genomes (from the genome organization to their annotation). However,

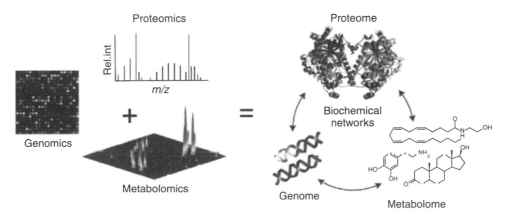

Figure 2.25. Integration of omics technology.

those genomic approaches quickly reach their limits when one looks at investigating the functional properties of genes in the wide context of a genome. Such limitation may be overcome; thanks to recent high throughput experimental progresses such as those obtained via metabolic and coexpression studies, which produce the so-called omics data. Therefore, integrating those data and the state-of-the-art computational genomic comparison is a natural evolution.

Angibaud et al. (2009) proposed a heuristic algorithm IISCS that incorporates omics knowledge into the IILCS heuristic, already known as accurate to compare genomes. When applied on bacteria, one emphasizes large functional units composed of several operons. This technique combined comparative genomics algorithms and *omics* data. This integrated genomics approach is not appropriate to predict operons as other standard methods, but emphasizes functional units (themselves composed by operons more or less conserved among species) and their respective organizations in terms of gene rearrangements. Beyond the simple help to the genome annotation, this information is useful to discriminate functional units conserved from one species to another that might be further investigated using systems biology techniques and other modeling approaches (Angibaud et al., 2009; Bayjanov et al., 2009).

In the current omics era, innovative high throughput technologies allow measuring temporal and conditional changes at various cellular levels. Individual analysis of each of these omics data undoubtedly results in interesting findings; it is only by integrating them a global insight into cellular behavior will be gained. A systems approach is thus predicated on data integration. However, because of the complexity of biological systems and the specificities of the data generating technologies (noisiness, heterogeneity, etc.), integrating omics data in an attempt to reconstruct signaling networks is not trivial. Developing its methodologies constitutes a major research challenge. Besides their intrinsic value toward health care, environment, and industry, prokaryotes are ideal model systems to further develop these methods because of their lower regulatory complexity compared with eukaryotes, and the ease with which they can be manipulated (De Keersmaecker et al., 2006).

Omics technology is widely used for diagnosis and drug discovery. Integration of omics technology (Fig. 2.26) would increase accuracy, reliability, and scope of omics-enabled application areas. Biomarkers are generally used for the detection of diseased condition (diagnosis), drug-induced toxicity, and drug efficacy. Omics technology offers very effective platform for biomarker discovery. Each omics technology represents certain aspect of cellular processes affected/altered in conditions such as disease and toxicity, and monitoring these changes forms the foundation for omics-technology-specific detection kit. Integration of these different aspects of cellular processes should form a better foundation for biomarker-based applications. Identification of a drug target is a very challenging task, which is directed toward finding a gene/protein that plays a critical role in disease mechanisms and has a control role in biochemical pathways. The results from biomarker analysis should be integrated with flux information derived from metabolomics experiment; this helps to identify the protein controlling the flux in the pathway and thus could be a candidate drug target. The presence of SNP in human genome is well established, which explains individual level differences toward drug response, disease susceptibility, and tolerance. Integration of knowledge gained from SNP analysis with identified drug targets would be a step toward implementation of personalized medicine, which would make therapy effective with minimal side effects.

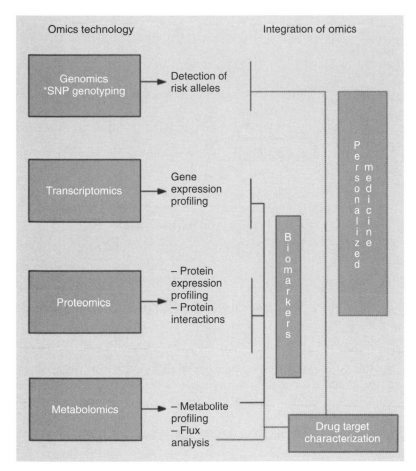

Figure 2.26. Integrating omics could create a definite result in biomarkers, drug target characterization, and personalized medicine.

The next-generation sequencing-based technology is the latest tool in the field of omics, with an ability to transform the way genomics and transcriptomics experiments were conducted till now. It has widespread applications; it can be used for transcriptomics, methylation analysis, genome resequencing, variations detection, among others. This technology generates huge data in very short time.

Omics technology provides a platform to monitor cellular process by high throughput monitoring of genes, proteins, metabolites. These technologies have potential for providing insights into changes happening at transcription (transcriptomics), translation (proteomics), and metabolism (metabolomics). Omics technologies have definitely added to our understanding of complex disease mechanism, and have thus expedited therapeutic efforts by providing potential drug targets, biomarkers (Fig. 2.27). However, full potential of omics technology has not been realized, because of factors such as lack of well-established standards, lack of integration of various omics technology, and lack of comprehensive reference omics profiles.

A technological advance in omics has surpassed its application, which can be attributed to evolving of newer platform before its potential was fully realized. Greater cooperation

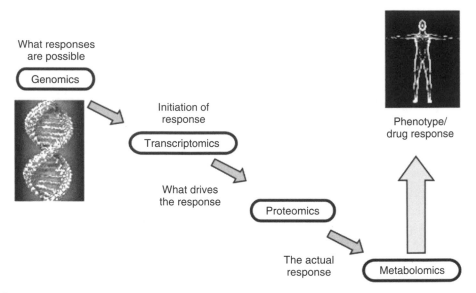

Figure 2.27. Drug response after an integrated omic research (http://sites.google.com/site/ prasadphapalenow/).

between scientific communities would be required to formulate data standards for seamless integration of data generated by different omics platforms and instruments.

2.7. BIOINFORMATICS IN MICROBIAL TECHNOLOGY

Biological data are being continuously produced at a phenomenal rate. As a result of this surge, computational analysis has become indispensable to biological research. Computational analysis of biological data helps experimentation in various ways, namely, hypothesis generation, hidden pattern extraction, and discovery of novel correlations. Bioinformatics is the most promising and fast growing sector in life sciences. Bioinformatics defines the application of computational technique to understand and organize the information associated with biological macromolecules (Luscombe et al., 2001). In general, the aims of bioinformatics are three-fold (Fig. 2.28). First, at its simplest bioinformatics organizes data in a way that allows researcher to access existing information and to submit new entries as they are produced, notable examples are biological sequence databanks such as NCBI, EBI, Swissport, and Ensembl. While data curation is an essential task, the information stored in the databases is essentially useless until analyzed. Thus, the purpose of bioinformatics extends much further. The second aim is to develop tools and resources that aid in the analysis of the data. For example, having sequenced a particular protein, it is of interest to compare it with previously characterized sequences. This needs more than just a simple text-based search, and programs such as FASTA and PSI-BLAST must constitute a biologically significant match. Development of such resources dictates expertise in computational theory as well as a thorough understanding of biology. The third aim is to use these tools to analyze the data and interpret the results in the biologically useful manner.

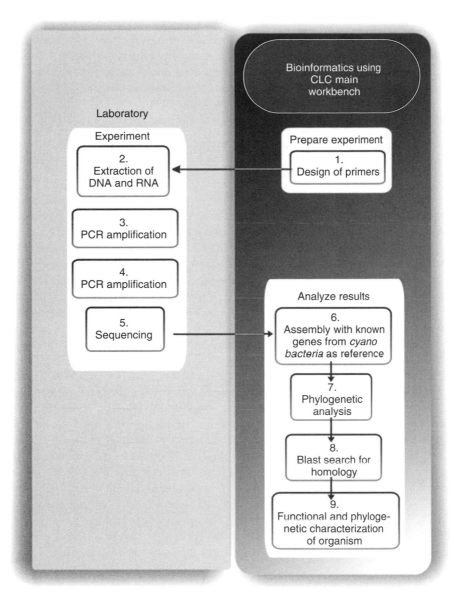

Figure 2.28. Typical bioinformatics research workflow (http://scicasts.com/resources/bioit-biotechnology/a-case-study-on-bioinformatics-in-microbial-research).

With the completion of human genome sequencing and the sequencing of various plants, animals, and microbes genomes, bioinformatics is poised to take up the challenges of the postgenomic era and the need for well-trained human resource is greater. Bioinformatics and sequence analysis allow us to develop hypotheses about the presence of a gene, and the nature and the role of proteins that the gene encodes. New insights into the molecular basis of the disease may come from investigating the function of homologs of a disease gene in model organisms. In this case, homology refers to the fact that two genes share a common evolutionary history. Equally interesting is the potential for uncovering evolutionary relationships and patterns between different forms of life. With the aid of nucleotide and protein sequences, it should be possible to find the ancestral ties between different organisms. So far it is known that closely related organisms have similar sequences and more distantly related organisms have more dissimilar sequences. Proteins that show a significant sequence conservation indicating a clear evolutionary relationship are said to form the same protein family. By studying protein folds and families, scientist are able to reconstruct the evolutionary relationship between two species and to estimate the line of divergence between two organisms since they shared a common ancestor.

The growing wealth of completely sequenced microbial genomes does not directly illuminate these natural ecosystems. Direct amplification or cloning from environmental samples of genes involved in metabolic processes of interest does allow us to investigate molecular processes at the genetic level among these poorly known systems. A major informatics challenge is to integrate data from the sequencing of multiple different genes from a mixture of unknown organisms and assemble a model of which genes co-occur within individual species. Dickerman and Tyler (2002) proposed to develop an informatics-based solution to this problem based on concordance of gene phylogenies and sequence attributes such as codon and oligonucleotide word frequencies. The system will take anonymous DNA sequences from multiple loci and estimate which ones come from the same species. A successful solution will enable a new level of sophistication in the way they can study the functional genomics of the largely unknown, unculturable microbes that dominate terrestrial, subterranean, and aquatic ecosystems (Dickerman and Tyler, 2002).

A self-organizing map (SOM) is an effective tool for clustering and visualizing high dimensional complex data on a two-dimensional map. Takashi et al. (2006) modified the conventional SOM to genome informatics, making the learning process and resulting map independent of the order of data input, and developed a novel bioinformatics tool for phylogenetic classification of sequence fragments obtained from pooled genome samples of microorganisms in environmental samples allowing visualization of microbial diversity and the relative abundance of microorganisms on a map. First, they constructed SOMs of tri- and tetranucleotide frequencies from a total of 3.3 Gb of sequences derived using 113 prokaryotic and 13 eukaryotic genomes, for which complete genome sequences are available. SOMs classified the 330,000 10 kb sequences from these genomes mainly according to species without information on the species. Importantly, classification was possible without orthologous sequence sets and thus was useful for studies of novel sequences from poorly characterized species such as those living only under extreme conditions and which have attracted wide scientific and industrial attention. Using the SOM method, sequences that were derived from a single genome but cloned independently in a metagenome library could be reassociated *in silico* (Takashi et al., 2006).

The Bioinformatics Links Directory features curated links to molecular resources, tools, and databases. The links listed in this directory are selected on the basis of recommendations from bioinformatics experts in the field. Table 2.4 contains links to microbial genome resources including genome sequences, annotations, and comparative analyses.

2.8. FUTURE CHALLENGES: THE BIOCHIPS

Technological advances in miniaturization have found a niche in biology and signal the beginning of a new revolution. A new revolution technology that may become promising

TABLE 2.4. Comprehensive Genomic Resource

Genomic Resource	Description
BAGET (http://archaea.u-psud.fr/bin/baget.dll)	BAGET (Bacterial and Archaeal Gene Exploration Tool) is a web service designed to facilitate extraction of specific gene and protein sequences from completely determined prokaryotic genomes. Query results can be exported as a rich text format file for printing, archival, or further analysis.
BROP (http://brop.org/)	The Bioinformatics Resource for Oral Pathogens (BROP) contains tools for genomics of oral pathogens including Genome Viewer, GOAL (genome-wide alignment), an oral pathogen microarray database, an entrez counter, oral pathogen-specific BLAST, and a codon usage database.
CMR (http://cmr.tigr.org/tigr-scripts/CMR/CmrHomePage.cgi)	The Comprehensive Microbial Resource (CMR) gives access to a central repository of the sequence and annotation of all complete public prokaryotic genomes as well as comparative genomics tools across all genomes in the database.
E. coli Genome Project (http://www.genome.wisc.edu/)	University of Wisconsin E. coli genome project site maintains and updates the annotated sequence of the E. coli K12 genome. It also has resources and tools such as ASAP for user input on functional annotation and characterization of E.coli genes.
E. coli Index (http://ecoli.bham.ac.uk/)	Comprehensive guide to information relating to E.coli; home of Echobase: a database of E. coli genes characterized since the completion of the genome.
GeneDB (http://www.genedb.org/)	The GeneDB project's primary goal is to develop and maintain curated database resources for three organisms: Schizosaccharomyces pombe, the kinetoplastid protozoa Leishmania major, and Trypanosoma brucei.
PseudoCAP (http://www.cmdr.ubc.ca/bobh/PAAP.html)	Pseudomonas aeruginosa Community Annotation Project (PseudoCAP); information and tools for genome analysis and annotation of P. aeruginosa.
Pseudomonas Genome Project (http://pseudomonas.com/)	This is a comprehensive database on all Pseudomonas species genomes providing primarily access to Pseudomonas aeruginosa genomic data and annotation. Its interface facilitates comparative analyses of genes, proteins, annotations, and gene orders, and contains a wealth of additional data including pathway-based, operon-based, protein-localization-based, gene-function-category- and ortholog/paralog-based information.
TIGR Microbial Database (http://www.tigr.org/tdb/mdb/mdbcomplete.html)	Table of complete and in-progress microbial genomes with links to publications.
TIGR Protist Gene Indices (http://www.tigr.org/tdb/tgi/protist.shtml)	Gene indices for various protists including Trichomonas vaginalis, Trypanosoma brucei, and Dictyostelium discoideum.

(continued)

TABLE 2.4. (*Continued*)

Genomic Resource	Description
Genome Annotation	
AHMII (http://www.wdcm.org/AHMII/ahmii.html)	Agent to Help Microbial Information Integration (AHMII) offers a search engine for particular strains present in culture collections and databases of bacteria, fungi, yeasts, and cell lines.
AMIGene (http://www.genoscope.cns.fr/agc/tools/amigene/Form/form.php)	Annotation of MIcrobial Genes (AMIGene) is a gene prediction server that can identify coding sequences in microbes.
BASys (http://wishart.biology.ualberta.ca/basys/)	BASys (Bacterial Annotation System) is a tool for automated annotation of bacterial genomic (chromosomal and plasmid) sequences including gene/protein names, GO functions, COG functions, possible paralogs and orthologs, molecular weights, isoelectric points, operon structures, subcellular localization, signal peptides, transmembrane regions, secondary structures, 3D structures, reactions, and pathways.
CGView Server (http://stothard.afns.ualberta.ca/cgview_server/)	The CGView Server generates graphical maps of circular genomes that can be used to visualize sequence conservation in the context of sequence features, imported analysis results, open reading frames, and base composition plots. The server uses BLAST to compare the primary sequence of up to three genomes or sequence sets, aiding in the identification of conserved genome segments, instances of horizontal transfer, or visualization of genome segments from newly obtained sequence reads.
CRISPRFinder (http://crispr.u-psud.fr/Server/CRISPRfinder.php)	Clustered Regularly Interspaced Short Palindromic Repeats (CRISPR) Finder detects this family of direct repeats found in the DNA of many bacteria and archaea.
DOE Joint Genome Institute Genome Portal (http://genome.jgi-psf.org/)	The Joint Genome Institute Genome Portal contains browsable and blastable genome assemblies for several organisms including pufferfish, frog, and sea squirt.
IBM Genome Annotation Page (http://cbcsrv.watson.ibm.com/Annotations/home.html)	IBM's Bio-Dictionary-based Annotations of Completed Genomes page lists annotations for over 75 complete genomes (archae, bacteria, eurkaryotes, and viruses). You can query these annotations at the sequence level as well as search/compare across genomes.
IGSA: Integrative Services for Genomic Analysis (http://isga.cgb.indiana.edu/)	IGSA is a web-based prokaryotic genome annotation server that provides intuitive web interfaces for biologist to customize their own annotation pipelines.

KAAS: KEGG Automatic Annotation Server
(http://www.genome.ad.jp/tools/kaas/)

KAAS (KEGG Automatic Annotation Server) provides functional annotation of genes by BLAST comparisons against the manually curated KEGG GENES database. The result contains KO (KEGG Orthology) assignments and automatically generated KEGG pathways.

Microbial Genome Viewer
(http://www.cmbi.ru.nl/MGV/)

Tool for visualization of microbial genomes. Chromosome wheels and linear genome maps with user specified features/color coding can be generated interactively. Graphics are created in SVG format.

Projector 2
(http://molgen.biol.rug.nl/websoftware/projector2/)

Projector 2 allows users to map completed portions of the genome sequence of an organism onto the finished (or unfinished) genome of a closely related species or strain. Using the related genome sequence as a template can facilitate sequence assembly and the sequencing of the remaining gaps.

YOGY
(http://www.sanger.ac.uk/PostGenomics/S_pombe/YOGY/)

Eukaryotic Orthology (YOGY) is a resource for retrieving orthologous proteins from nine eukaryotic organisms. Using a gene or protein identifier as a query, this database provides comprehensive, combined information on orthologs in other species using data from five independent resources: KOGs, Inparanoid, Homologene, OrthoMCL, and a table of curated orthologs between budding yeast and fission yeast. Associated gene ontology (GO) terms of orthologs can also be retrieved.

Functional Analysis

ATIVS (http://influenza.nhri.org.tw/ATIVS/)

Analytical Tool for Influenza Virus Surveillance (ATIVS) is a web server for analyzing serological data of all influenza viruses and provide interpretive summaries. ATIVS also compares the HA1 sequences of viruses to those of the reference vaccine strains to predict influenza A/H3N2 antigenic drift.

BAGEL
(http://bioinformatics.biol.rug.nl/websoftware/bagel/bagel_start.php)

BActeriocin GEnome mining tooL (BAGEL) identifies putative bacteriocin ORFs (antimicrobial peptides) based on a database containing information about known bacteriocins and adjacent genes involved in bacteriocin activity.

BAGEL2 (http://bagel2.molgenrug.nl/)

BActeriocin GEnome mining tooL (BAGEL2) identifies putative bacteriocins on the basis of conserved domains, physical properties, and genomic context. Improved genome mining capacity from BAGEL.

(continued)

TABLE 2.4. (*Continued*)

Genomic Resource	Description
CPA: Comparative Pathway Analyzer (https://www.cebitec.uni-bielefeld.de/groups/brf/software/cpa/index.html)	The Comparative Pathway Analyzer (CPA) is a web tool for comparative metabolic network analysis. The differences in metabolic reaction content between two sets of organisms are computed and displayed on KEGG pathway maps.
CRISPRcompar (http://crispr.u-psud.fr/CRISPRcompar/)	Clustered Regularly Interspaced Short Palindromic Repeats (CRISPR) elements are used in CRISPRcompar as a genetic marker for comparative and evolutionary analysis of closely related bacterial strains.
CVTree (http://tlife.fudan.edu.cn/cvtree)	Composition Vector Tree (CVTree) infers phylogenetic relationships between microbial organisms by comparing their proteomes using a composition vector approach.
EnteriX (http://globin.cse.psu.edu/enterix/)	EnteriX is a collection of tools for viewing pairwise and multiple alignments for bacterial genome sequences.
GeConT 2 (http://gecont.ibt.unam.mx/cgi-bin/gecont/gecont.pl)	The Gene Context Tool (GeConT) is a visualization tool for viewing the genomic context of a gene or group of genes, and their orthologous relationships, within any of the fully sequenced bacterial genomes. Sequence retrieval is also possible.
GFS (http://gfs.unc.edu/cgi-bin/WebObjects/GFSWeb)	Genome-based Fingerprint Scanning (GFS) takes as input an experimentally obtained peptide mass fingerprint, scans a genome sequence of interest, and outputs the most likely regions of the genome from which the mass fingerprint is derived.
G-language GAE (http://www.g-language.org/wiki/soap/)	G-language Genome Analysis Environment (G-language GAE) is a compendium of programs that focus on bacterial genome analysis. Included are programs for identification of binding sites, analysis of nucleotide composition bias, and visualization of genomic information.
Hhomp (http://toolkit.tuebingen.mpg.de/hhomp)	HHomp is a web server for prediction and classification of outer membrane proteins (OMPs). Beginning with sequence similarity of a protein to known OMPs, HHomp builds a hidden Markov model (HMM) and compares the input sequence to a database of OMPs by pairwise HMM comparison. The OMP database contains profile HMMs for over 20,000 putative OMP sequences.
Insignia (http://insignia.cbcb.umd.edu/)	Insignia provides a web interface for identifying unique genomic signatures from a database of all current bacterial and viral genomic sequences. Input is any set of target and background genomes.

Tool	Description
In silico experiments with complete bacterial genomes (http://insilico.ehu.es/)	Provides tools for theoretical PCR amplification, AFLP-PCR, and PFGE with all up-to-date public complete bacterial genomes (300+ genomes available).
Integrated Microbial Genomes (IMG) (http://img.jgi.doe.gov/)	The Integrated Microbial Genomes (IMG) system facilitates the comparison of genomes sequenced by the Joint Genome Institute (JGI). It can be searched using keywords or BLASTp, and the gene records displayed include biochemical properties, protein domains, chromosomal location and neighborhood and lists of paralogs and orthologs. One can easily build a list of genomes to be considered or excluded from the search and the Phylogenetic Profiler tool allows one to refine the selection by building a list of homologs either common to or excluded from specific organisms.
IslandPath (http://www.pathogenomics.sfu.ca/islandpath/)	IslandPath aids genomic island detection in prokaryotic genome sequences, using features such as dinucleotide bias, G + C, location of tRNA genes, annotations of mobility genes, etc. Genomic islands are defined here as genomic regions of potential horizontal origin.
JpHMM (http://jphmm.gobics.de/)	Jumping Profile Hidden Markov Model (jpHMM) takes a HIV-1 genome sequence and uses a precalculated multiple alignment of the major HIV-1 subtypes to predict the phylogenetic breakpoints and HIV subtype of the submitted sequence.
JproGO (http://www.jprogo.de/index.jsp)	JProGO is a tool for the functional interpretation of prokaryotic microarray data using Gene Ontology information.
JvirGel (http://www.jvirgel.de/)	JVirGel creates a virtual 2D gel based on theoretical molecular weights and calculated isoelectric points from a set of input proteins or proteomes.
mGenomeSubtractor (http://bioinfo-mml.sjtu.edu.cn/mGS/)	mGenomeSubtractor performs mpiBLAST-based comparison of reference bacterial genomes against multiple user-selected genomes. Such *in silico* subtractive hybridization also allows for definition of species-specific gene pools and can be used to develop genomic arrays.
MGIP (http://mgip.biology.gatech.edu/)	The Meningococcus Genome Informatics Platform (MGIP) is a suite of computational tools for the analysis of multilocus sequence typing data. Input DNA sequence files are analyzed to provide allele calls, and to characterize the specific sequence type and clonal complex of the *Neisseria meningitidis* strain.
Microbe Browser (http://microbe.vital-it.ch/)	The Microbe Browser is a web server providing comparative microbial genomics data integrated from GenBank, RefSeq, UniProt, InterPro, Gene Ontology, and the Orthologs Matrix Project (OMA) databases. Gene predictions based on five software packages are also displayed.

(continued)

131

TABLE 2.4. (*Continued*)

Genomic Resource	Description
Microbial Genome Database (http://mbgd.genome.ad.jp/)	MBGD is a database for comparative analysis of completely sequenced microbial genomes using ortholog identification, paralog clustering, motif analysis, and gene order comparison.
MicroFootPrinter (http://bio.cs.washington.edu/MicroFootPrinter.html)	MicroFootPrinter identifies the conserved motifs in regulatory regions of prokaryotic genomes using the phylogenetic footprinting program FootPrinter.
MIRU-VNTRplus (http://www.miru-vntrplus.org/)	MIRU-VNTRplus allows users to analyze genotyping data of their *Mycobacterium tuberculosis* strains either alone or in comparison with the reference DB of strains. The web server also includes tools to search for similar strains, phylogenetic analysis, and mapping of geographic information.
MLST (http://www.mlst.net/)	Multilocus Sequence Typing (MLST) is a nucleotide-sequence-based approach for the unambiguous characterization of isolates of bacteria and other organisms using the sequences of internal fragments of seven house-keeping genes.
MobilomeFINDER (http://mml.sjtu.edu.cn/MobilomeFINDER/)	MobilomeFINDER (Mobile Genome FINDER) is a tool for high throughput genomic island discovery.
NAST (http://greengenes.lbl.gov/cgi-bin/nph-NAST_align.cgi)	Nearest Alignment Space Termination (NAST) is a multiple sequence alignment server for comparative analysis of 16S rRNA gene sequences from bacteria and archaea.
Ogtree (http://bioalgorithm.life.nctu.edu.tw/OGtree/)	Overlapping genes (OG) in prokaryotic species are used in OGtree to construct genome phylogeny trees. Overlapping gene content and overlapping gene order of the whole genome is used for the distance-based method of tree construction.
Phydbac2 (http://igs-server.cnrs-mrs.fr/phydbac/)	Phylogenomic Display of Bacterial Genes (Phydbac2) is a tool to visualize and explore the phylogenomic profiles of bacterial protein sequences. It also allows the user to view sequence similarity across different organisms, access other genes with similar conservation profiles, and view genes that are found nearby a selected gene in multiple genomes.
ProdoNet (http://www.prodonet.tu-bs.de/)	A web-based application for the mapping of prokaryotic genes and the corresponding proteins to common gene regulatory and metabolic networks. Users input a list of genes from which shared operons, coexpressed genes, and shared regulators are detected. Common metabolic pathways are then viewed on KEGG maps.

PROFtmb
(http://cubic.bioc.columbia.edu/services/proftmb/)

PROFtmb predicts transmembrane beta-barrel (TMB) proteins in gram-negative bacteria. In addition to running your own predictions, you can also download predictions for all proteins in 78 gram-negative bacterial genomes.

RegAnalyst (http://www.nii.ac.in/ deepak/RegAnalyst/)

RegAnalyst is a web server for analysis of motifs, pathways, and network. It integrates MoPP (Motif Prediction Program), MyPatternFinder, and MycoRegDB (mycobacterial promoter and regulatory elements database).

RegPredict (http://regpredict.lbl.gov/)

RegPredict web server provides comparative genomics tools for reconstruction and analysis of microbial regulons using comparative genomics.

REPK
(http://rocaplab.ocean.washington.edu/tools/repk)

Restriction Endonuclease Picker (REPK) assists in the choice of restriction endonucleases for terminal restriction fragment length polymorphism (T-RFLP) by finding sets of four restriction endonucleases which together uniquely differentiate user-designated sequence groups.

sgTarget (http://www.ysbl.york.ac.uk/sgTarget/)

sgTarget is a structural genomics resource that helps to select and prioritize good targets for structure determination from a list of sequences. Target selection is based on multiple factors including homology searches and a range of physiochemical properties.

TargetRNA
(http://snowwhite.wellesley.edu/targetRNA)

TargetRNA is a web-based tool for identifying mRNA targets of small noncoding RNAs in bacterial species.

TiCo (http://tico.gobics.de/)

TIS Correction (TiCo) is a tool for improving predictions of prokaryotic Translation Initiation Sites (TIS). TiCo can be used to analyze and reannotate predictions obtained by the program GLIMMER.

VirHostnet
(http://pbildb1.univ-lyon1.fr/virhostnet/login.php)

Virus−Host Network is a knowledge-base system dedicated to the curation, the integration, the management and the analysis of virus−host molecular (mainly protein−protein) interaction networks as well as their functional annotation (molecular functions, cellular pathways, protein domains). VirHostNet contains high quality and up-to-date information gathered and curated from public databases.

Virus Genctyping Tools
(http://bioafrica.mrc.ac.za/rega-genotype/html/)

This web server facilitates high throughput virus genotyping by aligning query sequences to a predefined set of reference strains and then computing phylogenetic analysis.

133

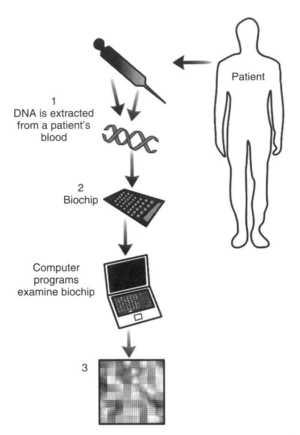

Figure 2.29. Biochip promises to be an important information tool to explore the molecular basis of a disease.

for research, diagnostics, and therapy enters into biology and medicine. Biochips containing microarrays of genetic information promise to be important research tools in the postgenomic era (Jain, 2001). The basic idea of the biochip technology is to convert the chemistry of life into a static form programmed to monitor genes, proteins, and relations between them. Biochip programmed by known sequences of DNA/RNA or proteins can recognize the real genes, mutations, and levels of expression. Biochip technology is a highly effective method that allows monitoring of thousands of genes/alleles at a time in computerized automatic operations with minimal volumes of necessary reagents. Biochips promise an important shift in molecular biology, DNA diagnostics and pharmacology, research in carcinogenesis and other diseases, and also the possibility of a holistic understanding of the world of biology. Development of high throughput "biochip" technologies has dramatically enhanced our ability to study biology and explore the molecular basis of a disease. Biochips enable massively parallel molecular analyses to be carried out in a miniaturized format with a very high throughput (Fig. 2.29).

Biochips are collections of miniaturized test sites (microarrays) arranged on a solid substrate onto which a large number of biomolecules are attached with high density. The word "biochip" derives from the computer term "chip." Although silicon surfaces bearing printed circuits can be used for DNA binding, the term biochip is now broadly used to describe all surfaces bearing microscopic spots, each one being formed by specific capture probes. The capture probes are chosen to complement the target sequence to be detected. Each capture probe will bind to its corresponding target sequence. Like a computer chip performing millions of mathematical operations in a few split seconds, a biochip allows for simultaneous analyses of thousands of biological reactions, such as decoding genes, in a few seconds. Biochip technologies can be applied to numerous fields including genomic, proteomic, and glycomic research, as well as pharmacology and toxicology. However, one of the most common applications is in the determination of gene expression in human cells and tissues. Global gene expression analysis has helped to identify important genes and signaling pathways in human malignant tumors.

Biochips are formed by *in situ* (on chip) synthesis of oligonucleotides (Braun et al., 2005), or peptide nucleic acids (PNAs), or spotting of DNA fragments. Hybridization of RNA- or DNA-derived samples on chips allows the monitoring of expression of mRNAs or the occurrence of polymorphisms in genomic DNA. Basic types of DNA chips are the sequencing chip, the expression chip, and chips for CGH. Advanced technologies used in automated microarray production are photolithography, mechanical microspotting, and ink jets. Bioelectronic microchips contain numerous electronically active microelectrodes with specific DNA capture probes linked to the electrodes through molecular wires. Several biosensors have been used in combination with biochips. The purpose of the chips is to detect many genes present in a sample in one assay rather than performing individual gene assays as is the practice, for example, in so-called multiwells, plates with 96 wells, where the reactions take place. The huge amount of information coming from the genome sequence and other research genome programs cannot be utilized to the full without the availability of methods such as biochips that enable these genes or specific DNA sequences to be detected in biological samples. DNA chip technology is an example of the enormous efforts undertaken in the genomic field in the last few years.

Use of protein array technology over conventional assay methods has advantages that include simultaneous detection of multiple analytes, reduction in sample and reagent volumes, and high output of test results. The susceptibility of ligands to denaturation, however, has impeded production of a stable, reproducible biochip platform, limiting most array assays to manual or, at most, semiautomated processing techniques. Such limitations may be overcome by novel biochip fabrication procedures (FitzGerald et al., 2005).

A design for a biochip memory device based on known materials and existing principles is presented in Fig. 2.29. The fabrication of this memory system relies on the self-assembly of the nucleic acid junction system, which acts as the scaffolding for a molecular wire consisting of polyacetylene-like units. A molecular switch to control current is described, which is based on the formation of a charge transfer complex. A molecular scale bit is presented, which is based on oxidation–reduction potentials of metal atoms or clusters. The readable "bit" which can be made of these components has a volume of 3×10^7 Å^3 and should operate at electronic speeds over short distances (Robinson and Seeman, 1987). After selection of a suitable biochip substrate, biochip surfaces were chemically modified and assessed to enable optimization of biochip fabrication procedures for different test panels. Each biochip has hundreds to thousands of gel drops on a glass, plastic, or membrane support, each about 100 μm in diameter. A segment of a DNA strand, protein, peptide, or antibody is inserted into each drop, tailoring it to recognize a specific biological agent or biochemical signature. These drops are in known positions, so, when a sample reacts, the reaction position can be detected, identifying the sample. The biochip system can identify infectious disease strains in less than 15 min when testing protein arrays and in less than 2 h when testing nucleic acid arrays.

In the next few years, biochips will enter into clinical medicine. Development of high throughput "biochip" technologies has dramatically enhanced our ability to study biology and explore the molecular basis of a disease. Biochips enable massively parallel molecular analyses to be carried out in a miniaturized format with a very high throughput (Kallioniemi, 2001). Each larger diagnostics laboratory will offer genetic tests with this method. The system can be used in hospitals and other laboratories as well as in the field. microSERS is a new biochip technology that uses surface-enhanced Raman scattering (SERS) microscopy for label-free transduction. The biochip itself comprises pixels of capture biomolecules immobilized on a SERS-active metal surface. Once the biochip has been exposed to the sample and the capture biomolecules have selectively bound their ligands, a Raman microscope is used to collect SERS fingerprints from the pixels on the chip. SERS, similar to other whole-organism fingerprinting techniques, is very specific. Initial studies have shown that the gram-positive *Listeria* and gram-negative *Legionella* bacteria, *Bacillus* spores, and *Cryptosporidium* oocysts can often be identified at the sub-species/strain level on the basis of SERS fingerprints collected from single organisms. Therefore, pathogens can be individually identified by microSERS, even when organisms that cross react with the capture biomolecules are present in a sample. Moreover, the SERS fingerprint reflects the physiological state of a bacterial cell, for example, when pathogenic *Listeria* and *Legionella* were cultured under conditions known to affect virulence, their SERS fingerprints changed significantly. Similarly, nonviable (e.g., heat or UV killed) microorganisms could be differentiated from their viable counterparts by SERS fingerprinting. Finally, microSERS is also capable of sensitive and highly specific detection of toxins (Grow et al., 2003).

Use of biological warfare (BW) agents has led to a growing interest in the rapid and sensitive detection of pathogens. Therefore, the development of field-usable detection devices for sensitive and selective detection of BW agents is an important issue. Song et al. (2005) reported a portable biochip system based on complementary metal oxide semiconductor (CMOS) technology that has great potential as a device for single bacteria detection. The possibility of single bacteria detection is reported using an immunoassay coupled to laser-induced fluorescence (LIF) detection. *Bacillus globigii* spores, which are a surrogate species for *B. anthracis* spores, were used as the test sample. Enzymatic amplification following immunocomplex formation allowed remarkably sensitive detection of *B. globigii* spores, and could preclude a complicated optical and instrumental system usually required for high sensitive detection. Atomic force microscopy (AFM) was employed to investigate whether *B. globigii* spores detected in the portable biochip system exist in a single cell or a multicellular form. It was found that *B. globigii* spores mostly exist in a multicellular form with a small minority of single cell form. Detection of aerosolized spores was achieved by coupling the miniature system to a portable bioaerosol sampler, and the performance of the antibody-based recognition and enzyme amplification method was evaluated by Stratis-Cullum et al. (2003).

Love et al. (2008) produced antibodies that would detect *B. anthracis* EA1 protein and intact spores with a high degree of specificity, but would not detect other *Bacillus* species. Existing monoclonal antibodies were evaluated and found to recognize *B. anthracis* EA1 and S-layer proteins from other closely related *Bacillus* species. The approach described can be used to generate specific antibodies to any desired target where homologous proteins also exist in closely related species, and demonstrates clear advantages to using recombinant technology to produce biological recognition elements for detection of biological threat agents.

Lee et al. (2008) reported a biochip, which could be a feasible tool for rapidly diagnosing mastitis causing pathogens in milk and providing information for a more effective treatment to cure mastitis. In this study, a biochip capable of detecting seven common species of mastitis causing pathogens, including *Corynebacterium bovis, Mycoplasma bovis, Staphylococcus aureus*, and the *Streptococcus* spp. *S. agalactiae, S. bovis, S. dysgalactiae*, and *S. uberis*, within 6 h, was developed. The technique is based on DNA amplification of genes specific to the target pathogens and consists of four basic steps: DNA extraction of bacteria, PCR, DNA hybridization, and colorimetric reaction.

Randox biochip-based Evidence Investigator™ system can measure the analytes in plasma and cerebrospinal fluid (CSF) from 60 patients suffering from Alzheimer's disease (AD), vascular dementia, frontotemporal dementia, dementia with Lewy bodies, or mild cognitive impairment, as well as from 20 cognitively healthy controls (Rosén et al., 2011). The aim was to test the analytical performance of two multiplex assays of cerebral biomarkers on a well-defined clinical material consisting of patients with various neurodegenerative diseases.

Lingxiang et al. (2010) established a biochip assay system for accurate detection of mycobacterial species including a biochip, sample preparation apparatus, hybridization instrument, chip washing machine, and laser confocal scanner equipped with interpretation software for automatic diagnosis. The biochip simultaneously identified 17 common mycobacterial species by targeting the differences in the 16S rRNA. The system was assessed with 64 reference strains and 296 *M. tuberculosis* and 243 nontuberculous mycobacterial isolates, as well as 138 otherbacteria and 195 sputum samples, and then compared to DNA sequencing.

The technology provides a point of care diagnostic system that would save time and money compared to current systems, which require sending samples to a centralized laboratory for confirmatory diagnosis. Various biochip technologies in cancer research, including analysis of disease predisposition by using SNP, global gene expression patterns by cDNA microarrays, concentrations, functional activities or interactions of proteins with proteomic biochips, and cell types or tissues, as well as clinical end points associated with molecular targets by using tissue microarrays. Novel protein biochips are under development in academic laboratories and emerging biotechnology companies to advance the pace and scope of scientific discovery. Biochips are a relevant topic for insurers partially already today, much more, however, in the future.

REFERENCES

Adams PS, Dolejsi MK, Hardin S, Mische S, Nanthakamur B, et al. 1996. DNA sequencing of a moderately difficult template: evaluation of the results from a *Thermus thermophilus* unknown test sample. *Biotechniques* **21**, 678.

Ahtiluoto S, Mannonen L, Paetau A, Vaheri A, Koskiniemi M, et al. 2000. *In situ* hybridization detection of human herpesvirus 6 in brain tissue from fatal encephalitis. *Pediatrics* **105**, 431–433.

Ahuja S, Ahuja S, Chen Q, Wahlgren M. 2006. Prediction of solubility on recombinant expression of Plasmodium falciparum erythrocyte membrane protein 1 domains in *Escherichia coli*. *Malar J* **5**, 52.

Albrecht M, Sharma CM, Reinhardt R, Vogel J, Rudel T. 2010. Deep sequencing-based discovery of the *Chlamydia trachomatis* transcriptome. *Nucleic Acids Res* **38**, 866–877.

Alm EJ, Huang KH, Price MN, Koche RP, Keller K, Dubchak IL, Arkin AP. 2005. The Microbes Online Web site for comparative genomics. *Genome Res* **15**, 1015–1022.

Angibaud S, Eveillard D, Fertin G, Rusu I. 2009. Comparing bacterial genomes by searching their common intervals. *Proceedings 1st Bioinformatics and Computational Biology Conference (BICoB 2009)*, vol. **5462**, pp. 102–113. Springer, New Orleans, USA.

Ansorge WJ. 2009. Next-generation DNA sequencing techniques. *Nat Biotechnol* **25**, 195–203.

Ansorge W, Rosenthal A, Sproat B, Schwager C, Stegemann J, Voss H. 1988. Non-radioactive automated sequencing of oligonucleotides by chemical degradation. *Nucleic Acids Res* **16**, 2203–2206.

Ansorge W, Sproat BS, Stegemann J, Schwager C. 1986. A non-radioactive automated method for DNA sequence determination. *J Biochem Biophys Methods* **13**, 315–323.

Ansorge W, Sproat B, Stegemann J, Schwager C, Zenke M. 1987. Automated DNA sequencing: ultrasensitive detection of fluorescent bands during electrophoresis. *Nucleic Acids Res* **15**, 4593–4602.

Ault GS, Ryschkewitsch CF, Stoner GL. 1994. Type-specific amplification of viral DNA using touchdown and hot start PCR. *J Virol Methods* **46**, 145–156.

Baek CH, Kim SI, KO YH, Chu KC. 2000. Polymerase chain reaction detection of *Mycobacterium tuberculosis* from fine needle aspirate for the diagnosis of cervical tuberculous lymphadenitis. *Laryngoscope* **110**, 30–34.

Baggerly KA, Morris JS, Edmonson SR, Coombes KR. 2005. Signal in noise:evaluating reported reproducibility of serum proteomic tests for ovarian cancer. *J Natl Cancer Inst* **97**, 307–309.

Baldi P, Long AD. 2001. A Bayesian framework for the analysis of microarray expression data: regularized t-test and statistical inferences of gene changes. *Bioinformatics* **17**, 509–519.

Barrett JC, Kawasaki ES. 2003. Microarrays: the use of oligonucleotides and cDNA for the analysis of gene expression. *Drug Discov Today* **8**, 134–141.

Barrett MT, Scheffer A, Ben-Dor A, Sampas N, Lipson D, et al. 2004. Comparative genomic hybridization using oligonucleotide microarrays and total genomic DNA. *Proc Natl Acad Sci U S A* **101**, 17765–17770.

Bauman JG, Wiegant J, Borst P, van Duijn P. 1980. A new method for fluorescence microscopical localization of specific DNA sequences by *in situ* hybridization of fluorochromelabelled RNA. *Exp Cell Res* **128**, 485–490.

Bayjanov JR, Wels M, Starrenburg M, van Hylckama Vlieg JE, Siezen RJ, Molenaar D. 2009. PanCGH: a genotype-calling algorithm for pangenome CGH data. *Bioinformatics* **25**, 309–314.

Berens C, Thain A, Schroeder R. 2001. A tetracycline-binding RNA aptamer. *Bioorg Med Chem* **9**, 2549–2556.

Bilello JA. 2005. The agony and ecstasy of "OMIC" technologies in drug development. *Curr Mol Med* **5**, 39–52.

Binder L, Vogel T, Hadary D. 1989, Chimeric bovine-human growth hormone prepared by recombinant DNA technology: binding properties and biological activity. *Mol Endocrinol* **3**, 923–930.

Bock LC, Griffin LC, Latham JA, Vermaas EH, Toole JJ. 1992. Selection of single-stranded DNA molecules that bind and inhibit human thrombin. *Nature* **355**, 564–566.

Boland EJ, Pillai A, Odom MW, Jagadeeswaran P. 1994. Automation of the Maxam-Gilbert chemical sequencing reactions. *Biotechniques* **16**, 1088–1095.

Braun V, Mahren S, Sauter A. 2005. Gene regulation by transmembrane signaling. *Biometals* **18**, 507–517.

Brenneis M, Soppa J. 2009. Regulation of translation in haloarchaea: 5′- and 3′-UTRs are essential and have to functionally interact *in vivo*. *Plos One* **4**, e4484.

Brenton JD, Aparicio SA, Caldas C. 2001. Molecular profiling of breast cancer: portraits but not physiognomy. *Breast Cancer Res* **3**, 77–80.

Brötz-Oesterhelt H, Bandow JE, Labischinski H. 2005. Bacterial proteomics and its role in antibacterial drug discovery. *Mass Spectrom Rev* **24**, 549–565.

Brown J, Horsley SW, Jung C, Saracoglu K, Janssen B, Brough M, Daschner M, Beedgen B, Kerkhoffs G, Eils R, Harris PC, Jauch A, Kearney L. 2000. Identification of a subtle t(16;19)(p13.3;p13.3) in an infant with multiple congenital abnormalities using a 12-colour multiplex FISH telomere assay, M-TEL. *Eur J Hum Genet* **8**, 903–910.

Bundy JG, Willey TL, Castell RS, Ellar DJ, Brindle KM. 2005. Discrimination of pathogenic clinical isolates and laboratory strains of *Bacillus cereus* by NMR-based metabolic profiling. *FEMS Microbiol Lett* **242**, 127–136.

Burgstaller P, Famulok M. 1996. Structural characterization of a flavin-specific RNA aptamer by chemical probing. *Bioorg Med Chem Lett* **6**, 1157–1162.

Callister SJ, McCue LA, Turse JE, Monroe ME, Auberry KJ, Smith RD, Adkins JN, Lipton MS. 2008. Comparative bacterial proteomics: analysis of the core genome concept. *PLoS ONE* **3**, e1542.

Carrington WA, Fogarty KE, Lifschitz L, Fay FS. 1995. Superresolution three-dimentional images of fluorescence in cells with minimal light exposure. *Science* **268**, 1483–1487.

Carson S, Cohen AS, Belenkii A, Ruiz-Martinez MC, Berka J, Karger BL. 1993. DNA sequencing by capillary electrophoresis: use of a two-laser-two-window intensified diode array detection system. *Anal Chem* **65**, 3219–3226.

Chakravarti B, Mallik B, Chakravarti DN. 2010. Proteomics and systems biology: application in drug discovery and development. *Methods Mol Biol* **662**, 3–28.

Chiocchia G, Smith KA. 1997. Highly sensitive method to detect mRNAs in individual cells by direct RT-PCR using *Tth* DNA polymerase. *Biotechniques* **22**, 312–318.

Church GM, Gilbert W. 1984. Genomic sequencing. *Proc Natl Acad Sci U S A* **81**, 1991–1995.

Clarke SC. 2002. Nucleotide sequence-based typing of bacteria and the impact of automation. *Bioessays* **24**, 858–862.

Croucher NJ, Fookes MC, Perkins TT, Turner DJ, Marguerat SB, et al. 2009. A simple method for directional transcriptome sequencing using illumine technology. *Nucleic Acids Res* **37**, e148.

Culhane AC, Perriere G, Higgins DG. 2003. Cross-platform comparison and visualisation of gene expression data using co-inertia analysis. *BMC Bioinformatics* **4**, 59.

Davidson BL, Breakefield XO. 2003. Viral vectors for gene delivery to the nervous system. *Nat Rev Neurosci* **4**, 353–364.

De Keersmaecker SC, Thijs IM, Vanderleyden J, Marchal K. 2006. Integration of omics data: how well does it work for bacteria? *Mol Microbiol* **62**, 1239–1250.

de Marco A. 2007. Protocol for preparing proteins with improved solubility by co-expressing with molecular chaperones in *Escherichia coli*. *Nat Protoc* **2**, 2632–2639.

de Marco A, Vigh L, Diamant S, Goloubinoff P. 2005. Native folding of aggregation-prone recombinant proteins in Escherichia coli by osmolytes, plasmid- or benzyl alcohol-overexpressed molecular chaperones. *Cell Stress Chaperones* **10**, 329–339.

de Souza GA, Arntzen MØ, Wiker HG. 2010. MSMSpdbb: providing protein databases of closely related organisms to improve proteomic characterization of prokaryotic microbes. *Bioinformatics* **26**, 698–699.

de Souza GA, Må H, Søfteland T, Saelensminde G, Prasad S, et al. 2008. High accuracy mass spectrometry analysis as a tool to verify and improve gene annotation using *Mycobacterium tuberculosis* as an example. *BMC Genomics* **2**, 316.

Deng Q, German I, Buchanan D, Kennedy RT. 2001. Retention and separation of adenosine and analogues by affinity chromatography with an aptamer stationary phase. *Anal Chem* **73**, 5415–5421.

Diamandis EP. 2004. Analysis of serum proteomic patterns for early cancer diagnosis: drawing attention to potential problems. *J Natl Cancer Inst* **96**, 353–356.

Diamandis EP. 2007. Oncopeptidomics: a useful approach for cancer diagnosis? *Clin Chem* **53**, 1004–1006.

Dickerman A, Tyler B. 2002. *A Bioinformatics Platform for Studying Metabolic Functions of Unculturable Microbes*. ACM DM Digital Library, UC Davis, CA.

Dolan M, Ally A, Purzycki MS, Gilbert W, Gillevet PM. 1995. Large-scale genomic sequencing: optimization of genomic chemicalsequencing reactions. *Biotechniques* **19**, 237–264.

Drake JW, Holland JJ. 1999. Mutation rates among RNA viruses. *Proc Natl Acad Sc U S A* **96**, 13910–13913.

Dutt A, Beroukhim R. 2007. Single nucleotide polymorphism array analysis of cancer. *Curr Opin Oncol* **19**, 43–49.

Ellington AD, Szostak JW. 1990. *In vitro* selection of RNA molecules that bind specific ligands. *Nature* **346**, 818–822.

Ellis DI, Goodacre R. 2006. Metabolic fingerprinting in disease diagnosis: biomedical applications of infrared and Raman spectroscopy. *Analyst* **131**, 875–885.

Englund PT. 1971. Analysis of nucleotide sequences at 3'-termini of duplex deoxyribonucleic acid with the use of the T4 deoxyribonucleic acid polymerase. *J Biol Chem* **246**, 3269–3276.

Englund PT. 1972. The 3'-terminal nucleotide sequences of T7 DNA. *J Mol Biol* **66**, 209–224.

Fauth C, Speicher MR. 2001. Classifying by colors: FISH-based genome analysis. *Cytogenet Cell Genet* **93**, 1–10.

Favis R, Barany F. 2000. Mutation detection in K-ras, BRCA1, BRCA2, and p53 using PCR/LDR and a universal DNA microarray. *Ann N Y Acad Sci* **906**, 39–43.

Favis R, Day JP, Gerry NP, Phelan C, Narod S, Barany F. 2000. Universal DNA array detection of small insertions and deletions in BRCA1 and BRCA2. *Nat Biotechnol* **18**, 561–564.

Feist AM, Henry CS, Reed JL, Krummenacker M, Joyce AR, Karp PD, Broadbelt LJ, Hatzimanikatis V, Palsson BO. 2007. A genome-scale metabolic reconstruction for *Escherichia coli* K-12MG1655 that accounts for 1260 ORFs and thermodynamic information. *Mol Syst Biol* **3**, 121.

Femino AM, Fay FS, Fogarty K, Singer RH. 1998. Visualization of single RNA transcripts *in situ*. *Science* **280**, 585–590.

Ferraboli S, Negri R, Di Mauro E, Barlati S. 1993. One-lane chemical sequencing of 3'-fluorescent-labeled DNA. *Anal Biochem* **214**, 566–570.

Ferrer-Miralles N, Domingo-Espín J, Corchero JL, Vázquez E, Villaverde A. 2009. Microbial factories for recombinant pharmaceuticals. *Microb Cell Fact* **8**, 17.

Fiehn O. 2002. Metabolomics the link between genotypes and phenotypes. *Plant Mol Biol* **46**, 155–171.

Fischer WB. 2003. Computational bioanalysis of proteins. *Anal Bioanal Chem* **375**, 23–25.

FitzGerald SP, Lamont JV, McConnell RI, Benchikh El O. 2005. Development of a high-throughput automated analyzer using biochip array technology. *Clin Chem* **51**, 1165–1176.

Fleischmann RD, Adams MD, White O, Clayton RA, Kirkness EF, et al. 1995. Whole-genome random sequencing and assembly of *Haemophilus influenzae* Rd. *Science* **268**, 496–512.

Forozan F, Karhu R, Kononen J, Kallioniemi A, Kallioniemi OP. 1997. Genome screening by comparative genomic hybridization. *Trends Genet* **13**, 405–409.

Fu X, Fu N, Guo S, Yan Z, Xu Y, et al. 2009. Estimating accuracy of RNA-Seq and microarrays with proteomics. *BMC Genomics* **10**, 161.

Fukuda S, Nakanishi Y, Chikayama E, Ohno H, Hino T, Kikuchiet J. 2009. Evaluation and characterization of bacterial metabolic dynamics with a novel profiling technique, real-time metabolotyping. *PLoS ONE* **4**, e4893.

Gall JG, Macgregor HC, Kidstone ME. 1969. Gene amplification in the oocytes of dytiscid water beetles. *Chromosoma (Berl.)* **26**, 169–187.

German JB, Hammock BD, Watkins SM. 2005. Metabolomics: building on century of biochemistry to guide human health. *Metabolomics* **1**, 3–9.

German JB, Roberts MA, Fay L, Watkins SM. 2002. Metabolomics and individual metabolic assessment: the next great challenge for nutrition. *J Nutr* **132**, 2486–2487.

German JB, Roberts MA, Watkins SM. 2003. Genomics and metabolomics as markers for the interaction of diet and health: lessons from lipids. *J Nutr* **133**, 2078S–2083S.

Goel MM, Ranjan V, Dhole TN, Srivastava AN, Mehrotra A, Kushwaha MR, Jain A. 2001. Polymerase chain reaction vs. conventional diagnosis in fine needle aspirates of tuberculous lymph nodes. *Acta Cytol* **45**, 333–340.

Golub TR, Slonim DK, Tamayo P, Huard C, Gaasenbeek M, et al. 1999. Molecular classification of cancer: class discovery and class prediction by gene expression monitoring. *Science* **286**, 531–537.

Gong G, Lee H, Kang GH, Shim YH, Huh J, Khang SK. 2002. Nested PCR for diagnosis of tuberculous lymphadenitis and PCR-SSCP for identification of rifampicin resistance in fine needle aspirates. *Diagn Cytopathol* **26**, 228–231.

Gonzalez O, Gronau S, Pfeiffer F, Mendoza E, Zimmer R, Oesterhelt D. 2009. Systems analysis of bioenergetics and growth of the extreme halophile *Halobacterium salinarum*. *PLoS Comput Biol* **5**, e1000332.

Gonzalez-Montalban N. 2007. Recombinant protein solubility-does more mean better. *Nat Biotechnol* **25**, 718–720.

Goszczynski B, McGhee JD. 1991. Resolution of sequencing ambiguities: a universal *Fok*I adapter permits Maxam±Gilbert re-sequencing of singlestranded phagemid DNA. *Gene* **104**, 71–74.

Graf A, Gasser B, Dragosits M, Sauer M, Leparc GG, et al. 2008. Novel insights into the unfolded protein response using Pichia pastoris specific DNA microarrays. *BMC Genomics* **9**, 390.

Griffin HG, Griffin AM. 1993. DNA sequencing. Recent innovations and future trends. *Appl Biochem Biotechnol* **38**, 147–159.

Grillot-Courvalin C, Goussard S, Huetz F, Ojcius DM, Courvalin P. 1998. Functional gene transfer from intracellular bacteria to mammalian cells. *Nat Biotechnol* **16**, 862–866.

Grow AE, Wood LL, Claycomb JL, Thompson PA. 2003. New biochip technology for label-free detection of pathogens and their toxins. *J Microbiol Methods* **53**, 221–233.

Grünenfelder B, Rummel G, Vohradsky J, Röder D, Langen H, Jenal U. 2001. Proteomic analysis of the bacterial cell cycle. *Proc Natl Acad Sci U S A* **98**, 4681–4686.

Guell M, van Noort V, Yus E, Chen WH, Leigh-Bell J, et al. 2009. Transcriptome complexity in a genome-reduced bacterium. *Science* **326**, 1268–1271.

Gupta N, Tanner S, Jaitly N, Adkins JN, Lipton M, Edwards R, et al. 2007. Whole proteome analysis of post-translational modifications: applications of mass-spectrometry for proteogenomic annotation. *Genome Res* **17**, 1362–1377.

Hacia JG. 1999. Resequencing and mutational analysis using oligonucleotide microarrays. *Nat Genet* **21**, 42–47.

Hacia JG, Brody LC, Chee MS, Fodor SP, Collins FS. 1996. Detection of heterozygous mutations in BRCA1 using high density oligonucleotide arrays and two-color fluorescence analysis. *Nat Genet* **14**, 441–447.

Harrigan GG, Goodacre R. 2003. *Metabolic Profiling: Its Role in Biomarker Discovery and Gene Function Analysis*, Kluwer Academic Publishers, Boston, MA.

Harris TD, Buzby PR, Babcock H, Beer E, Bowers J, et al. 2008. Single-molecule DNA sequencing of a viral genome. *Science* **320**, 106–109.

Hinton JC, Hautefort I, Eriksson S, Thompson A, Rhen M. 2004. Benefits and pitfalls of using microarrays to monitor bacterial gene expression during infection. *Curr Opin Microbiol* **7**, 277–282.

Homann M, Goringer HU. 2001. Uptake and intracellular transport of RNA Aptamers in African trypanosomes suggest therapeutic "Piggy-Back" approach. *Bioorg Med Chem* **9**, 2571–2580.

Hoque MO, Lee CC, Cairns P, Schoenberg M, Sidransky D. 2003. Genome-wide genetic characterization of bladder cancer: a comparison of high-density single-nucleotide polymorphism arrays and PCR-based microsatellite analysis. *Cancer Res* **63**, 2216–2222.

Hougaard DM, Hansen H, Larsson LI. 1997. Non-radioactive *in situ* hybridization for mRNA with emphasis on the use of oligodeoxynucleotide probes. *Histochem Cell Biol* **108**, 335–344.

Hughes TR, Mao M, Jones AR, Burchard J, Marton MJ, et al. 2001. Expression profiling using microarrays fabricated by an ink-jet oligonucleotide synthesizer. *Nat Biotechnol* **19**, 342–347.

Ishii N, Nakahigashi K, Baba T, Robert M, Soga T, et al. 2007. Multiple high-throughput analyses monitor the response of *E. coli* to perturbations. *Science* **316**, 593–597.

Isola NR, Allman SL, Golovlov VV, Chen CH. 1999. Chemical cleavage sequencing of DNA using matrix-assisted laser desorption/ionization time-of-flight mass spectrometry. *Anal Chem* **71**, 2266–2269.

Jain KK. 2001. Biochips for gene spotting. *Science* **294**, 621–623.

Jayasena SD. 1999. Aptamers: an emerging class of molecules that rival antibodies in diagnostics. *Clin Chem* **45**, 1628–1650.

Jeffreys AJ, Wislon V, Thein SL. 1985. Individual-specific 'fingerprints' of human DNA. *Nature* **316**, 76–79.

Jensen LJ, Julien P, Kuhn M, von Mering C, Muller J, et al. 2008. eggNOG: automated construction and annotation of orthologous groups of genes. *Nucleic Acids Res* **36**, D250–D254.

Johnson IS. 1983. Human insulin from recombinant DNA technology. *Science* **219**, 632–637.

Johnson RA, Wichern DW. 2002. *Applied Multivariate Statistical Analysis*, 5th ed., Prentice Hall, Upper Saddle River, NJ.

Joshi R, Janagama H, Dwivedi HP, Senthil Kumar TM, Jaykus LA, et al. 2009. Selection, characterization, and application of DNA aptamers for the capture and detection of *Salmonella enterica* serovar. *Mol Cell Probes* **23**, 20–28.

Jungblut PR. 2001. Proteome analysis of bacterial pathogens. *Microbes Infect* **3**, 831–840.

Kaddurah-Daouk R. 2006. Metabolic profiling of patients with schizophrenia. *PLoS Med* **3**, 363.

Kallioniemi OP. 2001. Biochip technologies in cancer research. *Ann Med* **33**, 142–147.

Kane MD, Jatkoe TA, Stumpf CR, Lu J, Thomas JD, Madore SJ. 2000. Assessment of the sensitivity and specificity of oligonucleotide (50mer) microarrays. *Nucleic Acids Res* **28**, 4552–4557.

Kanehisa M, Goto S, Hattori M, Aoki-Kinoshita KF, Itoh M, et al. 2006. From genomics to chemical genomics: new developments in KEGG. *Nucleic Acids Res* **34(Database issue)**, D354–357.

Karger AE, Harris JM, Gesteland RF. 1991. Multiwavelength fluorescence detection for DNA sequencing using capillary electrophoresis. *Nucleic Acids Res* **19**, 4955–4962.

Katayama S, Tomaru Y, Kasukawa T, Waki K, Nakanishi M, et al. 2005. Antisense transcription in the mammalian transcriptome' by the RIKEN Genome Exploration Research Group and Genome Science Group (Genome Network Project Core Group) and the FANTOM Consortium. *Science* **309**, 1564–1566.

Kieleczawa J, Dunn JJ, Studier FW. 1992. DNA sequencing by primer walking with strings of contiguous hexamers. *Science* **258**, 1787–1791.

Koide T, Reiss DJ, Bare JC, Pang WL, Facciotti MT, Schmid AK, Pan M, Marzolf B, Van PT, Lo FY, Pratap A, Deutsch EW, Peterson A, Martin D, Baliga NS. 2009. Prevalence of transcription promoters within archaeal operons and coding sequences. *Mol Syst Biol* **5**, 285.

Kothapalli R, Yoder SJ, Mane S, Loughran TP Jr. 2002. Microarray results: how accurate are they? *BMC Bioinformatics* **3**, 22.

Kotler LE, Zevinsonkin D, Sobolev IA, Beskin AD, Ulanovsky LE. 1993. DNA sequencing— modular primers assembled from a library of hexamers or pentamers. *Proc Natl Acad Sci U S A* **90**, 4241–4245.

Kuo WP, Jenssen TK, Butte AJ, Ohno-Machado L, Kohane IS. 2002. Analysis of matched mRNA measurements from two different microarray technologies. *Bioinformatics* **18**, 405–412.

Lander ES, Linton LM, Birren B, Nusbaum C, Zody MC, Baldwin J, et al. International Human Genome Sequencing Consortium. 2001. Initial sequencing and analysis of the human genome. *Nature* **409**, 860–921.

Lashkari DA, Hunicke-Smith SP, Norgren RM, Davis RW, Brennan T. 1995. An automated multiplex oligonucleotide synthesizer: development of high-throughput, low-cost DNA synthesis. *Proc Natl Acad Sci U S A* **92**, 7912–7915.

Lawrence JB, Singer RH. 1986. Intracellular localization of messenger RNAs for cytoskeletal proteins. *Cell* **45**, 407–415.

Le Tinevez R, Mishra RK, Toulme JJ. 1998. Selective inhibition of cell-free translation by oligonucleotides targeted to a mRNA hairpin structure. *Nucleic Acids Res* **26**, 2273–2278.

Lee DY, Bowen BP, Northern TR. 2010. Mass spectroscopy–based metabolomic analysis of metabolite–protein interaction and imaging. *Biotechniques* **49**, 557–565.

Lee SH, Jung BY, Rayamahji N, Lee HS, Jeon WJ, Choi KS, Kweon CH, Yoo HS. 2009. A multiplex real-time PCR for differential detection and quantification of *Salmonella* spp., *Salmonella enterica serovar Typhimurium* and Enteritidis in meats. *J Vet Sci* **10**, 43–51.

Lee H, Kim KN, Kee Chae Y. 2007. Reevaluating the capability of Taq DNA polymerase: long PCR amplification. *Protein Pept Lett* **14**, 321–323.

Lee KH, Lee J-W, Wang S-W, Liu L-Y, Lee M-F, et al. 2008. Development of a novel biochip for rapid multiplex detection of seven mastitis-causing pathogens in bovine milk samples. *J Vet Diagn Invest* **20**, 463–470.

Lein ES, Hawrylycz MJ, Ao N, Ayres M, Bensinger A, Bernard A, et al. 2007. Genome-wide atlas of gene expression in the adult mouse brain. *Nature* **445**, 168–176.

Leitch AR, Schwarzacher T, Jackson D, Leitch IJ. 1994. In Situ *Hybridization: A Practical Guide*, *RMS Microscopy Handbook*, No. 27, Bios Scientific@ Publishers Ltd, Oxford.

Leversha MA, Han J, Asga Z, Danila DC, Lin O, Gonzalez-Espinoza R, Anand A, Lilja H, Heller G, Fleisher M, Scher HI. 2009. Fluorescence *in situ* hybridisation analysis of circulating tumor cells in metastatic prostrate cancer. *Clin Cancer Res* **15**, 2091–2097.

Levsky JM, Braut SA, Singer RH. 2003. Single cell gene expression profiling by multiplexed expression fluorescence *in situ* hybridization: application to the analysis of cultured cells. In: Celis JE. eds., *Cell Biology: A Laboratory Handbook*, Academic Press, San Diego, California.

Li C, Wong WH. 2001. Model-based analysis of oligonucleotide arrays: model validation, design issues and standard error application. *Genome Biol* **2** (8), Research0032.

Lichter P, Cremer T, Borden J, Manuelidis L, Ward DC. 1988. Delineation of individual human chromosomes in metaphase and interphase cells by *in situ* suppression hybridization using recombinant DNA libraries. *Hum Genet* **80**, 224–234.

Lindon JC, Holmes E, Nicholson JK. 2004. Metabolomics technologies and their applications in physiological monitoring, drug safety assessment and disease diagnosis. *Biomarkers* **9**, 1–31.

Lingxiang Z, Guanglu J, Shengfen W, Can W, Qiang L, et al. 2010. Biochip system for rapid and accurate identification of mycobacterial species from isolates and sputum. *J Clin Microbiol* **48**, 3654–3660.

Lipshutz RJ, Fodor SP, Gingeras TR, Lockhart DJ. 1999. High density synthetic oligonucleotide arrays. *Nat Genet* **21**, 20–24.

Liu JM, Livny J, Lawrence MS, Kimball MD, Waldor MK, Camilli A. 2009. Experimental discovery of sRNAs in *Vibrio cholerae* by direct cloning, 5S/tRNA depletion and parallel sequencing. *Nucleic Acids Res* **37**, e46.

Lockhart DJ, Dong H, Byrne MC, Follettie MT, Gallo MV, et al. 1996. Expression monitoring by hybridization to high-density oligonucleotide arrays. *Nat Biotechnol* **14**, 1675–1680.

Lomedico PT. 1982. Use of recombinant DNA technology to program eukaryotic cells to synthesize rat proline: a rapid expression assay for cloned genes. *Proc Natl Acad Sci U S A* **79**, 5798–5802.

Lou X, Qian J, Xiao Y, Viel L, Gerdon AE, et al. 2009. Micromagnetic selection of aptamers in microfluidic channels. *Proc Natl Acad Sci U S A* **106**, 2989–2994.

Loughlin MF. 2007. Using 'omic' technology to target *Helicobacter pylori*. *Drug Discov* **2**, 1041–1051.

Loumaye E, Campbell R, Salat-Baroux J. 1995. Human follicle-stimulating hormone produced by recombinant DNA technology: a review for clinicians. *Hum Reprod Update* **1**, 188–199.

Love TE, Redmond C, Mayers CN. 2008. Real time detection of anthrax spores using highly specific anti-EA1 recombinant antibodies produced by competitive panning. *J Immunol Methods* **334**, 1–10.

Lu H, Zhu J, Zang Y, Ze Y, Quin J. 2006. Cloning, purification, and refolding of human paraoxonase-3 expressed in Escherichia coli and its characterization. *Protein Expr Purif* **46**, 92–99.

Luo B, Li W, Deng CH, Zheng FF, Sun XZ, et al. 2009. Utility of fluorescence *in situ* hybridisation in the diagnosis of upper urinary tract urothelial carcinomas. *Cancer Genet Cytogenet* **189**, 93–97.

Lupold SE, Hicke B, Lin Y, Coffey DS. 2002. Identification and characterization of nuclease-stabilized RNA molecules that bind human prostate cancer cells via the prostate-specific membrane antigen. *Cancer Res* **62**, 4029–4033.

Luscombe NM, Greenbaum D, Gerstein M. 2001. What is bioinformatics? A proposed definition and overview of the field. *Methods Inf Med* **40**, 346–358.

Mairal T, Ozalp VC, Lozano Sánchez P, Mir M, et al. 2008. Aptamers molecular tools tools for analytical applications. *Anal Bioanal Chem* **390**, 989–1007.

Manning MC, Patel K, Borchardt RT. 1989. Stability of protein pharmaceuticals. *Pharm Res* **6**, 903–918.

Manuelidis L, Langer-Safer PR, Ward DC. 1982. High-resolution mapping of satellite DNA using biotin-labeled DNA probes. *J Cell Biol* **95**, 619–625.

Marioni JC, Mason CE, Mane SM, Stephens M, Gilad Y. 2008. RNA-seq: an assessment of technical reproducibility and comparison with gene expression arrays. *Genome Res* **18**, 1509–1517.

Markowitz VM, Szeto E, Palaniappan K, Grechkin Y, et al. 2008. The integrated microbial genomes (IMG) system in 2007: data content and analysis tool extensions. *Nucleic Acids Res* **36**, D528–D533.

Martin-Gallardo A, McCombie WR, Gocayne JD. 1992. Automated DNA sequencing and analysis of 106 kilobases from human. *Nat Genet* **1**, 34–39.

Mattanovich D, Gasser B, Hohenblum H, Sauer M. 2004. Stress in recombinant protein producing yeasts. *J Biotechnol* **113**, 121–135.

Maxam AM. 1980. Sequencing the DNA of recombinant chromosomes. *Fed Proc* **39**, 2830–2836.

Maxam AM, Gilbert W. 1977. A new method for sequencing DNA. *Proc Natl Acad Sci U S A* **74**, 560–564.

Maxam AM, Gilbert W. 1980. Sequencing endlabeled DNA with base-specific chemical cleavages. *Methods Enzymol* **65**, 499–560.

McGrath PT, Lee H, Zhang L, Iniesta AA, et al. 2007. High-throughput identification of transcription start sites, conserved promoter motifs and predicted regulons. *Nat Biotechnol* **25**, 584–592.

Mecham BH, Klus GT, Strovel J, Augustus M, Byrne D, et al. 2004. Sequence-matched probes produce increased cross-platform consistency and more reproducible biological results in microarray-based gene expression measurements. *Nucleic Acids Res* **32**, e74.

Melin P. 2004. Proteomics as a tool to study microbial interactions. *Curr Proteomics* **1**, 27–34.

Metzker ML. 2010. Sequencing technologies—the next generation. *Nat Rev Genet* **11**, 31–46.

Meyer IM. 2008. Predicting novel RNA-RNA interactions. *Curr Opin Struct Biol* **18**, 387–393.

Min JJ, Gambhir SS. 2008. Molecular imaging of PET reporter gene expression. *Handb Exp Pharmacol* **185** (Pt 2), 277–303.

Mori H, Begley T. 2008. Metabolomic analysis of microorganisms. *Mol Biosyst* **4**, 108–109.

Morozova O, Hirst M, Marra MA. 2009. Applications of new sequencing technologies for transcriptome analysis. *Annu Rev Genomics Hum Genet* **10**, 135–151.

Mortazavi A, Williams BA, McCue K, Schaeffer L, Wold B. 2008. Mapping and quantifying mammalian transcriptomes by RNA-Seq. *Nat Methods* **5**, 621–628.

Mulkidjanian AY, Koonin EV, Makarova KS, Mekhedov SL, Sorokin A, et al. 2006. The cyanobacterial genome core and the origin of photosynthesis. *Proc Natl Acad Sci U S A* **103**, 13126–13131.

Myers TW, Gelfand DH. 1991. Reverse transcription and DNA amplification by a Thermus thermophilus DNA polymerase. *Biochemistry* **30**, 7661–7666.

Nagalakshmi U, Wang Z, Waern K, Shou C, Raha D, Gerstein M, Snyder M. 2008. The transcriptional landscape of the yeast genome defined by RNA sequencing. *Science* **320**, 1344–1349.

Nakamaye KL, Gish G, Eckstein F, Vosberg HP. 1988. Direct sequencing of polymerase chain reaction amplified DNA fragments through the incorporation of deoxynucleoside alpha-thiotriphosphates. *Nucleic Acids Res* **16**, 9947–9959.

Nederlof PM, Robinson D, Abuknesha R, Wiegant J, Hopman AH, et al. 1989. Three-color fluorescence *in situ* hybridization for the simultaneous detection of multiple nucleic acid sequences. *Cytometry* **10**, 20–27.

Nederlof PM, van der Flier S, Wiegant J, Raap AK, Tanke HJ, et al. 1990. Multiple fluorescence *in situ* hybridization. *Cytometry* **11**, 126–131.

Negri R, Costanzo G, Mauro ED. 1991. A single-reaction method for DNA sequence determination. *Anal Biochem* **197**, 389–395.

Nuovo GJ, Gallery F, MacConnell P. 1992. Detection of amplified HPV 6 and 11 DNA in vulvar lesions by hot start PCR *in situ* hybridization. *Mod Pathol* **5**, 444–448.

Oliver HF, Orsi RH, Ponnala L, Keich U, Wang W, et al. 2009. Deep RNA sequencing of L. monocytogenes reveals overlapping and extensive stationary phase and sigma B-dependent transcriptomes, including multiple highly transcribed noncoding RNAs. *BMC Genomics* **10**, 641.

Oliver SG, Winson MK, Kell DB, Baganz F. 1998. Systematic functional analysis of the yeast genome. *Trends Biotechnol* **16**, 373–378.

Ono N, Suzuki S, Furusawa C, Shimizu H, Yomo T. 2008. Quantitative expression analysis using oligonucleotide microarrays based on a physico-chemical model. *Bionetics '08. Proceedings of the 3rd International Conference on Bio-Inspired Models of Network, Information and Computing Systems*, Hyogo, Japan.

Onyango P. 2004. The role of emerging genomics and proteomics technologies in cancer drug target discovery. *Curr Cancer Drug Targets* **4**, 111–124.

Ornstein DL, Kashdan MA. 1985. Sequencing DNA using ^{35}S-labeling: a troubleshooting guide. *Biotechniques* **3**, 476–484.

Ozsolak F, Platt AR, Jones DR, Reifenberger JG, Sass LE, et al. 2009. Direct RNA sequencing. *Nature* **461**, 814–818.

Pachmann K. 1987. *In situ* hybridization with fluorochrome-labeled cloned DNA for quantitative determination of the homologous mRNA in individual cells. *J Mol Cell Immunol* **3**, 13–19.

Pahwa R, Hedau S, Jain S, Jain N, Arora VM, Kumar N, Das BC. 2005. Assessment of possible tuberculous lymphadenopathy by PCR compared to non-molecular methods. *J Med Microbiol* **54**, 873–878.

Pálffy R, Gardlík R, Hodosy J, Behuliak M, Resko P, et al. 2006. Bacteria in gene therapy: bactofection versus alternative gene therapy. *Gene Ther* **13**, 101–105.

Pan W, Craven RC, Qiu Q, Wilson CB, Wills JW, et al. 1995. Isolation of virus-neutralizing RNAs from a large pool of random sequences. *Proc Natl Acad Sci U S A* **92**, 11509–11513.

Pan Z, Trikalinos TA, Kavvoura FK, Lau J, Ioannidis JPA. 2005. Local literature bias in genetic epidemiology: an empirical evaluation of the Chinese literature. *PLoS Med* **2**, e334.

Passalacqua KD, Varadarajan A, Ondov BD, Okou DT, Zwick ME, Bergman NH. 2009. Structure and complexity of a bacterial transcriptome. *J Bacteriol* **191**, 3203–3211.

Pavski V, Le XC. 2001. Detection of human immunodeficiency virus type 1 reverse transcriptase using aptamers as probes in affinity capillary electrophoresis. *Anal Chem* **73**, 6070–6076.

Perkins TT, Kingsley RA, Fookes MC, Gardner PP, James KD, et al. 2009. A strand-specific RNASeq analysis of the transcriptome of the typhoid bacillus Salmonella typhi. *Plos Genet* **5**, e1000569.

Peterson JD, Umayam LA, Dickinson T, Hickey EK, White O. 2001. The comprehensive microbial resource. *Nucleic Acids Res* **29**, 123–125.

Petricoin EF, Ardekani AM, Hitt BA, Levine PJ, Fusaro VA, et al. 2002. Use of proteomic patterns in serum to identify ovarian cancer. *Lancet* **359**, 572–577.

Pileur F, Andreola ML, Dausse E, Michel J, Moreau S, et al. 2003. Selective inhibitory DNA aptamers of the human RNase H1. *Nucleic Acids Res* **31**, 5776–5788.

Pinkel D, Straume T, Gray JW. 1986. Cytogenetic analysis using quantitative, high-sensitivity, fluorescence hybridization. *Proc Natl Acad Sci U S A* **83**, 2934–2938.

Piper J, Fantes J, Gosden J, Ji LA. 1990. Automatic detection of fragile X chromosomes using an X centromere probe. *Cytometry* **11**, 73–79.

Piper J, Poggensee M, Hill W, Jensen R, Ji L, et al. 1994. Automatic fluorescence metaphase finder speeds translocation scoring in FISH painted chromosomes. *Cytometry* **16**, 7–16.

Plourde-Owobi L, Seguin D, Baudin MA, Moste C, Rokbi B. 2005. Molecular characterization of Clostridium tetani strains by pulsed-field gel electrophoresis and colony PCR. *Appl Environ Microbiol* **71**, 5604–5606.

Porro D, Sauer M, Branduardi P, Mattanovich D. 2005. Recombinant protein production in yeasts. *Mol Biotechnol* **31**, 245–259.

Potyrailo RA, Conrad RC, Ellington AD, Hieftje GM. 1998, Adapting selected nucleic acid ligands (aptamers) to biosensors. *Anal Chem* **70**, 3419–3425.

Power EGM. 1996. RAPD typing in microbiology; a technical review. *J Hosp Infect* **34**, 247–265.

Ravikirthi P, Suthers PF, Maranas CD. 2011. Construction of an *E. Coli* genome-scale atom mapping model for MFA calculations. *Biotechnol Bioeng* **108**, 1372–1382.

Raap AK. 1998. Advances in fluorescence *in situ* hybridization. *Mutat Res* **400**, 280–298.

Rahman M, Shimizu K. 2008. Altered acetate metabolism and biomass production in several *Escherichia coli* mutants lacking *rpoS*-dependent metabolic pathway genes. *Mol BioSyst* **4**, 160–169.

Ransohoff DF. 2005. Bias as a threat to the validity of cancer molecular-marker research. *Nat Rev Cancer* **5**, 142–149.

Rasmussen S, Nielsen HB, Jarmer H. 2009. The transcriptionally active regions in the genome of *Bacillus subtilis*. *Mol Microbiol* **73**, 1043–1057.

Redwan E, Matar SM, El-Aziz GA, Serour EA. 2008, Synthesis of the human insulin gene: protein expression, scaling up and bioactivity. *Prep Biochem Biotechnol* **38**, 24–39.

Revers RR, Henry R, Schmeiser L, Kolterman O, Cohen R, et al. 1984. The effects of biosynthetic human proinsulin on carbohydrate metabolism. *Diabetes* **33**, 762–770.

Richter M, Kube M, Bazylinski DA, Lombardot T, Glockner FO, et al. 2007. Comparative genome analysis of four magnetotactic bacteria reveals a complex set of group-specific genes implicated in magnetosome biomineralization and function. *J Bacteriol* **189**, 4899–4910.

Richterich P. 1989. Non-radioactive chemical sequencing of biotin labelled DNA. *Nucleic Acids Res* **17**, 2181–2186.

Risatti G, Holinka L, Lu Z, Kutish G, Callahan JD, et al. 2005. Diagnostic evaluation of a real-time reverse transcriptase PCR assay for detection of classical swine fever virus. *J Clin Microbiol* **43**, 468 471.

Robinson BH, Seeman NC. 1987. The design of a biochip: a self-assembling molecular-scale memory device. *Protein Eng* **1**, 295–300.

Rosén C, Mattsson N, Johansson PM, Andreasson U, Wallin A, et al. 2011. Discriminatory analysis of biochip derived protein patterns in CSF and plasma in neurodegenerative diseases. *Front Aging Neurosci* **3**, 1. DOI: 10.3389/fnagi.2011.00.

Rosenthal A, Sproat BS, Brown DM. 1990. A new guanine-specific reaction for chemical DNA sequencing using m-chloroperoxybenzoic acid. *Biochem Biophys Res Commun* **173**, 272–275.

Sachidanandam R, Weissman D, Schmidt SC, Kakol JM, Stein LD,et al. 2001. A map of human genome sequence variation containing 1.42 million single nucleotide polymorphisms. *Nature* **409**, 928–933.

Sambrook J, Russell DW. 2001. *Molecular Cloning: A Laboratory Manual*, 3rd ed., Cold Spring Harbor Laboratory Press, Cold Spring Harbor, NY.

Sanger F, Air GM, Barrell BG, Brown NL, Coulson AR,et al. 1977a. Nucleotide sequence of bacteriophage phi X174 DNA. *Nature* **265**, 687–695.

Sanger F, Nicklen S, Coulson AR. 1977b. DNA sequencing with chain terminating inhibitors. *Proc Natl Acad Sci U S A* **74**, 5463–5467.

Sanger F, Coulson AR. 1975. A rapid method for determining sequences in DNA by primed synthesis with DNA polymerase. *J Mol Biol* **94**, 441–448.

Sasidharan R, Agarwal A, Rozowsky J, Gerstein M. 2009a. An approach to compare genome tiling microarray and MPSS sequencing data for transcript mapping. *BMC Res Notes* **2**, 211.

Sasidharan R, Agarwal A, Rozowsky J, Gerstein M. 2009b. An approach to comparing tiling array and high throughput sequencing technologies for genomic transcript mapping. *BMC Res Notes* **2**, 150.

Sayers EW, Waring M J. 1993. Footprinting titration studies on the binding of echinomycin to DNA incapable of forming Hoogsteen base-pairs. *Biochemistry* **32**, 9094–9107.

Schneider A, Dessimoz C, Gonnet GH. 2007. OMA Browser–exploring orthologous relations across 352 complete genomes. *Bioinformatics* **23**, 2180–2182.

Schrock E, du Mamoir S, Veldman T, Schoell B, Wienberg J, et al. 1996. Multicolor spectral karyotyping of human chromosomes. *Science* **273**, 494–497.

Selinger DW, Cheung KJ, Mei R, Johansson EM, et al. 2000. RNA expression analysis using a 30 base pair resolution *Escherichia coli* genome array. *Nat Biotechnol* **18**, 1262–1268.

Shiaw-Lin W, Jeongkwon K, Russell WB, Lance L, et al. 2006. Dynamic profiling of the post-translational modifications and interaction partners of epidermal growth factor receptor signaling after stimulation by epidermal growth factor using extended range proteomic. *Mol Cell Proteomics* **5**, 1610–1627.

Siezen RJ, Wilson G, Todt T. 2010. Prokaryotic whole transcriptomic analysis: deep sequencing and tiling arrays. *Microbial Biotechnology* **3**, 125–130.

Siffroi JP, Dupuy O, Joye N, Le Bourhis C, Benzacken B, et al. 2000. Usefulness of fluorescence *in situ* hybridization for the diagnosis of turner mosaic fetuses with small ring X chromosomes fetal. *Diagn Ther* **15**, 229–233.

Singer RH, Ward DC. 1982. Actin gene expression visualized in chicken muscle tissue culture by using *in situ* hybridization with a biotinated nucleotide analog. *Proc Natl Acad Sci U S A* **79**, 7331–7335.

Smith K. Diggle MA, Clarke SC. 2004. Automation of a fluorescence-based multiplex PCR for the laboratory confirmation of common bacterial pathogens. *J Med Microbiol* **53** (Pt 2), 115–117.

Smith LM, Fung S, Hunkapiller MW, Hunkapiller TJ, Hood LE. 1985. The synthesis of oligonucleotides containing an aliphatic amino group at the 5′ terminus: synthesis of fluorescent DNA primers for use in DNA sequence analysis. *Nucleic Acids Res* **13**, 2399–2412.

Smith LM, Sanders JZ, Kaiser RJ, Hughes P, Dodd C, et al. 1986. Fluorescence detection in automated DNA sequence analysis. *Nature* **321**, 674–679.

Soga T, Ohashi Y, Ueno Y, Naraoka H, Tomita M, Nishioka T. 2003. Quantitative metabolome analysis using capillary electrophoresis mass spectrometry. *J Proteome Res* **2**, 488–494.

Song JM, Culha M, Kasili PM, Griffin GD, Vo-Dinh T. 2005. A compact CMOS biochip immunosensor towards the detection of a single bacteria. *Biosens Bioelectron* **20**, 2203–2209.

Sorek R, Cossart P. 2010. Prokaryotic transcriptomics: a new view on regulation, physiology and pathogenicity. *Nat Rev Genet* **11**, 9–16.

Speicher MR, Gwyn Ballard S, Ward DC. 1996. Karyotyping human chromosomes by combinatorial multi-fluor FISH. *Nat Genet* **12**, 368–375.

Stamm S, Longo FM. 1990. Direct sequencing of PCR products using the Maxam-Gilbert method. *Genet Anal Tech Appl* **7**, 142–143.

Steven A, Greenberg MD. 2001. DNA microarray gene expression analysis technology and its application to neurological disorders. *Neurology* **57**, 755–761.

Stoevesandt O, Taussig MJ, He M. 2009. Protein microarrays: high-throughput tools for proteomics. *Expert Rev Proteomics* **6**, 145–157.

Stratis-Cullum DN, Griffin GD, Mobley J, Vass AA, Vo-Dinh T. 2003. A miniature biochip system for detection of aerosolized Bacillus globigii spores. *Anal Chem* **75**, 275–280.

Studier FW. 1989. A strategy for high-volume sequencing of cosmid DNAs—random and directed priming with a library of oligonucleotides. *Proc Natl Acad Sci U S A* **86**, 6917–6921.

Suciu V, Botan E, Valent A, Chami L, Spatz A, Vielh P. 2008. The potential contribution of fluorescent *in situ* hybridization analysis to the cytopathological diagnosis of Merkel cell carcinoma. *Cytopathology* **19**, 48–51.

Sugimoto M, Hirayama A, Ishikawa T, Robert M, Baran R, et al. 2010. Differential metabolomics software for capillary electrophoresis-mass spectrometry data analysis. *Metabolomics* **6**, 27–41.

Swerdlow H, Gesteland R. 1990. Capillary gel electrophoresis for rapid, high resolution DNA sequencing. *Nucleic Acids Res* **18**, 1415–1419.

Tahara T, Kraus JP, Rosenberg LE. 1990. Direct DNA sequencing of PCR amplified genomic DNA by the Maxam-Gilbert method. *Biotechniques* **8**, 366–368.

Takashi A, Sujawara H, Kanaya S, Ikemura T. 2006. A novel bioinformatics tools for phylogenetic classification of genomic sequence fragments derived from the mixed genomes of uncultured environmental microbes. *Polar Biosci* **20**, 109–112.

Tanke HJ, Florijn RJ, Wiegant J, Raap AK, Vrolijk J. 1995. CCD microscopy and image analysis of cells and chromosomes stained by fluorescence *in situ* hybridization. *Histochem J* **27**, 4–14.

Tanke HJ, Wiegant J, van Gijlswijk RP, Bezrookove V, Pattenier H, et al. 1999. New strategy for multi-colour fluorescence *in situ* hybridisation: COBRA: COmbined Binary RAtio labelling. *Eur J Hum Genet* **7**, 2–11.

Tavazoie S, Hughes JD, Campbell MJ, Cho RJ, Church GM. 1999. Systematic determination of genetic network architecture. *Nat Genet* **22**, 281–285.

Tian J, Sang P, Gao P, Fu R, Yang D, et al. 2009. Optimization of a GC-MS metabolic fingerprint method and its application in characterizing engineered bacterial metabolic shift. *J Sep Sci* **32**, 2281–2288.

Tiefenbrunn AJ, Robinson AK, Kurnik PB, Ludbrook PA, Sobel BE. 1985. Clinical pharmacology in patients with evolving myocardial infarction of tissue-type plasminogen activator produced by recombinant DNA technology. *Circulation* **71**, 110–116.

Tirumalai RS, Chan KC, Prieto DA, Issaq HJ, Conrads TP, Veenstra TD. 2003. Characterization of the low molecular weight human serum proteome. *Mol Cell Proteomics* **2**, 1096–1103.

Tizard R, Cate RL, Ramachandran KL, Wysk M, Voyta JC, Murphy OJ, Bronstein I. 1990. Imaging of DNA sequences with chemiluminescence. *Proc Natl Acad Sci U S A* **87**, 4514–4518.

Toledo-Arana A, Dussurget O, Nikitas G, Sesto N, Guet- Revillet H, Balestrino D, et al. 2009. The Listeria transcriptional landscape from saprophytism to virulence. *Nature* **459**, 950–956.

Tsuchizaki N, Ishikawa J, Hotta K. 2000. Colony PCR for rapid detection of antibiotic resistance genes in MRSA and enterococci. *Jpn J Antibiot* **53**, 422–429.

Uchiyama I, Higuchi T, Kawai M. 2010. MBGD update 2010: toward a comprehensive resource for exploring microbial genome diversity. *Nucleic Acids Res* **38** (Suppl 1), D361–D365. doi: 10.1093/nar/gkp948.

Uchiyama I. 2003. MBGD: microbial genome database for comparative analysis. *Nucleic Acids Res* **31**, 58–62.

Uchiyama I. 2006. Hierarchical clustering algorithm for comprehensive orthologous-domain classification in multiple genomes. *Nucleic Acids Res* **34**, 647–658.

Uchiyama I. 2007. MBGD: a platform for microbial comparative genomics based on the automated construction of orthologous groups. *Nucleic Acids Res* **35**, D343–D346.

van Vliet AH. 2010. Next generation sequencing of microbial transcriptomes: challenges and opportunities. *FEMS Lett* **302**, 1–7.

van Vliet AH, Wren BW. 2009. New levels of sophistication in the transcriptional landscape of bacteria. *Genome Biol* **10**, 233.

Vassaux G, Nitcheu J, Jezzard S, Lemoine NR. 2006. Bacterial gene therapy strategie. *J Pathol* **208**, 290–298.

Van de Velde S, Delaive E, Dieu M, Carryn S, Van BF, et al. 2009. Isolation and 2-D-DIGE proteomic analysis of intracellular and extracellular forms of *Listeria monocytogenes*. *Proteomics* **9**, 5484–5496.

Venter JG, Adams MD, Myers EW, Li PW, Mural RJ, Sutton GG, et al. 2001. The sequence of the human genome. *Science* **291**, 1304–1351.

Venter JC, Smith HO, Hood L. 1996. A new strategy for genome sequencing. *Nature* **381**, 364–366.

Voss H, Wiemann S, Grothues D, Sensen C, Zimmermann J, Schwager C, Stegemann J, Erfle H, Rupp T, Ansorge W. 1993. Automated low-redundancy large-scale DNAsequencing by primer walking. *Biotechniques* **15**, 714–721.

Wada A. 1984. Automatic DNA sequencing. *Nature* **307**, 193.

Wada A, Yamamoto M, Soeda E. 1983. Automatic DNA sequencer: computer-programmed microchemical manipulator for the Maxam-Gilbert sequencing method. *Rev Sci Instrum* **54**, 1569–1572.

Wagstaff JL, Sadovskaya I, Vinogradov E, Jabbouri S, Mark J. Howard MJ. 2008. Poly-*N*-acetylglucosamine and poly(glycerol phosphate) teichoic acid identification from staphylococcal biofilm extracts using excitation sculptured TOCSYNMR. *Mol Biosyst* **4**, 170–174.

Wang Z, Gerstein M, Snyder M. 2009. RNA-Seq: a revolutionary tool for transcriptomics. *Nat Rev Genet* **10**, 57–63.

Waters LS, Storz G. 2009. Regulatory RNAs in bacteria. *Cell* **136**, 615–628.

Werner RG, Pommer CH. 1990. Successful development of recombinant DNA-derived pharmaceuticals. *Arzneimittelforschung* **40**, 1274–1283.

Wiegant J, Ried T, Nederlof PM, van der Ploeg M, Tanke HJ, Raap AK. 1991. *In situ* hybridization with fluoresceinated DNA. *Nucleic Acids Res* **19**, 3237–3241.

Wilhelm BT, Landry JR. 2009. RNA-Seq-quantitative measurement of expression through massively parallel RNA-sequencing. *Methods* **48**, 249–257.

Wilhelm BT, Marguerat S, Watt S, Schubert F, Wood V, et al. 2008. Dynamic repertoire of a eukaryotic transcriptome surveyed at single-nucleotide resolution. *Nature* **453**, 1239–1243.

Wishart DS, Tzur D, Knox C, Eisner R, Guo AC, et al. 2007. HMDB: the human metabolome database. *Nuceic Acids Res* **35**, 521–526.

Wittwer CT, Herrmann MG, Gundry CN, Elenitoba-Johnson KS. 2001. Real-time multiplex PCR assays. *Methods* **25**, 430–442.

Wurtzel O, Sapra R, Chen F, Zhu Y, Simmons BA, Sorek R. 2010. A single-base resolution map of an archaeal transcriptome. *Genome Res* **20**, 133–141.

Xi L, Lyons-Weiler J, Coello MC, Huang X, Gooding WE, et al. 2005. Prediction of lymph node metastasis by analysis of gene expression profiles in primary lung adenocarcinomas. *Clin Cancer Res* **11**, 4128–4135.

Yanagisawa K, Shyr Y, Baogang J Xu, Massion PP, Larsen PH, et al. 2003. Proteomic patterns of tumour subsets in non-small-cell lung cancer. *Lancet* **362**, 433–439.

Yoder-Himes DR, Chain PS, Zhu Y, Wurtzel O, Rubin EM, et al. 2009. Mapping the *Burkholderia cenocepacia* niche response via highthroughput sequencing. *Proc Natl Acad Sci U S A* **106**, 3976–3981.

Yuen T, Wurmbach E, Pfeffer RL, Ebersole BJ, Sealfon SC. 2002. Accuracy and calibration of commercial oligonucleotide and custom cDNA microarrays. *Nucleic Acids Res* **30**, e48.

Zhu M, Yang T, Wei S, DeWan T, Morell RJ, et al. 2003. Mutations in the gamma-actin gene (ACTG1) are associated with dominant progressive deafness (DFNA20/26). *Am J Hum Genet* **73**, 1082–1091.

3

MEDICAL MICROBIOLOGY AND MOLECULAR DIAGNOSTICS

3.1. PROLOGUE

Infectious diseases represent the second leading cause of death worldwide. The World Health Organization has estimated that each year, 3 million people die of tuberculosis, 0.5 million die of pertussis, and 25,000 die of typhoid. Diarrheal diseases, many of which are bacterial, the second leading cause of death in the world (after cardiovascular diseases), are killing 5 million people annually. Today, most microbial diseases of humans and their etiologic agents have been identified, although variants continue to evolve and sometimes new ones emerge. Parasitic infections remain a chronic health problem throughout the developing and underdeveloped world, and, unlike many viral or bacterial diseases, there are no successful vaccines for many of the parasitic diseases. Major advances in vaccinology over the last century resulted in the development of effective vaccines (e.g., pneumococcal polysaccharide vaccine, diphtheria toxoid, and tetanus toxoid), as well as of other vaccines (e.g., cholera, typhoid, and plague vaccines) that are less effective or have side effects. Another major advance was the discovery of antibiotics as powerful therapeutic tools. Their efficacy is reduced by the emergence of antibiotics-resistant bacteria. A major concern in public health remains the increase in the resistance of bacterial pathogens to available antibacterial agents coupled with an increase in virulence and persistence of some pathogenic strains. For example, multidrug-resistant clones of *Mycobacterium* have caused outbreaks of tuberculosis with no possibility of effective treatment, resulting in the death of the infected individuals.

Microbes: Concepts and Applications, First Edition. Prakash S. Bisen, Mousumi Debnath, Godavarthi B. K. S. Prasad
© 2012 Wiley-Blackwell. Published 2012 by John Wiley & Sons, Inc.

Many bacterial diseases can be viewed as a failure of the bacterium to adapt, since a well-adapted parasite ideally thrives in its host without causing significant damage. Relatively nonvirulent (i.e., well-adapted) microorganisms can cause disease under special conditions, for example, if they are present in unusually large numbers, if the host's defenses are impaired (e.g., AIDS and chemotherapy), or if anaerobic conditions exist. Despite advances in our knowledge of the etiology, treatment, and management of bacterial and viral infections in the community, they remain a major cause of morbidity and mortality worldwide. Dual viral infections are common, and a third of children have evidence of viral–bacterial co-infection. Pathogenic bacteria constitute only a small proportion of bacterial species; many nonpathogenic bacteria are beneficial to humans (i.e., intestinal flora produce vitamin K) and participate in essential processes such as nitrogen fixation, waste breakdown, food production, drug preparation, and environmental bioremediation. The understanding of the most intimate molecular mechanisms of the pathogen–host interaction might enable us to identify microbial functions required for pathogenesis or disease transmission that can be specifically targeted by new medicines and vaccines. This is possible only if enough basic knowledge is generated by research in the field of microbial pathogenesis.

3.2. MICROBIAL BIOLOGY

3.2.1. Morphology and Nature of Microorganisms

Microbes first appeared on earth about 3.5 billion years ago. They have a major role in sustaining life on our planet. Bacteria being prokaryotes, lack the complex cellular organization found in higher organisms, that is, no nuclear envelope and no specialized organelles. Yet, they engage in all basic life processes—transport of materials into and out of the cell, catabolism and anabolism of complex organic molecules, and the maintenance of structural integrity. The microbes exist everywhere in air, soil, rock, and water. Some live happily in searing heat, while others thrive in freezing cold. Some microbes need oxygen to live, but others do not. The microbes are found in plants and animals as well as in the human body. Some microbes may be pathogenic in humans, plants, and animals while others are essential for leading healthy life, and humans cannot exist without them. Indeed, the relationship between microbes and humans is delicate and complex. Most disease causing microbes belong to one of four major groups: bacteria, viruses, fungi, or protozoa.

Microbes come in a bewildering and exciting variety of size and shapes, with new ones being discovered all the time. The most common bacterial shapes are rod, cocci, and spiral; however, within each group are hundreds of unique variations. Rods may be long, short, thick, and thin with rounded or pointed ends. Cocci may be large, small, or oval shaped. Spiral-shaped bacteria may be with loose spirals or very tight spirals. Bacteria may exist as single cells or in groups such as chains, clusters, pairs, tetrads, octads, or as masses embedded in a capsule. Bacteria with different morphological shapes, namely, square bacteria, star-shaped bacteria, stalked bacteria, and budding bacteria do exist.

Viruses are obligate intracellular parasites and contain either DNA or RNA and a protein coat consisting of individual protein units. Viruses survive and multiply inside living cells by using the biosynthetic machinery of the host. Viruses can be classified into several morphological groups, namely, helical (e.g., bacteriophage M13), polyhedral/cubic (e.g., poliovirus), enveloped—may have poyhedral (e.g., herpes simplex virus, HSV) or helical (e.g., influenza virus) capsids—and complex (e.g., poxviruses).

Prions are particles that are transmissible and devoid of nucleic acid and contain exclusively a modified protein rich in beta-sheet content (PrPSc). The normal, cellular PrP (PrPC) is converted into PrPSc through a posttranslational process during which it acquires high beta-sheet content. The species of a particular prion is encoded by the sequence of the chromosomal PrP gene of the mammals in which it replicated last. The strain-specific properties of prions are embedded in the tertiary structure of PrPSc (Fig. 3.1).

Parasites are found worldwide and live inside the human body, feeding off either our own energy or the food. Once established in the body, parasites can settle in small niches, damaging the tissues that harbor them. Humans may host over hundred different types of parasites (roundworms, tapeworms, flukes, single-cell parasites, etc.).

The fungi are more evolutionarily advanced forms of microorganisms, and like plants and animals, fungi are classified as eukaryotes; that is, they have a diploid number of

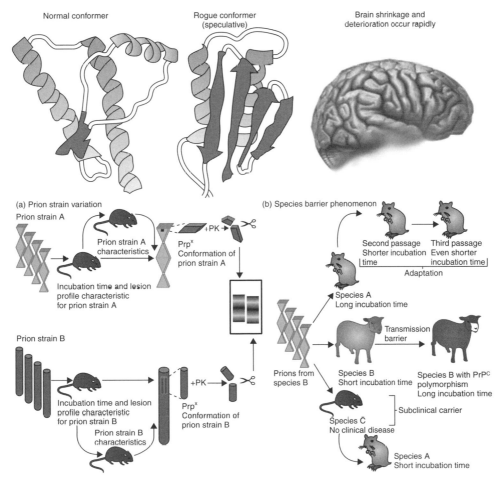

Figure 3.1. Structure and development cycle of prions.

chromosomes and a nuclear membrane and have sterols in their plasma membrane. Unlike these other groups, however, fungi possess chitin in their cell walls and are composed of filaments called *hyphae*. In addition to being filamentous, fungal cells often have multiple nuclei. Chitin is a long carbohydrate polymer that is also present in the exoskeletons of insects, spiders, and other arthropods. The chitin provides structural support and rigidity to the thin cells of the fungus, and makes fresh mushrooms crisp. Most members of the Fungi lack flagella. Fungi can be divided into two basic morphological forms, yeasts and hyphae. Yeasts are unicellular fungi, which reproduce asexually by blastoconidia formation (budding) or fission. Hyphae are multicellular fungi, which reproduce asexually and/or sexually.

3.2.2. Use of Bergey's Manual and Identibacter Interactus

Bergey's manual is one of the most authoritative resources in bacterial taxonomy. The five volume set is reorganized along phylogenetic lines and updated to reflect the current state of prokaryotic taxonomy. The first edition of *Bergey's Manual of Determinative Bacteriology* was initiated under the aegis of the Society of American Bacteriologists (now called the American Society for Microbiology) with the Editorial Board consisting of David H. Bergey (Chairman), Francis C. Harrison, Robert S. Breed, Bernard W. Hammer, and Frank M. Huntoon. Since the creation of the Bergey's Trust, the Trustees have published, successively, the fourth, fifth, sixth, seventh, eighth and ninth editions of the *Manual* and is updated from time to time. In 1977, the Trust published an abbreviated version of the eighth edition, called *The Shorter Bergey's Manual of Determinative Bacteriology*; this contained the outline classification of bacteria, descriptions of all genera and higher taxa, all keys and tables for the diagnosis of species, and all of the illustrations and two of the introductory chapters; however, it did not contain detailed species descriptions or most of the taxonomic comments. Through the years, the *Manual* has become a widely used international reference for bacterial taxonomy.

3.2.3. Bacterial Growth and Physiology

Bacterial growth is defined as the division of one bacterium into two cells, taking place by a process called *binary fission*. One initial bacterium cell eventually becomes millions. Each cell divides into two cells in minutes, making for the exponential rate at which the bacterial population increases. Binary fission is the asexual method of bacterial reproduction. The DNA of each cell first replicates then attaches itself to the plasma membrane. Once the DNA is attached to the membrane, the cell elongates by growing inward and splits into two separate entities, and this process is referred to as *cytokinesis*. Bacterial growth undergoes four stages. The first two stages are the lag phase, where growth is slow at first, bacteria become "acclimatized" to the new environmental conditions to which they have been introduced (pH, temperature, nutrients, etc.), and no significant increase in numbers with respect to time, and the log phase, where the number of bacteria doubles every few minutes. The final two stages of growth are the stationary phase, where growth stabilizes due to competition for nutrients among a high number of bacteria, and the death phase, where waste builds up and nutrients run out, allowing for the bacteria to die off (Fig. 3.2).

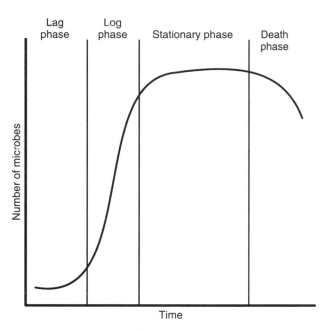

Figure 3.2. Bacterial growth pattern.

3.2.3.1. Bacterial Catabolism

Bacterial catabolism comprises the biochemical activities concerned with the net breakdown of complex substances to simpler substances by living cells. Substances with a high energy level are converted to substances of low energy content, and the organism utilizes a portion of the released energy for cellular processes. Endogenous catabolism relates to the slow breakdown of nonvital intracellular constituents to secure energy and replacement building blocks for the maintenance of the structural and functional integrity of the cell. This ordinarily occurs in the absence of an external supply of food. Exogenous catabolism refers to the degradation of externally available food. The principal reactions employed are dehydrogenation or oxygenation (either represents biological oxidation), hydrolysis, hydration, decarboxylation, and intermolecular transfer and substitution. The complete catabolism of organic substances results in the formation of carbon dioxide, water, and other inorganic compounds and is known as *mineralization*. Catabolic processes may degrade a substance only part way. The resulting intermediate compounds may be reutilized in biosynthetic processes, or they may accumulate intra- or extracellularly. Catabolism also implies a conversion of the chemical energy into a relatively few energy-rich compounds or "bonds," in which form it is biologically useful; also, part of the chemical energy is lost as heat.

Bacterial intermediary metabolism refers to the chemical steps involved in differing metabolic pathways. Normally, these intermediate compounds or precursors of subsequent products do not accumulate inside the bacterial cell and get transformed as rapidly as they are formed. The identification of such intermediary compounds, coenzymes, and enzymes catalyzing different steps of the pathway, and other details of the reaction, constitute the study of bacterial intermediary metabolism.

Many bacteria are able to grow and survive in the absence of oxygen and such anaerobic bacteria obtain energy by oxidizing suitable organic compounds by using another organic compound as an oxidizing agent in place of molecular oxygen. In most of the anaerobic fermentations, both the compounds oxidized and the compounds reduced (used as an oxidizing agent) are derived from a single substrate while, in other fermentations, one substrate is oxidized and the other is reduced. The substrates used by most of the bacteria include glucose and sucrose, polyalchohols such as mannitol, and salts of organic acids such as pyruvate and lactate.

3.2.3.2. Bacterial Anabolism

Bacterial anabolism comprises the physiological and biochemical activities concerned with the acquisition, synthesis, and organization of the numerous and varied chemical constituents of a bacterial cell. Clearly, when a cell grows and divides to form two cells, there exists twice the amount of cellular components that existed previously. These components are drawn, directly or indirectly, from the environment around the cell, and (usually) modified extensively in the growth processes when new cell material is formed (biosynthesis). This buildup, or synthesis, begins with a relatively small number of low molecular weight building blocks, which are either assimilated directly from the environment or produced by catabolism. By sequential and interrelated reactions, they are fashioned into different molecules (mostly of high molecular weight, and hence called *macromolecules*), for example, lipids, polysaccharides, proteins, and nucleic acids, and many of these molecules are, in turn, arranged into more complex arrays such as ribosomes, membranes, cell walls, and flagella. The anabolic products, of lower molecular weight, include pigments, vitamins, antibiotics, and coenzymes.

3.2.4. Antimicrobial Agents

Antimicrobial compounds interfere with growth and multiplication of microorganisms (bacteriostatic) or cause killing of the microbe (bactericidal). The broad spectrum antimicrobial compounds are active against a wide variety of microbes while the narrow spectrum is effective against few types of microorganisms. The *in vitro* antimicrobial activity is expressed as minimum inhibitory concentration (MIC), the lowest concentration that prevents visible growth of bacteria; minimum bactericidal concentration (MBC), the lowest concentration that kills the bacteria; or minimum antibiotic concentration (MAC), concentration that reduces the growth of an organism by factor of 10 (e.g., one log). An ideal antimicrobial agent exhibits powerful selective toxicity, slows down metabolism, good oral bioavailability, long elimination half life, no bacterial resistance or cross resistance, and no effect on the host immune system. The antimicrobial agents may be classified in many ways on the basis of the following criteria:

Mechanism of Action
- Inhibit cell wall synthesis: penicillin, cephalosporin, cycloserine, bacitracin, vancomycin, and clotrimazole.
- Inhibit cytoplasmic membrane function: polymyxins, amphotericin B, and nystatin.
- Inhibit protein synthesis: chloramphanicol, tetracycline, macrolides, and aminoglycosides.

- Interfere with intermediary metabolism: sulphonamides, trimethoprim, and sulphons.
- Affect nucleic acid metabolism and synthesis: quinolones, rifampicin, and acyclovir.

Chemical Structure

- Sulphonamides: sulphadimadine, sulphadiazine, sulphanilamide, sulphaquinoxalone.
- Diaminopyrimidines: trimethoprim, ormetoprim, and baquiloprim.
- Quinolones: nalidixic acid, enrofloxacin, difloxacin, and ciproflaxocin.
- β-Lactam antibiotics: penicillin G, ampicillin, cloxacillin, cephazolin, cephalexin.
- Aminoglycosides: streptomycin, gentamicin, amikacine, tobramycin.
- Tetracyclines: oxytetracycline, tetracycline, doxycyline, minocycline.
- Macrolide antibiotics: erythromycin and azithromycin.
- Nitrofuran—*derivatives*: nitrofurantoin and furazolidone.
- Nitroimidazole: metronidazole and imidazole.
- Polypeptide antibiotics: polymyxin B, colistin, and bacitricin.
- Polyene antibiotics: Nystatin and amphotericin B
- Imidazole derivatives: ketoconazole, fluconazole, and clotrimazole.

Therapeutic Applications (Type of Organism)

- Antibacterial: penicillin, aminoglycosides, tetracycline, and chloramphenicol.
- Antifungal: amphotericin B, griseofuvin, and ketoconazole.
- Antiviral: idoxuridine, vidarabine, zidovudine, and ribavirin.
- Antiprotozoal: metronidazole, quinapyramine, and diminazine.
- Anthelmintic: albendazole, levamisole, niclosamide, and praziquntel.
- Ectoparasiticide: cypermethrin, lindane, amitraz, and ethion.

Spectrum of Activity

- Narrow spectrum antimicrobials: penicillin G, streptomycin, erythromycin, and vancomycin.
- Broad spectrum antimicrobials: tetracycline, chloramphenicol, cephalexin, gentamycin, ampicllin.

Type of Action

- Bacteriostatic: sulphonamides, chloramphenicol, erythromycin, trimethoprim, clindamycin.
- Bactericidal: penicillin G, cephalexin, streptomycin, vancomycin and bacitracin.

Source

- Fungi: penicillin G, griseofulvin, and cephalexin.
- Actinomycetes: streptomycin, tetracycline, erythromycin, and chloramphenicol.
- Bacteria: polymyxin B, colistin, and bacitracin.
- Synthetic: sulphonamides, trimethoprim, quinolones, nitrofurans, and nitroimidazoles.

Antimicrobial resistance has emerged as one of the most important issues complicating the management of critically ill patients with infection. This is largely due to the increasing presence of pathogenic microorganisms with resistance to existing antimicrobial agents. Plants have an almost limitless ability to synthesize a variety of secondary metabolites—of which at least 12,000 have been isolated (tannins, terpenoids, alkaloids, and flavonoids), which have been found *in vitro* to have antimicrobial properties.

3.2.5. Bacterial Genetic Variations

Bacteria contain only one copy of genome (haploid). They have a circular chromosome with numerous extrachromosomal plasmids. Plasmids are small, circular pieces of DNA, and are independent from the main chromosome. They grow rapidly (double every 20–25 min, if conditions are favorable), and replicate independent of the main chromosome (have their own origin). Mutation produces genetic variation very rapidly in prokaryotes (bacteria) and viruses due to their short generation spans. Genetic variation, either by point mutations or gene rearrangement, is a fact of life in the microbial world. It not only allows pathogens to establish themselves in their chosen host, but also allows them to resist that host's subsequent attempts to evict them. They sometimes can actually take in DNA from other bacteria, which can integrate into their own circular DNA strands. Viruses inject their RNA strands into the host bacteria. Their genetic information becomes part of the host genetic code, where it is replicated along with the normal bacterial genes and forms more viruses. Sometimes, part of the bacterial DNA gets taken up into the newly formed viruses, which accounts for more genetic variation of the viruses. The sources of most of the genetic variation found in microbes include transformation (*Neisseria gonorrhoeae* pilin variation), transduction (*Vibrio cholera, Clostridium botulinum*), and conjugation (*Enterococcus faecium*). Genetic exchange plays a defining role in the evolution of many bacteria. The recent accumulation of nucleotide sequence data from multiple members of diverse bacterial genera has facilitated comparative studies that have revealed many features of this process. The transposable elements, which can move from place to place within chromosome or from plasmid to chromosome or vice versa by transposition in bacteria, generate a wide genetic variation. Some transposons are transcribed into RNA sequence, which yields reverse transcriptase enzyme. The reverse transcriptase converts RNA strand into a DNA strand, which is then inserted into the genome as a double-stranded DNA (dsDNA). There is evidence indicating that the extent and nature of recombination vary widely among microbiological species and often among lineages assigned to the same microbiological species (Didelot and Maiden, 2010).

3.2.6. Bacterial Pathogenecity

Pathogenicity is the ability of a microbe to cause disease following entry into the host. During the course of disease, bacterial pathogens have to respond and adapt to changing environmental conditions at the sites of infection. For this purpose, bacteria have developed complex regulatory processes that allow temporal and spatial control of virulence factor expression. Pathogenic bacteria may have one or several virulence factors. They may be common to all bacteria of a given genus or species, or they may be a characteristic of special pathogenic strains. Virulence factors that are common to the genus or species appear to give the organism a survival advantage; they may be selected during evolution. Those that are unique to some strains are usually acquired through mechanisms of genetic exchange (generally conjugation and transduction), and are not associated with specific survival advantage for the bacterium. For bacterial pathogens, successful *adherence* is usually a necessary prerequisite for virulence and even infection. Different pathogens exhibit different abilities of infecting a given host. For some pathogens, a very large number must be exposed to the host in order for infection to occur while for other pathogens relatively few individual microorganisms are necessary to induce disease. There are three general mechanisms by which a pathogen can induce damage to the host during infection, namely, *direct damage (minor), toxin-induced damage,* and *hypersensitivity.* *Toxins* are poisonous substances tend to be produced during a local infection, but are released and do their damage more systemically. There are about 220 bacterial *toxins* (exotoxins or endotoxins), which can *induce damage* through cardiovascular disturbance, destruction of blood vessels, disruption of the nervous system, plasma membrane disruption, inhibition of protein synthesis, or shock. About 40% of toxins disrupt plasma membranes. *Exotoxins* are proteins, often enzymes, produced inside cells and which do their damage only on release from the cell. Gram-positive rather than gram-negative bacteria tend to be producers of *exotoxins*. The endotoxins are relatively weak in terms of host damage induced, except when exposed to large quantities and/or include lipid A portion of lipopolysaccharide (LPS). Since LPS is found solely on gram-negative bacteria, *endotoxins* are also associated exclusively (or nearly so) with gram-negative bacteria. Qualitatively, all *endotoxins* produce the same symptoms. This is because the symptoms are a consequence of the body's response to the presence of *endotoxin* rather than *endotoxins* exerting some specific effect on the host. The host responses include fever, chills, weakness, aches, shock, death, septic shock, and so on. The number of pathogens required to cause disease (or, at least, infection) in half of the exposed hosts is called the ID_{50}. ID_{50} typically are one, that is, it usually takes far more than just exposure to a single organism for a host to become infected. The number of pathogens required to cause lethal disease in half of the exposed hosts is called an LD_{50}. The introduction of DNA techniques in the study of pathogenic microorganisms opened the way to the molecular analysis of virulence. The concomitant development of cell biology and the progressive application of its experimental approaches to the study of the interactions between microorganisms and host cells have made it possible, in this last decade of the millennium, to start elucidating the molecular and cellular mechanisms involved in the development of infectious diseases. The rapid and significant accumulation of knowledge in this domain has led to the establishment of molecular microbial pathogenesis as a new scientific discipline that is now studied as an independent subject in many universities. A clear indicator of the relevance of this emerging scientific field is that 64% of the 113 microbial genome sequencing projects currently recognized by the TIGR database concern infectious agents. Therefore, once again, the investigation of pathogenic microorganisms is driving the progress in life sciences (Mims, 1987).

3.2.7. Virus–Cell Interaction

Virus–cell interactions are fast processes; fusion of the viral envelope with cell membranes, for example. Viruses have to bind to cells in order to infect them. They require an intact cell to replicate and can direct the synthesis of hundreds to thousands of progeny viruses during a single cycle of infection. In contrast to other microorganisms, viruses do not replicate by binary fission; instead, the infecting particle must disassemble in order to direct synthesis of viral progeny. The interaction between a virus and its host cell begins with the attachment of the virus particle to specific receptors on the cell surface. Viral proteins that mediate the attachment include single-capsid components that extend from the virion surface, such as the attachment proteins of adenovirus, reovirus, and rotavirus; surface glycoproteins of enveloped viruses, such as influenza virus and HIV; and viral capsid proteins that form binding pockets that engage cellular receptors. Techniques such as X-ray crystallography are employed to characterize interactions between HIV and its cell-surface receptor. Adenovirus is the second virus to have its binding mechanism characterized at atomic resolution. Resistance to and recovery from viral infections will depend on the interactions that occur between viruses and host cell. The defenses mounted by the host may act directly on the virus or indirectly on virus replication by altering or killing the infected cell. The nonspecific host defenses function early in the encounter with virus to prevent or limit infection while the specific host defenses function after infection in recovery immunity to subsequent challenges. Fusion of the viral envelope with cell membranes, for example, is completed within 1 min.

3.3. INFECTION AND IMMUNITY

3.3.1. Overview of the Immune System

The immune response is how the body recognizes and responds to the entry of microbial pathogens and altered self-molecules such as tumor cells that appear foreign to the body. The immune system recognizes the pathogen as a whole or the substances known as *antigens* released by the pathogen and responds appropriately. Antigens are molecules found on the surface of pathogens, viruses, fungi, or bacteria. The immune system recognizes and destroys these antigens (Janeway, 1993).

Anatomically, the immune system is composed of specialized organs and several interdependent cell types that collectively protect the body from the invasion of foreign pathogens and altered self-cells. Many of these cell types have specialized functions and are specialized in engulfment of bacteria and other microbes, kill parasites, kill tumor cells, or viral-infected host cells. The immune system can be functionally divided into two wings, namely, innate immune system and acquired or adaptive immune system (Sompayrac, 1999).

The innate immune system is what we inherited by virtue of birth, which plays a key role in preventing the entry and destruction of foreign pathogens. The innate immune system is mainly composed of macrophages, polymorphonuclear cells (PMNs), natural killer (NK) cells, and acute phase proteins such as complement proteins, besides physical and mechanical barriers of the body. The cells of the innate immune system recognize pattern recognition receptors present on pathogens and act nonspecifically. The hallmarks of adaptive immune system are specificity and memory that makes it functionally different from the innate immune system. The adaptive immune system which is composed mainly of lymphocytes has two functional wings, namely, humoral and cell-mediated. The cell-mediated wing deals mainly with intracellular pathogens while the humoral wing deals with extracellular pathogens. The cells of the acquired immune system recognizes specific epitopes present on foreign invaders and mount effective immune response highly specific to a given pathogen or antigen. However, both wings of the immune system, namely, innate and acquired are highly interdependent on each other for mounting effective protective responses against pathogens (Janeway et al., 1999).

3.3.2. The Organs of the Immune System

3.3.2.1. Primary Lymphoid Organs

The primary lymphoid organs give rise to the birth of various cells associated with immune response. There are two primary lymphoid organs.

Bone Marrow. All cells of the immune system are initially derived from the bone marrow through a process called *hematopoiesis*. During hematopoiesis, bone-marrow-derived stem cells differentiate into mature cells either in bone marrow or into precursors of cells that migrate out of the bone marrow to continue their maturation elsewhere. The bone marrow generates B cells, NK cells, granulocytes, and immature thymocytes, besides red blood cells (RBCs) and platelets.

Thymus. Immature thymocytes, also known as *prothymocytes*, leave the bone marrow and migrate into the thymus. In the presence of thymic growth factors and stromal cells, the thymocytes get differentiated into T cells with specialized functions. Through a remarkable maturation process sometimes referred to as *thymic education*, T cells that are beneficial to the body are spared, while those T cells that might evoke a detrimental autoimmune response get eliminated. The mature T cells leave the thymus into the bloodstream.

3.3.2.2. Secondary Lymphoid Organs

The secondary lymphoid organs provide a battle ground where the interaction between cells of lymphoid system and microbial-derived antigens and activation of T and B cells take place resulting in the generation of appropriate immune responses.

Spleen. The spleen filters the bloodborne pathogens. It consists mainly of RBCs, macrophages, dendritic cells, B cells, T cells, and NK cells. In addition to direct capturing of foreign invaders (antigens) from the blood stream, migratory macrophages and antigen-presenting cells (APCs) bring antigens to the spleen via the bloodstream. An immune response is initiated when the macrophage or dendritic cells present the antigen to the appropriate B or T cells. This organ can be considered as a battle ground for cells of the immune system and foreign invaders. In the spleen, B cells get activated and differentiated to effector plasma cells in the presence of T-cell-derived cytokines and produce large amounts of specific antibodies. Also, the spleen serves as the site for destruction of aged RBCs.

Lymph Nodes. The lymph nodes filter tissueborne pathogens that enter the lymphatic system. Lymph nodes are found throughout the body and composed mainly of dendritic cells and macrophages, B cells, and T cells. The macrophages and dendritic cells dwelling in lymph nodes capture antigens, process them, and present these processed antigens to T- and B cells, consequently initiating an appropriate immune response.

Mucosa-Associated Lymphatic Tissue. This is less organized and consists of loosely organized clusters of lymphoid cells. These cells play a key role in dealing with mucosal pathogens.

3.3.3. Cells of the Immune System

3.3.3.1. Cells of Innate Immune System

GRANULOCYTES OR POLYMORPHONUCLEAR EUKOCYTES (PMNs). The PMNs consist of three cell types, namely, neutrophils, eosinophils, and basophils, based on their staining characteristics with certain dyes. These cells play a major role in the elimination of bacteria and parasites from the body. They engulf these foreign organisms and degrade them in an antigen nonspecific manner (Table 3.1).

MACROPHAGES. Macrophages are key cells in the regulation of adaptive immune responses. They are often referred to as *scavengers* or *antigen-presenting cells* because they pick up and ingest foreign materials and have the capability to present these antigens to T cells of the adaptive immune system. This is one of the important prerequisites in the initiation of adaptive immune response. Macrophages activated by certain cytokines exhibit increased levels of phagocytosis and also secrete large amounts of monokines required for the generation of inflammatory response.

NATURAL KILLER CELLS. These are similar to CD8+ cells and function as effector cells that directly kill viral-infected cells, most notably herpes- and cytomegalovirus-infected cells. NK cells also act on certain tumors such as melanomas and lymphomas. Unlike the CD8+ T cells, NK cells kill their targets without prior activation in an antigen nonspecific manner. However, in the presence of cytokines from CD4+ T cells, NK cells kill their tumor or viral-infected host cells more effectively.

DENDRITIC CELLS. Dendritic cells function as APCs and are considered superior to macrophages in antigen presentation to T-helper (TH) cells. These cells are found in the bloodstream and other tissues of the body such as skin and lungs. The dendritic cells capture antigen and bring it to the nearby lymphoid organs where an immune response is initiated (Banchereau et al., 2000). The dendritic cells are known to bind high amount of HIV, and are believed to serve as a reservoir of HIV that may be transmitted to CD4+ T cells.

3.3.3.2. Cells of Adaptive Immune System

T CELLS. T lymphocytes are divided into many subsets that are functionally and phenotypically different (Altman et al., 1996). The TH subset, also called the CD4+ T cell, is a pertinent coordinator of immune regulation (Constant and Bottomly, 1997). The main function of the TH cell is to augment immune responses by the secretion of specialized factors called *cytokines* that activate other cells of the immune system to fight off infection.

Another important subset of T cell is CD8+ T cell, which is involved in direct killing of certain tumor cells, viral-infected cells, and allogenic cells (Badovinac et al., 2000).

TABLE 3.1. Cells of the Immune System

Characteristics	Basophils and Mast Cells	Neutrophils	Eosionophils	Monocytes and Macrophages	Lymphocytes and Plasma Cells	Dendritic Cells
Structure						
% of WBCs in blood	Rare	50–70%	1–3%	1–6%	20–35%	NA
Subtypes and nicknames		Called *polys* or *segs* Immature forms called *bands* or *stabs*		Called the *mononuclear phagocyte system*	B lymphocytes Plasma cells T lymphocytes Cytotoxic T cells Helper T cells Natural killer cells Memory cells	Also called *Langerhans cells, veiled cells*
Primary function(s)	Release chemicals that mediate inflammation and allergic responses	Ingest and destroy invaders	Destroy invaders, particularly antibody-coated parasites	Ingest and destroy invaders Antigen presentation	Specific responses to invaders, including antibody production	Recognize pathogens and activate other immune cells by antigen presentation
Classification	Granulocytes	Granulocytes Phagocytes	Granulocytes Phagocytes Cytotoxic cells	Phagocytes Antigen-presenting cells	Cytotoxic cells (some types) Antigen-presenting cells	Antigen-presenting cells

Another important type of T cell is regulatory T cells, which play an important role in the regulation of immune responses. All types of T cells are found throughout the body and they often depend on the secondary lymphoid organs (the lymph nodes and spleen) as sites for activation to take place, but they are also found in other tissues of the body, most conspicuously the liver, lung, blood, and intestinal and reproductive tracts.

B Cells.   The major function of B lymphocytes is the production of specific antibodies in response to foreign proteins of bacteria, viruses, and parasites. Antibodies are specialized defense proteins that specifically recognize and bind to specific antigenic epitopes. Antibody binding to a foreign substance or antigen serves as a means of signaling phagocytic cells to engulf, kill, or remove that substance from the body.

3.3.4. Innate Immunity

Innate immunity provides the first line of defense against invading pathogens and refers to the basic resistance mechanisms that a species possesses. The basic attributes of the innate immune response are

- Pathogen nonspecific
- No generation of memory cells.

The innate defenses of the body include physical barriers, physiological barriers, chemical factors, and endocytic and phagocytic barriers.

3.3.4.1. Physical Barriers

1. *Skin*. No microbe can invade the skin when intact. The low pH of skin due to lactic and fatty acids is not congenial for the growth of microbes. The epidermis forms a thin outer layer with tightly packed epidermal cells and keratin (water proofing), and is completely renewed every 15–30 days. The dermis is a thicker inner layer that contains sebaceous glands associated with hair follicles. The sebum produced consists of lactic and fatty acids.

2. *Mucous Membranes*. These membranes that line the gastrointestinal (GI), urogenital, and respiratory tracts consist of ciliated epithelial cells, and collectively represent a surface area equivalent to that of a basketball court. Various secretions of the body such as saliva, tears, and mucous secretions along with cilia of epithelial cells play an important role in outward movement of microbial pathogens that gain entry through any of these entry points.

3.3.4.2. Physiological Barriers

1. *Temperature*. Normal body temperature inhibits the growth of most microorganisms, and elevated body temperature (fever) can have a direct effect on pathogenic microorganisms.

2. *pH*. Acidic pH of the stomach, skin, and vagina inhibits microbial colonization and growth.

3.3.4.3. Chemical Factors

A number of chemical factors also participate in defense against day-to-day encountered pathogens, which include fatty acids, lactic acid, lysozyme (able to cleave the peptidoglycan layer of the bacterial cell wall), cryptidins, α-defensins (produced at the base of crypts of small intestine—damage cell membranes), β-defensins (produced within skin, respiratory tract—also damages cell membranes), surfactant proteins A and D (present in lungs—function as opsonins, which enhance the efficiency of phagocytosis), interferons (group of proteins produced by cells following viral infection, which then bind to nearby cells and ignite mechanisms that inhibit viral replication), complement (a group of serum proteins that circulate in an inactive proenzyme state, which can be activated by a variety of specific and nonspecific molecules resulting in the destruction of pathogenic organisms).

3.3.4.4. Endocytic and Phagocytic Barriers

1. *Endocytosis*. It is a process by which pathogen-derived soluble antigens or macromolecules present in the extracellular tissue fluids get internalized by cells. Endocytosis takes place through pinocytosis or receptor-mediated endocytosis.
2. *Phagocytosis*. Ingestion of a particulate material including whole pathogenic microorganisms is referred to as *phagocytosis*. The plasma membrane expands around the particulate material to form large vesicles called *phagosomes* (10–20 times larger than the endosome). Once the particulate matter is ingested into phagosomes, the phagosomes fuse with lysosomes and the ingested material is then digested by a process similar to that seen in endocytosis. Only specialized cells are capable of phagocytosis, whereas endocytosis is carried out by virtually all cells.

Macrophages and neutrophils are considered *professional phagocytes* while fibroblasts and epithelial cells are known as *nonprofessional* phagocytes. The latter can be induced to become phagocytic under certain circumstances such as during intense inflammation.

3.3.5. Adaptive Immunity

Adaptive (acquired) immunity refers to antigen-specific defense mechanisms, and, unlike innate immunity, it is designed to react with a specific antigen. The immune system, like the nervous system, can remember. This is why, usually one siege of chicken pox confers immunity from subsequent infections.

3.3.5.1. Antigens

An antigen is any molecule that stimulates an immune response. Antigens are molecules on the surface of pathogens, viruses, fungi, or bacteria. Nonliving substances such as bacterial toxins, chemicals, drugs, and foreign particles may also function as antigens. Most antigens are proteins or polysaccharides, although small molecules coupled to carrier proteins (haptens) can also be antigenic (Benjamin et al., 1984). The segment of an antigenic molecule to which its cognate antibody binds is termed an *antigenic determinant* or *epitope*.

3.3.5.2. Generation of Adaptive Immunity

It is estimated that the human body has the ability to recognize 10^7 or more different epitopes and make up different antibodies, each with a unique specificity. In order to recognize this immense number of different epitopes, the body produces 2×10^{12} or more distinct clones of both B- and T lymphocytes, each with a unique B-cell receptor or T-cell receptor. The downside to the specificity of adaptive immunity is that only a few B- and T cells exist that recognize any given epitope. These few cells then get rapidly proliferated in order to mount an effective immune response against that particular antigenic epitope, and that typically takes several days.

Generation of adaptive immunity against an infectious agent involves the following steps:

- Antigen processing by APCs such as macrophages and dendritic cells and presentation of the processed antigen onto the surface of APCs by major histocompatibility (MHC) molecules to T cells. These APCs are described in Table 3.2.
- Interaction between APC and T cells and subsequent activation and proliferation of antigen-specific T lymphocytes.
- Activation and proliferation of antigen-specific B lymphocytes.
- Generation of plasma cells and secretion of antibody molecules.
- Generation of effector T cells such as CTLs.
- Production of activated macrophages.

Accordingly, there are two major branches of the adaptive immune responses: humoral and cell-mediated (Fig. 3.3).

1. *Humoral Immunity*. Humoral immunity or antibody-mediated immunity involves the production of antibody molecules by effector B cells in response to an antigen, and is effective against extracellular pathogens.
2. *Cell-Mediated Immunity* (CMI). CMI, which is mediated by T cells, is effective against intracellular pathogens and involves the production of CTLs, activated macrophages, activated NK cells, and so forth.

TABLE 3.2. Antigen-Presenting Cells (APCs)

Professional APCs	Nonprofessional APCs
Dendritic cells (several types)	Fibroblasts (skin)
Macrophages	Thymic epithelial cells
B cells	Thyroid epithelial cells
	Glial cells (brain)
	Pancreatic beta cells
	Vascular endothelial cells

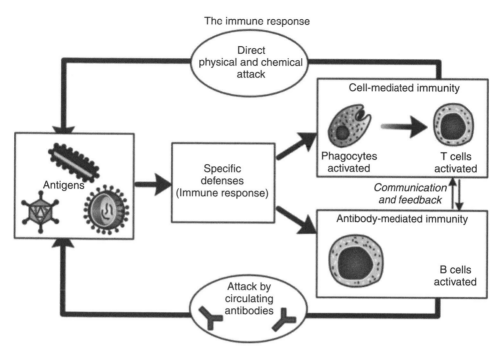

Figure 3.3. Generation of adaptive immune response.

3.3.5.3. Antigen Processing

Generation of specific adaptive immune response to an infectious agent requires antigen processing by APCs and its presentation on the surface of APCs by MHC molecules.

MAJOR HISTOCOMPATIBILITY COMPLEX. The MHC complex represents two major groups of molecules present on surfaces of cells. The class I MHC molecules are distributed on almost all nucleated cells and are involved in the presentation of endogenously derived antigens to CD8+ cells. The MHC I molecules present on the graft serve as targets in transplantation rejection. The class II MHC molecules, unlike MHC I molecules, have restricted cellular distribution and are confined to cells of the lymphoid system.

The genes representing MHC complex are located on chromosome 6 in humans. Class I molecules are encoded by the B, C, and A regions while class II molecules are encoded by the D region. A region between MHC I and MHC II encodes class III molecules such as specific complement components and other molecules that are not associated with antigen recognition. The class I and II regions are highly polymorphic (Guardiola et al., 1996).

MHC Class I Molecules. Class I molecules consist of two polypeptide chains: one larger chain consisting of about 350 amino acids (45 kDa) encoded by the gene in BCA region and another smaller chain, namely, β2-microglobulin of 12 kDa. The larger chain is folded into three identical domains known as α1, α2, and α3. The β2-microglobulin is noncovalently associated with the α3 domain of the larger chain. The α1 and α2 domains form antigen binding groove and consist of a β-pleated sheet at the bottom and two α helices on the sides. This groove can accommodate a peptide of about 10–13 amino acids length for presentation to a specific CTL (Buus, 1999).

MHC Class II Molecules. Class II molecules consist of two polypeptides, namely, α and β chains of about 230 and 240 amino acids long with 33 kDa and 28 kDa, respectively, and both the peptides are encoded by the D region. These polypeptides fold into two separate domains: α1 and α2 for the α-polypeptide and β1 and β2 for the β-polypeptide. The α1 and β1 domains together form a groove with β-pleated sheet on the bottom and two α-helices on the sides. This groove is capable of binding a small peptide of about 10 amino acids for presentation to specific TH cell (Kropshofer et al., 1999).

Class I versus Class II Molecules. Like all transmembrane proteins, the MHC molecules are synthesized by ribosomes on the rough endoplasmic reticulum (RER) and assembled within its lumen. Although class I and class II molecules appear somewhat structurally similar, they differ in their functions. Class I molecules present "endogenous" antigen while class II molecules present "exogenous" antigens to T cells. An endogenous antigen might be virus- or tumor-derived products. Exogenous antigens, in contrast, might be products of bacterial cells or viruses that are engulfed and processed by APCs. The TH cells, in turn, could activate appropriate B cells to produce antibody that would lead to the elimination of antigen.

PROCESSING OF ENDOGENOUS ANTIGENS (CYTOSOLIC PATHWAY). The APCs include macrophages, dendritic cells, and B cells. Dendritic cells are the most potent APCs as they express MHC II molecules and B-7 (required for generation of costimulus when APCs interact with T cells) constitutively, whereas macrophages and B cells require prior activation for the expression of B-7 on their cell surface. As with any other aged or altered self-protein, the endogenously derived viral antigens are degraded by proteosomes in the cytosol (Rock and Goldberg, 1999). These peptides from the cytosol are picked up by transmembrane TAP (transporters associated with antigen processing) proteins embedded in the membrane of the endoplasmic reticulum (ER) and are pumped in to the lumen of ER. The breakdown of proteins by proteosome and transport of peptides into the lumen of ER are energy-dependent processes. It is in the lumen of ER the peptides get associated with MHC I molecules and form a stable trimolecular complex. This complex then moves through the Golgi apparatus and is displayed on the surface of cells (Fig. 3.3).

PROCESSING OF EXOGENOUS ANTIGENS (ENDOCYTIC PATHWAY). The antigens derived from exogenous pathogens are degraded to smaller peptides of 10–13 amino acids by hydrolases in phagolysosome, and, unlike endogenous antigens, are not transported to the ER. The MHC II molecules synthesized by RER are stabilized by an "invariant" chain in the lumen of ER and this "ternary complex," through the Golgi apparatus, is transported to the lysosomes where exogenously derived degraded antigenic proteins are located. In the presence of lysosomal hydrolases, the "invariant" chain occupying the groove of MHC II is digested into smaller peptides and this frees the groove of MHC II for occupancy by the antigenic fragments in the phagolysosome. The MHC II-antigenic peptide complex moves to the plasma membrane and the complex is displayed at the cell surface (Fig. 3.4) (Reimann and Schirmbeck, 1999).

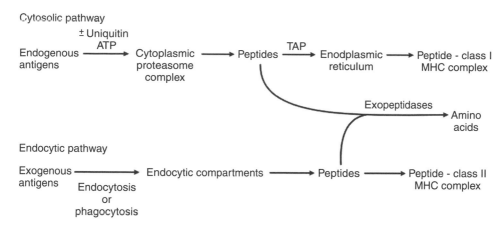

Figure 3.4. Outline of antigen processing pathways.

3.3.5.4. Humoral Immunity

The production of antibody involves three distinct phases:

1. *Induction Phase.* Antigen reacts with specific T and B cells.
2. *Expansion and Differentiation Phase.* Proliferation and differentiation of antigen-specific T- and B cells to a functional stage (i.e., effector cells).
3. *Effector Phase.* Secretion of antibodies by plasma cells that exert biological effects either independently or through complement or Fc (Fragment crystallized)-bearing cells (e.g., macrophages and other white blood cells).

Humoral immune response begins with the recognition of antigen by APCs followed by activation of T- and/or B cells (Agarwal et al., 1996). The interaction between antigen-specific TH cells (TH2) and antigen–MHC complex on the surface of APCs results in the activation of T cells and secretion of cytokines (Davis et al., 1998). These cytokines act on antigen specific "primed" B cells (B cells that have already encountered antigen), inducing B-cell proliferation and differentiation.

DIFFERENTIATION OF B LYMPHOCYTES. The antigen-dependent stages of B-lymphocyte activation, proliferation, and differentiation into effector plasma cells take place in secondary lymphoid organs, namely, spleen, lymph nodes, and other peripheral lymphoid tissues in the presence of T-cell cytokines. The activated B cell first differentiates into a larger B lymphoblast, which is devoid of all surface immunoglobulins (Igs). The B lymphoblast then differentiates into terminal plasma cell, which is, in essence, an antibody factory. The plasma cell secretes primarily IgM antibody during the initial "primary response." However, few B cells do not differentiate into plasma cells. Instead, these cells undergo secondary DNA rearrangements that place the constant region of the IgG, IgA, or IgE genes in conjunction with the VDJ (variable, diverse, and joining) genes and these cells remain as long-lived "memory cells." On subsequent encounter with antigen, these memory cells respond quickly and produce large amounts of IgG, IgA, or IgE antibodies, generating the "secondary response" (Ahmed and Gray, 1996).

TABLE 3.3. Features Differentiating Primary and Secondary Immune Responses

Primary Response	Secondary Response
Slow in onset	Rapid in onset
Low in magnitude	High in magnitude
Short lived	Long lived
IgM	IgG (or IgA, or IgE)

PRIMARY AND SECONDARY IMMUNE RESPONSES. Following exposure to an antigen for the first time, there is a lag of 7–10 days before specific antibody is produced. This antibody is IgM in nature, and, after a short time, the antibody level declines. These are the main characteristics of the primary response. On reexposure to the same antigen at a later date, there would be a rapid appearance of antibody, and in greater amount. It is of IgG class and remains detectable for a longer period. These are the features of the secondary response (Table 3.3).

STRUCTURE OF IMMUNOGLOBULINS. All immunoglobulins have a four chain structure as their basic unit. They are composed of two identical light chains (23 kDa) and two identical heavy chains (50–70 kDa) held together by interchain disulfide bonds. The amino acid sequence of N-terminal portion of each chain (110–115 amino acids), within a given class of immunoglobulin molecule, differs in antibodies of different antigenic specificities and this portion of immunoglobulin molecule is designated *variable portion*.

The amino acid sequence of C-terminal portion of each chain (110 amino acids in L chain and 330–440 amino acids in H chain) is constant in a given class of immunoglobulin molecule, irrespective of antibody specificity, and is designated *constant* region. The H chain contains a region that is rich in proline and cysteine and this region is called *hinge* region because proline induces some flexibility at this point. There are also intrachain disulfide bonds within each chain, which facilitate folding of the peptide chain into globular regions. These regions are called *domains* or *folds*. Each domain in an immunoglobulin molecule has an identical conformation of two β-sheets packed tightly against each other in a compressed antiparallel β-barrel. There are two domains in light chain (V_L and C_L) and four to five domains in heavy chain (V_H, C_{H1}–C_{H3} or C_{H4}). Immunogloblins contain a carbohydrate moiety attached generally to the C_{H2} domain. Although different immunoglobulins may differ structurally, they all are built from the same basic units.

Digestion with papain breaks the immunoglobulin molecule in the hinge region before the H–H interchain disulfide bond resulting in the formation of one Fc and two identical Fab (Fragment antigen binding) fragments that contain the light chain and the V_H and C_{H1} domains of the heavy chain. Each "Fab" fragment is monovalent, whereas the undigested immunoglobulin molecule is divalent. Treatment of immunoglobulins with pepsin results in cleavage of the heavy chain after the H–H interchain disulfide bonds resulting in a fragment that contains both antigen binding sites (Fab_2). The Fc region of the molecule is digested into small peptides by pepsin.

Most of the variability in variable regions of H and L chains is confined to three regions called the *hyper variable regions* or the *complementarily determining regions* (CDRs). The CDRs of one heavy and one light chain constitute "antigen binding" site of an antibody and interacts with antigenic epitopes. Antibodies with different antigenic specificities have different complementarity determining regions while antibodies of the exact same specificity have identical complementarity determining regions. The regions between the CDRs in the variable region are known as *framework regions*.

HEAVY CHAIN. The immunoglobulins can be divided into five major classes, based on differences in amino acid sequences in the constant region of the heavy chains. All immunoglobulins within a given class will have very similar heavy chain constant regions. The immunoglobulins of IgG and IgA are further divided into subclasses based on small differences in the amino acid sequences in the constant region of their heavy chains (Table 3.4).

LIGHT CHAIN. Immunoglobulin light chains are divided into two types based on differences in the amino acid sequence in the constant region of the light chain, namely, kappa light chains (60–65%) and lambda light chains (35–40%). The lambda light chains are further divided into four subtypes based on differences in the amino acid sequence in the constant region, namely., $\gamma 1$, $\gamma 2$, $\gamma 3$, and $\gamma 1$ (Fig. 3.5).

TABLE 3.4. Major Classes of Immunoglobulins

Immunoglobulin Class	Subclass	Nature of H Chain
IgG	IgG1	$\gamma 1$
	IgG2	$\gamma 2$
	IgG3	$\gamma 3$
	IgG4	$\gamma 4$
IgA	IgA1	$\alpha 1$
	IgA2	$\alpha 2$
IgM	—	μ
IgD	—	δ
IgE	—	ε

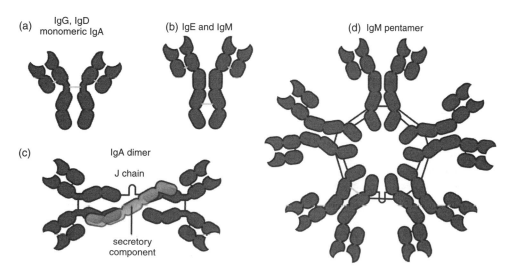

Figure 3.5. Structure of antibody classes.

The various class of immunoglobulins are described below.

IgG: IgG is a monomer and is the most abundant immunoglobulin and represents 75% of serum immunoglobulins. It is the only class of Ig that crosses the placenta except for IgG2. It can fix complement (except IgG4) and bind to Fc receptors (except IgG2 and IgG4) on blood cells including phagocytic cells. Hence, IgG is considered "opsonin."

IgM: IgM is the third most common serum Ig and exists generally as a pentamer (19S immunoglobulin), but it can also exist as a monomer. The monomeric form is present on the surface of B cells and mediates antigen recognition. The membrane-bound IgM contains an additional stretch of 20 amino acids at C-terminal end and is noncovalently associated with two additional membrane-bound proteins called Ig-alpha and Ig-beta with long cytoplasmic tails. These additional proteins function as signal transducing molecules on antigen interaction. IgM is the first Ig to be made by the fetus and the first Ig to be produced in primary immune response. The pentameric form contains an additional chain called J chain, which helps in polymerization of monomeric IgM. Single molecules of pentameric IgM can fix complement, whereas a minimum of two adjacent IgGs are required for complement fixation to take place. It is an efficient agglutinator.

IgA: IgA is the second most abundant serum immunoglobulin and exists as a monomer in serum and as a dimer in secretions. IgA is the major class of Ig in secretions—tears, saliva, and colostrums, and plays an important role in local (mucosal) immunity. The secretory IgA is associated with two other peptides, namely, a J chain and a secretory piece or T piece. The secretory piece helps IgA to be transported across mucosa and also protects it from degradation in the secretions. In general, IgA does not fix complement, unless aggregated.

IgD: IgD is primarily found on B-cell surfaces and found in low levels in serum. The function of serum IgD is uncertain while membrane-bound IgD functions as a receptor. IgD does not bind complement. The membrane bound IgD, like IgM, contains a hydrophobic stretch of amino acids and is associated with signal transducing Ig-alpha and Ig-beta chains.

IgE: IgE exists as a monomer and has an extra domain in the constant region. It is present in very low concentrations in serum as it can bind effectively to Fc receptors on basophils and mast cells even before interacting with antigens. It is the mediator of type I hypersensitivity (allergy) reactions and is elevated significantly in allergy and parasitic helminth infections. IgE does not fix complement, but eosinophils have Fc receptors for IgE and binding of eosinophils to IgE-coated helminths results in killing of the parasite by a process known as *antibody-dependent cell-mediated cytotoxicity* (ADCC).

GENERAL FUNCTIONS AND ANTIBODIES

1. *Antigen Binding*. Antibody binds to a specific antigenic determinant that is complementary to the CDR region. The primary function of immunoglobulins is antigen binding which may result in neutralization or elimination from the host.
2. *Effector Functions*. Frequently the binding of an antibody to an antigen may not result in elimination; rather, the antigen–antibody complexes may bind to phagocytic cells via Fc receptors and get phagocytosed. Antibody is thus considered "opsonin". The effector functions of immunoglobulins are as follows:
   Complement Activation. Antibody with bound antigen can fix and activate complement proteins by "classical pathway." The complement activation leads to

lysis or breakdown of bacterial and other pathogens and release of biologically active substances that mediate inflammation.

Binding to Various Cell Types. Antibodies may bind to phagocytic cells, lymphocytes, platelets, mast cells, and basophils which have Fc receptors. The binding of free antibodies or antigen–antibody complexes to these cell types result in a variety of effector functions such as "opsonization," mast cell degranulation (IgE) and ADCC. Some immunoglobulins such as IgG bind to receptors on placental trophoblasts, which facilitate transfer of the IgG across the placenta and these results in transplacental transfer of maternal immunity to the fetus and newborn.

ANTIGEN–ANTIBODY INTERACTION. The hypervariable regions of heavy and light chains of Fab portion of the antibody form antigen combining site of an antibody. The CDRs of one H chain and one L chain constitute one antigen binding site of an antibody. The antigenic epitopes nestles in a cleft formed by the combining site of the antibody and held together by the noncovalent bonds, namely, hydrogen bonds, electrostatic bonds, van der Waals forces, and hydrophobic bonds. Multiple bonding between the antigen and the antibody ensures that the antigen will be bound tightly to the antibody, although the interaction is reversible. The antigen–antibody interactions are affected by factors such as affinity and avidity of the antibody toward the antigen.

Specificity refers to the ability of an individual antibody to bind to only one antigenic determinant or the ability of a population of antibodies to react with only one antigen. Antibodies can distinguish differences in the primary structure of an antigen, isomeric forms of an antigen, secondary and tertiary structure of an antigen, and thus highly specific for a given epitope.

Cross reactivity refers to the ability of an individual antibody to react with more than one antigenic determinant or the ability of a population of antibodies to react with more than one antigen. Cross reactions arise because of shared epitopes.

3.3.5.5. Complement System

Historically, the term complement (C) was used to refer to a heat-labile serum component that was able to lyse bacteria (activity is destroyed (inactivated) by heating the serum at 56 °C for 30 min). Complement can opsonize bacteria for enhanced phagocytosis; it can recruit and activate various cells including PMNs and macrophages; it can participate in the regulation of antibody responses; and it can aid in the clearance of immune complexes and apoptotic cells. Complement can also have detrimental effects for the host; it contributes to inflammation and tissue damage and it can trigger anaphylaxis.

Complement comprises over 20–25 different serum proteins that are produced by a variety of cells including hepatocytes, macrophages, and gut epithelial cells. Almost all complement proteins but for Factor D exist as proenzymes that, when activated, cleave one or more other complement proteins. On cleavage, some of the complement breakdown products activate cells, increase vascular permeability, or opsonize bacteria.

The complement proteins are activated by three different pathways: *classical, alternative*, and *mannan-binding* (Fig. 3.6).

CLASSICAL PATHWAY
C1 Activation. C1, a multisubunit protein containing three different proteins (C1q, C1r, and C1s), binds to the Fc region of IgG and IgM antibody molecules that have

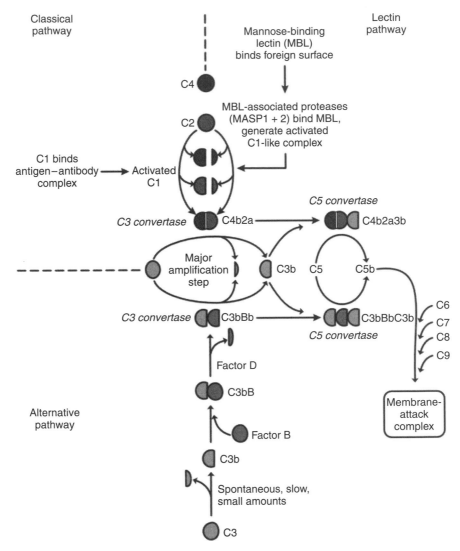

Figure 3.6. Complement activation pathways.

interacted with antigen. C1 binding does not occur to antibodies that have not complexed with antigen and binding requires calcium and magnesium ions. The binding of C1 to an antibody is via C1q, and C1q must cross-link at least two antibody molecules before it is firmly fixed. The binding of C1q results in the activation of C1r, which in turn activates C1s. The result is the formation of an activated "C1qrs," which is an enzyme that cleaves C4 into two fragments C4a and C4b.

C4 and C2 Activation (Generation of C3 Convertase). The bigger C4b fragment binds to the membrane and the C4a fragment is released into the microenvironment. Activated "C1qrs" also cleaves C2 into smaller C2a and bigger C2b fragments. C2a binds to the membrane in association with C4b, and C2b is released into the medium. The resulting C4bC2a complex functions as a C3 convertase, which cleaves C3 into C3a and C3b.

C3 Activation (Generation of C5 Convertase). The bigger C3b fragment binds to the membrane in association with C4b and C2a, and C3a is released. The resulting C4b2a3b is a C5 convertase and cleaves C5 into C5a and C5b.

The breakdown products of the classical pathway have potent biological activities that contribute to host defenses while some may exhibit detrimental effects if produced in an unregulated manner. Factors regulating the classical pathway are shown in Table 3.5.

The C1-INH deficiencies are associated with the development of hereditary angioedema, a hereditary disorder.

LECTIN PATHWAY. The lectin pathway is very similar to the classical pathway. It is initiated by the binding of mannose-binding lectin (MBL) to bacterial surfaces with mannose-containing polysaccharides (mannans). Binding of MBL to a pathogen results in the association of two serine proteases, MASP-1 and MASP-2 (MBL-associated serine proteases). MASP-1 and MASP-2 are similar to C1r and C1s, respectively, and MBL is similar to C1q. Formation of the MBL/MASP-1/MASP-2 trimolecular complex results in the activation of MASPs and subsequent cleavage of C4 into C4a and C4b. The C4b fragment binds to the membrane and the C4a fragment is released into the microenvironment. Activated MASPs also cleave C2 into C2a and C2b. C2a binds to the membrane in association with C4b and C2b is released into the microenvironment. The resulting C4bC2a complex is a C3 convertase, which cleaves C3 into C3a and C3b. C3b binds to the membrane in association with C4b and C2a and C3a is released into the microenvironment. The resulting C4bC2aC3b is a C5 convertase. The generation of C5 convertase is the end of the lectin pathway. The biological activities and the regulatory proteins of the lectin pathway are the same as those of the classical pathway.

ALTERNATIVE PATHWAY. The alternative pathway begins with the activation of C3 and requires Factors B and D and Mg^{2+} cation, all present in normal serum. In serum there is low level spontaneous hydrolysis of C3 to produce C3a and C3b. Activators of the alternate pathway are components on the surface of pathogens, and include LPS of gram-negative bacteria and the cell walls of some bacteria and yeasts. Thus, when C3b binds to an activator surface, the C3 convertase formed will be stable and continue to generate additional C3a and C3b by cleavage of C3. Once C3b is associated with a cell surface, Factor B will bind to it and becomes susceptible to cleavage by Factor D. The resulting C3bBb complex is a C3 convertase that will continue to generate more C3b, thus amplifying C3b production.

TABLE 3.5. Regulation of the Classical Pathway

Component	Regulation
C1	C1-INH; dissociates C1r and C1s from C1q
C3a	C3a inactivator (C3a-INA; carboxypeptidase B); inactivates C3a
C3b	Factors H and I; Factor H facilitates the degradation of C3b by Factor I
C4a	C3 INA
C4b	C4 binding protein (C4-BP) and Factor I; C4-BP facilitates degradation of C4b by Factor I; C4-BP also prevents association of C2a with C4b thus blocking the formation of C3 convertase

Some of the C3b generated by the stabilized C3 convertase on the activator surface associate with the C3bBb complex to form a C3bBbC3b complex. This is the C5 convertase of the alternative pathway. The generation of C5 convertase is the end of the alternative pathway. A deficiency of C3 results in an increased susceptibility to these organisms. The alternative pathway provides a means of nonspecific resistance against infection without the participation of antibodies, and hence provides a first line of defense against a number of infectious agents. Many gram-negative and some gram-positive bacteria, certain viruses, parasites, heterologous red cells, aggregated immunoglobulins (particularly, IgA), and some other proteins (e.g., proteases and clotting pathway products) can activate the alternative pathway. One protein, cobra venom factor (CVF), has been extensively studied for its ability to activate this pathway.

MEMBRANE ATTACK COMPLEX. C5 convertase from the classical (C4b2a3b), lectin (C4b2a3b), or alternative (C3bBb3b) pathway cleaves C5 into C5a and C5b. C5a remains in the fluid phase and the C5b rapidly associates with C6 and C7 and inserts into the membrane. Subsequently C8 binds, followed by several molecules of C9. The C9 molecules form a pore on the membrane through which the cellular contents leak and lysis occurs. The complex consisting of C5bC6C7C8C9 is referred to as the *membrane attack complex* (MAC).

C5a generated in the lytic pathway has several potent biological activities. It is the most potent anaphylotoxin. In addition, it is a chemotactic factor for neutrophils and stimulates the respiratory burst in them, and it stimulates inflammatory cytokine production by macrophages. Its activities are controlled by inactivation by carboxypeptidase B (C3-INA).

Some of the C5b67 complex formed can dissociate from the membrane and enter the fluid phase. If this were to occur it could then bind to other nearby cells and lead to their lysis. The damage to bystander cells is prevented by Protein S (vitronectin). Protein S binds to soluble C5b67 and prevents its binding to other cells.

BIOLOGICALLY ACTIVE PRODUCTS OF COMPLEMENT ACTIVATION. Activation of complement results in the production of several biologically active molecules that contribute to resistance, anaphylotoxis, and inflammation.

Anaphylotoxins. C4a, C3a, and C5a (in increasing order of activity) are all anaphylotoxins that cause basophil/mast cell degranulation and smooth muscle contraction. Undesirable effects of these peptides are controlled by carboxypeptidase B (C3a-INA).

Chemotactic Factors. C5a and MAC (C5b67) are both chemotactic. C5a is also a potent activator of neutrophils, basophils, and macrophages, and causes induction of adhesion molecules on vascular endothelial cells.

Opsonins. C3b and C4b on the surface of microorganisms attach to C-receptor (CR1) on phagocytic cells and promote phagocytosis.

Other Biologically Active Products of C Activation. Degradation products of C3 (iC3b, C3d, and C3e) also bind to different cells by distinct receptors and modulate their functions. In summary, the complement system takes part in both specific and nonspecific resistance and generates a number of products of biological and pathophysiological significance (Table 3.6).

TABLE 3.6. Functions of Complement Activation Products and their Regulatory Factors

Fragment	Activity	Effect	Control Factor(s)
C2a	Prokinin, accumulation of fluids	Edema	C1-INH
C3a	Basophil and mast cell degranulation, enhanced vascular permeability, smooth muscle contraction	Anaphylaxis	C3a-INA
C3b	Opsonin, phagocyte activation	Phagocytosis	Factors H and I
C4a	Basophil and mast cell degranulation, enhanced vascular permeability, smooth muscle contraction	Anaphylaxis (least potent)	C3a-INA
C4b	Opsonin	Phagocytosis	C4-BP and Factor I
C5a	Basophil and mast cell degranulation, enhanced vascular permeability, smooth muscle contraction	Anaphylaxis (most potent)	C3a-INA
	Chemotaxis, stimulation of respiratory burst, activation of phagocytes, stimulation of inflammatory cytokines	Inflammation	
C5bC6C7	Chemotaxis	Inflammation	Protein S (vitronectin)
	Attaches to other membranes	Tissue damage	

There are known genetic deficiencies of most individual complement components, but C3 deficiency is most serious and fatal. Complement deficiencies also occur in immune complex diseases and acute and chronic bacterial, viral, and parasitic infections (Table 3.7).

3.3.5.6. Cell-Mediated Immunity

The second arm of the adaptive immune response is referred to as *cell-mediated immunity* (CMI). As the name implies, the immunity is mediated mainly by cells and includes phagocytosis by phagocytes, direct cell killing by CTLs, and direct cell killing by NK cells. These responses are especially effective in destroying intracellular bacteria, eliminating virus-infected host cells, and tumor cell destruction.

MACROPHAGE ACTIVATION. While the production of antibody through the humoral immune response can effectively lead to the elimination of a variety of pathogens, bacteria that have evolved to invade and multiply within phagocytic cells of the immune response pose a different threat (Fig. 3.7).

CELL-MEDIATED CYTOTOXICITY. The second half of the cell-mediated immune response is involved in the rejection of foreign grafts and the elimination of tumors and virus-infected cells. The responses are mediated by CTLs, NK cells, and K cells. Each of these effector cells recognizes its target by different means as described below.

T cells are primed in the thymus, where they undergo two selection processes. The first negative selection process eliminates those T cells that can recognize the MHC molecules associated with self-products. Then, a positive selection process begins whereby T cells that can recognize MHC molecules complexed with foreign peptides are allowed to

TABLE 3.7. Complement Deficiencies and Diseases

Pathway/Component		Disease	Mechanism
Classical pathway	C1-INH	Hereditary angioedema	Overproduction of C2b (prokinin)
	C1, C2, and C4	Predisposition to SLE	Opsonization of immune complexes helps keep them soluble; deficiency results in increased precipitation in tissues and inflammation
Lectin pathway	MBL	Susceptibility to bacterial infections in infants or immunosuppressed	Inability to initiate the lectin pathway
Alternative pathway	Factors B or D	Susceptibility to pyogenic (pus-forming) bacterial infections	Lack of sufficient opsonization of bacteria
	C3	Susceptibility to bacterial infections	Lack of opsonization and inability to utilize the membrane attack pathway
	C5, C6, C7, C8, and C9	Susceptibility to gram-negative infections	Inability to attack the outer membrane of gram-negative bacteria
	Properdin (X-linked)	Susceptibility to meningococcal meningitis	Lack of opsonization of bacteria
	Factor H or I	C3 deficiency and susceptibility to bacterial infections	Uncontrolled activation of C3 via alternative pathway resulting in depletion of C3

pass out of the thymus. The large number of CD molecules (cluster of differentiation molecules) on the surfaces of lymphocytes allows huge variability in the forms of the receptors. There are more than 160 CD molecules, each of which is a different chemical molecule that coats the surface. Every T and B cell has about 10^5 molecules on its surface. B cells have on their surface major CD molecules such as CD21, CD35, CD40, and CD45 in addition to other non-CD molecules. The major CD molecules on T cells include CD2, CD3, CD4, CD8, CD28, CD45R, and other non-CD molecules on their surfaces.

The large number of molecules on the surfaces of lymphocytes allows huge variability in the forms of the receptors. They are produced with random configurations on their surfaces. There are some 10^{18} different structurally different receptors (Hauser, 1995). Essentially, an antigen may find a near-perfect fit with a very small number of lymphocytes, perhaps as few as one (Germain and Štefanová, 1999).

CTLs (CD8+) do their work by releasing lymphotoxins, which cause cell lysis. TH cells (CD4+) serve as managers, directing the immune response. They secrete

Extracellular microorganisms

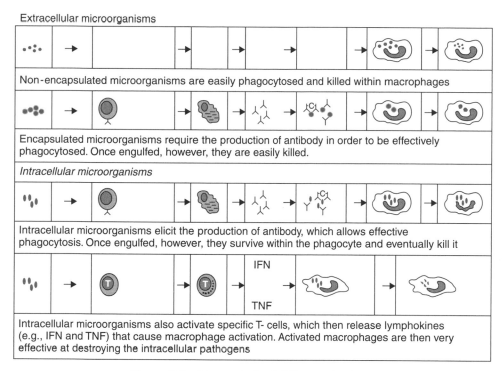

Figure 3.7. Phygocytosis of different microbes.

lymphokines that stimulate cytotoxic T and B cells to grow and divide, attract neutrophils, and enhance the ability of macrophages to engulf and destroy microbes. T regulatory cells inhibit the production of CTLs once they are unneeded, lest they cause more damage than necessary. Memory T cells are programmed to recognize and respond to a pathogen once it has invaded and been repelled (Jamieson and Ahmed, 1989; Dutton et al., 1998).

Cytotoxic T Lymphocytes. CTLs, like other T cells are self-MHC-restricted (Fig. 3.8). Briefly, CTLs recognize the antigen presented by MHC I molecules and once recognition is successful, the CTL "programs" the target cell for self-destruction. First, CTLs release perforin in the space between the CTL and its target. In the presence of calcium ions, perforin polymerizes, forming channels in the target cell's membrane. The CTL may also release various granzymes that pass through the polyperforin channels, causing target cell lysis. The CTL may also release lymphokines and/or cytokines that interact with specific receptors on the target cell surface, causing internal responses that lead to destruction of the target cell. The CTLs also express FAS ligand which induces apoptosis of target cell on interaction with FAS. CTLs principally act to eliminate endogenous antigens. The memory T cells last for several years in certain viral infections (Farber, 2000).

NK Cells. NK cells are "large granular lymphocytes." These cells are generally nonspecific, MHC-unrestricted cells involved primarily in the elimination of neoplastic or tumor cells. The precise mechanism by which they recognize their target cells is not

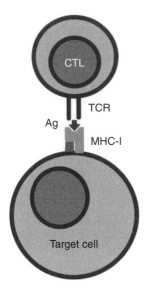

Figure 3.8. Interaction of CTL with target cell.

clear. The receptors present on NK cells distinguish host cells with abnormal expression of MHC and perhaps other molecules. Once the target cell is recognized, killing occurs in a manner similar to that produced by the CTL.

Antibody-Dependent Cell-Mediated Cytotoxicity (ADCC). Cells containing immunoglobulin Fc receptors on their surface are involved in a process known as *antibody-dependent cell-mediated cytotoxicity*. ADCC occurs as a consequence of antibody being bound to a target cell surface via specific antigenic determinants expressed by the target cell. Once bound, the Fc portion of the immunoglobulin can be recognized by the cells bearing Fc receptors, namely, macrophages, neutrophils, monocytes, and NK cells; killing then ensues by a mechanism similar to that employed by CTLs (Ferguson et al., 1999).

Cytokines. Cytokines are secreted *de novo* by a variety of cells including T cells in response to antigenic stimuli, and usually act locally, at very low concentrations. The cytokines regulate and mediate immunity, inflammation, and hematopoiesis. Cytokines bind to specific receptors on target cells, and then signal the cell via second messengers, often tyrosine kinases, to alter cellular activity. Cytokines are often produced in cascades. Actions of cytokines include (i) up- or downregulation of the expressin of membrane proteins including MHC, (ii) secretion of effector molecules, and cell adhesion molecules (CAMs) expression, (iii) cellular proliferation, (iv) chemotaxis of neutrophils, monocytes, and T cells, (v) cellular differentiation, (vi) inflammation, and (vii) death of tumor cells. The sources and functions of various cytokines are listed out in Table 3.8.

3.3.6. Hypersensitivity

Occasionally, the immune system responds inappropriately to the presence of antigen and these responses are referred to as *hypersensitivities*. Hypersensitivities are divided

TABLE 3.8. Cytokines, Sources, Functions on β Target Cells

Cytokine	Secreted by	Targets and Effects
Some Cytokines of Innate Immunity		
Interleukin 1 (IL-1)	Monocytes, macrophages, endothelial cells, epithelial cells	Vasculature (inflammation); hypothalamus (fever); liver (induction of acute phase proteins)
Tumor necrosis factor α (TNF α)	Macrophages	Vasculature (inflammation); liver (induction of acute phase proteins); loss of muscle, body fat (cachexia); induction of death in many cells types; neutrophils activation
Interleukin 12 (IL-12)	Macrophages, dendritic cells	NK cells; influences adaptive immunity (promotes T_H1 subset)
Interleukin 6 (IL-6)	Macrophages, endothelial cells	Liver (induces acute phase proteins); influences adaptive immunity (proliferation and antibody secretion of B cell lineage)
Interferon α (IFN-α) (this is a family of molecules)	Macrophages	Induces an antiviral state in most nucleated cells; increases MHC class I expression; activates NK cells
Some Cytokines of Adaptive Immunity		
Interleukin 2 (IL-2)	T cells	T-cell proliferation; can promote AICD, NK cell activation and proliferation; B-cell proliferation
Interleukin 4 (IL-4)	TH2 cells; mast cells	Promotes TH2 differentiation; isotype switch to IgE
Interleukin 5 (IL-5)	TH2 cells	Eosinophil activation and generation
Interleukin 25 (IL-25)	Unknown	Induces secretion of TH2 cytokine profile
Transforming growth factor β (TGF-β)	T cells, macrophages, other cell types	Inhibits T-cell proliferation and effector functions; inhibits B-cell proliferation, promotes isotype switch to IgE; inhibits macrophages
Interferon γ (IFN-γ)	TH1 cells, CD8+ cells, NK cells	Activates macrophages; increases expression of MHC class I and class II molecules; increases antigen presentation

into four categories on the basis of the type of response generated and the causative agent.

- Type I: *immediate hypersensitivity*, mediated by IgE class of antibodies.
- Type II: *cytotoxic hypersensitivity*, mediated by IgG or IgM class of antibodies through ADCC.
- Type III: *immune complex hypersensitivity*, mediated by antibody–antigen complexes.
- Type IV: *delayed hypersensitivity*, mediated by TH1 cells.

3.3.6.1. Type I Reaction

Initial exposure to an antigen produces Age class of antibody response. Immunoglobulin Age binds very specifically to receptors on the surface of mast cells, which remain circulating. On subsequent exposure, the antigen cross-links Age on mast cells causing the cells to degranulate and release large amounts of histamine, lipid mediators, and chemotactic factors that cause smooth muscle contraction, vasodilation, increased vascular permeability, bronco constriction, and edema. These reactions occur very suddenly, and may lead to death. Examples of Type I hypersensitivities include allergies to penicillin, insect bites, and molds. If the specific antigen in question is injected intradermally and the patient is sensitive, a specific reaction known as *wheal and flare* can be observed within 15 min.

3.3.6.2. Type II Reaction

Type II hypersensitivity involves IgG- or IgM-mediated cytotoxicity. In addition, Type II hypersensitivity may also involve complement that binds to cell-bound antibody. When antibodies react with a host cell surface, cell lysis may occur. The examples of Type II hypersensitivity are as follows:

> *Autoimmune hemolytic anemia*—drugs such as penicillin become attached to the surface of RBCs and act as hapten for the production of antibody, which then binds the RBC surface leading to lysis of RBCs.
>
> *Goodpasture's syndrome*—IgG antibodies react against glomerular basement membrane surfaces, which can lead to kidney destruction.

3.3.6.3. Type III Reaction

Type III hypersensitivity involves circulating antibody bound to an antigen. These circulating antigen–antibody complexes may get deposited on tissues and this may lead to complement activation, leading to tissue damage. This type of hypersensitivity develops as a result of systematic exposure to an antigen and is dependent on the type of antigen and antibody and the size of the AG–AB complex. The circulating immune complexes that are too small fail to activate complement system; complexes that are too large are removed by the reticuloendothelial system; and intermediate complexes may get lodged in the glomerulus activating the complement system leading to kidney damage. An example of Type III hypersensitivity is serum sickness, a systemic condition that may develop when a patient is injected with a large amount of antibodies, for example, antitoxin raised in an animal. The antitoxin antibodies react with the toxin-forming immune complexes that get deposited in tissues. Type III hypersensitivities can also be ascertained by intradermal injection of the antigen, followed by the observance of an "Arthus" reaction (swelling and redness at the site of injection) after a few hours.

3.3.6.4. Type IV Reaction

Type IV or delayed hypersensitivity, as the name implies, develops several hours/days following exposure to an antigen. Initial exposure to a cellular antigen produces a cell-mediated response. Memory T cells respond on second exposure to the same antigen leading to activation of macrophages and destruction of both specific (MT) and nonspecific (LM) microorganisms.

3.3.7. Immunity in Viral Infections

3.3.7.1. Humoral Components

NONSPECIFIC COMPONENTS. A number of humoral components of the nonspecific immune system function in resistance to viral infections (Gupta et al., 1999).

Interferons. Interferon is one of the first lines of defense against viruses because it is induced early after virus infection before any of the other defense mechanisms appear (e.g., antibody and Tc cells).

There are three types of interferon, IFN-α (also known as *leukocyte interferon*), IFN-β (also known as *fibroblast interferon*), and IFN-γ (also known as *immune interferon*). IFN-α and IFN-β are also referred to as *Type I interferon* and IFN-γ as *Type II*. There are approximately 20 subtypes of IFN-α, but only one IFN-β and IFN-γ (Table 3.9). Transcription of the IFN genes occurs only after exposure of cells to an appropriate inducer such as viruses, double-stranded RNA (dsRNA), LPS. Among the viruses, the RNA viruses are the best inducers while DNA viruses are poor IFN inducers (Holland, 1992).

Complement. Most viruses do not fix complement by the alternative route. However, the interaction of a complement-fixing antibody with a virus-infected cell or with an enveloped virus can result in the lysis of the cell or virus. Thus, by interfacing with the specific immune system, complement also plays a role in resistance to viral infections.

SPECIFIC RESPONSES. Virus-specific antibody can directly neutralize virus infectivity by preventing the attachment of virus to receptors on host cells or entry of the virus into the cell (Dimmock, 1993, Doherty et al., 2000). Antibodies can also prevent uncoating of virus by interfering with the interaction of viral proteins involved in uncoating. Complement fixing antibodies can assist in the lysis of viral infected cells or enveloped viruses. Antibodies can also act as opsonins and augment phagocytosis of viruses either by promoting their uptake via Fc or C3b receptors or by agglutinating viruses to make them more easily phagocytosed (Evans et al., 1999). Antibody-coated virus infected cells can be killed by K cells thereby preventing the spread of the infection (Stevenson and Doherty, 1998; Edwards and Dimmock, 2000). Table 3.10 shows in brief the various antiviral activities of antibodies.

Opsonization of viruses with antibody can enhance their uptake by phagocytic cells (Table 3.10). If the virus is able to survive in the phagocyte, this allows for the spread of virus infection. HIV is an example of viruses that can survive in macrophages.

TABLE 3.9. Types and Properties of Interferons

Property	Type		
	Alpha (α)	Beta (β)	Gamma (γ)
Genes	>20	1	1
Inducers	Viruses (RNA > DNA) dsRNA	Viruses (RNA > DNA) dsRNA	Antigens, mitogens
Principal source	Leukocytes, epithelium	Fibroblasts	Lymphocytes

TABLE 3.10. Mediation of Antiviral Activities of Antibodies

Target	Agent	Mechanism
Free virus	Antibody alone	Blocks binding to cell
		Blocks uncoating of virus
	Antibody + complement	Damage to virus envelope
		Opsonization of virus
Virus-infected cell	Antibody bound to infected cells	ADCC by K cells, NK cells, and/or macrophages

3.3.7.2. Cellular Components

In addition to the barriers and humoral components involved in resistance to and recovery from viral infections, there are several different cells that play a role in our antiviral defenses.

NONSPECIFIC
Macrophages. Macrophages contribute to antiviral defenses in a number of ways.

- *Intrinsic antiviral activity*—Macrophages help limit viral infections by virtue of their intrinsic ability to prevent replication of viruses. However, some viruses are able to replicate or at least survive in macrophages and thus can be spread by macrophages.
- *Extrinsic antiviral activity*—Macrophages are also able to recognize virus-infected cells and to kill them. Thus, macrophages also contribute to antiviral defenses by virtue of their cytotoxic activity.
- *ADCC*—Virus-infected cells that are coated with IgG antibodies can be killed by macrophages by ADCC.
- *IFN production*—Macrophages are a source of IFN.

NK Cells. NK cells act by recognizing and killing virus-infected cells. The recognition of virus-infected cells is not MHC-restricted or antigen-specific. Thus, NK cells will kill cells infected with many different viruses. NK cells can also mediate ADCC and can kill virus-infected cells by this mechanism. The activities of NK cells can be enhanced by IFN-γ and IL-2.

SPECIFIC. T cells play a major role in recovery from viral infections (Butz and Bevan, 1998). CTLs generated in response to viral antigens on infected cells can kill the infected cells, thereby preventing the spread of infection and these T-cell responses are governed by host genetic factors (Doherty et al., 1978). Human memory CTL response specific for certain viruses such as influenza A virus is broad and multispecific (Gianfrani et al., 2000). In addition, lymphokines secreted by T cells can recruit and activate macrophages and NK cells, thereby mobilizing a concerted attack in the virus.

3.3.7.3. Immunosuppression

Many viruses are able to suppress immune responses and thereby overcome or minimize host defenses (Pircher et al., 1990; Borrow and Shaw, 1998; Tortorella et al., 2000). The best example is HIV which infects the CD4+ cells, thereby destroying the specific

TABLE 3.11. Virus-Mediated Immunosuppression

Host Defense Affected	Virus	Virus Product	Mechanism
Interferon	EBV	EBERS (small RNAs)	Blocks protein kinase activation
	Vaccinia	eIF-2alpha homolog	Prevents phosphorylation of eIF-2alpha by protein kinase
Antibody	HSV-1	gE/gI	Binds Fc-γ and blocks function
Cytokines	Myxoma	IFN-γ receptor homolog	Competes for IFN-γ and blocks function
	Shope fibroma virus	TNF receptor	Competes for TNF and blocks function
	EBV	IL-10 homolog	Reduces IFN-gamma function
MHC class I	CMV	Early protein	Prevents transport of peptide-loaded MHC
	Adenovirus	E3	Blocks transport of MHC to surface
Apoptosis	Adenovirus	14.7K	Inhibits capsases
	EBV	Bcl-2 homolog	Antiapoptotic

immune system. Other viruses (e.g., measles virus) can also infect lymphocytes and affect host immune responses (Haydon and Woolhouse, 1998; Table 3.11).

3.4. BACTERIAL PATHOGENS AND ASSOCIATED DISEASES

An estimated 30% of bacteria are disease causing pathogens. According to epidemiological estimates, infectious diseases caused by microbes are responsible for more deaths worldwide than any other single cause. As per the estimates, more than half the people who ever lived died from smallpox (caused by a virus) or malaria (caused by a protozoan) at the beginning of the twentieth century. Bacteria have been the cause of some of the most deadly and widespread diseases in human history. Bacterial diseases such as pneumonia, tuberculosis, typhoid, plague, diphtheria, cholera, and dysentery have taken a huge toll on humanity. The most common fatal bacterial diseases include respiratory infections, with tuberculosis alone killing about 3 million people a year, in tropical countries (Spratt and Maiden, 1999). Pathogenic bacteria can enter the body through a variety of means, including inhalation into the nose and lungs, ingestion through food, or transmission through sexual contact. Accordingly, the human diseases caused by pathogenic bacteria may be categorized under the following groups:

- Airborne bacterial diseases
- Foodborne and waterborne bacterial diseases
- Soilborne bacterial diseases
- Arthropodborne bacterial diseases
- Sexually transmitted bacterial diseases
- Other bacterial diseases.

3.4.1. Airborne Bacterial Infections

3.4.1.1. Streptococcal Infections

Streptococci constitute a large number of species with diverse biological properties, and hemolytic streptococci, in particular, those of group A (*Streptococcus pyogenes*), are by far the most frequent streptococcal pathogens found in man. Group A organisms produce diseases with extremely varied symptomatology (Lancefield, 1962). This group A infections may lead to rheumatic fever and acute glomerulonephritis. Infections in newborns and urogenital tract infections in women are the most serious forms of disease produced by group B streptococci (*Streptococcus agalactiae*). Group D organisms are very common in human and are normally found in the GI and genitourinary tracts, and on the skin. Under specific conditions, group D streptococci (*Enterococcus faecalis*) cause subacute or acute systemic infections, or localized septic processes. Groups C, G, and F streptococci frequently colonize the pharynx and sporadically produce upper respiratory tract (URT) diseases, and they can provoke localized sepsis at other sites. Group A streptococci are transmitted from person to person through air when there is close contact between individuals. The diagnosis of streptococcal infections based on clinical symptoms alone is highly unreliable, especially in the case of upper respiratory infections. Microbiological confirmation of the clinical diagnosis is, therefore, essential (Fig. 3.9). The methods, based on procedures that were introduced several decades ago, entail sampling of the material, cultivation on blood–agar plates overnight, identification of streptococcal colonies, and determination of the serological group. One of the sensitive techniques includes coagglutination method, the results of which are available in five minutes after the specimen has been taken.

DIPHTHERIA. Diphtheria is an acute contagious respiratory disease caused by *Corynebacterium diphtheriae*, which is a gram-positive, nonmotile, aerobic bacterium forming irregular rod-shaped or club-shaped cell during growth. The pathogen enters the body via the respiratory passage with cells lodging in the throat and tonsils. The inflammatory response of throat tissues to the pathogen leads to the formation of a characteristic lesion referred to as *pseudomembrane*.

Certain strains of *C. diphtheriae* secrete powerful exotoxin called *diphtheria toxin* that inhibits eukaryotic protein synthesis. The diphtheria toxin, if absorbed into the circulatory system and distributed throughout the body, damages the peripheral nerves, heart, and kidneys. Both culture of the bacterium and toxin detection assay are required for confirmation of the diagnosis (Fig. 3.10).

WHOOPING COUGH (PERTUSSIS). Whooping cough or pertussis is a highly contagious, primarily childhood respiratory disease caused by *Bordetella pertussis*, which is a small gram-negative, aerobic coccobacillus. The spasmodic coughing gives the disease its name, because a whooping sound results from the patient inhaling in deep breaths to obtain sufficient air. It is a serious disease that can cause permanent disability in infants, and even death. When an infected person sneezes or coughs, tiny droplets containing the bacteria move through air, and the disease is easily spread from person to person and incubates 7–14 days in the body of the victim. Once inside the URT, the pathogens adhere to the ciliated epithelial cells and secrete several toxins, the most important being the pertussis toxin. The initial diagnosis is usually based on the symptoms. Serological testing also

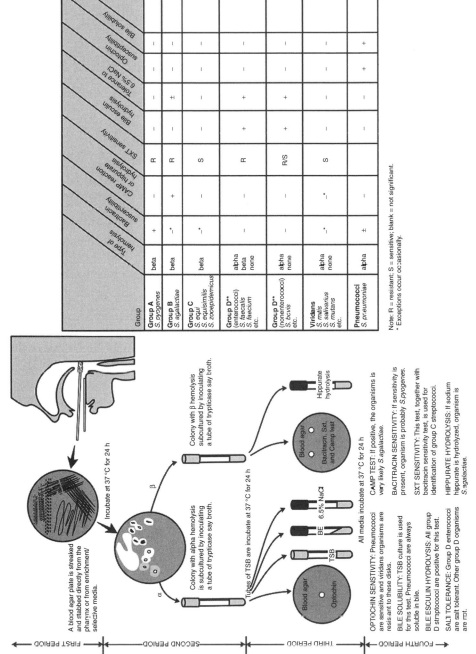

Figure 3.9. Streptococci: microbiological methods for isolation and identification.

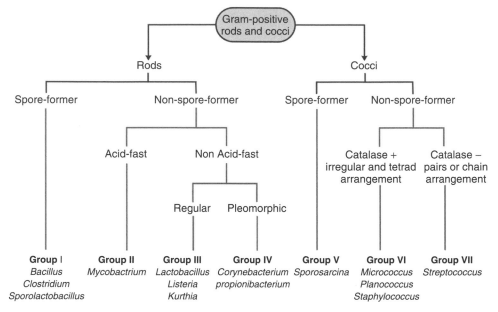

Figure 3.10. Microbiological examination for separation of gram-positive rods and cocci.

resulted in a significant increase in diagnostic yield compared to culture alone. PCR is a useful technique, but validity of results must be assured by careful control. The disease can be prevented by vaccinating children by DPT-vaccine when they are 2–3 months old. Antibiotics such as erythromycin, tetracycline, ampicillin, or chloramphenicol is the choice of drugs.

3.4.1.2. Meningococcal Infections

Neisseria meningitidis is the cause of adult bacterial meningitis and the clinical manifestations of the disease can be quite varied, ranging from transient fever and bacteremia to fulminant disease with death ensuing within hours of the onset of clinical symptoms. The bacteria are transmitted from person to person through droplets of respiratory or throat secretions. Close and prolonged contact such as kissing, sneezing, or coughing on someone, or living in close quarters (such as a dormitory, sharing eating, or drinking utensils) with an infected person facilitates the spread of the disease. The average incubation period is 4 days, but can range between 2 and 10 days. Group A meningococcus accounts for an estimated 80–85% of all cases in the meningitis belt, with epidemics occurring at intervals of 7–14 years. The most common symptoms are a stiff neck, high fever, sensitivity to light, confusion, headache, and vomiting. Even when the disease is diagnosed early and adequate treatment is started, 5–10% of patients die, typically within 24–48 h after the onset of symptoms. Bacterial meningitis may result in brain damage, hearing loss, or a learning disability in 10–20% of survivors. The gold standard for the diagnosis of systemic meningococcal infection is the isolation of *N. meningitidis* from a usually sterile body fluid such as blood or cerebrospinal fluid (CSF), or less commonly synovial, pleural, or pericardial fluid (Zhou and Spratt, 1992). PCR-based technique with high sensitivity and specificity for diagnosis of meningococcal meningitis is available. Several vaccines are available to control the disease.

3.4.1.3. Haemophilus influenzae Infections

Haemophilus influenzae, a gram-negative coccobacillus, is divided into unencapsulated and encapsulated strains. The latter are further classified into serotypes, with the *H. influenzae* serotype b being the most pathogenic to humans, responsible for respiratory infections, ocular infection, sepsis, and meningitis (Alamgir et al., 2000). Infected children start showing symptoms of meningitis after an incubation period of about 2–4 days and clinical manifestations tend to evolve rapidly. When *H. influenzae* infects the larynx, trachea, and bronchial tree, it leads to mucosal edema and thick exudate; when it invades the lungs, it leads to bronchopneumonia. Another notable species is *Haemophilus ducreyi*, the causative agent of chancroid. Rapid serodiagnostic methods such as latex agglutination test and enzyme immunoassays (EIAs) are available for diagnosis of *H. influenzae* infections. A highly sensitive and specific EIA for the detection of *H. influenzae* serotype b antigens in body fluids and broth cultures was developed, with a polyclonal antibody directed against polyribose phosphate as the solid-phase reagent and a biotinylated monoclonal antibody directed against *H. influenzae* type b outer membrane protein as the liquid-phase reagent (Bednarek et al., 1991).

3.4.1.4. Tuberculosis

Tuberculosis is a chronic infection caused by the bacterium *Mycobacterium tuberculosis* (Garg et al., 2006; Tiwari et al., 2007b). It usually involves the lungs, but other organs of the body can also be involved. Tuberculosis is the second most infectious disease and is the leading cause of death in the tropical world. It was first isolated in 1882 by a German physician named Robert Koch who received the Nobel Prize for this discovery. A person can become infected with tuberculosis bacteria when he or she inhales minute particles of infected sputum from air. Only about 10% of those infected with TB develop the disease. About 15% of people with the disease develop TB in an organ other than the lung, such as the lymph nodes, GI tract, and bones and joints. A delayed or missed diagnosis of active infection and prevalence of drug-resistant strains of *M. tuberculosis* are major causes of transmission and mortality. Studies indicate that there exists a close association between TB and AIDS (Kartikeyan et al., 2007). Therefore, spread of AIDS among the people is resulting in dramatic increases in TB. The identification of *M. tuberculosis* in secretions or tissues from the patient is the mainstay of the diagnosis of tuberculosis. Detection of *M. tuberculosis* by microscopy is difficult in specimens containing fewer than 104 bacteria per milliliter. Several serological- and molecular-based diagnostic tests are available for diagnosis of tuberculosis (Zacharia et al., 2010). Serological tests based on *M. tuberculosis*-specific lipids appeared highly promising for diagnosis of tuberculosis (Bisen et al., 2010). *M. tuberculosis*-derived glycolipids when interchelated into liposomes diagnosed tuberculosis infection with 94% sensitivity and 98.3% specificity. On the basis of this approach, a rapid card test, namely, "TB screen test" was designed for the diagnosis of pulmonary and extrapulmonary tuberculosis. The "TB screen test" is found more economical and rapid (4 min) than other currently available products with superior sensitivity, specificity, and simplicity (Tiwari et al.,

2005). The "TB card test" based on antigen detection in biological fluids offers the advantage of diagnosing current infections (Tiwari et al., 2007a, b). These tests proved to be simple, sensitive, specific, cost-effective, and highly rapid. Isoniazid and rifampin are the keystones of treatment, but, because of increasing resistance to them, pyrazinamide and either streptomycin sulfate or ethambutol HCL is added to regimens. Six months is the minimum acceptable duration of treatment for all adults and children with culture-positive TB. It has been estimated that one in seven cases of tuberculosis is resistant to drugs and resistance arises when patients fail to complete their drug therapy, lasting six months or longer. National and International Health agencies call for aggressive intervention to prevent the further spread of drug-resistant TB. Directly observed therapy (DOT) is being used—that is, the actual, documented observation of the patient when he or she takes the medicine and this method has been shown to reduce the likelihood of treatment failures (Fig. 3.11).

3.4.1.5. Pneumococcal Pneumonia

Pneumococcal pneumonia is an infection in the lungs caused by gram-positive diplococci called *Streptococcus pneumoniae*. *S. pneumoniae*, also called *pneumococcus*, can infect the URTs of adults and children and can spread to the blood, lungs, middle ear, or nervous system. Pneumococcal pneumonia mainly causes illness in children younger than 2 years and in adults 65 years of age or older. The pneumococcus is acquired by aerosol or through inhalation, leading to colonization of the nasopharynx. At any given time, the noses and throats of up to 70% of healthy people contain pneumococcus. Children, generally, harbor more of the bacteria than adults do. In about 30% of people with pneumococcal pneumonia, the bacteria invade the blood stream from the lungs. The incidence of disease increases strongly in association with a viral illness, such as influenza, parainfluenza, respiratory syncytial virus (RSV), adenovirus, or human metapneumovirus. *Mycoplasma pneumoniae* may also cause severe secondary pneumonia in immunocompromised patients. The diagnosis is based on physical examination, radiologic findings, and microbiologic diagnosis. However, etiologic diagnosis using traditional culture methods is difficult to obtain. The pneumococcal vaccine is the only way to prevent getting pneumococcal pneumonia.

3.4.1.6. Legionellosis

Legionella are gram-negative, rod-shaped bacteria that are ubiquitous in freshwater. The most common sources of infection are showers, taps, and spray heads and spa baths, and transmission occurs through aerosolization. The incubation period is 2–6 days. The symptoms include high fever, chills, headache, severe muscular ache, cough, fever, chest pain, breathlessness, diarrhea, vomiting, and confusion/delirium. Isolation of *Legionella* from respiratory secretions, lung tissue, pleural fluid, or a normally sterile site remains an important method for diagnosis. Urinary antigen assay and culture of respiratory secretions on selective media are the preferred diagnostic tests for Legionnaires' disease.

3.4.2. Foodborne and Waterborne Diseases

The two most common types of foodborne illness are intoxication and infection. A foodborne infection is caused by the ingestion of food containing pathogenic microorganisms (i.e., bacteria, virus, or parasite), which must multiply within the GI tract, producing widespread inflammation. The foodborne illness may be due to *Campylobacter jejuni, C. botulinum, Clostridium perfringens, Cyclospora cayetanensis, Escherichia coli* (O157:H7), *Shigella, Salmonella*, and *Staphylococcus aureus. C. jejuni* can be transmitted to humans via unpasteurized milk, contaminated water, and raw or undercooked meats, poultry, and shellfish. *C. botulinum*, which causes botulism, is the most deadly of all food pathogens. Toxins, not bacteria, cause illness. Bacteria that produce toxins include *S. aureus* and *C. botulinum*. Symptoms may include nausea, vomiting, diarrhea, abdominal pain or distress, and fever (Fig. 3.12).

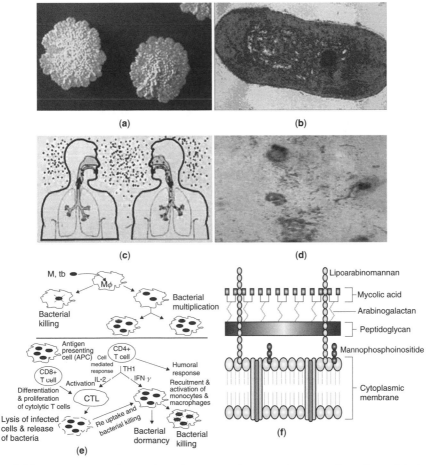

Figure 3.11. (a) Growth of *M. tuberculosis* on LJ (Lowenstein–Jensen) medium, (b) electron micrograph of *M. tuberculosis*, (c) transmission of tuberculosis infection from person to person through aerosol liberated from TB +ve individual inhaled by the normal individual and acquires infection; (d) acid-fast stained smear showing pink colored bacilli of *M. tuberculosis*; (e) flow chart progression of *M. tuberculosis* bacillus after inhalation, entry into macrophages, and activation of immune response; (f) *M. tuberculosis* (acid fast) cell wall; (g) schematic representation of principle of TB antigen detection (TB/M card test) kit (Tiwari et al., 2007a, b). (*See insert for color representation of the figure*).

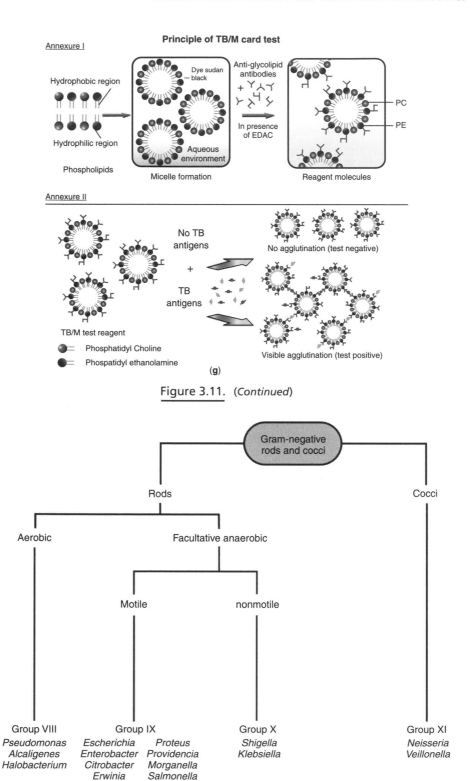

Figure 3.11. (*Continued*)

Figure 3.12. Microbiological examination outline for gram-negative rods and cocci.

3.4.2.1. Botulism

Botulism is caused by the ingestion of food containing the neurotoxins produced by *C. botulinum*, an anaerobic spore-forming gram-positive bacterium. On the basis of specificity, seven types (type A, B, C, D, E, F, and G) of these neurotoxins have been recognized. The neurotoxin of *C. botulinum* is a protein and a dose as low as 0.01 mg is said to be fatal to humans. A and B producing strains are often found on fruits and vegetables and honey. Symptoms appear 12–36 h after ingestion and include nausea and vomiting (B and E); visual impairment—blurred vision, ptosis, dilated pupils; loss of mouth and throat function (A and B)—dry mouth, throat, tongue, sore throat; fatigue and loss of coordination; respiratory impairment; abdominal pain; and either diarrhea or constipation; in general, cranial nerve is first affected and then descend. Sixty to seventy percent cases of botulism die. Neurotoxin (BoTox) is water soluble; produced as a single polypeptide of 150,000 MW (progenitor), cleaved by a protease to form two polypeptides (H and L chains), which then become S–S bonded. All serotypes block the exocytotic release of acetylcholine from synaptic vesicles at peripheral motor nerve terminals.

3.4.2.2. Staphylococcal Food Poisoning

Cold meat, poultry, and prepared foods such as custards, trifles, and cream products are prone to *S. aureus* contamination. Staphylococcal food poisoning is due to enterotoxin B, and the symptoms are characterized by severe vomiting, diarrhea, abdominal pain, and cramps (Rasmussen et al., 2011). Recovery usually occurs within 6–24 h.

3.4.2.3. Clostridial Food Poisoning

The disease produced by *C. perfringens* is not as severe as botulism. Spores are found in soil, nonpotable water, unprocessed foods, and in the intestinal tract of animals and humans. Meat and poultry are frequently contaminated with these spores from one or more sources during processing. Spores of some strains are so heat resistant that they survive boiling for four or more hours (Paredes-Sabja and Sarker, 2009). Symptoms occur within 8–24 h after consuming the contaminated food, which include acute abdominal pain and diarrhea. Nausea, vomiting, and fever are less common. Recovery is usually within 1–2 days, but symptoms may persist for 1 or 2 weeks (Paredes-Sabja and Sarker, 2009).

3.4.2.4. Typhoid Fever

Typhoid fever is caused by *Salmonella typhi*. The organisms gain entry into the body through contaminated water supplies and food material. The bacteria enter the lymphatic system in the Peyer's patch of the intestine forming ulcer, and disseminate through the circulatory system. It invades the mononuclear phagocyte system and continues to multiply within the phagocytic blood cells. The organism grows and causes damage to the liver and gallbladder, and sometimes to the kidneys, spleen, and lungs. The symptoms of typhoid include high fever (104 °C), headache, and weakness, and abdominal pain, ulceration, blood vessel hemorrhaging, and rash. The symptoms develop in a stepwise manner in 3 weeks span and begin to decline after the third week. Chlorination of water supply is able to kill the bacteria.

3.4.2.5. Salmonellosis

Salmonellosis is a form of food infection that may result when foods containing *Salmonella* bacteria are consumed. Once they gain entry, the bacteria may continue to live and grow in the intestine, set up an infection and cause illness. The possibility and severity of the illness depends in large part on the size of the dose, the resistance of the host, and the specific strain of *Salmonella* causing the illness. *Salmonella* bacteria are readily destroyed by cooking and do not grow at refrigerator or freezer temperatures. Symptoms of salmonella infection include diarrhea, abdominal pain, nausea, chills, fever, and vomiting. These usually occur within 12–36 h after consuming contaminated food and may last up to 2–7 days. Arthritis symptoms may follow 3–4 weeks after the onset of acute symptoms (Fig. 3.13).

3.4.2.6. Shigellosis

Shigellae are gram-negative, nonmotile, nonspore-forming, rod-shaped bacteria, very closely related to *E. coli* and cause Shigellosis. People infected with *Shigella* develop diarrhea, fever, and stomach cramps starting a day or two after they are exposed to the bacterium. Shigella was discovered over 100 years ago by the Japanese microbiologist, Shiga, after whom the genus is named. There are four species of *Shigella: S. boydii*,

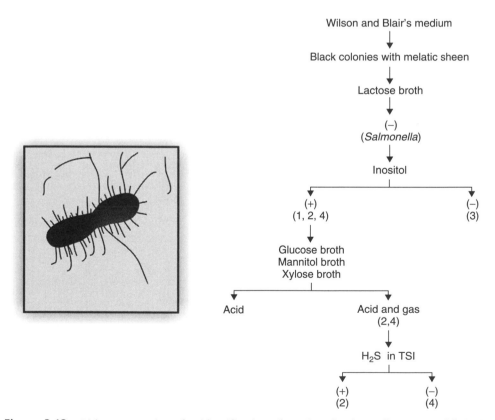

Figure 3.13. Dichotomous chart for identification of species of Salmonella: 1. *S. typhii*, 2. *S. paratyphii*, 3. *S. typhimurium*, 4. *S. enteritidis*.

S. dysenteriae, S. flexneri, and S. sonnei. S. sonnei, also known as *Group D Shigella*, accounts for over two-thirds of the shigellosis in the United States. *S. flexneri*, or Group B Shigella, accounts for almost all of the rest. *Shigella* are present in the diarrheal stools of infected persons while they are ill and for a week or two afterward. Diagnosis of *Shigella* is by identifying the bacteria in the stool of an infected person. Shigellosis can usually be treated with antibiotics such as ampicillin, trimethoprim/sulfamethoxazole, nalidixic acid, and the fluoroquinolone, ciprofloxacin (Figs. 3.14 and 3.15).

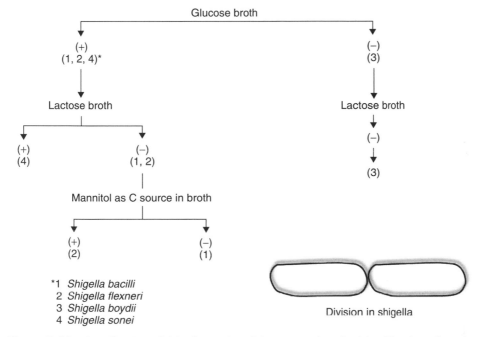

Figure 3.14. Identification of shigella species: dichotomous chart for identification of species.

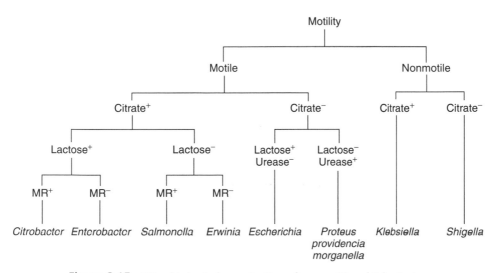

Figure 3.15. Microbiological examination of groups IX and X bacteria.

3.4.2.7. Cholera

Cholera is a severe waterborne diarrheal disease caused by *V. cholerae*, a gram-negative curved bacillus bacterium, transmitted almost exclusively via contaminated water and food. *V. cholerae* and *V. parahaemolyticus* are pathogens of humans. Both produce diarrhea, but in ways that are entirely different. *V. parahaemolyticus* primarily affects the colon while *V. cholerae* is noninvasive and affects the small intestine by secreting an enterotoxin.

   V. cholerae produces cholera toxin, the model for enterotoxins, whose action on the mucosal epithelium is responsible for the characteristic diarrhea of the disease cholera. The clinical description of cholera begins with sudden onset of massive diarrhea. This results from the activity of the cholera enterotoxin, which activates the adenylate cyclase enzyme in the intestinal cells, converting them into pumps that extract water and electrolytes from blood and tissues and pump it into the lumen of the intestine. This loss of fluid leads to dehydration, anuria, acidosis, and shock. The watery diarrhea is speckled with flakes of mucus and epithelial cells and contains enormous numbers of vibrios. The loss of potassium ions may result in cardiac complications and circulatory failure. The immediate treatment of the disease is the oral rehydration therapy with NaCl plus glucose to estimate water uptake by the intestine. Antibiotics such as tetracyclin, trimethoprim-sulfamethoxazole, or ciprofloxacin are commonly used for treatment. (Fig. 3.16).

3.4.2.8. Escherichia coli

Many strains of *E. coli* live peacefully in the gut, helping to keep the growth of more harmful microorganisms in check (Chao and Cox, 1983). However, one strain, *E. coli*

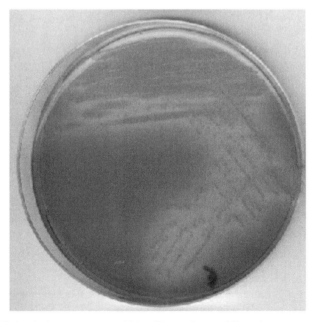

Figure 3.16. *Vibrio cholerae* on TCBS (thiosulfate–citrate–bile salt–sucrose agar plate). (*See insert for color representation of the figure*).

O157:H7, causes a distinctive and sometimes deadly disease. The pathogenic *E. coli* are divided into four groups according to their action in the body: enteropathogenic (causes severe infantile diarrhea), enterotoxigenic (causes diarrhea, severe cholera-like illness), enteroinvasive (dysentery-like illness), and enterohemorrhagic bloody diarrhea and colitis (Guttman, 1997).

Symptoms begin with nonbloody diarrhea 1–5 days after the uptake of contaminated food, and progress to bloody diarrhea, severe abdominal pain, and moderate dehydration. In young children, hemolytic uremic syndrome (HUS) is a serious complication that can lead to renal failure and death. In adults, the complications include thrombocytopenic purpura, characterized by cerebral nervous system deterioration, seizures, and strokes. Ground beef is the food most associated with *E. coli* O157:H7. Since *E. coli* is always present in feces, they are used as a marker for the fecal pollution of water, milk, and food (Table 3.12; Gross and Siegel, 1981).

3.4.2.9. Camphylobacteriosis and Helicobacteriosis

Campylobacteriosis is caused by consuming food or water contaminated with *C. jejuni*. It generally inhabits intestinal tracts of healthy animals (especially chickens) and in untreated surface water. Raw and inadequately cooked foods and nonchlorinated water are the main sources of human infection. The organism grows best in a reduced oxygen environment, and is easily killed by heat. Diarrhea, nausea, abdominal cramps, muscle pain, headache, and fever are common symptoms. Onset usually occurs 2–5 days

TABLE 3.12. Summary of Results (Expected) of Morphological and Biochemical Reactions of *Escherichia coli*

Test	Result
Morphological colony	*Circular, raised and smooth, medium sized*
1 Flagella stain	Positive
2 Capsular stain	Negative
3 Gram stain	Negative
4 Endospore	Negative
5 Motility	Positive
Biochemical	
1 Catalase	Positive
2 Oxidase	Negative
3 Gas production from glucose	Positive
4 Acid from mannitol and sucrose	Positive
5 Gelatin hydrolysis	Negative
6 H_2S from TSI	Negative
7 Urease	Negative
8 IMViC reactions	
(a) Indole production	Positive
(b) Methyl Red test	Positive
(c) Voges–Proskauer Test	Negative
(d) Citrate utilization	Negative
9 Nitrate reduction	Positive

Abbreviation: TSI, triple sugar iron.

after consuming contaminated food. Duration is 2–7 days, but can be weeks with such complications as urinary tract infections and reactive arthritis.

Infection of gastric epithelial cells with *Helicobacter pylori* (Salaun et al., 1998) induces strong proinflammatory responses by activating nuclear transcription factors NF-KB and AP-1. Several reports indicate that multiple bacterial factors and cellular molecules are involved in this signaling (Backert and Naumann, 2010).

3.4.3. Soilborne Bacterial Infections

3.4.3.1. Anthrax

Anthrax is primarily a disease of domesticated and wild animals. Humans become infected when brought into contact with infected animals, which includes their flesh, bones, hides, hair, and excrement. *Bacillus anthracis* is a very large, gram-positive, spore-forming rod-shaped bacterium, with $1-1.2$ μm width $\times 3-5$ μm length. Genotypically and phenotypically, it is very similar to *Bacillus cereus*, which is found in soil habitats around the world, and to *Bacillus thuringiensis*, the pathogen for larvae of Lepidoptera. The three species have the same cellular size and morphology, and form oval spores located centrally in a nonswollen sporangium.

The most common form of the disease is cutaneous anthrax in humans. The spores germinate, and a characteristic gelatinous edema develops at the site, which develops into papule within 12–36 h after infection. The papule changes rapidly to a vesicle, then to a pustule, and finally into a necrotic ulcer from which infection may disseminate, giving rise to septicemia. Another form of the disease, inhalation anthrax (Woolsorters' disease), results most commonly from inhalation of spore-containing dust. The disease begins abruptly with high fever and chest pain. It progresses rapidly to a systemic hemorrhagic pathology and is often fatal.

GI anthrax is analogous to cutaneous anthrax but occurs on the intestinal mucosa. The bacteria spread from the mucosal lesion to the lymphatic system. Intestinal anthrax results from the ingestion of poorly cooked meat from infected animals. The virulence of *B. anthracis* is attributable to three bacterial components: (i) capsular material composed of poly-D-glutamate polypeptide; (ii) EF (edema factor) component of exotoxin; and (iii) LF (lethal factor) component of exotoxin. Anthrax spores may survive in the soil, water, and on surfaces for many years. Anthrax spores are killed by boiling at 100 °C for 30 min. The possibility of creating aerosols containing anthrax spores has made *B. anthracis* a chosen weapon of bioterrorism.

3.4.3.2. Tetanus

Tetanus is a highly fatal disease of humans caused by *Clostridium tetani*, a motile anaerobic, gram-negative, spore-forming, and soil-inhabiting bacterium. The organism is found in soil, especially in heavily manured soils, and in the intestinal tracts and feces of various animals. Carrier rates in humans vary from 0% to 25%, and the organism is thought to be a transient member of the intestinal flora whose presence depends on ingestion. The pathogen enters the human body through a soil-contaminated wound. The disease stems from a potent neurotoxin (tetanus toxin) that causes uncontrolled stimulation of skeletal muscles. In the beginning, the toxin causes cramping and twisting in skeletal muscles surrounding the wound and tightness of the jaw muscles. Later on, there is trismus resulting in inability to open the mouth. Tetanus toxin is one of the three most poisonous substances known, the other two being the toxins of botulism and diphtheria. The bacterium synthesizes the tetanus toxin as a single 150 kDa polypeptide that can be separated into two polypeptide domains or fragments. The B domain is responsible for binding of the toxin to host cell neurons, and the A fragment is responsible for the neurotoxicity of the toxin.

3.4.3.3. Gas Gangrene

C. perfringens, which produces a huge array of invasins and exotoxins, causes wound and surgical infections that lead to gas gangrene, in addition to severe uterine infections. *Clostridial hemolysins* and extracellular enzymes such as proteases, lipases, collagenase and hyaluronidase contribute to the invasive process. *C. perfringens* also produces an enterotoxin and is an important cause of food poisoning. Most of the clostridia are saprophytes, but a few are pathogenic to humans, primarily *C. perfringens, C. difficile*, and *C. tetani*. Those that are pathogens have primarily a saprophytic existence in nature and, in a sense, are opportunistic pathogens. *C. tetani* and *C. botulinum* produce the most potent biological toxins known to affect humans. *C. perfringens* is classified into five types (A–E) on the basis of its ability to produce one or more of the major lethal toxins, alpha, beta, epsilon, and iota (α, β, ε, and ι). Enterotoxin (CPE) producing (cpe ǀ) *C. perfringens* type A is reported continuously as one of the most common food poisoning agents worldwide. Gas gangrene is marked by a high fever, brownish pus, gas bubbles under the skin, skin discoloration, and a foul odor. It can be fatal if not treated immediately.

3.4.3.4. Leptospirosis

Leptospirosis is sometimes referred to as *swineherd's disease, swamp fever*, or *mud fever*. Leptospirosis, caused by pathogenic spirochetes of the genus *Leptospira*, is considered the most common zoonosis and has been recognized as a reemerging infectious disease among animals and humans. The spirochetes are finely coiled, thin, motile, obligate, slow-growing anaerobes, and occur in warmer climates. The organism enters the body when mucous membranes or abraded skin come in contact with contaminated environmental sources and the infection often leads to renal and hepatic dysfunction. Leptospirosis in humans is characterized by an acute febrile illness followed by mild self-limiting sequelae, but at times causes more severe and often fatal, multiorgan involvement. The clinical manifestations are highly variable and may include (i) a mild, influenza-like illness; (ii) Weil's syndrome characterized by jaundice, renal failure, hemorrhage, and myocarditis with arrhythmias; (iii) meningitis/meningo-encephalitis; (iv) pulmonary hemorrhage with respiratory failure. Leptospirosis responds to antibiotics such as doxycycline or penicillin and ampicillin.

3.4.3.5. Listeriosis

Listeriosis is a dangerous infection caused by consuming food contaminated with bacteria called *Listeria monocytogenes*, small gram-positive rods and nonspore forming.

The *L. monocytogene* infections are common in wild and domesticated animals, and are found in soil and water. Raw milk or products made from raw milk may carry these bacteria. The bacteria most often cause a GI illness. In some cases, it may lead to septicemia or meningitis. The bacteria may cross the placenta and infect the developing baby. Infections in late pregnancy may lead to stillbirth or death of the infant within a few hours of birth. About half of infants infected at or near term will die.

3.4.4. Arthropodborne Bacterial Infections

3.4.4.1. Plague

Plague causes more human deaths than any other infectious diseases except malaria and tuberculosis. It is caused by an aerobic gram-negative rod-shaped bacterium *Yersinia pestis*. Rats are the primary reservoir of the bacterium and fleas serve as intermediate hosts spreading the disease to mammals. *Y. pestis* multiply in flea's intestine and can be transmitted to a healthy animal (including man) in the next bite (Fig. 3.17).

There are three forms of plague:

Bubonic, which causes the tonsils, adenoids, spleen, and thymus to become inflamed. Symptoms include fever, aches, chills, and tender lymph glands.

Septicemic, in which bacteria multiply in the blood. It causes fever, chills, shock, and bleeding under the skin or other organs.

Pneumonic, in which the bacteria enter the lungs and cause pneumonia. People with the infection can spread this form to others.

Pneumonic plague is the most contagious form because it can spread from person to person in airborne droplets, and this type could be a biowarfare agent. Symptoms usually begin within 2–6 days after exposure to the plague bacteria. If diagnosed in time, plague is treatable with antibiotics. The drug of choice is streptomycin and there is no vaccine for plague.

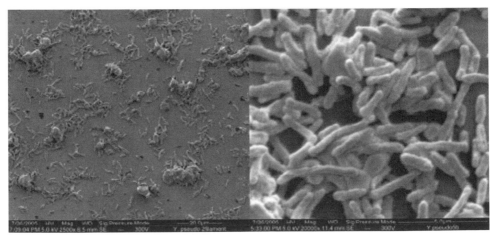

Figure 3.17. Scanning electron micrograph of *Yersinia* sp. *Source:* Courtesy of Dr. G.P. Rai, DRDE, Gwalior.

3.4.4.2. Lyme Disease

Lyme disease is caused by the bacterium *Borrelia burgdorferi* and is spread by ticks. Lyme disease is not contagious, but affects different parts of the body in varying degrees as it progresses. It can affect the nervous system, joints, skin, and heart. The early symptoms include circular rash after 1–2 weeks of infection, which looks like a bull's eye. It can affect the nervous system, causing facial paralysis, or tingling and numbness in the arms and legs. The clinical diagnosis is based on physical examination and medical history. Laboratory tests such as enzyme-linked immunosorbent assay (ELISA), Western blot, and PCR help to confirm the disease.

3.4.4.3. Rocky Mt. Spotted Fever

Rocky Mountain spotted fever (RMSF) is a disease caused by *Rickettsia rickettsii*, and is usually transmitted by ticks. The bacteria are unlikely to be transmitted to a person by a tick that is attached for less than 20 h. If a tick is infected with *R. rickettsii*, the longer that tick stays attached, the more likely it is to transmit the bacteria through its mouth into the person's blood. Usually, several hours are needed for this to take place. Treatment involves careful removal of the tick from the skin and antibiotics to eliminate the infection. Doxycycline or tetracycline is the drugs of choice for both confirmed and suspected cases. Chloramphenicol may be used in pregnant women. Complications are rare, but can include nerve damage, paralysis, hearing loss, and, rarely, death in unattended cases.

3.4.4.4. Epidemic Typhus

Epidemic typhus is a severe acute disease with prolonged high fever up to $40\,^{\circ}$C, intractable headache, and a pink-to-purple raised rash, due to infection with a microorganism called *Rickettsia prowazekii*. It is transmitted by the human body louse (*Pediculus humanus corporis*). People become infected when infected lice feces or crushed infected body lice are rubbed into small cuts on the skin. Symptoms include fever, headache, chills, tiredness, and muscle aches. About half of people who are infected develop a flat red rash that begins on the back, chest, and stomach, and then spreads to the rest of the body except for the face, palms, and soles. Other symptoms may include vomiting, eye sensitivity to light, and confusion. In severe cases, complications may include kidney failure and brain inflammation (encephalitis). The neurologic features gave the disease its name from the Greek word "typhos," which means smoke, cloud, and stupor arising from fever. The diagnosis of epidemic typhus fever is based on signs and symptoms of illness, as well as laboratory blood tests. *R. prowazekii* is considered category B bioterrorism agent, because it is stable in dried louse feces and can be transmitted through aerosols.

3.4.5. Sexually Transmitted Bacterial Infections

3.4.5.1. Syphilis

Syphilis is a contagious sexually transmitted disease (STD) caused by a spirochete, namely, *Treponema pallidum*, which is about 10–5 μm in length and about 0.15 μm in diameter. The pathogen enters the body through mucous membranes or minor breaks or abrasions of the skin. Once in the body, it migrates to the regional lymph nodes and rapidly spreads throughout the body. Syphilis occurs in three distinct stages. In primary stage, the pathogen multiplies developing a characteristic lesion called *chancre* formed within 3–6 weeks. The secondary stage begins 6–8 weeks after appearance of the lesion. During this stage, there is appearance of cutaneous lesions involving mucus membrane, such as on lips, tongue, throat, penis, angina, and other body surfaces. The tertiary stage occurs several years after the primary stage. Although, this stage is asymptomatic, it can cause damage of any organ of the body especially damage to aorta resulting in death of the patient or paralysis or may cause change in personality. The syphilis-infected pregnant woman gives birth to still baby who is mentally retarded and neurologically impaired. The diagnosis is confirmed by microscopy. Serological tests proved to be more sensitive. Penicillin is highly effective in syphilis treatment; the primary and secondary stages can be controlled by a single injection of penicillin G, and, in tertiary stage of syphilis, penicillin treatment must be extended for a longer period.

3.4.5.2. Gonorrhea

Gonorrhea is an acute, infectious, STD of the mucous membranes of the genitourinary tract, eye, rectum, and throat. It is caused by the gram-negative, encapsulated Diplococcus, *N. gonorrhoeae*. It may affect urogenital, anorectal, pharyngeal, and conjunctival areas. The most common site of *N. gonorrhoeae* infection is the urogenital tract (Stern et al., 1986; Stern and Meyer, 1987). Men with this infection may experience dysuria with penile discharge and swollen testicles, and women may have mild vaginal mucopurulent discharge, severe pelvic pain, swelling, abnormal menstrual bleeding, or no symptoms. The organisms are transmitted through sexual intercourse; symptoms usually appear within 2–10 days after sexual contact with an infected partner; however, symptoms can appear up to 30 days after sexual contact. A gonorrhea culture is done on a sample of body fluid collected from the potentially infected area, such as the cervix, urethra, eye, rectum, or throat. Serological and nucleic acid-based tests with superior sensitivity and specificity are also available. *N. gonorrhoea* is susceptible to almost all classes of antibiotics. If left untreated, *N. gonorrhoea* infections can disseminate to other areas of the body, which commonly causes synovium and skin infections.

3.4.5.3. Chlamydia

Chlamydia is a curable STD caused by bacteria called *Chlamydia trachomatis*. It is estimated that chlamydia is the most prevalent STD with 3–4 million new cases each year (Barry et al., 1999). Chlamydia infection is very common among young adults and teenagers. Women who have chlamydia may experience lower abdominal pain or pain during intercourse, bleeding during intercourse, and bleeding between menstrual periods. Men may experience burning and itching around the opening of the penis and/or pain and swelling in the testicles. Symptoms of chlamydia may appear within 1–3 weeks after being infected. Pregnant woman with chlamydia are at risk of spontaneous abortion. In men, untreated chlamydia can affect the testicles and ultimately lead to infertility. Chlamydia can also cause salpingitis, which is a painful inflammation of the fallopian tubes. Chlamydia infection can affect sperm function and male fertility. It is the most common cause of inflammation in reproductive organs (Hackstadt, 1999). The two most commonly used antibiotics are azithromycin and doxycycline.

3.4.5.4. Chancroid

Chancroid is a sexually transmitted genital ulcer disease caused by a gram-negative bacterium *H. ducreyi*. Chancroid is a local infection, which does not become systemic. Chancroid may also lead to swelling, tenderness, and inflammation of the lymph nodes in the groin, a side effect not associated with syphilis. The open sore of chancroid resembles the sore present in syphilis. After first being infected, an incubation period of up to 2 weeks will be followed by a small bump that becomes an ulcer a day later. The ulcer can range in size from 1/8 of an inch to 2 inches in diameter. The center of the ulcer has a gray, yellowish pus-like fluid, and it easily bleeds if knocked or injured. Fifty percent of male patients have a single ulcer. Female patients typically have four or five ulcers. In men, the ulcers are usually on the glands of the penis and in women they appear on the labia minora. Chancroid is diagnosed by Gram stain of culture, biopsy, or microscopic examination of a smear sample of the sore. Diagnosis is confirmed by culturing *H. ducreyi* with results available within 4 h. Standard treatment is 2 weeks of regular doses of one of the following antibiotics: erythromycin and trimethoprin. A single dose of axithromycin or ceftriaxone can also be used to treat chancroid.

3.4.6. Other Important Bacterial Diseases

3.4.6.1. Leprosy

Leprosy is caused mainly by *Mycobacterium leprae*, a rod-shaped "acid-fast" bacillus that is an obligate intracellular bacterium. The bacteria take an extremely long time to reproduce inside of cells (about 12–14 days as compared to minutes to hours for most bacteria). The bacteria grow very well in the body's macrophages and Schwann cells (cells that cover and protect nerve axons). *M. leprae* cannot be cultured on artificial media. *M. leprae* is genetically related to *M. tuberculosis* and other mycobacteria that infect humans. The early signs and symptoms of leprosy occur slowly (usually over years). Numbness and loss of temperature sensation (cannot sense very hot or cold temperatures) are some of the first symptoms. As the disease progresses, the sensations of touch, then pain, and eventually deep pressure are decreased or lost. The majority of cases of leprosy are diagnosed on the basis of hypopigmented patches of skin or reddish skin patches with loss of sensation, thickened peripheral nerves, or both clinical findings together often comprise the clinical diagnosis. Skin smears or biopsy material that show acid-fast bacilli with the Ziehl–Neelsen stain or the Fite stain (biopsy) can diagnose multibacillary leprosy, or, if bacteria are absent, diagnose paucibacillary leprosy. Other specialized tests include lepromin test, phenolic glycolipid-1 test, and PCR and immunodiagnostic tests. In general, paucibacillary leprosy is treated with two antibiotics, dapsone and rifampicin, while multibacillary leprosy is treated with the same two plus a third antibiotic, clofazimine. Usually, the antibiotics are given for at least 6–12 months or more. WHO suggested that single-dose treatment of patients with only one skin lesion with rifampicin, minocycline (Minocin), or ofloxacin (Floxin) is effective. There is no commercially available vaccine to prevent leprosy.

3.4.6.2. *Staphylococcal Infections*

Over 30 different types of *Staphylococci* can infect humans, but most infections are caused by *S. aureus*. *Staphylococci* can be found normally in the nose and on the skin of 20–30% of healthy adults. The staph infection often begins with a small cut, abrasion, or crack, which gets infected by the *Staphylococcus* bacteria. Staph infections range from a simple boil to antibiotic-resistant infections to flesh-eating infections. Some more conditions caused by the staph bacteria include folliculitis (an infection of the hair follicles that causes small white head boils on the skin), boils (localized skin infections that run deeper within hair follicles, which leave swollen, red, painful lump in the skin), styes or hordeolum (infection of the follicle surrounding the eyelashes, which causes pimple-like sores on eyelids), impetigo (red sores around mouth, which develop into blisters), and abscess (an infection of skin that causes localized collection of pus accompanied by inflammation) (Figs. 3.18 and 3.19).

3.4.6.3. *Pseudomonas aeruginosa Infections*

Pseudomonas aeruginosa which is primarily a nosocomial pathogen is a gram-negative bacterium measuring 0.5–0.8 μm by 1.5–3.0 μm. *P. aeruginosa* is a free-living bacterium, commonly found in soil and water. It causes urinary tract infections, respiratory system infections, dermatitis, soft tissue infections, bacteremia, bone and joint infections, gastrointestinal infections, and a variety of systemic infection, particularly in patients with severe burns and cancer and patients who are immunosuppressed. The most serious complication of cystic fibrosis is respiratory tract infection by the ubiquitous bacterium *P. aeruginosa*. It produces two extracellular protein toxins, exoenzyme S and exotoxin A. Exoenzyme S has the characteristic subunit structure of the A component of a bacterial toxin, and it has ADP-ribosylating activity (for a variety of eucaryotic proteins) characteristic of many bacterial exotoxins. Exoenzyme S is produced by bacteria growing in burned tissue and may be detected in the blood before the bacteria are colonized. Exotoxin A appears to mediate both local and systemic disease processes caused by *P. aeruginosa* (Figs. 3.20 and 3.21 and Table 3.13).

3.4.7. Actinomycetes Mycetoma

Mycetoma is a chronic infection of the skin, subcutaneous tissue, and sometimes bone, characterized by discharging sinuses filled with organisms. The most common

Figure 3.18. EM of *Staphylococci* and coagulase test. (*See insert for color representation of the figure*).

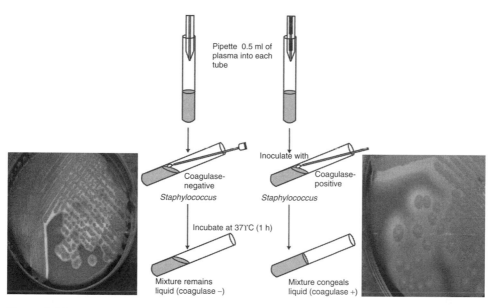

Figure 3.19. Blood agar plates and hemolysis, steps in the coagulase test, and enlarged view of hemolysis. (*See insert for color representation of the figure*).

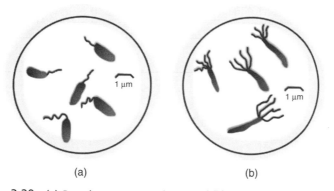

(a)   (b)

Figure 3.20. (a) *Pseudomonas aeruginosa* and (b) *Pseudomonas fluorescens*.

actinomycetes to cause mycetoma with black grains, which is present worldwide, is *Madurella mycetomatis*; the mycetoma with white grains having worldwide distribution is *Pseudallescheria boydii*; mycetoma with white/yellow grains are *Actinomadura madurae, Nocardia asteroids*; and mycetoma with brown or red grains include *Actinomadura pelletieri* (Africa) and *Streptomcyes somaliensis* (North Africa, Middle East). The mycetoma generally presents as a single lesion on an exposed site and may persist for years. Two-thirds arise on the foot. The diagnosis of mycetoma depends on identifying grains. These are obtained using a needle and syringe to extract material from a soft part of the lesion under the skin or by collecting pus. The color of the grains may suggest the likely diagnosis; black grains suggest a fungal infection, minute white grains suggest nocardia, and red grains are due to *A. pelletieri*. Larger white grains or yellow-white grains may be fungal or actinomycotic in origin. Microscopy using potassium hydroxide

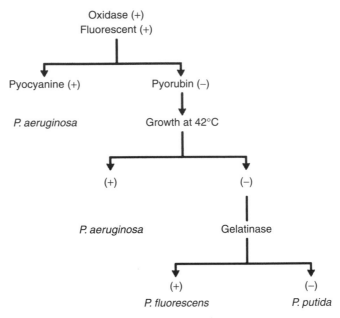

Figure 3.21. Dichotomy chart for identification of *Pseudomonas* sp.

TABLE 3.13. Summary of Results (Expected) of Morphological and Biochemical Reactions of *Pseudomonas* sp

S. No.	Tests	Results
Morphological		
1	Colony morphology	Smooth, round
2	Flagella stain	Positive
3	Gram stain	Negative
4	Endospore stain	Negative
5	Motility	Positive
6	Pigment production	Positive
Biochemical		
1	Catalase	Positive
2	Oxidase	Positive
3	Gas production in glucose broth	Negative
4	Gelatinase	Negative
5	Caseinase	Negative
6	Oxidation/fermentation reactions	Oxidation
7	Amylase	Negative
8	H_2S production in TSI	Negative
9	Nitrate reduction	Negative
10	Growth at 42 °C	Positive

(KOH) confirms the diagnosis and type of mycetoma. Actinomycetoma responds well to treatment with appropriate antibiotics, but they are required for months or years. The sinuses dry up, swelling and tenderness improve, and the grains disappear. Deformity may persist. Single or combination of streptomycin injections, oral cotrimoxasole, amikacin are in practice.

3.4.7.1. Actinomycosis

Actinomycosis is a chronic suppurative and granulomatous disease of the cervico-facial, thoracic, or abdominal areas.

The most common cause of actinomycosis is the organism *Actinomyces israelii* which infects both man and animals. In cattle, the disease is called "lumpy jaw" because of the huge abscess formed in the angle of the jaw. In humans, *A. israelii* is an endogenous organism that can be isolated from the mouths of healthy people. Frequently, the infected patient has a tooth abscess or a tooth extraction and the endogenous organism becomes established in the traumatized tissue and causes a suppurative infection. These abscesses are not confined to the jaw and may also be found in the thoracic area and abdomen. The patient usually presents with a pus-draining lesion, so the pus will be the clinical material sent to the laboratory. This diagnosis can be made on the hospital floor. When the vial of pus is rotated, the yellow sulfur granules, characteristic of this organism, can be seen with the naked eye. These granules can also be seen by running sterile water over the gauze used to cover the lesion. The water washes away the purulent material leaving the golden granules on the gauze. This organism, which occurs worldwide, can be seen histologically as "sulfur granules" surrounded by PMNs forming the purulent tissue reaction. The organism is a gram-positive rod that frequently branches. The laboratory must specifically be instructed to culture for this anaerobic organism. These lesions must be surgically drained prior to antibiotic therapy and the drug of choice is large doses of penicillin.

Using a modified Fite-Faraco stain, a "sulphur granule" is shown in the middle of the image (Fig. 3.22). These granules actually represent colonies of *A. israelii*, a gram-positive, anaerobic filamentous bacteria.

3.4.7.2. Nocardiosis

Nocardiosis primarily presents as a pulmonary disease or brain abscess in the United States. In Latin America, it is more frequently seen as the cause of a subcutaneous infection, with or without draining abscesses. It can even present as a lesion in the chest wall that drains onto the surface of the body similar to actinomycosis. Brain abscesses are frequent secondary lesions.

The most common species of *Nocardia* that cause disease in humans are *N. brasiliensis* and *N. asteroides*. These are soil organisms, which can also be found endogenously in the sputum of apparently healthy people. *N. asteroides* is usually the etiologic agent of pulmonary nocardiosis while *N. brasiliensis* is frequently the cause of subcutaneous lesions (Fig. 3.23). The material sent to the laboratory, depending on the presentation of the disease, is sputum, pus, or biopsy material. These organisms rarely form granules. *Nocardia* are aerobic, gram-positive rods, and stain partially acid-fast (i.e., the acid-fast staining is not uniform). There are no serological tests, and the drug of choice is Bactrim (trimethoprim plus sulfamethoxazole). *Nocardia* grow readily on most bacteriologic and TB media. The geographic distribution of these organisms is worldwide.

3.4.7.3. Streptomycosis

Streptomyces spp. are gram-positive aerobic actinomycetes known for their production of antimicrobial substances. Though they seldom cause human disease, infections can manifest as localized, chronic suppurative lesions of the skin.

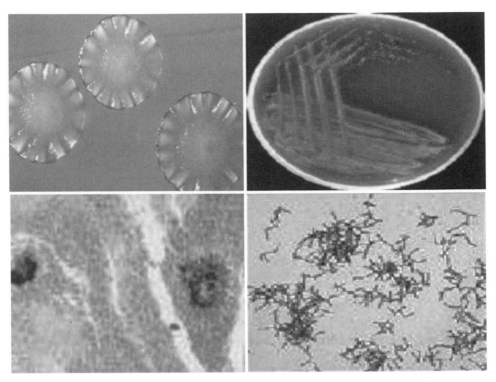

Figure 3.22. *Actinomyces* spp. are gram-positive, fungus-like bacteria (hematoxylin and eosin stain). (*See insert for color representation of the figure*).

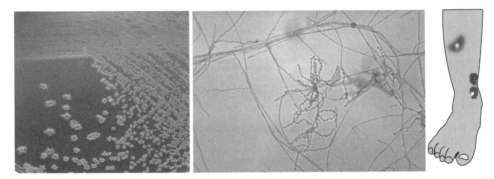

Figure 3.23. Gram-positive aerobic *Nocardia asteroides* slide culture reveals chain of spores amongst aerial mycelia and pathological symptoms form. (*See insert for color representation of the figure*).

The streptomyces species usually cause the disease entity known as *mycetoma* (fungus tumor). These infections are usually subcutaneous, but they can penetrate deeper and invade the bone. Some species produce a protease that inhibits macrophages. Material sent to the laboratory is pus or skin biopsy. The streptomycetes are aerobic like *Nocardia*, and can grow on both bacterial and fungal (Sabouraud) media. They produce a chalky aerial mycelium with much branching (Fig. 3.24). It is important to know that the suspected organism, that is, actinomycetes, unlike most bacterial pathogens

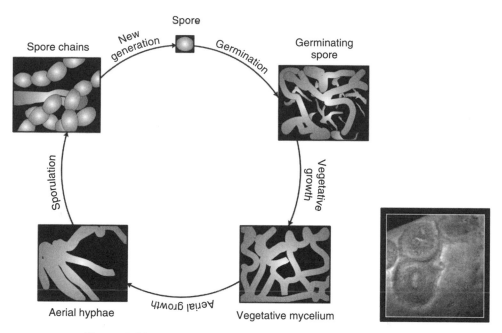

Figure 3.24. Life cycle of *Streptomyces* and mycetoma infection.

that grow out overnight, take a longer period to become visible on the culture plates (48–72 h). The various species of streptomyces produce granules of different size, texture, and color. These granules, along with colonial growth and biochemical tests, allow the bacteriologist or mycologist to identify each species. The organisms are found worldwide. There are no serological tests, and the drugs of choice are the combination of sulfamethoxazole/trimethoprim or amphotericin B. In the tropics this disease may go undiagnosed or untreated for so long that surgical amputation may be the only effective treatment.

3.5. VIRAL PATHOGENS AND ASSOCIATED DISEASES

Viruses may be classified into different groups on the basis of pathogenicity to humans, transmissibility to the community, and whether effective prophylaxis is available or not. Viruses that cause severe human disease (Schneider-Schaulies, 2000) present a serious hazard to laboratory workers, for example, HIV, HBV, Japanese B encephalitis virus, Hantaviruses, Rift Valley fever virus, yellow fever (YF) virus, and rabies virus. There are virus groups that cause severe human disease that present a high risk of spread to the community and there is usually no effective prophylaxis or treatment, for example, Lassa fever virus, filoviruses, smallpox virus, and Crimean–Congo hemorrhagic fever virus. The other category includes those that may present a risk of spread to the community but there is usually effective prophylaxis or treatment available, for example, Herpesviruses, ortho and paramyxoviruses, picornaviruses, and adenoviruses, and those that may cause human disease that may be a hazard to laboratory workers but is unlikely to spread to the community (Mateu, 1995; Alcami and Koszinowski, 2000).

3.5.1. Herpes Simplex Viruses

HSVs are dsDNA-enveloped viruses and contain four basic structures—the envelope, tegument, nucleocapsid (NC), and a DNA-containing core. The genome of HSV-1 and HSV-2 shares 50–70% homology. Humans are the only natural host to HSV, the virus is spread by contact, and the usual site for implantation is the skin or mucous membrane. HSV is able to escape the immune response and persists indefinitely in a latent state in certain tissues. Appropriate techniques reveal virus in 50% of normal human trigeminal ganglia and, to a lesser extent, in cervical, sacral, and vagal ganglia. HSV is spread by contact, as the virus is shed in saliva, tears, genital, and other secretions. The risk of infection to a nonimmune individual in contact with contaminated secretions can be as high as 80%. HSV infection leads to a variety of clinical manifestations, which includes acute gingivostomatitis, herpes labialis (cold sore), ocular herpes, herpes genitalis, other forms of cutaneous herpes, meningitis, encephalitis, and neonatal herpes. Laboratory diagnosis is based on light microscopy and electron microscopy. Direct antigen detection by immunohistology is more sensitive and specific than that by light and electron microscopy (90% sensitive, 90% specific), but cannot match virus culture. HSV-1 and HSV-2 are among the easiest viruses to cultivate. Serological tests based on antigen and antibody detection are highly promising. Acyclovir is the drug of choice for most situations at present.

3.5.2. Varicella Zoster Virus

Varicella zoster virus (VZV) is a dsDNA-enveloped virus and is one of the classic diseases of childhood, with the highest prevalence occurring in the 4–10 years age group.

Varicella is highly communicable, with an attack rate of 90% in close contacts. Little is known about the route and the source of transmission of the virus. After an incubation period of 14 days, the virus arrives at its main target organ, the skin. The skin lesions progress rapidly through the stages of macules to papules to vesicles, which rapidly break down with crust formation. Hemorrhagic symptoms sometimes occur during the course of varicella and usually present 2–3 days after the onset of the rash (Fig. 3.25). Viral pneumonia is the most serious complication of varicella and this condition may be difficult to distinguish from bacterial pneumonia unless a biopsy is taken.

Herpes zoster mainly affects a single dermatome of the skin. Postherpetic neuralgia is the most common and important complication of herpes zoster. The clinical presentations of varicella or zoster are so characteristic that laboratory confirmation is rarely required. This remains the definitive method for diagnosing VZV infections. Cytology, immunofluorescence, and molecular methods based on PCR are available, which distinguish between HSV and VZV infection.

3.5.3. Cytomegalovirus

Cytomegalovirus is a dsDNA-enveloped virus belonging to herpesvirus family. CMV is one of the most successful human pathogens; it can be transmitted vertically or horizontally usually with little effect on the host. Transmission may occur *in utero*, perinatally, or postnatally. Once infected, the person carries the virus for life, which may be activated from time to time, during which infectious virions appear in the urine and the saliva.

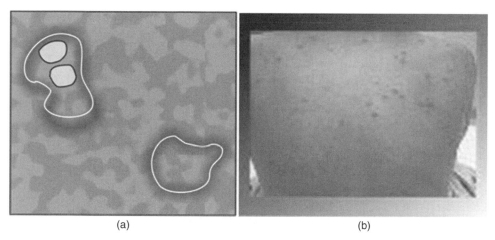

(a)   (b)

Figure 3.25.   (a) Diagrammatic electronmicrograph of VZV particles; (b) vesicular rash of chick-enpox. (*See insert for color representation of the figure*).

CMV is the second commonest cause of mental retardation after Down's syndrome and causes more cases of congenital damage. The classical presentation of cytomegalic inclusion disease is IUGR, jaundice, hepatosplenomegaly, thrombocytopenia, and encephalitis with or without microcephaly. Diagnosis is based on histopathology, and rapid serological and molecular methods based on PCR are available for sensitive diagnosis.

3.5.4. Epstein–Barr Virus

Epstein–Barr virus (EBV) is a dsDNA-enveloped virus belonging to herpesvirus family. EBV causes infectious mononucleosis and is found to be associated with Burkitt's lymphoma and nasopharyngeal carcinoma.

3.5.5. Human Herpes Virus

Human herpes virus (HHV) belongs to the betaherpesvirus subfamily of herpesviruses; it is a dsDNA-enveloped virus of icosahedral symmetry and is the causative agent for roseala infantum (Fig. 3.26). There are two variants (HHV-6A and HHV-6B) with a high degree of homology to each other (95–99%) but with different biological properties, for example, cell tropism and tissue distribution. Children older than 4 months acquire HHV-6 rapidly, as the protective effects of maternal antibody wears off. Association of HHV-6 with febrile convulsions, meningitis, and encephalitis has been documented. Serology is highly promising and used to make a diagnosis in a routine laboratory. Both IgM and IgG can be demonstrated by immunofluorescence.

3.5.6. Hepatitis

3.5.6.1. Hepatitis A

Hepatitis A virus (HAV) is a picornavirus containing single-stranded RNA (ssRNA). HAV is a fecal–oral pathogen and transmission is particularly associated with fecally

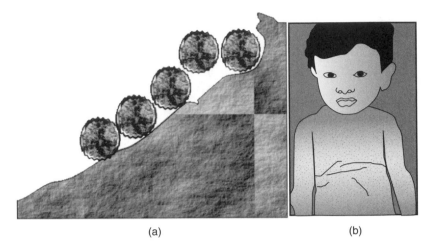

(a)   (b)

Figure 3.26. (a) HHV-6 particles on the surface of lymphocyte; (b) roseala infantum.

contaminated food and water. The pathological changes are common to all types of viral hepatitis, with parenchymal cell necrosis and histiocytic periportal inflammation. Following an incubation period of around 4 weeks, there is an acute onset of nonspecific symptoms such as fever, chills, headache, fatigue, malaise, and aches and pains. Complications of hepatitis A include prolonged cholestatic jaundice characterized by elevated levels of serum transaminase levels. No carrier state exists for hepatitis A and the mortality rate is very low.

The virus can be detected in the feces up to 2 weeks before the appearance of jaundice and up to 2 weeks afterward. HAV can also be detected in the serum, saliva, urine, and semen. A variety of ELISAs and molecular diagnostic tests are available for detection of Hepatitis A in serum.

3.5.6.2. Hepatitis B

Hepatitis B is responsible for 1.5 million deaths per year. Around 40% of chronically infected individuals will die as a result of this infection. Blood and blood products are the main routes through which the virus is transmitted. The virus is also found in semen, vaginal discharges, breast milk, and serous exudates such as the CSF, and these have been implicated as possible vehicles of transmission. Hepatitis B infection is reported to be associated with certain HLA (human leukocyte antigen) haplotypes (Almarri and Batchelor, 1994). Hepatitis B virus (HBV) is a dsDNA virus, and has not yet been possible to propagate the virus in cell culture. The genome comprises of circular DNA and the major protein is HBsAg. The core protein is the major component of the NC and HBeAg may be generated from the core protein by proteolytic cleavage. The incubation period for hepatitis B is 6 weeks to 6 months, that is, 40–180 days (average of 90 days or 3 months). As with hepatitis A, the clinical picture is very variable, although the disease is on the whole more severe than hepatitis A. The carrier state is defined as persistence of the HBsAg in the circulation for more than 6 months. The carrier state may persist lifelong and may be associated with mild liver damage varying from minor changes in liver function to chronic active hepatitis and cirrhosis and hepatocellular carcinoma.

It takes about 4–5 years for cirrhosis to develop. Some carriers may go on to develop hepatocellular carcinoma. The first serological marker to appear after infection is HBsAg, and this occurs 2–8 weeks before biochemical evidence of liver dysfunction or the onset of jaundice. The HBsAg persists throughout the course of the illness. Next to appear is the viral DNA polymerase and the e antigen. The e antigen is a distinct soluble antigen, which is located within the core and correlates closely with the number of virus particles and the relative infectivity. In general, detection of HBsAg is used for the diagnosis of acute infection and the screening for carrier (infectious stage) status. Antibodies to specific HBcAg indicate past infection and are of value in monitoring progress (Chisari, 1995).

Two types of interferons, namely, α interferon (Intron A) and Peg interferon (Pegasys) are in use against Hepatitis B infection. Prevention by vaccination is the best approach.

3.5.6.3. Hepatitis C

Hepatitis C genome resembled that of a flavivirus and the mode of HCV (hepatitis C virus) transmission is similar to that of HBV. HCV infection is particularly associated with high alanine transminase levels. In general, hepatitis C is a slowly progressing disease than hepatitis B. It takes 10 years to develop chronic hepatitis, 20 years to develop cirrhosis, and 30 years to develop hepatocellular carcinoma. Immunodiagnosis based on antigen detection by a variety of immunodiagnostic assays and molecular diagnosis based on PCR are widely in use.

3.5.6.4. Hepatitis E

Hepatitis E disease is primarily associated with the ingestion of fecally contaminated drinking water. Hepatitis E was first documented in New Delhi in 1955 when 29,000 cases of icteric hepatitis occurred, following the contamination of the city's drinking water. Hepatitis E is caused by a calicivirus-like, unenveloped RNA virus, which is highly labile and sensitive and which is not yet cultured. Clinically, the disease is similar to hepatitis A, but carries a higher mortality, especially among pregnant women. HEV (hepatitis E virus) has moderately higher incubation period than HAV (2–9 weeks, mean = 6 weeks) Similar to HAV, the disease is usually self-limiting. The clinical attack rate is from 0.7% to 10%. Virus particles are present in the feces just before the onset of jaundice and the appearance of antibodies. Hepatitis E accounts for 12–50% of clinical hepatitis in the Indian subcontinent.

3.5.7. Respiratory Syncytial Virus

RSV is a major respiratory pathogen of young children and belongs to the family of paramyxoviruses. RSV is distributed globally and is the most common cause of severe lower respiratory disease in young infants. It is responsible for 50–90% of cases of bronchiolitis, 5–40% of pneumonias and bronchitis, and less than 10% of croups in young children. RSV is an ssRNA-enveloped virus, and envelope has two proteins and two glycoproteins, M protein, a 22–24K protein of unknown function, and F (fusion) and G (glycoprotein) appears to effect virus attachment. The virus enters the body usually through the eye or nose, rarely through the mouth. The virus then spreads along the epithelium of the respiratory tract, mostly by cell to cell transfer. The incubation period for RSV is usually 3–6 days but may vary from 2–8 days. As the virus spreads to the lower respiratory tract, it may produce bronchiolitis and/or pneumonia. Early in bronchiolitis, a peribronchiolar inflammation with lymphocytes occurs, which progresses to the characteristic necrosis and sloughing of the bronchiolar epithelium. This sloughed necrotic material may plug the bronchioles resulting in an obstruction to the flow of air, the hallmark of bronchiolitis. Air may be trapped distal to the sites of occlusion, causing the characteristic hyperinflation of bronchiolitis, which, when absorbed, results in multiple areas of focal atelectasis. No one appears to escape infection with RSV, almost all infections are acquired during the first 3 years of life, and thus virtually all adults possess specific antibody. Apnoea is a frequent complication of RSV infection, occurring in approximately 20% of cases. RSV infection may be diagnosed by cell culture techniques or by the identification of viral antigen through rapid diagnostic techniques.

3.5.8. Influenza Viruses

Influenza viruses belong to orthomyxovirus group in the family of Orthomyxoviridae (Lamb and Krug, 2001). It is an ssRNA-enveloped virus and 80–120 nm in diameter. The RNA is associated with a nucleoprotein and is segmented with eight RNA fragments (seven for influenza C). There are four antigens, namely, the hemagglutinin (HA), neuraminidase (NA), NC, the matrix (M), and the NC proteins that are well characterized. The nucleoprotein is a type-specific antigen that occurs in three forms, A, B, and C, which provides the basis for the classification of human influenza viruses (Deitsch et al., 1997; Colman, 1998). The matrix protein (M) surrounds the NC proteins and makes up 35–45% of the particle mass. The HA mediates the attachment of the virus to the cellular receptor and is made up of two subunits, HA1 and HA2 (Röhm et al., 1996). On the basis of their NC and M protein antigens, the influenza viruses are divided into three distinct immunological types (A, B, and C). Influenza A viruses also occur in pigs, birds, and horses (Couceiro et al., 1993, Drescher and Aron, 1999). However, only humans are infected by influenza B and C. The antigenic differences of the HA and the NA antigens of influenza A viruses provide the basis for their classification into subtypes. The virus changes the antigenic composition frequently and this phenomenon is called *antigenic shift*. The HA antigen is always involved in antigenic shift as also the NA. Lesser antigenic changes in the virus are known as antigenic drift. Antigenic drift is thought to arise through natural mutation, and selection of new strains takes place by antibody pressure in an immune or partially immune population (Bush et al., 1999).

The onset of infection is abrupt with a marked fever, headache, photophobia, shivering, a dry cough, malaise, myalgia, and a dry tickling throat. Fever is continuous and lasts around 3 days. Influenza B infection is similar to influenza A, but infection with influenza C is usually subclinical or very mild in nature. The clinical complications include pneumonia, myositis and myoglobinuria, and Reye's syndrome.

Nasopharyngeal aspirates are the specimens of diagnostic choice, and, for the presence of viral particles (influenza A and B antigens), the specimen can be examined by indirect immunofluorescence. Other sensitive tests include enzyme immune assays and PCR-based tests. At present, treatment of influenza is entirely symptomatic (Cox and Subbarao, 2000). Vaccines against influenza have been around.

3.5.9. Parainfluenza Viruses

Parinfluenza virus contains one molecule of ss RNA of negative sense, although some contain positive sense strands. The virus is homologous to mumps virus, HA and NA present, unlike RSV. The virus belongs to the family Paramyxoviridae, which contains five subtypes: parainfluenzavirus 1, 2, 3, 4a, and 4b. Parinfluenza virus subtypes 1 and 3 belong to the paramyxovirus genus while the subtypes 2, 4a, and 4b belong to the rubullavirus genus along with mumps. The virions are 120–300 nm in diameter with helical NCs (Wright and Webster, 2001). Parainfluenza virus multiplication in the host occurs throughout the tracheobronchial tree, inducing the production of mucus. As a result, the vocal cords of the larynx get swollen, causing obstruction to the inflow of air. In adults, the virus is responsible for inflammation in the upper parts of the respiratory tract while, in infants and young children, the bronchi, bronchioles, and lungs may also get occasionally involved. About 80% of patients exhibit cough and runny nose 1–3 days before the onset of the cough.

Croup is a well-defined, easily recognized clinical entity. The best diagnostic specimens are nasopharyngeal secretions or saline mouth gargles. Virus identification by microscopy, virus isolation techniques, and immune and molecular diagnostic approaches are available.

3.5.10. Adenoviruses

Adenoviruses are widely responsible for acute URT infections, that is "colds." About 51 human adenovirus serotypes have been reported which are further divided into seven subgroups based on their capacity to agglutinate erythrocytes. Adenovirus predominantly infects the respiratory tract, the GI tract, and the eye. Adenoviruses are associated with a wide variety of clinical syndromes such as pharyngitis, pneumonia, acute hemorrhagic cystitis, and meningitis. The diagnosis is generally based on virus isolation from infected body fluids and secretions, throat swabs, and urine. Serological and molecular diagnosis holds great promise.

3.5.11. Rhinoviruses

Rhinovirus infections occur worldwide and are responsible for common colds. Rhinovirus is a naked ssRNA virus and there are about 100 serotypes. Rhinoviruses are responsible for 30% of common colds, coronaviruses for 10%, and other viruses include adenoviruses, enteroviruses, RSV, influenza, and parainfluenza. Rhinovirus appears to be transmitted mainly by the aerosol route.

The clinical symptoms of common cold include nasal obstruction, sneezing, sore throat, and cough. There is little fever and systemic reactions are uncommon. Illness may last for a week or more. Rhinovirus infections may induce the onset of asthmatic attacks in atopic individuals. Usually, a common cold does not require laboratory investigation. If required, the diagnosis is generally made by the isolation of the virus in a sensitive cell culture. Nasal washings are the best diagnostic specimens. Because of the multiplicity of serotypes, it would be very difficult to develop an effective vaccine against rhinoviruses.

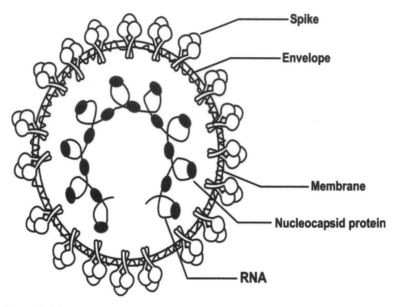

Figure 3.27. Diagrammatic illustration of the genome of corona virus particle.

3.5.12. Coronaviruses

Human coronaviruses are ssRNA-enveloped viruses with characteristic club-shaped projections of 20 nm (Fig. 3.27). Human coronaviruses are responsible for 10–30% of all common colds. Coronavirus-like particles are also often seen in the feces of children and adults suffering from diarrhea. The diagnostic approaches include direct detection of virus and serological tests.

3.5.13. SARS Virus

SARS virus outbreak was reported in China in 2003. It is a novel coronavirus with genome size of 29,000 bases. The incubation period is about 6 days, and children are rarely affected by this virus. The initial diagnosis of SARS is clinical as per the guidelines issued by the WHO. RT-PCR is the mainstay of diagnosis of SARS infection and the diagnostic specimens include throat swabs, trachael aspirates, and feces. Serological diagnosis is often the only means of confirming the diagnosis of SARS.

3.5.14. Diarrheal Viruses

Rotaviruses and adenoviruses are regularly involved in endemic diarrhea. Rotaviruses are by far the largest cause of diarrhea in children, accounting for 50–80% of cases in developing countries. Rotavirus is a dsRNA virus with cubical symmetry. It was previously thought that rotaviruses were by far the most common cause of viral gastroenteritis. It is thought that noroviruses could account for up to 90% of all cases of gastroenteritis. Fever and vomiting are the common symptoms of diarrhea. The site of infection is principally the upper small bowel. Astrovirus is unique among human viruses and is found so far only in the gut. They are five serotypes of human astroviruses and may be found in the stool in large numbers.

Norwalk virus was the first fastidious enteric virus and is the most common viral gastroenteritis agent, well surpassing rotaviruses. Laboratory diagnostic approaches include RT-PCR and serological tests based on specific antigen and antibody detections.

3.5.15. Calicivirus (Sapovirus)

Calicivirus, generally, infects pigs, cats, sea-lions, and fur-seals, but may be found in humans as well. The human calicivirus is morphologically indistinguishable but antigenically distinguishable from animal strains and contains a single positive stranded RNA genome. The virus shows icosahedral symmetry, the face and the apices of the icosahedron are represented by cupped hollows, of which 32 cup-shaped depressions are on the surface. Vomiting being a more prominent feature than diarrhea, and infection by human caliciviruses resemble that by Norwalk agents.

3.5.16. Enterovirus

Enterovirus belongs to the family Picornaviridae and over 72 serotypes of enterovirus have been isolated from humans. Enteroviruses enter the body through the alimentary tract where they multiply, and may spread to other organs causing severe disease that is typical of individual enterovirus types. The human enteroviruses belonging to this category include the polioviruses, coxsackieviruses, enterocytopathic human orphan viruses, and enteroviruses. Incubation period widely varies from 2 to 40 days. It contains ssRNA, and is a naked virus with icosahedral symmetry. There is a close homology among human enteroviruses and also with rhinoviruses.

3.5.17. Poliomyelitis

Poliovirus was first identified in 1909, and humans are the only natural hosts for polioviruses. Polioviruses are disseminated globally. Almost all population have antibodies to all three strains of the virus before 5 years of age. The incubation period is usually 7–14 days. Following ingestion, the virus multiplies in the oropharyngeal and intestinal mucosa and enters the lymphatic system. Virus isolation and serological techniques are the mainstay of diagnosis. No specific treatment is available except supportive measures in paralytic poliomyelitis. However, it is possible to prevent the disease through vaccination. Protection is required against all three types of poliovirus. There are two vaccines available: the formalin-inactivated Salk vaccine and the live-attenuated Sabin vaccine. Poliovirus was targeted for eradication by the WHO and has been eradicated from most regions of the world except the Indian subcontinent and subSaharan Africa.

3.5.18. Coxsackie

Coxsackie viruses are divided into two groups on the basis of the lesions observed in suckling mice. Group A viruses produce a diffuse myositis with acute inflammation and necrosis of fibers of voluntary muscles, and produce a flaccid paralysis. Group B viruses produce focal areas of degeneration in the brain and necrosis in the skeletal muscles, and produce a spastic paralysis. The coxsackieviruses can produce a remarkable variety of diseases. Still, a number of group A viruses have not definitely been implicated as causative agents of any human disease. Some syndromes are almost caused exclusively by group A viruses (herpangina and hand-foot-mouth disease), and some others by group B (epidemic pleurodynia and myocarditis of the newborn).

3.5.19. Echoviruses

The first echoviruses are picornaviruses isolated from the GI tract. Similar to polioviruses, the mouth is the portal of entry of the viruses although they can gain entry through the respiratory route. The incubation period varies between 2 and 7 days and recovery from infection is accompanied by the development of lifelong immunity. The diseases caused by echo viruses include aseptic meningitis, encephalitis, paralysis, enteritis, pleurodynia, and myocarditis.

The diagnosis of enterovirus-associated myocarditis is usually made on laboratory tests. Virus isolation or serological tests can be used. PCR assays are becoming increasingly used for the detection and identification of enteroviruses. Vaccination is not available against coxsackie or echoviruses.

3.5.20. Mumps

Mumps is one of the commonly acquired viral diseases of childhood and is the most common cause of aseptic meningitis. The virus is transmitted by aerosol droplets or by direct contact. The virus multiplies in the epithelium of the URT or the GI tract or eye, and quickly spreads to the local lymphoid tissue and to distant sites in the body. The parotid gland is commonly infected but so may the CNS (central nervous system), testis or epididymis, pancreas, and ovary. The incubation period of the virus is 16–18 days but may vary from 14–25 days. Parotid swelling develops in 95% of those with clinical illness and virus shedding into the saliva starts a couple of days before the onset of parotitis and ends 7–8 days later. Serological diagnosis is based on the detection of virus-specific antibodies.

3.5.21. Measles Virus

Measles is one of the typical viral diseases of childhood and often leads to severe complications that may be fatal. The infection is acquired through the URT or conjunctiva. The virus contains ssRNA and belongs to the family of Paramyxoviruses. Envelope consists of HA protein (Liebert et al., 1994). The incubation period is 10–11 days, and the distinctive maculopapular rash (Fig. 3.28) appears about 4 days after exposure and starts behind the ears and on the forehead. From here the rash spreads to involve the whole body (Griffin, 2001).

The majority of cases develop pneumonia. Occasionally, marked hepatosplenomegaly, hyperaesthesia, numbness, and paraesthesia are also found. In the majority of patients, measles is an acute self-limiting disease that will run its course without the need for specific treatment. However, it is far more serious in the immunocompromised, the

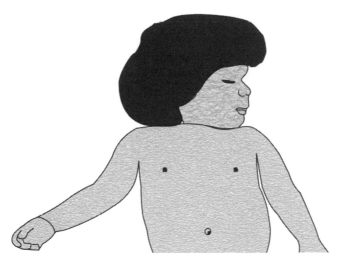

Figure 3.28. Morbilliform rash of measles.

undernourished, and children with chronic debilitating diseases. The symptoms of acute measles are so distinctive that laboratory diagnosis is seldom required. However, as the vaccination program progressed, atypical forms of measles have emerged and laboratory diagnosis may be required. Production of multinucleate giant cells with inclusion bodies is pathognomonic for measles, and definitive diagnosis of measles infection can be made by immunological approaches.

3.5.22. Rubella

Rubella virus is an ssRNA-enveloped virus with worldwide distribution. The incubation period is 13–20 days, and the onset is abrupt in children with the appearance of maculopapular rash (Fig. 3.29), which lasts for not more than 3 days.

The common complication being joint involvement; fingers, wrists knees, and ankles are most frequently affected. Complications are reported in about 60% of adult females. Similar complications may be caused by parvovirus, arboviruses, and enteroviruses. Patients are potentially infectious over a prolonged period and therefore rapid diagnosis of rubella is essential. Virus can be isolated from nasopharyngeal secretions a week before the onset of the rash, but as rubella antibodies develop the viraemia ceases. Antibody-based diagnosis is the mainstay of diagnosis of rubella infection.

3.5.23. Parvovirus B

Human parvovirus is an ssDNA, naked icosahedral virus. The common manifestation is a mild febrile illness with a maculopapular rash. The differential diagnosis of erythema infectiosum includes all diseases where a maculopapular rash may be present, for example, rubella, enteroviruses, arboviruses, streptococcal infection, and allergy. Rubella causes the greatest problems as the two viruses may circulate together. Therefore, a definitive diagnosis can be made only by serological tests.

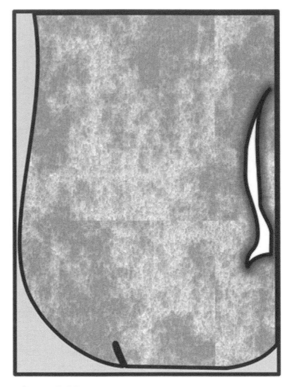

Figure 3.29. Maculopapular rash found in rubella.

3.5.24. Human Immunodeficiency Viruses

Human immunodeficiency virus was isolated by Montagnier and Gallo in 1984, and, in 1986, HIV-2 was isolated from West Africa, which did not cross-react serologically with HIV-1 (Fig. 3.30). There are reports about the existence of other strains of this virus (Gao et al., 1998; McCutchan et al., 1999). HIV virus is an enveloped RNA virus (Crandall, 1999). The genome consists of 9200 nucleotides (HIV-1); gag core proteins, namely, p15, p17, and p24; pol—p16 (protease), p31 (integrase/endonuclease) (Cornelissen et al., 1996); env–gp160 (gp120: outer membrane part, gp41: transmembrane part) (Fig. 3.30). The HIV envelope proteins, glycoprotein 120 (gp120) and glycoprotein 41 (gp41), play crucial roles in HIV entry (Caffrey, 2011).

The first step of infection is the binding of gp120 to the CD4 receptor of the cell, which is followed by the fusion of the virus and cell membrane and is mediated by the gp41 molecule. The virus then penetrates into the cell, and uncoating, reverse transcription, provirus synthesis, and integration take place. This is followed by the synthesis and maturation of virus progeny (Schols and De Clercq, 1996).

Chemokine receptors such as CXCR4 and CCR5 are essential coreceptors for HIV-1 entry. Activated cells that become infected with HIV produce viruses immediately and die within 1–2 days. The immunosuppression seen in AIDS is due to the depletion of T4 helper lymphocytes. The time required to complete a single HIV life cycle is approximately 1.5 days. The diagnosis of AIDS is established with the appearance of opportunistic infections, or of certain neoplasms, such as Kaposi's sarcoma, primary

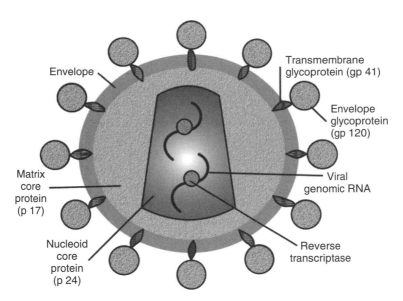

Figure 3.30. Schematic representation of an HIV particle.

lymphoma of the brain, and other non Hodgkin's lymphomas. The diagnosis of AIDS is also established by the finding of a CD4 count of less than 200 cells per cubic millimeter. In all stages of HIV infection, characteristic skin manifestations include oral hairy leucoplakia and an itching maculopapular eruption. A high percentage of HIV-infected patients show neurological changes and the spectrum of symptoms ranges from slight neuropsychological abnormalities to organic psychosis and complete dementia (Kimata et al., 1999). It is now thought that up to 95–99% of all HIV-infected persons eventually develop AIDS (Back et al., 1996). HIV is transmitted by sexual contact, blood and blood products, or from the mother to the newborn infant. It appears that infectivity is correlated with viraemia and is greater very early after infection before the appearance of antibodies and later in the course after the development of symptomatic immunodeficiency (Kartikeyan et al., 2007).

The diagnosis of HIV infection is usually based on serological tests. ELISA followed by Western blot is the most frequently used method for screening of blood samples for HIV antibody. HIV antigen can be detected early in the course of HIV infection and is undetectable during the latent period (antigen–antibody complexes are present) but becomes detectable during the final stages of infection (Coffin, 1995; Fig. 3.31).

Zidovudine (AZT) was the first antiviral agent used for the treatment of HIV and was introduced in 1987. Combination therapy with two nucleoside analogs was found to be better than monotherapy with one alone. A further breakthrough occurred with the introduction of HIV protease inhibitors, which were specifically designed against HIV protease and were shown to be the most potent anti-HIV effect to date. Compliance is a major issue when therapy is expected to be lifelong.

The main types of approaches for development of AIDS vaccine included live-attenuated viruses, inactivated viruses, live-recombinant viruses, synthetic peptides, recombinant DNA products (gp120, gp160), and native envelope and/or core proteins (Singh and Bisen, 2006). However, HIV vaccine is unlikely to be available in the near future.

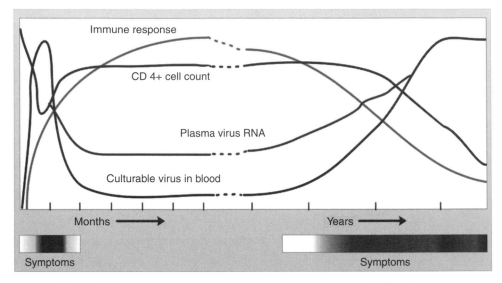

Figure 3.31. Course of immune response during the course of HIV infection.

3.5.25. Human T Cell Lymphotropic Viruses (HTLV)

HTLV-I infection occurs in clusters in certain geographic locations around the world. Transmission through breast milk is implicated as the major route for the maintenance of infection in high prevalence areas. HTLV, an ssRNA-enveloped virus, belongs to the oncornavirus subfamily of retroviruses. Both HTLV-I and HTLV-II are tropic for CD4+ lymphocytes but other cells can also be infected. In contrast to HIV, the cellular receptor is not known. HTLV-I infection leads to two kinds of disease manifestation: adult T-cell leukemia and tropical spastic paraparesis. Laboratory diagnosis rests mainly based on the detection of antibodies. Western blot using recombinant HTLV antigens is used to distinguish between HTLV-I and HTLV-II. PCR can also be used to detect HTLV from peripheral blood mononuclear cells and can distinguish between HTLV-1 and HTLV-II. A combination of interferon-α and zidovudine had been reported to be effective in treating the infection (Larsen et al., 2000).

3.5.26. Papillomaviruses

The human papillomavirus (HPV) is a dsDNA virus and is linked with genital cancers. The genome of the virus consists of about 7800 bp and exists in a supercoiled circular form. Not much is known about the proteins produced by the papillomaviruses. HPVs have not been grown *in vitro*. At least 57 HPV types are known based on DNA homology (Van Doorslaer et al., 2011). HPVs infect and replicate in squamous epithelium on both keratinized and nonkeratinized (mucosal) cells. Most people are infected with the common HPV types 1, 2, 3, and 4. HPVs are responsible for the numerous types of warts such as common hand warts, juvenile warts, and genital warts. The association between certain strains of HPV and cervical cancer is well documented and it is the most common virus associated with cancer worldwide.

The appearance of warts varies from the scaly flat lesions on the epithelium of individuals with EV, to the ace-to-white flat CIN lesion of the cervix. The appearance of wart lesions is a good indication followed by detection of the common HPV antigen in the tissues, which confirms the virus. Vaccines have been shown to offer 100% protection against the development of precancerous cervical cancer lesions and genital warts caused by the HPV.

3.5.27. Polyomaviruses

Polyomaviruses are dsDNA viruses and belong to the family Papovaviridae. The genome consists of about 5000 bp and both strands of DNA are used for transcription. There is no evidence for the existence of animal reservoirs, and the exact route of transmission is still unknown. Although, the genome of the virus was detected in several tumors, no evidence of association is seen with any tumors.

3.5.28. Rabies

Rabies virus infects the CNS of all warm-blooded animals, where the resultant disease is almost invariably fatal. It is an ssRNA-enveloped virus with helical symmetry.

The natural hosts of the virus are unknown and are thought to be bats or rodents. The commonest mode of transmission in man is by the bite of a rabid animal or the contamination of scratch wounds by virus-infected saliva. The incubation period is highly variable, ranging from seven days to several years. The diagnosis of animal and human rabies can be made by four methods: (i) histopathology, (ii) virus cultivation, (iii) serology, and (iv) virus antigen detection. Several types of live- and killed-attenuated vaccines are available (Fig. 3.32).

3.5.29. Smallpox

Poxviruses are brick-shaped dsDNA viruses, and are the largest viruses known. Poxviruses have a biconcave "nucleoid" and two lateral bodies. Inside the cell, the

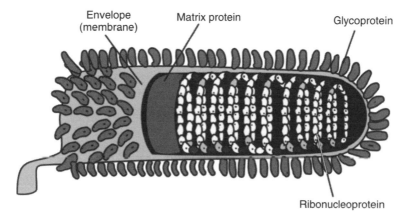

Figure 3.32. Schematic diagram of rabies virus particles.

virion often has a double membrane. About 100 polypeptides have been demonstrated in vaccinia. Poxviruses are very easy to isolate and will grow in a variety of cell cultures.

Smallpox was transmitted by a respiratory route. Variola major caused severe infections with 20–50% mortality and variola minor with <1% mortality. The vaccine was highly effective. If administered during the incubation period, the vaccine either prevents or reduces the severity of clinical symptoms. Smallpox was eradicated and the world was declared to be free of smallpox in 1980. Availability of the most effective vaccine is one of the prime reasons for smallpox elimination from the globe.

3.5.30. Monkeypox

The pathogenesis and clinical features of monkeypox is the same as that of smallpox. The mortality in human monkeypox is appreciable, being in the order of 10%. One important difference between human monkeypox and smallpox is the lower capacity for human spread.

3.5.31. Vaccinia

Vaccination with vaccinia in certain cases is associated with complications that ranged from mild reactions to fatal encephalitis. Vaccinia is being used as a vector for immunization against other viruses.

3.5.32. Cowpox

Cowpox is a relatively unimportant zoonosis and has been described in humans, cows, and cats. Infection in humans often produces a localized lesion which is similar to that caused by vaccination. In humans, the lesions are restricted to the hands, but may be transferred to the face in some cases. EM is generally used for the diagnosis of infection.

3.5.33. Arthropodborne Viruses

Arthopodborne viruses (Arboviruses) are viruses that could be transmitted to humans via an insect (arthropod) vector (Scott et al., 1994). In general, arboviruses belong to three families:

1. *Toga viruses*: genera alphaviruses, for example, EEE, WEE, VEE.
2. *Buna viruses*: for example, Sicilian sandfly fever, Rift Valley fever, and Crimean–Congo hemorrhagic fever.
3. *Flavi viruses*: for example, St. Louis encephalitis, Japanese encephlitis (JE), YF, and dengue.

3.5.33.1. Toga Viruses

There are four toga virus groups belonging to the family Togaviridae, namely, alphaviruses, rubiviruses, pestiviruses, and arteriviruses. The *Alphavirus* genus has 25 members, of which 11 are recognized to be pathogenic to humans. All alphaviruses are transmitted by mosquitoes and are widely distributed. The *Rubivirus* genus has only one member, rubella. There are three viruses in *Pestivirus* genus, which are of veterinary significance, namely, bovine diarrhea virus, hog cholera virus, and border disease virus. The *Arterivirus* genus contains equine arteritis virus. All are ssRNA-enveloped viruses with icosahedral symmetry.

3.5.33.2. Chikungunya Virus

The best example of an *Alphavirus* is chikungunya virus. There are two possible modes of virus transmission: the first mode is when humans come in contact with monkeys, and the second is through infected mosquitoes. The clinical disease is a flu-like fever of acute onset, followed by a pharyngitis, maculopapular rash, and arthritis. Minor hemorrhagic manifestations may be seen. Diagnosis of infection is possible by serological and molecular approaches.

3.5.33.3. Bunyaviruses

Bunyaviruses are enveloped ssRNA viruses and comprise more than 200 viruses. Viruses of this group are thought to be transmitted by arthropods; mostly by mosquito vectors, but occasionally phlebotomine sandflies, midges, and ticks may also carry the infection. Vertebrate reservoirs have been demonstrated for some viruses, and humans are not known to be a natural host or reservoir for any of these viruses (exception sandfly fever). Diagnosis of bunyavirus infections is usually made by serological and molecular approaches.

3.5.33.4. Flaviviruses

Flaviviruses are ssRNA-enveloped arthropodborne viruses, which produce encephalitis. Transmitted by mosquito and ticks, produce a broad spectrum of clinical responses in humans ranging from asymptomatic infection to fulminant encephalitis or hemorrhagic fever. The various types of flaviviruses are as follows:

St Louis Encephalitis. The incubation period is 21 days. Patients who are symptomatic will usually present with or progress to one of the three syndromes: (i) febrile headache, (ii) aseptic meningitis, and (iii) encephalitis. The laboratory diagnosis is usually made by serology. Treatment is supportive and no vaccine is available.

Japanese Encephalitis. JE is a major public health problem in Asia. The transmission cycle in nature involves the Culex and Aedes mosquitoes, and domestic animals, birds, bats, and reptiles. Diagnosis is usually made by serological or molecular means. No specific treatment is available.

Murray Valley Encephalitis. This virus closely resembles JE clinically, and is confined to Australia and New Guinea, where it is an important cause of epidemic encephalitis periodically. The diagnosis is made by serology and no specific treatment or vaccine is available.

West Nile Fever. West Nile virus is transmitted by Culex mosquitoes. Diagnosis is usually made by serological approaches, although the virus can be isolated from blood in tissue culture. Neither vaccine nor specific therapy is available. It is difficult to distinguish clinically West Nile, dengue, and chikungunya.

Ilheus Virus. This virus is found in Latin America where it causes a febrile illness with arthralgia. Occasionally, mild encephalitis is seen. The virus is often confused with dengue, St Louis encephalitis, YF, and influenza viruses.

Yellow Fever. The incubation period varies from 3 to 6 days, and the hemorrhagic manifestations may vary from petechial lesions to epitaxis, bleeding gums, and GI hemorrhage (black vomit of YF). Fifty percent of patients with frank YF will develop fatal disease characterized by severe hemorrhagic manifestations, oliguria, and hypotension. YF can be diagnosed serologically or by virus isolation.

Kyasanur Forest Disease. This is a tick-borne disease closely related to the tick-borne encephalitis complex, and is geographically restricted to Karnataka State in India. Hemorrhagic fever and meningoencephalitis may be seen. The case-fatality rate is 5%.

Dengue. Several thousands of dengue cases are reported every year in tropical and subtropical areas of the world. Several serologically distinguishable types of dengue viruses are known to exist in different geographic zones. The virus resides in the salivary glands of the mosquito from where it can be transmitted to healthy individuals. The virus has an incubation period of 2–7 days in humans. The virus spreads to the liver, spleen, bone marrow, and lymph nodes, and involvement of other organs such as the heart, lungs, and GI tract have also been reported.

The WHO recommended the following criteria for the diagnosis of dengue shock syndrome (DSS). They are fever, hemorrhagic manifestations including at least a positive tourniquet test, enlarged liver, and thrombocytopenia ($\leq$100,000 ul). Diagnosis of DSS is made when there is a circulatory failure. The pathogenesis of DSS is attributed both to virus virulence and immunopathological mechanisms. Diagnosis is made by serological, RT-PCR, and virus isolation methods. There is no specific antiviral treatment or vaccine available for dengue virus.

Orbiviruses. Orbiviruses are insect-borne viruses, primarily of veterinary importance. Orbiviruses contain dsRNA. A maculopapular rash may be seen in a minority of patients. A more severe clinical picture may be seen in children, who may develop hemorrhagic manifestations including severe gastrointestinal bleeding (GI) and disseminated intravascular coagulation (DIC). Aseptic meningitis or encephalitis may be seen. Colorado tick fever (CTF) may be diagnosed by virus isolation and serological diagnosis.

3.5.34. Lymphocytic Choriomenigitis Virus

Lymphocytic choriomenigitis virus (LCMV) rarely infects humans. The disease is generally mild and is manifested usually as a form of aseptic meningitis (lymphocytic meningitis) or an influenza-like illness. Very rarely, severe or fatal disease is seen with hemorrhagic manifestations. LCMV may be pathogenic for the fetus.

3.5.35. Filovirus

Filovirus causes the most severe form of viral hemorrhagic fever known, with 60–90% mortality. It is an enveloped RNA virus of unique morphology, and forms tubular structures of 80 nm diameter and up to 10,000 nm length. The onset of illness is sudden and marked by fever, chills, headache, myalgia, and anorexia. Virological diagnosis is readily achieved by virus isolation. Diagnosis is usually made by the direct detection of viral antigen by ELISA or viral-RNA by PCR. No vaccine is available. The mortality is extremely high, being in the order of 60–90%.

3.5.36. Hantaviruses

Hantaviruses are responsible for causing hemorrhagic fever with renal syndrome (HFRS) or, more recently, hantavirus disease (HVD). Hantavirus is a member of the Bunyaviridae; it is an enveloped ssRNA virus, with virion measuring 98 nm in diameter, and the surface has characteristic square grid-like structure. At least six serotypes are known. Damage to capillaries and small vessel walls, resulting in vasodilation and congestion with hemorrhages are reported in hantavirus infection. Damage is most severe during the hypotensive and the oliguric phases, and the kidney is usually involved resulting in nephritis. Diagnosis of HVD is usually made on the clinical findings and serology.

3.6. PRIONS

The subacute spongiform encephalopathies are associated with unconventional agents whose purification and exact physical identification are yet to be achieved. During the 1970s, the major thrust of the research was in purification of the infectious agent. The diseases associated with the unconventional agents in humans include kuru, Creutzfeldt–Jacob disease (CJD), and Gerstmann–Straussler syndrome (GSS).

3.6.1. Kuru

Kuru is most commonly seen in women. Kuru is probably transmitted through cannibalism. With the cessation of cannibalism in the 1950s, the incidence of kuru gradually disappeared, first among children, and successively thereafter among adolescents and young adults.

3.6.2. Creutzfeldt–Jacob Disease (CJD)

CJD has a worldwide distribution and had been reported in 50 countries. Up to 15% of all cases of CJD have a family history of disease consistent with autosomal dominant transmission. CJD presents as a subacute dementia, evolving over weeks to several months, and is accompanied by pyramidal, extrapyramidal, and cerebellar signs. The precise mechanism of spread in familial cases is uncertain. What is known is that the sequence of the PrP gene is changed in affected individuals, which may render more susceptible to infection by the agent. Alzheimer's disease shares a number of clinical and pathological features in common with CJD. Pathologically, there is a spectrum of amyloid deposition which overlaps in both diseases. The natural mechanism of spread and the reservoir of CJD virus remain unknown at the present.

3.6.3. Gerstmann–Straussler Syndrome (GSS)

GSS is a very rare, usually familial, neurodegenerative disease that affects patients in the third to the seventh decade of life. The exact incidence of GSS is unknown, but is estimated to be between 1 and 10 per 100 million. Familial cases are associated with autosomal dominance inheritance in families. A change in codon 102 from proline to leucine was found in the PrP gene of affected individuals. Therefore, it appears that this genetic change is essential for the acquisition of the disease.

The long incubation period of prions sets them apart from most other viral infections. The prions appear to replicate continuously from inoculation and manifest clinically when they gain access to the CNS. The complete absence of a detectable host immune response is something interesting. Current evidence suggests that modified PrP protein is an integral, if not the sole component of prion, and it is not clear whether there is any nucleic acid component.

There are no laboratory tests available for definitive diagnosis of prion diseases. Confirmatory diagnosis is made on histopathological changes in the CNS by light microscopy or electron microscopy. Forty-three years have passed since it was first proposed that a protein could be the sole component of the infectious agent responsible for the enigmatic prion diseases. Many discoveries have strongly supported the prion hypothesis, but only recently has this once heretical hypothesis been widely accepted by the scientific community. In the past 3 years, researchers have achieved the "Holy Grail" demonstration that infectious material can be generated *in vitro* using completely defined components. These breakthroughs have proven that a misfolded protein is the active component of the infectious agent, and that propagation of the disease and its unique features depend on the self-replication of the infectious folding of the prion protein. In spite of these important discoveries, it remains unclear whether another molecule besides the misfolded prion protein might be an essential element of the infectious agent. Future research promises to reveal many more intriguing features about the rogue prions (Soto, 2010).

3.7. PARASITIC INFECTIONS

Parasitic infections are caused by worms, microscopic protozoa, skin parasites, and so forth. GI parasites can reduce food absorption by causing inflammation of the intestinal wall. Food might also get stuck resulting in excessive toxins, smelly farts, bad breath, and bloating. Infected body is filled with poisons and harmful stress hormones. Sweating and urinating are quick ways to excrete poisons. Some bloodsucking worms leave open wounds resulting in darker feces. The loss of blood can cause iron deficiency, anemia, and dizziness. Other symptoms caused by parasitic infections include anorexia, chills, colitis, coughing, diarrhea, dysuria, fatigue, fever, headache, hematochezia, hemoptysis, immunodeficiency, itching, jaundice, joint pain, memory loss, mental problems, muscle spasms and pains, vomiting and nausea, rash, rectal hemorrhage, rectal prolapse, shortness of breath, stomach pain, swelling, sweating and grinding teeth while sleeping, and vaginitis. The main sources of parasite infections include drinking water, food, and arthropod vectors.

3.7.1. Nematoda

3.7.1.1. Roundworms

Roundworms have small teeth for biting and oral suckers to hold on to tissues. The length of roundworms may range from a few millimeters to about 30 cm. The most common human roundworms are as follows:

Anisakis. These are parasitic roundworms (2 cm long) living in the stomach. Anisakiasis is acquired by eating uncooked fish or squid infected with *Anisakis simplex* (the most common Anisakis species). The parasite is resistant to gastric acid and survives burrowed into the gastric wall. Larval *A. simplex* grows into a reproducing adult only in marine mammals. In humans it cannot survive and dies within a few weeks. But the short time that it lives, it causes stomach pain and nausea. Anisakiasis can be diagnosed gastroscopically, and responds to albendazole therapy.

Ascaris lumbricoides (*Giant Roundworm*). It is the most common parasitic worm in humans. Adult females are 20–35 cm long and 3–6 mm in diameter. Male worms are a little smaller reaching 15–30 cm length and 2–4 mm width. According to some estimates, 25% of humans are infected with *A. lumbricoides*. The life cycle of *A. lumbricoides* takes about 3 months. Adults live up to 2 years and a female produces about 200,000 microscopic eggs per day. The eggs are very resistant to chemicals, extreme temperatures, and other rough conditions, and can survive for months.

Ascariasis symptoms include blockage of the biliary tract, diarrhea, fever, nausea, obstruction of the bowel (which can be fatal), stomachache, slower growing of a child or a teen, vomiting, and weakness. Diagnosis is done by examining feces for the presence of *A. lumbricoides* eggs. Some common drugs are albendazole, ivermectin, nitazoxanide, and mebendazole.

Enterobius vermicularis (Pinworm). Human pinworm, *E. vermicularis*, is the most common parasitic worm infection in the United States and Western Europe. *E. vermicularis* does not need an intermediate host to complete its life cycle. Humans get infected by accidentally swallowing or inhaling microscopic pinworm eggs. Adults are white, thin worms. Males are 0.2 mm thick and 2–5 mm long, whereas females are 0.5 mm thick and 8–13 mm long. Life expectancy for males is 7 weeks, whereas females live up to 5–13 weeks. The males usually die after the pinworms have mated in the last part of the small intestine, the ileum. The gravid (pregnant) female resides at the beginning of the large intestine, colon, eating whatever food passes through the intestinal tract. Female pinworm reaches fertility within 4 weeks. She swims at the rate of 12 cm/h toward the rectum. During sleep when body temperature is low and there is less movement, the female pushes out from the anus and lays eggs on the outside skin. The eggs get stuck on the skin, underwear, or bedding, and become infective within a few hours. Eggs survive up to 3 weeks on clothing, sheets, or other objects. After the female has laid 11,000–16,000 eggs, it dies.

Diagnosis is made by identifying pinworms or their eggs. Worms can sometimes be seen on the skin around the anus 2–3 h after falling asleep. Some common drugs against pinworms are albendazole and mebendazole.

3.7.1.2. Hookworms

Hookworms are the second most common human worms (the most common is *A. lumbricoides*). There are thousands of hookworm species but only two of them target humans. *Necator americanus* (necatoriasis) and *Ancylostoma duodenale* (ancylostomiasis) infect over one billion people around the globe, mostly in tropical and subtropical climates.

N. americanus is gray-pink in color. Male is 5–9 mm and female 10 mm long and about 0.5 mm thick. Usually, they live a few years but can live up to 15 years. Females produce up to 10,000 eggs per day. *N. americanus* is very similar to *A. duodenale*. *A. duodenale* males are 5–10 mm and females 10 mm or more in length and 0.5 mm thick. They live only about 6 months. Females produce up to 30,000 eggs per day.

Hookworms can cause symptoms such as anemia and protein deficiency caused by blood loss, constipation, congestive heart failure, decreased rate of growth and mental development in children (caused by protein and iron deficiency), diarrhea, dizziness, dyspnea (difficulty to breath), excessive cough during larvae migration, fatigue (tiredness), fever, indigestion (food digestion not taking place properly), loss of appetite, nausea, rash after larval invasion, sore and itchy feet after larval invasion, stomach and chest pain, vomiting, and weight loss.

Diagnosis of an intestinal hookworm infection is done by microscopic identification of hookworm eggs present in a stool sample. Hookworm infection is rarely lethal. Infection is usually treated for 1–3 days. The drugs against hookworms are benzimidazoles such as albendazole and mebendazole.

3.7.1.3. Loa loa (Eye Worm)

Loa loa is a thread-like worm that lives under the skin and is often spotted migrating in the eye. The life cycle starts with microfilariae (prelarval eggs) released by the female worm. The microfilariae remain in the peripheral blood during daytime, but during the night they reside in the lungs. When a noninfected deer fly takes a blood meal containing microfilariae, the microfilariae migrate from the fly's midgut to the hemocoel and eventually to the thoracic muscles. There the microfilariae develop into first stage and eventually into third-stage (infective) filarial larvae in about 15 days. The infective larvae migrate to the fly's proboscis and invade another human during the next blood meal. The larva enters the subcutaneous layer where they mature into adults in one year time. The cycle is completed, when male and female mate and release microfilariae into the bloodstream. *L. loa* adults live up to 17 years.

Adult *L. loa* male is about 30–34 mm long and 0.35–0.42 mm thick, and the female is about 40–70 mm long and 0.5 mm thick. Loiasis can be asymptomatic and is often the cause of eosinophilia. Other symptoms include arthritis (joint pain), colonic lesion (damaged large intestine), inflammation, swelling and accumulation of fluid in testicles, lymphadenitis (infection of the lymph glands), glomerulonephritis (kidney disease), and peripheral neuropathy and retinopathy.

Diagnosis can be made by analyzing blood under a microscope to find *L. loa* microfilariae. Microfilariae have also been found in urine, sputum, and spinal fluids. Identification of an adult worm is possible from a tissue sample collected during a subcutaneous biopsy. Sometimes adult worms are seen migrating across the eye, but the short time (often only 15 min) for the worm's passage through the conjunctiva makes this observation less used. Calabar swellings are also of diagnostic value.

Loiasis is usually treated with diethylcarbamazine (DEC) or sometimes with ivermectin. Both drugs have severe side effects which include death. DEC is effective against
microfilariae and less effective against adult worms.

3.7.1.4. Strongyloides stercoralis

Strongyloides stercoralis is an intestinal parasite (2 mm long) responsible for strongyloidiasis. Unlike most other parasites, *S. stercoralis* has a heterogonic life cycle. In
addition to the parasitic life cycle in humans, it can live freely in the soil and reproduces
without a host. *S. stercoralis* can infect the same host over and over without any intermediate host. Diagnosis is by microscopic examinations of duodenal or stool samples
for the presence of the parasite. Strongyloidiasis is treated with ivermectin.

3.7.1.5. Trichuris trichiura

The human whipworm is called *Trichuris trichiura* and causes trichuriasis. Whipworm
more likely infects people who already have other parasites. Adult female is 35–50 mm
long, whereas male is about 30–45 mm long. Adult females live about 5 years. Small
numbers of whipworms might not cause any symptoms. But, if there are hundreds of
worms, there may be bloody diarrhea and anemia due to severe vitamin and iron loss.
Trichuriasis is diagnosed by detection of eggs. Mebendazole and albendazole are the
drugs of choice.

3.7.1.6. Wuchereria bancrofti

Lymphatic filariasis is a parasitic disease caused by thread-like worms, *Wuchereria bancrofti*. The parasite is carried from person to person by mosquitoes. The life cycle of
W. bancrofti starts, when a male and a female mate inside lymphatic vessels of an
infected human. The female releases thousands of microfilariae into the bloodstream.
When ingested by the mosquito, the microfilariae migrate through the midgut eventually
reaching the thoracic muscles. Within 1–2 weeks they mature into first-stage larvae and
eventually into infective third-stage larvae, which migrate through the hemocoel to the
mosquito's prosbocis. When the mosquito bites another person, the larvae are injected
into the human skin. They migrate to the lymph vessels and mature into adults within
6 months. Adult females can live up to 7 years. An adult female *W. bancrofti* is about
80–100 mm long and 0.24–0.30 mm in diameter, whereas a male is about 40 mm long
and 0.1 mm in diameter. A microfilaria is about 240–300 μm long and 7.5–10 μm thick.
It is sheathed and has nocturnal periodicity.

W. bancrofti infection is usually asymptomatic. Some people can develop lymphedema
and swelling, which is prevalent in the legs, but sometimes also in the arms, genitalia, and
breasts. Over time, the disease causes thickening and hardening of the skin, a condition
called *elephantiasis*. Filarial infection might also lead to pulmonary tropical eosinophilia
syndrome, characterized by cough, shortness of breath, and wheezing.

Diagnosis of lymphatic filariasis is traditionally done by microscopic examination of
nocturnal blood samples. The blood can also be tested for the presence of parasite-specific
antibodies. A new method of a highly sensitive "card test" has been developed to detect

antigens without laboratory equipment using finger-prick blood droplets taken anytime of the day. Molecular diagnosis by PCR is also possible.

DEC is the drug of choice and is associated with some side effects, which include dizziness, fever, headache, nausea, and muscle and joint pain. According to recent reports *Wolbachia* bacteria live in symbiosis with *W. bancrofti*. The bacteria live inside the parasite. If the bacteria are killed with antibiotics, such as tetracycline, *W. bancrofti* dies too.

3.7.2. Trematoda

Trematoda class or "trematodes" are commonly known as flukes. Flukes are flat worms. Parasitic flukes live in the intestine, tissue, or in the blood. Their life cycle begins when molluscs such as snails get infected with fluke larvae. The first-stage larvae are called *miracidia*. They have tail-like structures, *cilia*, for moving and finding molluscs. Depending on the fluke species, the larva goes through different developmental stages, which are miracidium, sporocyst, redia, cercaria, mesocercaria, and metacercaria. Adulthood is reached inside the final host, humans. Adults reproduce either sexually or asexually. Eggs exit the body with the feces and infect new molluscs. Flukes cause diarrhea, inflammation, ulcers, allergic and inflammation reactions, among other symptoms.

3.7.2.1. Fasciola Hepatica

Fasciola hepatica is a parasitic liver fluke that lives in the liver. In addition to humans, it infects cows and sheep. Adults reach a length of 3 cm and a width of 1 cm. The life cycle of *F. hepatica* starts when a female lays eggs in the liver of an infected human. Immature eggs are discharged in the biliary ducts and taken out in the feces. If landed in water, the eggs become embryonated and develop larvae called *miracidia*. A miracidium invades an aquatic snail and develops into cercaria, a larva that is capable of swimming with its large tail. The cercaria exits and finds aquatic vegetation where it forms a cyst called *metacercaria*. When a human eats the raw freshwater plant containing the cyst, the metacercaria excysts in duodenum and then penetrates the intestinal wall and gets into the peritoneal cavity. The larva spends a few weeks just browsing and eating the liver, then it relocates to the bile duct where it begins its final stage and becomes an adult. It takes about 3 months for the metacercaria to develop into an adult. Adult females can produce up to 25,000 eggs per day.

The adults in the biliary ducts cause liver inflammation and obstruction of the biliary fluid. Symptoms include diarrhea, eosinophilia, fever, nausea, stomachache, vomiting, among others. Adult *F. hepatica* is identified from eggs in a stool sample. Early stage of the infection can be diagnosed by antibody detection in the blood. Fascioliasis responds to triclabendazole drug.

Fasciolopsis buski is the largest intestinal fluke in humans. It causes a parasitic disease called *fasciolopsiasis* and is commonly known as the giant intestinal fluke.

3.7.2.2. Paragonimus westermani

About 22 million people have been found to be infected with *Paragonimus westermani*, the human lung fluke. Humans get infected with the disease, paragonimiasis, by eating raw crabs or fish that are carrying the parasite. Adult lung flukes are 4–6 mm wide, 3–5 mm thick, and 7–12 mm long. Life cycle of a lung fluke begins when the female lays eggs that are carried out from the human lungs in the sputum by the motion of microvilli. Then the eggs travel through the GI tract and are passed out of the body along with stool. If the feces end up in water, then after two weeks larvae called *miracidia* hatch and start to grow. A miracidium finds a snail and penetrates its skin. In 3–5 months miracidium develops further and produces another larval form called *cercaria*. The cercaria crawls out of the snail to find fresh water crayfish (a lobster-like creature) or crabs. It finds its way to the muscles of the crab and starts forming a cyst. Within 2 months it transforms into metacercaria, which is the resting form of cercaria. If humans eat this infected crab raw, the metacercaria cyst gets into the stomach. Once inside the duodenum, the metacercaria excysts and penetrates the intestinal wall, and continues through the abdominal wall and diaphragm into the lungs where it forms a capsule and develops into an adult. Sometimes lung fluke larvae accidentally travel to the brain or other organs and reproduce there. But, because the passage of eggs from the brain is blocked, the life cycle does not take place. If the worm reaches the spinal cord instead of the lungs, the host might become paralyzed, and, if it infects the heart, the host might die. Lung flukes cause pain and severe cough. Paragonimiasis diagnosis is done by examining the sputum for the presence of lung fluke eggs. Feces can be examined, too. Alternatively, X rays and biopsies can be taken. Paragonimiasis responds well to praziquantel.

3.7.2.3. Schistosoma

Human schistosomiasis is caused by blood flukes, namely, *Schistosoma mansoni, Schistosoma haematobium*, and *Schistosoma japonicum*, and two geographically localized species *Schistosoma intercalatum* and *Schistosoma mekongi*. More than 200 million people are infected worldwide by Schistosomiasis. *Schistosoma* requires two hosts to complete its life cycle. Depending on the species, the parasite sheds the eggs either in the feces or urine of an infected human. If the feces end up in water, larvae called *miracidia* hatch up and enter freshwater snails where they transform into sporocysts. They multiply asexually producing hundreds of cercariae that exit the snail and start waiting in the water. When they find humans nearby, the circariae quickly attach to the skin with suckers. As they penetrate the skin they transform into schistosomulae. Each schistosomula stays a few days in the skin and then enters the bloodstream through dermal lymphatic vessels or blood venules. In humans *Schistosoma* reaches fertility in 6–8 weeks. Adult blood flukes are 1–2 cm long and can live for many years. Females lay eggs on the endothelial lining of the venous capillary walls at the rate of 300–3000 eggs per day depending on the *Schistosoma* species. Most eggs, however, travel to the lumen of the intestinal tract (*S. japonicum* and *S. mansoni*) and of the ureters and bladder (*S. haematobium*), thus exiting the body in the feces or urine. But each species has a preferred location; for example, *S. japonicum* resides more frequently in the veins that drain the small intestine, *S. mansoni* is found more often in the veins that drain the large intestine, *S. haematobium* is usually seen in the venous plexus of the bladder, but can also be found in the rectal venules.

The first symptoms are a rash or itch during the first few days. Within 2 months chills, cough, diarrhea, fatigue, fever, and muscle aches can occur. Usually, however, during the first few weeks, schistosomiasis is asymptomatic. After years of infection eggs inflame organs such as the liver, bladder, and lungs. If eggs end up in the brain or spinal cord, they can cause paralysis, seizures, or inflammation of the spinal cord. Diagnosis is done from a stool or urine sample by microscopic examination. Tissue biopsy (bladder or rectal biopsy) can also be used in finding eggs, if stool or urine samples are negative. Schistosomiasis is treated with praziquantel.

3.7.3. Cestoda

Diphyllobothrium latum, the fish tapeworm, is the largest tapeworm in humans. It causes a parasitic infection called *diphyllobothriasis*, which is acquired by eating raw fish infected with the parasite. One adult tapeworm can shed up to million eggs per day. It can grow over 10 m long and live up to 20 years. Diphyllobothriasis is usually asymptomatic but can last for decades. In some cases, it causes severe vitamin B_{12} deficiency because *D. latum* can absorb most of the B_{12} intake. In some cases it can lead to neurological symptoms. Diphyllobothriasis symptoms include constipation, diarrhea, fatigue, obstruction of the bowel, and pernicious anemia (caused by vitamin B_{12} deficiency), which can lead to subacute combined degeneration of spinal cord, stomach pain or discomfort, vomiting, and weight loss. Diphyllobothriasis is diagnosed by examining the stool sample to find eggs or sometimes proglottids. Identification by microscopy is restricted to a genus level, and species can be identified by performing PCR on eggs. Diphyllobothriasis is treated by injecting diatrizoic acid into the duodenal wall in order to have the worms dissociated from intestinal wall followed by treatment with niclosamide or praziquantel.

3.7.3.1. Hymenolepis nana

Hymenolepis nana, also known as *dwarf tapeworm*, is the most common tapeworm that infect humans. The disease, hymenolepiasis is found worldwide. *H. nana* is the only human tapeworm that does not necessarily need an intermediate host to complete its life cycle (although *Taenia solium* can autoinfect and cause cysticercosis). Hymenolepiasis is usually asymptomatic in adults. But prolonged infection or multiple of tapeworms especially in children can cause more severe symptoms. Hymenolepiasis symptoms sometimes include anal itching, bloody diarrhea, diarrhea, headache, increased appetite, insomnia, loss of appetite, muscle spasms, nausea, nervousness, seizures, stomachache, vomiting, weakness, and weight loss. Diagnosis is made by identifying tapeworm eggs in stool. Alternatively, an adult worm can be identified during endoscopic examination. Hymenolepiasis is treated with a single dose of praziquantel, which has an efficacy of 96%. If praziquantel is not available, niclosamide or albendazole can be used instead.

3.7.3.2. Taenia saginata

Taenia saginata (Fig. 3.33) is a large tapeworm that causes an infection called taeniasis. It is commonly known as the beef tapeworm or cattle tapeworm because it uses cows as intermediate hosts. Humans are the only definitive hosts. *T. saginata* can live up to 25 years. It can grow up to 5 m, but, in some cases, can reach lengths of over 20 m (coiled in the intestinal tract). The disease is often asymptomatic. Taeniasis caused by *T. saginata* (Fig. 3.33) is more noticeable than taeniasis caused by *T. solium* (Fig. 3.34), although *T. solium* is, overall, more dangerous because of the risk of cysticercosis. Heavy infection of

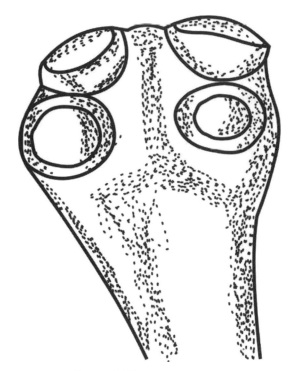

Figure 3.33. *Taenia saginata.*

T. saginata can cause some of the symptoms like allergic reactions, chronic indigestion, constipation, diarrhea, dizziness, headache, loss of appetite, nausea, obstruction of the bowel, stomachache, and weight loss.

Diagnosis is by identifying eggs or proglottids. During the first 3 months antibody detection methods can be used to find *T. saginata* from a blood sample. All *Taenia* species have similar eggs so identification can only be done at the genus level. Diagnosis can also be done during an endoscopic examination. Treatment is traditionally done with praziquantel. Alternatively, niclosamide can be used. Both drugs have some side effects (especially praziquantel) that are similar to actual tapeworm infection symptoms.

3.7.3.3. Taenia solium

The pork tapeworm, *T. solium* (7 m), is the most harmful tapeworm in humans (Fig. 3.34). *T. solium* infection is acquired either from human feces containing *T. solium* eggs or from uncooked pork that contains larval cysts. The larvae mature into adults in the small intestine and if eggs are ingested, the resulting disease is called cysticercosis. When uncooked pork is consumed, the dormant larva excysts in the bowel (Fig. 3.35). The flat body of an adult *T. solium* consists mostly of segments, proglottids. Pork tapeworm is attached to the intestinal wall with its head, the scolex. Taeniasis diagnosis is made by an endoscopic examination or by finding segments (or eggs) from the feces. Taeniasis is usually treated with niclosamide.

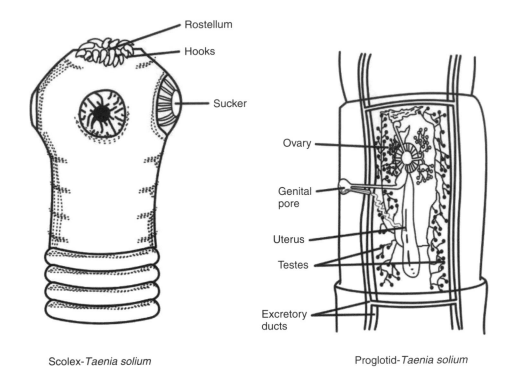

Scolex-*Taenia solium*

Proglotid-*Taenia solium*

Figure 3.34. *Taenia solium.*

3.7.3.4. Cysticercosis

About 60% of patients with cysticercosis have cysticerci in the CNS, which is called *neurocysticercosis*. Cysticerci molt into adults only in the intestine. Immune system does not recognize the cysts. The adults can live in the tissue for many years without causing any symptoms but generate inflammatory response. Symptoms include muscle spasms, dizziness, headaches, and so on. In few cases, lack of attention, confusion, and difficulty with balance may occur. Major cysticerci infections can lead to a sudden death. Fibrosis may take place in the infected organs. Vital functions of the organ may be lost. Cysticercosis diagnosis is done by magnetic resonance imaging scans or X rays. The cysts resemble tumors, so diagnosis is not foolproof. Cysticercosis responds to albendazole when used in combination with anti-inflammatory drugs.

3.7.4. Protozoan Parasites

Protozoa are unicellular organisms and are 10–100 μm long. The most common protozoan parasites (Tibayrenc et al., 1990) in humans are discussed in the following sections.

3.7.4.1. Balantidium coli

Balantidium coli is the largest protozoan parasite in humans living in the GI tract. It belongs to the Ciliophora phylum and is the only protozoan ciliate to infect humans.

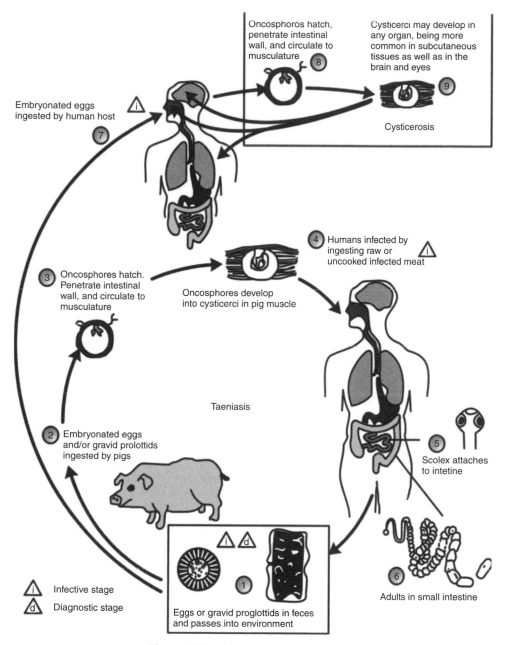

Oncosphoros hatch, penetrate intestinal wall, and circulate to musculature ⑧

Cysticerci may develop in any organ, being more common in subcutaneous tissues as well as in the brain and eyes ⑨

Cysticerosis

Embryonated eggs ingested by human host ⚠ⁱ ⑦

④ Humans infected by ingesting raw or uncooked infected meat ⚠ⁱ

③ Oncosphores hatch. Penetrate intestinal wall, and circulate to musculature

Oncosphores develop into cysticerci in pig muscle

Taeniasis

② Embryonated eggs and/or gravid prolottids ingested by pigs

⑤ Scolex attaches to intetine

⚠ⁱ Infective stage
⚠ᵈ Diagnostic stage

⚠ⁱ ⚠ᵈ

① Eggs or gravid proglottids in feces and passes into environment

⑥ Adults in small intestine

Figure 3.35. Life cycle of *Taenia solium*.

It goes through two development phases: a cyst and a trophozoite (Fig. 3.36). Trophozoites are 0.03–0.15 mm long and 0.025–0.12 mm wide with both a micronucleus and a macronucleus. Trophozoites live in the cecum and the colon of the large intestine.

About 1% of the world's population is infected with balantidiasis. People who raise pigs have bigger risk of getting infected with balantidiasis.

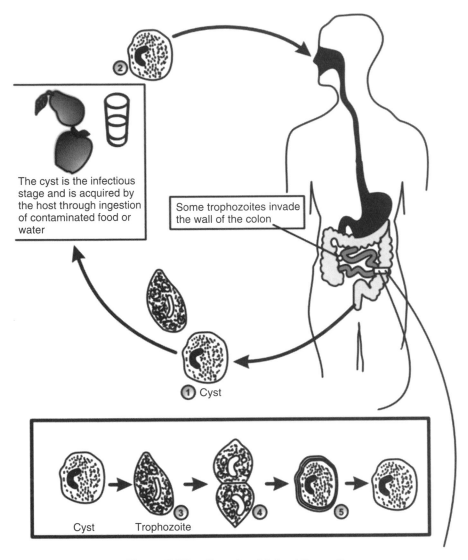

The cyst is the infectious stage and is acquired by the host through ingestion of contaminated food or water

Some trophozoites invade the wall of the colon

① Cyst

Cyst   Trophozoite

Figure 3.36.   Life cycle of *Balantidium coli.*

Balantidiasis is often asymptomatic. But in some cases the patient might have diarrhea, weight loss, and dysentery. Dysentery is an inflammatory disorder of the intestine, particularly of the colon, which causes severe diarrhea containing blood and/or mucus in the feces with stomach pain and fever. Untreated dysentery cases can be fatal. Diagnosis can be made by finding trophozoites from a stool or tissue sample (collected during endoscopy). Balantidiasis is treated with tetracycline. Iodoquinol and metronidazole can also be used.

3.7.4.2. Entamoeba histolytica

Entamoeba histolytica infects the large intestine and causes internal inflammation. About 50 million people are infected worldwide, mostly in tropical countries in areas of poor sanitation. In humans *E. histolytica* lives and multiplics as a trophozoitc. Trophozoites are oblong and about 15–20 μm in length. The trophozoites encyst and exit the body in the feces. Both cysts and trophozoites may be seen in the feces. The life cycle of *E. histolytica* does not require any intermediate host. Another human can get infected by ingesting fecally contaminated water, food, or hands. If the cysts survive the acidic environment of stomach, they transform back into trophozoites in the small intestine. Trophozoites migrate to the large intestine where they live and multiply by binary fission. Cysts are usually confined to firm stool, whereas trophozoites are found in loose stool. Only cysts can survive longer periods and infect other humans (Fig. 3.37).

Most *E. histolytica* infections are asymptomatic and trophozoites live in the intestinal lumen feeding on surrounding nutrients. About 10–20% of the infections develop into amebiasis, which is responsible for 70,000 deaths each year.

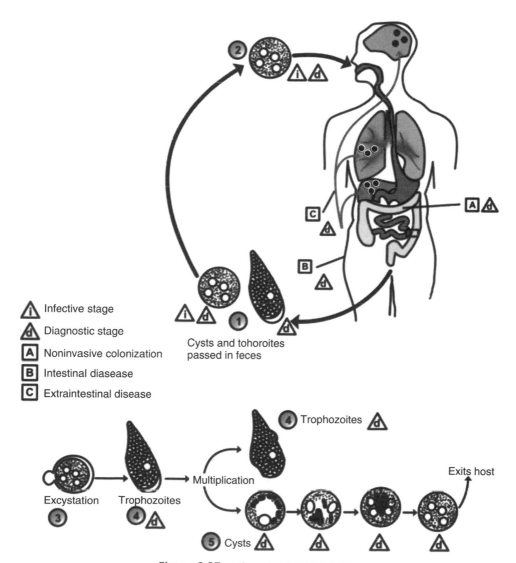

Figure 3.37. Life cycle of *E. histolytica*.

Minor infections (luminal amebiasis) can cause symptoms that include gas (flatulence), intermittent constipation, loose stools, stomachache, and stomach cramping. Severe infections inflame the mucosa of the large intestine causing amoebic dysentery. The parasites can also penetrate the intestinal wall and travel to organs such as the liver via blood stream causing extraintestinal amebiasis. Symptoms of these severe infections include anemia, appendicitis bloody diarrhea, fatigue, fever, flatulence, genital and skin lesions, liver abscesses, painful defecation (passage of the stool), peritonitis (inflammation of the peritoneum which is the thin membrane that lines the abdominal wall), pleuropulmonary abscesses, and toxic megacolon (dilated colon).

Amebiasis is diagnosed by detection of cysts in the stool sample. Trophozoites can be identified under a microscope from biopsy samples taken during colonoscopy or surgery. *E. histolytica* should be differentiated from the nonpathogenic *Entamoeba dispar*. The two are morphologically similar. Differentiation is based on immunologic or isoenzymatic analysis or molecular methods. The preferred drugs are metronidazole or tinidazole immediately followed with paromomycin, diloxanide furoate, or iodoquinol. Asymptomatic intestinal amebiasis is treated with paromomycin, diloxanide furoate, or iodoquinol.

3.7.4.3. Giardia intestinalis

Giardia intestinalis also known as *Giardia lamblia* or *Giardia duodenalis* causes giardiasis in the small intestine. *G. intestinalis* lives as active trophozoites in the small intestine. Some trophozoites encyst into cysts, which are released in a bowel movement (Fig. 3.38). The cyst has a protective shell and it can survive in the environment for many weeks. Each cyst releases two trophozoites in the small intestine. The trophozoites encyst as they move toward the colon. Cysts are found more often in firm stool, whereas both trophozoites and cysts are present in loose stool. Common giardiasis symptoms include bloating, bad breath and farts, dehydration, diarrhea, fatigue, greasy floating stools, loss of appetite, malabsorption (in chronic giardiasis), nausea, stomachache and cramps, vomiting (rarely), weakness, and weight loss. Diarrheal can be fatal. Another not so recognizable effect is the lack of vitamin B_{12}.

G. intestinalis trophozoites are pear shaped and $10-20$ μm long. Other characteristics include flagella, median bodies, sucking disks, and two big nuclei. *G. intestinalis* cysts are oval to ellipsoid and $8-19$ μm long. Immature cysts have two nuclei, whereas mature cysts have four. Diagnosis is by microscopy. Trophozoites can also be found from duodenal fluid or from biopsies taken during endoscopy. Giardiasis treatment is accomplished with antimicrobial drugs such as metronidazole, nitazoxanide, tinidazole, paromomycin, quinacrine, and furazolidone.

3.7.4.4. Leishmania

Leishmaniasis is spread by sandflies and about 12 million people are infected worldwide. It causes the most parasitic deaths after malaria. The two most common forms are visceral and cutaneous leishmaniasis. There are about 500,000 new visceral leishmaniasis cases each year. More than 90% of the visceral leishmaniasis infections take place in Bangladesh, Brazil, India, Nepal, and Sudan. The disease develops a few months after the sandfly bite and causes anemia, enlarged liver and spleen, fever, weaker inflammatory response, and weight loss (Coffman and Beebe, 1998).

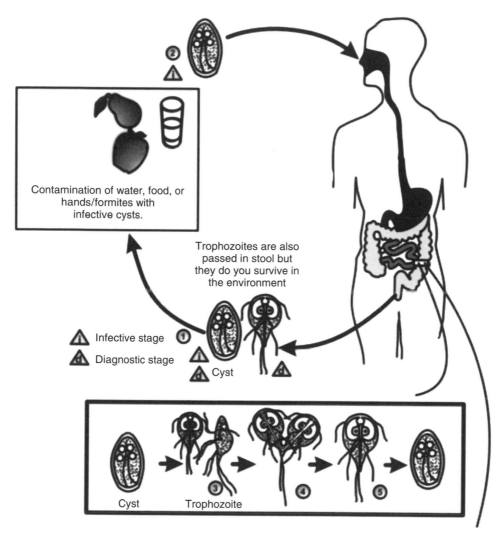

Figure 3.38. Life cycle of *G. intestinalis.*

There are about 1.5 million new cutaneous leishmaniasis cases each year. The disease causes skin sores a few weeks after the bite. Some sores are covered by crust tissue. Nearby lymph glands might swell up, if the sore is located further from the heart than the glands. Over 90% of the cutaneous leishmaniasis infections occur in Afghanistan, Algeria, Brazil, Iran, Iraq, Peru, Saudi Arabia, and Syria.

Leishmania parasite has two separate life-forms (Fig. 3.39). In humans it mainly lives as amastigote, which does not have flagella to move around with. Inside insects Leishmania appears as promastigote, which has a flagellum and is able to move. When a sandfly infected with Leishmania feeds, the parasites are transmitted to the biting area. Sandflies are smaller than regular mosquitoes and fly very quietly. Their bite might be painless and might go unnoticed. Leishmania resides in macrophage (Menon and Bretscher, 1998).

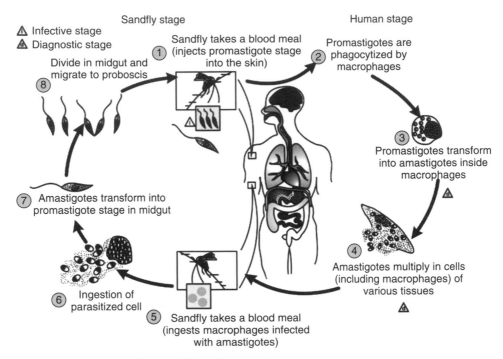

Figure 3.39. Life cycle of *Leishmania parasite*.

Diagnosis is done by doing lab tests. Depending on the particular Leishmania species and the location of the infection, different treatment is used. There is a new drug sodium stibogluconate, which might become the preferred medicine in the future.

3.7.4.5. *Plasmodium falciparum (Malaria)*

Plasmodium falciparum is responsible for 85% of malarial cases. The three less common and less dangerous *Plasmodium* species are: *P. ovale, P. malariae,* and *P. vivax.* Over 200 million people are affected by malaria annually, and it is the deadliest parasitic disease killing over 1 million people each year (Babiker and Walliker, 1997). About 90% of the deaths occur south of the Sahara desert and most are under 5-year-old children.

LIFE CYCLE. Malaria is carried by *Anopheles* mosquitoes. Of the over 400 *Anopheles* species, only 30–40 can transmit malaria (Babiker and Walliker, 1997). The infection starts when a female mosquito injects "sporozoites" into human skin while taking a blood meal. A sporozoite travels into the liver where it invades a liver cell. It matures into a "schizont" (mother cell), which produces 30,000–40,000 "merozoites" (daughter cells) within 6 days. The merozoites burst out and invade RBCs. Within 2 days one merozoite transforms into a trophozoite, then into a schizont, and finally 8–24 new merozoites burst out from the schizont and the red cell as they rupture. Then the merozoites invade new red cells (Fig. 3.40). *P. falciparum* can prevent an infected red cell from entering the spleen (the organ where old and damaged red cells are destroyed) by sending adhesive proteins to the cell membrane of the red cell. The proteins make the red cell to stick to small blood vessel walls. This poses a threat for the human host since the clustered

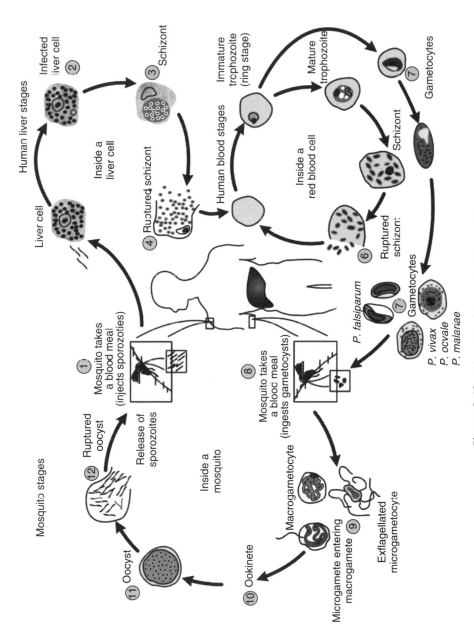

Figure 3.40. Life cycle of the malarial parasite.

Human liver stages

Infected liver cell ②

③ Schizont

Liver cell

Inside a liver cell

④ Ruptured schizont

Immature trophozoite (ring stage)

Mature trophozoite

Gametocytes ⑦

Human blood stages

Schizont

Inside a red blood cell

⑥ Ruptured schizont

Gametocytes

⑦ Gametocytes

P. falciparum

P. vivax
P. ovale
P. malariae

① Mosquito takes a blood meal (injects sporozoites)

⑧ Mosquito takes a blood meal (ingests gametocysts)

Mosquito stages

⑫ Ruptured oocyst

Release of sporozoites

Inside a mosquito

Macrogametocyte

⑪ Oocyst

⑩ Ookinete

Microgamete entering macrogamete ⑨

Exflagellated microgametocyte

red cells might create a blockage in the circulation system (Cohen and Lambert, 1982).

A merozoite can also develop into a "gametocyte," which is the stage that can infect a mosquito. There are two kinds of gametocytes: males (microgametes) and females (macrogametes). They get ingested by a mosquito. Inside the mosquito's midgut, male and female gametocytes merge into "zygotes," which then develop into "ookinetes." The motile ookinetes penetrate the midgut wall and develop into "oocysts." The cysts eventually release sporozoites, which migrate into the salivary glands where they get injected into humans. The development inside a mosquito takes about two weeks and only after that time can the mosquito transmit the disease. *P. falciparum* cannot complete its life cycle at temperatures below 20 °C (Barnwell, 1999).

SYMPTOMS. Following infected mosquito bite, symptoms usually begin within 10–30 days; in some cases, within 7 days. Malaria can be uncomplicated or severe. Symptoms of uncomplicated malaria might include chills, diarrhea, fever, general discomfort, headache, muscle pain, nausea, sweating, vomiting, and weakness. Other less common manifestations include enlargement of the spleen or liver, high body temperature, increased breathing frequency, mild anemia, mild jaundice, and so on. Symptoms of severe malaria might include breathing difficulties, coma, confusion, death, focal neurologic signs, seizures, and severe anemia.

DIAGNOSIS. Malaria is usually diagnosed by examining the blood sample under a microscope. There are also test kits that detect antigens of *P. falciparum* in the patient's blood. These immunologic tests are known as *rapid diagnostic tests* (*RDTs*). RDTs can detect two different malarial antigens, one for *P. falciparum* and the other is found for all four human malarial species. Antigenic variation is one of the major problems in serological diagnosis (Iqbal et al., 1993; Brannan et al., 1994).

P. falciparum and *P. vivax* have been confirmed to be resistant (in some areas) to many antimalarial drugs (Anderson et al., 2000). For example, chloroquine-resistant strain of *P. falciparum* has spread to most endemic areas (Scherf et al., 1998). The drugs that are usually recommended by national malaria control programs include artemesinin-containing combination treatments (for example, artemether–lumefantrine and artesunate–amodiaquine), atovaquone–proguanil, chloroquine, doxycycline, mefloquine, quinine, and sulfadoxine–pyrimethamine. Primaquine is used as an adjunct against certain *Plasmodium* variants (Rich et al., 2000). It is active against the dormant liver forms and is not recommended for people who are deficient in glucose-6-phosphate dehydrogenase or for pregnant women.

3.7.4.6. Toxoplasma gondii

Toxoplasma gondii causes a disease called *toxoplasmosis*, which is found all over the world. Some estimates suggest that over 30% of human population is infected and the disease is usually asymptomatic. *T. gondii* is known to change the host's behavior. The parasite is shown to make rats fearless near cats. When the rat is eaten by the cat it gets inside the primary host. There have been a few studies with humans, too. There appear to be strong correlation between schizophrenia and toxoplasmosis. There are reports that women with toxoplasmosis are more likely to cheat their husbands. Men with the parasite have shown to be more aggressive.

LIFE CYCLE. The life cycle of *T. gondii* (Fig. 3.41) starts, when oocysts exit the primary host (cat) in the feces. Millions of oocysts are shed for as long as 3 weeks after

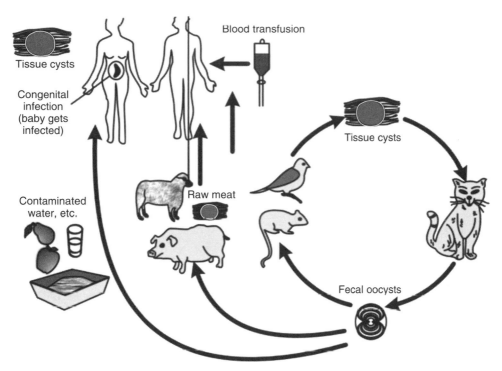

Figure 3.41. Life cycle of *toxoplasma gondii*.

infection. Oocysts sporulate and become infective within a few days in the environment. The oocysts are found only in the feces of domestic and wild cats. Birds, humans, and other intermediate hosts get infected after ingesting water or food contaminated with the cat feces. (The primary hosts, cats, can get infected at this point, too.) In the gut oocysts transform into tachyzoites, which are about 4–8 μm long and 2–3 μm wide. They travel to other parts of the body via bloodstream and further develop into tissue cyst, bradyzoites, in muscle and neural tissues. Cysts are about 5–50 μm in diameter. They are commonly found in skeletal muscles, brain, myocardium, and eyes where they can remain for many decades. If a cat (or a human) eats the intermediate host, the tissue cysts get ingested.

SYMPTOMS. About 10–20% of infected people develop sore lymph nodes, muscle pain, and other minor symptoms that last for several weeks and then subside (acute toxoplasmosis). Symptoms of acute ocular toxoplasmosis include blurred or reduced vision, eye pain, redness of the eye, sensitive to light, tearing of the eyes.

DIAGNOSIS. Toxoplasmosis diagnosis is generally made by serologic tests. Direct observation of the parasite is possible in CSF, stained tissue sections, or other biopsy samples. Prenatal diagnosis is done to monitor mother-to-child transmission. Ocular toxoplasmosis has also been reported. Serologic tests (antibody based) are not reliable in immunosuppressed patients.

Toxoplasmosis can be treated (not completely) with combinations of pyrimethamine with either trisulfapyrimidines or sulfadiazine plus folinic acid in the form of leucovorin calcium to protect the bone marrow from the toxic effects of pyrimethamine. Treatment of persons with ocular disease depends on the size of the eye lesion.

3.7.4.7. Trypanosoma brucei

Sleeping sickness, African trypanosomiasis, is a deadly blood disease caused by two variates of *Trypanosoma brucei* and transmitted by tsetse fly (Osinski, 1980; Aslam and Turner, 1992). *T. brucei rhodesiense* causes East African trypanosomiasis. *T. brucei gambiense* causes West African trypanosomiasis (also known as *Gambian sleeping sickness*). The two subspecies do not overlap in geographic distribution (Barry and Turner, 1991; Barry, 1997). They infect humans and tsetse flies (*Glossina* genus) and less than 1% of tsetse flies carry the parasite. *T. brucei* needs two hosts to live and reproduce. Its life cycle starts, when an infected tsetse fly bites the human skin (Fig. 3.42). While it is feeding on blood, metacyclic trypomastigotes are transmitted to the skin from the salivary glands of the fly. The parasites get into the bloodstream by entering lymphatic or blood vessels. They travel in different body fluids (such as blood, lymphatic, or spinal fluid), transform into bloodstream trypomastigotes, and multiply by binary fission. The disease can be spread by another tsetse fly that drinks the infected blood. Inside the fly the life cycle takes about 3 weeks. Ingested bloodstream trypomastigotes transform into procyclic trypomastigotes in the fly's midgut and multiply. They transform into epimastigotes, migrate to the salivary glands, then transform into metacyclic trypomastigotes, and multiply once again by binary fission.

T. brucei is not killed by the immune system because it has a glycoprotein (VSG) coating that makes its cell membrane very thick and hard to recognize (Barbet and Kamper, 1993). It also changes frequently its structure to always keep ahead of the immune response (Gray, 1965; Augur et al., 1989). *T. b. rhodesiense* and *T. b. gambiense* are indistinguishable under a microscope. A trypomastigote is 14–33 µm long and has a tiny kinetoplast located at the posterior end, a centrally located nucleus, an undulating membrane, and a flagellum. Trypomastigotes are the only stage found in patients (Fig. 3.43). Humans are the main host for *T. b. gambiense*, but it is sometimes found in animals as well. Wild animals are the main reservoir of *T. b. rhodesiense*.

African trypanosomiasis has three symptomatic stages, the last one being the most dangerous eventually leading to death, if left untreated. In 1–3 weeks after the bite, a chancre (a red sore skin lesion) can develop on the bite area. Several weeks or months later, *Trypanosoma* parasites in the blood, spinal, and lymphatic fluid (hemolymphatic stage) can cause anemia, cardiac dysfunction, pruritus (itching), fatigue, fever, headache, muscle and joint pain, renal failure, skin rash, splenomegaly, swelling of the lymph nodes (most prominently in the back of the neck and in the groin), hands and face, thrombocytopenia (low level of platelets, thrombocytes), and weight loss. The disease reaches its final stage when the parasites get through the blood–brain barrier entering the brain. The CNS involvement can occur as early as within a month in some cases. This meningoencephalitic stage (inflammation of the CNS) causes symptoms such as confusion and abnormal behavior, insomnia (sleeping troubles), personality changes, somnolence (extreme fatigue) coma, and death. Diagnosis is made by examining blood, lymph node or tissue aspirates (fluid suction), bone marrow, chancre, or CSF under a microscope.

Treatment is based on symptoms and laboratory results. The drug of choice depends on the infecting species and the stage of infection. Pentamidine isethionate and suramin are usually used for treating the hemolymphatic stage of West and East African trypanosomiasis, respectively. Melarsoprol is used for late disease with CNS involvement (infections by *T.b. gambiense* or *T. b. rhodiense*).

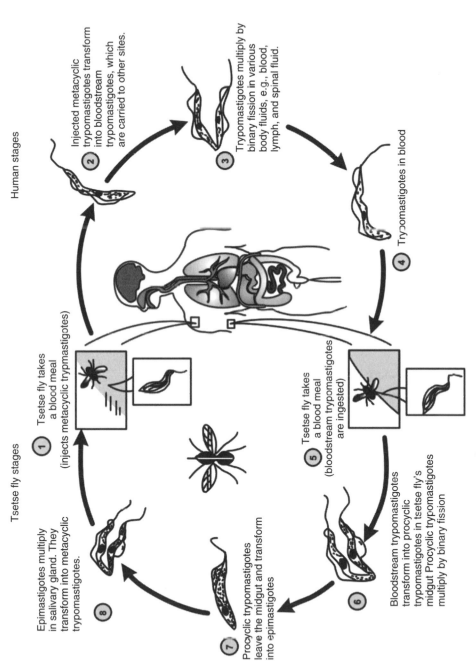

Human stages

② Injected metacyclic trypomastigotes transform into bloodstream trypomastigotes, which are carried to other sites.

③ Trypomastigotes multiply by binary fission in various body fluids, e.g., blood, lymph, and spinal fluid.

④ Trypomastigotes in blood

Tsetse fly stages

① Tsetse fly takes a blood meal (injects metacyclic trypmastigotes)

⑤ Tsetse fly takes a blood meal (bloodstream trypomastigotes are ingested)

⑧ Epimastigotes multiply in salivary gland. They transform into metacyclic trypomastigotes.

⑦ Procyclic trypomastigotes leave the midgut and transform into epimastigotes

⑥ Bloodstream trypomastigotes transform into procyclic trypomastigotes in tsetse fly's midgut Procyclic trypomastigotes multiply by binary fission

Figure 3.42. Life cycle of *Trypanosoma brucei*.

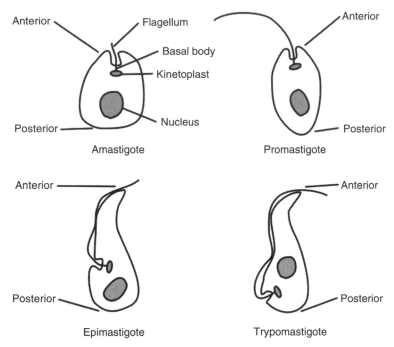

Figure 3.43. Major stages of *T. brucei*.

3.7.5. Skin Parasites

3.7.5.1. Cimex lectularius

Bedbugs are skin parasites that live inside or near beds and feed on blood during the night. The most common bedbug species is *Cimex lectularius*. One bedbug can leave a number of sore red spots, if it has to drill many holes before finding a blood vein. The preferred treatment for an apartment is to use heat. High temperature of at least 45 °C is applied to troubled areas.

3.7.5.2. Dermatobia hominis

The human botfly, *Dermatobia hominis*, belongs to the Oestroidea family. There are about 150 botfly species but only *D. hominis* uses humans as a host. The larvae of these huge hairy flies are parasitic living inside the skin. Human botfly lays eggs on the skin. When a larva inside the egg detects warmth, it hatches and penetrates the skin. It develops deep inside the skin and breaths through a special tube that has an opening at the wound spot. After feeding for 2 months, the botfly comes out and drops off. On the soil it will take another week to become an adult botfly.

3.7.5.3. Sarcoptes scabiei

Sarcoptes scabiei is a skin parasite causing severe itching and infections. There are at least 300 million cases every year worldwide. *S. scabiei* goes through four stages in its life cycle: egg, larva, nymph, and adult. Female *S. scabiei* is 0.25–0.35 mm wide and 0.30–0.45 mm long. Males are about half of that size. The movement of *S. scabiei* and eggs inside the tunnels causes local inflammation. This allergic reaction causes very intense rashes. Scabies spreads easily through skin-to-skin contact. Diagnosis is done by finding mites or their tunnels. Living mites can be identified under a microscope. Treatment includes use of Sulfuric soap for a few days. To get relief from itching during the treatment period antihistamines such as chlorpheniramine is used.

3.8. FUNGAL PATHOGEN

Fungi are parasites or saprophytes and human fungal infections are common and generally mild. However, in very sick or otherwise immune suppressed people, fungi can sometimes cause serious diseases. Fungi are identified and classified according to their appearance by microscopy and in culture, and by the method of reproduction, which may be sexual or asexual. Growing fungi have branched filaments called *hyphae*, which make up the mycelium. Some fungi are compartmented by septae. Yeasts form a subtype of fungus characterized by clusters of round or oval cells. These bud out similar cells from their surface to divide and propagate. The human fungal infections may be categorized into superficial, subcutaneous, and systemic fungal infections.

3.8.1. Superficial Fungal Infections

These affect the outer layers of the skin, the nails, and hair (Fig. 3.44). The main groups of fungi causing superficial fungal infections are dermatophytes, yeasts, and molds. The dermatophytes are so well adapted to living on human skin that they provoke minimal inflammatory reaction in the host, for example, *Tinea rubrum, Trichophyton interdigitale, Trichophyton tonsurans, Microsporum audouinii, Trichophyton violaceum, Microsporum ferrugineum, Trichophyton schoenleinii*, and *Trichophyton megninii*. The presence of a dermatophyte infection is confirmed by microscopy and culture of skin scrapings.

Candida refers to a group of yeasts that commonly infect the skin. Candida is a normal inhabitant of the human digestive tract from early infancy, where it lives without causing any disease most of the time. However, in immunocompromised subjects, the organism can cause infection of the mucosa (the lining of the mouth, anus, and genitals), the skin, and rarely, deep-seated infection. The most common *Candida* (*C*) species to result in candidiasis is *C. albicans, C. tropicalis, C. parapsilosis, C. glabrata*, and *C. guilliermondii*.

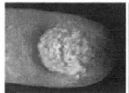 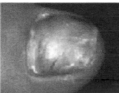  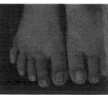

Figure 3.44. (a) Candida nail infection, (b) nail infection due to *Microsporum canis* (rare), (c) *Aspergillus flavus*, (d) *Trichophyton rubrum*.

Microscopy and culture of skin swabs and scrapings are the diagnostic approaches for candidal infections. Candida can live on a mucosal surface.

Mold infections can occasionally infect the skin and nails (Fig. 3.44) and cause indolent infections in immunocompromised individuals, especially the elderly. Mold infections originate from soil. The responsible organisms include *Scopulariopsis brevicaulis, Fusarium* spp., *Aspergillus* spp., *Alternaria* spp., *Acremonium* spp.*Scytalidinum dimidiatum (Hendersonula toruloides)*, and *Scytalidinium hyalinum*. They can be mild or severe. *Scopulariopsis brevicaulis* and *Scytalidinum dimidiatum* are the most likely causes of skin infection.

3.8.1.1. *Aspergillus* sp

Mold infections of the finger and toenails can be indistinguishable from other types of onchomycoses. However, unlike dermatophyte infections, molds frequently result in paronchia (inflamed nail folds). One or more toenails may be infected, or the mold may simply be a contaminant. The surrounding skin is often dry and may itch. The appearance of the nail may include brownish dull discoloration of the nail, which starts at one edge, streaked and pitted nail plate, or complete nail destruction. Mold infections are diagnosed by microscopy and culture of skin scrapings and/or nail clippings. Mold nail infections are notoriously difficult to clear with currently available medications, which may be required for longer courses or in combination with other topical and oral antifungal agents.

3.8.1.2. *Chromoblastomycosis*

Chromoblastomycosis is a chronic fungal infection in which there are raised crusted lesions affecting the skin and subcutaneous tissue. It usually affects the limbs. Chromoblastomycosis may be due to several fungi found in soil, wood, and decaying plant material. The most common organisms are *Phialophora verrucosa, Fonsecaea pedrosi, Fonsecaea compacta, Cladosporium carrionii*, and *Rhinocladiella aquaspersa*.

Chromoblastomycosis generally presents as a single lesion on an exposed site such as the foot or hand. It grows very slowly: only about 2 mm per year. Eventually, a warty dry nodule or plaque develops (Fig. 3.45). It may cause no discomfort but is frequently very itchy. Microscopy and culture of scrapings or pus swabs suggest the diagnosis. There may be typical thick-walled dark-brown cells on skin biopsy confirming the presence of a dematiaceous fungus. It is dark colored due to melanin on the walls of the organism. Clusters of characteristic thick-walled brown "sclerotic" (hard) cells are seen on microscopy. Treatment is difficult and prolonged. It may include posaconazole or voriconazole, flucytosine, and thiobendazole.

3.8.2. Systemic Fungal Infections

Systemic mycoses are fungal infections affecting internal organs. In the right circumstances, the fungi enter the body via the lungs, through the gut, paranasal sinuses, or skin. The fungi can then spread via the bloodstream to multiple organs including the skin, often causing multiple organs to fail and eventually resulting in the death of the patient. Patients who are immunocompromised are more predisposed to systemic mycoses, but

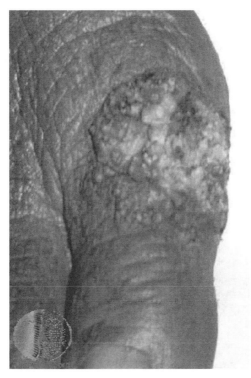

Figure 3.45. Chromoblastomycosis.

they can develop in otherwise healthy patients. Systemic mycoses can be split between two main varieties, endemic respiratory infections and opportunistic infections.

Fungi that can cause systemic infection in people with normal immune function as well as those who are immunocompromised, include *Histoplasma capsulatum* (histoplasmosis), *Coccidioides immitis* (coccidioidomycosis), *Blastomyces dermatitidis, Paracoccidioides brasiliensis* (paracoccidiodomycosis), and *Penicillium marneffei* (penicilliosis). Fungi that result in systemic infection only in immunocompromised or sick people include *Candida* species (Candidiasis), *Aspergillus* species (Aspergillosis, Fig. 3.46), *Cryptococcus* (Cryptococcosis), and *Zygomycetes* (Zygomycosis). These fungi are found in or on normal skin, decaying vegetable matter, and bird droppings, but not exclusively. They are present throughout the world.

The clinical features of the illness depend on the specific infection, and fungal infections in people with normal immune function may result in very minor symptoms or none at all. General symptoms may include fever, cough, and loss of appetite.

3.8.2.1. Lungs

Invasive lung mycoses typically result in a progressive dry cough, shortness of breath, pain when taking a deep breath, and fever. These symptoms may progress to the point of life threatening acute respiratory distress syndrome. Hemoptysis (the coughing up of blood) is also sometimes seen, particularly if inflammation of the large airways is present.

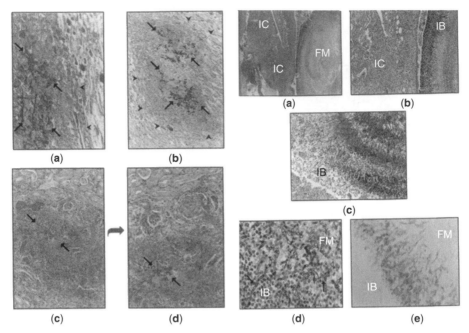

Figure 3.46. Systemic infection of *Aspergillus fumigates*: (1) Periodic acid Shiff staining of positive fungal hyphae in kidney. (2) Fungal mass (FM) in pelvis of ureter, IB denotes peripheral area of fungal mass, and IC denotes inflammatory catico-medullary area (stains used are haematoxylin and eosin, Celestin Blue, antibody against crude metabolic antigens of *Aspergillus fumigatus*; Raghuwanshi and Bisen, 2002). (*See insert for color representation of the figure*).

3.8.2.2. Bone

Bone infection can develop and spread through the blood or rarely via direct spread from an overlying ulcer and infected skin. Fever and pain in the affected bone are the cardinal symptoms. This would usually be investigated with X rays, CT scanning, and aspiration (suction for sampling) via a needle to obtain a specimen for culture.

3.8.2.3. Brain

Systemic mycoses with brain involvement are associated with high mortality. Symptoms that are suggestive of brain involvement include headache, seizures, and deficits in normal brain control over movement or sensation. These are confirmed by CT or MRI brain scans. Inflammation of the brain (meningitis) may be seen, particularly with candidiasis and cryptococcosis. Symptoms usually include stiff neck and irritability.

3.8.2.4. Eyes

Almost any of the eye structures may be infected by mycoses. *Candida* and *Cryptococcus* are the most commonly infecting mycoses. The symptoms depend on which part of the eye is infected, but may include visual blurring, dark or black images floating in the visual field, pain, and a red eye.

3.8.2.5. Skin

Skin changes of wider degree may be seen in systemic mycoses and the nature of the skin lesions depend partly on the nature of fungus involved.

The most reliable tests to confirm an infection are skin biopsy for histological analysis and fungal culture. Systemic antifungal medications include amphotericin b, fluconazole, itraconazole, voriconazole, and caspofungin. The prognosis depends on the patient's immune function and the extent of infection when treatment is started. Unfortunately, despite treatment, many patients die of their infection.

3.8.2.6. Inhaled Fungal Infection

Although uncommon, some may infect healthy individuals. The result is most often a mild infection and long lasting resistance to further attack, but occasionally these infections are more serious and chronic (especially in the immune suppressed). The systemic fungal infections include histoplasmosis. Histoplasmosis mainly affects the lungs with most patients often showing minimal or no symptoms. There are several clinical presentations of histoplasmosis.

Laboratory and radiological studies are performed to confirm the diagnosis of histoplasmosis.

> *Sputum cultures*—positive yields found in 10–15% of patients with acute pulmonary histoplasmosis and in 60% of patients with chronic pulmonary histoplasmosis.
>
> *Blood cultures*—positive results found in 50–90% of patients with progressive disseminated histoplasmosis.
>
> *Serologic testing* may indicate spreading of the disease.
>
> *Chest X ray and CT scanning.*
>
> *Tissue biopsy.*

Amphotericin B is an effective drug for severe cases of acute pulmonary histoplasmosis, chronic pulmonary histoplasmosis, and all forms of disseminated histoplasmosis.

3.8.2.7. Opportunistic Infections

The "opportunists" that infect immunocompromised include *Aspergillus* sp. (aspergillosis), fungi of the order Mucorales (mucormycosis), *Cryptococcus neoformans* and *Cryptococcus gattii* (cryptococcosis), *Trichosporon beigelii*, and *P. boydii* (Romani, 2011).

3.9. MICROBIAL DIAGNOSTICS

3.9.1. Immunodiagnostic Methods

Immunodiagnosis is based on the detection of antigen or antibody or both either in body fluids or tissues. Several techniques are available for detection of antigens and antibodies. They include low sensitive and high specific techniques such as precipitation-based tests, agglutination-based tests, fluorescent-marker-based tests, and enzyme immune assays.

3.9.1.1. Eastern Blotting

A unique liposome-based immunodiagnostic kit based on immunoblotting of antiglycolipid antibodies and glycolipid antigen in patients' body fluids, for early diagnosis of pulmonary as well as extrapulmonary tuberculosis caused by *M. tuberculosis* strain H37Rv (Fig. 3.47) has been developed first of its kind by employing a fourth-generation blotting "eastern blotting" (Bisen et al., 2003; Tiwari et al., 2005; Bisen and Tiwari, 2011a,b). US patent 7,888,037 B2 (February 2011); Japanese patent 4,601,628 (2006-553765) (January 2011).

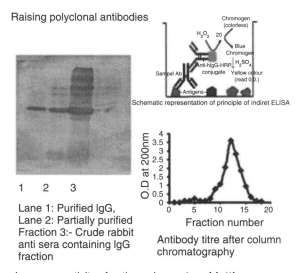

Raising polyclonal antibodies

Schematic representation of principle of indiret ELISA

- Polyclonal antibodies were rasied against the glycolipid antigens in rabbits.
- Priming was done with Ag + CFA.
- Booster dose was done with Ag + IFA.
- The antibody titre was assayed by ELISA technique.

1   2   3

Lane 1: Purified IgG,
Lane 2: Partially purified
Fraction 3:- Crude rabbit anti sera containing IgG fraction

Antibody titre after column chromatography

Immunoreactivity of antigens by **eastern blotting**

- The antibody raised was incubated on the glycolipid chromatographed TLC and followed by washing with PBS and incubated with anti rabbit IgG (HRP) conjugated.

- Substrate (DAB+H_2O_2) was added and bands were visualized.

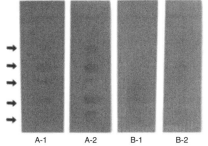

A-1   A-2   B-1   B-2

A-1. Immunoreactivity with sera of active pulmonary TB on aTLC plate;
A-2. Immunoreactivity with extra - pulmonary
B-1, Immunoreactivity BCG vaccinated
B-2, Immunoreactivity BCG unvaccinated healthy humans for cross reactivity study.

Figure 3.47. Immunoblotting and raising polyclonal antibodies against *Mycobacterium tuberculosis*. (*See insert for color representation of the figure*).

3.9.1.2. Agglutination Tests

The interaction between particulate antigens and specific antibodies generates agglutination, which is visible to the naked eye. The antibodies interacting with particulate antigens are generally called *agglutinins*. Pentavalent IgM, due to its high valence, is a good agglutinin than other classes of antibodies. Agglutination tests can be used either in a qualitative manner or in a semiquantitative manner to assay for the presence of an antigen or an antibody. The antibody is mixed with the particulate antigen and a positive test is indicated by the agglutination of the particulate antigen. For example, a patient's RBCs can be mixed with antibody to a blood group antigen to determine a person's blood type.

For semiquantitative purposes, serial dilutions of the serum are employed and then a fixed number of RBCs or bacteria or other such particulate antigens are added. Then the highest dilution of the serum that gives visible agglutination is determined. The reciprocal of the highest dilution of serum that gives visible agglutination is called *antibody titer*.

Agglutination is at its maximum when antibody and antigen are at optimal concentrations. At extremely high concentrations or extremely low concentrations of antibody, no agglutination will be visualized and this is due to prozone effect. Lack of agglutination in the prozone is due to excessive antibodies resulting in very small complexes that do not clump to form visible agglutination.

3.9.1.3. Precipitation Tests

Precipitation tests can be used either for qualitative detection or semiquantitative detection of an antigen or an antibody. The precipitation tests are categorized as follows:

Double Immunodiffusion. Antigens and antibodies placed in adjacent wells of an agar gel diffuse radially and form precipitin lines when they interact with each other. If more than one precipitin line appears, it indicates the presence of more than one antigen/antibody interactions. The application of this technique includes qualitative detection of antigens or antibodies, determination of antibody titer in immune sera, and study of cross-reactivity or homology between antigen preparations.

Radial Immunodiffusion. Radial immunodiffusion is commonly used for the determination of immunoglobulin levels in patient samples. In this test, the anti-antibody is embedded in the gel bed and different dilutions of the serum being studied are placed in holes punched into the antibody containing agar. As the serum immunoglobulin diffuses into the gel, it reacts with the anti-antibody and when the equivalence point is reached a ring of precipitation is formed. The diameter of the ring is proportional to the log of the concentration of antigen, since the amount of antibody is constant. Thus, by running different concentrations of standard immunoglobulins one can generate a standard cure from which one can quantitate the amount of a given immunoglobulin in an unknown sample. Thus, this is a quantitative test.

Immunoelectrophoresis. Immunoelectrophoresis is used for the qualitative analysis of complex mixtures of antigens. In this technique, a complex mixture of antigens is electrophoresed so that the antigens get separated according to their charge. After electrophoresis, a trough is cut in the gel (parallel to electrophoretic run) and antibodies are added. As the antibodies diffuse into the agar, precipitin lines are produced in the form of arc.

Countercurrent Electrophoresis. In this test the antigen and antibody are placed in wells punched out of an agar gel, and the antigen and antibody are electrophoresed into each other where they form a precipitation line. This test only works if the agar bed possesses endosmotic property. This test is similar to that of double immunodiffusion, and the major advantage is its speed.

3.9.1.4. Enzyme Immunoassay

The EIA methods hold great promise for application in diagnosis of a wide variety of pathogens. In this assay, one of the components (antigen or antibody) is adsorbed onto a solid surface (polyvinyl chloride or polystyrene). This is then used to "capture" the relevant antigen or antibody in the test solution and the complex is detected by means of an enzyme-labeled antibody or antigen. The degradation of the enzyme substrate, measured photometrically, is proportional to the concentration of the unknown "antibody" or "antigen" in the test solution. The EIAs are highly sensitive but low specificity techniques. These assays employ an enzyme as a marker. The basic prerequisite is the enzyme should generate a colored product, which is quantifiable. The assay demands removal of unbound antigen or antibody from antigen–antibody complexes. The enzymes that are commonly used as labels include peroxidase and alkaline phosphatase. The EIAs are categorized on the basis of labeling component (antigen or antibody). These techniques detect antigens derived from pathogens or specific antibodies mounted against pathogens at nanogram/pictogram levels.

Competitive ELISA. In this assay, the antigen is labeled; by using known amounts of enzyme-labeled standard antigen, one can generate a standard curve relating enzyme activity to bound antigen. From this standard curve, one can determine the amount of an antigen in an unknown sample. The amount of labeled antigen required to saturate all antigen binding sites of antibodies (fixed quantity) is determined first. In the presence of a test sample containing specific antigen, the amount of standard (labeled) antigen displaced is worked out, which is directly proportional to the antigen present in the test sample.

Indirect ELISA. Noncompetitive ELISA is used for the measurement of antibodies. The solid surface is coated with the antigen and is used for the detection of antibody in the unknown sample. Following incubation with test sample, the amount of labeled second antibody (anti-antibody) bound is related to the amount of antibody in the unknown sample. This assay is commonly employed for the measurement of antibodies of the IgE class or any class directed against particular allergens or antigens. The amount of labeled second antibody that binds is proportional to the amount of antigen that bound to the first antibody.

Sandwich ELISA. This assay is used for detection of antigen. Specific antibody is first immobilized on the surface. The antigen in the test sample bound to immobilized antibody is detected by adding first antibody coupled to an enzyme. The antigen is sandwiched between two antibodies.

3.9.1.5. Flow Cytometry

Flow cytometry is commonly used in the clinical laboratory to identify and enumerate cells bearing a particular antigen. Cells in suspension are labeled with a fluorescent tag.

The cells are then analyzed on the flow cytometer. In a flow cytometer, the cells exit a flow cell and are illuminated with a laser beam. The amount of laser light that is scattered off the cells as they pass through the laser can be measured, which gives information concerning the size of the cells. In addition, the laser can excite the fluorochrome on the cells and the fluorescent light emitted by the cells can be measured by one or more detectors. In a one-parameter histogram, increasing amount of fluorescence (e.g., green fluorescence) is plotted on the x axis and the number of cells exhibiting that amount of fluorescence is plotted on the y axis. The fraction of cells that are fluorescent can be determined by integrating the area under the curve. In a two-parameter histogram, the x axis is one parameter (e.g., red fluorescence) and the y axis is the second parameter (e.g., green fluorescence). The number of cells is indicated by the contour and the intensity of the color.

3.9.1.6. Western Blotting

Blotting refers to the transfer of biological samples from a gel to a membrane and their subsequent detection on the surface of the membrane. Western blotting (also called *immunoblotting*) is now a routine laboratory technique for protein analysis. The first step in a Western blotting technique is separation of the macromolecules using polyacrylamide gel electrophoresis (PAGE). After electrophoresis, the separated proteins are transferred or blotted onto a nitrocellulose or polyvinylidene difluoride (PVDF) membrane electrophoretically. Electrophoretic transfer of proteins involves placing a protein-containing polyacrylamide gel in direct contact with a piece of nitrocellulose or other suitable membrane and "sandwiching" this between two electrodes submerged in a conducting solution. When an electric field is applied, the proteins move out of the polyacrylamide gel and onto the surface of the membrane, where the proteins become tightly bound. The result is a membrane with a copy of the protein profile that was originally in the polyacrylamide gel.

The membrane is then blocked to prevent any nonspecific binding of antibodies to the surface of the membrane. A variety of blocking buffers ranging from milk or normal serum to highly purified proteins are available to block free sites on a membrane. The transferred protein is made to react with an enzyme-labeled antibody as a probe. An appropriate insoluble substrate (DAB for peroxidase) is then added to the membrane, which results into a chromogenic precipitate on the membrane (Raghuwanshi and Bisen, 2002). The most sensitive detection methods use a chemiluminescent substrate that, when combined with the enzyme, produces light as a by-product. The light output can be captured by appropriate means for chemiluminescent detection. Alternatively, fluorescently tagged antibodies can also be used, which are directly detected with the aid of a fluorescence imaging system. Whatever system is used, the intensity of the signal should correlate with the abundance of the antigen on the membrane (Figs. 3.48 and 3.49).

In the indirect detection method, a primary antibody is added first to bind to the antigen. This is followed by a labeled secondary antibody that is directed against the primary antibody. Labels include biotin, fluorescent probes such as fluorescein or rhodamine, and enzyme conjugates such as horseradish peroxidase or alkaline phosphatase. The indirect method offers many advantages over the direct method, the primary one includes amplification of signal.

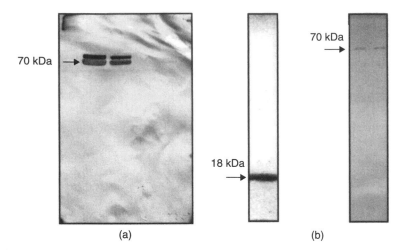

Figure 3.48. (a) Western blot showing production of antibodies raised against doublet of 70–72 kDa protein in rabbit. (b) Western blot showing production of antibodies against 18 and 70 kDa proteins in *Aspergillus fumigatus* antigens (Raghuwanshi and Bisen, 2002). (*See insert for color representation of the figure*).

Monoclonal antibodies directed against the major subtypes of the influenza virus, as well as the various serotypes of *Salmonella*, are commonly used in speciation. Specific antigenic proteins may be detected by antibodies directed against these proteins in immunoblot methods.

3.9.1.7. Tests for Cell-Associated Antigens

IMMUNOFLUORESCENCE. Immunofluorescence is a technique whereby an antibody labeled with a fluorescent molecule (fluorescein or rhodamine or one of many other fluorescent dyes) is used to detect the presence of an antigen in or on a cell or tissue by the fluorescence emitted by the bound antibody (Fig. 3.50).

> *Direct Immunofluorescence*. In direct immunofluorescence, the antibody specific to the antigen is directly tagged with the fluorochrome.
>
> *Indirect Immunofluorescence*. In indirect immunofluorescence, the antibody specific to the antigen is unlabeled and a second anti-immunoglobulin antibody directed toward the first antibody is tagged with the fluorochrome. Indirect fluorescence is more sensitive than direct immunofluorescence, since there is amplification of the signal.

3.9.2. Molecular Diagnostic Approaches

Molecular diagnostics has become an indispensable tool in clinical laboratories for early and sensitive diagnosis of microbial disorders. The application of molecular testing methods in the clinical laboratory has dramatically improved our ability to diagnose microbial diseases. Microbial DNA/RNA extracted from a clinical specimen may be analyzed

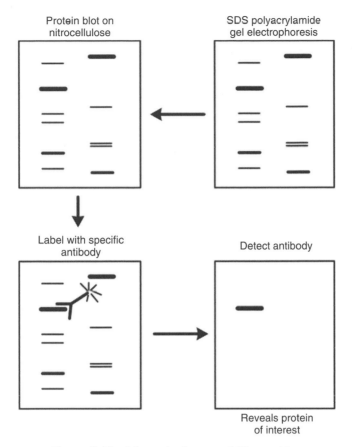

Figure 3.49. Schematic diagram of Western blot.

for the presence of various organism-specific nucleic acid sequences regardless of the physiological requirements or viability of the organism. Molecular methods hold more significance in instances of limited specimen volume. Even from low volume specimens, enough DNA/RNA can be made to allow performance of numerous molecular assays. The basic application is to determine changes in sequence or expression levels in crucial genes involved in a disease. Transciptomics (the study of gene expression), proteomics (the study of protein expression), and metabolomics hold great deal of interest in understanding a disease process and development of diagnostics (Figs. 3.51 and 3.52). Nucleic acid techniques such as plasmid profiling, various methods for generating restriction fragment length polymorphisms, and the PCR are making increasing inroads into clinical microbiology laboratories (Debnath Mousumi and Bisen, 2010). Some of the techniques are discussed below:

1. Plasmid profile analysis was among the earliest nucleic-acid-based techniques applied to the diagnosis of infectious diseases, and has been proved to be useful in numerous investigations. This method has also been widely utilized for tracking antimicrobial resistance.

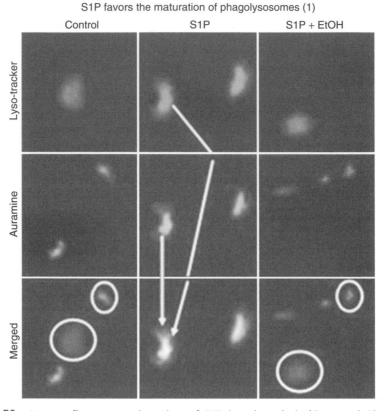

Figure 3.50. Immunoflouresence detection of S1P in tuberculosis (Garg and Bisen, 2004). (*See insert for color representation of the figure*).

2. The PCR facilitates several-fold amplification of extremely small quantities of DNA. PCR technology facilitates the detection of DNA or RNA (or both) of pathogenic organisms and, as such, is the basis for a broad range of clinical diagnostic tests for detection of a wide variety of pathogens. Detection of infectious organisms and detection of genetic variations, including mutations, in human genes are two major areas of applications in relation to microbes. PCR-based diagnostic tests are reported for detecting and/or quantifying several microbial pathogens, namely, HIV-1, HBV and HCV, HPV, *C. trachomatis, N. gonorrhoeae*, cytomegalovirus, and *M. tuberculosis*. PCR-based tests have several advantages over serological methods, particularly in terms of sensitivity. Earlier detection of infection can mean earlier treatment and an earlier return to good health.

3. Restriction enzyme pattern facilitates application of the procedure in identification of various microbial pathogens (Xu et al., 1999). This technique is widely used in studying the tuberculosis osocomial outbreak in HIV-positive populations.

4. Ribotyping has been shown to have both taxonomic and epidemiological value. Restriction patterns can be obtained by hybridizing Southern-transferred DNA fragments with labeled bacterial ribosomal operon(s), which encode for 16S and/or 23S rRNA. Ribotyping assays have been used to differentiate bacterial strains in

dTHP1 proteome

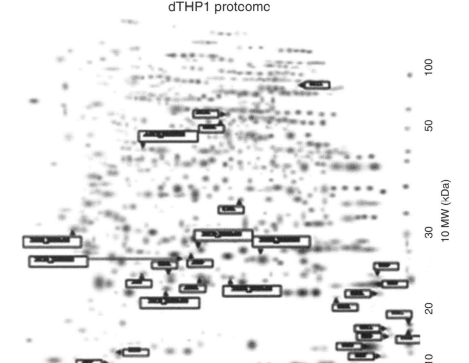

Figure 3.51. Proteomic approach in S1P analysis in tuberculosis (Garg and Bisen, 2004).

different serotypes and to determine the serotype(s) most frequently involved in outbreaks.

5. RAPD can be used to differentiate strains within species, various serotypes within strains/species, and subtypes within a given serotype. It is, therefore, useful for determining whether two isolates of same species are epidemiologically related or not.

6. Phage analysis may be used to type bacterial strains within a given species. However, this approach is labor-intensive and requires the maintenance of bacteriophage panels for a wide variety of bacteria in the laboratory. Moreover, bacteriophage profiles may fail to identify isolates, are often difficult to interpret, and may give poor reproducibility.

7. Another commonly used molecular diagnostic method is RT-PCR. It is the most sensitive technique that is currently available for mRNA detection and quantitation. Compared to the two other commonly used techniques for quantifying mRNA levels, northern blot analysis and RNase protection assay, RT-PCR, can be used to quantify mRNA levels from much lesser samples. In fact, this technique is sensitive enough to enable quantitation of RNA from a single cell. Detection of PCR products on a real-time basis has led to widespread adoption of real-time RT-PCR as the

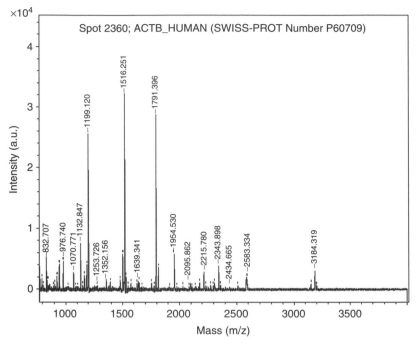

Figure 3.52. MALDI analysis of S1P in tuberculosis (Garg and Bisen, 2004).

method of choice for quantitating changes in gene expression. Investigations have been successfully employed in cancer biology (Fig. 3.53) using immunohistochemistry and reverse transcriptase-polymerase in oral cancer tissue showing frequent overexpression of survivin and p53 in oral squamous cell carcinoma (OSCC) and oral premalignant lesions (Khan and Bisen, 2007; Khan et al, 2009, 2010a, 2010b, 2011).

8. DNA microarrays offer the latest technological advancement for multigene detection and diagnostics. They were conceived originally to examine gene expression for large numbers of genes, but have also been applied to genotyping and diagnostics. It can be used to distinguish between DNA sequences that differ by as little as a single nucleotide polymorphism (SNP). DNA microarrays can be used to detect multiple pathogens based on differences in 16S rDNA sequences. For example, nucleic acids can be extracted from a sample and 16S rDNA sequences can be amplified by PCR using universal 16S primers. The resulting PCR products can be hybridized to an array consisting of many oligonucleotide probes, which can be designed to detect and characterize pathogens by taxonomy (e.g., Gram type), by genus, or by species, if sufficient discriminatory sequences are available.

9. Biosensors as discussed earlier in the book, is also used as a diagnostic tool. Electrochemical DNA biosensors exploit the affinity of ssDNA for complementary strands of DNA, and are used in the detection of specific sequences of DNA with a view toward developing portable analytic device. Electrochemical DNA biosensor is proposed as a fast and easy screening method of specific DNA sequences in human, viral, and bacterial nucleic acids. Sequence-specific electrochemical biosensing of

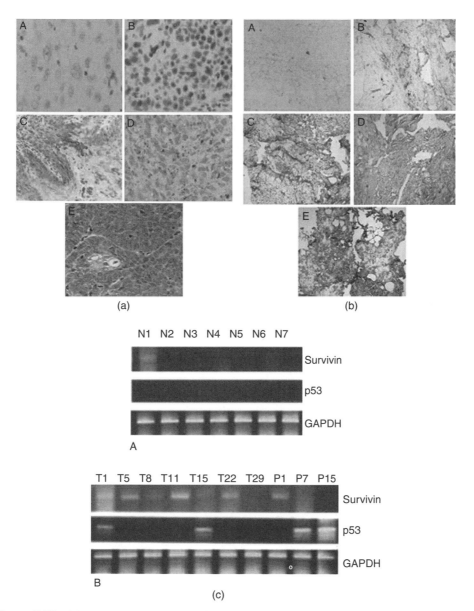

Figure 3.53. (a) Representative results of immunohistochemical staining for oral squamous cell carcinoma (OSCC) survivin expression. (b) Representative results of immunohistochemical staining for p53. (c) Representative RT-PCR results of survivin and p53 expression. (*See insert for color representation of the figure*).

M. tuberculosis DNA for detection and discrimination of herpes simplex I and type II viruses from PCR amplified DNA is already available.

10. Also discussed earlier in detail is another tool "proteomics," a technology to analyze gene products, proteins, on a large scale. Proteomics uses various technologies such as one- and two-dimensional gel electrophoresis, X-ray crystallography, and nuclear magnetic resonance, MALDI-TOF, and so on. Mass-spectrometry-based

proteomics is an indispensable tool for systems biology. One of the most promising developments to come from the study of microbial genome sequences and proteins has been the identification of potential new drugs.

The antigenic epitopes carried to the immune system on the surface of liposomes, grafted with PEG, not only showed very high immune response but also prolonged the persistence of immune response considerably against HIV infection. Free antigenic epitopes, control liposomes without epitopes, and PEG grafted liposomes without epitopes did not develop antibodies to the antigenic epitope (Fig. 3.54). These stealth liposomes

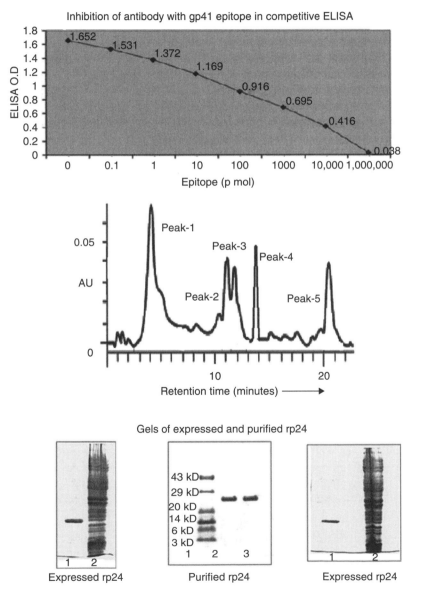

Figure 3.54. Strategy for raising HIV-1 serodiagnostic ability and immunogenicity of synthetic gp41 epitope of HIV-1.

open the door to the researchers to employ this concept to develop more effective vaccines (Singh and Bisen, 2006; Singh et al., 2007).

A good example of this is the identification of new drugs to target and inactivate the HIV-1 protease. Proteomics facilitates identification of specific protein biomarkers facilitating diagnosis of the disease (Windle et al., 2010).

It is clear that the molecular approaches allow for the rapid and accurate identification of the etiologic agent, identification of drug-resistant pathogens in a time substantially shorter than traditional methods. This facilitates earlier initiation of a focused antimicrobial regimen(s) and decreases the likelihood of disease progression.

3.10. FUTURE CHALLENGES: PROMISES OF PHARMACOGENOMICS

Microbes outnumber all other species on the planet and most microbes do not cause disease. Many of the microorganisms, for example, the typhoid bacillus, gonococcus, tubercle bacillus, and treponema of syphilis, are adapted exclusively to humans; while others, like *Salmonella typhimurium*, can cause disease in humans, animals, birds, and reptiles. The major distinction between commensal, opportunistic, and pathogenic microbes is that pathogenic microbes have evolved ability to breach anatomic barriers that generally restrict other microbes. Although lot of advances have been made in understanding the virulence factors of microbial pathogens, host-microbial interactions, and pathogenesis of many of the microbial infections, we have to go a long way in understanding the basis of drug resistance and means to tackle it. Prevalence of multidrug-resistant strains of *M. tuberculosis* in the community poses a serious challenge to public health authorities. Also, extremely alarming is the progressive emergence of drug resistance to vancomycin, the last resort for treating nosocomial infections with enterococci and staphylococci resistant to the usual antibiotics. This situation, together with the constant emergence of new infectious agents, and the reemergence of others that were thought to have been banished forever, constitutes a clear threat that infectious diseases will once again become the principal cause of death and morbidity. The only way in which we can successfully deal with this situation is to move from empiricism to rationalism in the design and development of new weapons in the war against infection. Availability of a suitable laboratory test for early diagnosis of microbial disease facilitates containment of disease at individual and community level. Diagnostics significantly affect therapeutic management of human diseases. There are a myriad of ways diagnostics influence patient care: assessing disease risk sooner; diagnosing disease earlier, long before symptoms occur; targeting disease more specifically, with often less invasive treatments; estimating prognosis more accurately and managing chronic disease more effectively. Identification of the molecular mechanisms of host–pathogen interaction holds promise for significant developments in the identification of new therapeutic strategies to combat infectious diseases, a process that will also benefit parallel improvements in molecular diagnostics, biomarker identification, and drug discovery. With the advent of PCR, the molecular testing field has been dominated by technology platforms sprung from this basic invention, which played a pivotal role in human genome sequencing. Later, the invention of real-time PCR, development of novel PCR-alternative methods such as nucleic-acid-sequence-based amplification (NASBA), transcription-based amplification (TMA), branched DNA method, strand displacement amplification, and isothermal amplification methods contributed significantly to sensitivity and rapidity of diagnosis. Thus, the addition of molecular detection methods to the microbiology laboratory has resolved many of the problems, particularly those associated with limitation of diagnostic material, sensitivity, and specificity of diagnosis. Despite enormous advances in the detection of infectious agents by amplification methods, there are also limitations that must be addressed. The assays should be validated in terms of analytic sensitivity and specificity, and clinical sensitivity and specificity since nonspecificity is one of the major problems encountered with PCR-based diagnostics. Also, the sensitivity of molecular tests could lead to potential cross-contamination or assay inhibition from sample preparation. The method and timing of sampling and the source of the specimen also have a huge impact on results. The inherent high cost of molecular diagnostic tests is one of the acceptance barriers, especially when the outgoing method is inexpensive.

The completion of the Human Genome Project has dramatically increased the potential of pharmacogenomics to individualize antimicrobial therapies based on genetic constitution. The current therapeutic drug monitoring guided by microbial genotyping and phenotyping appears to be significantly enhanced by incorporation of host genomics. Pharmacogenomics will provide a powerful tool to investigate variable responses to antimicrobial therapies. Pharmacogenomics focuses on differences among several drugs or compounds with regard to a generic set of expressed or nonexpressed genes. The genetic factors of the individual have been reported to have a significant impact on the efficacy of the antimicrobial drug following administration *in vivo*. Anti-TB drug (ATD)-related hepatotoxicity is a worldwide serious medical problem among TB patients. Association of genetic polymorphisms in drug-metabolizing enzymes with anti-TB drugs has been reported. Apart from acting on the bacteria, isoniazid, the principal ATD, is also metabolized by human enzymes to generate toxic chemicals that might cause hepatotoxicity. It has been proposed that the production and elimination of the toxic metabolites depends on the activities of several enzymes, such as N-acetyl transferase 2 (NAT2), cytochrome P450 oxidase (CYP2E1), and glutathione S-transferase (GSTM1). There is now evidence that DNA sequence variations or polymorphisms at these loci (NAT2, CYP2E1, and GSTM1) could modulate the activities of these enzymes and, hence, the risk of hepatotoxicity. Since the prevalence of polymorphisms is different in worldwide populations, the risk of ATD hepatotoxicity varies in the populations. Thus, the knowledge of polymorphisms at these loci, before medication, may be useful in evaluating risk and controlling ATD hepatotoxicity.

Pharmacogenomics will also provide the foundation for antiretroviral treatment individualization based on race/ethnicity and gender. Therapeutic management of HIV infection has also been characterized by substantially differing response rates and adverse effects among AIDS patients. To date, few antiretrovirals appear to have a clear genotype–phenotype correlation. However, such correlations have been demonstrated for CYP2B6 and efavirenz disposition, HLA-B5701 and abacavir hypersensitivity, and UGT1A1 and atazanavir hyperbilirubinemia. Socially, the genetic profiling of pharmacogenomics also generated public anxiety over ethics concern of privacy and discrimination. Although progresses have been made in recent years to address some of the above issues, it is a long way before these molecular diagnostics penetrate tertiary health centers. The combination of pharmacology with genomics allows physicians to predict the efficacy of an antimicrobial drug response based on a person's genetic makeup, that is, tailored or personalized medicine. The science of pharmacogenomics is still in its infancy and the clinical implications of pharmacogenomic advances need to be interpreted with caution until further confirmation is available.

REFERENCES

Agarwal A, Sarkar S, Nazbal C, Balasundaram G, Rao KVSB. 1996. Cell responses to a peptide epitope. I. The cellular basis for restricted recognition. *J Immunol* **157**, 2779–2788.

Ahmed R, Gray D. 1996. Immunological memory and protective immunity: understanding their relation. *Science* **272**, 54–60.

Alamgir ASM, Matsuzaki Y, Hongo S, Tsuchiya E, Sugawara K, Muraki Y, Nakamura K. 2000. Phylogenetic analysis of influenza C virus non-structural (NS) protein genes and identification of the NS2 protein. *J Gen Virol* **81**, 1933–1940.

Alcami A, Koszinowski UH. 2000. Viral mechanisms of immune evasion. *Trends Microbiol* **8**, 410–418.

Almarri A, Batchelor JR. 1994. HLA and hepatitis B infection. *Lancet* **344**, 1194–1195.

Altman JD, Moss PAH, Goulder PJR, Barouch DH, McHeyzer-Williams MG, et al. 1996. Phenotypic analysis of antigen-specific T lymphocytes. *Science* **274**, 94–96.

Anderson TJC, Haubold B, Williams JT, Estrada-Franco JG, Richardson L, et al. 2000. Microsatellite markers reveal a spectrum of population structures in the malaria parasite *Plasmodium falciparum*. *Mol Biol Evol* **17**, 1467–1482.

Aslam N, Turner CMR. 1992. The relationship of variable antigen expression and population growth rates in *Trypanosoma brucei*. *Parasitol Res* **78**, 661–664.

Augur Z, Abiri D, van der Ploeg LHT. 1989. Ordered appearance of antigenic variants of African trypanosomes explained in a mathematical model based on a stochastic switch process and immune-selection against putative switch intermediates. *Proc Natl Acad Sci U S A* **86**, 9626–9630.

Babiker HA, Walliker D. 1997. Current views on the population structure of *Plasmodium falciparum*: implications for control. *Parasitol Today* **13**, 262–267.

Back NK, Nijhuis M, Keulen W, Boucher CA, Oude Essink BO, et al. 1996. Reduced replication of 3TC-resistant HIV-1 variants in primary cells due to a processivity defect of the reverse transcriptase enzyme. *EMBO J* **15**, 4040–4049.

Backert S, Naumann M. 2010. What a disorder: proinflammatory signaling pathways induced by *Helicobacter pylori*. *Trends Microbiol* **18**, 479–486.

Badovinac VP, Tvinnereim AR, Harty JT. 2000. Regulation of antigen-specific CD8$^+$ T cell homeostasis by perforin and interferon-γ. *Science* **290**, 1354–1357.

Banchereau J, Briere F, Caux C, Davoust J, Lebecque S, et al. 2000. Immunobiology of dendritic cells. *Ann Rev Immunol* **18**, 767–811.

Barbet AF, Kamper SM. 1993. The importance of mosaic genes to trypanosome survival. *Parasitol Today* **9**, 63–66.

Barnwell JW. 1999. Malaria: a new escape and evasion tactic. *Nature* **398**, 562–563.

Barry JD. 1997. The relative significance of mechanisms of antigenic variation in African trypanosomes. *Parasitol Today* **13**, 212–218.

Barry JD, Peeling RW, Brunham RC. 1999. Analysis of the original antigenic sin antibody response to the major outer membrane protein of *Chlamydia trachomatis*. *J Infect Dis* **179**, 180–186.

Barry JD, Turner CMR. 1991. The dynamics of antigenic variation and growth of African trypanosomes. *Parasitol Today* **7**, 207–211.

Bednarek MA, Sauma SY, Gammon MC, Porter G, Tamhankar S, et al. 1991. The minimum peptide epitope from the influenza virus matrix protein: extra and intracellular loading of HLA-A2. *J Immunol* **147**, 4047–4053.

Benjamin DC, Berzofsky JA, East IJ, Gurd FRN, Hannum C, et al. 1984. The antigenic structure of proteins: a reappraisal. *Ann Rev Immunol* **2**, 67–101.

Bisen PS, Tiwari RP. 2011a. Diagnostic kit for detecting pulmonary and extra-pulmonary tuberculosis. US Patent Published No US7, 888, 037 B2, Feb 15, 2011.

Bisen PS, Tiwari RP. 2011b. A diagnostic kit for detecting pulmonary and extra-pulmonary tuberculosis. Japanese Patent Published No 4601628, (2006-553765) Jan 31, 2011.

Bisen PS, Garg SK, Tiwari RP, Tagore PRN, Chandra R, et al. 2003. Analysis of shotgun expression library of *Mycobacterium tuberculosis* genome for immunodominant polypeptides: potential use in serodiagnosis. *Clin Diagn Lab Immunol* **10**, 1051–1058.

Bisen PS, Prasad GBKS, Zacharia A, Jadon N, Dubey R, Tiwari RP. 2010. Liposomes in diagnosis of tuberculosis. In: Anninos P, Rossi M, Pham TD, Falugi C, Bussing A, Koukkou M, eds., *Advances in Biomedical Research*, pp. 381–400. WSEAS Press, Cambridge University, Cambridge. February 23–25, 2010. www.wseas.org.

Borrow P, Shaw GM. 1998. Cytotoxic T-lymphocyte escape viral variants: how important are they in viral evasion of immune clearance *in vivo*? *Immunol Rev* **164**, 37–51.

Brannan LR, Turner CMR, Phillips RS. 1994. Malaria parasites undergo antigenic variation at high rates *in vivo*. *Proc Royal Soc London Series B Biol Sci* **256**, 71–75.

Bush RM, Bender CA, Subbarao K, Cox NJ, Fitch WM. 1999. Predicting the evolution of influenza A. *Science* **286**, 1921–1925.

Butz EA, Bevan MJ. 1998. Massive expansion of antigen-specific CD8+ T cells during an acute virus infection. *Immunity* **8**, 167–175.

Buus S. 1999. Description and prediction of peptide-MHC binding: the 'human MHC project'. *Curr Opin Immunol* **11**, 209–213.

Caffrey M. 2011. HIV envelope: challenges and opportunities for development of entry inhibitors Trends *Microbiol* **19**, 191–197.

Chao L, Cox EC. 1983. Competition between high and low mutating strains of *Escherichia coli*. *Evolution* **37**, 125–134.

Chisari FV. 1995. Hepatitis B virus immunopathogenesis. *Ann Rev Immunol* **13**, 29–60. 76.

Coffin JM. 1995. HIV population dynamics *in vivo*: implications for genetic variation, pathogenesis, and therapy. *Science* **267**, 483–489.

Coffman RL, Beebe AM. 1998. Genetic control of the T cell response to *Leishmania major* infection. In: Gupta S, Sher A, Ahmed R, eds., *Mechanisms of Lymphocyte Activation and Immune Regulation VII: Molecular Determinants of Microbial Immunity*, pp. 61–66. Plenum Press, New York.

Cohen S, Lambert PH. 1982. Malaria. In: Cohen S, Warren D, eds., *Immunology of Parasitic Infections*, 2nd ed., pp. 422–438. Blackwell, London.

Colman PM. 1998. Structure and function of the neuraminidase. In: Nicholson KG, Webster RG, Hay AJ, eds., *Textbook of Influenza*, pp. 65–73. Blackwell Science, Oxford.

Constant SL, Bottomly K. 1997. Induction of Th1 and Th2 CD4+ T cell responses: the alternative approaches. *Ann Rev Immunol* **15**, 297–322.

Cornelissen M, Kampinga G, Zorgdrager F, Goudsmit J. 1996. Human immunodeficiency virus type 1 subtypes defined by *env* show high frequency of recombinant *gag* genes. *J Virol* **70**, 8209–8212.

Couceiro JN, Paulson JC, Baum LG. 1993. Influenza virus strains selectively recognize sialy-loligosaccharides on human respiratory epithelium: the role of the host cell in selection of hemagglutinin receptor specificity. *Virus Res* **29**, 155–165.

Cox NJ, Subbarao K. 2000. Global epidemiology of influenza: past and present. *Ann Rev Med* **51**, 407–421.

Crandall KA, ed., 1999. *The Evolution of HIV*, Johns Hopkins University Press, Baltimore, MD.

Davis MM, Boniface JJ, Reich Z, Lyons D, Hampl J, et al. 1998. Ligand recognition by αβ T cell receptors. *Ann Rev Immunol* **16**, 523–544.

Debnath Mousumi PGBKS, Bisen PS. 2010. Impact of HGP on molecular diagnostics. *Molecular Diagnostics: Promises and Possibilities*, pp. 85–96, Springer, Dordrecht.

Deitsch KW, Moxon ER, Wellems TE. 1997. Shared themes of antigenic variation and virulence in bacterial, protozoal, and fungal infections. *Microbiol Mol Biol Rev* **61**, 281–293.

Didelot X, Maiden MC. 2010. Impact of recombination on bacterial evolution. *Trends Microbiol* **18**, 315–322.

Dimmock NJ. 1993. Neutralization of animal viruses. *Curr Topics Microbiol Immunol* **183**, 1–149.

Doherty PC, Biddison WE, Bennink JR, Knowles BB. 1978. Cytotoxic T-cell responses in mice infected with influenza and vaccinia viruses vary in magnitude with H-2 genotype. *J Exp Med* **148**, 534–543.

Doherty PC, Christensen JP. 2000. Accessing complexity: the dynamics of virus-specific T cell responses. *Ann Rev Immunol* **18**, 561–592.

Drescher J, Aron R. 1999. Influence of the amino acid differences between the hemagglutinin HA1 domains of influenza virus H1N1 strains on their reaction with antibody. *J Med Virol* **57**, 397–404.

Dutton RW, Bradley LM, Swain SL. 1998. T cell memory. *Ann Rev Immunol* **16**, 201–223.

Edwards MJ, Dimmock NJ. 2000. Two influenza A virus-specific Fabs neutralize by inhibiting virus attachment to target cells, while neutralization by their IgGs is complex and occurs simultaneously through fusion inhibition and attachment inhibition. *Virology* **278**, 423–435.

Evans DT, O'Connor DH, Jing P, Dzuris JL, Sidney J, et al. 1999. Virus-specific cytotoxic T-lymphocyte responses select for amino-acid variation in simian immunodeficiency virus Env and Nef. *Nat Med* **5**, 1270–1276.

Farber DL. 2000. T cell memory: heterogeneity and mechanisms. *Clin Immunol* **95**, 173–181.

Ferguson N, Anderson R, Gupta S. 1999. The effect of antibody-dependent enhancement on the transmission dynamics and persistence of multiple-strain pathogens. *Proc Natl Acad Sci U S A* **96**, 790–794.

Gao P, Robertson DL, Carruthers CD, Li Y, Bailes E, Kosrikis LG, et al. 1998. An isolate of human immunodeficiency virus type 1 originally classified as subtype I represents a complex mosaic comprising three different group M subtypes (A, G, and I). *J Virol* **72**, 10234–10241.

Garg SK, Bisen PS. 2004. Proteomic analysis for the identification of human macrophage molecule/s involved in Mycobacterial killing: Sphingosine 1-Phosphate as a novel enhancer of phospholipase-D mediated antimycobacterial activity (unpublished data). PhD thesis, Bundelkhand University, Jhansi, UP, India.

Garg SK, Santucci MB, Panitti M, Pucillo L, Bocchino M, et al. 2006. Does sphingosine 1-phosphate play a protective role in the course of pulmonary tuberculosis? *Clin Immunol* **121**, 260–264.

Germain RN, Štefanová I. 1999. The dynamics of T cell receptor signaling: complex orchestration and the key roles of tempo and cooperation. *Ann Rev Immunol* **17**, 467–522.

Gianfrani C, Oseroff C, Sidney J, Chesnut RW, Sette A. 2000. Human memory CTL response specific for influenza A virus is broad and multispecific. *Human Immunol* **61**, 438–452.

Gray AR. 1965. Antigenic variation in a strain of *Trypanosoma brucei* transmitted by *Glossina morsitans* and *G. palpalis*. *J Gen Microbiol* **41**, 195–214.

Griffin DE. 2001. Measles virus. In: Knipe DM, Howley PM, eds., *Fields Virology*, 4th ed., pp. 1401–1441. Lippincott-Raven, Philadelphia.

Gross MD, Siegel EC. 1981. Incidence of mutator strains in *Escherichia coli* and coliforms in nature. *Mutat Res* **91**, 107–110.

Guardiola J, Maffei A, Lauster R, Mitchison NA, Accolla RS, Sartoris S. 1996. Functional significance of polymorphism among MHC class II gene promoters. *Tissue Antigens* **48**, 615–625.

Gupta S, Maiden MCJ, Anderson RM. 1999. Population structure of pathogens: the role of immune selection. *Parasitol Today* **15**, 497–501.

Guttman DS. 1997. Recombination and clonality in natural populations of *Escherichia coli*. *Trends Ecol Evol* **12**, 16–22.

Hackstadt T. 1999. Cell biology. In: Stephens RS, ed., *Chlamydia: Intracellular Biology, Pathogenesis, and Immunity*, pp. 101–138. American Society for Microbiology, Washington, DC.

Hauser SL. 1995. T-cell receptor genes. *Ann N Y Acad Sci* **756**, 233–240.

Haydon DT, Woolhouse MEJ. 1998. Immune avoidance strategies in RNA viruses: fitness continuums arising from trade-offs between immunogenicity and antigenic variability. *J Theor Biol* **193**, 601–612.

Holland JJ, ed., 1992. *Genetic Diversity of RNA Viruses*, Springer-Verlag, New York.

Iqbal J, Perlmann P, Berzins K. 1993. Serologic diversity of antigens expressed on the surface of *Plasmodium falciparum* infected erythrocytes in Punjab (Pakistan). *Trans R Soc Trop Med Hyg* **87**, 583–588.

Jamieson BD, Ahmed R. 1989. T cell memory: long-term persistence of virus-specific cytotoxic T cells. *J Exp Med* **169**, 1993–2005.

Janeway CA Jr. 1993. How the immune system recognizes invaders. *Sci Am* **269**, 73–79.

Janeway CA Jr, Travers P, Capra JD, Walport MJ. 1999. *Immunobiology: The Immune System in Health and Disease*, Garland Publishers, New York.

Kartikeyan S, Bharmal RN, Tiwari RP, Bisen PS. 2007. Fundamentals of Immunity. *HIV and AIDS Basic Elements and Priorities*, pp. 27–37, Springer, Dordrecht.

Khan Z, Bisen PS. 2007. Tumor control by manipulation of the human anti-apoptotic survivin gene. PhD thesis, Bundelkhand University, Jhansi, UP, India.

Khan Z, Khan N, Tiwari RP, Patro IK, Prasad GBKS, Bisen PS. 2010a. Down-regulation of survivin by oxaliplatin diminishes radioresistance of head and neck squamous carcinoma cells. *Radiother Oncol* **96**, 267–273. http://dx.doi.org/. doi: 10. 1016/j. radonc.2010.06.005, 16 July 2010.

Khan Z, Khan N, Varma AK, Tiwari RP, Mouhamad S, Prasad GBKS, Bisen PS. 2010b. Oxaliplatin mediated inhibition of survivin increase sensitivity of head and neck squamous cell carcinoma cell lines to paclitaxel. *Curr Cancer Drug Target PMID* **10**, 660–669. 20578991 BSP/CCDT/E-Pub/00034.

Khan Z, Khan N, Tiwari RP, Sah NK, Prasad GBKS, Bisen PS. 2011. Biology of Cox-2: An application in cancer therapeutics. *Curr Drug Target* **12**, 1082–1093.

Khan Z, Tiwari RP, Mulherkar Rita SNK, Prasad GBKS, Shrivastava BR, Bisen PS. 2009. Detection of survivin and p53 in human oral cancer: correlation with clinicopathological findings. *Head Neck* **31**, 1039–1048. www.interscience.wiley.com. doi: 10.1002/hed.21071, 1st April 2009.

Kimata JT, Kuller L., Anderson DB, Dailey P, Overbaugh J. 1999. Emerging cytopathic and antigenic simian immunodeficiency virus variants influence AIDS progression. *Nat Med* **5**, 535–541.

Kropshofer H, Hämmerling GJ, Vogt AB. 1999. The impact of the non-classical MHC proteins HLA-DM and HLA-DO on loading of MHC class II molecules. *Immunol Rev* **172**, 267–278.

Lamb RA, Krug RM. 2001. *Orthomyxoviridae:* the viruses and their replication. In: Knipe DM, Howley PM, eds., *Fields Virology*, 4th ed., pp. 1487–1531. Lippincott-Raven, Philadelphia, PA.

Lancefield RC. 1962. Current knowledge of type-specific M antigens of group A streptococci. *J Immunol* **89**, 307–313.

Larsen O, Andersson S, da Silva Z, Hedegaard K, Sandstrom A, et al. 2000. Prevalences of HTLV-1 infection and associated risk determinants in an urban population in Guinea–Bissau, West Africa. *J Acquir Immune Defic Syndr* **25**, 157–163.

Liebert UG, Flanagan SG, Loffler S, Baczko K, ter Meulen V, Rima BK. 1994. Antigenic determinants of measles virus hemagglutinin associated with neurovirulence. *J Virol* **68**, 1486–1493.

Mateu MG. 1995. Antibody recognition of picornaviruses and escape from neutralization: a structural view. *Virus Res* **38**, 1–24.

McCutchan FE. 1999. Global diversity in HIV. In: Crandall KA, ed., *The Evolution of HIV*, pp. 41–101. Johns Hopkins University Press, Baltimore, MD.

Menon JN, Bretscher PA. 1998. Parasite dose determines the Th1/Th2 nature of the response to *Leishmania major* independently of infection route and strain of host or parasite. *Eur J Immunol* **28**, 4020–4028.

Mims CA. 1987. *The Pathogenesis of Infectious Disease*, 3d ed. Academic Press, London.

Osinski RJ. 1980. Antigenic variation in trypanosomes: a computer analysis of variant order. *Parasitology* **80**, 343–357.

Paredes-Sabja D, Sarker MR. 2009. *Clostridium perfringens* sporulation and its relevance to pathogenesis. *Future Microbiol* **4**, 519–525.

Pircher H, Moskophidis D, Rohrer U, Burki K, Hengartner H, Zinkernagel RM. 1990. Viral escape by selection of cytotoxic T cell–resistant virus variants *in vivo*. *Nature* **346**, 629–633.

Raghuwanshi SK, Bisen PS. 2002. Characterization of Aspergillus fumigates antigens with respect to diagnosis and pathogenesis (unpublished data). PhD thesis, Barkatullah University, Bhopal.

Rasmussen RV, Fowler VG Jr, Skov R, Bruun NE (2011. Future challenges and treatment of Staphylococcus aureus bacteremia with emphasis on MRSA. *Future Microbiol* **6**, 43–56.

Reimann J, Schirmbeck R. 1999. Alternative pathways for processing exogenous and endogenous antigens that can generate peptides for MHC class I–restricted presentation. *Immunol Rev* **172**, 131–152.

Rich SM, Ferreira MU, Ayala FJ. 2000. The origin of antigenic diversity in *Plasmodium falciparum*. *Parasitol Today* **16**, 390–396.

Rock KL, Goldberg AL. 1999. Degradation of cell proteins and the generation of MHC class I–presented peptides. *Ann Rev Immunol* **17**, 739–779.

Röhm C, Zhou N, Süss J, MacKenzie J, Webster RG. 1996. Characterization of a novel influenza hemagglutinin, H15: criteria for determination of influenza A subtypes. *Virology* **217**, 508–516.

Romani L. 2011. Immunity to fungal infections. *Nat Rev Immunol* **11**, 275–288.

Salaun L, Audibert C, Le Lay G, Burucoa C, Fauchere JL, Picard B. 1998. Panmictic structure of *Helicobacter pylori* demonstrated by the comparative study of six genetic markers. *FEMS Microbiol Lett* **161**, 231–239.

Scherf A, Hernandez-Rivas R, Buffet P, Bottius E, Benatar C, Pouvelle B, Gysin J, Lanzer M. 1998. Antigenic variation in malaria: *in situ* switching, relaxed and mutually exclusive transcription of var genes during intra-erythrocytic development in *Plasmodium falciparum*. *EMBO J* **17**, 5418–5426.

Schneider-Schaulies J. 2000. Cellular receptors for viruses: links to tropism and pathogenesis. *J Gen Virol* **81**, 1413–1429.

Schols D, De Clercq E. 1996. Human immunodeficiency virus type 1 gp120 induces energy in human peripheral blood lymphocytes by inducing interleukin-10 production. *J Virol* **70**, 4953–4960.

Scott TW, Weaver SC, Mallampalli VL. 1994. Evolution of mosquito-borne viruses. In: Morse SS, ed., *Evolutionary Biology of Viruses*, pp. 293–324. Raven Press, New York.

Singh SK, Bisen PS. 2006. Adjuvanticity of stealth liposomes on the immunogenicity of synthetic gp41 epitope of HIV-1. *Vaccine* **24**, 4161–4166.

Singh SK, Sah NK, Bisen PS. 2007. A synthetic gag-p24 epitope chemically coupled to BSA through a deca alanine peptide enhances HIV-1 serodiagnostic ability by several fold. *AIDS Res Hum Retroviruses* **23**, 153–160.

Sompayrac LM. 1999. *How the Immune System Works*, Blackwell Science, Malden, MA.

Soto C. 2010. Prion hypothesis: the end of the controversy? *Trends Biochem Sci* doi: 10.1016/j.tibs.2010.11.001.

Spratt BG, Maiden MCJ. 1999. Bacterial population genetics, evolution and epidemiology. *Philos Trans R Soc Lond BBiol Sci* **354**, 701–710.

Stern A, Meyer TF. 1987. Common mechanism controlling phase and antigenic variation in pathogenic neisseriae. *Mol Microbiol* **1**, 5–12.

Stern A, Brown M, Nickel P, Meyer TF. 1986. Opacity genes in *Neisseria gonorrhoeae*: control of phase and antigenic variation. *Cell* **47**, 61–71.

Stevenson PG, Doherty PC. 1998. Cell-mediated immune response to influenza virus. In: Nicholson KG, Webster RG, Hay AJ, eds., *Textbook of Influenza*, pp. 278–287. Blackwell Science, Oxford.

Tibayrenc M, Kjellberg F, Ayala FJ. 1990. A clonal theory of parasitic protozoa: the population genetic structure of *Entamoeba, Giardia, Leishmania* and *Trypanosomes*, and its medical and taxonomic consequences. *Proc Natl Acad Sci U S A* **87**, 2414–2418.

Tiwari RP, Bharmal RN, Kartikeyan S, Bisen PS. 2007a. Development of a rapid Liposomal agglutination card test for the detection of glycolipid antigen in patients with meningeal, pulmonary and other extra-pulmonary tuberculosis. *Int J Tuberc Lung Dis* **11**, 1143–1151.

Tiwari RP, Hattikudur NS, Bharmal RN, Kartikeyan S, Deshmukh NN, Bisen PS. 2007b. Modern approaches to a rapid diagnosis of tuberculosis: promises and challenges. *Tuberculosis* **87**, 193–201.

Tiwari RP, Tiwari D, Garg SK, Chandra R, Bisen PS. 2005. Glycolipids of *Mycobacterium tuberculosis* strain H37Rv are potential serological markers for diagnosis of active tuberculosis. *Clin Diagn Lab Immunol* **12**, 465–473.

Tortorella D, Gewurz BE, Furman MH, Schust DJ, Ploegh HL. 2000. Viral subversion of the immune system. *Ann Rev Immunol* **18**, 861–926.

Van Doorslaer K, Bernard HU, Chen Z, de Villiers EM, Hausen HZ, Burk RD. 2011. Papilloma viruses: evolution, Linnaean taxonomy and current nomenclature. *Trends Microbiol* **19**, 49–50.

Windle HJ, Brown PA, Kelleher DP. 2010. Proteomics of bacterial pathogenicity: therapeutic implications. *Proteomics Clin Appl* **4**, 215–227.

Wright PF, Webster RG. 2001. Orthomyxoviruses. In: Knipe DM, Howley PM, eds., *Fields Virology*, 4th ed., pp. 1533–1579. Lippincott-Raven, Philadelphia, PA.

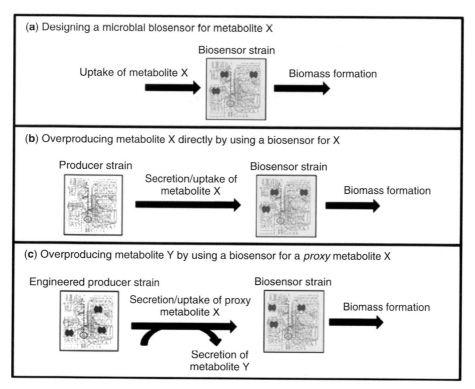

Figure 1.2. Concept of microbial biosensor (Tepper and Shlomi, 2010).

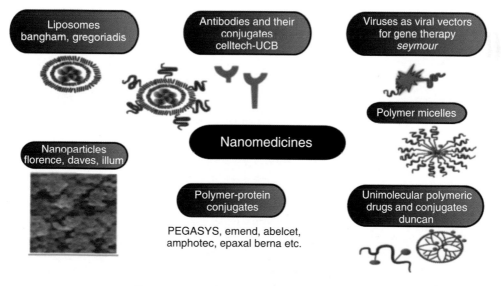

Figure 1.5. Various types of nanomedicines.

Microbes: Concepts and Applications, First Edition. Prakash S. Bisen, Mousumi Debnath, Godavarthi B. K. S. Prasad
© 2012 Wiley-Blackwell. Published 2012 by John Wiley & Sons, Inc.

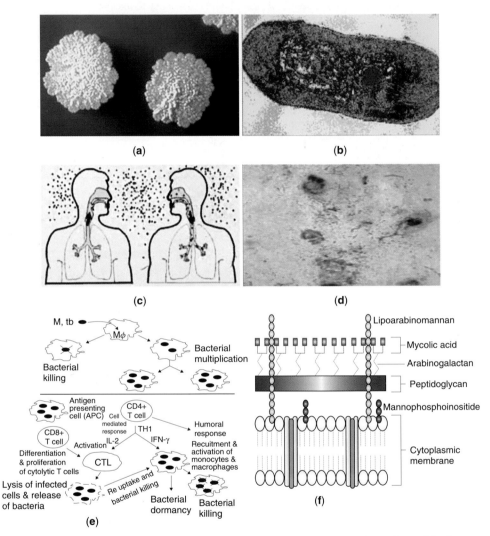

Figure 3.11. (a) Growth of *M. tuberculosis* on LJ (Lowenstein–Jensen) medium, (b) electron micrograph of *M. tuberculosis*, (c) transmission of tuberculosis infection from person to person through aerosol liberated from TB +ve individual inhaled by the normal individual and acquires infection; (d) acid-fast stained smear showing pink colored bacilli of *M. tuberculosis*; (e) flow chart progression of *M. tuberculosis* bacillus after inhalation, entry into macrophages, and activation of immune response; (f) *M. tuberculosis* (acid fast) cell wall; (g) schematic representation of principle of TB antigen detection (TB/M card test) kit (Tiwari et al., 2007a, b).

Principle of TB/M card test

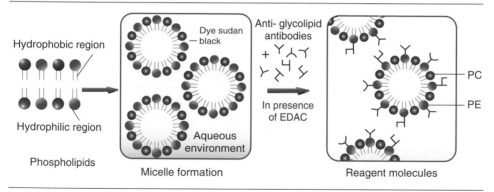

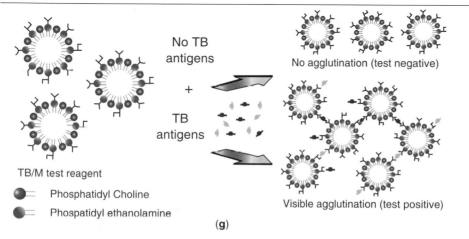

(g)

Figure 3.11. (*Continued*)

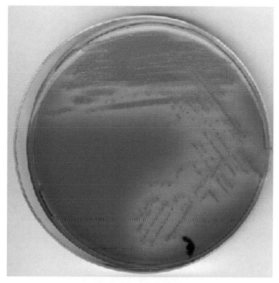

Figure 3.16. *Vibrio cholerae* on TCBS (thiosulfate–citrate–bile salt–sucrose agar plate).

Figure 3.18. EM of *Staphylococci* and coagulase test.

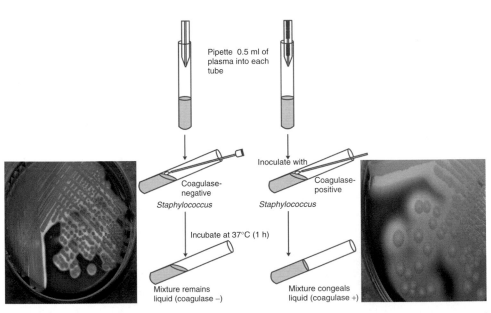

Figure 3.19. Blood agar plates and hemolysis, steps in the coagulase test, and enlarged view of hemolysis.

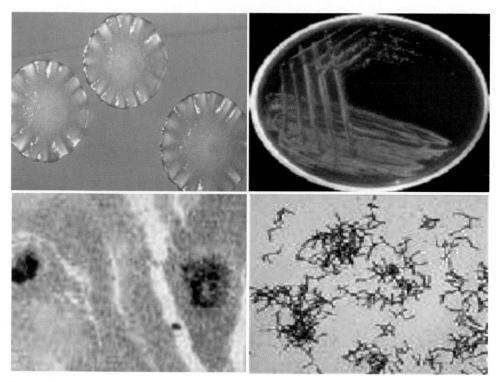

Figure 3.22. *Actinomyces* spp. are gram-positive, fungus-like bacteria (hematoxylin and eosin stain).

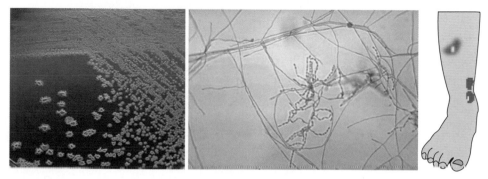

Figure 3.23. Gram-positive aerobic *Nocardia asteroides* slide culture reveals chain of spores amongst aerial mycelia and pathological symptoms form.

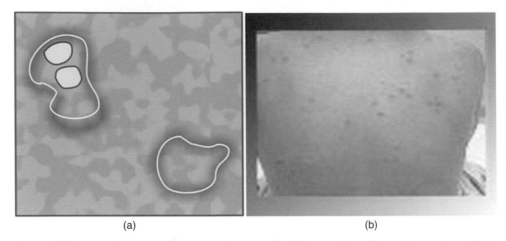

(a)

(a)   (b)

Figure 3.25. (a) Diagrammatic electronmicrograph of VZV particles; (b) vesicular rash of chick-enpox.

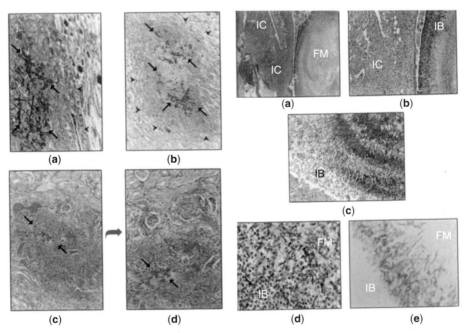

Figure 3.46. Systemic infection of *Aspergillus fumigates*: (1) Periodic acid Shiff staining of positive fungal hyphae in kidney. (2) Fungal mass (FM) in pelvis of ureter, IB denotes peripheral area of fungal mass, and IC denotes inflammatory catico-medullary area (stains used are haematoxylin and eosin, Celestin Blue, antibody against crude metabolic antigens of *Aspergillus fumigatus*; Raghuwanshi and Bisen, 2002).

Raising polyclonal antibodies

- Polyclonal antibodies were rasied against the glycolipid antigens in rabbits.
- Priming was done with Ag + CFA.
- Booster dose was done with Ag + IFA.
- The antibody titre was assayed by ELISA technique.

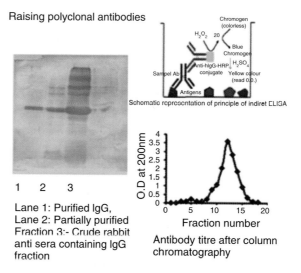

Schematic representation of principle of indiret ELISA

Lane 1: Purified IgG,
Lane 2: Partially purified
Fraction 3:- Crude rabbit anti sera containing IgG fraction

Antibody titre after column chromatography

Immunoreactivity of antigens by eastern blotting

- The antibody raised was incubated on the glycolipid chromatographed TLC and followed by washing with PBS and incubated with anti rabbit IgG (HRP) conjugated.

- Substrate (DAB+H$_2$O$_2$) was added and bands were visualized.

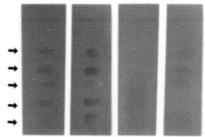

A-1. Immunoreactivity with sera of active pulmonary TB on aTLC plate;
A-2. Immunoreactivity with extra - pulmonary
B-1, Immunoreactivity BCG vaccinated
B-2, Immunoreactivity BCG unvaccinated healthy humans for cross reactivity study.

Figure 3.47. Immunoblotting and raising polyclonal antibodies against *Mycobacterium tuberculosis*.

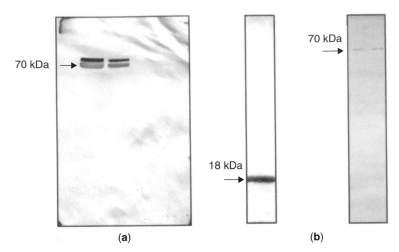

(a)   (b)

Figure 3.48. (a) Western blot showing production of antibodies raised against doublet of 70–72 kDa protein in rabbit. (b) Western blot showing production of antibodies against 18 and 70 kDa proteins in *Aspergillus fumigatus* antigens (Raghuwanshi and Bisen, 2002).

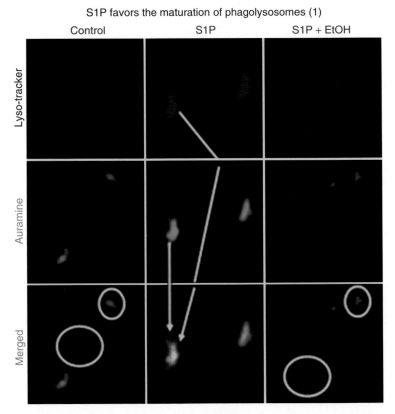

Figure 3.50. Immunoflouresence detection of S1P in tuberculosis (Garg and Bisen, 2004).

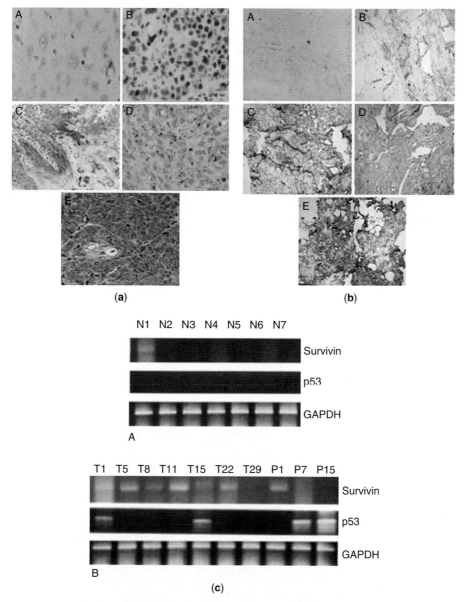

Figure 3.53. (a) Representative results of immunohistochemical staining for oral squamous cell carcinoma (OSCC) survivin expression. (b) Representative results of immunohistochemical staining for p53. (c) Representative RT-PCR results of survivin and p53 expression.

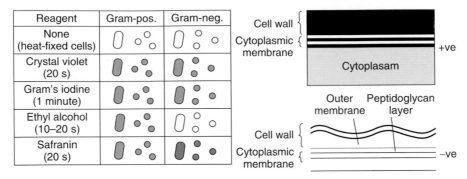

Reagent	Gram-pos.	Gram-neg.
None (heat-fixed cells)		
Crystal violet (20 s)		
Gram's iodine (1 minute)		
Ethyl alcohol (10–20 s)		
Safranin (20 s)		

Figure 4.5. Gram staining procedure and cell wall structure.

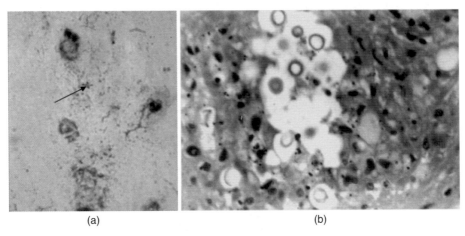

(a)   (b)

Figure 4.6. (a) Acid-fast-stained smear showing pink colored bacilli of *M. tuberculosis* (Tiwari et al., 2005). (b) Lymph node biopsy showing many large, encapsulated yeast cells of *C. neoformans*. Haematoxylin and eosin stain; original magnification 400× (Khan et al., 2003).

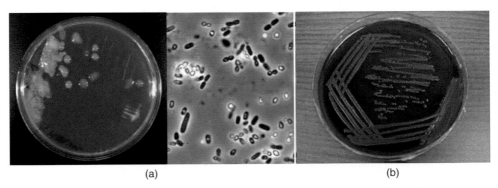

(a)   (b)

Figure 4.7. (a) Isolation of *Azotobacter* spp. from soil sample. (b) *Escherichia coli* produces red colonies on MacConkey agar.

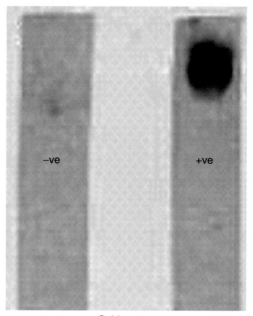

Oxidase test

Figure 4.10. Oxidase test specimen.

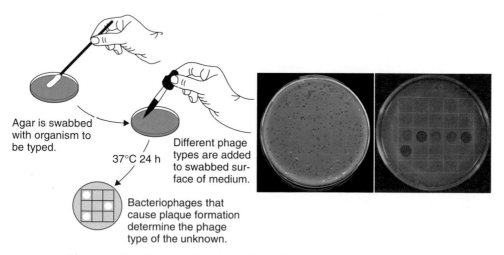

Agar is swabbed with organism to be typed.

37°C 24 h

Different phage types are added to swabbed surface of medium.

Bacteriophages that cause plaque formation determine the phage type of the unknown.

Figure 4.23. Phage typing plaque formation for *Staphylococcus aureus*.

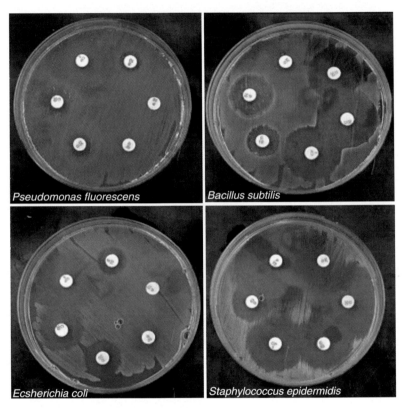

Figure 5.10. Antibiotic sensitivity test.

Xu J, Mitchell TG, Vilgalys R. 1999. PCR-restriction fragment length polymorphism (RFLP) analyses reveal both extensive clonality and local genetic differences in *Candida albicans*. *Mol Ecol* **8**, 59–73.

Zacharia A, Jadaun N, Dubey R, Prasad GBKS, Bisen PS. 2010. Culture filtrate antigens in tuberculosis diagnosis. In: Anninos P, Rossi M, Pham TD, Falugi C, Bussing A, Koukkou M, eds., *Advances in Biomedical Research*, pp. 417–425. WSEAS Press, Cambridge University, Cambridge. February 23–25, 2010. www.wseas.org.

Zhou J, Spratt BG. 1992. Sequence diversity within the *argF, fbp* and *recA* genes of natural isolates of *Neisseria meningitides* —interspecies recombination within the *argF* gene. *Mol Microbiol* **6**, 2135–2146.

4

IDENTIFICATION AND CLASSIFICATION OF MICROBES

4.1. PROLOGUE

The future of microbial taxonomy is being moulded by the work carried out today. The question of whether or not microorganisms are clustered in groups with certain likenesses is pivotal. If microbial diversity is organized in some way, then taxonomy merely needs to find a way to reflect that diversity. If microorganisms do not form groups with meaningful traits and evolutionary history, then organizing microbial diversity will be more difficult, if not impossible (http://www.envismadrasuniv.org). It is also possible that some microorganisms fit nicely into clusters while others do not. In this event, microbial taxonomy may have to embrace multiple approaches to classification. Moving forward in the fields of microbial taxonomy and systematics, research needs to address more fully the issues of clustering, community genomics, and the development of novel approaches for classification and characterization.

The correct identification of microorganisms is of fundamental importance to microbial systematists as well as to scientists involved in many other areas of applied research and industry (e.g., agriculture, clinical microbiology, and food production). Increased use of automation and user-friendly software makes these technologies more widely available. In all, the detection of infectious agents at the nucleic acid level represents a true synthesis of clinical chemistry and clinical microbiology techniques (Cook et al., 2003; Cohen and Smith, 1964; Tang et al., 1997). Accurate identification requires a sound classification or system of ordering organisms into groups, as well as an unequivocal nomenclature for naming them (Truper and Schleifer, 2006).

Microbes: Concepts and Applications, First Edition. Prakash S. Bisen, Mousumi Debnath, Godavarthi B. K. S. Prasad
© 2012 Wiley-Blackwell. Published 2012 by John Wiley & Sons, Inc.

4.2. PRINCIPLES OF TAXONOMY

The art of biological classification is known as *taxonomy*. Taxonomy is a particular classification arranged in a hierarchical structure. Typically, this is organized by supertype–subtype relationships, also called *generalization–specialization relationships*, or, less formally, *parent–child relationships*. In such an inheritance relationship, the subtype by definition has the same properties, behaviors, and constraints as the supertype plus one or more additional properties, behaviors, or constraints. Bacterial taxonomy is divisible into three parts: (i) classification, (ii) nomenclature, (iii) identification (Fig. 4.1; Cowan, 1965; Young et al., 1992). It has two functions: the first is to describe as completely as possible the basic taxonomic units, or species; the second, to devise an appropriate way of arranging and cataloging these units. The notion of species consists of assemblage of individuals that share a high degree of phenotypic similarity, coupled with an appreciable dissimilarity from other assemblage of the same general kind. Every assemblage of individuals shows some degree of internal phenotypic diversity because of genetic variation. Ideally, species should be characterized by complete description of their phenotypes and genotypes (Doolittle and Zhaxybayeya, 2009).

There are rules for nomenclature but none for classification or identification. Both classification and identification depend on characterization of the bacterium, but each makes different use of the individual feature. The aim of "identification" is to equate the properties of a pure culture with those of a well-characterized and accepted species. When identification in this sense cannot be accomplished, the aim of identification must shift to characterization of a new species, that is, to a new description (Truper and Schleifer, 2006). In classification, equal weight is given to each independent character; some as important, others less so. Systematics is the study of multiple items, units, or individuals with the aim of finding common factors and differences; lines of cleavage are made so that the like fall on the same side of the dividing line, and the unlike on the other. Biological systematics bears the special name taxonomy, and the subject can be subdivided into three sections:

1. *Classification*. The orderly arrangement of units into groups of larger units. A simple analogy can be found in a pack of cards; the individual cards can first be sorted by color, then into suits. Within each suit the cards can be arranged in a numerical sequence, and the face cards placed in some order of seniority.

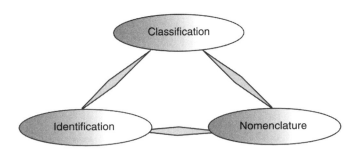

Figure 4.1. Three facets of taxonomy.

2. *Nomenclature*. The naming of the units defined and delineated by the classification. In the example of cards, the face cards are given names and more than one name, for example, jack or knave, may be given to the same card.

3. *Identification*. It can be done through various methods either by physical methods or by methods based on phylogeny.

These three facets, or the trinity that is taxonomy, are to some extent interdependent, but in an orthodox scheme they are considered in the order given above. It is arguable whether the hen or the egg came first, but, since the end of the nineteenth century, bacteriological ethics have demanded that we should not name a bacterium before we have allotted it to a unit in an orderly classificatory system.

Numerical taxonomy typically invokes a number of these criteria at once. The reason for this is that if only one criterion was invoked at a time there would be a huge number of taxonomic groups, each consisting of only one of a few microorganisms. The purpose of grouping would be lost. By invoking several criteria at a time, fewer groups consisting of larger number of microorganisms result. The groupings result from the similarities of the members with respect to the various criteria. A so-called similarity coefficient can be calculated. At some imposed threshold value, microorganisms are placed in the same group.

A well-known example of taxonomic characterization is the kingdom, division, class, family, genus, species, and strain divisions. Such a "classical" bacterial organization, which is typified by the *Bergey's Manual of Determinative Bacteriology*, is based on metabolic, immunological, and structural characteristics. Strains, for example, are all descended from the same organism, but differ in an aspect such as the antigenic character of a surface molecule.

Microbial taxonomy can create much order from the plethora of microorganisms. For example, the American Type Culture Collection maintains the following, which are based on taxonomic characterization (the numbers in brackets indicate the number of individual organisms in the particular category): algae (120), bacteria (14,400), fungi (20,200), yeast (4300), protozoa (1090), animal viruses (1350), plant viruses (590), and bacterial viruses (400). The actual number of microorganisms in each category will continue to change as new microbes are isolated and classified. The general structure, however, of this classical, so-called phenetic system will remain the same.

The classification of bacteria is based not only on how they look but also on what they can do. These molecular techniques for characterizing bacterial genotypes provide a possible basis for defining a bacterial species (Table 4.1). Molecular microbial taxonomy relies on the generation and inheritance of genetic mutations that is the replacement of a nucleotide building block of a gene by another nucleotide. Sometimes the mutation confers no advantage to the microorganism and so is not maintained in subsequent generations. Sometimes the mutation has an adverse effect, and so is actively suppressed or changed. But sometimes the mutation is advantageous for the microorganism, such a mutation will be maintained in succeeding generations.

Because mutations occur randomly, the divergence of two initially genetically similar microorganisms will occur slowly over evolutionary time (millions of years). By sequencing a target region of genetic material, the relatedness or dissimilarity of microorganisms can be determined. When enough microorganisms have been sequenced, relationships can be established and a dendrogram is constructed.

TABLE 4.1. Molecular Techniques for Characterizing Bacterial Genotypes

Approach	Target	Outcome Experiment	Main Limitations
Sequencing of rRNA			
16S rRNA gene sequencing	16S rRNA gene	16S rRNA gene sequence collection	Bias in NA extraction, PCR and cloning; laborious
RT-PCR	mRNA	Specific gene expression	Bias in NA extraction and PCR
Fingerprinting DDGE, TGGE, TTGE, T-RFLP, SSCP	16s rRNA gene	Diversity profiles	Bias in NA extraction and PCR
Non-16S rRNA gene fingerprinting	GC fractionation of DNA, cellular fatty acid	Diversity profiles	16S rRNA approaches required for identification
Quantification of 16S rRNA and its Encoding Genes			
Dot blot hybridization	16S rRNA	Relative abundance of 16s rRNA	Laborious at species level; requires 16S rRNA gene sequence data
qReal-time PCR	16S rRNA gene	Relative abundance of 16s rRNA	Laborious and expensive in early stages of development
FISH	16S rRNA	Enumeration of bacterial populations	Laborious at species level; requires 16S rRNA gene sequence data
DNA Microarray Technology			
Diversity arrays	16S rRNA genes; antibiotic resistance genes	Diversity profiles	Laborious; expensive; in early stages of development
DNA microarray	mRNA	Transcriptional fingerprints	Bias in NA extraction and NA labeling; expensive
Parallel sequencing technologies	Hundreds to thousands of genes	Sequence of microbial genome	Cost; sequences to compare and identify correctly

For a meaningful genetic categorization, the target of the comparative sequencing must be carefully chosen. Molecular microbial taxonomy of bacteria relies on the sequence of ribonucleic acid (RNA), dubbed 16S RNA, that is present in a subunit of prokaryotic ribosomes. Ribosomes are complexes that are involved in the manufacture of proteins using messenger RNA as the blueprint. Given the vital function of the 16S RNA, any mutation tends to have a meaningful, often deleterious, effect on the functioning of the RNA. Hence, the evolution (or change) in the 16S RNA has been very slow, making it a good molecule to compare microorganisms that are billions of years old.

A rigid and complex set of rules governs the biological nomenclature; the rules are designed to keep nomenclature as stable as possible. According to the conventional binomial system of nomenclature, every biological species bears a Latinized name that consists of two words. The first word indicates the taxonomic group of the higher order of immediately higher order or genus to which the species belongs and the second word identifies it as a particular species of that genus. The first name is capitalized and the whole phrase is italized: *Escherichia* (generic name) *coli* (specific name). In contexts, in which no confusion is possible, the generic name is often abbreviated to its initial letter *E. coli*.

In bacterial taxonomy new species is named; a particular strain is designated as a type strain. In the taxonomic treatment of a biological group, the individual species are grouped in a series of categories of successive higher order: genus, family, order, class, and division (or phylum). Such an arrangement is known as a hierarchical one.

Molecular microbial taxonomy has been possible because of the development of polymerase chain reaction (PCR) technique. Using this technique, a small amount of genetic material can be amplified to detectable quantities.

The use of the chain reaction has produced a so-called bacterial phylogenetic tree. The structure of the tree is even now evolving. But the current view has the tree consisting of three main branches. One branch consists of the bacteria. There are some 11 distinct groups within the bacterial branch. Three examples are the green nonsulfur bacteria, Gram-positive bacteria, and cyanobacteria. Suzuki and Yamasato (1994) have divided bacteria into 12 groups that have been defined on the basis of ribosomal RNA analysis (16S rRNA). Most groups (similar to phylum's) contain a variety of physiological and morphological types of bacteria. This reinforces the idea that phenotypic characteristics are inadequate to define evolutionary relationships between microbial species. The second branch of the evolutionary tree consists of the Archaea, which are thought to have been very ancient bacteria that diverged from both bacteria and eukaryotic organisms billions of years ago. Evidence to date places the Archaea a bit closer on the tree to bacteria than to the final branch (the Eukarya). There are three main groups in the archaea: halophiles (salt-loving), methanogens, and the extreme thermophiles (heat loving) (Madigan et al., 2003). This last group is composed of extreme thermopiles that require elemental sulfur for optimal growth. For most members, the sulfur serves as an electron acceptor in anaerobic respiration. Evolution of the eukaryotic line was characterized by periods of rapid evolution interspersed with eras of slow evolution. The accumulation of O_2 in the atmosphere about 1.5 billion years ago seems to correspond to a period of rapid evolution.

Finally, the third branch consists of the Eucarya, or the eukaryotic organisms. Eucarya includes organisms as diverse as fungi, plants, slime molds, and animals (including humans).

Small subunit ribosomal DNA sequences were determined for 17 strains belonging to the genera *Alteromonas, Shewanella, Vibrio*, and *Pseudomonas*, and their sequences were analyzed by phylogenetic methods (Gauthier et al., 1995). The resulting data confirmed the existence of the genera *Shewanella* and *Moritella*, but suggested that the genus *Alteromonas* should be split into two genera. They proposed that a new genus, *Pseudoalteromonas*, should be created to accommodate 11 species that were previously *Alteromonas* species, including *Pseudoalteromonas atlantica* comb. nov., *Pseudoalteromonas aurantia* comb. nov., *Pseudoalteromonas carrageenovoa* comb. nov.,

Pseudoalteromonas citrea comb. nov., *Pseudoalteromonas denitrificans comb.nov.*, *Pseudoalteromonas espejiana* comb. nov., *Pseudoalteromonas haloplanktis* comb. nov. (with two subspecies, *Pseudoalteromonas haloplanktis haloplanktis* comb. nov. and *Pseudoalteromonas haloplanktis tetraodonis* comb. nov.), *Pseudoalteromonas luteoviolacea* comb. nov., *Pseudoalteromonas nigrifaciens* comb. nov., *Pseudoalteromonas rubra* comb. nov., and *Pseudoalteromonas undina* comb. nov., and one species that previously was placed in the genus *Pseudomonas, Pseudoalteromonas piscicida* comb. nov.

In conventional taxonomy, some characteristics are given special emphasis. These include the Gram stain, cell morphology, and the presence of cell structures such as endospores. In numerical taxonomy, all phenotypic characteristics are given equal weight in classifying strains. *Bergey's Manual of Systematic Bacteriology* contains the phenotypic characteristics used to classify bacteria by conventional taxonomy, and keys that can be used to identify unknown strains from their phenotypic characters. Some analyses of nucleic acids have been used in conventional taxonomy; these include measurement of DNA base composition and nucleic acid hybridization.

4.2.1. Strategies Used to Identify Microbes

Over the past century, microbiologists have searched for more rapid and efficient means of microbial identification. The identification and differentiation of microorganisms has principally relied on microbial morphology and growth variables. Advances in molecular biology over the past 10 years have opened new avenues for microbial identification and characterization.

The traditional methods of microbial identification rely solely on the phenotypic characteristics of the organism. Bacterial fermentation, fungal conidiogenesis, parasitic morphology, and viral cytopathic effects are a few phenotypic characteristics commonly used. Some phenotypic characteristics are sensitive enough for strain characterization; these include isoenzyme profile, antibiotic susceptibility profile, and chromatographic analysis of cellular fatty acids. However, most phenotypic variables commonly observed in the microbiology laboratory are not sensitive enough for strain differentiation. When methods for microbial genome analysis became available, a new frontier in microbial identification and characterization was opened.

Early DNA hybridization studies were used to demonstrate relatedness among bacteria. This understanding of nucleic acid hybridization chemistry made possible the nucleic acid probe technology (Tang et al., 1997). Advances in plasmid and bacteriophage recovery and analysis have made possible plasmid profiling and bacteriophage typing, respectively. Both have been proved to be powerful tools for the epidemiologist investigating the source and mode of transmission of infectious diseases. These technologies, however, like the determination of phenotypic variables, are limited by microbial recovery and growth.

Nucleic acid amplification technology has opened new avenues of microbial detection and characterization, such that growth is no longer required for microbial identification. In this respect, molecular methods have surpassed traditional methods of detection for many fastidious organisms. The PCR and other recently developed amplification techniques have simplified and accelerated the *in vitro* process of nucleic acid amplification. The amplified products, known as amplicons, may be characterized by various methods, including nucleic acid probe hybridization, analysis of fragments after restriction endonuclease digestion, or direct sequence analysis. Rapid techniques of nucleic

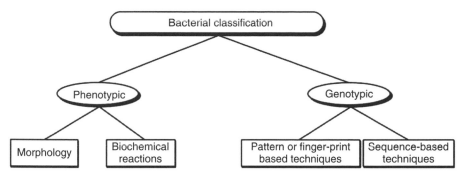

Figure 4.2. Basis of bacterial classification.

acid amplification and characterization have significantly broadened the microbiologist's diagnostic arsenal (Tang et al., 2004).

4.2.2. Methods for Bacterial Identification

Methods of bacterial identification can be broadly delimited to genotypic techniques based on profiling an organism's genetic material (primarily its DNA) and phenotypic techniques based on profiling either an organism's metabolic attributes or some aspect of its chemical composition (Fig. 4.2). Genotypic techniques have the advantage over phenotypic methods that they are independent of the physiological state of an organism; they are not influenced by the composition of the growth medium or by the organism's phase of growth.

Phenotypic techniques, however, can yield more direct functional information that reveals what metabolic activities are taking place to aid the survival, growth, and development of the organism. These may be embodied, for example, in a microbe's adaptive ability to grow on a certain substrate, or in the degree to which it is resistant to a cohort of antibiotics. Genotypic and phenotypic approaches are complementary and use different techniques. However, this division is historical; we predict that as molecular-based identification matures, there will be more and more overlap in the information obtained using different methodologies.

Genotypic microbial identification methods can be broken into two broad categories: pattern- or fingerprint-based techniques and sequence-based techniques. Pattern-based techniques typically use a systematic method to produce a series of fragments from an organism's chromosomal DNA. These fragments are then separated by size to generate a profile or fingerprint that is unique to that organism and its very close relatives. With enough of this information, one can create a library, or database, of fingerprints from known organisms, to which test organisms can be compared. When the profiles of two organisms match, they can be considered very closely related, usually at the strain or species level.

4.3. USING PHENOTYPIC CHARACTERISTICS TO IDENTIFY MICROBES

Phenotypic characters of bacteria include morphology and biochemical reactions carried out by bacteria whose results can be viewed (Table 4.2). Morphological characteristics

TABLE 4.2. Methods Used to Characterize Genotypic and Phenotypic Characteristics

Method	Comments
Phenotypic characteristics	Most of these methods do not require sophisticated equipment and can be done easily anywhere in the world.
Microscopic morphology	Size, shape, and staining characteristics such as Gram stain can give suggestive information as to the identification of an organism. Further testing is needed to confirm the organism.
Metabolic differences	Culture characteristics can give suggestive information. A battery of biochemical tests can be used to confirm the identification.
Serology	Proteins and polysaccharides that make up a prokaryote are sometimes characteristics enough to be considered as an identifying marker. These can be detected using specific antibodies.
Fatty acid analysis	Cellular fatty acid composition can be used as an identifying marker and is analyzed by gas chromatography.
Genotypic characteristics	These methods are increasingly being used to identify microorganisms.
Nucleic acid hybridization	Probes can be used to identify prokaryotes grown in cultures. In some cases, the method is sensitive enough to detect the organism directly in a specimen.
Amplifying specific DNA sequences using PCR	Even an organism that occurs in very low number in a mixed culture can be identified.
Sequencing rRNA genes	This method requires amplifying, cloning and then sequencing rRNA genes, but it can also be used to identify unculturable organisms.

include colony morphology such as color, size, shape, opacity, elevation, margin surface texture, consistency, and so on. These characters are observed after the incubation period on the cultures on the solid media. In liquid cultures, we can observe the pellicle formation and sediment formation. Biochemical characteristics include enzyme production, utilization of a particular sugar, aerobic or anaerobic reactions, and so on.

Limited information exists on the phenotypic characteristics of bacteria found in biofilm. Both wet-mounted and properly stained bacterial cell suspensions can yield a great deal of information. These simple tests can indicate the Gram reaction of the organism; whether it is acid-fast; its motility; the arrangement of its flagella; the presence of spores, capsules, and inclusion bodies; and, of course, its shape. This information can often allow identification of an organism to the genus level or minimize the possibility that it belongs to one or another group. Colony characteristics and pigmentation are also quite helpful. For example, colonies of several *Porphyromonas* species auto fluoresce under long-wavelength ultraviolet light, and *Proteus* species swarm on appropriate media.

A primary distinguishing characteristic is whether an organism grows aerobically, anaerobically, facultatively (i.e., in the presence or absence of oxygen), or microaerobically (i.e., in the presence of a less than atmospheric partial pressure of oxygen). The proper atmospheric conditions are essential for isolating and identifying bacteria. Other important growth assessments include the incubation temperature, pH, nutrients required, and resistance to antibiotics. For example, one diarrheal disease agent, *Campylobacter jejuni*, grows well at 42 °C in the presence of several antibiotics; another, *Yersinia enterocolitica*, grows better than most other bacteria at 4 °C. *Legionella, Haemophilus*, and some other pathogens require specific growth factors, whereas *E. coli* and most other Enterobacteriaceae can grow on minimal media. Most bacteria are identified and classified largely on the basis of their reactions in a series of biochemical tests. Some tests are routinely used for many groups of bacteria (oxidase and nitrate reduction, amino acid degrading enzymes; fermentation or utilization of carbohydrates); others are restricted to a single family, genus, or species (coagulase test for staphylococci, pyrrolidonyl arylamidase test for gram-positive cocci).

Both the number of tests needed and the actual tests used for identification vary from one group of organisms to another. Therefore, the lengths to which a laboratory should go in detecting and identifying organisms must be decided in each laboratory on the basis of its function, the type of population it serves, and its resources. Clinical laboratories today base the extent of their work on the clinical relevance of an isolate to the particular patient from which it originated, the public health significance of complete identification, and the overall cost-benefit analysis of their procedures. For example, the Centers for Disease Control and Prevention (CDC) reference laboratory uses at least 46 tests to identify members of the Enterobacteriaceae, whereas most clinical laboratories, using commercial identification kits or simple rapid tests, identify isolates with far fewer criteria.

4.3.1. Microscopic Morphology

An important initial step in identifying a microorganism is to determine its shape, size, and staining characteristics. Microscopic examination gives information very quickly and is sometimes enough to make a presumptive identification.

Bacteria are unicellular microorganisms found in every habitat on earth. Nearly all have cell walls composed of peptidoglycan and reproduce by binary fission (cloning of cells). Although many of these microbes are harmless or beneficial to humans, others are pathogenic, causing infectious diseases.

4.3.1.1. Size and Shape

The size and shape of microorganisms can be readily determined by microscopic examination of the wet mount. On the basis of size and shape, one can readily decide whether the organism in question is a prokaryote, fungus, or protozoan. In a clinical laboratory, it can sometimes provide all information needed for diagnosis of certain infections. For example, a wet mount of vaginal secretion is routinely used to diagnose infections caused by yeast or by protozoans (*Trichomonas*).

4.3.1.2. Identification of Bacteria

Some of the first steps in identifying a bacterium include examination of

1. The shape of the individual bacterium
2. Whether the bacteria exist in specific groupings
3. The colony morphology (the appearance of a "colony"; a group of millions of bacteria that arose from one single parent cell).

TABLE 4.3. Bacterial Types According to the Shapes

Bacteria	Shape
Bacillus	Rod
Coccus	Spherical
Spirillum	Spiral
Coccobacilli	Elongated coccal form
Filamentous	Bacilli that occur as long threads
Vibrios	Short, slightly curved rods
Fusiform	Bacilli with tapered ends

Bacterial Shapes: Most bacteria are classified according to their shapes listed in Table 4.3. Thus, we find a range of varied bacteria ranging from rod shaped, spherical, spiral, elongated, bacilli, to curved bacteria (Fig. 4.3).

4.3.1.3. Prokaryote Arrangement of Cells

Bacteria sometimes occur in groups rather than singly, and the single cell's shape influences cell arrangements that they form as the bacterial cells divide. Bacilli divide along a single axis, and are sometimes seen in pairs or chains. Since they only divide along one axis, bacilli will not be found in clusters, such as those formed by staphylococcal bacteria. Cocci divide on one or more planes, producing cells in pairs (diplococci), chains (streptococci), packets (sarcinae), clusters (staphylococci), and so forth. (Fig. 4.4). Size, shape, and arrangement of cells are often the first clues in the identification of bacteria. However, since there are many "look-alikes," methods other than microscopy must be used to determine the genus and species of an organism.

Bacterial populations grow extremely fast when they are supplied with nutrients and environmental conditions that allow them to thrive. Through this growth, different types of bacteria will sometimes produce colonies that are distinctive in appearance. Some colonies may be colored, some are circular in shape, while others are irregular. The characteristics of a colony (shape, size, color, etc.) are termed the *colony morphology*.

4.3.1.4. Gram Staining

Effective utilization and understanding of the clinical bacteriology laboratory can greatly aid in the diagnosis of infectious diseases. Although described more than a century ago, the Gram stain remains the most frequently used rapid diagnostic test, and, in conjunction with various biochemical tests, is the cornerstone of clinical laboratories. First described by Danish pathologist Christian Gram in 1884 and later slightly modified, the Gram stain easily divides bacteria into two groups, gram positive and gram negative, on the basis of their cell wall and cell membrane permeability to organic solvents (Steinbach and Shetty, 2001). The test was originally developed by Christian Gram in 1884, but was modified by Hucker (1921). The modified procedure provided greater reagent stability and better differentiation of organisms. Other modifications have been specifically developed for staining anaerobes and for weakly staining gram-negative bacilli (*Legionella* spp., *Campylobacter* spp., *Bacteroides* spp. *Fusobacterium* spp., and *Brucella* spp.) by using a carbol-fuchsin or a basic fuchsin counterstain.

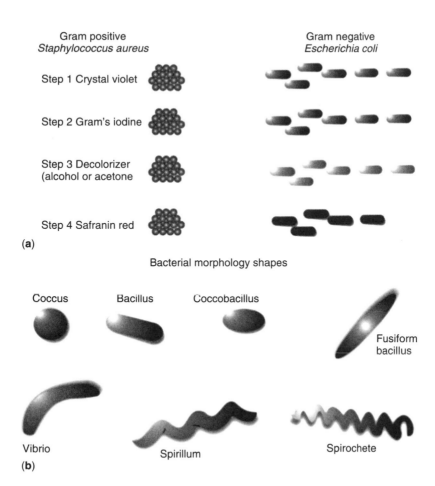

Gram positive
Staphylococcus aureus

Gram negative
Escherichia coli

Step 1 Crystal violet

Step 2 Gram's iodine

Step 3 Decolorizer
(alcohol or acetone

Step 4 Safranin red

(a)

Bacterial morphology shapes

Coccus   Bacillus   Coccobacillus   Fusiform bacillus

Vibrio   Spirillum   Spirochete

(b)

Figure 4.3. Different bacterial morphologies.

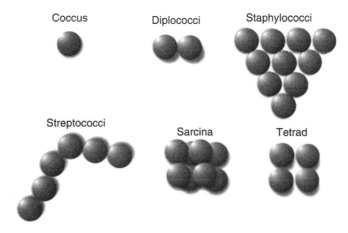

Coccus   Diplococci   Staphylococci

Streptococci   Sarcina   Tetrad

Figure 4.4. Various morphology of bacterial coccus.

Interpretation of Gram-stained smears involves consideration of staining characteristics and cell size, shape, and arrangement. These characteristics may be results of influence by a number of variables including culture age, media, incubation atmosphere, staining methods, and the presence of inhibitory substances. Similar considerations apply to the interpretation of smears from clinical specimens, and additional factors include different host cell types and possible phagocytosis. Gram stain permits the separation of all bacteria into two large groups, those which retain the primary dye (gram positive) and those that take the color of the counterstain (gram negative). The primary dye is crystal violet and the secondary dye is usually either safranin O or basic fuchsin. Some of the more common formulations include saturated crystal violet (approximately 1%), Hucker's crystal violet, and 2% alcoholic crystal violet. The Gram stain is one of the most useful methods in identifying bacteria. In this procedure, the bacteria are stained and examined under the microscope. Organisms are judged to be gram positive if they retain the crystal violet after decolorization and will appear purple. Gram-negative organisms are decolorized and appear pink to red because they take up the safranine counterstain.

The Gram stain classifies bacteria (Fig. 4.5) phenotypically based on differences in cell wall thickness with differing glycosaminopeptide and lipoprotein compositions: Gram-positive bacteria have a peptidoglycan layer 10–15 times thicker than gram-negative bacteria. The cell wall, synonymous with the peptidoglycan layer, is a rigid framework of cross-linked peptidoglycan forming the outermost component of the cell. The more complex gram-negative bacteria also have an outer membrane beyond the peptidoglycan layer, which consists of lipopolysaccharide (endotoxin), lipoprotein, and phospholipids. In some gram-negative species there also exists a periplasmic space between the outer membrane and the inner cytoplasmic membrane with β-lactamases that degrade β-lactam antibiotics (Steinbach and Shetty, 2001).

Tips for Success: Perform Gram stains only on fresh cultures (24 h incubation). Using older cultures often yields unexpected results. Be sure to stop the application of decolorizer immediately after it runs clear to avoid overdecolorizing. Always run positive and negative controls from known cultures.

Procedure

1. Place a loopful of sterile distilled water onto a microscope slide.
2. Touch an isolated colony with inoculating loop and swirl it in the drop of water on the slide.

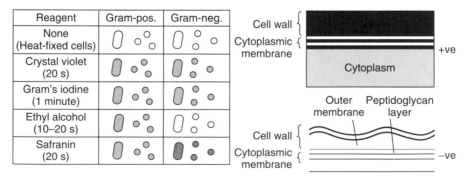

Figure 4.5. Gram staining procedure and cell wall structure. (*See insert for color representation of the figure*).

3. Let the smear air dry at room temperature.

4. Heat fix the smear by waving the slide over a flame, being careful not to overheat.

5. Flood the slide with crystal violet, and let stand for 1 min.

6. Wash the slide briefly with cold water.

7. Flood the slide with Gram's iodine; let stand for 1 min; wash off with water.

8. Decolorize until the solvent flows colorlessly from the slide.

9. Flood the slide with safranine; let stand for 30 s; wash off with water.

10. Blot the slide dry with bulbous paper.

11. Examine the slide under the microscope for Gram reaction (100× oil-immersion objective).

12. Gram-positive organisms will appear purple, whereas gram-negative organisms will appear pink.

In a clinical laboratory, the gram stain of a specimen by itself is generally not sensitive and specific enough to diagnose the cause of most infection, but it is still a useful tool. The clinician can see the Gram reaction, the shape, and the arrangement of the bacteria and whether the organism appears to be growing as a pure culture or with other bacteria or cell of the host. However, most medically important bacteria do not have distinctive shapes or staining characteristics and usually cannot be identified by Gram stain alone. For example, *Streptococcus pyogenes*, which causes strep throat, cannot be distinguished microscopically from other streptococci that are part of the normal flora of the throat. A Gram stain of a stool specimen cannot distinguish *Salmonella* species from *E. coli*. These organisms must be generally isolated in pure cultures and tested for the biochemical attributes to provide precise identification.

In certain cases, the Gram stain gives enough information to start appropriate antimicrobial therapy while awaiting more accurate information. For example, if a urine sample from an otherwise healthy woman reveals more than the permissible level of negative rods per oil-immersion field, the clinician will suspect a urinary tract infection caused by *E. coli*, the common cause of such infection. Likewise, a Gram stain of sputum showing numerous white blood cells and gram-positive encapsulated diplococci is highly suggestive of *Streptococcus pneumoniae*, an organism that causes pneumonia. In certain cases, the result of a Gram stain is enough to accurately diagnose the infection. For example, the presence of gram-negative diplococci clustered in the white blood cells in a sample of urethral secretion from a male is considered diagnostic for gonorrhea, the sexually transmitted disease by *Neisseria gonorrhoeae*. This diagnosis can be made because this microorganism is the only gram-negative dipllococcus found inhabiting the normal sterile urethra of a male.

4.3.1.5. Special Strains

Certain microorganisms have unique characteristics that can be detected with special staining procedures. For example, *Filobasidiellla neoformans* is one of the few types of yeast that produce a capsule. Thus, a capsule strain of the cerebospinal fluid that shows the presence of encapsulated yeast is diagnostic of the cryptococcal meningitis (Fig. 4.6). Members of the genus *Mycobacterium* are some of the few microorganisms

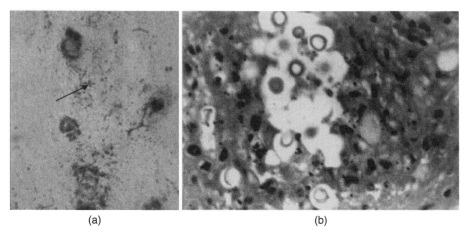

(a)   (b)

Figure 4.6. (a) Acid-fast-stained smear showing pink colored bacilli of *M. tuberculosis* (Tiwari et al., 2005). (b) Lymph node biopsy showing many large, encapsulated yeast cells of *C. neoformans*. Haematoxylin and eosin stain; original magnification 400× (Khan et al., 2003). (*See insert for color representation of the figure*).

that are acid-fast. If the patient has symptoms of tuberculosis, then an acid-fast stain is usually employed on a sample of their sputum to determine *Mycobacterium tuberculosis* infection.

4.3.2. Metabolic Differences

The identification of most prokaryotes relies on analyzing their metabolic capabilities such as the types of sugars utilized or the end products produced. In some cases these characteristics are revealed by the growth and colony morphology on the cultivation media, but most often they are demonstrated using biochemical tests.

4.3.2.1. Culture Characteristics

Microorganisms can be grown in a pure culture. They are the easiest to identify, because it is possible to obtain high number of a single type of microorganisms. Even the colony morphology can give initial clues to the identity of the organisms. For example, colonies of the *Streptococci* are generally fairly small relative to many other bacteria such as *Staphylococci*. Colonies of *Serratia marcescens* are often red when incubated at 22 °C owing to the production of pigments. *Pseudomonas aeruginosa* often produces a soluable pigment, which discolors the growth medium. In addition, cultures of *P. aeruginosa* have a distinct fruity odor.

The use of selective and differential media in the isolation process can provide additional information that helps to identify an organism. For example, if a soil sample is plated onto a medium that lacks a nitrogen source and is then incubated aerobically, any resulting colonies are likely members of the genus *Azotobacter*. The ability to fix nitrogen under aerobic condition is an identifying characteristic of these bacteria.

In clinical laboratories, where rapid but accurate diagnosis is essential, specimens are plated onto media, specially designed to provide important clues as to identify the disease causing organism. For example, a specimen taken by swabbing the throat of a patient, complaining of a sore throat, can be inoculated onto blood agar, a nutritionally rich medium containing red blood cells. This differential medium enables to detect the characteristic β-hemolytic colonies of *Streptococcus pyrogenes*. Urine sample collected from a patient suspected of having a urinary tract infection is plated onto MacConkey agar, which is both selective and differential. MacConkey agar has bile salts, which inhibit the growth of most nonintestinal organisms, and lactose along with a pH indicator, which differentiates lactose fermenting organisms. *E.coli*, the most common causative organism of urinary tract infections, forms characteristic pink colonies on MacConkey agar because of its ability to ferment lactose. Other bacteria can also grow and ferment lactose on this medium; however, colony appearance alone is not capable enough to conclusively identify *E. coli* (Fig. 4.7).

4.3.2.2. Streak Method

In order to identify bacteria, it is necessary to obtain a pure culture. This is done by using the streak plate method (Fig. 4.8). Bacterial cells are spread over the surface of an agar plate in a continuous dilution, so that the cells will be separated from each other. When the plate is incubated, the individual cells will grow into colonies originated from a single cell.

Tips for Success: Obtaining well-isolated colonies takes practice. While streaking a section of the plate, try to keep the inoculating loop in contact with the surface of the agar at all times.

Procedure

1. Heat sterilize an inoculating loop.
2. For a liquid culture, dip the loop into the broth, or, for solid media, lightly touch a colony with the loop.
3. Using the loop, spread the culture over the surface of one quadrant of the plate as shown in Figure 4.8.

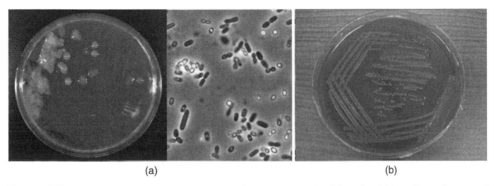

(a)   (b)

Figure 4.7. (a) Isolation of *Azotobacter* spp. from soil sample. (b) *Escherichia coli* produces red colonies on MacConkey agar. (*See insert for color representation of the figure*).

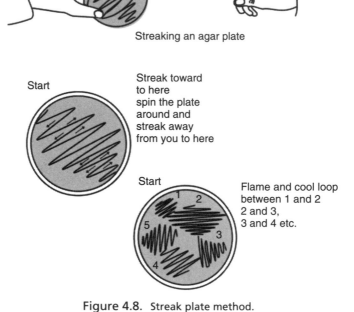

Figure 4.8. Streak plate method.

4. Sterilize the inoculating loop again.
5. Continue the streak into the next quadrant.
6. Repeat steps 3–5 until the pattern is complete.
7. Incubate the plate.
8. Observe for the presence of well-isolated colonies.

4.3.2.3. Biochemical Characteristics

Biochemical test-based identification systems are familiar to most microbiologists and require little training to operate. Systems range from strip cards for specific groups of bacteria (e.g., coryneforms, bacillus, and enterics) to large plate arrays that may be automatically scanned for changes due to pH shifts or redox reactions. The strength of identification in enterics is generally quite good, and the ease of use and cost per sample for identification is considerably less than that for DNA sequencing, but higher than that for FAME (fatty acid methyl ester) analysis (Cook et al., 2003; O'Hara, 2005). The use of these systems depends on the choice of the correct "card" or "strip" of wells of reagents. This is typically done using information such as that gained from the Gram stain (a prerequisite step not involved in the other two major technologies). One problem with most biochemical test systems, however, is that these systems are geared to the clinical market, and, as a result, are limited in the number of environmental species they can identify.

CATALASE TEST. It is a test to detect the presence of the catalase enzyme. Most organisms possess this enzyme, which is capable of breaking down hydrogen peroxide. Organisms containing the catalase enzyme will form oxygen bubbles (Fig. 4.9) when exposed to hydrogen peroxide.

Tips for Success: Do not use a nichrome loop as this will give a false-positive reaction. Because hydrogen peroxide is unstable, always check the expiration date on the solution. Do not perform the catalase test on blood agar, because blood cells in the agar contain the catalase enzyme. Always run positive and negative controls from known cultures.

Procedure

1. Place a drop of 3% hydrogen peroxide onto a clean microscope slide.
2. Touch an isolated colony with an inoculating loop.
3. Place the loop, carrying some of the isolate, into the drop of hydrogen peroxide.
4. Observe the slide for the evolution of bubbles.
5. The reaction is positive if oxygen bubbles form rapidly.

OXIDASE TEST. It is a test to detect the presence of the enzyme cytochrome oxidase. In the presence of this enzyme, the oxidase reagent (*N,N,N',N'*-tetramethyl-*p*-phenylenediamine dihydrochloride) is oxidized and turns from colorless to purple (Fig. 4.10).

Tips for Success: Perform the oxidase test only on fresh cultures. Using older cultures may yield unexpected results. The color change should occur within the first 30 s. Disregard any color changes after this time period. Do not use a nichrome loop as this may lead to a false-positive result. Always run positive and negative controls from known cultures.

Procedure

1. Place a piece of filter paper on a clean microscope slide.
2. Place two to three drops of oxidase reagent onto the filter paper.
3. Touch an isolated colony with a wooden applicator stick.
4. Place the end of the stick, carrying some of the isolate, onto the reagent saturated filter paper.
5. Observe for the appearance of dark purple color.
6. The reaction is positive if the smear turns purple within 10–30 s.

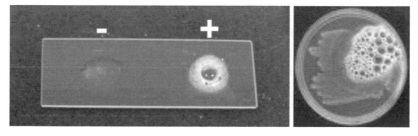

Figure 4.9. Catalase test specimen.

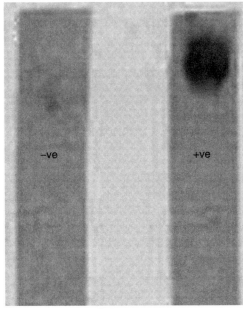

Oxidase test

Figure 4.10.   Oxidase test specimen. (*See insert for color representation of the figure*).

COAGULASE TEST. This test is used to differentiate *Staphylococcus aureus* from other staphylococci species. Coagulase is an enzyme produced by *S. aureus*, which converts fibrinogen to fibrin.*S. aureus* will coagulate rabbit plasma when incubated in its presence.

Tips for Success: Weakly coagulase-positive strains may require overnight incubation for a positive result to be observed. Always run positive and negative controls from known cultures. Be sure to inoculate the unknown and control tubes with approximately the same amount of inoculum. Refrigerate the reconstituted rabbit plasma when not in use.

Procedure

1. Add 0.5 ml reconstituted lyophilized rabbit coagulase plasma with EDTA to a sterile 13 × 100 mm tube.

2. Touch an isolated colony with an inoculating loop.

3. Place the loop, carrying some of the isolate, into the tube containing the rabbit plasma, and mix thoroughly.

4. Incubate the tube at 35 °C for 6 h.

5. Observe the tube for clot formation.

6. If no clotting is observed, incubate again for 24 h and observe (Fig. 4.11).

7. The reaction is positive if clotting is observed. (The plasma will gel to a viscosity where it will not flow down the tube when tilted 45 °.)

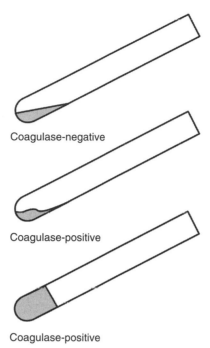

Coagulase-negative

Coagulase-positive

Coagulase-positive

Figure 4.11. Coagulase test specimen.

MOTILITY TEST. This test is used to determine the presence of flagella (external appendages used for movement). Bacteria that are motile have flagella, where as non-motile do not.

Tips for Success: Be sure that the microscope is located on a solid surface free from vibration. It is helpful to reduce the illumination, and increase contrast to help view the bacteria, as they are not stained in this procedure. Motility by direct microscopic observation can be difficult for an amateur microbiologist. Always run positive and negative controls from known cultures. Do not mistake Brownian motion or fluid movement under the cover slip for motility. Motility agar may be used as an alternative to this procedure.

Procedure

1. Inoculate an isolated colony into tryptic soy broth or brain heart infusion broth.
2. Incubate the culture for 24 h (Fig. 4.12).
3. Place a drop of the broth onto a clean microscope slide, and cover with a cover slip.
4. Observe microscopically for motility.
5. A positive test result is indicated by individual bacterial cells moving in random directions.

Figure 4.12. Motility test specimen.

HEMOLYTIC TEST. In this procedure, the unknown bacteria are subcultured onto blood agar and observed for their ability to lyse red blood cells. Blood agar is composed of a nutrient agar base enriched with 5–10% sheep, rabbit, or horse blood. Bacteria that can completely lyse the blood cells, causing a clearing around the colony, are called β *hemolytic*. Bacteria that can partially break down the blood cells, causing a green or brown discoloration of the agar around the colony, are called α *hemolytic*. Bacteria that cannot lyse the cells, causing no change in the agar, are called γ *hemolytic*.

Tips for Success: Avoid contamination by heat sterilizing the inoculating loop just before streaking plates. After sterilizing the loop, touch it to a portion of the agar where no growth is present to cool the loop (this avoids "hot looping" the colony chosen for isolation). After incubation, it is helpful to make observations with proper back-lighting. Always run positive and negative controls from known strains.

Procedure

1. Touch an isolated colony with a sterile inoculating loop.
2. Using the loop, streak for isolation onto a blood agar plate.
3. Incubate the plate for 24–48 h.
4. Observe for the presence of a zone of clearing or discoloration around isolated colonies.

TRIPLE SUGAR IRON TEST. Triple sugar iron agar is used for a dual purpose. It can indicate both carbohydrate fermentation and hydrogen sulfide production. This test is generally used to help identify enteric bacteria. In this procedure, bacteria can be differentiated based on their pattern of utilization of several nutrients.

There are two parts to the test, the slant and the butt:

- Alkaline/alkaline—A red slant and red butt indicates the absence of carbohydrate fermentation.
- Alkaline/acid—A yellow butt and red slant indicates glucose fermentation only.
- Acid/acid—A yellow butt and yellow slant indicates lactose and/or sucrose and glucose fermentation.
- H_2S positive—A black butt indicates the presence of hydrogen sulfide production.

Tips for Success: Examine the tubes at no earlier or later than 18–24 h; otherwise, a false-positive or false-negative reaction may be observed. If the butt of the slant is H_2S positive, and obscures the acid–alkaline reaction, record the butt as acid. Always run positive and negative controls from known strains.

Procedure
1. Touch an isolated colony with a sterile needle.
2. Using the needle, stab the butt, and withdraw the needle along the same line.
3. Touch the same isolated colony with a sterile loop.
4. Streak the surface of the slant.
5. Incubate the tube for 18–24 h.
6. Observe for carbohydrate fermentation, and hydrogen sulfide production indicated by a change in the color of the media.

4.3.3. Serology

The protein and polysaccharides that make up a bacterium are sometimes characteristic enough to be considered identifying markers. The most useful of these are the molecules that make up surface structures including the cell wall, glycocalyx, flagella, and pili.

For example, some species of *Streptococcus* contain a unique carbohydrate molecule as part of their cell wall, which can be used to distinguish them from other species. These carbohydrates, as well as any distinct protein or polysaccharide, can be detected using techniques that rely on the specificity of interaction between antibodies and antigens. Methods that exploit such interactions are called *serology*.

Highly specific identification of microorganisms can be obtained by serological techniques. *In vitro* (that is, outside the body and in an artificial environment, such as a test tube), antigens and antibodies react together in certain visible ways. The chemical composition of antigens differs, and, therefore, the reactions are highly specific; that is, each antigen provokes an antibody response with only that antibody. When it provokes an antibody response, the antigen is known as an *immunogen*.

The cell wall of gram-negative bacteria consists of several layers of various polysaccharides. The periplasm contains peptidoglycan, a copolymer of polysaccharide and short peptides, and a class of β-glucans. In gram-negative bacilli, the carbohydrate antigens within the wall of the organism are called *somatic* (associated with the soma, that is, the body of the cell) or *O antigens* (Figs. 4.13 and 4.14). Each species has a different array of O antigens that can be detected in serological tests. In a similar manner, those bacilli that are motile also contain characteristic flagellar protein components called *H antigens* (H is from the German word hauch, which refers to motility). In streptococci, the carbohydrate wall antigens are used to group the organisms by alphabetic designations A through V. Many bacteria also contain antigenic carbohydrate capsules that can be used for identification, the primary example being the pneumococci, whose capsules permit them to be differentiated into more than 80 different types. Exotoxins and other protein metabolites of bacterial cells are also antigenic. The interaction of an antibody with an antigen may be demonstrated in several ways. Examples of these are latex agglutination, coagglutination, and enzyme-linked assays.

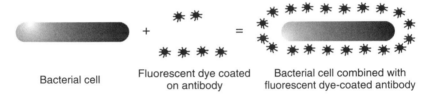

Figure 4.13. Enzyme-linked fluorescent dye assay.

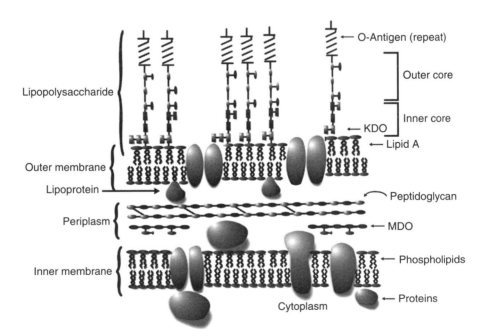

Figure 4.14. Structure of the cell wall of *E. coli*.

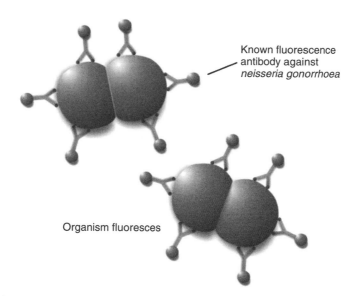

Known fluorescence antibody against *neisseria gonorrhoea*

Organism fluoresces

Figure 4.15. Direct fluorescent antibody test for *Neisseria gonorrhoeae.*

These tests depend on linking the antibody to a particle or an enzyme in order for a positive reaction to be observed. The fluorescent antibody test is similar to the enzyme immunoassay except that the antibody is linked to a dye that fluoresces when it is reviewed microscopically under an ultraviolet light source. Fluorescent antibody tests can provide a rapid diagnosis of infections caused by pathogens that are difficult to grow in culture, or that grow slowly. Thus, they have become popular for detecting organisms such as *Legionella pneumophilia* (the agent of Legionnaires disease), *Bordetella pertussis, Chlamydia trachomatis*, and several viruses, directly in patient specimens. A portion of the specimen dried on a microscope slide is treated with the fluorescent antibody reagent, rinsed to remove unbound antibody, and then viewed under a fluorescence microscope with an ultraviolet light source (Figs. 4.13 and 4.15).

In a positive test, bacteria or viral inclusions fluoresce apple green. This test is used in a similar way to identify microorganisms isolated on culture plates or in cell cultures. Bacterial agglutination test is a simpler test, which detects O and H antigens of gram-negative enteric bacilli, (usually *Salmonella* and *Shigella* species and *E. coli*). When the unknown organism isolated in a culture is mixed with an antiserum (prepared in animals) that contains antibodies specific for its antigenic makeup, agglutination (clumping) of the bacteria occurs. If the antiserum does not contain specific antibodies, no clumping is seen. A control test in which saline is substituted for the antiserum must always be included to be certain that the organism does not clump in the absence of the antibodies.

Commercially available antibodies are routinely used to specifically identify antigenic proteins from a wide variety of organisms. In some instances, the test may be used to identify only the genus and species of an organism. Examples of this include the cryptococcal antigen agglutination assay and the exoantigen assay for *Histoplasma capsulatum*. Other immunoassays are designed to subtype microbes. Monoclonal antibodies directed against the major subtypes of the influenza virus, as well as the various serotypes of *Salmonella*, are commonly used in speciation. Specific antigenic proteins may be detected by antibodies directed against these proteins in immunoblot methods.

Electrophoretic typing techniques have been used to examine outer membrane proteins, whole-cell lysates, and particular enzymes. Several electrophoretic methods are available to examine the protein profile of an organism. Generally, outer membrane proteins and proteins from cell lysates are examined by sodium dodecyl sulfate–polyacrylamide gel electrophoresis. This technique denatures the proteins and separates them on the basis of molecular mass. The protein profile may be used to compare strains.

Nondenaturing conditions are used for the electrophoretic separation of active enzymes. Multilocus enzyme electrophoresis is the typing technique based on the electrophoretic pattern of several constitutive enzymes. Differences in electrophoretic migration of functionally similar enzymes (e.g., lactate dehydrogenase isoenzymes) represent different alleles. These differences or similarities, especially when numerous enzymes are examined, may be used to exclude or infer relatedness. The absence of a particular protein may simply reflect downregulation of that particular gene product, rather than the loss of that particular gene. Additionally, the electrophoretic migration of proteins is dependent on molecular mass, net protein charge, or both. Mutations that do not alter these characteristics will not be detected.

4.3.4. Fatty Acid Analysis (FAME)

Another popular method of bacterial classification is through characterization of the types and proportions of fatty acids present in the cytoplasmic membrane and outer membrane. This technique is nicknamed as FAME. The fatty acid composition of prokyotes can be highly variable including differences in fatty acid length, the presence or absence of a double bond, rings, branched chains, or hydroxyl groups. The fatty acid profile can help to identify a particular bacterial species.

FAME is in widespread use in clinical, public health, and food and water inspection laboratories, where the identification of pathogens and other bacterial hazards needs to be done on a routine basis. A FAME can be created by an alkali-catalyzed reaction between fats or fatty acids and methanol (Fig. 4.16). The molecules present in biodiesel are primarily FAMEs, usually obtained from vegetable oils by transesterification.

Every microorganism has its specific FAME profile (microbial fingerprinting); therefore, it can be used as a tool for microbial source tracking (MST). The types and proportions of fatty acids present in the cytoplasmic membrane and outer membrane (gram-negative) lipids of cells are major phenotypic trains.

Clinical analysis can determine the lengths, bonds, rings, and branches of the FAME. To perform this analysis, a bacterial culture is taken, and the fatty acids are extracted and used to form methyl esters. The volatile derivatives are then introduced into a gas

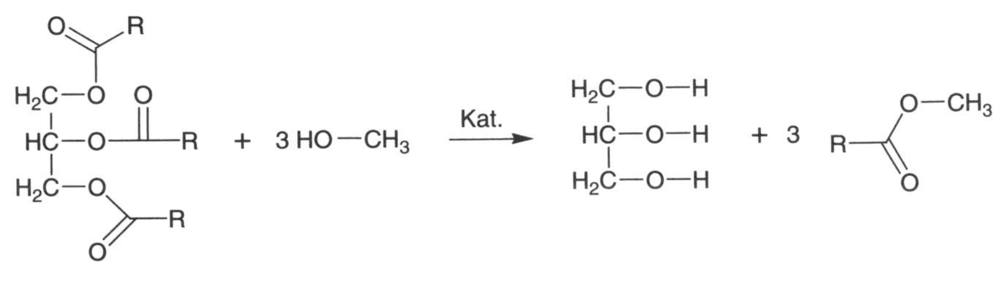

Figure 4.16. Fatty acid methyl ester is formed by an alkali-catalyzed reaction.

chromatagraph, and the patterns of the peaks help to identify the organism (Olsen et al., 1990). This is widely used in characterizing new species of bacteria, and is useful for identifying pathogenic strains.

More than 300 fatty acids and related compounds are found in bacteria. The wealth of information contained in these compounds is both in the qualitative differences (usually at the genus level) and quantitative differences (commonly at the species level). As the biochemical pathways for creating fatty acids are known, various relationships can be established. Thus, $16:0 \rightarrow 16:1$ through the action of a desaturase enzyme, and is a mole-for-mole conversion. Following this, as the bacterial cell becomes physiologically mature, the shift of $16:1 \rightarrow 17:0$ cyclopropane is again a mole-for-mole conversion.

This information suggests that use of the cells in an actively growing stage minimizes the differences between cultures. Use of a $24 + 2$ h culture and harvesting from a rapidly growing quadrant of a quadrant streak plate reduce the differences. Controlled growth temperature and use of standardized commercially available media also contribute to the reproducibility of the fatty acid profile. Branched chain fatty acids (iso and anti-iso acids) are common in many gram-positive bacteria, while gram-negative bacteria are composed predominantly of straight chain fatty acids. The presence of lipopolysaccharides (LPS) in gram-negative bacteria gives rise to the presence of hydroxy fatty acids in those genera. Thus, the presence of 10:0 3OH, 12:0 3OH, and/or 14:0 3OH fatty acids indicates that the organism is gram-negative, and, conversely, the absence of the LPS and hydroxy fatty acids indicates that the organism is gram-positive (Kunitsky et al., 2005). As a result, it is not necessary to perform the traditional Gram stain prior to FAME analysis. Fatty acid profiles are quite unique for *Bacillus anthracis*, compared with other bacillus species (Song et al., 2000).

As bacteria frequently exchange plasmids, the system would not work well if such changes did cause alterations in the fatty acid composition. Similarly, treatment with ultraviolet light (a frame-shift mutagen) or point-mutagens such as nitrosoguanidine and ethyl methanesulfonate at levels that kill 99.999% of the cells and create large numbers of auxotrophic and/or motility mutants did not affect the fatty acid profile, as long as the growth rate was relatively normal. This suggests that the fatty acid composition is highly conserved genetically and that significant changes take place only over considerable periods. As a result, the same genus and species of bacteria from anywhere in the world will have highly similar fatty acid profiles as long as the ecological niche is similar. The adaptation to different ecological niches over long periods provides information vital to strain tracking by fatty acid profiling.

4.4. USING GENOTYPIC CHARACTER TO IDENTIFY MICROBES

Recent years have seen an explosion in the development and application of molecular tools for identifying microbes and analyzing their activity (Amor et al., 2007). The tools that have been developed for identifying microbes and analyzing their activity can be divided into those based on nucleic acids and other macromolecules and approaches directed at analyzing the activity of complete cells. The nucleic-acid-based tools are more frequently used because of the high throughput potential provided by using PCR amplification or *ex situ* or *in situ* hybridization with DNA, RNA, or even peptide nucleic acid probes. These methods involve the study of the microbial DNA, the chromosome and plasmid, their composition, homology, and presence or absence of specific genes. Application of genome-scale analysis such as DNA microarray technology has revolutionized multiple scientific disciplines. Diagnostic evaluation using genotypic methods—such as PCR of the species-specific ligase and glycopeptide resistance genes helps to identify four *Enterococcus* species and 16S RNA sequencing, the "gold standard" for identification of enterococci—confirmed the results obtained by the FT-IR classification (Kirschner et al., 2001).

Approaches based on complete or partial genomes include DNA arrays that can be used in comparative genomics or genome-wide expression profiling (de Vos, 2001). These omics approaches have now become feasible for probiotic bacteria after the recent realization of the complete genome sequences of human isolates of *Bifidobacterium longum* (Schell et al., 2002) and *Lactobacillus plantarum* (Kleerebezem et al., 2003).

4.4.1. Nucleic Acid Probes to Detect Specific Nucleotide Sequence

DNA gene probes may become extremely useful in studying gene transfer and adaptation mechanisms in natural bacterial communities, and in the laboratory. This technology allows the detection of specific gene sequence(s) in bacterial species, and can be used to find and monitor recombinant DNA clones in microorganisms being considered for release into the natural environment. It may provide a new generation of highly specific tests that offer advantages over the classical approaches for identifying specific organisms (Trevors, 1985). Single-stranded DNA from an organism of interest is allowed to attach itself to a membrane. A single-stranded DNA probe binds to its immobilized complementary strand. This binding can be detected by labeling the probes with radioisotopes or with nonradioactive reporter molecules, such as the biotin–streptavidin–enzyme complex (http://www.ilri.org; Fig. 4.17).

To adapt DNA probe methodology for use in soils, the following features of a protocol were needed to be improved or developed: (i) A procedure was needed, which would allow processing of more samples simultaneously and in a shorter period for analysis of the number of treatments and replicates needed for ecological studies. (ii) the isolated DNA had to be of sufficient purity and size for use in experiments involving digestion with restriction endonucleases, transfer to cellulose nitrate membranes, and hybridization to DNA probes. If contaminants are not removed, reduction in the efficiency of digestion by restriction endonucleases and the specificity of hybridization will be seen. (iii) It was also necessary to develop probes both sensitive and specific enough to detect the presence of a particular sequence of low frequency in the complex mixture of DNAs isolated from the soil bacterial community. The standard method of labeling probes by nick translation did not appear to be sensitive or specific enough for probing natural populations.

A probe is a single-stranded nucleic acid that has been labeled with a detectable tag, such as radioisotope or a fluorescent dye. It is complementary to the sequence of interest. Floresecent *in situ* hybridization is increasingly used to observe and identify intact microorganisms in environmental samples and clinical samples. By using a probe that binds to certain ribosomal RNA (rRNA) sequences, either species specific or groups of related organism can be identified, and characteristics of rRNA can be studied that make it ideal for classification.

Nucleic acid probing is based on two major techniques: dot blot hybridization and whole-cell *in situ* hybridization. Dot blot hybridization is an *ex situ* technique in which total RNA is extracted from the sample and is immobilized on a membrane together

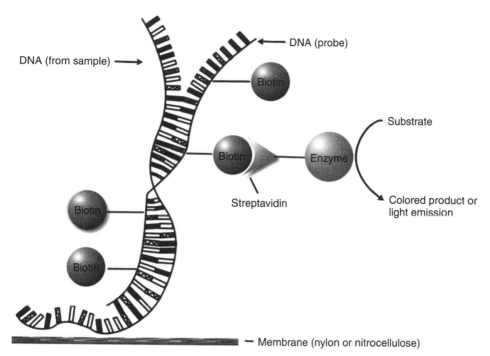

Figure 4.17. Nucleic acid hybridization process.

with a series of RNAs of reference strains. Subsequently, the membrane is hybridized with a radioactively labeled probe, and after stringent washing, the amount of target rRNA is quantified. Because cellular rRNA content is dependent on the physiological activity of the cells, no direct measure of the cell counts can be obtained. In contrast to dot blot hybridization, fluorescent *in situ* hybridization (FISH) is applied to morphologically intact cells and thus provides a quantitative measure of the target organism. The listed probes can all be used for dot blot hybridizations, but for application in FISH, specific validation is required. Some regions of the rRNA are not accessible because of their secondary structure and protection in the ribosome. Hence, the number of validated FISH probes is much lesser than that of the probes suitable for *ex situ* analysis.

4.4.2. Amplifying Specific DNA Sequences Using PCR

The PCR can be used to amplify a specific nucleotide present in nearly any environment. This includes DNA in samples such as body, fluids, soil, food, and water. This technique can be used to detect organisms that are present in extremely small numbers as well as those that cannot be grown in cultures. The most commonly used DNA sequence for bacterial phylogenetics is the highly conserved 16S rRNA gene sequence (Fig. 4.18), and primers have been designed to selectively amplify bacterial 16S rRNA genes (Weisburg et al., 1991).

To use PCR to detect a microbe of interest, a sample is first treated to release and denature the DNA. Specific primers and other ingredients are then added to the denatured DNA

A simplified map of the 16S rRNA molecule

Primer	Primer sequence	Primer length	E.coli position
Univ Bact F	5′ GAG TTT GAT YMT GGC TC	17-mer	9–25
Univ Back R	5′ GYT ACC TTG TTA CGA CTT	18-mer	1509–1492

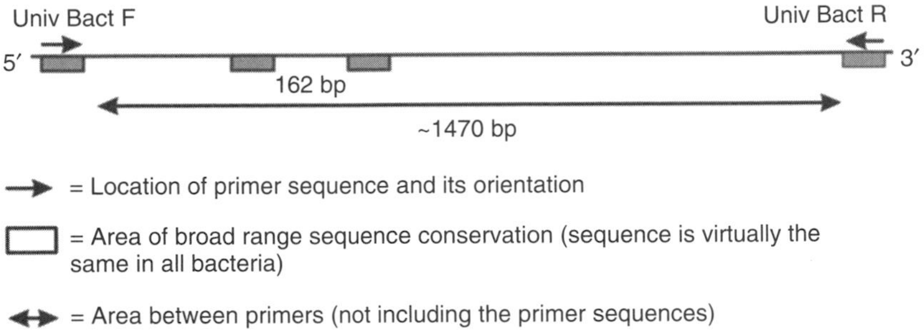

→ = Location of primer sequence and its orientation

▭ = Area of broad range sequence conservation (sequence is virtually the same in all bacteria)

◄► = Area between primers (not including the primer sequences)

Figure 4.18. A simplified map of the 16S rRNA molecule.

forming the components of the PCR reaction. Some information about the nucleotide sequence of the organism must be known in order to select the appropriate primers. After approximately 30 cycles of PCR, the DNA region flanked by the primer will be amplified a billion-fold. In most of the cases the results in a sufficient quantity for the amplified fragment can be readily visible as a discrete band on the gel after staining with ethidium bromide.

In such situations the DNA markers most commonly used have been restriction fragment length polymorphisms (RFLPs). Fragments are usually generated by frequent-cutting enzymes and separated by conventional agarose gel electrophoresis, but occasionally rare-cutting enzymes are used and larger fragments are separated by pulsed-field gel electrophoresis (PFGE). RFLPs have been used successfully to generate numerous microbial typing systems, but for some organisms discrimination is suboptimal because there is a tendency for one or two genetic types to predominate amongst an apparently heterogeneous population. Better discrimination between isolates can be achieved by the secondary step of Southern blot hybridization with radiolabeled probes recognizing repetitive DNA sequences. However, this adds a rather laborious, expensive second step, which is incompatible with large-scale epidemiological studies.

The PCR profiles obtained were unique for unrelated strains, whereas similar patterns were observed for epidemiologically related strains isolated from members of the same family. In some studies, such as that carried out on human herpes virus 6 with primers from known viral DNA sequences, the amplified products were analyzed by a combination of Southern blot hybridization, digestion with restriction endonucleases, and partial nucleotide sequencing. For many organisms genetic maps are not available and relatively little is known of their molecular biology.

4.4.3. Sequencing Ribosomal RNA Genes

Full and partial 16S rRNA gene sequencing methods have emerged as useful tools for identifying phenotypically aberrant microorganisms. Hence, 16S rRNA gene sequencing

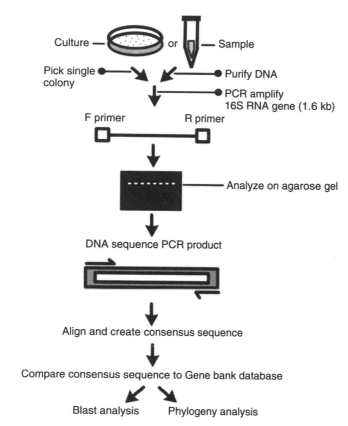

Figure 4.19. Bacterial identification by 16S rRNA gene sequencing.

is also performed (Fig. 4.19). In a particular case it was found that all three patients had endocarditis, and conventional methods identified isolates from patients A, B, and C as a *Facklamia* sp., *Eubacterium tenue*, and a *Bifidobacterium* sp. But when 16S rRNA gene sequencing was performed, the isolates were identified as *Enterococcus faecalis, Cardiobacterium valvarum,* and *Streptococcus mutans*, respectively (Petti et al., 2005).

Technologist bias or inexperience with an unusual phenotype or isolate may similarly compromise identification when results of biochemical tests are interpreted to fit expectations. Although not perfect, genotypic identification of microorganisms by 16S rRNA gene sequencing has emerged as a more objective, accurate, and reliable method for bacterial identification, with the added capability of defining taxonomical relationships among bacteria (Clarridge, 2004; Yuan et al., 2005).

Phenotypic methods have numerous strengths but often fail because the phenotype is inherently mutable and subject to biases of interpretation. 16S rRNA gene sequencing is a more accurate and objective method for identification of microorganisms (Bosshard et al., 2004; Fredericks and Relman, 1999; Han and Falsen, 2005) with particular utility in the clinical laboratory. It also reduces the interpretive bias and shows the need for a "pretest" probability regarding a microorganism's classification to direct workup and database selection. Medical technologists may pursue an erroneous identification algorithm based on their phenotypic "intuition," such that when unusual microorganisms are encountered, they are made to "fit" with technologist expectations, or when common microorganisms with atypical phenotypes are encountered, they are made to "fit" characteristics of extremely unusual pathogens. Conventional automated identification systems often rely on technologist's interpretations of a microorganism's Gram stain morphology (e.g., RapID ANA) or oxidase result (e.g., Biolog) for selecting the correct reference database. This case series demonstrates that seemingly simple biochemical or Gram reactions are not unquestionably foolproof and may lead to inappropriate use of comparative databases. Such exhaustive phenotypic testing potentially delays turnaround time without the added benefit of accuracy.

The nucleotide sequence of the rRNA may be used to identify prokaryotes, particularly those that are difficult or currently impossible to grown cultures. The prokaryotic 70S ribosome, which plays an indispensable role in protein synthesis, is composed of proteins and three different rRNAs (5S, 16S, and 23S). Because of its highly constrained and essential function, the nucleotide sequence changes that can occur in the rRNAs, yet, still allow the ribosome to operate. This is why it is proved to be so important in classification and, more recently, in identification.

Of the different rRNAs, the 16S molecule has proved most useful in taxonomy because of its moderate size ($\sim$1500) nucleotides. The 5S molecules lacks the critical amount of information because of its small size (120 nucleotides), whereas the larger size of 23S molecule ($\sim$3000 nucleotides) has made it more difficult to sequence in the past.

Some regions in the prokaryotes are virtually the same in all prokaryotes, whereas others are variable. It is the variable region that is used to identify an organism. Once the nucleotide sequence is determined, it can be compared with 16S region of known organisms by searching the extensive database of rRNA sequences that exists. For example, the Ribosomal Database Project (RDP) contains a large collection of such sequences, now numbering over 100,000. The RDP can be assessed electronically (http://rdp.cme.msu.edu/html/) and besides sequences contains phylogenetic tutorials, reference citations, previews of new release of sequences, and a host of other features.

The methods for obtaining rRNA sequences and generating phylogenetic trees are now quite routine (Fig. 4.20). Newly generated sequences can be compared with sequences in the RDP and other genetic databases such as Gen Bank(USA), DDBS(Japan), or EMBL(Germany). Then, using a treeing algorithm, a phylogentic tree is produced describing the evolutionary information inherent in the sequences.

The separation of the microorganisms is typically represented by what is known as a dendrogram. Essentially, a dendrogram appears as a tree oriented on a horizontal axis. The dendrogram becomes increasingly specialized. The similarity coefficient increases as the dendrogram moves from the left to the right. The right-hand side consists of branches of the tree. Each branch contains a group of microorganisms.

The dendrogram depiction of relationships can also be used for another type of microbial taxonomy. In the second type of taxonomy, the criterion used is the shared evolutionary heritage. This heritage can be determined at the genetic level. This is termed *molecular taxonomy*.

To begin with the process, the PCR is used to amplify the gene encoding 16S rRNA from the genomic DNA. Following this, the PCR product is sequenced by the dideoxy DNA sequencing method. Using PCR primers complementary to the conserved sequences in the small unit of rRNAs, only a tiny amount of the cell material can yield a huge amount of DNA product for sequencing purposes. Once sequencing is done, it is ready for computer analysis.

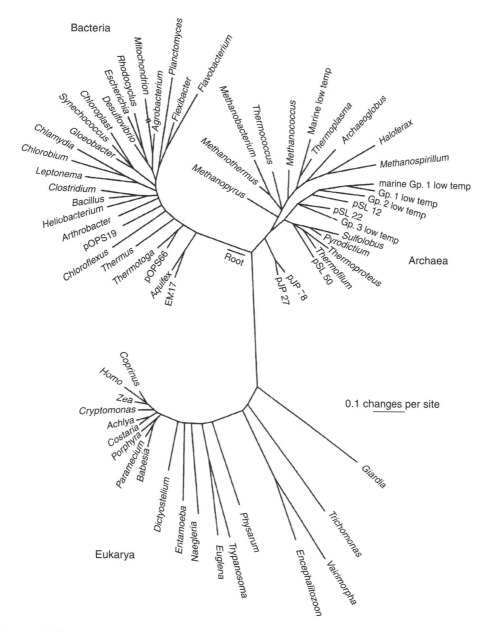

Figure 4.20. Sequencing 16S rDNA from individual organisms has given us the ability to map phylogenetic relationships between single-celled organisms based on the similarity of their 16S genes.

Several different algorithms for sequence analysis and phylogenetic tree formation are available for comparative ribosomal sequencing (Fig. 4.20).

However, regardless of which program to be used, the raw data must first be aligned with the previous aligned sequences using a sequence editor. Not all the rRNAs are exactly the same length. Thus, during alignment, gaps can be inserted wherever necessary in regions where one sequence can be shorter than the other. The aligned sequences are then imported into a treeing programme and comparative analysis is done. Two widely used treeing algorithms are distance and parsimony. Using distance method, sequences are aligned and then an evolutionary distance (ED) is calculated by having the computer record every position in the dataset in which there is a difference in the sequence. From these dataset a data matrix can be constructed, which shows the ED between two sequences in the dataset. Following this, a statistical correlation is factored into the ED, which considers the possibility that more than one change can occur at a given site. Once this is accounted for, a phylogenetic tree is generated in which the lengths of the lines in the tree are proportional to the EDs (Fig. 4.20).

4.5. CHARACTERIZING STRAIN DIFFERENCES

If a species of bacteria is isolated and cultivated in the laboratory it is known as a *strain*. A single isolate with distinctive characteristic[s] may also represent a strain. Members of the same species that have small differences between them can be distinguished by additional methods. These species are then subdivided into subspecies, subgroups, biotypes, serotypes, variants, and so on. Methods of bacterial strain identification can be broadly delimited to genotypic techniques based on profiling an organism's genetic material (primarily its DNA) and phenotypic techniques based on profiling either an organism's metabolic attributes or some aspect of its chemical composition (Table 4.2). Genotypic techniques have the advantage over phenotypic methods that they are independent of the physiological state of an organism; they are not influenced by the composition of the growth medium or by the organism's phase of growth. The process of differentiating strains based on their phenotypic and genotypic differences is known as *typing*. These typing methods are useful to understand typability, reproducibility, discriminatory power, ease of performance, and ease of interpretation. Two methods of typing are found. Phenotypic techniques detect characteristics expressed by the microorganism (such as shape, size, staining properties, biochemical properties, and antigenic properties that can be measured without reference to the genome) and genotypic techniques involve direct DNA-based analysis of chromosomal or extrachromosomal genetic elements.

Molecular diagnostics provides outstanding tools for detection, identification and characterization of microbial strains. The application of these and other related techniques, along with the development of molecular markers for bacterial strains, greatly facilitates understanding of the ecological interactions of microbial strains, their roles, succession, competition, and prevalence in food fermentations, and allows the correlation of these features to desirable quality attributes of the final product. Several strains of microorganisms have been selected or genetically modified to increase the efficiency with which they produce enzymes (Table 4.2).

4.5.1. Phenotypic Typing Methods

Traditional methods for microbial identification require the recognition of differences in morphology, growth, enzymatic activity, and metabolism to define genera and species. Phenotypic identification often suggests unusual organisms not typically associated with the submitted clinical diagnosis. Phenotypic profiles including Gram stain results, colony morphologies, growth requirements, and enzymatic and/or metabolic activities are generated, but these characteristics are not static and can change with stress or evolution.

Thus, when common microorganisms present with uncommon phenotypes, when unusual microorganisms are not present in reference databases, or when databases are out of date, reliance on phenotypes can compromise accurate identification.

4.5.1.1. Biochemical Typing

Traditional microbial identification methods typically rely on phenotypes such as morphological features, growth variables, and biochemical utilization of organic substrates. The biological profile of an organism is termed a *biogram*. The determination of relatedness of different organisms on the basis of their biograms is termed *biotyping*. Investigators must determine which profile variables have the greatest differentiating capabilities for a given organism (Kilian et al., 1979; Maslow et al., 1993). For example, Gram stain characteristics, indole positivity, and the ability to grow on MacConkey medium do not aid in the differentiation of nonentero hemorrhagic *E. coli* from *E. coli* O157:H7. However, sorbitol fermentation has proven to be an extremely useful characteristic of the biochemical profile used to differentiate these strains.

Biograms that are identical have been used to infer relatedness between strains in epidemiological investigations (Martin de Nicolas et al., 1995). The biograms of organisms are not entirely stable, and several isotypes may exist from a single isolate. Biograms may be influenced by genetic regulation, technical manipulation, and the gain or loss of plasmids. In many instances, biotyping is used in conjunction with other methods to more accurately profile microorganisms.

Biotyping makes use of the pattern of metabolic activities expressed by an isolate, colonial morphology, and environmental tolerances. Strains are referred to as *biotypes*. Biochemical tests are used to identify many bacteria and also used to distinguish strains. If the biochemical variation is uncommon; it can be used for tracing the source of certain disease outbreaks. A strain has a characteristic biochemical pattern (Fig. 4.21) and is called a *biovar* or *biotype* (Tester et al., 2009). They performed Western blot analysis of the H-type BSE zebu (Charly-04) with (a) a core-binding antibody (Sha31), (b) an amino-terminal binding antibody (12B2), and (c) a carboxy-terminal binding antibody (SAF84). Samples are assigned to the lanes as follows: negative control (N), L-type BSE (L), C-type BSE (C) and for the zebu medulla oblongata (lane 1, 15 mg tissue equivalent), cerebellar cortex (lane 2, 15 mg), hippocampus (lane 4, 0.75 mg), piriform lobe (lane 5, 15 mg), basal ganglia (lane 7, 1.5 mg), frontal cortex (lane 8, 15 mg), occipital cortex (lane 9, 15 mg), and temporal cortex (lane 10, 15 mg). The dashed line indicates the molecular mass of the unglycosylated C-type PrP and helps to visualize differences compared to the H-type BSE zebu. The same samples, but deglycosylated are shown in (d) with a carboxy-terminal binding antibody (SAF84). A molecular mass marker (in kDa) is indicated on the left in Figure 4.21.

Biotyping may be performed manually or using automated systems. Sugar fermentation, amino acid decarboxylation/deamination, standard enzymatic tests (such as IMViC, citrate, and urease), tolerance to pH, chemicals and dyes, hydrolysis of compounds, hemagglutination, and hemolysis are some examples of biotyping methods. They offer some advantages as most strains are typeable. The techniques are reproducible with relatively ease in performance and interpretation. But the main disadvantage is that they have a poor discriminatory power. Variation in gene expression is the most common reason for isolates that represent single strain to differ in one or more biochemical reactions. Point mutation too contributes to this problem.

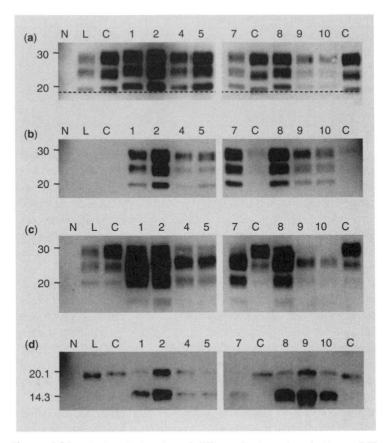

Figure 4.21. Biochemical typing of different brain regions in H-type BSE.

4.5.1.2. Serological Typing

Serological typing or serotyping is based on the fact that strains of same species can differ in the antigenic determinants expressed on the cell surface (Fig. 4.22). Surface structures such as lipopolysaccharides, membrane proteins, capsular polysaccharides, flagella, and fimbriae exhibit antigenic variations. Strains differentiated by antigenic differences are known as *serotypes*. Serotyping is used for several gram-negative and gram-positive bacteria identification.

Serotyping is performed using several serological tests such as bacterial agglutination, latex agglutination, coagglutination, and fluorescent and enzyme labeling assays. Most strains are typeable. They have good reproducibility and ease of interpretation though some have ease of performance. But they have some disadvantages. Some autoagglutinable (rough) strains are untypeable. Some methods of serotyping are technically demanding. There is dependency on a good quality reagent from commercial sources. In-house preparation of reagents is a difficult process. Serotyping has a poor discriminatory power due to large number of serotypes, cross reaction of antigens, and untypeable nature of some strains.

The invention of serological typing concerns a method for typing antibodies in a sample liquid by means of type-specific antigens and in particular a method for typing

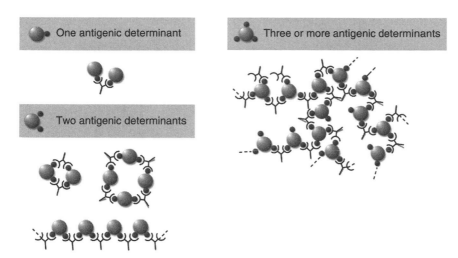

Figure 4.22. Serotyping distinguished by antigenic determinants on the surface of the cell.

antibodies to the hepatitis C virus and peptide antigens suitable for this. A further possibility of serological type differentiation of infections with the HCV types 1, 2, and 3 can be carried out by means of an indirect ELISA using peptide antigens of the amino acid regions. For this, type-specific peptide antigens can be immobilized separately according to their type in individual wells of a microtiter plate and each was contacted with separate aliquots of a plasma sample from HCV-infected blood donors. The typing was carried out according to the reactivity of the serum sample with the individual peptide antigens. However, this method is relatively inaccurate and, moreover, does not allow the determination of individual viral subtypes, that is, individual virus strains whose immunogenicity differs only to a slight extent.

4.5.1.3. Genomic Typing

Currently, genomic typing of microorganisms is widely used in several major fields of microbiological research (Table 4.4). Taxonomy, research aimed at elucidation of evolutionary dynamics or phylogenetic relationships, population genetics of microorganisms, and microbial epidemiology all rely on genetic typing data for discrimination between genotypes. Apart from being an essential component of these fundamental sciences, microbial typing clearly affects several areas of applied microbiological research. The epidemiological investigation of outbreaks of infectious diseases and the measurement of genetic diversity in relation to relevant biological properties such as pathogenicity, drug resistance, and biodegradation capacities are obvious examples. The diversity among nucleic acid molecules provides the basic information of genomic typing. However, researchers in various disciplines tend to use different vocabularies, a wide variety of different experimental methods to monitor genetic variation, and sometimes widely differing modes of data processing and interpretation (van Belkum et al., 2001).

In a unique example cited by Arostegui et al. (2000), it was shown that minor histocompatibility antigen (HA-1) genomic typing by RSCA (reference-strand-mediated conformation analysis) is easy to perform and that could be used as a routine typing

TABLE 4.4. Survey of the Characteristics of Several Currently Used Microbial Typing Methods

Typing Method	Typeability	Reproducibility	Discriminatory Power	Ease of Performance	Ease of Interpretation	General Availability	Cost
Phenotypic							
P1. Antimicrobial susceptibility	Good	Good	Poor	Excellent	Excellent	Excellent	Low
P2. Manual biotyping	Good	Poor	Poor	Excellent	Excellent	Excellent	Low
P3. Automated biotyping	Good	Good	Poor	Good	Good	Variable	Medium
P4. Serotyping	Variable	Good	Variable	Good	Good	Variable	Medium
P5. Bacteriophage typing	Variable	Fair	Variable	Poor	Poor	Excellent	Medium
P6. PAGE	Excellent	Good	Good	Excellent	Fair	Good	Medium
P7. Immunoblotting	Excellent	Good	Good/excellent	Good	Fair	Variable	Medium
P8. MLEE	Excellent	Excellent	Good	Good	Excellent	Variable	High
Genotypic							
G1. Plasmid profiles	Variable	Fair	Variable	Fair	Good	Excellent	Medium
G2. Plasmid REA	Variable	Excellent	Good	Good	Excellent	Excellent	Medium
G3. Chromosomal REA	Excellent	Variable	Variable	Good	Fair	Variable	Medium
G4. Ribotyping	Excellent	Excellent	Good	Good	Good	Variable	High

method for The Kidd (JK) blood group system that is clinically important in transfusion medicine. In another example (Aarts et al., 1999), the genetic relationship between isolates of Listeria monocytogenes belonging to different serotypes was determined and the suitability of automated laser fluorescent analysis (ALFA) of amplified fragment length polymorphism (AFLP) fingerprints was assessed by genomic typing of 106 L.

4.5.1.4. Phage Typing

Phage typing is a method used for detecting single strains of bacteria. It is used to trace the source of outbreaks of infections. The viruses that infect bacteria are called *bacteriophages* ("phages" for short), and some of these can infect only a single strain of bacteria. These phages are used to identify different strains of bacteria within a single species. They help to characterize bacteria, extending to strain differences, by demonstration of susceptibility to one or more (a spectrum) races of bacteriophage; widely applied to staphylococci and typhoid bacilli, for epidemiological purposes. Phage typing requires the use of a standard collection of dissimilar phages. In the process of developing a phage typing set, numerous phages are first isolated and tests are undertaken to determine if they are different and useful in delineating the types of organisms under study.

For many years phage typing has been a useful epidemiologic tool for studying outbreaks of *Salmonella typhi* and *Salmonella typhimurium* (Fig. 4.23). Ten types of phages (podoviruses) were found to be morphologically identical to *Salmonella* phage P22. Two phages are siphoviruses and identical to flagella-specific phage chi (Eisenstark et al., 2009). This system was particularly useful for differentiating a group of animal strains that had a number of diverse phage types. Strains can be characterized by their pattern of resistance or susceptibility to a standard set of bacteriophages. This relies on the presence or absence of particular receptors on the bacterial surface, which are used by the virus to bind to the bacterial wall. This method is used to type isolates of *S. aureus* and *Salmonella* sp. Such stains are referred to as *phage types*. The susceptibility of an organism to a particular type of phage can be readily demonstrated in the laboratory.

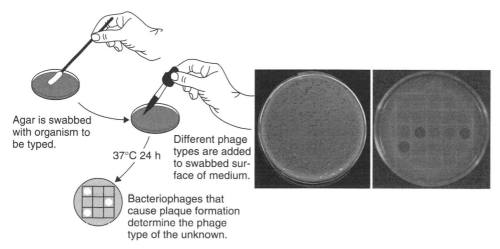

Figure 4.23. Phage typing plaque formation for *Staphylococcus aureus*. (*See insert for color representation of the figure*).

Firstly, a culture of the test organism is inoculated into melted, cooled nutrient agar and poured onto the surface of an agar plate thus creating a uniform layer of cells, then drops of different types of bacteriophage are carefully placed on the surface of the agar. During incubation the bacteria will multiply, forming a visible haze of cells. A clear zone will be formed at each spot where bacteriophage has been added in case the organism is susceptible to the type of phage. The pattern of clearing indicates the susceptibility to different phage and can be compared to determine the strain differences. Using phages to differentiate bacteria is justified the term *phage typing*.

Phage typing can be extremely important in many health situations because it can identify random, unrelated organisms as well as the isolates that are actually responsible for a given problem. Aside from relating an organism to an outbreak, this laboratory method can also be used for surveillance, assessing strain distribution, and ascertaining the effectiveness of therapeutic measures. This technique has a fair amount of reproducibility, discriminatory power, and ease of interpretation. But this technique also requires maintenance of biologically active phages, and hence is available only at reference centers. Even for the experienced worker, the technique is demanding. Many strains are nontypeable.

4.5.1.5. Antibiograms

An antibiogram is the result of a laboratory testing for the sensitivity of an isolated bacterial strain to different antibiotics. It is by definition an *in vitro* sensitivity. In clinical practice, antibiotics are most frequently prescribed on the basis of general guidelines and knowledge about sensitivity, for example, uncomplicated urinary tract infections can be treated with a first generation quinolone. This is because *E. coli* is the most likely causative pathogen, and it is known to be sensitive to quinolone treatment. Infections that are not acquired in the hospital, are called *community acquired* infections (Pitout Johann et al., 2005).

However, many bacteria are known to be resistant to several classes of antibiotics, and treatment is not so straightforward. This is especially the case in vulnerable patients, such as patients in the intensive care unit. When these patients develop "hospital-acquired" or "nosocomial" pneumonia, more hardy bacteria such as *P. aeruginosa* are potentially involved. Treatment is then generally started on the basis of surveillance data about the local pathogens probably involved. This first treatment, based on statistical information about former patients and aimed at a large group of potentially involved microbes, is called *empirical treatment*.

Before starting this treatment, the physician will collect a sample from a suspected contaminated compartment: a blood sample when bacteria possibly have invaded the bloodstream, a sputum sample in the case of ventilator associated pneumonia, and a urine sample in the case of a urinary tract infection. These samples are transferred to the microbiology laboratory, which examines the sample under the microscope, and tries to culture the bacteria (Fig. 4.24). This can help in the diagnosis.

Once a culture is established, there are two possible ways to get an antibiogram:

- *Semiquantitative based on diffusion* (Kirby–Bauer method). Small discs containing different antibiotics, or impregnated paper discs, are dropped in different zones of the culture on an agar plate, which is a nutrient-rich environment in which bacteria can grow. The antibiotic will diffuse in the area surrounding each tablet, and a disc

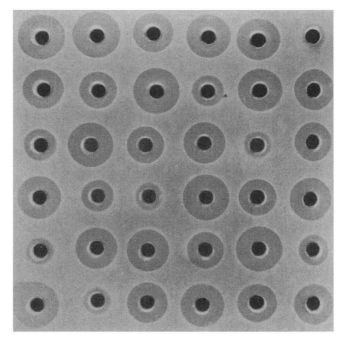

Figure 4.24. Antibiograms of *Klebsiella pneumoniae* isolates, a streptomycin assay plate using the agar diffusion techniques.

of bacterial lysis will become visible. Since the concentration of the antibiotic was the highest at the center, and the lowest at the edge of this zone, the diameter is suggestive of the minimum inhibitory concentration, or MIC, (conversion of the diameter in millimeter to the MIC, in μg/ml, is based on known linear regression curves).

- *Quantitative based on dilution*. A dilution series of antibiotics is established (this is a series of reaction vials with progressively lower concentrations of antibiotic substance). The last vial in which no bacteria grow contains the antibiotic at the minimal inhibiting concentration.

Once the MIC is calculated, it can be compared to known values for a given bacterium and antibiotic; for example, an MIC > 0.06 μg/ml may be interpreted as a penicillin-resistant *S. pneumoniae*. Such information may be useful to the clinician, who can change the empirical treatment to a more custom-tailored treatment that is directed only at the causative bacterium.

Antibiograms are an important resource for healthcare professionals involved in deciding and prescribing empiric antibiotic therapy. Appropriate empiric therapy is essential in attempting to treat infections correctly and quickly in an effort to decrease mortality. The use of antibiograms is also helpful in identifying trends in antibiotic resistance. Basic components of an antibiogram include antibiotics tested, organisms tested, number of isolates for each organism, percentage susceptibility data for each drug/pathogen combination, specimen sites notations (e.g. blood, urine, and catheters), and specific area or unit being tested.

It is important to tailor antibiotics as soon as sensitivities are known. This is the best way to avoid drug resistance and new/emerging organisms that are resistant. The goal to minimizing infection is to prescribe broad-spectrum antibiotics based on unit specific antibiograms.

The susceptibility or resistance of an organism to a possibly toxic agent forms the basis of the following typing techniques. The antibiogram is the susceptibility profile of an organism to a variety of antimicrobial agents, whereas the resistogram is the susceptibility profile to dyes and heavy metals. Bacteriocin typing is the susceptibility of the isolate to various bacteriocins, that is, toxins that are produced by a collected set of producer strains. These three techniques are limited by the number of agents tested per organism.

By far, the antibiogram is the most commonly used susceptibility/resistance typing technique, most probably because the data required for antibiogram analysis are available routinely from the antimicrobial susceptibility testing laboratory. Antibiograms have been used successfully to demonstrate relatedness with limitations (Thurm and Gericke, 1994). Organisms with similar antibiograms may be related, such is not necessarily the case. The antibiogram of an organism is not always constant. Selective pressure from antimicrobial therapy may alter an organism's antimicrobial susceptibility profile in such a way that related organisms show different resistance profiles. These alterations may result from chromosomal point mutations or from the gain or loss of extrachromosomal DNA such as plasmids or transposons.

This typing technique involves comparison of different isolates to a set of antibiotics. Isolates differing in their susceptibilities are considered as different strains. The identification of new or unusual pattern of antibiotic resistance among isolates cultured from multiple patients is often the first indication of an outbreak. The technique has ease of performance and interpretation with a fair amount of reproducibility.

As a consequence of various genetic mechanisms, different strains may develop similar resistance pattern, thus reducing the discriminating power. The susceptibility pattern of isolates taken over a period of time that represents the same strain may differ for one or more antibiotics due to acquisition of resistance.

4.5.1.6. Protein Typing

Protein typing relies on major or minor differences in the range of proteins made by different strains. Variations in the types and structures of the proteins expressed by bacteria can be detected by several methods. The proteins, glycoproteins, or polysaccharides are extracted from a culture of the strain, separated by sodium dodecyl sulfate-polyacrylamide gel electrophoresis and stained to compare with those of other strains. More-similar organisms display more-similar protein patterns. In another method termed *immunoblotting*, the electrophoresed products are transferred to nitrocellulose membrane and then exposed to antisera raised against specific strain. The bound antibodies are then detected by enzyme-labeled anti-immunoglobulins. These methods are currently employed for epidemiological studies of *S. aureus* and *Clostridium difficile*. All strains are typeable and techniques have good reproducibility and ease of interpretation. Yet, as the patterns detected are very complex, comparisons among multiple strains are difficult and the interpretation becomes difficult. Methods employed are technically demanding and equipment required is costly and hence are not available in all laboratories.

4.5.1.7. Multilocus Enzyme Electrophoresis (MLEE)

Here, the isolates are analyzed for differences in the eletrophoretic mobilities of a set of metabolic enzymes. Cell extracts containing soluble enzymes are electrophoresed in starch gels. Variations in the electrophoretic mobility of an enzyme, referred to as *electromorph*, typically reflect amino acid substitution that alter the charge of the protein. But this method is only moderately discriminatory for the epidemiological analysis of clinical isolates. It requires techniques and equipment that are not available in most laboratories.

4.5.2. Molecular Typing Methods

Genotypic characterization is becoming a more widely practised and standard method for characterizing and identifying bacteria. The technique is universally applicable as all bacterial genera and species become uniformly defined according to the genotypic uniqueness. The results of the phenotypic tests will correlate with the genotypic characteristics and bring about accurate and useful identification of an organism. Several molecular typing techniques have been developed during the past decade for the identification and classification of bacteria at or near the strain level. The most powerful of these are genetic-based molecular methods known as *DNA fingerprinting techniques*, for example, PFGE of rare-cutting restriction fragments, ribotyping, randomly amplified polymorphic DNA (RAPD), and AFLP, which have been extensively applied for the infraspecific identification and genotyping (McCartney, 2002). Basically, these methods rely on the detection of DNA polymorphisms between species or strains and differ in their dynamic range of taxonomic discriminatory power, reproducibility, ease of interpretation, and standardization.

4.5.2.1. Plasmid Analysis

The number and sizes of plasmids carried by an isolate can be determined by preparing a plasmid extract and subjecting it to gel electrophoresis. But reproducibility of this method suffers due to the existence of plasmid in different molecular forms such as supercoiled, nicked, or linear, each of which migrates differently on electrophoresis. Since plasmids can be spontaneously lost or readily acquired, related strains can exhibit different plasmid profiles. Clinical isolates lacking plasmids are untypeable. Those strains with one or two plasmids provide a poor discriminatory power.

4.5.2.2. Restriction Endonuclease Analysis (REA) of Chromosomal DNA

A restriction endonuclease enzymatically cuts DNA at a specific nucleotide recognition sequence (Fig. 4.25). The number and sizes of restriction fragments are influenced by the recognition sequence of enzyme and composition of DNA. Bacterial DNA is digested with endonucleases that have relatively frequent restriction sites, thereby generating hundreds of fragments ranging from approximately 0.5 to 50 kb in length. Such fragments can be separated by size using agarose gel electrophoresis. The pattern is stained by ethidium bromide and examined under UV light. Different strains of the same species have different REA (restriction endonuclease analysis) profiles because of variations in

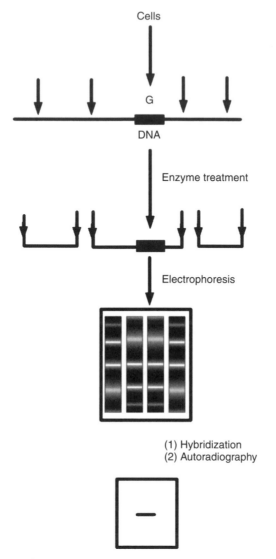

Figure 4.25. Restriction endonuclease analysis.

their DNA sequences. The complex profile consists of hundreds of bands that may be unresolved or overlapping, thus making comparison difficult. The pattern may consist of bands generated from digestion of plasmids too. These reduce the ease of interpretation and discriminatory power.

4.5.2.3. Pulsed-Field Gel Electrophoresis (PFGE) of Chromosomal DNA

PFGE is a technique that overcomes the limitations of REA. It is a variation of agarose gel electrophoresis in which the orientation of the electric field across the gel is periodically changed. This modification enables large fragments to be effectively separated by their size. RFLP analysis of bacterial DNA involves the digestion of genomic DNA with

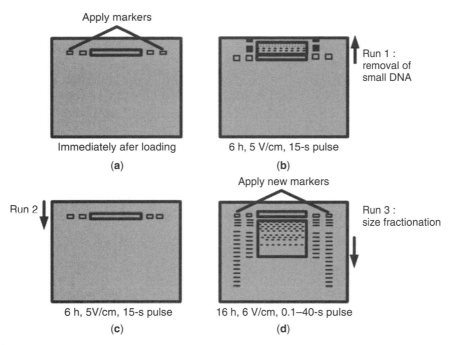

Figure 4.26. Isolation of insert DNA from agarose blocks by PFGE (Osoegawa et al., 2001).

rare-cutting restriction enzymes to yield a relatively few large fragments. The restriction fragments are then size-fractionated using PFGE, which allows separation of large genomic fragments. The generated DNA fingerprint obtained depends on the specificity of the restriction enzyme used and the sequence of the bacterial genome and is therefore characteristic of a particular species or strain of bacteria (Fig. 4.26).

This fingerprint represents the complete genome and thus can detect specific changes (DNA deletion, insertions, or rearrangements) within a particular strain over time. Its high discriminatory power has been reported for the differentiation between strains of important probiotic bacteria, such as *B. longum* and *B. animalis, Lactobacillus casei* and *Lactobacillus rhamnosus, Lactobacillus acidophilus* complex, *Lactobacillus helveticus,* and *Lactobacillus johnsonii*. Amor et al. (2007) reported a new approach combining RFLP with DNA fragment sizing by flow cytometry for bacterial strain identification. DNA fragment sizing by flow cytometry was found to be faster and more sensitive than PFGE, and this technique is also amenable to automation.

4.5.2.4. Ribotyping

Ribotyping is a variation of the conventional RFLP analysis (Fig. 4.27). It combines Southern hybridization of the DNA fingerprints, generated from the electrophoretic analysis of genomic DNA digests, with rDNA-targeted probing. The probes used in ribotyping vary from partial sequences of the rDNA genes or the intergenic spacer regions to the whole rDNA operon. Ribotyping has been used to characterize strains of *Lactobacillus* and *Bifidobacterium* from commercial products as well as from human fecal samples

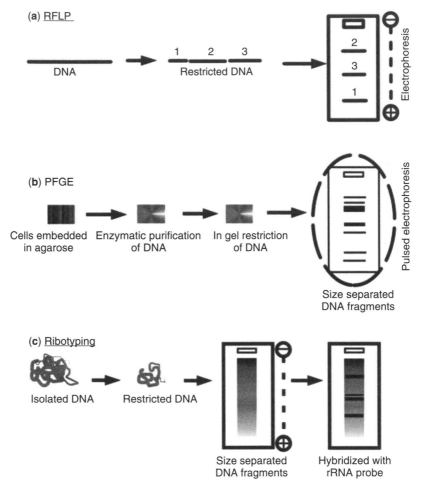

Figure 4.27. A ribotype is essentially an RFLP but differs from PFGE and RFLP (O'Sullivan, 2000).

(Giraffa et al., 2000; Zhong et al., 1998). However, ribotyping provides high discriminatory power at the species and subspecies level rather than at the strain level. PFGE was shown to be more discriminatory in typing closely related *Lb. casei* and *Lb. rhamnosus*, as well as *Lb. johnsonii* strains than either ribotyping or RAPD analysis (Ventura and Zink, 2002; Ventura et al., 2003).

4.5.2.5. Randomly Amplified Polymorphic DNA

Arbitrary amplification, also known as RAPD, has been widely reported as a rapid, sensitive, and inexpensive method for genetic typing of different strains of LAB and bifidobacteria. This PCR-based technique makes use of arbitrary primers that are able to bind under low stringency to a number of partially or perfectly complementary sequences of unknown location in the genome of an organism. If binding sites occur in a spacing and orientation that allow amplification of DNA fragments, fingerprint patterns are generated that are specific to each strain (O'Sullivan, 2000). RAPD profiling has been applied to distinguish between strains of *Bifidobacterium* and between strains of the *Lb. acidophilus* group and related strains. Several factors have been reported to influence the reproducibility and discriminatory power of the RAPD fingerprints, that is, annealing temperature, DNA template purity and concentration, and primer combinations. The use of five single-primer reactions under optimized conditions improved the resolution and accuracy of the RAPD method for the characterization of dairy-related bifidobacteria including *B. adolescentis, B. animalis, B. bifidum, B. breve, B. infantis*, and *B. longum* (Vincent et al., 1998).

4.5.2.6. Amplified Restriction Length Polymorphism

AFLP combines the power of RFLP with the flexibility of PCR-based methods by ligating primer-recognition sequences (adaptors) to the digested DNA (Fig. 4.28). Total genomic DNA is digested using two restriction enzymes, one with an average cutting frequency and a second with a higher cutting frequency. Double-stranded nucleotide adapters are usually ligated to the DNA fragments serving as primer binding sites for PCR amplification. The use of PCR primers complementary to the adapter and the restriction site sequence yields strain-specific amplification patterns. At present, AFLP has mostly been employed in clinical studies, but its successful application for strain typing of the *Lb. acidophilus* group and *Lb. johnsonii* isolates has also been reported (Gancheva et al., 1999; Ventura and Zink, 2002).

4.5.2.7. Other PCR Approaches

PCR-based approaches other than RAPD and AFLP have been used for molecular typing, such as amplified ribosomal DNA restriction analysis, ARDRA; Figure 4.28 (Roy et al., 2001; Ventura et al., 2001). Repetitive extragenic palindromic PCR (Rep-PCR) (Ventura et al., 2003) and triplicate arbitrary primed PCR (TAP-PCR) have been shown to offer a high discriminatory power for the identification.

4.5.2.8. Southern Blot Analysis of RFLPs

In contrast to REA of DNA, Southern blot analyses detect only the particular restriction fragment. The DNA is digested by endonuclease, the fragments are separated by gel electrophoresis, and the fragments are transferred to nitrocellulose membranes (Fig. 4.29). The fragments containing specific sequences are then detected by labeled DNA probes. Variations in the number and sizes of the fragments detected are referred to as *restriction fragment length polymorphism* (RFLP).

4.6. CLASSIFICATION OF MICROBES ON THE BASIS OF PHENOTYPIC CHARACTERISTICS

Microbial taxonomy is a means by which microorganisms can be grouped together. Organisms having similarities with respect to the criteria used are in the same group, and are separated from other groups of microorganisms that have different characteristics.

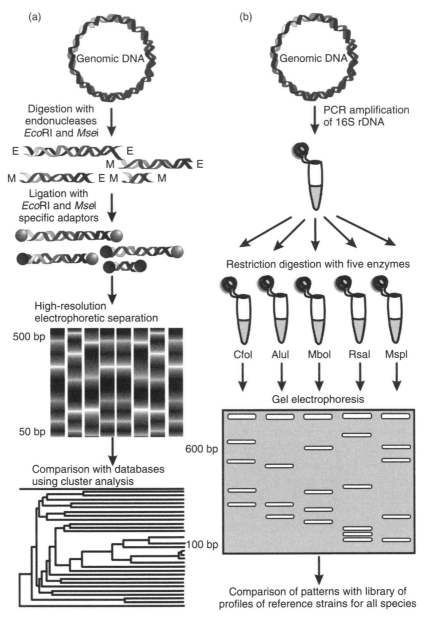

Figure 4.28. The genotypic identification of *Acinetobacter* species can be achieved by whole-genome fingerprinting, restriction analysis of a particular DNA sequence, and DNA-sequence determination. (a) AFLP comprises the following steps: digestion of cellular DNA with two restriction enzymes (a frequent and a rare cutter); ligation of adaptors to the restriction fragments; selective amplification of the fragments; electrophoresis of fragments on a sequencing machine; and visualization of the restriction profiles. The profiles are compared by cluster analysis to those of a reference library of AFLP fingerprints using dedicated software. (b) ARDRA comprises amplification of 16S rDNA sequences, followed by restriction of the amplified fragments with five restriction enzymes and electrophoresis. The resulting profiles are compared with a library of restriction profiles (Dijkshoorn et al., 1996, 2007).

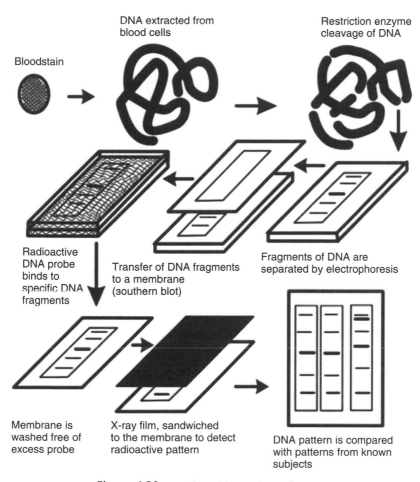

Figure 4.29. Southern blot analysis of RFLPs.

There are a number of taxonomic criteria that can be used. For example, numerical taxonomy differentiates microorganisms, typically bacteria, based on their phenotypic characteristics. Phenotypes are the appearance of the microbes or the manifestation of the genetic character of the microbes. Examples of phenotypic characteristics include the Gram stain reaction, shape of the bacterium, size of the bacterium, whether or not the bacterium can propel itself along, the capability of the microbes to grow in the presence or absence of oxygen, types of nutrients used, chemistry of the surface of the bacterium, and the reaction of the immune system to the bacterium.

Bacterial taxonomy relies on phenotypic characteristics to classify organisms, and is useful for the practical identification of unknown strains. The primary taxonomic unit is the species, which is defined by the phenotypic characteristics of a collection of similar strains. Culture collections contain type strains to serve as standards of the characteristics attributed to a particular species (Clarridge, 2004).

Microorganisms can be classified or distinguished from one another, by the ability to (i) grow on different substrates and/or production of different end products, (ii) produce specific enzymes, (iii) use oxygen, or (iv) be motile. For example, certain microbes can use different carbohydrates as sources of energy and/or carbon. Because such variability exists in carbohydrate utilization between different microbes, this can aid in the group, genus, or species identification.

4.6.1. Carbohydrate Utilization

Carbohydrates can be fermented or oxidized via aerobic or anaerobic respiration. When a carbohydrate is fermented, acid, and sometimes gas, is produced. The most common end product of carbohydrate fermentation is lactic acid (lactate). Other acids include formic and acetic acid. Microbes including *Streptococcus* and *Lactobacillus* can produce lactic acid, formic acid, and/or ethanol. Enteric organisms can produce lactic acid, formic acid, and succinic acid, as well as ethanol and gasses CO_2 and H_2. Acid production can be detected by addition of a pH indicator, such as phenol red (PR) or bromo cresol purple to the medium. PR turns yellow in acidic conditions (slightly under pH 7.0) while at neutral or basic pH it is red. Gas production can be monitored by addition of a small tube, called *a Durham tube*, which has been inverted in the carbohydrate-containing growth medium to trap gas bubbles (Fig. 4.30).

4.6.2. Enzyme Production

The ability of an organism to produce or express a particular enzyme can be assayed (tested) in the laboratory using simple biochemical reagents and substrates. Several of these tests require only small amounts of cellular material and may take only a few minutes to interpret. Subsets of microbes can then be classified into groups based on their ability to produce a positive or negative reaction. The results of biochemical tests are used with other data, including the Gram reaction, to identify an organism by name. These tests play critical roles in the swift identification of bacteria causing infection.

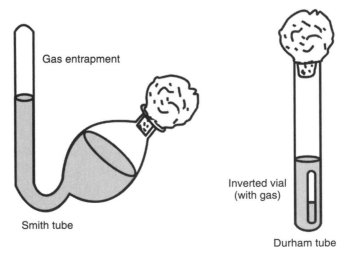

Figure 4.30. Gas entrapment vessel.

4.6.3. Mode of Growth

A very important requirement for cellular growth and metabolism is oxygen. Microbes can be grouped into five different categories based on their requirement for molecular oxygen. Strict or obligate aerobes must have O_2 for growth because O_2 is used as the terminal electron acceptor for oxidative phosphorylation. In contrast, strict or obligate anaerobes cannot grow in the presence of O_2 and may be killed by trace amounts of O_2. Microaerophilic organisms need a small amount of O_2 for growth but too much O_2 will kill them. A subset of anaerobes can tolerate O_2 but they do not use it. They are called *aerotolerant anaerobes*. Facultative anaerobes grow best when O_2 is available, but they can also grow, though not as well, if O_2 is not present (Fig. 4.31).

4.6.4. Motility

Microbes move in a variety of ways that are roughly analogous to our walking, running, swimming, or crawling movements. Many organisms can swim in liquid (such as lakes or oceans and blood system). Most swimmers move by rotation of one or more external

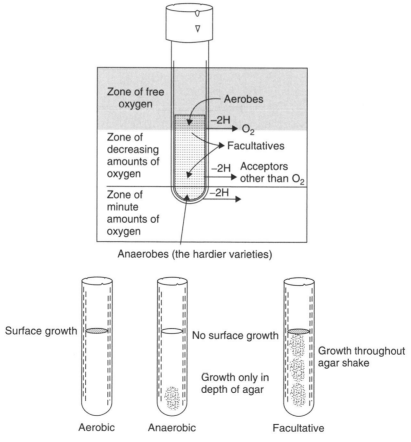

Figure 4.31. Oxygen requirement of microorganisms.

appendages called *flagella*. This is a complex process that is very well studied. The helical shape of the flagellum is reminiscent of a motorboat propeller; movement results from rotating one or more flagella. Flagella can rotate very fast when the cell is in liquid medium, allowing for rapid swimming. When cells with flagella are in more viscous (thicker) medium, such as broth containing some agar, the rate of swimming is reduced. Some examples of swimmers include *E. coli, S. typhimurium*, and *B. pertussis*. Some swimmers can also swarm. Swarming occurs when a swimmer cell enters a viscous environment or a solid surface. The swimmer then increases in length and makes a large number of flagella. Some examples include *Proteus mirabilis* and *S. marcescens*. Twitching motility is a slow movement of cells over a surface via type IV pili. The cell throws out a pilus which attaches to a surface and is subsequently depolymerized so that the cell moves toward the attachment point. Pathogens, such a *Neisseria* and *Pseudomonas*, use twitching motility, as does the soil bacterium *Myxococcus xanthus*.

4.6.5. Antigen and Phage Susceptibility

Cell wall (O), flagellar (H), and capsular (K) antigens are used to aid in classifying certain organisms at the species level, to serotype strains of medically important species for epidemiologic purposes, or to identify serotypes of public health importance. Serotyping is also sometimes used to distinguish strains of exceptional virulence or public health importance, for example, with *V. cholerae* (O1 is the pandemic strain) and *E. coli* (enterotoxigenic, enteroinvasive, enterohemorrhagic, and enteropathogenic serotypes).

Phage typing (determining the susceptibility pattern of an isolate to a set of specific bacteriophages) has been used primarily as an aid in epidemiologic surveillance of diseases caused by *S. aureus, Mycobacteria, P. aeruginosa, V. cholerae,* and *S. typhi* (Baron, 1996). Susceptibility to bacteriocins has also been used as an epidemiologic strain marker. In most cases, phage and bacteriocin typing have been supplemented by molecular methods.

Bacteriophages, viruses that infect and lyse bacteria, are often specific for strains within a species. A collection of bacteriophages, many of which often infect similar bacteria, is termed a *panel*. When a bacterial isolate is exposed to a panel of bacteriophages, a profile is generated on the basis of bacteriophages capable of infecting and lysing the bacteria. The bacteriophage profile may be used to type bacterial strains within a given species. The more closely related the bacterial strains, the greater the similarity of the bacteriophage profiles. Bacteriophage profiles have been used successfully to type various organisms associated with epidemic outbreaks (Tang et al., 1997). However, this typing method is labor-intensive and requires the maintenance of bacteriophage panels for a wide variety of bacteria. Additionally, bacteriophage profiles may fail to identify isolates, are often difficult to interpret, and may give poor reproducibility (Bannerman et al., 1995).

4.7. CLASSIFICATION OF MICROBES ON THE BASIS OF GENOTYPIC CHARACTERS

Genotypic identification is emerging as an alternative or complement to establish phenotypic methods. Characterization of organisms can also be done utilizing the genotypic properties. As discussed earlier, several kinds of analysis performed on isolated nucleic acids furnish information about the genotype, the analysis of the base composition of DNA, the study of chemical hybridization between nucleic acids isolated from different organisms, and the sequencing of nucleic acids. 16S rRNA-sequence-based methods, DNA base ratio, and DNA hybridization offer a viable option for rapid and reliable identification.

4.7.1. DNA Base Ratio (G + C Ratio)

DNA base composition can only prove that organisms are unrelated. The ratio of bases in DNA can vary over a wide range. If two organisms have different DNA base compositions, they are not related. However, organisms with identical base ratios are not necessarily related, because the nucleotide sequences in the two organisms could be completely different.

In molecular biology, GC content (guanine–cytosine content) is the percentage of nitrogenous bases on a DNA molecule, which are either guanine or cytosine (from a possibility of four different ones, also including adenine and thymine). This may refer to a specific fragment of DNA or RNA, or that of the whole genome. When it refers to a fragment of the genetic material, it may denote the GC content of part of a gene (domain), single gene, group of genes (or gene clusters), or even a noncoding region. G (guanine) and C (cytosine) undergo a specific hydrogen bonding, whereas A (adenine) bonds specifically with T (thymine).

The GC pair is bound by three hydrogen bonds, while AT pairs are bound by two hydrogen bonds. DNA with high GC content is more stable than that with low GC content, but, contrary to popular belief, the hydrogen bonds do not stabilize the DNA significantly and stabilization is mainly due to stacking interactions. In spite of the higher thermostability conferred to the genetic material, it is envisaged that cells with DNA with high GC content undergo autolysis, thereby reducing the longevity of the cell *per se*. Owing to the robustness endowed to the genetic materials in high GC organisms, it was commonly believed that the GC content played a vital part in adaptation temperatures, a hypothesis that has recently been refuted.

In PCR experiments, the GC content of primers are used to predict their annealing temperature to the template DNA. A higher GC content level indicates a higher melting temperature.

GC content is usually expressed as a percentage value, but sometimes as a ratio (called G + C *ratio* or *GC ratio*). GC content percentage is calculated as

$$\frac{G + C}{A + T + G + C} \times 100$$

whereas the AT/GC ratio is calculated as

$$\frac{A + T}{G + C}$$

The GC content percentages as well as GC ratio can be measured by several means, but one of the simplest methods is to measure what is called the melting temperature of the DNA double helix using spectrophotometry. The absorbance of DNA at a wavelength of 260 nm increases fairly sharply when the double-stranded DNA separates into two single strands when sufficiently heated. The most commonly used protocol for determining GC ratios uses flow cytometry for large number of samples.

GC content is found to be variable with different organisms, the process of which is envisaged to be contributed by variation in selection, mutational bias, and biased recombination-associated DNA repair. The species problem in prokaryotic taxonomy has led to various suggestions in classifying bacteria, and the ad hoc committee on reconciliation of approaches to bacterial systematics has recommended the use of GC ratios in higher level hierarchical classification.

For example, the *Actinobacteria* are characterized as "high GC content bacteria." In *Streptomyces coelicolor*, GC content is 72%. The GC content of yeast (*Saccharomyces cerevisiae*) is 38%, and that of another common model organism Thale Cress (*Arabidopsis thaliana*) is 36%. Because of the nature of the genetic code, it is virtually impossible for an organism to have a genome with a GC content approaching either 0% or 100%. A species with an extremely low GC content is *Plasmodium falciparum* (GC% = ~20%), and it is usually common to refer to such examples as being AT-rich instead of GC-poor.

Physical methods of analysis also provide an indication of the molecular homogeneity of a DNA sample. If every molecule of DNA had the same GC content, both the thermal transition in a melting curve and the band position will be same.

The GC content is often measured by determining the temperature at which the double-stranded DNA denatures (Fig. 4.32). Because three hydrogen bonds occur between G and C base pairs, and only two hydrogen bonds between A and T, high GC content melts at a higher temperature.

The temperature at which the double-stranded DNA melts can be readily determined by monitoring the absorbance of UV light by the solution of DNA as it is heated. The absorbance readily increases as double-stranded DNA denatures. In a typical melting curve (Fig. 4.33), the increase in UV absorbance can be measured as the temperature increases. This tracks the unwinding and denaturation of DNA. The melting point (T_m) is the temperature at which half the DNA is unwound.

DNA that consists entirely of AT base pairs melts at about $70\,^{\circ}$C and DNA that has only GC base pairs melts at over $100\,^{\circ}$C. The T_m of any DNA molecule can be calculated if one knows the base composition. The simplest formulas just take the overall composition into account and they are not very accurate. More accurate formula will use the stacking interactions of each base pair to predict the melting temperature. The GC content varies among the different kinds of bacteria, with numbers ranging from 28% to 78%. Organisms that are related by other criteria have similar or identical DNA base composition. Thus, if the GC content of two organisms differs by more than a small percent, they cannot be closely related. However, similarity does not necessarily mean that the organism is related, since many arrangements of the bases are possible. The genome size and the actual nucleotide sequences also differ greatly.

4.7.2. DNA Hybridization

DNA–DNA hybridization generally refers to a molecular biology technique that measures the degree of genetic similarity between pools of DNA sequences. It is usually used to determine the genetic distance between two species. When several species are compared that way, the similarity values allow the species to be arranged in a phylogenetic tree; it is therefore one possible approach to carrying out molecular systematics.

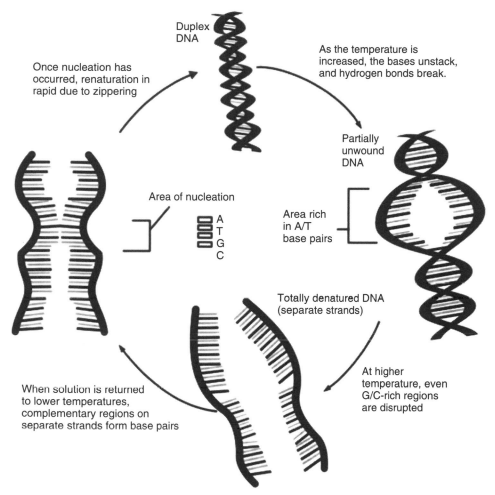

Once nucleation has occurred, renaturation in rapid due to zippering

Duplex DNA

As the temperature is increased, the bases unstack, and hydrogen bonds break.

Partially unwound DNA

Area of nucleation

A
T
G
C

Area rich in A/T base pairs

Totally denatured DNA (separate strands)

When solution is returned to lower temperatures, complementary regions on separate strands form base pairs

At higher temperature, even G/C-rich regions are disrupted

Figure 4.32. Effect of raising the temperature of DNA.

Hybridization between the total DNA of two organisms is useful for detecting relationships between closely related organisms. The extent of nucleotide sequence similarity between two organisms can be determined by measuring how completely single strands of their DNA will hybridize to one another. Just as two complementary strands of DNA from one organism will base pair or anneal, so will the similar DNA of the different organism. The degree of hybridization will reflect the degree of sequence similarity. DNA from organisms that share many sequences will hybridize more completely the DNA from those that do not.

On rapid cooling of the solution of thermally denatured DNA, the single strands remain separated. However, if the solution is held at a temperature from 10 to $30\,^{\circ}$C below the T_m value, specific reassociation (annealing) of the complementary strands to form double-stranded molecules occurs. There is always random pairing, but since a randomly matched duplex contains many mismatched base pairs, its thermal stability is low and its strands separate very rapidly at temperatures near the T_m. In contrast, pairing of the complementary strands forms duplexes that are quite stable because each base

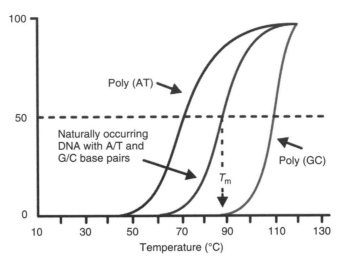

Figure 4.33. Melting curve of DNA.

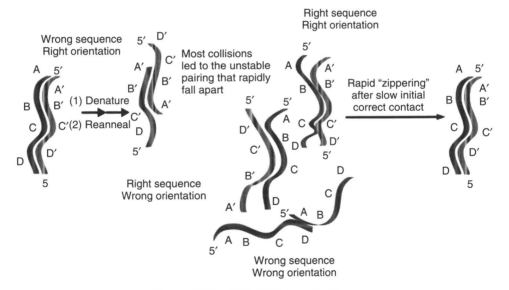

Figure 4.34. DNA–DNA hybridization.

participates in interstrand hydrogen bonding. Thus, at temperatures near the T_m, only duplexes between the strands with high degree of complementarity persist; the closer that the temperature of incubation is to the T_m, the more stringent is the requirement of base pairing.

Shortly after the discovery of this phenomenon, it was shown that when DNA preparations from two related strains of bacteria are mixed and treated in this manner, hybrid DNA molecules are formed (Fig. 4.34). The discovery of the reassociation of stranded DNA molecules from different biological sources to from hybrid duplexes laid the foundations of an entirely new approach to the study of genetic relatedness in bacteria.

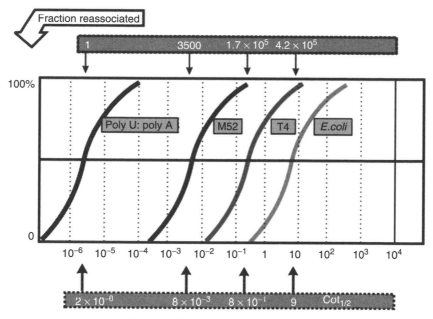

Figure 4.35. Rate of reassociation.

In vitro experiments of DNA–DNA associations permit an assessment of the overall degree of genetic homology between the bacteria. Since duplexes can also be formed between single-stranded DNA and complementary RNA strands, analogous DNA–RNA reassociations can be performed. If the RNA preparations consist of either tRNAs or rRNAs, such experiments permit an assessment of the genetic homology between two bacteria with respect to specific, relatively small segments of chromosome: those that code the base sequences either of the tRNAs or of the rRNA. The range of organisms among which genetic homology is detectable can be greatly extended by parallel studies on DNA–rRNA reassociation, because the relatively small portion of the bacterial genome that codes for the rRNA has a much more conserved sequence than the bulk of the chromosomal DNA. As a result, it is frequently possible to detect the DNA–rRNA reassociation, relatively high homology, between the genomes of the two bacteria which shows no specific homology by DNA–DNA reassociation. The rate of the reassociation is inversely proportion to the length of the reassociating DNA (Fig. 4.35).

In a bacterial group, the value of nucleic acid reassociation studies is directly related to the number of strains and species that have been compared. Extensive comparative data has been available for several major bacterial groups.

Whole genomic DNA–DNA hybridization has been a cornerstone of bacterial species determination but is not widely used because it is not easily implemented. Cluster analysis of the hybridization profiles revealed taxonomic relationships between bacterial strains tested at species to strain level resolution, suggesting that this approach is useful for the identification of bacteria as well as determining the genetic distance among bacteria. Since arrays can contain thousands of DNA spots, a single array has the potential for broad identification capacity.

4.7.3. Nucleotide Sequence Analysis

Genotype information at highest precision may be determined as DNA (or RNA) nucleotide-base sequences. RNAs are often sequenced either by converting the RNAs into DNA or by sequencing the DNA gene that gives rise to the RNA. By using PCR to amplify a known DNA segment and automated techniques to sequence the amplified product, it is possible to compare multiple isolates.

One is the analysis of the base composition of DNA, that is, to determine the mole percentage of guanine and cytosine in DNA (%G + C). The second is to determine the degree of similarity between two DNA samples by hybridization between DNA and DNA or DNA and RNA. The basis of this test is that the degree of hybridization would be an indication of the degree of relationship (homology). The relative percentage of guanine and cytosine $(G + C/A + T + G + C) \times 100$ varies widely with different bacteria. The composition of chromosomal DNA is a fixed property of each cell and is independent of age and other external influences. The percentage $(G + C)$ of chromosomal DNA can be determined by extracting DNA from cells by rupturing carefully. The DNA is then purified to remove nonchromosomal DNA.

Since no preparation shows absolute molecular homogeneity, the $G + C$ content is always a mean value and represents the peak in the normal distribution curve. Each bacterial species have DNA with a characteristic mean $G + C$ content; this can be considered one of the important specific characters. Mean DNA base composition is a character of taxonomic value among bacteria, since the range for the group as a whole is so wide.

The base composition can then be determined either by subjecting the purified DNA to increasing temperature and determining the increase in hypochromicity or by centrifugation of the DNA in cesium chloride density gradients. The basis of the first method, that is, the melting point method, is that when double-stranded DNA is subjected to increasing temperature, the two DNA strands separate at a characteristic temperature. The melting temperature depends on the $G + C$ content of the DNA. The higher the $G + C$ content, the higher will be the melting point.

The mean temperature at which thermal denaturation of DNA occurs is called *the melting point* (T_m), and this is determined by noting the change in optical density of DNA solution at 260 nm during the heating period. From the melting point, the mole percentage $(G + C)$ can be calculated as $\%G + C = T_m \times 63.54/0.47$.

The percentage $(G + C)$ composition can also be calculated by determining the relative rate of sedimentation in a cesium chloride solution. DNA preparation when subjected to high gravitational force (as in an ultracentrifuge) in a heavy salt solution will sediment at a region in the centrifuge tube where its density is equal to the density of the medium. By this method, DNA samples that are heterogeneous can also be separated simultaneously. The buoyant density is very characteristic of each type of DNA and is dependent on the percentage GC content, From the buoyant density one can calculate the percentage GC content by using the empirical formula $P = 1.660 + 0.00098\ (\%GC)$ g cm^3

A third method of determining the percentage $(G + C)$ is by the controlled hydrolysis of DNA with acids and separating and measuring the nucleotides by chromatography. This method is laborious but simple. The base composition of DNA from a variety of organisms are determined by these procedures. The genetic relatedness can also be determined by measuring the extent of hybridization between denatured DNA molecules between single-stranded DNA and RNA species. The degree of homology is determined by mixing two kinds of single-stranded DNA or single-stranded DNA with RNA under appropriate conditions, and then measuring the extent to which they associate to form double-stranded structures. This can be precisely measured by making either the DNA or RNA radioactive. The degree of relatedness of different bacteria as determined by DNA–RNA hybridization. Although genetic relatedness can be determined by DNA–RNA hybridization, the DNA–DNA hybridization is more accurate, provided precautions are taken to ensure that hybridization between two strands is uniform. The technique is advantageous as it can be applied on all strains; results are reproducible with ease in interpretation. But the process requires costly reagents and equipment besides being labor-intensive.

Early in the chemical study of DNA preparation from different organisms and subsequent work has revealed that the base composition of DNA is a character of profound taxonomic importance, particularly among microorganisms.

4.7.4. Comparing the Sequence of 16S Ribosomal Nucleic Acid

Many of the modern molecular tools are based on 16S ribosomal DNA sequence (Amor et al., 2007), complete or partial genomes or specific fluorescent probes, that monitor the physiological activity of microbial cells (Table 4.2). The tools that have been developed for identifying microbes and analyzing their activity can be divided into those based on nucleic acids and other macromolecules and approaches directed at analyzing the activity of complete cells. The nucleic-acid-based tools are more frequently used because of the high throughput potential provided by using PCR amplification or *ex situ* or *in situ* hybridization with DNA, RNA, or even peptide nucleic acid probes. Notably, these include 16S rDNA sequences that can be used to place diagnostics into a phylogenetic framework and can be linked to databases providing up to 100,000 sequences (Amann and Ludwig, 2000). These 16S rDNA-based methodologies are robust and superior to traditional methods based on phenotypic approaches, which are often unreliable and lack the resolving power to analyze the microbial composition and activity of bacterial populations. In addition, a panoply of approaches that are based on DNA sequences other than rDNAs have been applied frequently to probiotic bacteria. These have been shown to be particularly useful for strain identification.

A promising method for simultaneous and selective detection of both culturable and nonculturable bacteria of defined taxonomic groups is the amplification of 16S ribosomal DNA (rDNA) or rRNA sequences using PCR. Sequence comparisons of small subunit rRNA have been used as a source for determining phylogenetic and evolutionary relationships among organisms of the three kingdoms Archaea, Eucarya, and Bacteria. The present compilation of complete genes for the small subunit rRNA contains over 2200 16S and 16S-like sequences. The 16S rRNAs are highly conserved, sharing common three-dimensional structural elements of similar function. The primary structures are well investigated and conserved, and variable regions have been determined. Primers located in highly conserved regions have been published, allowing the amplification of 16S rDNA and subsequent sequence analysis. Certain signatures in the nucleotide sequence can be unique for particular phylogenetic groups, offering the opportunity to design genus-specific probes, whereas the variable regions can be used to assign organisms to lower taxonomic groups (Mehling et al., 1995). The determination of full-length 16S rDNA sequences, as opposed to partial gene sequences, of streptomycete and some other actinomycete strains has provided data that may be useful in elucidating taxonomic levels or detecting chimeric PCR products. The design of PCR primers with potential for the differentiation of strains at the genus, species, and strain levels was made possible by sequence analysis of the complete 16S rDNA sequences. The possible combinations of genus- and strain-specific primers permit diverse assays, such as multiplex PCR or PCR with nested primers, lessening the likelihood of false-positive identification of streptomycetes and thus increasing the fidelity of the assay.

DNA-based technology for the identification of bacteria typically uses only the 16S rRNA gene as the basis for identification. This technique has the advantage of being able to identify difficult to cultivate strains, and is growth and operator independent. As the 16S rRNA gene is highly conserved at the species level, speciation is commonly quite good, but as a result, subspecies and strain level differences are not shown. Some problems with the 16S rRNA technology are that it requires a high level of technical proficiency, and the costs per sample, as well as equipment costs are high. As a result, the technology is not well suited for routine microbial quality control (QC), but rather is best used for direct product failures (Sutton and Cundell, 2004). Technology that uses information from both the 16S rRNA and 23S rRNA genes is also used in pharmaceutical QC, but primarily to aid in strain tracking.

Sequence comparisons of small subunit rRNA have been used as a source for determining phylogenetic and evolutionary relationships among organisms of the three kingdoms Archaea, Eucarya, and Bacteria. The present compilation of complete genes for the small subunit rRNA contains over 2200 16S and 16S-like sequences (Gutell et al., 1994). The 16S rRNAs are highly conserved, sharing common three-dimensional structural elements of similar function. To facilitate the differential identification of the genus *Streptomyces*, the 16S rRNA genes of 17 actinomycetes were sequenced and screened for the existence of streptomycete-specific signatures. The 16S rDNA of the *Streptomyces* strains and *Amycolatopsis orientalis* subsp *lurida* exhibited 95–100% similarity, while that of the 16S rDNA of *Adnoplanes utahensis* showed only 88% similarity to the streptomycete 16S rDNAs. Potential genus-specific sequences were found in regions located around nucleotide positions 120,800 and 1100. Several sets of primers derived from these characteristic regions were investigated as to their specificity in PCR-mediated amplifications. Most sets allowed selective amplification of the streptomycete rDNA sequences studied. RFLPs in the 16S rDNA permitted all strains to be distinguished (Mehling et al., 1995).

Over the last decade, hybridizations with rRNA-targeted probes have provided a unique insight into the structure and spatiotemporal dynamics of complex microbial communities. Nucleic acid probes can be designed to specifically target taxonomic groups at different levels of specificity (from species to domain) by virtue of variable evolutionary conservation of the rRNA molecules. Appropriate software environments such as the ARB package, a software environment for sequence data (http://www.arb-home.de/) and availability of large databases (http://rdp.cme.msu.edu/html/), or the online resource for oligonucleotide probes probeBase (http://www.microbial-ecology.de/probebase/index.html) offer powerful platforms for a rapid probe design and *in silico* specificity profiling. Oligonucleotide probes that are complementary to regions of 16S or 23S rRNA have been successfully used for the identification of LAB, and, hence, they offer the potential to be used as reliable and rapid diagnostic tools.

4.8. FUTURE CHALLENGES: APTAMERS FOR DETECTION OF PATHOGENS

Aptamers are specific nucleic acid sequences that can bind to a wide range of non-nucleic acid targets with high affinity and specificity. Aptamers, simply described as chemical antibodies, are synthetic oligonucleotide ligands or peptides that can be isolated *in vitro* against diverse targets including toxins, bacterial and viral proteins, virus-infected cells, cancer cells, and whole pathogenic microorganisms. Aptamers assume a defined three-dimensional structure and generally bind functional sites on their respective targets. They possess the molecular recognition properties of monoclonal antibodies in terms of their high affinity and specificity. The applications of aptamers range from diagnostics and biosensing, target validation, targeted drug delivery, therapeutics, templates for rational drug design to biochemical screening of small-molecule lead compounds (Khati, 2010).

These molecules are identified and selected through an *in vitro* process called *SELEX* (*systematic evolution of ligands by exponential enrichment*). Proteins are the most common targets in aptamer selection. In diagnostic and detection assays, aptamers represent an alternative to antibodies as recognition agents. Cellular detection is a promising area in aptamer research. One of its principal advantages is the ability to target and specifically differentiate microbial strains without having previous knowledge of the membrane molecules or structural changes present in that particular microorganism (Torres-Chavolla and Alocilja, 2009).

So far, several bioanalytical methods have used nucleic acid probes to detect specific sequences in RNA or DNA targets through hybridization. More recently, specific nucleic acids, aptamers, selected from random sequence pools have been shown to bind non-nucleic acid targets, such as small molecules or proteins. The development of *in vitro* selection and amplification techniques has allowed the identification of specific aptamers, which bind to the target molecules with high affinity. Many small organic molecules with molecular weights from 100 to 10,000 Da have been shown to be good targets for selection.

Aptamers can be selected against difficult target haptens, such as toxins or prions. The selected aptamers can bind to their targets with high affinity and even discriminate between closely related targets (Fig. 4.36). Aptamers can thus be considered as

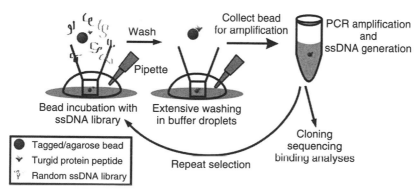

Figure 4.36. An efficient and easy-to-execute single microbead SELEX approach is developed to generate high affinity ssDNA aptamers against botulinum neurotoxin (Jeffrey et al., 2008).

a valid alternative to antibodies or other biomimetic receptors, for the development of biosensors and other analytical methods. Aptamers are generally produced by the SELEX process, which, starting from large libraries of oligonucleotides, allows the isolation of large amounts of functional nucleic acids by an iterative process of *in vitro* selection and subsequent amplification through PCR. Aptamers are suitable for electrochemical detection through several strategies and for detection of pathogens. An aptasensor consists of immobilization onto the surface of a sensor of specific nucleic acids, selected from random sequence pools. The selected aptamers bind to their targets with high affinity and even discriminate between closely related targets. This is due to the adaptive recognition: aptamers, unstructured in origin, fold on associating with their ligands into molecular architectures in which the ligand becomes an intrinsic part of the nucleic acid structure. The coupling of reliable electrochemical techniques with micro- or nanoarray of electrodes offers the possibility of monitoring multiple affinity reaction, since on each electrode surface will be immobilized a specific probe (that identifies a microbial species or the expression of a particular toxin).

Currently, aptamers can be identified to most proteins, including blood-clotting factors, cell-surface receptors, and transcription factors. Chemical modifications to the oligonucleotides enhance their pharmacokinetics and pharmacodynamics, thus extending their therapeutic potential. Several aptamers have entered the clinical pipeline for applications and diseases such as macular degeneration, coronary artery bypass graft surgery, and various types of cancer. Furthermore, the functional repertoire of aptamers has expanded with the descriptions of multivalent agonistic aptamers and aptamers–siRNA chimeras. The first aptamer-based clinical drugs have recently entered service. Meanwhile, active research programmes have identified a wide range of antiviral aptamers that could form the basis for future therapeutics (Thiel and Giangrande, 2009).

REFERENCES

Aarts HJM, Hakemulder LE, Van Hoef AMA. 1999. Genomic typing of *Listeria monocytogenes* strains by automated laser fluorescence analysis of amplified fragment length polymorphism fingerprint patterns. *Int J Food Microbiol* **49**, 95–102.

Amann R, Ludwig W. 2000. Ribosomal RNA-targeted nucleic acid probes for studies in microbial ecology. *FEMS Microbiol Rev* **24**, 555–565.

Amor KB, Vaughan EE, de Vos WM. 2007. Advanced molecular tools for the identification of lactic acid bacteria. *J Nutr* **137**, 741S–747S.

Arostegui JI, Gallardo D, Rodriguez-Luaces M, Querol S, Madrigal JA, Garcia-Lopez J, Granena A. 2000. Genomic typing of minor histocompatibility antigen HA-1 by reference strand mediated conformation analysis (RSCA). *Tissue Antigens* **56**, 69–76.

Bannerman TL, Hancock GA, Tenover FC, Miller JM. 1995. Pulsed-field gel electrophoresis as a replacement for bacteriophage typing of Staphylococcus aureus. *J Clin Microbiol* **33**, 551–555.

Baron EJ. 1996. Classification. In: Baron S, et al., eds., *Baron's Medical Microbiology*, 4th ed., University of Texas Medical Branch, Galveston.

van Belkum A, Struelens M, de Visser A, Verbrugh H, Tibayrenc M. 2001. Role of genomic typing in taxonomy, evolutionary genetics, and microbial epidemiology. *Clin Microbiol Rev* **14** (3), 547–560.

Bosshard PP, Abels S, Altwegg M, Bottger EC, Zbinden R. 2004. Comparison of conventional and molecular methods for identification of aerobic catalase-negative gram-positive cocci in the clinical laboratory. *J Clin Microbiol* **42**, 2065–2073.

Clarridge J. 2004. Impact of 16S rRNA gene sequence analysis for identification of bacteria on clinical microbiology and infectious diseases. *Clin Microbiol Rev* **17**, 840–862.

Cohen JO, Smith PB. 1964. Serological typing of *Staphylococcus aureus* II. Typing by slide agglutination and comparison with phage typing. *J Bacteriol* **88**, 1364–1371.

Cook VJ, Turenne CY, Wolfe J, Pauls R, Kabani A. 2003. Conventional methods versus 16S ribosomal DNA sequencing for identification of nontuberculous mycobacteria: cost analysis. *J Clin Microbiol* **41**, 1010–1015.

Cowan ST. 1965. Principles and practice of bacterial taxonomy—a forward look. *J Gen Microbiol* **39**, 143–153.

Dijkshoorn L, Aucken H, Gerner-Smidt P, Janssen P, Kaufmann ME, Garaizar J, Ursing J, Pitt TL. 1996. Comparison of outbreak and nonoutbreak Acinetobacter baumannii strains by genotypic and phenotypic methods. *J Clin Microbiol* **34**, 1519–1525.

Dijkshoorn L, Nemec A, Seifert H. 2007. An increasing threat in the hospital: multidrug resistant Acinetobacter baumannii. *Nat Rev Microbiol* **5**, 939–951.

Doolittle WF, Zhaxybayeya O. 2009. On the origin of prokaryotic species. *Genome Res* **19**, 744–756.

Eisenstark A, Rabsch W, Ackermann HW. 2009. Morphology of *Salmonella typhimurium* typing phages of the Lilleengen set. *Can J Microbiol* **55**, 1403–1405.

Fredericks DN, Relman DA. 1999. Application of polymerase chain reaction to the diagnosis of infectious diseases. *Clin Infect Dis* **29**, 475–488.

Gancheva A, Pot B, Vanhonacker K, Hoste B, Kersters K. 1999. A polyphasic approach towards the identification of strains belonging to *Lactobacillus acidophilus* and related species. *Syst Appl Microbiol* **22**, 573–585.

Gauthier G, Gauthier M, Christen R. 1995. Phylogenetic analysis of the gene r a alteromonas, shewanella, and moritella using genes coding for small-subunit rRNA sequences and division of the genus alteromonas into two genera, alteromonas (Emended) and pseudoalteromonas gen. nov., and proposal of twelve new species combinations. *Int J Syst Bacteriol* **45**, 755–761.

Giraffa G, Gatti M, Rossetti L, Senini L, Neviani E. 2000. Molecular diversity within *Lactobacillus helveticus* as revealed by genotypic characterization. *Appl Environ Microbiol* **66**, 1259–1265.

Gutell R, Larsen N, Woese C. 1994. Lessons from an evolving rRNA: 16s and 23s rRNA structures from a comparative perspective. *Microbiol Rev* **58**, 10–26.

Han XY, Falsen E. 2005. Characterization of oral strains of *Cardiobacterium valvarum* and emended description of the organism. *J Clin Microbiol* **43**, 2370–2374.

Hucker GJ. 1921. A new modification and application of the gram stain. *J Bacteriol* **6**, 395–397.

Jeffrey B, Tok H, Fische NO. 2008. Single microbead SELEX for efficient ssDNA aptamer generation against botulinum neurotoxin. *Chem Commun* **16**, 1883–1885.

Khan ZU, Al-Anezi AA, Chandy R, Xu J. 2003. Disseminated cryptococcosis in an AIDS patient caused by a canavanine-resistant strain of *Cryptococcus neoformans* var. *grubii*. *J Med Microbiol* **52**, 271–275.

Khati I. 2010. The future of aptamers in medicine. *J Clin Pathol* **63** (6), 480–487.

Kilian M, Sorensen I, Frederiksen W. 1979. Biochemical characteristics of 130 recent isolates from *Haemophilus influenzae* meningitis. *J Clin Microbiol* **9**, 409–412.

Kirschner C, Maquelin K, Pina P, Ngo Thi NA, Choo-Smith LP, Sockalingum GD, Sandt C, Ami D, Orsini F, Doglia SM, Allouch P, Mainfait M, Puppels GJ, Naumann D. 2001. Classification and identification of enterococci: a comparative phenotypic, genotypic, and vibrational spectroscopic study. *J Clin Microbiol* **39**, 1763–1770.

Kleerebezem M, Boekhorst J, van Kranenburg R, Molenaar D, Kuipers OP, Leer R, Tarchini R, Peters SA, Sandbrink HM, Fiers MW, Stiekema W, Lankhorst RM, Bron PA, Hoffer SM, Groot

MN, Kerkhoven R, de Vries M, Ursing B, de Vos WM, Siezen RJ. 2003. Complete genome sequence of *Lactobacillus plantarum* WCFS1. *Proc Natl Acad Sci U S A* **100**, 1990–1995.

Kunitsky C, Osterhout G, Sasser M. 2005. Identification of microorganisms using fatty acid methyl ester (FAME) analysis and the MIDI Sherlock Microbial Identification System. *Encyclopedia of Rapid Microbiological Methods*, vol. **3**, 1–18.

Madigan M, Martinko J, Parker J. 2003. *Brock Biology of Microorganism*, 10th ed., Pearson Education, Upper Saddle River, NJ.

Martin de Nicolas MM, Vindel A, Offez-Nieto JA. 1995. Epidemiological typing of clinically significant strains of coagulase-negative staphylococci. *J Hosp Infect* **29**, 35–43.

Maslow JN, Brecher SM, Adams KS, Durbin A, Loring S, Arbeit RD. 1993. Relationship between indole production and differentiation of *Klebsiella* species: indole-positive and negative isolates of *Klebsiella* determined to be clonal. *J Clin Microbiol* **31**, 2000–2003.

McCartney A. 2002. Application of molecular biological methods for studying probiotics and the gut flora. *Br J Nutr* **88**, S29–S37.

Mehling A, Wehmeier UF, Piepersberg W. 1995. Nucleotide sequences of streptomycete 16s ribosomal DNA: towards a specific identification system for streptomycetes using PCR. *Microbiology* **141**, 2139–2147.

O'Hara CM. 2005. Manual and automated instrumentation for identification of Enterobacteriaceae and other aerobic Gram-negative bacilli. *J Clin Microbiol* **18**, 147–162.

Olsen WP, Groves MJ, Klegerman ME. 1990. Identifying bacterial contaminants in a pharmaceutical manufacturing facility by gas chromatographic fatty acid analysis. *Pharm Technol* **14**, 32–36.

Osoegawa K, Mammoser AG, Wu C, Frengen E, Zeng C, Catanese JJ, de Jong PJ. 2001. A bacterial artificial chromosome library for sequencing the complete human genome. *Genome Res* **11**, 483–496.

O'Sullivan DJ. 2000. Methods for analysis of the intestinal microflora. *Curr Issues Intest Microbiol* **1**, 39–50.

Petti CA, Polage CR, Schreckenberger P. 2005. The role of 16S rRNA gene sequencing in identification of microorganisms misidentified by conventional methods. *J Clin Microbiol* **43**, 6123–6125.

Pitout Johann DD, Nordmann P, Laupland KB, Poirel L. 2005. Emergence of enterobacteriaceae producing extended-spectrum β-lactamases (ESBLs) in the community. *J Antimicrob Chemother* **56**, 52–59.

Roy D, Sirois S, Vincent D. 2001. Molecular discrimination of lactobacilli used as starter and probiotic cultures by amplified ribosomal DNA restriction analysis. *Curr Microbiol* **42**, 282–289.

Schell MA, Karmirantzou M, Snel B, Vilanova D, Berger B, Pessi G, Zwahlen MC, Desiere F, Bork P, Delley M, Pridmore RD, Arigoni F. 2002. The genome sequence of *Bifidobacterium longum* reflects its adaptation to the human gastrointestinal tract. *Proc Natl Acad Sci U S A* **99**, 14422–14427.

Song Y, Yang R, Guo Z, Zhang M, Wang X, Zhou F. 2000. Distinctness of spore and vegetative cellular fatty acid profiles of some aerobic endosporeforming bacteria. *J Microbiol Methods* **39**, 225–241.

Steinbach WJ, Shetty AK. 2001. Use of the diagnostic bacteriology laboratory: a practical review for the clinician. *Postgrad Med J* **77**, 148–156.

Sutton SVW, Cundell AM. 2004. Microbial identification in the pharmaceutical industry. *Pharmacopeial Forum* **30**, 1884–1894.

Suzuki T, Yamasato K. 1994. Phylogeny of spore-forming lactic acid bacteria based on 16S rRNA gene sequences. *FEMS Microbiol Lett* **115**, 13–17.

Tang P, Louie M, Richardson SE, Smieja M, et al. Ontario Laboratory Working Group for the Rapid Diagnosis of Emerging Infections. 2004. Interpretation of diagnostic laboratory tests for severe acute respiratory syndrome: the Toronto experience. *CMAJ* **170**, 47–54

Tang YW, Procop GW, Persing DH. 1997. Molecular diagnostics of infectious diseases. *Clin Chem* **43**, 2021–2038.

Tester S, Juillerat V, Doherr MG, Haase B, Polak M, et al. 2009. Biochemical typing of pathological prion protein in aging cattle with BSE. *J Virol* **6**, 64.

Thiel KW, Giangrande PH. 2009. Therapeutic applications of DNA and RNA aptamers. *Oligonucleotides* **19**, 209–222.

Thurm V, Gericke B. 1994. Identification of infant food as a vehicle in a nosocomial outbreak of *Citrobacter freundii*: epidemiological subtyping by allozyme, whole-cell protein, and antibiotic resistance. *J Appl Bacteriol* **76**, 553–558.

Tiwari RP, Tiwari D, Garg SK, Chandra R, Bisen PS. 2005. Glycolipids of *Mycobacterium tuberculosis* Strain H37Rv Are Potential Serological Markers For Diagnosis of Active Tuberculosis. *Clinical and Diagnostic Laboratory Immunology* **12**, 465–473.

Torres-Chavolla E, Alocilja EC. 2009. Aptasensors for detection of microbial and viral pathogens. *Biosens Bioelectron* **24**, 3175–3182.

Trevors JT. 1985. DNA probes for the detection of specific genes in bacteria isolated from the environment. *Trends Biotechnol* **3**, 291–293.

Truper HG, Schleifer KH. 2006. Prokaryote characterization and identification. In: Dworkin M, Falkow S, Rosenberg E, Schleifer KH, Stackebrandt E. eds., vol. 1, *The Prokaryotes: A Handbook on the Biology of Bacteria*, 3rd ed., pp. 58–79. Springer, New York.

Ventura M, Elli M, Reniero R, Zink R. 2001. Molecular microbial analysis of *Bifidobacterium* isolates from different environments by the species-specific amplified ribosomal DNA restriction analysis (ARDRA). *FEMS Microbiol Ecol* **36**, 113–121.

Ventura M, Meylan V, Zink R. 2003. Identification and tracing of *Bifidobacterium* species by use of enterobacterial repetitive intergenic consensus sequences. *Appl Environ Microbiol* **69**, 4296–4301.

Ventura M, Zink R. 2002. Specific identification and molecular typing analysis of *Lactobacillus johnsonii* by using PCR-based methods and pulsed-field gel electrophoresis. *FEMS Microbiol Lett* **217**, 141–154.

Vincent D, Roy D, Mondou F, Dery C. 1998. Characterization of bifidobacteria by random DNA amplification. *Int J Food Microbiol* **43**, 185–193.

de Vos WM. 2001. Advances in genomics for microbial food fermentations and safety. *Curr Opin Biotechnol* **12**, 493–498.

Weisburg WG, Barns SM, Pelletier DA, Lane DJ. 1991. 16S ribosomal DNA amplification for phylogenetic study. *J Bacteriol* **173**, 697–703.

Young JM, Takikawa Y, Gardan L, Stead DE. 1992. Changing concepts in the taxonomy of plant pathogenic bacteria. *Annu Rev Phytopathol* **30**, 67–105.

Yuan S, Astion ML, Schapiro J, Limaye AP. 2005. Clinical impact associated with corrected results in clinical microbiology testing. *J Clin Microbiol* **43**, 2188–2193.

Zhong W, Millsap K, Bialkowska-Hobrzanska H, Reid G. 1998. Differentiation of *Lactobacillus* species by molecular typing. *Appl Environ Microbiol* **64**, 2418–2423.

DIVERSITY OF MICROORGANISMS

5.1. PROLOGUE

Microbes are the most abundant and diverse organisms on Earth. The diverse groups of organism and physiological reactions that form the core of the biogeochemical cycles are of immense importance in understanding and managing global environmental problems. Thousands of microbial species in soil and ocean samples have been cataloged. These extremely high levels of biodiversity may help ensure the stability of ecosystem processes in the face of environmental change. The abundance and abilities of organism on Earth modify the direction of flow and size of reservoir of many chemical elements. The existence of a vast diversity of genotypes relates to the functional and phenotypic diversity. Microbial phenotypes are, therefore, ancient in terms of the geological timescale and have been maintained through stabilizing selection (Fenchel and Finlay, 2006).

5.2. PHYSIOLOGICAL DIVERSITY OF MICROORGANISMS

Microbial organisms are pervasive, ubiquitous, and essential components of all ecosystems. The geochemical composition of Earth's biosphere has been molded largely by microbial activities. Many of these new microbes are widespread and abundant among contemporary microbiota and fall within novel divisions that branch deep within the tree of life (Delong and Pace, 2001). The breadth and extent of extant microbial diversity

Microbes: Concepts and Applications, First Edition. Prakash S. Bisen, Mousumi Debnath, Godavarthi B. K. S. Prasad
© 2012 Wiley-Blackwell. Published 2012 by John Wiley & Sons, Inc.

has become much clearer. The independent development of microbial biology encouraged the birth of some fields, a prime example being molecular biology. Meanwhile, however, appreciation of ecology and evolution in microbial systems has lagged far behind parallel developments in mainstream biology. Metabolic diversity allows to differentiate the microorganisms present in diverse environments into distinct categories as chemolithotrophs, chemoorganotrophs, phototrophs, and lactic acid bacteria. These organisms are further classified according to their nature of aerobic or anaerobic mode of respiration and for their special properties (Fig. 5.1).

5.2.1. Anaerobic Chemotrophs

Chemotrophs are organisms that can obtain energy by the oxidation of inorganic compounds. They are generally defined by their ability to grow in mineral media and to derive their cell carbon from CO_2. Many chemolithotrophs are commonly referred to as *autotrophs*. Anaerobic chemotrophs probably appeared subsequently followed by the evolution of chlorophyll-containing anaerobic photoautotrophs. These organisms require temperature above $50\,^{\circ}C$ for growth and use of sulfur as electron acceptor (except *Sulfolobus* that uses oxygen or ferric iron). These microorganisms more likely developed at temperatures below $110\,^{\circ}C$ and at water depths of about 1000 m. Under such environmental conditions, the microorganisms would have been anaerobic chemotrophs, metabolizing in a reducing environment and obtaining their energy and nutrients from the hydrothermal fluids. *Micrococus* is a methanogen and an anaerobic chemolithotroph (Fig. 5.2). The anaerobic electron transport chains are not restricted to facultative anaerobes, although they have evolved the ability to respond efficiently to the availability of various electron acceptors and the changes in the redox potential of the medium. Strict anaerobes cannot grow in the presence of oxygen. Moreover, they are restricted to certain biotope.

5.2.2. Anoxygenic Phototrophs

Anoxygenic phototrophs inhabit a variety of extreme environments, including thermal, polar, hypersaline, acidic, and alkaline aquatic and terrestrial habitats. Anoxygenic phototrophs have photosynthetic pigments called *bacteriochlorophylls* (similar to chlorophyll found in eukaryotes). Bacteriochlorophyll *a* and *b* have maximum wavelength absorption at 775 and 790 nm, respectively. However, these pigments were found to absorb longer wavelengths (many in the infrared spectrum) *in vivo*. Unlike oxygenic phototrophs, anoxygenic photosynthesis only functions using a single photosystem. This restricts them to cyclic electron flow only, and they are therefore unable to produce O_2 from the oxidization of H_2O. The anoxygenic phototrophs fall into five phylogenetic groups: the purple sulfur bacteria (*Chromatium*, Fig. 5.3), purple nonsulfur bacteria (*Rhodobacter*), green sulfur bacteria (*Chlorobium*), green nonsulfur bacteria (*Chloroflexus*), and heliobacteria. Anoxygenic phototrophs such as *Rhodobacter rubrum* can contain several bacteriochlorophylls, and most purple bacteria have bacteriochlorophyll *a*, which absorbs maximally between 800 and 925 nm. Organisms with many different types of chlorophylls are at an advantage because they can use more of the energy of the electromagnetic spectrum.

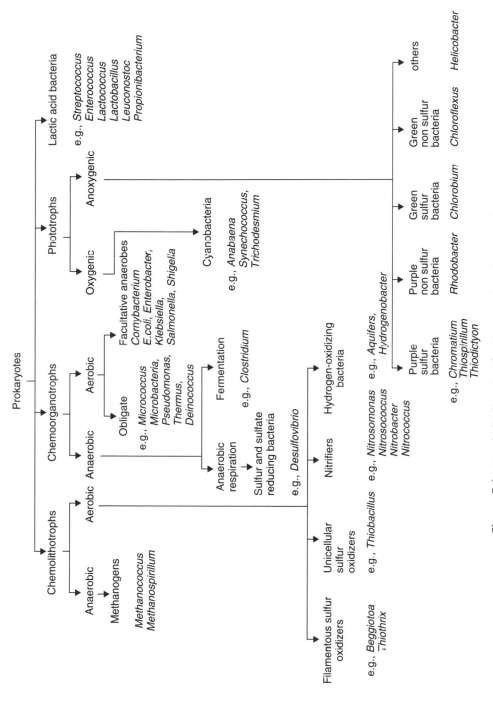

Figure 5.1. Metabolic diversity of prokaryotes in a diverse environment.

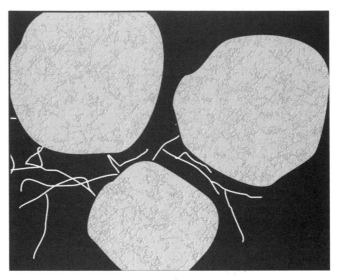

Figure 5.2. Electron microscopic view of *Methanococcus*, an anaerobic chemolithotroph.

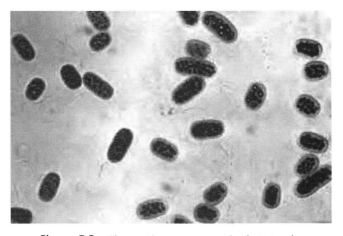

Figure 5.3. *Chromatium*: anoxygeneic phototroph.

5.2.3. Oxygenic Phototrophs

Phototrophs are the organisms (usually plants) that carry out photosynthesis to acquire energy. They use the energy from sunlight to convert carbon dioxide and water into organic materials to be utilized in cellular functions such as biosynthesis and respiration. Oxygenic phototrophs represent a group of photosynthetic, mostly photolysis-mediated oxygen-evolving monerans (prokaryotes). *Anabeana* is one such organism representing this class (Fig. 5.4). These are the only organisms performing oxygenic photosynthesis that can also fix nitrogen. These organisms are among the oldest organisms known, dating back to the early Precambrian period of 3.6×10^9 years ago and probably played a crucial role in the evolution of higher plants.

Figure 5.4. *Anabaena torulosa* with dark colored akinetes and colorless differentiated heterocyst.

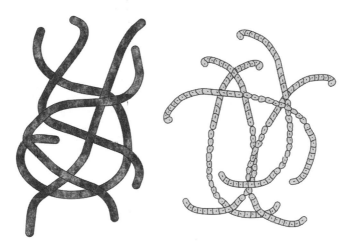

Figure 5.5. *Beggiatoa.*

Photosynthesis directly or indirectly drives the biogeochemical cycles in all extant ecosystems of the planet. Even hydrothermal vent communities, which use inorganic electron donors of geothermal origin and assimilate CO_2 by chemolithoautotrophy (rather than photoautotrophy), still depend on the molecular O_2 generated by oxygenic phototrophs outside these systems (Jannasch, 1989).

5.2.4. Aerobic Chemolithotrophs

Chemolithotrophs are organisms that obtain their energy from the oxidation of inorganic compounds. Aerobic chemolithotrophs oxidize inorganic compounds such as NH_4, H_2S, and iron (Fe). *Beggiatoa* (Fig. 5.5) and *Thiothrix* are some of the representatives of aerobic chemolithotrophs. Important marine examples include the nitrifying bacteria, sulfide-oxidizing bacteria, and methanotrophs (Munn, 2004). The aerobic bacteria capable of obtaining energy from the oxidation of H_2 form a heterogeneous group that includes both facultative and obligate chemolithotrophs and representatives of both gram-negative and gram-positive genera. H_2-oxidizing aerobes inhabit such diverse biotypes as soil, oceans, and hot springs. The oxidation of H_2 in these bacteria is catalyzed by [NiFe] metalloenzymes called *hydrogenases*. The hydrogenases studied so far belong to two families: dimeric, membrane-bound enzymes (MBH (membrane-bound hydrogenase)) coupled to electron transport chains and tetrameric, cytoplasmic NAD-reducing enzymes (SH). Ni^{2+} is an essential component of the active site contained in the large

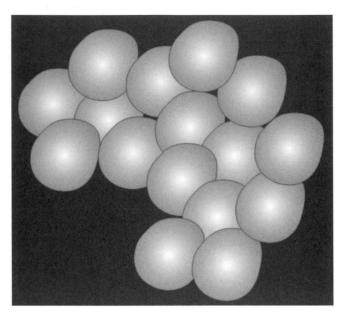

Figure 5.6. *Micrococcus.*

subunit of the MBH enzymes (Friedrich and Schwartz, 1993). Nitrifying bacteria are aerobic chemolithotrophs using carbon dioxide as a carbon source and oxidizing reduced inorganic nitrogen compounds to obtain energy (Carter et al., 2008).

5.2.5. Aerobic Chemoorganotrophs

Aerobic chemoorganotrophs form long filaments (trichomes). Some of them are marine bacteria that attach to solid substrates by a holdfast and have a complex lifestyle in which dispersal is by the formation of gonidia. Chemoorganotrophs obtain their energy from organic compounds such as glucose. *Micrococcus* (Fig. 5.6), *Microbacter, Pseudomonas, Thermus*, and *Deinonococcus* are representatives of this group.

   Anaerobically respiring chemoorganotrophs use sulfur compounds instead of O_2 as their terminal electron acceptor. Generally, their metabolism produces hydrogen sulfide (H_2S), which has a rotten egg smell. H_2S also reacts with iron, which is corrosive, and forms a black product (iron sulfide). Anaerobic chemoorganotrophs consequently are responsible for significant iron corrosion under anaerobic conditions (iron pipes). H_2S reacting with iron also turns anaerobic environments black, such as mud. Aerobic chemoorganotrophs are the most important of bacteria. Many pathogens are aerobic chemorganotrophs. Aerobic chemorganotrophs can be divided, based on their oxygen requirements, into obligate aerobes and facultative anaerobes. The Sphingomonadaceae comprises of genera *Erythrobacter, Erythromicrobium*, and *Porphyrobacter*, which are aerobic chemorganotrophs, although their cells contain bacteriochlorophyll *a* and carotenoids (Folkow, 2006).

5.3. THRIVING IN TERRESTRIAL ENVIRONMENT

Most of the terrestrial life-forms of organisms are able to thrive well in intolerably hostile environments (Michael, 1998). The majority of known extremophiles are varieties of archaea and bacteria. They are classified, according to the conditions in which they exist, as thermophiles, hyperthermophiles, psychrophiles, halophiles, acidophiles, alkaliphiles, barophiles, and endoliths. These categories are not mutually exclusive, so that, for example, some endoliths are also thermophiles. The resistance of some bacterial cells to environmental destruction is impressive. Some bacteria form resistant cells called *endospores*. The original cell replicates its chromosome, and one copy becomes surrounded by a durable wall. The outer cell disintegrates, but the endospore it contains survives all sorts of trauma, including lack of nutrients and water, extreme heat or cold, and most poisons. Unfortunately, boiling water is not hot enough to kill most endospores in a reasonable length of time. Endospores may remain dormant for centuries (Campbell, 1993).

Chemoautotrophs, which obtain their energy from the oxidation of inorganic chemicals, might be particularly suited to alien environments. *Thiobacillus*, for example, makes a living by oxidizing sulfur to sulfuric acid, while other types oxidize hydrogen to water or nitrites to nitrates (Fig. 5.7; Suzuki et al., 1999). *Ferrobacillus* is especially interesting because the energy obtained by oxidizing ferrous iron to ferric is very small, so that this organism must have evolved some ingenious way of boosting its low grade energy input to the high levels needed to synthesize essential chemicals such as adenosine triphosphate. Certain species can tolerate wide extremes of acidity and alkalinity. *Thiobacillus* will grow in solutions containing 3% sulfuric acid, while other types of bacteria have been found in the saturated brine of the Great Salt Lake in Utah and in an icy pool in Antarctica containing 33% calcium chloride. Some bacteria have survived

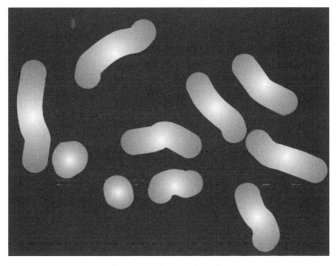

Figure 5.7. *Thiobacillus.*

enormous pressures, up to 10 tons/cm^2, while low pressures appear to pose no threat to them at all, provided that liquid water is present. However, some species of bacteria, such as *Deinococcus radiodurans*, have been found growing in the radioactive cooling ponds where fuel cans from nuclear reactors are stored; however, they are not tolerant of highly ionizing radiation. For this reason alone, the notion of panspermia is not easy to uphold. The information voyage begins where bacteria can only grow at temperatures well above the known range for terrestrial living organisms. These "hyperthermophiles," whose optimum temperature range is around and in some cases even above the boiling point of water, are found in the turbulent zones of deep-sea hydrothermal vents. Furthermore, they were found to belong to a distinctive group of look alike bacteria named *archaebacteria* (or more recently *archaea*); however, since "The Outer Reaches of Life" was penned, other hyperthermophilic bacteria have been found that cohabitate with the hyperthermophilic archaea. Also, undiscovered at the time was the astonishing finding, in a submarine hydrothermal vent, of a "nanoorganism" growing on the surface of a normal-sized archaea.

Microbial life exists in extreme terrestrial environments where temperatures, pressures, or other factors would have been foreseen as unbearable to any kind of life as we know it. Bacteria, however, have been found in such environments and have established exotic, or some might say, eccentric ways of coping with extreme conditions.

5.3.1. Microbial Population Counts in Soil

The organic matter in soil exists mainly as humus or is partially decomposed (plant and animal tissues). It is also prepared artificially as farmyard manure, green manure/green leaf manure, compost, vermicompost, biofertilizers, and so on. On the whole, the balanced availability of both inorganic and organic matter in the soil determines the soil fertility. Indirectly, these organic and inorganic matter help in the proliferation of various qualitative microflora that play a vital role in maintaining the nutritional balance of soil. Thus, microorganisms have a great role to play in determining soil fertility, for without a proper distribution of microflora, no soil can support plant growth which speaks of its fertility. Microorganisms in soil affect the fertility of soil by means of physical or chemical changes. Soil structure depends on the association between mineral soil particles (sand, silt, and clay) and organic matter, in which aggregates of different size and stability are formed. Soil organic matter has been considered an important indicator of soil quality because it is a nutrient sink and source and it not only enhances soil physical and chemical properties and promotes biological activity but also maintains environmental quality (Astaraei, 2008).

An active microbial population performs many beneficial activities such as organic matter decomposition, nutrient availability and recycling, and pathogen suppression (Sylvia et al., 1997). Thus, turfgrasses grown in root zones containing lower microbial populations may be less healthy and possibly more easily affected by some soil-borne turfgrass pathogens such as *Pythium* spp. and *Rhizoctonia*, resulting in an overall lower quality turfgrass (Hodges, 1990; Couch, 1995).

Mancino et al. (1993) monitored microbial properties of a 5-year-old sand peat moss root zone and found that microbial populations in the thatch of a sand-based root zone planted to creeping bentgrass were much higher than those of the underlying rootzone. They reported that a mature (older than 5 years) putting green supported a relatively large microbial population ($> 10^7$ cfu/g of soil), which is similar to populations found in some native soils.

All the factors that affect soil formation also affect the microbes. These are climate, parent material and texture (especially clay), vegetation, topography, and time. For example, the microbial status of a dryland wheat field in the Palouse region is quite different on the north-facing slope than on the south-facing slope. Both long-term and short-term factors have shaped the soils differently, particularly the influence of moisture on vegetation growth and the resulting soil organic matter levels. Other factors include the chemical status of the soil (pH, salt levels, redox potential) and physical conditions (water potential, temperature, soil structure); soil organisms, especially microbes, influence the same factors that influence them. Our management actions dramatically affect soil microbes. These include the choice of crop, rotation, and tillage, as well as the use of irrigation, fertilization, or fumigation. Research is increasingly pointing to certain practices that almost always enhance the microbial status, such as the use of composts and cover crops. Of all factors, soil organic matter can be considered the primary determinant of microbial status. It is the "fuel" for the soil microbes. The organic matter is often described as consisting of three components, each differing in composition and function.

The biologically active fraction is the smallest and turns over every 1–5 years. It is highly important for microbes, nutrient cycling, and disease suppression. The protected fraction is intermediate in size, turns over every 5–30 years, and plays a key role in soil structure and water relations. The stable fraction is the largest, turns over every 50–10,000 years, and contributes to the cation exchange capacity, other chemical properties, color, and microaggregation. A number of researchers are looking for analytical tests beyond simple soil organic matter to better understand the more active fraction and the role it plays in supporting soil biology. Particulate organic matter (POM) is one of these. Microbial biomass carbon generally makes up 1–5% of the soil carbon. Let us imagine that your soil has 3.5% organic matter as reported from a laboratory test. Because soil organic matter contains carbon, nitrogen, oxygen, phosphorus, and so on, divide the organic matter by 1.75 to determine the percentage of soil organic carbon (organic matter is about 57% C). So this soil has 2% soil C. Then multiply that by the approximate weight of the top 6–7 in of soil, called the *acre furrow slice*, which is 2 million lb. So, 2% soil C × 2 million lb = 40,000 lb of organic carbon per acre. It is referred to as *organic* since some soil may contain significant amounts of inorganic carbon in carbonate compounds that are not part of the soil organic matter equation. Then, if 3% of the organic carbon is in the microbes, that equals about 1200 lb of microbial carbon per acre, which is similar to the weight of a cow gazing in a 1-acre field.

5.3.1.1. Measuring Soil Microbes

There are lots of microbes in a small amount of soil. Microbial numbers can be extremely variable, making measurements difficult. Most researchers feel that there need to be at least a 10-fold difference in microbial number to be meaningful because of variability in measuring organisms. Sometimes we see "desirable" levels of microbes proposed. For example, a productive agricultural soil might be expected to contain 10–100 million bacteria per gram of soil. So if you compared two soils, one with 15 million bacteria and the other with 90 million, they have a similar number of bacteria and would perform the same if all else was equal.

There are several common techniques for measuring the number or biomass of soil microbes. The most basic is the direct count, where microbes are extracted from soil in a liquid, sometimes dyed to illustrate particular groups, and then counted on a slide under the microscope. This method tends to overestimate since dead organisms cannot always be distinguished from live ones. Also, sampling error can be magnified by the small sample size and the potential for uneven microbial distribution in the extract (Granatstein and Bezdicek, 2003).

5.3.2. Bacteria that Form a Resting Stage

In certain species of the lower bacteria, under certain circumstances, changes take place in the protoplasm, which result in the formation of bodies called *spores* to which the vital activities of the original bacteria are transferred. Spore formation occurs chiefly among the bacilli and in some spirilla. The commencement of spore formation in a bacterium is indicated by the appearance in the protoplasm of a minute highly refractile granule (or by a number of minute highly refractile granules scattered throughout the protoplasm, which gradually coalesce) unstained by the ordinary methods. The granule increases in size and assumes a round, oval, or short rod-shaped form, always shorter but often broader than the original bacterium. In the process of spore formation, the rest of the bacterial protoplasm may remain unchanged in both appearance and staining power for a considerable time (*Bacillus tetani*) or, on the other hand, may soon lose its power of staining and ultimately disappear, leaving the spore in the remains of the envelope (*Bacillus anthracis*). This method of spore formation is called *endogenous* (Fig. 5.8). Bacterial spores are always nonmotile. The spore may be in the center of the bacterium, at one extremity, or at a short distance from one extremity. In structure, the spore consists of a mass of protoplasm surrounded by a dense membrane.

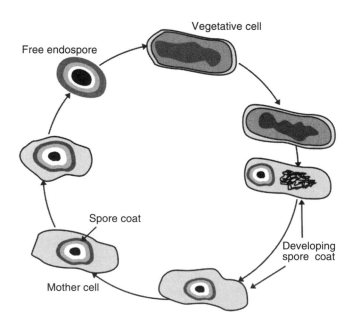

Figure 5.8. Endospore formation.

It is important to note that in bacteria, spore formation is rarely, if ever, to be considered as a method of multiplication. In at least the great majority of cases, only one spore is formed from one bacterium and only one bacterium in the first instance from one spore. Sporulation is looked upon as a resting stage of a bacterium and should be contrasted with the stage when active multiplication takes place. The latter is usually referred to as the *vegetative stage* of the bacterium. Regarding the significance of spore formation in bacteria, there has been some difference of opinion. According to one view, it may be regarded as representing the highest stage in the vital activity of a bacterium. There is thus an alternation between the vegetative and spore stages, the occurrence of the latter being necessary to the maintenance of the species in its greatest vitality. Such a rejuvenescence, as it were, through sporulation, is known in many algae. In many cases, for instance, spore formation only occurs at temperatures especially favorable for growth and multiplication. There is often a temperature below which, while vegetative growth still takes place, sporulation will not occur, and in the case of *B. anthracis*, if the organism be kept at a temperature above the limit at which it grows best, not only are no spores formed but also the species may lose the power of sporulation. Furthermore, in the case of bacteria preferring the presence of oxygen for their growth, an abundant supply of this gas may favor sporulation. Most bacteriologists are, however, of the opinion that when a bacterium forms a spore, it only does so when its surroundings, especially its food supply, become unfavorable for vegetative growth; it then remains in this condition until it is placed in more suitable surroundings. Such an occurrence would be analogous to what takes place under similar conditions in many of the protozoa. Often sporulation can be prevented from taking place for an indefinite time if a bacterium is constantly supplied with fresh food (the other conditions of life being equal). The presence of substances excreted by the bacteria themselves plays, however, a more important part in making the surroundings unfavorable than the mere exhaustion of the food supply. A living spore will always develop into a vegetative form if placed in a fresh food supply. With regard to the rapid formation of spores when the conditions are favorable for vegetative growth, it must be borne in mind that in such circumstances, the conditions may really very quickly become unfavorable for continuance of growth since not only will the food supply around the growing bacteria be rapidly exhausted but also the excretion of effete and inimical matters will be all the more rapid.

Bacteria used under these conditions lend themselves to investigation in the same manner as enzymes or other catalytic systems, and it has been possible to show that the bacterial suspensions have many factors in common with enzyme systems. The resting organism technique enables important chemical distinctions to be drawn between various bacteria. For instance, in contrast to *Escherichia coli* or *Bacillus pyocyaneus, Bacillus alcaligenes* is practically incapable of activating nitrates, chlorates, and some of the sugars. Such distinctions become of considerable value when taken in conjunction with the growth characteristics of these organisms. The easy manipulation of resting bacteria and the fact that conditions can be well controlled make it possible to obtain reproducible and consistent results.

5.3.3. Bacteria that Associate with Plants

Plants support a diverse array of bacteria, including parasites, mutualists, and commensals on and around their roots (Fig. 5.9), in the vasculature, and on aerial tissues. These microbes have a profound influence on plant health and productivity. Bacteria physically

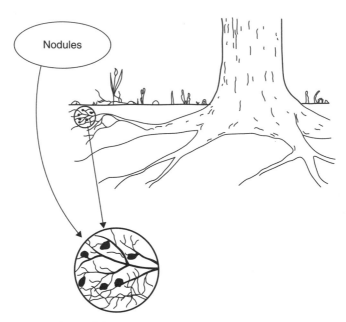

Figure 5.9. Interaction of microbes with plants.

interact with surfaces to form complex multicellular and often multispecies assemblies, including biofilms and smaller aggregates. There is growing appreciation that the intensity, duration, and outcome of plant–microbe interactions are significantly influenced by the conformation of adherent microbial populations (Danhorn and Fuqua, 2007). Owing to their small size, commensal bacteria are ubiquitous and grow on plants exactly as they will grow on any other surface. However, their growth can be increased by warmth and sweat, and large populations of these organisms in humans are the cause of body odor.

The presence of Fe-oxidizing bacteria in the rhizosphere of four different species of wetland plants was investigated in a diverse wetland environment that had Fe(II) concentrations ranging from tens to hundreds of micromoles per liter and a pH range of 3.5–6.8. Enrichments for neutrophilic putatively lithotrophic Fe-oxidizing bacteria were successful on roots from all four species; acidophilic Fe-oxidizing bacteria were enriched only on roots from plants whose root systems were exposed to soil solutions with a pH of less than 4 (Emerson et al., 1999). In *Sagittaria australis*, there was a positive correlation ($p < 0.01$) between cell numbers and the total amount of Fe present; the same correlation was not found for *Leersia oryzoides*. These results present the first evidence for culturable Fe-oxidizing bacteria associated with Fe-plaque in the rhizosphere.

Microorganisms interact with plants because plants offer a wide diversity of habitats including the phyllosphere (aerial plant part), the rhizosphere (zone of influence of the root system), and the endosphere (internal transport system) (Lindow et al., 2002; Lynch, 1990). Interactions of epiphytes, rhizophytes, or endophytes may be detrimental or beneficial for either the microorganism or the plant and may be classified as neutralism, commensalism, synergism, mutualism, amensalism, competition, or parasitism (Montesinos, 2003).

Plant-associated microorganisms play essential roles in agriculture and food safety and contribute to the environmental equilibrium. Their study has been classically based on cultivation-dependent methods, which often recover only 0.01–10% of direct counts. However, studies based on molecular analysis have estimated more than 4000 species per gram of soil. Most of these microorganisms are probably noncultivatable, such as the plant symbiotic vesicular arbuscular mycorrhiza (VAM), endomycorrhizae common in many angiosperms and gymnosperms, or are transiently in a viable but noncultivatable state (Colwell and Grimes, 2000).

Microbial composition and population levels in the rhizosphere and phyllosphere are determined mainly by carbon sources released as root exudates and often change with cultivars of the same plant species. The degree of change in microbial composition of the phyllosphere or rhizosphere is associated even with the type of genetic modification in some transgenic plants. The high proportion of ampicillin-resistant bacteria occurring naturally in agricultural fields fertilized often with manure may put into perspective the potential impact of the antibiotic-selectable-marker acquisition by corn-associated microbiota. More than 10,000 species of fungi, bacteria, and viruses are estimated to cause plant diseases, several of which are of major economic importance (Montesinos, 2003).

Most plant-associated microorganisms whose genomes have already been sequenced are bacteria, including *Agrobacterium tumefaciens, Pseudomonas syringae* pv. *tomato, Xanthomonas campestris* pv. *campestris, Ralstonia solanacearum, Xanthomonas axonopodis* pv. *citri*, and *Xylella fastidiosa; Bradyrhizobium japonicum, Sinorhizobium meliloti*, and *Mesorhizobium loti; Clavibacter michiganensis* subsp. *sepedonicus; Azotobacter; Bacillus thuringiensis; Pseudomonas fluorescens; Rhizobium leguminosarum;* and fungal pathogens including *Aspergillus* spp., *Botryotinia fuckeliana, Fusarium* spp., and *Mycospherella graminicola*.

The easy commercial exchange of agricultural products (fresh products and plant material) has caused the highest spread of plant pathogens and diseases ever observed. Quarantine systems to prevent the introduction of new plant pathogens in protected areas and eradication measures to avoid epidemic spread are the first protection barriers. However, many pathogens remain latent in the plant material and in very low numbers, and methods with high sensitivity, specificity, and reliability are required. New methods of plant pathogen analysis have evolved parallel to the advancement of methods of analysis in human and veterinary microbiology (Montesinos, 2003).

5.3.4. Isolation of Antibiotic-Producing Bacteria from Soil

Serious infections caused by bacteria that have become resistant to commonly used antibiotics have become a major global health care problem in the twenty-first century (Alanis, 2005). Antibiotics are the best known products of actinomycete. Over 5000 antibiotics have been identified from the cultures of gram-positive and gram-negative organisms, as well as filamentous fungi, but only about 100 antibiotics have been commercially used to treat human, animal, and plant diseases. The genus, *Streptomyces*, is responsible for the formation of more than 60% of known antibiotics, while a further 15% are produced by a number of related actinomycetes, *Micromonospora, Actinomadura, Streptoverticillium*, and *Thermoactinomycetes* (Waksman, 1954).

Antibiotics, because of their industrial importance, are the best known products of actinomycetes. The actinomycetes produce an enormous variety of bioactive molecules, for example, antimicrobial compounds. One of the first antibiotics used is streptomycin produced by *Streptomycin griseus*. Bacteria, such as *Pseudomonas aeruginosa*, are common environmental organisms, which act as opportunistic pathogens in clinical cases where the defense system of the patient is compromised. In addition, other intrinsically antibiotic-resistant organisms such as *Stenotrophomonas maltophilia* are emerging as opportunistic pathogens.

5.3.4.1. Method of Isolation

Ceylan et al. (2008) collected soil samples from various locations in Mugla province from 2006 to 2007. They selected several diverse habitats in different areas for the isolation of *Streptomyces* strains. These habitats included the rhizosphere of plants, agricultural soil, preserved areas, and forest soils. The samples were taken up to a depth of 20 cm after removing approximately 3 cm of the soil surface. The samples were placed in polyethylene (PE) bags, closed tightly, and stored in a refrigerator. Korn-Wendisch and Kutzner (1992) devised a method for screening for the isolation of *Streptomyces*.

Soil is pretreated with $CaCO_3$ (10:1 w/w), incubated at 37 °C for 4 days, and suspended in sterile Ringer solution (1/4 strength). Test tubes containing 10^{-2} dilution of the samples were placed in a water bath at 45 °C for 16 h. The spores would separate from the vegetative cells, and then the dilutions were inoculated on the surface of the actinomycete isolation agar plates. The plates were incubated at 28 °C until the sporulation of *Streptomyces* colonies occurred. *Streptomyces* colonies (where the mycelium remained intact and the aerial mycelium and long spore chains were abundant) were then removed and transferred to the yeast extract–malt extract agar (ISP2) slants. Pure cultures were obtained from selected colonies for repeated subculturing. After antimicrobial activity screening, the isolated *Streptomyces* strains were maintained as suspensions of spores and mycelial fragments in 10% glycerol (v/v) at −20 °C.

5.3.4.2. Characterization of the Isolates

Streptomyces colonies can be characterized morphologically and physiologically following the international guidelines of the International Streptomyces Project (ISP) (Shirling and Gottlieb, 1966):

General morphology is determined by plating on oatmeal agar and incubating in the dark at 28 °C for 21 days and then by performing a direct light microscopic examination of the surface of the crosshatched cultures. Colors are determined according to the scale adopted by Prauser (1964). Melanin reactions are detected by growing the isolates on peptone-yeast extract-iron agar (ISP 6) medium (Shirling and Gottlieb, 1966). Williams et al. (1983a, 1983b) determined some diagnostic characters of highly active *Streptomyces* strains following the directions given in *Bergey's Manual of Systematic Bacteriology*.

In Vitro Screening of Isolates for Antagonism. Madigan et al. (1997) prepared the balanced sensitivity medium (BSM, (Difco, 1863) plates and inoculated *Streptomyces* by a single streak of inoculum in the center of the petri dish. After 4 days of incubation at 28 °C, the plates were seeded with test organisms by a single streak at 90° angle to the *Streptomyces* strains, and the microbial interactions were analyzed by determination of the size of the inhibition zone (Fig. 5.10).

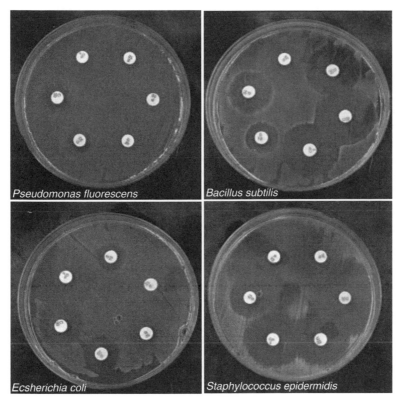

Figure 5.10. Antibiotic sensitivity test. (*See insert for color representation of the figure*).

5.4. AQUATIC ENVIRONMENT

In reality, all microorganisms require an aquatic environment. Aquatic environment comprises 70% of Earth's surface, most of the area occupied by oceans. There are also a broad spectrum of other aquatic environment including estuaries, harbors, major river system, lakes, wetland, streams, springs, and aquifers. Microorganisms are the primary producers in that environment and are responsible for approximately half of all primary production on Earth. They are also primary consumers. The microbiota that inhibits aquatic environment include bacteria, virus, fungi, algae, and other microfauna. Identifying the microbial composition of the aquatic environment and determining the physiological activity of the earth component are the first steps in understanding the ecosystem as a whole. In general, methods used in monitoring microbes in aquatic environment are similar to those used for most other environments.

5.4.1. Sheathed Bacteria

Sheathed bacteria are bacteria that grow as long filaments, whose exterior is covered by a layer known as a *sheath*. The bacteria are able to grow and divide within the sheath. Examples of sheathed bacteria include *Leptothrix discophora* (also known as *iron bacteria*) and *Sphaerotilus natans*.

Sheathed bacteria are common among the bacterial communities in water and soil. In these environments, the sheath is often coated with precipitates of elements in the water or soil environment, such as oxides of iron and manganese. The elements are unstable in solution and thus will readily come out of solution when presented with an appropriate site.

The sheath that covers the bacteria can be of varied construction. Much of the structural information has been gleaned from the observation of thin slices of sample using the transmission electron microscope. The sheath surrounding *Leptothrix* species is glycocalyx-like in appearance. Often the deposition of metals within the sheath network produces areas where the material has crystallized. In contrast, the sheath of *S. natans* presents the "railroad track" appearance, which is typical of a biological membrane consisting of two layers of lipid molecules.

Electron microscopic studies of *Leptothrix* species (Fig. 5.11) have shown that the bacterium is intimately connected with the overlying sheath. The connections consist of protuberances that are found all over the surface of the bacterium. In contrast, *S. natans* is not connected with the overlying sheath. Both *Leptothrix* and *S. natans* (Fig. 5.12) can exist independent of the sheath. Bacteria in both genera have a life cycle that includes a free-swimming form (called a *swarmer cell*) that is not sheathed. The free-swimming forms have flagella at one end, which propels the cells along. When encased in the sheath, the bacteria are referred to as *sheathed* or *resting bacteria*.

Bacterial sheaths are produced when the bacteria are in an aquatic or a soil environment that contains high amounts of organic matter. The sheath may serve to provide protection to the bacteria in these environments; also, the ability of metallic compounds to precipitate on the sheath may provide the bacteria with a ready supply of such inorganic nutrients. For example, *Leptothrix* is able to utilize the manganese contained in the manganese oxide precipitate on the sheath. Sheaths may also help the bacteria survive over a wide range of temperature and pH, by providing a relatively inert barrier to the external environment.

5.4.2. Prosthecated Bacteria

Prosthecate bacteria are a nonphylogenetically related group of gram-negative bacteria that possess appendages, termed *prosthecae*. These cellular appendages are neither pili

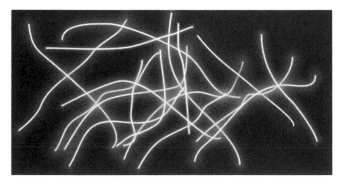

Figure 5.11. *Leptothrix*.

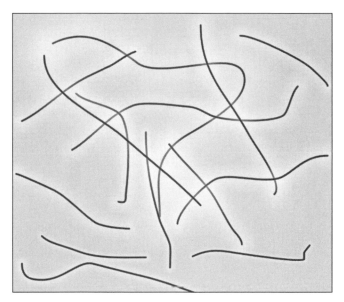

Figure 5.12. *Sphaerotilus natans.*

nor flagella, as they are extensions of the cellular membrane and contain cytosol. One notable group of prosthecates is the genus *Caulobacter* (Fig. 5.13).

Prosthecates are generally chemoorganotrophic aerobes that can grow in nutrient-poor habitats, being able to survive at nutrient levels on the order of parts per million, for which reason they are often found in aquatic habitats. These bacteria will attach to surfaces with their prosthecae, allowing a greater surface area to take up nutrients (and release waste products). Some prosthecates will grow in nutrient-poor soils as aerobic heterotrophs.

Although the overall role of the heterotrophic bacterial community is known, the importance of specific bacterial populations is poorly understood. In evaluating the significance of bacteria in a habitat, it is desirable to enumerate individual genera and species since numerically important organisms may play very important roles. In addition, correlation of fluctuating bacterial numbers with changing conditions in a habitat may suggest the environmental variables that affect or are affected by the heterotrophic community. Unfortunately, most heterotrophic bacteria in aquatic habitats are morphologically nondescript and must be obtained in pure culture for positive identification. Thus, enumeration of specific genera in mixed populations relies on the ability of the investigator to quantitatively cultivate the organisms of interest.

Badger et al. (2006) reported on the dimorphic prosthecate bacteria (DPB), *Hyphomonas neptunium*. This is an α-proteobacteria that reproduce in an asymmetric manner rather than by binary fission and is of interest as simple models of development (Fig. 5.14). Before this work, the only member of this group for which genome sequence was available was the model freshwater organism *Caulobacter crescentus*. The genome sequence of *H. neptunium*, a marine member of the DPB that differs from *C. crescentus* in that *H. neptunium* uses its stalk as a reproductive structure. Genome analysis indicates that this organism shares more genes with *C. crescentus* than it does with *Silicibacter pomeroyi* (a closer relative according to 16S rRNA phylogeny), that it relies on a

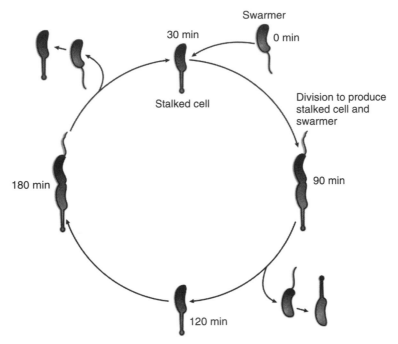

Figure 5.13. Life cycle of *Caulobacter*.

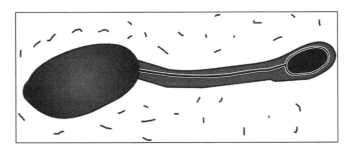

Figure 5.14. *Hyphomonas neptunium*.

heterotrophic strategy utilizing a wide range of substrates, that its cell cycle is likely to be regulated in a manner similar to that of *C. crescentus*, and that the outer membrane components of *H. neptunium* and *C. crescentus* are remarkably similar. *H. neptunium* swarmer cells are highly motile via a single polar flagellum. With the exception of cheY and cheR, genes required for chemotaxis were absent in the *H. neptunium* genome. Consistent with this observation, *H. neptunium* swarmer cells did not respond to any chemotactic stimuli that were tested, suggesting that *H. neptunium* motility is a random dispersal mechanism for swarmer cells rather than a stimulus-controlled navigation system for locating specific environments. In addition to providing insights into bacterial development, the *H. neptunium* genome will provide an important resource for the study of other interesting biological processes including chromosome segregation, polar growth, and cell aging.

5.4.3. Bacteria that Derive their Nutrients from Other Organisms

Bacteria are consistently associated with the body surfaces of animals. There are many more bacterial cells on the surface of a human (including the gastrointestinal tract) than there are human cells that make up an animal. Bacteria and other microbes that are consistently associated with an animal are called the *normal flora*, or more properly the *indigenous microbiota*, of the animal. These bacteria have a full range of symbiotic interactions with their animal hosts.

5.4.3.1. Types of Symbiotic Associations

In biology, symbiosis is defined as *life together*, that is, those two organisms live in an association with one another. Thus, there are at least three types of relationships based on the quality of the relationship for each member of the symbiotic association.

MUTUALISM. Both members of the association benefit. For humans, one classic mutualistic association is that of the lactic acid bacteria that live on the vaginal epithelium of a woman. The bacteria are provided a habitat with constant temperature and supply of nutrients (glycogen) in exchange for the production of lactic acid, which protects the vagina from colonization and disease caused by yeast and other potentially harmful microbes

COMMENSALISM. There is no apparent benefit or harm to either member of the association. A problem with commensal relationships is that if you look at one long enough and hard enough, you often discover that at least one member is being helped or harmed during the association. Consider our relationship with *Staphylococcus epidermidis*, a consistent inhabitant of the skin of humans. Probably, the bacterium produces lactic acid that protects the skin from colonization by harmful microbes that are less acid tolerant. Other metabolites that are produced by the bacteria are an important cause of body odor (good or bad, depending on your personal point of view) and possibly associated with certain skin cancers. "Commensalism" works best when the relationship between two organisms is unknown and not obvious.

Bacteria exhibit different modes of nutrition. On this basis, broadly two types of bacteria can be recognized: autotrophic bacteria and heterotrophic bacteria. The photoautotrophic bacteria possess photosynthetic pigments in membrane-bound lamellae (or thylakoids) and utilize solar energy. The bacterial photosynthesis is different from that of green plants since here water is not used as a hydrogen donor. Hence, oxygen is not released as a by-product. For this reason, the process is described as anoxygenic photosynthesis, where bacteria that produce organic compounds from inorganic raw materials utilizing energy liberated from the oxidation of inorganic substances. Following are the common types of chemoautotrophic bacteria that are unable to manufacture their own organic food and hence are dependent on external source. These bacteria can be distinguished into two groups, saprophytic and parasitic bacteria.

Saprophytic bacteria obtain their nutritional requirements from dead organic matter. They break down the complex organic matter into a simple soluble form by secreting exogenous enzymes and subsequently absorb the simple nutrients and assimilate and release energy. These bacteria have a significant role in the ecosystem, functioning as decomposers. The aerobic breakdown of organic matter is called *decay* or *decomposition*.

It is usually complete and not accompanied by the release of foul gases. Anaerobic breakdown of organic matter is called *fermentation*. It is usually incomplete and is always accompanied by the release of foul gases. Anaerobic breakdown of proteins is called *putrefaction*. Parasitic bacteria are those that occur in the body of animals and plants, obtaining their organic food from them. Most of these bacteria are pathogenic, causing serious diseases in the host organisms by either exploiting them or releasing poisonous secretions called *toxins*.

5.4.4. Bacteria that Move by Unusual Mechanisms

Magnetotactic bacteria of strain Mar 1–83 show unusual mechanisms. When swimming in an applied magnetic field, they do not move as a homogeneous cell suspension, but aggregate in distinct wavelike structures (Fig. 5.15) called the *magnetosome chains*. The waves remained stable during forward movement. The number of cells per wave ranged from a few cells in permanent lateral contact to hundreds of bacteria moving visibly within a wave. Wave formation required a horizontal and vertical magnetic component. Electron microscopy indicated at least three distinct parallel chains of magnetosomes inside the bacterium. The cellular magnetic dipole moment was determined, and cell to cell magnetic interaction was ruled out as the sole mechanism that induced wave formation and kept waves stable (Spormann, 1987). The mechanisms responsible for bacterial gliding motility have been a mystery for almost 200 years. Gliding bacteria move actively over surfaces by a process that does not involve flagella. Gliding bacteria are phylogenetically diverse and are abundant in many environments. It is suggested that more than one mechanism is needed to explain all forms of bacterial gliding motility. The *Myxococcus xanthus* "social gliding motility" and *Synechocystis* gliding are similar to the bacterial "twitching motility" and rely on type IV pilus extension and retraction for cell movement (Fig. 5.15). In contrast, gliding of filamentous cyanobacteria, mycoplasmas, members of the Cytophaga–Flavobacterium group, and "adventurous gliding" of *M. xanthus* do not appear to involve pili. The mechanisms of movement used by these bacteria are still a

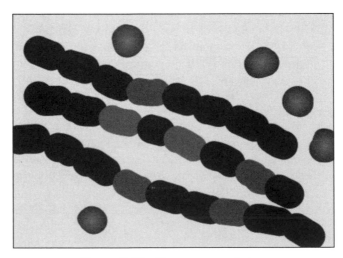

Figure 5.15. Magnetosome chains.

matter of speculation. Genetic, biochemical, ultrastructural, and behavioral studies are providing insight into the machineries employed by these diverse bacteria, which enable them to glide over surfaces (McBride, 2001).

5.4.5. Bacteria that Form Storage Granules

In many bacteria the cytoplasm contains large granules of storage materials. These storage polymers have a high molecular weight and are therefore osmotically inert. They can be stored without increasing the osmolarity of the cytoplasm. Storage polymers often occur in the form of granules within the cytoplasm and may be surrounded by a layer (Fig. 5.16). They serve as stores of energy or organic compounds in the form of readily metabolizable substrates. Three different types of storage granules are found in the cytoplasm: polymetaphosphate granules, poly-β-hydroxybutyrate granules, and polyglucan granules. Some bacterial species are mobile and possess locomotory organelles, namely, flagella. Those that do are able to taste their environment and respond to specific chemical foodstuffs or toxic materials and move toward or away from them (chemotaxis).

Flagella are embedded in the cell membrane, extend through the cell envelope, and project as a long strand. Flagella consist of a number of proteins, including flagellin. They move the cell by rotating with a propeller-like action. Axial filaments in spirochetes have a similar function to flagella. Binding proteins in the periplasmic space or cell membrane bind food sources (such as sugars and amino acids), causing methylation of other cell membrane proteins that in turn affect the movement of the cell by flagella. Permeases are proteins that then transport these foodstuffs through the cell membrane. Energy and carbon sources can then be stored when necessary in cytoplasmic "storage granules," which consist of glycogen, polyhydroxybutyrate, or polyphosphate. Some bacteria produce intracellular nutrient storage granules, such as glycogen, polyphosphate, sulfur, or

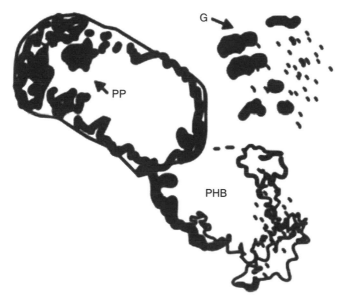

Figure 5.16. Thin section of *Pseudomonas pseudoflava* showing polyphosphate (PP) (volutin) granules, poly-β-hydroxybutyrate (PHB), and glycogenlike granules (G).

polyhydroxyalkanoates (PHAs). These granules enable bacteria to store compounds for later use. Certain bacterial species, such as the photosynthetic cyanobacteria, produce internal gas vesicles, which they use to regulate their buoyancy, thus allowing them to move up or down into water layers with different light intensities and nutrient levels.

5.4.6. Bacteriological Examination of Water—Qualitative Tests

Various techniques have been devised to permit the analysis of the structure and function of microorganisms. Some techniques are qualitative in their intent. That is, they provide a "yes or no" answer. Other techniques are quantitative in their intent. These techniques provide numerical information about a sample.

Assessing the growth of a bacterial sample provides examples of both types of analysis techniques. An example of a qualitative technique would be the growth of a bacterial sample on a solid growth medium to solely assess whether the bacteria in the sample are living or dead. An example of a quantitative technique is the use of a solid growth medium to calculate the actual number of living bacteria in a sample.

Microscopic observation of microorganisms can reveal a wealth of qualitative information. The observation of a suspension of bacteria on a microscopic slide (the wet mount) reveals whether the bacteria are capable of self-propelled motion. Microorganisms, particularly bacteria, can be applied to a slide as the so-called smear, which is then allowed to dry on the slide (Fig. 5.17). The dried bacteria can be stained to reveal, for example, whether they retain the primary stain in the Gram staining protocol (gram positive) or whether that stain is washed out of the bacteria and a secondary stain is retained (gram negative). Examination of such smears will also reveal the shape, size,

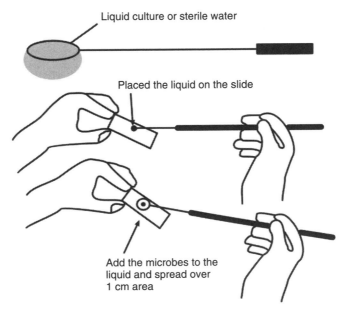

Liquid culture or sterile water

Placed the liquid on the slide

Add the microbes to the liquid and spread over 1 cm area

Air dry or heat gently. When dry, briefly heat fix the cells to the slide.

Figure 5.17. Method of making a bacterial smear.

and arrangement (single, in pairs, in chains, in clusters) of the bacteria. These qualitative attributes are important in categorizing bacteria.

Microscopy can be extended to provide qualitative information. The incorporation of antibodies to specific components of the sample can be used to calculate the proportion of the samples in a population that possesses the target of interest. Fluorescent-labeled antibodies, or antibodies combined with a dark appearing molecule such as ferritin, are useful in such studies. The scanning confocal microscope is proving to be tremendously useful in this regard. The optics of the microscope allows visual data to be obtained at various depths through a sample (typically the sample is an adherent population of microorganisms). These optical thin sections can be reconstructed via computer imaging to produce a three-dimensional image of the specimen. The use of fluorescent-tagged antibodies allows the location of protein within the living biofilm to be assessed.

The self-propelled movement of living microorganisms, a behavior that is termed *motility*, can also provide quantitative information. For example, recording the moving cells is used to determine their speed of movement and whether the presence of a compound acts as an attractant or a repellent to the microbes.

Bacterial growth is another area that can yield qualitative or quantitative information. Water analysis for the bacterium *E. coli* provides an example. A specialized growth medium allows the growth of only *E. coli*. Another constituent of the growth medium is utilized by the growing bacteria to produce a by-product that fluoresces when exposed to ultraviolet (UV) light. If the medium is dispensed in bottles, the presence of growing *E. coli* can be detected by the development of fluorescence. However, if the medium is dispensed in smaller volumes in a gridlike pattern, then the number of areas of the grid that are positive for growth can be related to a mathematical formula to produce the most probable number (MPN) of living *E. coli* in the water sample. Viable bacterial counts can be determined for many other bacteria by several other means. The ability of bacteria to grow or not to grow on a media containing controlled amounts and types of compounds yields quantitative information about the nutritional requirements of the microbes.

The advent of molecular techniques has expanded the repertoire of quantitative information that can be obtained. For example, a technique involving reporter genes can show whether a particular gene is active and can indicate the number of copies of the gene product that is manufactured. Gene probes have also been tagged to fluorescent or radioactive labels to provide information as to where in a population a certain metabolic activity is occurring and the course of the activity over time.

Many other qualitative and quantitative techniques exist in microbiological analysis. A few examples include immunoelectrophoresis, immunoelectron microscopy, biochemical dissection of metabolic pathways, the molecular construction of cell walls and other components of microorganisms, and mutational analysis. The scope of these techniques is ever expanding.

5.4.7. Membrane Filter Method

The membrane filter (MF) technique is an effective, accepted technique for testing fluid samples for microbiological contamination. It involves less preparation than many traditional methods and is one of a few methods that will allow the isolation and enumeration of microorganisms. The MF technique also provides information on the presence or absence of microbes within 24 h. The MF technique was introduced in the late 1950s as

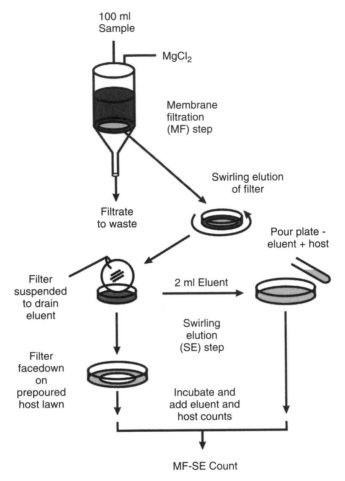

100 ml
Sample

MgCl$_2$

Membrane
filtration
(MF) step

Filtrate
to waste

Swirling elution
of filter

Pour plate -
eluent + host

Filter
suspended
to drain
eluent

2 ml Eluent

Swirling
elution
(SE) step

Filter
facedown
on
prepoured
host lawn

Incubate and
add eluent and
host counts

MF-SE Count

Figure 5.18. Microfiltration method.

an alternative to the MPN procedure for microbiological analysis of water samples. The MF technique offers the advantage of isolating discrete colonies of bacteria, whereas the MPN procedure only indicates the presence or absence of an approximate number or organisms. The various steps are described below (Fig. 5.18).

1. Collect the sample and make any necessary dilutions.
2. Select the appropriate nutrient or culture medium. Dispense the broth into a sterile petri dish, evenly saturating the absorbent pad.
3. Flame the forceps, and remove the membrane from the sterile package.
4. Place the MF into the funnel assembly.
5. Flame the pouring lip of the sample container and pour the sample into the funnel.
6. Turn on the vacuum and allow the sample to draw completely through the filter.
7. Rinse funnel with sterile buffered water. Turn on vacuum and allow the liquid to draw completely through the filter.
8. Flame the forceps and remove the MF from the funnel.

9. Place the MF on the prepared petri dish.
10. Incubate at the proper temperature and for the appropriate time period.
11. Count the colonies under 10–15× magnification.
12. Confirm the colonies, and report the results.

When the number of indicator organisms in water is very low, direct inoculation on solid media is not practicable and other methods must be used by which large volumes can be examined. Other methods of examination include membrane filtration and multiple tube tests. MF techniques are widely used for the enumeration of bacteria from water sample. In this method, the bacterial cells are filtered through a membrane as the sample passes through it. The cells along with the membrane are placed on a suitable solid medium. On incubation, these cells produce visible colonies that can be counted.

5.4.8. Standard Plate Count—Quantitative Test

This determines the total number of bacteria in a specified amount of milk, generally a milliliter (ml). This is used for grading of milk. Under aseptic conditions, 1 ml of milk is added to 99 ml of distilled water or buffer. The 1-ml and one-tenth ml samples are then transferred to sterile petri dishes, and 1:100 and 1:1000 dilutions of the milk were prepared, respectively. Other dilutions may also be prepared successively (Fig. 5.19).

A growth medium such as plate count agar or tryptone glucose yeast extract agar is then prepared, and the milk samples are mixed with the medium. The dishes are incubated at 37 °C for 24–48 h. The plates are then placed on a counting device such as a Quebec colony counter and the number of bacterial colonies is recorded. The colony count falling between 30 and 300 is selected and multiplied by the reciprocal of the dilution factor to obtain the bacterial count per milliliter of milk. If 248 colonies appeared on the 1:100 plate and 16 colonies on the 1:1000 plate, 248 would be selected and multiplied by 100, giving 24,800 total bacteria per milliliter of milk sample.

The standard plate count (SPC), also referred to as the *aerobic plate count* or the *total viable count*, is one of the most common tests applied to indicate the microbiological quality of food. The significance of SPCs, however, varies markedly according to the type of food product and the processing it has received. When SPC testing is applied on a regular basis, it can be a useful means of observing trends by comparing SPC results over time.

Three levels of SPCs are listed below based on food type and the processing/handling the food.

Level 1 applies to ready to eat foods in which all components of the food have been cooked in the manufacturing process/preparation of the final food product and, as such, microbial counts should be low.

Level 2 applies to ready to eat foods that contain some components that have been cooked and then further handled (stored, sliced, or mixed) before preparation of the final food or where no cooking process has been used.

Level 3 is where SPCs are not applicable. This applies to foods such as fresh fruits and vegetables (including salad vegetables), fermented foods, and foods incorporating these (such as sandwiches and filled rolls). It would be expected that these foods would have an inherent high plate count because of the normal microbial flora present.

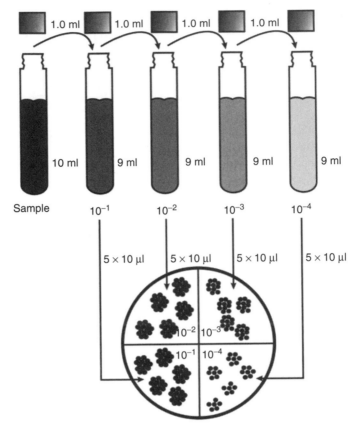

Figure 5.19. Standard plate count after dilution of the samples.

Note: An examination of the microbiological quality of a food should not be based on SPCs alone. The significance of high (unsatisfactory) SPCs cannot truly be made without identifying the microorganisms that predominate or without other microbiological testing.

5.5. ANIMALS AS HABITAT

Animals are always used by microbes as a habitat. There are various microorganisms that totally depend on animals and plants for their nutritional aspects. Microbes are everywhere in the biosphere, and their presence invariably affects the environment that they are growing in. The effect of microorganisms on their environment can be beneficial, harmful, or unapparent with regard to human measure or observation. The beneficial effects of microbes derive from their metabolic activities in the environment, their associations with plants and animals, and their use in food production and biotechnological processes. Bacteria differ dramatically with respect to the conditions that are necessary for their optimal growth. In terms of nutritional needs, all cells require sources of carbon, nitrogen, sulfur, phosphorus, and numerous inorganic salts.

5.5.1. Bacteria that Inhabit the Skin

The skin is an extremely effective barrier to microorganisms. Besides preventing the cells of our body from escaping, it also prevents the entry of microorganisms. Skin contains several layers of tightly packed, heavily keratinized cells (keratin is a fibrous protein that gives toughness to skin, hair, and nails). It is very difficult for most organisms to squeeze in between skin cells. Skin cells are also continually being shed and replaced by new ones, thus removing any microorganisms attached to them. The skin surface is also very hydrophobic and dry, which prevents the growth of many microorganisms. Sebaceous glands are present throughout the skin and they secrete hydrophobic oils that further repel water and microorganisms. The oil also helps keep the skin supple and flexible, preventing cracking that might allow microbial access to internal layers. Finally, melanin in the skin also helps reduce the harmful impact of UV light by absorbing it. UV light can be damaging to all cells, including cells of the immune system. A large number of bacteria can be found underneath the toenail also (Fig. 5.20). Bacteria living on the skin surface, including staphylococcal types that typically induce inflammation below the skin, actually prevent excessive inflammation after injury to the skin.

The adult human is covered with approximately 2 m^2 of skin. The density and composition of the normal flora of the skin varies with anatomical locale. The high moisture content of the axilla, groin, and areas between the toes supports the activity and growth of relatively high densities of bacterial cells, but the density of bacterial populations at most other sites is fairly low, generally in hundreds or thousands per square centimeter. Most bacteria on the skin are sequestered in sweat glands.

Propionibacterium acnes live in hair follicles, the tiny pores in our skin from which hairs sprout (Fig. 5.21). When these pores become blocked, the bacteria can multiply and contribute to the inflammation we call *acne*. But the bacteria can also cause inflammation inside our tissues, leading to tissue damage.

The "hygiene hypothesis" emerged in the late 1980s to explain why allergies like hay fever and eczema were less common in children from large families who were exposed

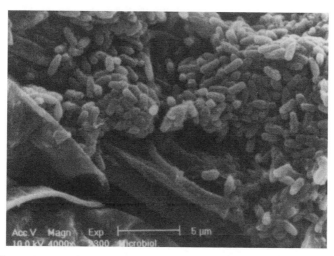

Figure 5.20. High magnification reveals a host of bacteria underneath a human toenail.

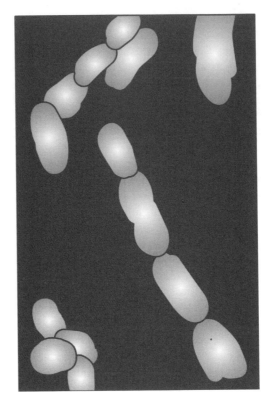

Figure 5.21. *Propionibacterium acnes.*

to more infectious agents. The theory suggests that a lack of early childhood exposure to infectious agents and microorganisms changes how the immune system reacts to bacterial threats. Bacterial skin infections are very common, and they can range from merely annoying to deadly. Most bacterial infections of the skin are caused by two bacteria, *Staphylococcus aureus* and a form of *Streptococcus*.

5.5.2. Bacteria that Inhabit the Mucous Membrane

Humans are colonized by normal flora at the moment of birth and passage through the birth canal. In uterus, the fetus is sterile, but as the mother's water breaks and the birth process begins, so does the colonization of the body surfaces. Handling and feeding the infant after birth leads to establishment of a stable normal flora on the skin, oral cavity, and intestinal tract in about 48 h.

Bacteria are found at various anatomical locations in adults (Table 5.1). The skin microbes found in the most superficial layers of the epidermis and the upper parts of the hair follicles are gram-positive cocci (*S. epidermidis* and *Micrococcus* sp. and corynebacteria such as *Propionibacterium* sp.). These are generally nonpathogenic and considered to be commensals, although mutualistic and parasitic roles have been assigned to them; for example, staphylococci and propionibacteria produce fatty acids that inhibit the growth of fungi and yeast on the skin. But, if *P. acnes*, a normal inhabitant of the skin, becomes trapped in hair follicle, it may grow rapidly and cause inflammation and acne.

TABLE 5.1. Predominant Bacteria at Various Anatomical Locations in Adults

Anatomical Location	Predominant Bacteria
Skin	Staphylococci and corynebacteria
Mucous membrane	Staphylococci and lactic acid bacteria
Teeth	Staphylococci and lactobacilli
Nasal membranes	Staphylococci and corynebacteria
Pharynx	Staphylococci, *Neisseria*, gram-negative rods and cocci
Stomach	*Helicobacter pylori* (up to 50%)
Small intestine	Lactics, enterics, enterococci, bifidobacteria
Anterior urethra	Staphylococci, corynebacteria, enterics
Vagina	Lactic acid bacteria during childbearing

Sometimes potentially pathogenic *S. aureus* is found on the face and hands of individuals who are nasal carriers. This is because the face and hands are likely to become inoculated with the bacteria on the nasal membranes. Such individuals may autoinoculate themselves with the pathogen or spread it to other individuals or food.

Studies have indicated that the attachment of bacteria to mucosal surfaces is the initial event in the pathogenesis of most infectious diseases due to bacteria in animals and humans. An understanding of the mechanisms of attachment and a definition of the adhesive molecules on the surfaces of bacteria (adhesins) as well as those on host cell membranes (receptors) have suggested new approaches to the prevention of serious bacterial infections: (i) application of purified adhesion or receptor materials or their analogs as competitive inhibitors of bacterial adherence, (ii) administration of sublethal concentrations of antibiotics that suppress the formation and expression of bacterial adhesins, and (iii) development of vaccines against bacterial surface components involved in adhesion to mucosal surfaces. Progress has already been made in the development of antiadhesive vaccines directed against the fimbrial adhesins of several human bacterial pathogens (Beachey, 1981).

5.5.3. Obligate Intracellular Parasites

Obligate intracellular parasites are parasitic microorganisms that cannot reproduce outside their host cell, forcing the host to assist in their reproduction. Obligate intracellular parasites of humans include the following:

- Viruses
- Certain bacteria, including
  - *Chlamydia* and closely related species
  - *Rickettsia*
  - *Coxiella*
  - Certain species of *Mycobacterium* such as *Mycobacterium leprae*
- Certain protozoa, including
  - *Plasmodia* species
  - *Leishmania* sp
  - *Toxoplasma gondii*
  - *Trypanosoma cruzi*

The mitochondria in eukaryotic cells may also have originally been such parasites but ended up forming a mutualistic relationship. Other organisms invariably cause disease in humans, such as obligate intracellular parasites that are able to grow and reproduce only within the cells of other organisms. Still, infections with intracellular bacteria may be asymptomatic, such as during the incubation period. An example of intracellular bacteria is *Rickettsia*. One species of *Rickettsia* causes typhus, while another causes Rocky Mountain spotted fever. Chlamydia, another phylum of obligate intracellular parasites, contains species that can cause pneumonia or urinary tract infection and may be involved in coronary heart disease. *Mycobacterium* and *Brucella* can exist intracellularly, although they are facultative (not obligate intracellular parasites).

The phylogeny of obligate intracellular coccoid parasites of acanthamoebae isolated from the nasal mucosa of humans was analyzed by the rRNA approach. The primary structures of the 16S and 23S rRNA molecules of one strain were determined in almost full length. *In situ* hybridization with a horseradish-peroxidase-labeled oligonucleotide probe targeted to a unique signature site undoubtedly correlated the retrieved 16S rRNA sequence to the respective intracellular parasite. This probe also hybridized with the second strain, suggesting a close relationship between the two intracellular parasites. Comparative sequence analysis demonstrated a distinct relationship with the genus *Chlamydia*. With 16S rRNA similarities of 86–87% to the hitherto-sequenced *Chlamydia* species, the intracellular parasites are likely not new species of this genus but representatives of another genus in the family Chlamydiaceae. Consequently, it was proposed to provisionally classify the endoparasite of *Acanthamoeba* sp. strain Bn9 as "*Candidatus Parachlamydia acanthamoebae*." From an epidemiological perspective, small amebae could be environmental reservoirs and vectors for a variety of potentially pathogenic bacteria including the members of Chlamydiaceae (Amann et al., 1997).

5.6. ARCHAEA IN EXTREME ENVIRONMENTS

The domain Archaea was not recognized as a major domain of life until quite recently. Until the twentieth century, most biologists considered all living things to be classifiable as either a plant or an animal. But in the 1950s and 1960s, most biologists came to the realization that this system failed to accommodate the fungi, protists, and bacteria. Archaea are microbes that mostly live in extreme environments. These are called *extremophiles*. Other archaea species are not extremophiles and live in ordinary temperatures and salinities.

Archaea and bacteria are quite similar in size and shape, although a few archaea have very unusual shapes, such as the flat and square cells of *Haloquadra walsbyi* (Fig. 5.22). Despite this visual similarity to bacteria, archaea possess genes and several metabolic pathways that are more closely related to those of eukaryotes, notably, the enzymes involved in transcription and translation. The archaea exploit a much greater variety of sources of energy than eukaryotes, ranging from familiar organic compounds such as sugars to using ammonia, metal ions, or even hydrogen gas as nutrients. Salt-tolerant archaea (the Halobacteria) use sunlight as an energy source and other species of archaea fix carbon; however, unlike plants and cyanobacteria, no species of archaea is known to do both. Archaea reproduce asexually and divide by binary fission, fragmentation, or budding; in contrast to bacteria and eukaryotes, no known species form spores. Archaea were first detected in extreme environments, such as volcanic hot springs. It is argued

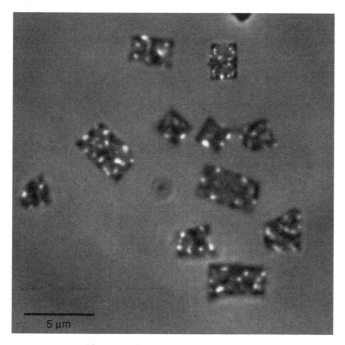

Figure 5.22. *Haloquadra walsbyi.*

that the bacteria, archaea, and eukaryotes represent separate lines of descent that diverged early on from an ancestral colony of organisms. A few biologists, however, argue that the Archaea and Eukaryota arose from a group of bacteria. It is possible that the last common ancestor of the bacteria and archaea was a thermophile, which raises the possibility that lower temperatures are "extreme environments" in archaeal terms, and organisms that live in cooler environments appeared only later. Since the archaea and bacteria are no more related to each other than they are to eukaryotes, the term *prokaryote*'s only surviving meaning is "not a eukaryote," limiting its value. This resistance to extreme environments has made archaea the focus of speculation about the possible properties of extraterrestrial life.

5.6.1. Extreme Halophiles

Halophiles are extremophile organisms that thrive in environments with very high concentration of salt. The name comes from the Greek for "salt-loving." While the term is perhaps most often applied to some halophiles classified into the Archaea domain, there are also bacterial halophiles and some Eukaryota, such as the alga *Dunaliella salina* (Fig. 5.23). Some well-known species give off a red color from carotenoid compounds. Such species contain the photosynthetic pigment bacteriorhodopsin. Halophiles are categorized as slight, moderate, and extreme by the extent of their halotolerance. Halophiles can be found anywhere with a concentration of salt five times greater than the salt concentration of the ocean, such as the Great Salt Lake in Utah, Owens Lake in California, the Dead Sea, and in evaporation ponds.

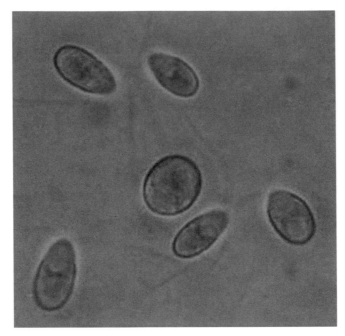

Figure 5.23. *Dunaliella salina.*

High salinity represents an extreme environment that relatively few organisms have been able to adapt to and occupy. Most halophilic and all halotolerant organisms expend energy to exclude salt from their cytoplasm to avoid protein aggregation ("salting out"). In order to survive the high salinities, halophiles use two differing strategies to prevent desiccation through the osmotic movement of water out of their cytoplasm. Both strategies work by increasing the internal osmolarity of the cell. In the first strategy (that is used by the majority of bacteria, some archaea, yeasts, algae, and fungi), organic compounds are accumulated in the cytoplasm—these osmoprotectants are known as *compatible solutes*. These can be synthesized or accumulated from the environment. The second, more radical, adaptation involves the selective influx of potassium (K^+) ions into the cytoplasm. This adaptation is restricted to the moderately halophilic bacterial order Halanerobiales, the extremely halophilic archaeal family Halobacteriaceae, and the extremely halophilic bacterium *Salinibacter ruber*. The presence of this adaptation in three distinct evolutionary lineages suggests convergent evolution of this strategy, it being unlikely to be an ancient characteristic retained in only scattered groups or through massive lateral gene transfer. The primary reason for this is that the entire intracellular machinery (enzymes, structural proteins, etc.) must be adapted to high salt levels, whereas in the compatible solute adaptation, little or no adjustment is required to intracellular macromolecules. In fact, the compatible solutes often act as more general stress protectants and as just osmoprotectants.

The extreme halophiles or haloarchaea (often known as *halobacteria*), a group of archaea, require at least a 2 M salt concentration and are usually found in saturated solutions (about 36% w/v salts). These are the primary inhabitants of salt lakes, inland seas, and evaporating ponds of seawater, such as the Dead Sea and solar salterns, where they tint the water column and sediments bright colors. In other words, they will most likely perish if they are exposed to anything other than a very high concentration salt conditioned environment. These prokaryotes require salt for growth. The high concentration of NaCl in their environment limits the availability of oxygen for respiration. Their cellular machinery is adapted to high salt concentrations by having charged amino acids on their surfaces, allowing the retention of water molecules around these components. They are heterotrophs that normally respire by aerobic means. Most halophiles are unable to survive outside their high salt native environment. Indeed, many cells are so fragile that when placed in distilled water, they immediately lyse from the change in osmotic conditions.

Haloarchaea, and particularly, the family Halobacteriaceae are members of the domain Archaea and comprise the majority of the prokaryotic population. There are currently 15 recognized genera in the family. The domain Bacteria (mainly *Salinibacter ruber*) can comprise up to 25% of the prokaryotic community but comprises more commonly a much lower percentage of the overall population.

A comparatively wide range of taxa have been isolated from saltern crystallizer ponds, including members of the following genera: *Haloferax, Halogeometricum, Halococcus, Haloterrigena, Halorubrum, Haloarcula*, and *Halobacterium* (Oren, 2002). However, the viable counts in these cultivation studies have been small when compared to total counts, and the numerical significance of these isolates has been unclear. Only recently it has become possible to determine the identities and relative abundances of organisms in natural populations, typically using polymerase chain reaction (PCR)-based strategies that target 16S small subunit ribosomal ribonucleic acid (16S rRNA) genes. While comparatively few studies of this type have been performed, results from these suggest that some of the most readily isolated and studied genera may not in fact be significant in the *in situ* community. This is seen in cases such as the genus *Haloarcula*, which is estimated to make up less than 0.1% of the *in situ* community but commonly appears in isolation studies.

5.6.2. Extreme Thermophiles

A thermophile is a type of extremophilic organism that thrives at relatively high temperatures, between 45 and 80 °C (113 and 176 °F, respectively). Many thermophiles are archaea. Extreme thermophiles are critters that live in some of the most unwelcoming environments on the planet. Archaea such as *Sulfolobus acidocaldarius* live in hot springs and geysers where the water temperature can be up to 100 °C and the water is filled with sulfuric acid (Fig. 5.24). *Chlororflexus aurantiacus* can carry out photosynthesis at over 60 °C. *Pyrococcus furiosus* lives in boiling-hot undersea mud surrounding volcanic islands where there is virtually no oxygen. Close by, living in the tissues of large tube-dwelling worms, archaea are also found. In fact, the entire ecosystem down there depends on them. Their membranes and proteins are unusually stable at these extremely high temperatures. Thus, many important biotechnological processes utilize thermophilic enzymes because of their ability to withstand intense heat.

Many of the hyperthermophilic archaea require elemental sulfur for growth. Some anaerobes use sulfur instead of oxygen as an electron acceptor during cellular respiration. Some are lithotrophs that oxidize sulfur to sulfuric acid to use as an energy source, thus requiring the microorganism to be adapted to very low pH conditions. These organisms are inhabitants of hot sulfur-rich environments. Often these organisms are colored,

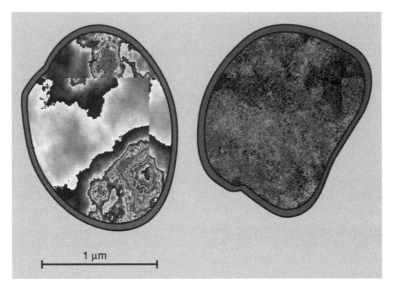

Figure 5.24. *Sulfolobus acidocaldarius.*

because of the presence of photosynthetic pigments. Examples of thermophilic species are *Thermus aquaticus* and *Thermococcus litoralis*.

5.6.3. Thermophilic Extreme Acidophiles

Acidophiles are not just present in exotic environments such as the Yellowstone National Park or deep-sea hydrothermal vents. Genera such as *Acidithiobacillus* and *Leptospirillum* bacteria and *Thermoplasmales* archaea are present in syntrophic relationships in the more mundane environments of concrete sewer pipes and implicated in the heavy-metal-containing sulfurous river waters.

The use of acidophilic organisms in mining is a nascent technique for extracting trace metals through bioleaching and offers solutions for the phenomenon of acid mine drainage (AMD) in mining spoils. On exposure to oxygen (O_2) and water (H_2O), metal sulfides undergo oxidation to produce a metal-rich acidic effluent. If the pH is low enough to overcome the natural buffering capacity of the surrounding rocks ("calcium carbonate equivalent" or "acid-neutralizing capacity"), the surrounding area may become acidic, as well as contaminated with high levels of heavy metals. Although acidophiles have an important place in the iron and sulfur biogeochemical cycles, strongly acidic environments are overwhelmingly anthropogenic in cause, primarily created after the cessation of pyritic (iron disulfide or FeS_2) mining operations.

In a comparative study of continuous flow iron-oxidation reactors, moderate thermophiles did not produce higher rates of ferric iron than the mesophile *Thiobacillus ferrooxidans* (Fig. 5.25) but iron oxidation was less sensitive to inhibition by chloride in a vessel containing a thermophile than in a vessel operating with the mesophile. Iron oxidation during autotrophic growth of moderately thermophilic acidophiles and the rapid dissolution of mineral sulfides during the autotrophic growth of both the moderate and the extreme thermophiles were demonstrated, thus considerably increasing the potential

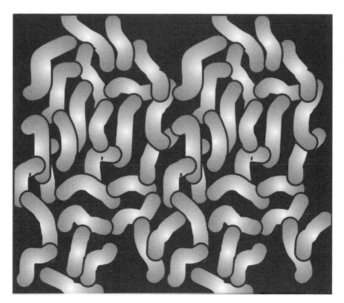

Figure 5.25. *T. ferrooxidans.*

industrial significance of these bacteria. The yield of soluble copper from a chalcopy-rite concentrate was shown to increase with temperature from relatively low yields with the mesophile *T. ferrooxidans* through moderate yields with the moderately thermophilic bacteria to almost complete mineral solubilization with the newly isolated *Sulfolobus* strains.

5.7. BIOGEOCHEMICAL CYCLES

In ecology and Earth science, a biogeochemical cycle or nutrient cycle is a pathway by which a chemical element or molecule moves through both biotic (biosphere) and abiotic (lithosphere, atmosphere, and hydrosphere) compartments of the Earth (Fig. 5.26). Separate biogeochemical cycles can be identified for each chemical element, such as the nitrogen (N), phosphorus (P), and carbon (C) cycles. However, through chemical transformations, elements combine to form compounds, and the biogeochemical cycle of each element must also be considered in relation to the biogeochemical cycles of other elements.

The element is recycled, although in some cycles there may be places (called *reservoirs*) where the element is accumulated or held for a long period (such as an ocean or lake for water). Water, for example, is always recycled through the water cycle. Water undergoes evaporation, condensation, and precipitation, falling back to Earth clean and fresh. Elements, chemical compounds, and other forms of matter are passed from one organism to another and from one part of the biosphere to another through the biogeochemical cycles.

Life on Earth is inextricably linked to climate through a variety of interacting cycles and feedback loops. There has been a growing awareness of the extent to which human activities, such as deforestation and fossil fuel burning, have directly or indirectly

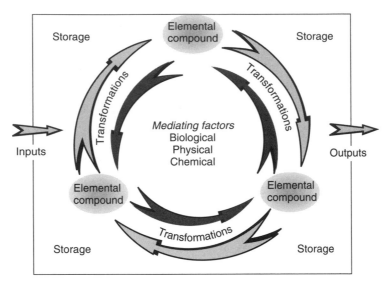

Figure 5.26. Biogeochemical cycle.

modified the biogeochemical and physical processes involved in determining the Earth's climate. These changes in atmospheric processes can disturb a variety of the ecosystem services that humanity depends on. In addition to helping maintain relative climate stability and a self-cleansing oxidizing environment, these services include protection from most of the sun's harmful UV rays, mediation of runoff and evapotranspiration (which affects the quantity and quality of freshwater supplies and helps control floods and droughts), and regulation of nutrient cycling.

The transport and transformation of substances in the environment, through life, air, sea, land, and ice, occurs via the biogeochemical cycles. These global cycles include the circulation of certain elements, or nutrients, on which life and the Earth's climate depend. One way in which climate influences life is by regulating the flow of substances through these biogeochemical cycles, in part through atmospheric circulation; water vapor is one such substance. It is critical for the survival and health of human beings and ecological systems and is part of the climatic state. When water vapor condenses to form clouds, more of the sun's rays are reflected back into the atmosphere, usually cooling the climate. Conversely, water vapor is also an important greenhouse gas in the atmosphere, trapping heat in the infrared part of the spectrum in the lower atmosphere. The water or hydrologic cycle intersects with most of the other element cycles, including the cycles of carbon (Fig. 5.27), nitrogen (Fig. 5.28), sulfur (Fig. 5.29), and phosphorus, as well as the sedimentary cycle. The processes involving each one of these elements may be strongly coupled with that of other elements and, ultimately, with important regional and global-scale climatic or ecological processes.

Managing and finding solutions to many of the important environmental problems facing humanity begin with understanding and integrating biogeochemical cycles and the scales at which they operate. Examples of these links include world climate and the potential threat of global climate change; agricultural productivity and its strong reliance on climatic factors, including temperature and precipitation, and the availability of nutrients; the cleansing of toxics in soils and streams through precipitation and runoff; acid

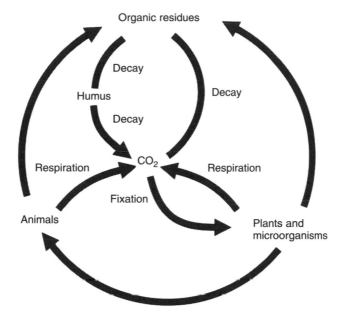

Figure 5.27. Carbon cycle.

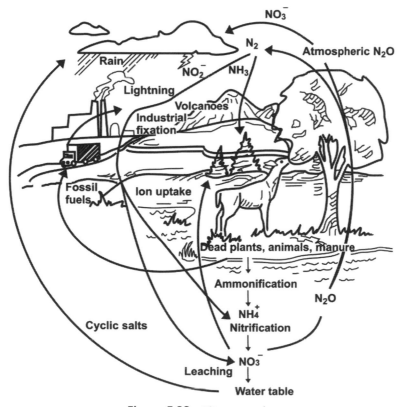

Figure 5.28. Nitrogen cycle.

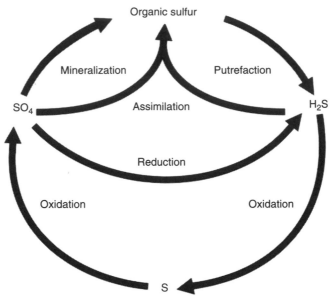

Figure 5.29. Sulfur cycle.

precipitation and the perturbation of ecosystem processes; the depletion of stratospheric ozone and its potential threat to human health and the food chain; and the often destructive interaction with natural cycles of other man-made compounds such as pesticides and synthetic hormones.

5.7.1. Various Cycles

Biogeochemical cycling refers to the complex series of process by which chemical elements and compounds move between different sources and sinks. This interdisciplinary study involves transitions between atmospheric gases, soils, living organisms, oceans, and geological formations. The global pressure of the human population is significantly altering the natural course of many of Earth's biogeochemical cycles, throwing important nutrients and compounds out of balance with the usual climatic and biological feedback mechanisms. These imbalances contribute to marine dead zones, climate change, ocean acidification, and many other environmental problems.

More than 30 chemical elements are cycled through the environment by biogeochemical cycles. There are six important biogeochemical cycles that transport carbon, hydrogen, oxygen, nitrogen, sulfur, and phosphorus. These six elements comprise the bulk of atoms in living things. Carbon, the most abundant element in the human body, is not the most common element in the crust.

5.7.2. Consequences of Biogeochemical Cycles Gone Wild

Human technology and population growth can both directly and indirectly disturb the biosphere. The key question now is can humans cause global climate change? The human population has experienced phenomenal exponential growth since the industrial revolution. Modern agriculture and medicine have increased growth rates in the human population, resulting in over 90 million people added each year. The United Nations estimates indicate that a human population of 10 billion may exist by the end of the twenty-first century. Since evolving around 200,000 years ago, the human species has proliferated and spread over the Earth. Beginning in 1650, the slow population increases of the human species has been exponential. New technologies for hunting and farming have facilitated this expansion. It took 1800 years to reach a total population of 1 billion, but only 130 years to reach 2 billion, and a mere 45 years to reach 4 billion.

The 1% of the atmosphere (trace gases) that is neither nitrogen nor oxygen plays an important role in global climate and in shielding the Earth surface from solar radiation. Agricultural and industrial gases may affect the atmosphere's ability to protect as well as alter the world's climate.

Carbon dioxide has many sources (cellular respiration and the burning of wood or fossil fuels such as coal or petroleum). There are two main sinks for carbon dioxide: plants and the oceans. Plants convert carbon dioxide into organic molecules by photosynthesis. Oceans form calcium carbonate and over long periods, store it as limestone. Since the industrial revolution, carbon dioxide levels in the atmosphere have increased. This increase has rapidly accelerated during the past 40 years. The average temperature of the Earth has risen by 0.5 °C over the past 100 years. Although a long-term rise of 2° would seem minor, this is thought sufficient to completely melt the glacial ice caps in Antarctica and Greenland, causing global sea levels to rise 100 m. This can alter climate patterns such as rainfall, ocean currents, and climate zones. Climate changes can have biological (such as causing migrations) as well as geopolitical and economic consequences.

Earth's climate fluctuates on both short- and long-term timescales. There have been periods in Earth's history with higher average annual temperatures than the present, and vice versa, when glaciers covered extensive expanses of the northern and/or southern hemispheres. We are currently between ice ages, the last of which ended nearly 10,000 years ago. Climate fluctuations have left evidence in the distribution of fossils, living forms and their close relatives, and locations of certain types of sedimentary rocks. Studies of fossils and the sedimentary rocks have led to estimates of temperature. Paleoecology is the branch of science that deals with such data in an attempt to reconstruct the environment of the distant (and not so distant) past.

5.8. ENVIRONMENTAL INFLUENCE AND CONTROL OF MICROBIAL GROWTH

Control of microbial growth, as used here, means to inhibit or prevent growth of microorganisms. This control is affected in two basic ways: by killing microorganisms and by inhibiting the growth of microorganisms. Control of growth usually involves the use of physical or chemical agents that either kill or prevent the growth of microorganisms. Agents that kill cells are called *cidal agents*; agents which inhibit the growth of cells (without killing them) are referred to as *static agents*. Thus, the term *bactericidal* refers to killing bacteria, and *bacteriostatic* refers to inhibiting the growth of bacterial cells. A bactericide kills bacteria, a fungicide kills fungi, and so on. In microbiology, sterilization refers to the complete destruction or elimination of all viable organisms in or on a substance being sterilized. There are no degrees of sterilization: an object or substance is either sterile or not. Sterilization procedures involve the use of heat, radiation, or chemicals or the physical removal of cells.

Physical and chemical properties of the medium carrying the organism (i.e., environment) have a profound influence on the rate and efficiency of destruction of microorganisms by chemical agents; for example, the effect of heat is greater in acidic medium than in alkaline medium and the consistency of medium greatly influences the interaction of a chemical agent with the microorganism. High concentration of carbohydrates increases the thermal resistance or heat stability of organisms. Presence of organic matter reduces the efficiency of an antimicrobial agent by inactivating it or protecting the organism from the chemical agent by masking it. Organic matter added to a disinfectant may have any one of the following outcomes:

1. Combination of disinfectant and organic matter to form a nonbactericidal produce/compound.
2. Combination of the disinfectant with organic material to form precipitate, thus forming a physical barrier between disinfectant and the microorganism.
3. Accumulation of organic matter on microbial cell surface to provide a coating/covering that will impair the contact disinfectant and the microbial cell. If components such as serum or yeast extract are added to system, more disinfectant will be required for the same effect.

Medium of growth of microorganisms is very important because one medium may support the growth of a particular type of microorganism, while another medium may not support the growth of the same type. Death may be the consequence of the growth medium. Bacteriostasis may be mistaken for bacterial action. Some compounds are bactericidal at higher concentrations. It may be necessary to add compounds that will neutralize the cidal or static effect of disinfectant in the medium, otherwise there will be no growth.

5.8.1. Temperature

Temperature is probably the most important environmental factor affecting growth. If temperature is too hot or too cold, microorganisms will not grow. The minimum and maximum temperatures for microbial growth vary widely among microorganisms and are usually a reflection of the temperature range and average temperature of their habitat. Temperature affects living organisms in two opposing ways:

1. As temperature increases, chemical and enzymatic reactions precede at a faster rate and the growth rate increases.
2. Above a certain temperature, proteins are irreversibly damaged.

Therefore, as temperature is increased within a certain range, growth and metabolic activity increase up to a point where inactivation reactions set in. Above this temperature, cell functions fall sharply to zero. Each microorganism thus has

- Minimum temperature below which no growth occurs.
- Optimum temperature at which growth is most rapid.
- Maximum temperature above which growth is not possible.

The optimum temperature is always closer to the maximum rather than the minimum. These three temperatures, called the *cardinal temperatures*, are usually characteristic of each type of organism. However, they are not completely fixed because they can be modified by other environmental factors especially the chemical composition of the medium.

5.8.1.1. Factors Affecting Maximum and Minimum Temperatures

The maximum growth temperature usually reflects the inactivation of one or more key proteins in the cell. The factors affecting minimum temperatures are less clear. It may result from the "freezing" of the cytoplasmic membrane, impairing its ability to transport nutrients or form proton gradients. Experiments have shown that adjustment in membrane lipid composition can cause changes in minimum temperature. Cardinal temperatures vary greatly throughout the microbial world.

Temperature classes of microorganisms (Fig. 5.30) can be broadly categorized under four broad categories:

1. Psychrophiles—these have low temperature optima.
2. Mesophiles—these have midrange temperature optima.
3. Thermophiles—these have high temperature optima.
4. Hyperthermophiles–these have very high temperature optima.

The various temperature classes of microorganisms have fixed temperature for growth (Table 5.2).

Optimal temperatures vary from $40\,^{\circ}C$ to higher than $100\,^{\circ}C$. Growth of different bacteria can range from below freezing temperature to above boiling temperature, though, no one organism can grow over this whole range (Table 5.2). Most bacteria have a temperature range of about $30\,^{\circ}C$, although some have broader ranges than others.

5.8.2. Osmotic Pressure

Osmotic pressure is another limiting factor in the growth of bacteria. Bacteria contain about 80–90% water; they require moisture to grow because they obtain most of their

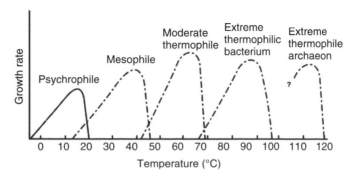

Figure 5.30. Growth rate of the various classes of bacteria classified on the basis of their growth in different temperature ranges.

TABLE 5.2. Terms Used to Describe Microorganisms in Relation to
Temperature Requirements for Growth

| Group | Temperature for Growth (°C) | | | Comments |
	Minimum	Optimum	Maximum	
Psychrophiles	Below 0	10–15	Below 20	Grow best at relatively low temperature
Psychrotrophs	0	15–30	Above 25	Able to grow at low temperature but prefer moderate temperature
Mesophiles	10–15	30–40	Below 45	Most bacteria, especially those living in association with warm-blooded animals
Thermophiles	45	50–85	Above 100 (boiling)	Among all thermophiles is wide variation in optimum and maximum temperatures

TABLE 5.3. Limiting Water Activities (A_w) for Growth of
Certain Prokaryotes

Organism	Minimum A_w for Growth
Caulobacter	1.00
Spirillum	1.00
Pseudomonas	0.91
Salmonella/E. coli	0.91
Lactobacillus	0.90
Bacillus	0.90
Staphylococcus	0.85
Halococcus	0.75

nutrients from their aqueous environment. Cell walls protect prokaryotes against changes in osmotic pressure over a wide range. However, sufficiently hypertonic media at concentrations greater than those inside the cell (such as 20% sucrose) cause water loss from the cell by osmosis. Fluid leaves the bacteria causing the cell to contract, which, in turn, causes the cell membrane to separate from the overlying cell wall. This process of cell shrinkage is called *plasmolysis*. Because plasmolysis inhibits bacterial cell growth, the addition of salts or other solutes to a solution inhibits food spoilage by bacteria, as occurs when meat or fish is salted.

Osmolarity is inversely related to water activity (A_w), which is more like a measure of the concentration of water (H_2O) in a solution. Osmolarity is determined by solute concentration in the environment. Increased solute concentration means increased osmolarity and decreased A_w. Some of the limiting water activities (A_w) for growth of certain prokaryotes are shown in Table 5.3. Some types of bacteria, called *extreme or obligate halophiles*, are adapted to and require high salt concentrations, such as found in the Dead Sea, where salt concentrations can reach 30%. Facultative halophiles do not require high salt environments to survive but are capable of tolerating these conditions. Halophiles can grow in salt concentrations up to 2%, a level that would inhibit the growth of other bacteria. However, some facultative halophiles, such as *Halobacterium halobium* grow in salt lakes, salt flats, and other environments where the concentration of salt is up to seven times greater than that of the oceans. When bacteria are placed in hypotonic media with concentrations weaker than the inside of the cell, water tends to enter by osmosis.

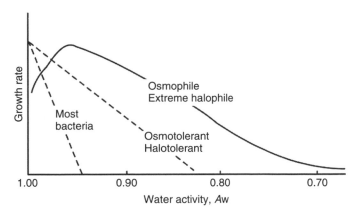

Figure 5.31. Growth rate versus osmolarity for different classes of prokaryotes.

The accumulation of water causes the cell to swell and then to burst, a process called *osmotic lysis*.

When the osmolarity of the different prokaryotes was plotted, various types of growth curve patterns were found (Fig. 5.31). From left to right, the graph shows the growth rates of a normal (nonhalophile) such as *E. coli* or *Pseudomonas*, a halotolerant bacterium such as *S. aureus*, and an extreme halophile such as the archaea *Halococcus*. Note that a true halophile grows best at salt concentrations where most bacteria are inhibited.

The oxidation of elemental sulfur by *Thiobacillus thiooxidans* was studied by Suzuki et al. (1999) at pH 2.3, 4.5, and 7.0 in the presence of different concentrations of various anions (sulfate, phosphate, chloride, nitrate, and fluoride) and cations (potassium, sodium, lithium, rubidium, and cesium). The results agree with the expected response of this acidophilic bacterium to charge neutralization of colloids by ions, pH-dependent membrane permeability of ions, and osmotic pressure.

5.8.3. UV Light

UV light is a nonionizing short-wavelength radiation that falls between 4 and 400 nm in the visible spectrum. Most bacteria are killed by the effects of UV light (Fig. 5.32), and it is routinely used to sterilize surfaces, such as work areas of transfer hoods used for the inoculation of cultures. The primary lethal effects of UV light are due to its mutagenic properties. When DNA absorbs UV light, pyrimidine dimers are produced because of the formation of a covalent bond between two adjacent thymine or cytosine molecules in a strand of DNA. These dimers deform the DNA molecule, so that DNA polymerase is unable to replicate the strand of DNA past the site of dimer formation and genes can no longer be transcribed. Cells have evolved various repair mechanisms to deal with the damage (including photoreactivation, nucleotide excision repair, and the SOS system). However, if sufficiently large numbers of dimers form in the DNA, the systems cannot cope and begin making errors by inserting incorrect bases for the damaged bases, eventually resulting in cell death.

UV light destroys microorganisms as described under "sterilization." Many spoilage organisms are readily killed by irradiation. In some parts of Europe, fruits and vegetables

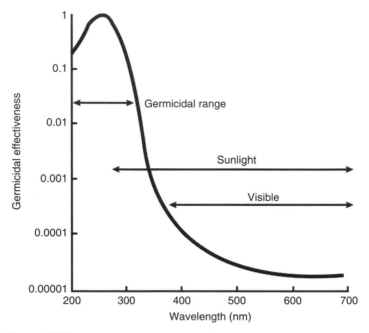

Figure 5.32. Germicidal effectiveness of various wavelengths of light.

are irradiated to increase their shelf life up to 500%. The practice has not been accepted in the United States. UV light can be used to pasteurize fruit juices by passing the juice over a high intensity UV light source. Free *Legionella* are destroyed easily by UV light and by a range of chemical biocides (Fig. 5.32).

UV systems for water treatment are available for personal, residential, and commercial applications and may be used to control bacteria, viruses, and protozoan cysts. The FDA has approved irradiation of poultry and pork to control pathogens, as well as foods such as fruits, vegetables, and grains to control insects and spices, seasonings, and dry enzymes used in food processing to control microorganisms. Food products are treated by subjecting them to radiation from radioactive sources, which kills significant numbers of insects, pathogenic bacteria, and parasites. According to the FDA, irradiation neither makes food radioactive nor noticeably changes taste, texture, or appearance of food. Irradiation of food products to control foodborne disease in humans has been generally endorsed by the United Nations World Health Organization (WHO) and the American Medical Association. Two important disease-causing bacteria that can be controlled by irradiation are *E. coli* O157:H7 and *Salmonella* species.

5.8.4. pH

Bacteria need a physiological pH inside their cells, just like all other living organisms (Table 5.4). Their ability to survive in extreme pH (either high or low) depends on their ability to correct for the difference between inside and out. One example of a bacterium that can live in acidic environments is *Helicobacter pylori*, which live in the stomach. It produces high amounts of urease, an enzyme that degrades urea and by doing so decreases the acidity (raises the pH). Imagine the bacteria produce a "cloud" of neutral

TABLE 5.4. Minimum, Maximum, and Optimum pH for Growth of Certain Prokaryotes

Organism	Minimum pH	Optimum pH	Maximum pH
Thiobacillus thiooxidans	0.5	2.0–2.8	4.0–6.0
Sulfolobus acidocaldarius	1.0	2.0–3.0	5.0
Bacillus acidocaldarius	2.0	4.0	6.0
Zymomonas lindneri	3.5	5.5–6.0	7.5
Lactobacillus acidophilus	4.0–4.6	5.8–6.6	6.8
Staphylococcus aureus	4.2	7.0–7.5	9.3
Escherichia coli	4.4	6.0–7.0	9.0
Clostridium sporogenes	5.0–5.8	6.0–7.6	8.5–9.0
Erwinia carotovora	5.6	7.1	9.3
Pseudomonas aeruginosa	5.6	6.6–7.0	8.0
Thiobacillus novellus	5.7	7.0	9.0
Streptococcus pneumoniae	6.5	7.8	8.3
Nitrobacter sp.	6.6	7.6–8.6	10.0

pH around them to protect them from the acidic environment. There are other bacteria that are specialized to live in basic pH, for instance, near black smokers, geological fountains of minerals that shoot highly alkaline minerals into the ocean. In conclusion, which pH is lethal for the bacteria depends on the species. Their defense is to keep the protons or OH^- ions out. Would they not succeed, their proteins would rapidly denature, which is the lethal toxicity of nonphysiological pH. pH affects bacteria the same way it affects all living things.

Extremes of pH affect the function of enzyme systems by denaturing them. However, bacteria become adapted over time to their surroundings. Just like we have enzymes that are adapted to the pH of our stomachs (very acidic) or to the small intestine (basic), bacteria that live in acid conditions are adapted to them. If they are moved to an environment that is neutral or basic, they will probably die. Bacteria that are human or animal pathogens are generally adapted to a pH of about 7.4, which is slightly basic.

Each organism has a pH range within which growth is possible, and most have well-defined pH optima (Fig. 5.33). Most natural environments have pH values between 5 and 9, and most organisms have pH optima in this range. Very few species can grow at pH values below 2 or above 10. Organisms capable of living at low pH levels are called *acidophiles*. Those capable of living at very high pH levels are called *alkaliphiles*.

As a group, fungi tend to be more acid tolerant than bacteria. Many grow optimally at pH 5 or below, and a few grow well at high pH values. Some bacteria are also acidophilic, and some cannot grow at neutral pH (obligate acidophiles). These include several species of *Thiobacillus* and several genera of the archaea, including *Sulfolobus*. A high concentration of hydrogen ions is required for the stability of the cytoplasmic membrane of obligate acidophiles. This dissolves if pH is raised to neutrality. Despite the pH requirements of particular organisms for growth, the optimal growth pH represents the pH of the extracellular environment only. The intracellular pH must remain near neutrality to prevent destruction of acid- or alkali-labile macromolecules in the cell. In extreme acidophiles and extreme alkalinophiles, the intracellular pH may vary by several units from neutrality.

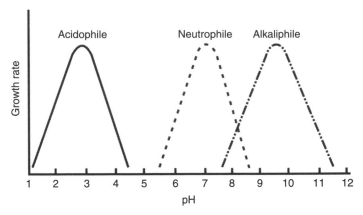

Figure 5.33. Growth rate versus pH for three environmental classes of prokaryotes. Most free-living bacteria grow over a pH range of about 3 units. Note the symmetry of the curves below and above the optimum pH for growth.

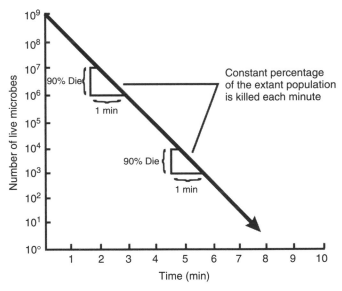

Figure 5.34. Oligodynamic property.

5.8.5. Oligodynamic Action

The inhibition or killing of microorganisms using very small amounts of a chemical substance is oligodynamic. The activity of heavy metals on microorganisms is termed *oligodynamic action* (Fig. 5.34). Metals such as as silver, mercury, and copper are used. Mercury is an older antiseptic used as mercuric chloride ($HgCl_2$). In products such as mercurochrome, merthiolate, and metaphen, mercury is combined with organic carrier compounds, which reduces its toxicity to skin. Copper is particularly active against algae. It is used as copper sulfate in swimming pools and municipal water supplies. Silver, as silver nitrate, is used as an antiseptic and disinfectant. One drop of 1% $AgNO_3$ solution is placed in the eyes of newborn to protect against infection by the gonococcus, *Neisseria gonorrhoeae*. Silver nitrate can also be combined with an antimicrobial drug for use in treatment of burns. Ethylene oxide has been used to sterilize many substances, including spices, biological preparations, soil, plastics, certain medical preparations, and contaminated laboratory equipment. It has also been used to decontaminate certain spacecraft components. Beside its good penetration, this gas has a wide spectrum of activity (meaning it is effective against many different kinds of microbes). It is also effective at relatively low temperatures, and it does not damage materials exposed to it. One advantage is its relatively slow action on microorganisms. The mode of action of ethylene oxide is believed to be alkylation reactions with organic compounds such as enzymes and other proteins. Alkylation consists of the replacement of an active hydrogen atom in an organic compound with an alkyl group. In this reaction, the ring in the ethylene oxide molecule splits and attaches itself where the hydrogen was originally. The reaction would, for instance, inactivate an enzyme. Ethylene oxide is usually measured in terms of milligrams of the pure gas per liter of space. For sterilization, concentrations of 450/1000 mg of gas per liter are necessary. Concentrations of 500 mg/l are generally effective in 4 h at about 130 °F (58 °C) and relative humidity of about 40%.

β-Propiolactone is a colorless liquid at room temperature with a high boiling point (162.3 °C). It has a sweet but irritating odor. Although unstable at room temperature, it may be stored at 4 °C (refrigerated) for months without undergoing deterioration. Although β-propiolactone is not flammable and does not penetrate as well as ethylene oxide, it is much more active than ethylene oxide. Aqueous solutions effectively inactivate some viruses, including polio and rabies, and also kill bacteria and bacterial spores. The vapor in a concentration of about 1.5 mg of the lactone per liter of air with a high relative humidity (75–80%) at about 25 °C kills spores in a few minutes. β-Propiolactone is not a substitute for ethylene oxide (because of its low penetrating power) but is used in place of formaldehyde for surface disinfection or sterilization. It is used to sterilize large enclosed spaces. Advantages of β-propiolactone over formaldehyde include more rapid antimicrobial action and faster removal from enclosure after application. Its mode of action is thought to be similar to that of ethylene oxide.

Formaldehyde is a gas that is stable only at high concentrations and elevated temperatures. At room temperature it polymerizes, forming a solid substance. The important polymer is paraformaldehyde, a colorless solid that rapidly yields formaldehyde on heating. Formaldehyde is also marketed in aqueous solution as formalin, which contains 37–40% formaldehyde. Formalin and paraformaldehyde are the two main sources of formaldehyde for gaseous sterilization. Vaporization of formaldehyde from either of these sources into an enclosed space for an adequate time will affect sterilization. Humidity and temperature have a pronounced effect on the microbicidal action of formaldehyde; in order to sterilize an enclosure, the temperature should be at about room temperature (22 °C) and the relative humidity between 60% and 80%. One disadvantage is the limited ability of these vapors to penetrate covered surfaces.

5.9. MICROORGANISMS AND ORGANIC POLLUTANTS

Many factors control biodegradability of a contaminant in the environment. Before attempting to employ bioremediation technology, one needs to conduct a thorough characterization of the environment where the contaminant exists, including the microbiology, geochemistry, mineralogy, geophysics, and hydrology of the system; many chemical contaminants in the environment can be readily degraded because of their structural similarity to naturally occurring organic carbon, and the amounts added may exceed the carrying capacity of the environment. Many pollutants are currently or were in the past used as pesticides. Others are used in industrial processes and in the production of a range of goods such as solvents, polyvinyl chloride, and pharmaceuticals. Microbes remove ammonia from wastewater treatment systems, clean up toxic waste, and reduce phosphorus content in lakes, all benefits that could be hindered or lost with the presence of nanoparticles.

Beneficial soil bacteria cannot tolerate silver, copper oxide, and zinc oxide nanoparticles, also used in sunscreens and other products. Damage to the bacteria, which clean up organic pollutants, occurred at very low levels of exposure, equivalent to two drops in an Olympic-size swimming pool. Researchers warned these particles could be toxic to aquatic life.

5.9.1. Environmental Law

Environmental law is a complex and interlocking body of treaties, conventions, statutes, regulations, and common law that, very broadly, operate to regulate the interaction of humanity and the rest of the biophysical or natural environment toward the purpose of reducing the impacts of human activity both on the natural environment and on humanity itself. The topic may be divided into two major areas: pollution control and remediation and resource conservation and management. Laws dealing with pollution are often media limited, that is, pertain only to a single environmental medium, such as air, water (whether surface water, groundwater, or oceans), soil, and control both emissions of pollutants into the medium and liability for exceeding permitted emissions and responsibility for cleanup. Laws regarding resource conservation and management generally focus on a single resource, for example, natural resources such as forests, mineral deposits, or animal species, or more intangible resources such as scenic areas or sites of high archeological value and provide guidelines and limitations for the conservation, disturbance, and use of those resources. These areas are not mutually exclusive; for example, laws governing water pollution in lakes and rivers may also conserve the recreational value of such waterbodies. Furthermore, many laws that are not exclusively "environmental" nonetheless include significant environmental components and integrate environmental policy decisions. Municipal, state, and national laws regarding development, land use, and infrastructure are examples.

Environmental law draws from and is influenced by principles of environmentalism, including ecology, conservation, stewardship, responsibility, and sustainability. Pollution control laws generally are intended (often with varying degrees of emphasis) to protect and preserve both the natural environment and human health. Resource conservation and management laws generally balance (again, often with varying degrees of emphasis) the benefits of preservation and economic exploitation of resources. From an economic perspective, environmental laws may be understood as concerned with the prevention of present and future externalities and the preservation of common resources from individual exhaustion. The limitations and expenses that such laws may impose on commerce, and the often unquantifiable (nonmonetized) benefit of environmental protection, have generated and continue to generate significant controversy.

In international law, a distinction is often made between hard and soft law. Hard international law generally refers to agreements or principles that are directly enforceable by a national or international body. Soft international law refers to agreements or principles that are meant to influence individual nations to respect certain norms or incorporate them into national law. Although these agreements sometimes oblige countries to adopt implementing legislation, they are not usually enforceable on their own in a court.

If a treaty or convention does not specify an international forum that has subject-matter jurisdiction, often the only place to bring a suit with respect to that treaty is the member state's domestic court system. This presents at least two additional hurdles. If the member state being sued does not have domestic implementing legislation in place to hear the dispute, there will be no forum available. Even in the event that the domestic legislation provides for such suits, since the judges who decide the case are residents of the country against which it is brought, potential conflicts of interest arise.

Only nations are bound by treaties and conventions. In international forums, such as the International Court of Justice (ICJ), countries must consent to being sued. Thus, it is often impossible to sue a country. The final question in the jurisdictional arena is who may bring a suit. Often, only countries may sue countries; individual citizens and nongovernmental organizations (NGOs) cannot. This has huge repercussions. First, the environmental harm must be large and notorious for a country to notice. Second, for a country to have a stake in the outcome of the subject matter, some harm may have to cross the borders of the violating country into the country that is suing. Finally, even if transboundary harm does exist, the issue of causation, especially in the environmental field, is often impossible to prove with any certainty.

The enforcement issue is one where advocates for a safer environment often find themselves stymied. Even if a treaty or convention provides for specific substantive measures to be taken by a country (many treaties merely provide "frameworks"), specifies a forum for dispute resolution, and authorizes sanctions for noncompliance, international law remains largely unenforceable. A country cannot be forced to do what it is not willing to do. One can sanction the country, order damages, restrict trade, or, most frequently, publicize noncompliance, but beyond that, if a country will not comply, there is very little to be done.

5.9.2. Process of Biodegradation

The process of biodegradation may be different for different substances, but, in general, biodegradable substances will be decomposed into carbon dioxide, methane, and water as the final products. Biodegradation is the chemical breakdown of materials by a physiological environment. The term is often used in relation to ecology, waste management, and environmental remediation (bioremediation). Organic material can be degraded aerobically with oxygen or anaerobically without oxygen. A term related to biodegradation is *biomineralization*, in which organic matter is converted into minerals. Biosurfactant, an extracellular surfactant secreted by microorganisms, enhances the biodegradation process.

Biodegradable matter is generally organic material, such as plant and animal matter and other substances originating from living organisms, or artificial materials that are similar enough to plant and animal matter to be put to use by microorganisms. Some microorganisms have the astonishing, naturally occurring, microbial catabolic diversity to degrade, transform, or accumulate a huge range of compounds including hydrocarbons (e.g., oil), polychlorinated biphenyls (PCBs), polyaromatic hydrocarbons (PAHs), pharmaceutical substances, radionuclides, and metals. Major methodological breakthroughs in microbial biodegradation have enabled detailed genomic, metagenomic, proteomic, bioinformatic, and other high throughput analyses of environmentally relevant microorganisms, providing unprecedented insights into key biodegradative pathways and the ability of microorganisms to adapt to changing environmental conditions. Biodegradation can be measured in a number of ways. The activity of aerobic microbes can be measured by the amount of oxygen they consume or the amount of carbon dioxide they produce. It can be measured by anaerobic microbes and the amount of methane or alloy that they may be able to produce. In formal scientific literature, the process is termed *bioremediation*.

Biodegradable plastics are of two main types in the market: hydro-biodegradable plastics (HBPs) and oxo-biodegradable plastics (OBPs). Both have to undergo chemical degradation by hydrolysis and oxidation respectively. This results in their physical disintegration and a drastic reduction in their molecular weight. These smaller, lower molecular weight fragments are then amenable to biodegradation (Fig. 5.35).

OBPs are made by adding a small proportion of compounds of specific transition metals (iron, manganese, cobalt, and nickel are commonly used) into the normal production of polyolefins such as PE, polypropylene (PP), and polystyrene (PS). The additives act as catalysts to speed up the normal oxidative degradation, increasing the overall process by up to several orders of magnitude (factors of 10). It generally takes a small amount of catalyst, relative to the amount of material being catalyzed, to perform the required function. Also, by definition, the catalyst is not consumed in the reaction.

The products of the catalyzed oxidative degradation of the polyolefins are precisely the same as those for conventional polyolefins because, other than a small amount of additive present, the plastics are conventional polyolefins. Many commercially useful hydrocarbons (e.g., cooking oils, polyolefins, many other plastics) contain small amounts of additives called *antioxidants* that prevent oxidative degradation during storage and use. Antioxidants function by "deactivating" the free radicals that cause degradation. Lifetime (shelf life+use life) is controlled by antioxidant level, and the rate of degradation after disposal is controlled by the amount and nature of the catalyst.

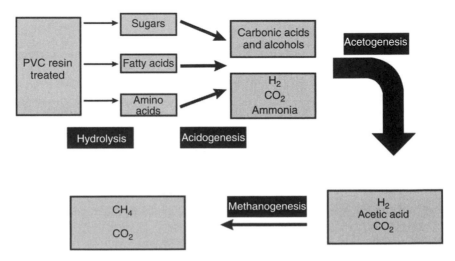

Figure 5.35. Biodegradation of PVC. resin.

Since there are no existing corresponding standards that can be used directly in reference to plastics that enter the environment in ways other than compost, that is, as terrestrial or marine litter or in landfills, OBP technology is often attacked by the HBP industry as unable to live up to the standards (which are actually the standards for composting). It has to be understood that composting and biodegradation are not identical. OBPs can, however, be tested according to ASTM D6954, and (as from January 1, 2010) UAE 5009:2009. HBPs tend to degrade and biodegrade somewhat more quickly than OBP, but they have to be collected and put into an industrial composting unit. The end result is the same—both are converted to carbon dioxide (CO_2), water (H_2O), and biomass. OBPs are generally less expensive, possess better physical properties, and can be made with current plastic processing equipment. HBP emits methane in anaerobic conditions, but OBP does not.

Polyesters play a predominant role in HBPs because of their potentially hydrolyzable ester bonds. HBPs can be made from agricultural resources, such as corn, wheat, sugar cane, fossil (petroleum-based) resources, or a blend of the two. Some of the commonly used polymers include PHAs, PHBV (polyhydroxybutyrate-valerate), PLA (polylactic acid), PCL (polycaprolactone), PVA (polyvinyl alcohol), and PET (polyethylene terephthalate). It would be misleading to call these resources "renewable" because the agricultural production process burns significant amounts of hydrocarbons and emits significant amounts of CO_2. OBPs (such as normal plastics) are made from a by-product of oil or natural gas, which would be produced whether or not the by-product were used to make plastic.

HBP technology claims to be biodegradable by meeting the ASTM D6400-04 and EN 13432 standards. However, these two commonly quoted standards are related to the performance of plastics in a commercially managed compost environment. They are not biodegradation standards. Both were developed for hydro-biodegradable polymers where the mechanism including biodegradation is based on reaction with water and state that for a product to be compostable, the following criteria need to be met:

Disintegration—the ability to fragment into nondistinguishable pieces after screening and safely support bioassimilation and microbial growth.

Inherent biodegradation—conversion of carbon to carbon dioxide to the level of 60% and 90% over a period of 180 days for ASTM D6400-04 and EN 13432, respectively. There is therefore little or no carbon left for the benefit of the soil, but the CO_2 emitted to atmosphere contributes to climate change.

Safety—that there is no evidence of any ecotoxicity in finished compost and soils can support plant growth.

Toxicity—that heavy metal concentrations are less than 50% of the regulated values in soil amendments.

5.9.3. Relationship between Contaminant Structure, Toxicity, and Biodegradability

Some types of contaminant structure can lead to low degradation even if the structure is similar to that of naturally occurring molecules (Fig. 5.36). Microbial degradation is thought to be the primary mechanism for PAH degradation in soil and is dominated by members of the *Sphingomonas, Burkholderia, Pseudomonas*, and *Mycobacterium* taxonomic groups. Microbial communities that are capable of digesting specific compounds

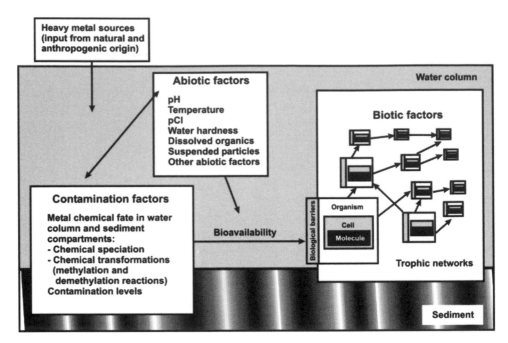

Figure 5.36. Ecotoxicological approach to metal bioaccumulation and transfers at ecosystem level.

will proliferate exponentially in response to digestible contamination. However, some higher order PAHs are not accessible to microbes as food due to their absorption inside organic particles or location in small pores that are inaccessible to bacteria. The degree to which these "etches" accelerate degradation is unclear, and the addition of nitrogen- and phosphorus-containing soil amendments is the best way to facilitate the growth of biofilms, given supportive pH conditions. The presence of lower molecular weight PAHs supports microbial communities that may be needed for the metabolization of higher molecular weight PAHs, which puts older, more weathered sites at a disadvantage in the absence of amendments.

Actions and interactions between the three fundamental sets of ecotoxicological factors contamination, abiotic, and biotic factors are depicted in Fig. 5.36 (Boudou and Ribeyre, 1997).

Pesticides that reach the soil are acted on by several physical, chemical, and biological forces. Although physical and chemical forces are acting on or degrading the pesticides to some extent, microorganisms play a major role in the degradation of pesticides. Many soil microorganisms have the ability to act on pesticides and convert them into simpler nontoxic compounds. This process of degradation of pesticides and conversion into nontoxic compounds by microorganisms is known as *biodegradation*. Not all pesticides reaching the soil are biodegradable and such chemicals that show complete resistance to biodegradation are called *recalcitrant*.

5.9.3.1. Conversion of the Pesticide Molecule to a Toxic Compound

Detoxification is not synonymous with degradation because a single change in the side chain of a complex molecule may render the chemical nontoxic. It is the conversion of nontoxic substrate into a toxic molecule, for example, the herbicide, 4-(2,4-dichlorophenoxy)butyric acid (2,4-DB) and the insecticide phorate are transformed and activated microbiologically in soil to give metabolites that are toxic to weeds and insects. Organisms must have the necessary catabolic activity required for degradation of contaminants at a fast rate to bring down the concentration of contaminant, the target contaminant must be bioavailable, the soil conditions must be congenial for microbial/plant growth and enzymatic activity, and the cost of bioremediation must be less than other technologies of removal of contaminants.

5.9.4. Environmental Factors

The physicochemical factors of natural environment determine the rates of microbial growth and the nature and size of the indigenous population. High growth rates require suitable control of nutrients and physical and chemical conditions of the environment. Environmental temperature is one of the most important factors affecting the growth rate of microbes. There is a minimum temperature below which growth does not occur. As the temperature increases above the minimum, rate of growth increases in accordance with the laws governing the effect of temperature on the chemical reactions that make up growth. These reactions are mostly enzyme catalyzed. However, a point called the *optimum temperature* is reached, during which there is also a very rapid increase in the rate of inactivation of heat-sensitive cell components, such as enzymes, ribosomes, DNA, membranes. Above an optimum temperature, this heat denaturation will occur so rapidly that there is a corresponding rapid drop in the rate of growth, determining a maximum temperature of growth for that particular microorganism. Most microbes are capable of growth in a temperature range of 20–30 °C. Most microorganisms have a growth optimum between 20 and 40 °C and are called *mesophilic*. Those inhabiting cold environments such as polar areas can grow at much lower temperatures. There are three environmental factors affecting the interwoven chemical and microbial breakdown of the organic matter.

1. Oxygen
2. Moisture
3. Temperature.

5.9.4.1. Oxygen

Composting can be defined in terms of availability of oxygen. Aerobic decomposition means that the active microbes in the heap require oxygen, while in anaerobic decomposition, the active microbes do not require oxygen to live and grow. Temperature, moisture content, the size of bacterial populations, and availability of nutrients limit and determine how much oxygen the heap uses.

5.9.4.2. Moisture

The amount of moisture in the heap should be as high as possible, while still allowing air to filter into the pore spaces for the benefit of aerobic bacteria. Individual bacteria hold various percentages of moisture in compost and determine the amount of water that can be added. For example, woody and fibrous material, such as bark, sawdust, wood chips, hay, and straw, have the capacity to hold up to 75–85% of moisture. "Green manure," such as lawn clipping and vegetable trimming, are able to hold 50–60% moisture.

5.9.4.3. Temperature

Temperature is an important factor in biological composting of heap. Low outside temperatures during winter slow the decomposition process, while warmer temperatures speed it up. During the warmer months of the year, intense microbial activity inside the heap caused composting to proceed at extremely high temperatures. The microbes that decompose the raw materials fall into basically two categories: mesospheric, those that live and grow at temperatures of 50–113 °F (10–45 °C), and thermophillic, those that thrive in temperatures of 113–158 °F (45–70 °C). The initial heat heap phase that most garden compost goes through is thermophillic. The organic material gets dehydrated very quickly in this phase and is kept aerated and moistened. The high temperatures are beneficial to the gardener because they kill weed seeds and germs that could be detrimental to vegetation. The next phase holds at 100 °F for a while and different microbes predominate. Then finally, the ambient phase occurs where the pleasant earthly odor originates and material has produced compost.

5.9.5. Biodegradation of Organic Pollutants

Hypersaline environments are important for both surface extension and ecological significance. However, little information is available on the biodegradation of organic pollutants by halophilic microorganisms in such environments. In addition, it is estimated that 5% of industrial effluents are saline and hypersaline. Conventional nonextremophilic microorganisms are unable to efficiently perform the removal of organic pollutants at high salt concentrations. Halophilic microorganisms are metabolically different and are adapted to extreme salinity; these microorganisms are good candidates for the bioremediation of hypersaline environments and treatment of saline effluents. A diversity of contaminating compounds are susceptible to be degraded by halotolerant and halophilic bacteria. Significant research efforts are still necessary to estimate the true potential of these microorganisms to be applied in environmental processes and in the remediation of contaminated hypersaline ecosystems (Borgne et al., 2008).

Interest in the microbial biodegradation of pollutants has intensified as humanity strives to find sustainable ways to clean up contaminated environments (Bagchi et al., 1985; Singh et al., 1989a, 1989b, 1993). These bioremediation and biotransformation methods endeavor to harness the astonishing, naturally occurring ability of microbial xenobiotic metabolism to degrade, transform, or accumulate a huge range of compounds including hydrocarbons (e.g., oil), PCBs, PAHs, heterocyclic compounds (such as pyridine or quinoline), pharmaceutical substances, radionuclides, and metals (Gothalwal and Bisen, 1998; Bisen and Khare, 1991; Sharma and Bisen, 1992; Sharma et al., 2001). Major methodological breakthroughs in recent years have enabled detailed genomic, metagenomic, proteomic, bioinformatic, and other high throughput analyses of environmentally relevant microorganisms providing unprecedented insights into key biodegradative pathways and the ability of organisms to adapt to changing environmental conditions (Table 5.5).

TABLE 5.5. Degradation of Pollutants by Dehalogenases

Mechanism of Dehalogenation	Pollutant	Source
Reductive	3-Chlorobenzoate	*Desulfomonile tiedjei, Flavobacterium*
	Perchloroethylene (PCE)	
	3-Chloro-4-hydroxybenzoate	
Oxygenolytic	Pneumocystis pneumonia (PCP),	*Burkholderia cepacia, Flavobacterium* sp.
	Chloroalkanes	
	4-Chlorophenylacetate	
Hydrolytic	Haloaromatic compounds	*Arthrobacter, Pseudomonas*
	Haloalkanes	*Pseudomonas, Hyphomicrobium, Methylobacterium*
	Haloacid compounds	*Pseudomonas, Alcaligenes*
Haloalcohol	C-2, C-3, bromo, chloro alcohols	*Flavobacterium, Pseudomonas, Arthrobacter*
Dehydroalogenation	Lindane	*Pseudomonas paucimobilis*

Several strategies may be adopted to efficiently improve the process and help natural biodegradation processes work faster. These biological methodologies, collectively indicated as *bioremediation*, and usually considered environment-friendly treatments constitute essentially a managed or spontaneous process mediated by living organisms (mainly microorganisms), which degrade or transform contaminants to less toxic or nontoxic products, with mitigation or elimination of environmental contamination. The use of plants (phytoremediation) and microorganisms associated with them has also been considered as an appealing technology to degrade, contain, or render harmless contaminants in soil or groundwater.

Intra-, ecto-, and extracellular enzymes are, ultimately, the main effectors of pollutant transformation during plant and/or microbial processes. Enzymes perform a wide range of very important functions in Nature. They are highly specific and efficient, guiding the biochemistry of life with great precision and fidelity. This fidelity is essential in the cells of living organisms, and a multitude of mechanisms have evolved for controlling the activity of these enzymes. Enzymes play a key role in harvesting energy from the sun via photosynthesis, perform a wide range of metabolic functions throughout every living cell in the bodies of plants and animals, and are in fact really the catalysts of all biological processes constituting life on Earth.

Many enzymes of both microbial or plant origin have been recognized to be able to transform pollutants at a detectable rate and are potentially suitable to restore polluted environment (Fig. 5.37). The main enzymatic classes involved in such a process are hydrolases, dehalogenases, and oxidoreductases. Amide, ester, and peptide bonds undergo hydrolytic cleavage by amidases, esterases, and proteases, respectively, in several xenobiotic compounds and may lead to products with little or no toxicity. Hydrolases, responsible for the cleavage of pesticides, are among the best studied groups of enzymes. Most of these hydrolases are extracellular enzymes, except for the cell-wall-bound enzymes of *Penicillium* and *Arthrobacter* sp., which hydrolyze barban and propham, respectively.

Evidence have been provided that lytic enzymes produced by many fungi and bacteria are strictly involved in the microbes' antagonism toward plant pathogens and

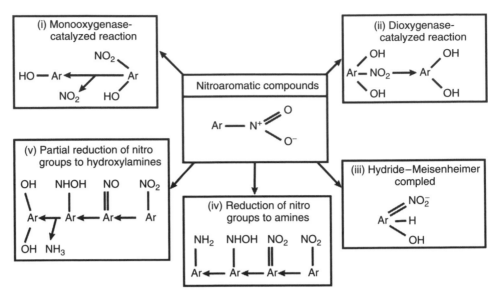

Figure 5.37. Microbial strategies for remediation of nitroaromatic compounds by aerobes (i, ii, iii, and iv) and anaerobes (v) (Kulkarni and Chaudhari, 2007).

pests. Exoenzymes possessing chitinolytic, glucanolytic, cellulolytic, or proteolytic activities can be used individually or in combination to provide an enzymatic basis for a number of processes, which ultimately lead to a biocontrol effect. Very interesting is the finding that selected strains of *Trichoderma* when introduced in rhizosphere soil rapidly catabolized phytotoxic concentrations (10 mM) of cyanide through the action of two cyanide-catabolizing enzymes, formamide hydrolyase and rhodanese (Lynch, 1990).

The elimination of a wide range of pollutants and wastes from the environment is an absolute requirement to promote a sustainable development of our society with low environmental impact. Biological processes play a major role in the removal of contaminants, and they take advantage of the astonishing catabolic versatility of microorganisms to degrade/convert such compounds. New methodological breakthroughs in sequencing, genomics, proteomics, bioinformatics, and imaging are producing vast amounts of information. In the field of environmental microbiology, genome-based global studies open a new era, providing unprecedented *in silico* views of metabolic and regulatory networks, as well as clues to the evolution of degradation pathways and to the molecular adaptation strategies to changing environmental conditions. Functional genomic and metagenomic approaches are increasing our understanding of the relative importance of different pathways and regulatory networks to carbon flux in particular environments and for particular compounds and they will certainly accelerate the development of bioremediation technologies and biotransformation processes (Fig. 5.38).

5.9.6. Bioremediation

This process is defined as the use of living organisms for the recovery/cleaning up of a contaminated medium (soil, sediment, air, water). The process of bioremediation

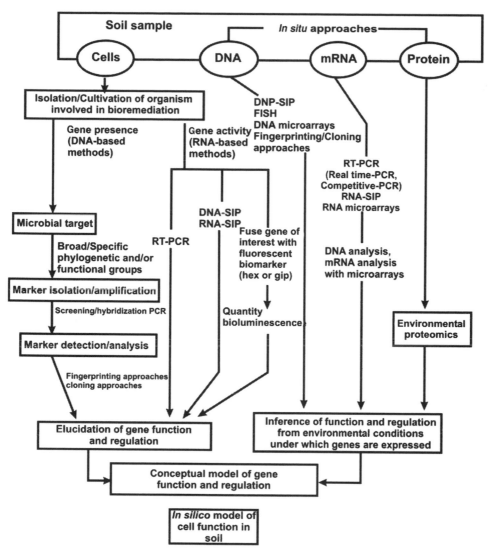

Figure 5.38. Different genomic approaches available to examine microbial communities of polluted soils. *Source:* After Andreoni and Gianfreda (2007).

might involve introduction of new organisms to a site or adjustment of environmental conditions to enhance degradation rates of indigenous fauna. Bioremediation can be applied to recover brownfields for development and for preparing contaminated industrial effluents before discharge into waterways. Bioremediation technologies are also applied to contaminated wastewater, groundwater, surface water, soil, sediments, and air where there has been either accidental or intentional release of pollutants or chemicals that pose a risk to human, animal, or ecosystem health.

Environmental biotechnology is not a new field; composting and wastewater treatments are familiar examples of old environmental biotechnologies. However, studies in molecular biology and ecology offer opportunities for more efficient biological processes. Notable accomplishments of these studies include the clean up of polluted water and land areas (Bisen et al., 1987; Bisen and Khare, 1991; Mehta et al., 1992; Gothalwal and Bisen, 1993). Bioremediation is defined as the process where organic wastes are biologically degraded under controlled conditions to an innocuous state or to levels below concentration limits established by regulatory authorities (Mueller et al., 1996). Bioremediation is the use of living organisms, primarily microorganisms, to degrade the environmental contaminants into less toxic forms. It uses naturally occurring bacteria and fungi or plants to degrade or detoxify substances hazardous to human health and/or the environment. The microorganisms may be indigenous to a contaminated area or they may be isolated from elsewhere and brought to the contaminated site. Contaminants are transformed by living organisms through reactions that take place as a part of their metabolic processes. Biodegradation of a compound is often a result of the actions of multiple organisms. Microorganisms are imported to a contaminated site to enhance degradation by a process known as *bioaugmentation*. For bioremediation to be effective, microorganisms must enzymatically attack the pollutants and convert them to harmless products. As bioremediation can be effective only where environmental conditions permit microbial growth and activity, its application often involves the manipulation of environmental parameters to allow microbial growth and degradation to proceed at a faster rate. Like other technologies, bioremediation has its limitations. Some contaminants, such as chlorinated organic or high aromatic hydrocarbons, are resistant to microbial attack. It is not easy to predict the rates of cleanup for a bioremediation exercise, as the cleanup operation is either slow or absent. There are no rules to predict if a contaminant can be degraded. Bioremediation techniques are typically more economical than traditional methods such as incineration, and some pollutants can be treated on site, thus reducing exposure risks for cleanup personnel and potentially wider exposure risks as a result of transportation accidents. Since bioremediation is based on natural attenuation, public consider it more acceptable than other technologies. Most bioremediation systems are run under aerobic conditions, but running a system under anaerobic conditions may permit microbial organisms to degrade otherwise recalcitrant molecules (Colberg and Young, 1995).

Bioremediation technologies can be generally classified as *in situ* or *ex situ*. *In situ* bioremediation involves treating the contaminated material at the site, while *ex situ* involves the removal of the contaminated material to be treated elsewhere (Fig. 5.39). Some examples of bioremediation technologies are bioventing, landfarming, bioreactor, composting, bioaugmentation, rhizofiltration, and biostimulation. Bioremediation can occur on its own (natural attenuation or intrinsic bioremediation) or can be spurred on via the addition of fertilizers to increase the bioavailability within the medium (biostimulation). Advancements have also proven successful via the addition of matched microbe strains to the medium to enhance the resident microbe population's ability to break down contaminants. Microorganisms performing the function of bioremediation are known as *bioremediators*.

5.10. MICROORGANISMS AND METAL POLLUTANTS

High concentrations of heavy metals in soil have an adverse effect on microorganisms and microbial processes. Among soil microorganisms, mycorrhizal fungi are the only ones providing a direct link between soil and roots and can therefore be of great importance in heavy metal availability and toxicity to plants (Bisen et al., 1995, 1996 and Dev et al.,

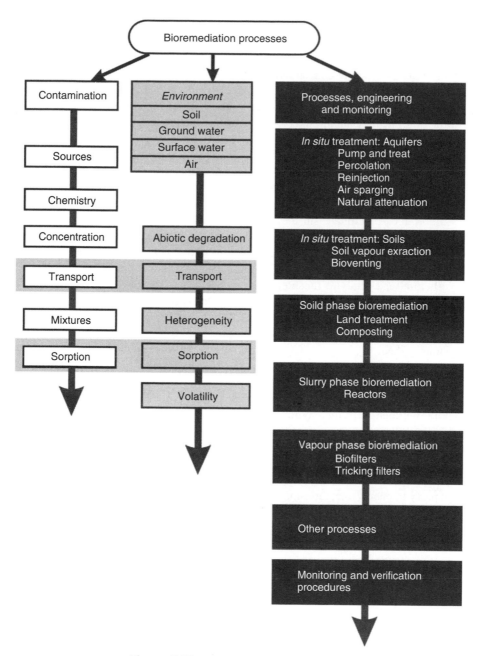

Figure 5.39. Bioremediation technology.

1997). The mycorrhizal fungi can be possibly used as bioremediation agents in polluted soils or as bioindicators of pollution (Leyval et al., 1997). Heavy-metal-contaminated soil problem has already become the global focus. With the increasing attention to environment protection, scientists began to search new methods for remediation of contaminated soil (Gothalwal and Bisen, 1998, 2001; Singh et al., 1993).

5.10.1. Metals Defined

Many base metals and a few precious metals, as well as some metalloids, can be enzymatically or nonenzymatically concentrated and dispersed by microbes in their environment. Some of these activities are or have a potential to be commercially exploited. Microbes encounter metals and metalloids of various kinds in the environment, and it is, therefore, not surprising that they should interact with them, sometimes to their benefit and at other times to their detriment. Of particular practical interest are the base metals including vanadium, chromium, manganese, iron, cobalt, nickel, copper, zinc, molybdenum, silver, cadmium, and lead; the precious metals gold and silver; and the metalloids arsenic, selenium, and antimony. In Nature, these metals and metalloids exist as mostly cations, oxyanions, or both in aqueous solution; as salts or oxides in crystalline (mineral) form; or as amorphous precipitates in the insoluble form. A few metals, such as iron, copper, and gold, may also exist in the metallic state in Nature, but the first two of these only very rarely. All microbes, whether prokaryotic or eukaryotic, require metal species for structural functions and/or catalytic functions. The alkali metals Ca and Mg serve both structural and catalytic functions. The metals V, Cr, Mn, Fe, Co, Ni, Cu, Zn, Mo, and W and the metalloid Se may participate in catalytic functions. For such uses, low environmental concentrations are sufficient.

Some prokaryotes can use metal species that can exist in more than one oxidation state, among them Cr, Mn, Fe, Co, Cu, As, and Se, as electron donors or acceptors in their energy metabolism. For this function, the metals or metalloids must occur in sufficiently high concentrations locally to meet the organisms' demand. Microbial interactions with small quantities of metals or metalloids do not exert a major impact on metal or metalloid distribution in the environment, whereas interactions with larger quantities of metals or metalloids, as are required in energy metabolism for instance, have a noticeable impact. Some of the latter type of interactions are or have a potential to be commercially exploited.

A number of microbes are able to use some metals or metalloids as electron donors or acceptors in energy metabolism, including eubacteria and archaea. Depending on the element, the metal species may be in simple ionic form or oxyanionic form (Bisen and Khare, 1991; Sharma et al., 2001; Sharma and Bisen, 1992; Singh et al., 1989b, 1993). As energy sources, oxidizable metals or metalloids may satisfy the entire energy demand of an organism (chemolithotrophs). For example, the eubacteria *T. ferrooxidans* and *Leptospirillum ferrooxidans* and the archaea *Acidianus brierleyi* and *S. acidocaldarius* are able to obtain all their energy for growth from the oxidation.

Miles underground, microbes survive without oxygen or sunlight by feeding on metals such as iron and manganese. One of these microorganisms, *Geobacter metallireducens*, has an unusual survival tactic for life in the underworld. It uses a sensor to "sniff out" metals. If metal is not nearby, *G. metallireducens* can spontaneously grow flagella, whiplike cellular propellers, to find new energy sources. Previously, researchers believed *G. metallireducens* was immotile. By studying the genome sequence of another metal-eating microbe of the same genus, scientists have now discovered that *G. metallireducens* contains genes for flagella. The flagella genes surprised the scientists who had never seen *G. metallireducens* swim. Looking back, the researchers realized that in the past motility experiments, the organism had only been grown on soluble metals, which are easy to work with in the laboratory. After performing experiments with insoluble metals, such as iron oxide, the team found that *G. metallireducens* indeed grows flagella and swims.

The microbe's metal diet has made it intriguing to researchers. In addition to using iron, the organism uses metals such as plutonium and uranium to metabolize food. *G. metallireducens* consumes these radioactive elements and essentially eats away at the contaminants. In the case of uranium, it changes the metal from a soluble to an insoluble form. The insoluble uranium drops out of the groundwater, thus decontaminating streams and drinking water. It remains in the soil and could then be extracted.

5.10.2. Metal Toxicity Effect on Microbial Cells

The environmental and microbiological factors that can influence heavy metal toxicity are discussed with a view to understanding the mechanisms of microbial metal tolerance. It is apparent that metal toxicity can be heavily influenced by environmental conditions. Binding of metals to organic materials, precipitation, complexation, and ionic interactions are all important phenomena that must be considered carefully in laboratory and field studies. It is also obvious that microbes possess a range of tolerance mechanisms, most featuring some kind of detoxification. Many of these detoxification mechanisms occur widely in the microbial world and are not specific to only microbes growing in metal-contaminated environments (Gothalwal and Bisen, 2001; Gadd and Griffiths, 1978).

Toxic effects of heavy metals on soil microorganisms have been extensively studied in the past, and almost every group of organisms has been studied in this respect. Fungi and bacteria constitute the main components of the soil microbial biomass. It has often been stated that fungi are more tolerant to heavy metals as a group than bacteria. High loads of metals disturb tree growth and induce effects in the mycorrhiza fungi. Thus, to compare heavy metal effects on the saprotrophic part of the fungal and bacterial communities, experiments should be performed without the involvement of plants. Furthermore, measurements of activity would be the most direct way of comparing these two groups of microorganisms since this is a more sensitive measure than biomass measurements. Bacterial activity and growth rate have usually been determined *in situ* by the thymidine or leucine incorporation technique. Both methods were initially developed for use in aquatic environments but are nowadays frequently used in terrestrial habitat. In a similar way, fungal activity and growth rate in aquatic habitats have been estimated by the acetate-in-ergosterol incorporation technique. However, this method has only recently been applied to the soil environment.

Anaerobic degradation of sulfate laden organics involves competitive interactions among various groups of bacteria including methane-producing bacteria (MPB) and sulfate-reducing bacteria (SRB) for substrate utilization with inherent generation of hydrogen sulfide. Microorganisms including MPB and SRB require various metals including Fe, Ni, Zn, and Co for enzymatic activities and growth. Sulfide affects process performance and stability of high rate anaerobic treatment systems by its direct toxicity to microorganisms and indirectly through precipitation of key metals. Metals form complexes in natural as well as artificial aquatic systems including anaerobic treatment processes. Natural organic material and biogenic ligands chelate metals; in some cases, this chelation increases the bioavailability of nutrient metals, whereas in others, it decreases the toxicity of metals by making them less bioavailable.

Damage is caused through a variety of mechanisms, but most toxins do so by increasing oxidative stress, poisoning enzymes, directly damaging DNA or cellular membranes, or acting as endocrine disrupters. For example, the toxic metal cadmium increases oxidative damage by both causing the formation of free radicals such as H_2O_2, O_2, and OH

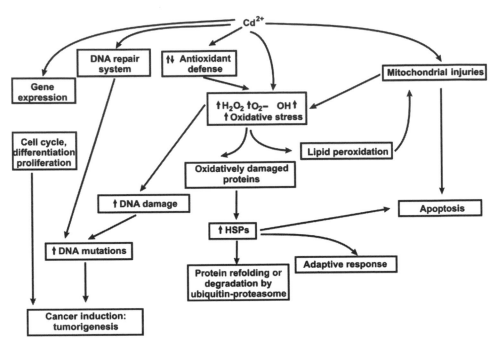

Figure 5.40. General scheme of biological consequence of cadmium intoxication in cells.

and directly poisoning several enzymes that reduce oxidative stress, including catalase (CAT), glutathione reductase (GR), and the most abundant cellular antioxidant, glutathione (GSH). This is a fairly common mechanism for heavy metal toxicity. Cadmium interferes with various important mechanisms such as gene expression, cell cycle, differentiation, and proliferation (Fig. 5.40). Cadmium gives rise to oxidative damage affecting DNA, proteins, and membrane lipids. The induction of oxidative damage is associated with mitochondrial dysfunction, deregulation of intracellular antioxidants, and apoptosis. Oxidative stress on proteins induces HSPs (heat shock proteins), associated with an adaptive response, initiation of protein refolding, and/or degradation by ubiquitine proteasome. Oxidative damage to DNA leads to mutations and induction of cancer. The inhibition of some DNA repair pathways contributes to the rise in mutations and cancer.

5.10.3. Mechanism of Microbial Metal Resistance and Detoxification

Microbial cells have resistances to essentially all the toxic heavy metals in the periodic table. In bacterial cells, the genetic determinants of these resistances are frequently found on small extrachromosomal plasmids and transposons. Sometimes the resistances are associated with detoxifying enzymes. This is true for the $Hg^{2+} \rightarrow Hg^{0}$ reductase, the $As^{3+} \rightarrow As^{5+}$ oxidase, and the $Cr^{6+} \rightarrow Cr^{3+}$ reductase. In other cases, such as in case of As^{5+}, Ag^{+}, and Cd^{2+}, no change in redox state occurs but, rather, uptake and transport differences accompany resistance determinants (Jeffery and Simon, 1984).

Some metals, including iron, zinc, copper, manganese, are micronutrients used in the redox processes, in the regulation of the osmotic pressure, and also as enzyme components. Other metals are not essential. However, even essential metals such as zinc

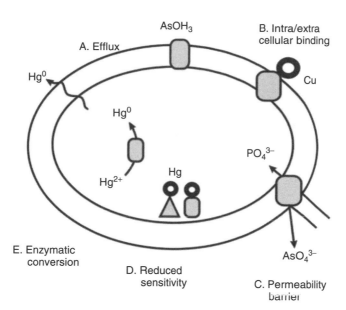

Figure 5.41. Methods of metal resistance: (A) metals can be effluxed out of the cell; (B) metals are bound either inside or outside of the cell; (C) reduced uptake because of more specific uptake systems; (D) expression of less sensitive proteins; and (E) enzymatic conversion of the metal (Dopson et al., 2003).

and copper are toxic at high concentrations. The effects of high metal concentration are DNA and membrane damage and loss of enzyme function. To protect themselves from toxic concentrations of metals, bacteria utilize a variety of resistance mechanisms that involve permeability barriers, intra- and extracellular sequestration, efflux pumps, enzymatic detoxification, and reduction.

Even though microbes require a number of metals, most metals are nonessential, have no nutrient value, and will become toxic to microbes if they accumulate above normal physiological concentrations by the action of unspecific, constitutively expressed transport systems. At high intracellular levels, both essential and nonessential metals can cause damage.

Many microorganisms demonstrate resistance to metals in water, soil and industrial waste. Genes located on chromosomes, plasmids, or transposons encode specific resistance to a variety of metal ions. There are five basic mechanisms that convey an increased level of cellular resistance to metals (Fig. 5.41): active transport efflux of the toxic metal from the cell, enzymic conversion, intra- or extracellular sequestration, exclusion by a permeability barrier, and reduction in sensitivity of cellular targets.

Bacterial plasmids contain genetic determinants for resistance systems for Hg^{2+} (and organomercurials), Cd^{2+}, AsO_2, AsO_4^{3-}, CrO_4^{2-}, TeO_3^{2-}, Cu^{2+}, Ag^+, Co^{2+}, Pb^{2+}, and other metals of environmental concern. In some cases, there is the potential for using genetically engineered microbes for bioremediation. Recombinant DNA analysis has been applied to mercury, cadmium, zinc, cobalt, arsenic, chromate, tellurium, and copper resistance systems. The eight mercury resistance systems that have been sequenced all contain the gene for mercuric reductase, the enzyme that converts toxic Hg^{2+} ions to less toxic volatile metallic Hg. Four of these systems also determine the enzyme organomercurial lyase, which cuts the HgC bond and thus detoxifies methylmercury and phenylmercury. Two sequenced Cd^{2+} resistance determinants govern cellular efflux of Cd^{2+}, assuring a low level of intracellular Cd^{2+}, making them not an obvious candidate for bioremediation. Cadmium accumulation by bacterial metallothionein or phytochelatin is a potentially useful process, but only preliminary reports have appeared on bacteria producing polythiol polypeptides. For arsenic resistance, a unique efflux ATPase maintains low intracellular As levels. A bacterial AsO_2^- oxidase has been reported, which has the potential to convert more toxic As(III) into less toxic As(V), but this system has not been studied in recent years. For chromate, resistance results from reduced cellular uptake. However, both soluble and membrane-bound Cr(VI) reductase bacterial activities convert more toxic Cr(VI) to less toxic Cr(III) in different bacteria (Silver, 1992).

Heavy metals are naturally present in some ecosystems, their industrial use leads to serious environmental problems. The use of metal-resistance bacteria can help to remove metal from contaminated environments. Understanding the regulation of heavy metal resistance could be useful for biological waste treatment and estimating the impact that industrial activity may have on natural ecosystems.

5.10.4. Metal–Microbe Interaction

Microorganisms play an important role in the environmental fate of toxic metals and radionucleotides with a multiplicity of mechanisms effecting transformation between soluble and insoluble forms. Although some heavy metals are essential trace elements, most can be, at high concentrations, toxic to all branches of life, including microbes, by forming complex compounds within the cell. Because heavy metals are increasingly found in microbial habitats due to natural and environmental processes, microbes have evolved several mechanisms to tolerate the presence of heavy metals. These mechanisms include the efflux of metal ions outside the cell, accumulation and complexation of metal ions inside the cell, and reduction of the heavy metal ions to a less toxic state. Because the intake and subsequent efflux of heavy metal ions by microbes usually includes a redox reaction involving the metal, bacteria that are resistant to and grow on metals also play an important role in the biogeochemical cycling of those metal ions. The metal-accumulating capacity of microorganisms can be exploited to remove, concentrate, and recover metals from mine tailings and industrial effluents. Although metals cannot be broken down into nontoxic components such as organic compounds, bioremediation can be used to stabilize, extract, or reduce the toxicity of soil and groundwater contaminated by AMD. Several reports of aerobic bacteria accumulating metals such as Ag, Co, Cu, Cr, and Ni are available.

Metal ions can interact with microorganisms via a range of mechanisms. For example, metals can adsorb directly to an organism's cell wall or can react with microbial by-products, such as extracellular polymers (Ruggiero et al., 2000, 2004). A classic system of microbial–metal interaction involves low molecular weight organic ligands (siderophores) that are excreted and used by plants and microbes to acquire iron (Fig. 5.42).

All microorganisms, except the lactobacilli, have nutritional requirements for Fe(III) that are not met in aqueous aerobic environments. Under those conditions, Fe(III) has a low solubility (at neutral pH, typical concentrations are about 10–18 M) so that its bioavailability is limited. In order to acquire sufficient iron, bacteria synthesize and secrete siderophores that can chelate Fe(III) and carry it into the cell via specific high affinity uptake receptors. The siderophores are typically multidentate, oxygen-donor ligands

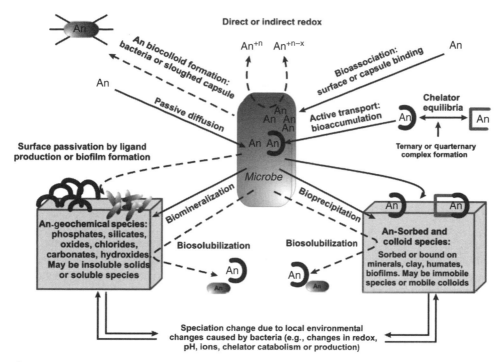

Figure 5.42. Microbial interactions with actinides in the environment. All processes could increase or decrease solubility and/or mobility depending on specific bacteria and numerous environmental and chemical factors (Ruggiero et al., 2004). An, actinide species.

that usually have hydroxamate, catecholate, or carboxylate moieties. Although they are designed to have an extremely high affinity for Fe^{3+}, siderophores can bind other "hard" ions, such as Al(III), Zn(II), Ga(III), Cr(III), and Pu(III,IV) (Neu, 2000).

All processes could increase or decrease solubility and/or mobility depending on specific bacteria and numerous environmental and chemical factors.

Essential elements, such as K, Ca, Mn, Mg, Cu, Zn, Fe, Co, and those with no essential biological function, such as Cs, Cd, Pb, Al, Sn, Hg, can be accumulated by microorganisms by nonspecific physiochemical interactions and specific mechanisms of sequestration and transport (Singh et al., 1989a, 1989b; Sharma et al., 2001; Sharma and Bisen, 1992). Presently, metal-accumulating bacteria have shown potential for material science. Biomimetics is the area of research dealing with material science and engineering through biology. Bacteria are involved as workers in the living factory, and a plethora of novel nanostructure particles with unexpected properties are produced in the living factory, which have applications in biomedical sciences, optics, magnetics, mechanics, catalysis, and energy science.

5.10.5. Microbial Approaches in Remediation of Metal-Contaminated Soil and Sediments

Biological processes have been used for some inorganic materials, such as metals, to lower radioactivity and to remediate organic contaminants. Heavy metal contamination has become a worldwide problem by disturbing the normal functions of rivers and lakes. Sediment, as the largest storage and resources of heavy metal, plays a rather important role in metal transformations. With metal contamination the usual challenge is to accumulate the metal into harvestable plant parts, which must then be disposed of in a hazardous waste landfill before or after incineration to reduce the plant to ash. Two exceptions are mercury and selenium, which can be released as volatile elements directly from plants into the atmosphere. The concept and practice of using plants and microorganisms to remediate contaminated soil have developed over the past 30 years (Pilon-Smits and Pilon, 2002).

The *in situ* remediation of sediment aims at increasing the stabilization of some metals such as the mobile and the exchangeable fractions, whereas the *ex situ* remediation mainly aims at removing those potentially mobile metals, such as the manganese oxides and the organic matter fraction. The pH and organic matter can directly change metals distribution in sediment; however, oxidation–reduction potential (ORP), mainly through changing the pH values, indirectly alters metals distribution. Mainly ascribed to their simple operation mode, low costs, and fast remediation effects, *in situ* remediation technologies, especially being fit for slight pollution sediment, are applied widely. However, for avoiding metal secondary pollution from sediment release, *ex situ* remediation should be the hot point in future research (Peng et al., 2009).

The idea of bioremediation has become popular with the onset of the twenty-first century. Microbes are often used to remedy environmental problems found in soil, water, and sediments. Table 5.6 states the essential factors for microbial bioremediation.

In principle, genetically engineered plants and microorganisms can greatly enhance the potential range of bioremediation. For example, bacterial enzymes engineered into plants can speed up the breakdown of TNT and other explosives. With transgenic poplar trees carrying a bacterial gene, methylmercury may be converted to elemental mercury, which is released into the atmosphere at an extreme dilution. However, concern about release of such organisms into the environment has limited actual field applications.

Although microorganisms cannot destroy metals, they can alter their chemical properties via a surprising array of mechanisms, some of which can be used to treat metal contamination (Bisen et al., 1987; Bisen and Mathur, 1993). In some cases, these processes involve highly specific biochemical pathways that have evolved to protect the microbial cell from toxic heavy metals; a good example of such pathway is the microbial reduction of mercury. Because these detoxification mechanisms are very specific, the biochemical components that recognize and detoxify the target metals may also prove useful in the design of biosensors for "bioavailable" concentrations of toxic metals. In other examples, microbes can produce new mineral phases via nonspecific mechanisms that

TABLE 5.6. Essential Factors for Microbial Bioremediation

Factor	Desired Conditions
Microbial population	Suitable kinds of organisms that can biodegrade all the contaminants
Oxygen	Enough to support aerobic biodegradation (about 2% oxygen in the gas phase or 0.4 mg/l in the soil water)
Water	Soil moisture should be 50–70% of the water holding capacity of the soil
Nutrients	Nitrogen, phosphorus, sulfur, and other nutrients to support good microbial growth
Temperature	Appropriate temperatures for microbial growth (0–40 °C)
pH	Best range is from 6.5 to 7.5

result in the entrapment of toxic metals within the soil or sediments. Other mechanisms of potential commercial importance rely on the production of biogenic ligands that can complex metals, resulting in their mobilization from contaminated soils. The mobilized metals can then be pumped out of the soil or sediment and trapped in a bioreactor on the surface.

5.11. ENVIRONMENTALLY TRANSMITTED PATHOGENS

Pathogens are agents (e.g., bacteria, virus, and protozoa) that can cause a disease. While natural environments are filled with millions of different microorganisms, most of these are not pathogens. Most environmental microorganisms are free living (do not require a host to function), as they are able to utilize the carbon and nitrogen they encounter in soil and water. Pathogens generally originate from an infected host and are transmitted in the environment. For a pathogen to function fully, it must encounter a host where it can grow and reproduce. Inside the host, the pathogen encounters the appropriate environment (with required nutrients, optimum pH and temperature, to mention a few) that will sustain its growth and reproduction. Environmentally transmitted pathogens are microorganisms that spend most of their life outside a host, but when introduced to a host, they can (not always) cause disease. Primarily while in the environment, pathogens are just able to survive, although growth and reproduction is possible. Pathogens can be transmitted in all types of environments including water, soil, air, and food. While the direct detection of pathogens in the environment will provide evidence of their presence, this is a difficult task.

Pathogens can spread from person to person in a number of ways (Fig. 5.43). Not all pathogens use all the available routes. For example, the influenza virus is transmitted from person to person through air, typically via sneezing or coughing. But the virus is not transmitted via water. In contrast, *E. coli* is readily transmitted via water, food, and blood but is not readily transmitted via air or the bite of an insect. While routes of transmission vary for different pathogens, a given pathogen will use a given route of transmission. This has been used in the weaponization of pathogens. The best known example is anthrax. The bacterium that causes anthrax, *B. anthracis*, can form an environmentally hardy form called a *spore*. The spore is very small and light. It can float on currents of air and can be breathed into the lungs, where the bacteria resume growth and swiftly cause a serious and often fatal form of anthrax. As demonstrated in the United States in the last few months of 2001, anthrax spores are easily sent through the mail to targets. As well, the powdery spores can be released from an aircraft. Over a major urban center, modeling studies have indicated that the resulting casualties could number in the hundreds of thousands.

Contamination of water by pathogens is another insidious route of disease spread. Water remains crystal clear until there are millions of bacteria present in each milliliter. Viruses, which are much smaller, can be present in even higher numbers without affecting the appearance of the liquid. Thus, water can be easily laced with enough pathogens to cause illness.

5.11.1. Transmission of Pathogen

Microorganisms that cause disease in humans and other species are known as *pathogens* (Fig. 5.44). The transmission of pathogens to a human or other host can occur in a

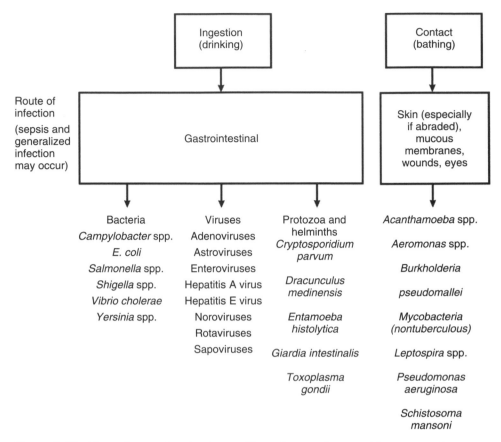

Figure 5.43. The various routes of entry used by pathogens for spread from person to person.

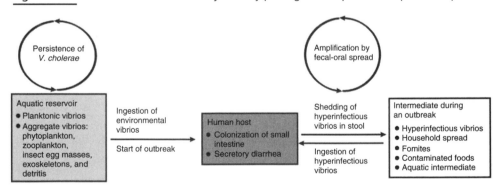

Figure 5.44. Transmission of *Vibrio cholerae*.

number of ways, depending on the microorganism. A common route is via water. The ingestion of contaminated water introduces microbes into the digestive system. Intestinal upsets can result. As well, an organism may be capable of entering the cells that line the digestive tract, gaining entry into the bloodstream. From there, an infection can become widely dispersed. A prominent example of a waterborne pathogen is *Vibrio cholerae*, the bacterium that causes cholera. Contamination of drinking water by this bacterium is still at epidemic proportions in some areas of the world. Toxigenic strains of *V. cholerae* persist in aquatic environments alongside nontoxigenic strains, aided by biofilm formation on biological surfaces and the use of chitin as a carbon and nitrogen source. On ingestion of these aquatic-environment-adapted bacteria via contaminated food or water, toxigenic strains colonize the small intestine, multiply, secrete cholera toxin, and are shed back into the environment by the host in secretory diarrhea. The stool-shed pathogens are in a transient hyperinfectious state that serves to amplify the outbreak through transmission to subsequent hosts (Nelson et al., 2009).

Pathogens can also be transmitted via air. Viruses and bacterial spores are light enough to be lifted off by breeze. These agents can subsequently be inhaled, resulting in lung infections. An example of such a virus is the Hantavirus. A particularly prominent bacterial example is the spore form of the anthrax-causing bacterium *B. anthracis*. The latter has also been identified as a bioweapon used by terrorists that can, as exemplified in a 2001 terrorist attack on the United States, be transmitted in mail that when opened or touched can result in cutaneous or inhalation anthrax. Still other microbial pathogens are transmitted from one human to another via body fluids such as blood. This route is utilized by a number of viruses. The most publicized example is the transmission of the human immunodeficiency virus (HIV). HIV is generally regarded to be the cause of the acquired immunodeficiency syndrome. As well, viruses that cause hemorrhagic fever (e.g., Ebola virus) are transmitted through blood. If precautions are not taken when handling patients, the caregiver can become infected.

Transmission of pathogens can occur directly, as in the above mechanisms. As well, transmission can be indirect. An intermediate host that harbors the microorganism can transfer the microbes to humans via a bite or by other contact. *Coxiella burnetii*, the bacterium that causes Q fever, is transmitted to humans by the handling of animals such as sheep. As another example, the trypanosome parasite that causes sleeping sickness enters the bloodstream by the bite of a female mosquito that acts as a vector for the transmission of the parasite. Finally, some viruses are able to transmit infection over long periods by becoming latent in the host. More specifically, the genetic material of viruses such as the hepatitis viruses and the herpes virus can integrate and be carried for decades in the host genome before the symptoms of infections appear.

5.11.1.1. Bacteria

Bacteria are single-celled microorganisms that can exist either as independent (free-living) organisms or as parasites (dependent on another organism for life). The term bacteria was devised in the nineteenth century by the German botanist Ferdinand Cohn (1828–1898) who based it on the Greek bakterion meaning a small rod or staff. In 1853, Cohn categorized bacteria as one of three types of microorganisms: bacteria (short rods), bacilli (longer rods), and spirilla (spiral forms). The term bacteria was preceded in the seventeenth century by the microscopic animalcules described by Antony van Leeuwenhoek (1632–1723). There are typically 40 million bacterial cells in a gram of soil, and a million bacterial cells in a milliliter of freshwater; in all, there are approximately five nonillion (5×10^{30}) bacteria on Earth, forming much of the world's biomass. Bacteria are vital in recycling nutrients, with many steps in nutrient cycles depending on these organisms, such as the fixation of nitrogen from the atmosphere and putrefaction. However, most bacteria have not been characterized, and only about half of the phyla of bacteria have species that can be grown in the laboratory.

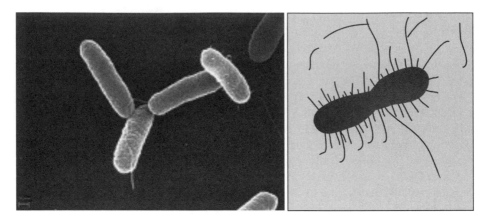

Figure 5.45. *Salmonella.*

SALMONELLA. *Salmonella* is a genus of rod-shaped, gram-negative, non-spore-forming, predominantly motile enterobacteria with diameters around 0.7–1.5 μm, lengths from 2 to 5 μm, and flagella which project in all directions (i.e., peritrichous; Fig. 5.45). They are chemoorganotrophs, obtaining their energy from oxidation and reduction reactions using organic sources, and are facultative anaerobes. Most species produce hydrogen sulfide that can readily be detected by growing them on media containing ferrous sulfate, such as TSI (triple sugar iron). Most isolates exist in two phases: a motile phase I and a nonmotile phase II. Cultures that are nonmotile on primary culture may be switched to the motile phase using a Craigie tube. *Salmonella* are closely related to the *Escherichia* genus and found worldwide in cold- and warm-blooded animals (including humans) and in the environment. They cause illnesses such as typhoid fever, paratyphoid fever, and the foodborne illness salmonellosis. *Salmonella* infections are zoonotic and can be transferred between humans and nonhuman animals. Many infections are due to ingestion of contaminated food. A distinction is made between enteritis *Salmonella* and typhoid/paratyphoid *Salmonella*, where the latter, because of a special virulence factor and a capsule protein (virulence antigen), can cause serious illness, such as *Salmonella enterica* subsp.*enterica* serovar Typhi, or *Salmonella typhi*). *S. typhi* is adapted to humans and does not occur in animals.

The bacteria were first isolated by Theobald Smith in 1885 from pigs. The genus name *Salmonella* was derived from the last name of D. E. Salmon. Salmonellosis (gastroenteritis characterized by nausea, vomiting, and diarrhea) is the most common disease caused by the organism. Abdominal cramping also may occur. Salmonellosis thus produces the symptoms that are commonly referred to as *food poisoning*. Although food poisoning is usually a mild disease, nausea, vomiting, and diarrhea can lead to dehydration and even death (about 500 deaths per year in the United States). It is important to note that many other organisms (e.g., viruses, *E. coli, Shigella*) and toxins (e.g., botulism, mushroom toxin, pesticides) can produce food poisoning symptoms. However, over 1.4 million cases of salmonellosis occur per year in the United States, and the rest of industrialized countries have similar high rates. Countries with poor sanitation have a much higher incidence of salmonellosis.

Typhoid fever occurs when some of the *Salmonella* organisms (often identified as *S. typhi*) are not killed by the normal human immune defenses (macrophage cells) after they enter the gastrointestinal tract. *Salmonella* then survive and grow in the human spleen, liver, and other organs and may reach the blood (bacteremia). *Salmonella* can be shed from the liver to the gallbladder, where they can continue to survive and be secreted in the patient's feces for up to a year. Symptoms include high fever with temperatures up to 104 °F, sweating, inflammation of the stomach and intestines, and diarrhea. Symptoms usually resolve, but many patients become *Salmonella* carriers. Approximately half of patients develop slow heartbeat (bradycardia), and about 30% of patients get flat, slightly raised red or rose spots on the chest and abdomen. Typhoid fever is also referred to as *enteric fever*.

Paratyphoid fever, also termed *enteric fever*, has symptoms like typhoid, but it is usually not as severe. Subtypes are A, B, and C and vary by having small changes in symptoms, such as more rose spots (A), occurring in conjunction with herpes labialis and gastroenteritis (B), and rarely, with septicemia and abscesses (C).

E. COLI.   *E. coli* is a gram-negative rod-shaped bacterium (Fig. 5.46) that is commonly found in the lower intestine of warm-blooded organisms (endotherms). Most *E. coli* strains are harmless, but some, such as the serotype O157:H7, can cause serious food poisoning in humans and are occasionally responsible for product recalls. The harmless strains are part of the normal flora of the gut and can benefit their hosts by producing vitamin K_2 and by preventing the establishment of pathogenic bacteria within the intestine. *E. coli* are not always confined to the intestine, and their ability to survive for brief periods outside the body makes them an ideal indicator organism to test environmental samples for fecal contamination.

The bacteria can also be grown easily and their genetics is comparatively simple and can be easily manipulated or duplicated through a process of metagenics, making them one of the best studied prokaryotic model organisms and an important species in biotechnology and microbiology. *E. coli* was discovered by the German pediatrician and bacteriologist Theodor Escherich in 1885 and is now classified as part of the Enterobacteriaceae family of Gammaproteobacteria.

E. coli normally colonizes an infant's gastrointestinal tract within 40 h of birth, arriving with food or water or from the individuals handling the child. In the bowel, it adheres

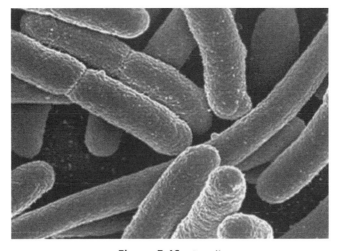

Figure 5.46.   *E. coli.*

to the mucous membrane of the large intestine. It is the primary facultative anaerobe of the human gastrointestinal tract (facultative anaerobes are organisms that can grow in either the presence or the absence of oxygen). As long as these bacteria do not acquire genetic elements encoding for virulence factors, they remain benign commensals.

Nonpathogenic *E. coli* strain Nissle 1917, also known as *Mutaflor*, is used as a probiotic agent in medicine, mainly for the treatment of various gastroenterological diseases, including inflammatory bowel disease.

E. coli O157:H7 was first recognized as a cause of foodborne illness in 1982, when an outbreak of severe bloody diarrhea was traced to contaminate undercooked hamburgers. The organism is found in the intestines of cattle but does not cause illness in cattle.

SHIGELLA. *Shigella* is a genus of bacteria that are a major cause of diarrhea and dysentery, diarrhea with blood and mucus in the stools, throughout the world. The bacteria are transmitted by ingestion of contaminated food or water, or through person-to-person contact. In the body, they can invade and destroy the cells lining the large intestine, causing mucosal ulceration and bloody diarrhea. Apart from diarrhea, symptoms of *Shigella* infection include fever, abdominal cramps, and rectal pain. Most patients recover without complications within 7 days. Shigellosis can be treated with antibiotics, although some strains have developed drug resistance.

Shigella infection is typically via ingestion (fecal-oral contamination); depending on the age and condition of the host, as few as 100 bacterial cells can be enough to cause an infection. *Shigella* causes dysentery that results in the destruction of the epithelial cells of the intestinal mucosa in the cecum and rectum. Some strains produce enterotoxin and Shiga toxin, similar to the verotoxin of *E. coli* O157:H7. Both Shiga toxin and verotoxin are associated with causing hemolytic uremic syndrome.

Shigella invade the host through epithelial cells of the large intestine. With a type III secretion system acting as a biological syringe, the bacterium injects IpaD protein into the host cell, triggering bacterial invasion and the subsequent lysis of vacuolar membranes by the IpaB and IpaC proteins (Fig. 5.47). The bacterium utilizes a mechanism for its motility by which its IcsA protein triggers actin polymerization in the host cell (via N-WASP recruitment of Arp2/3 complexes) in a "rocket" propulsion fashion for cell to cell spread. The most common symptoms are diarrhea, fever, nausea, vomiting, stomach cramps, flatulence, and constipation. The stool may contain blood, mucus, or pus. In rare cases, young children may have seizures. Symptoms can take as long as a week to show up, but most often begin 2–4 days after ingestion. Symptoms usually last for several days but can last for weeks. *Shigella* is implicated as one of the pathogenic causes of reactive arthritis worldwide. Severe dysentery can be treated with ampicillin, TMP-SMX (trimethoprim/sulfamethoxazole), or fluoroquinolones such as ciprofloxacin and with, of course, rehydration. Each of the *Shigella* genomes include a virulence plasmid that encodes conserved primary virulence determinants. The *Shigella* chromosomes share most of their genes with that of *E. coli* K12 strain MG1655.

Shigella species are negative for motility and are nonlactose fermenters (however, *Shigella sonnei* can ferment lactose). They typically do not produce gas from carbohydrates (with the exception of certain strains of *Shigella flexneri*) and tend to be overall biochemically inert. *Shigella* should also be urea hydrolysis negative. When inoculated to a TSI slant, they react as follows: K/A, gas, H_2S. Indole reactions are mixed, positive and negative, with the exception of *S. sonnei*, which is always indole negative.

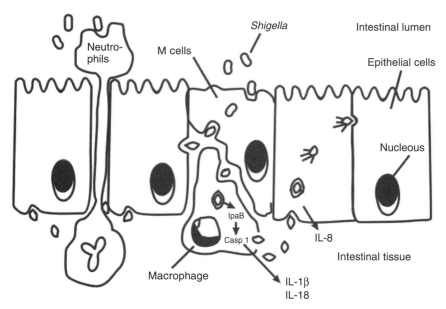

Figure 5.47. Mechanism of *Shigella* invading the host through the epithelial cells of the large intestine.

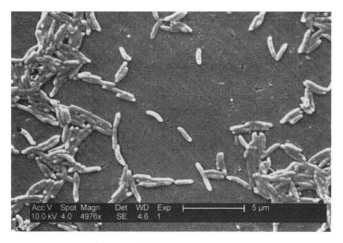

Figure 5.48. *Campylobacter.*

CAMPYLOBACTER. *Campylobacter* is a genus of bacteria that are gram negative, spiral, and microaerophilic (Fig. 5.48). Motile with either uni- or bipolar flagella, the organisms have a characteristic spiral/corkscrew appearance and are oxidase positive. *Campylobacter jejuni* is now recognized as one of the main causes of bacterial foodborne disease in many developed countries. At least a dozen species of *Campylobacter* have been implicated in human disease, with *C. jejuni* and *C. coli* being the most common. *C. fetus* is a cause of spontaneous abortions in cattle and sheep, as well as an opportunistic pathogen in humans. Campylobacteriosis is an infection caused by *Campylobacter*. The common routes of transmission are fecal-oral, ingestion of contaminated food or water, and intake of raw meat. It produces an inflammatory, sometimes bloody, diarrhea, periodontitis or dysentery syndrome, mostly including cramps, fever, and pain. The infection is usually self-limiting and in most cases, symptomatic treatment by reposition of liquid and electrolyte replacement is enough in human infections. The use of antibiotics, on the other hand, is controversial. It lasts for 5–7 days with constant pain and diarrhea. Species of *Campylobacter* include *C. coli, C. concisus, C. curvus, C. fetus, C. gracilis, C. helveticus, C. hominis, C. hyointestinalis, C. insulaenigrae, C. jejuni, C. lanienae, C. lari, C. mucosalis, C. rectus, C. showae, C. sputorum*, and *C. upsaliensis*.

Campylobacter is found in the intestines of many wild and domestic animals. The bacteria are passed in their feces, which can lead to infection in humans via contaminated food and meat (especially chicken), water taken from contaminated sources (streams or rivers near where animals graze), and milk products that have not been pasteurized. Bacteria can be transmitted from person to person when someone comes into contact with the fecal matter from an infected person, especially a child in diapers. Household pets can carry and transmit the bacteria to their owners. Once inside the human digestive system, *Campylobacter* infects and attacks the lining of both the small and large intestines.

The bacteria also can affect other parts of the body. In some cases, particularly in very young patients and those with chronic illnesses or a weak immune system, the bacteria can get into the bloodstream (condition called *bacteremia*).

YERSINIA. *Yersinia* is a genus of bacteria in the family Enterobacteriaceae. *Yersinia* are gram-negative rod-shaped bacteria, a few micrometers long and fractions of a micrometer in diameter, and facultative anaerobes (Fig. 5.49). Some members of *Yersinia* are pathogenic in humans; in particular, *Y. pestis* is the causative agent of plague. Rodents are the natural reservoirs of *Yersinia*; less frequently other mammals serve as the host. Infection may occur either through blood (in the case of *Y. pestis*) or in an alimentary fashion, occasionally via consumption of food products (especially vegetables, milk-derived

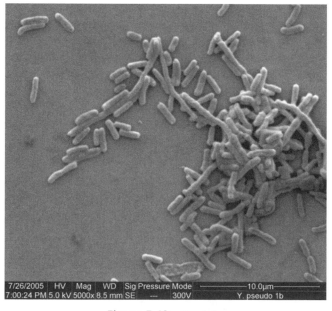

Figure 5.49. *Yersinia*.

products, and meat) contaminated with infected urine or feces. The genus is named after A. E. J. Yersin, a Swiss bacteriologist, who codiscovered the *Y. pestis* bacterium as the causative agent of bubonic plague along with Kitasato Shibasaburō. The special genus *Yersinia* has been recognized since 1971, mainly for taxonomic reasons.

Speculations exist as to whether or not certain *Yersinia* can also be spread via protozoonotic mechanisms since *Yersinia* are known to be facultative intracellular parasites; studies and discussions of the possibility of ameba-vectored (through the cyst form of the protozoan) *Yersinia* propagation and proliferation are now in progress. An interesting feature peculiar to some of the *Yersinia* bacteria is the ability to not only survive but also actively proliferate at temperatures as low as $1-4\,°C$ (e.g., on cut salads and other food products in a refrigerator). *Yersinia* are relatively quickly inactivated by oxidizing agents such as hydrogen peroxide and potassium permanganate solutions. *Y. pestis* is the causative agent of plague. The disease caused by *Y. enterocolitica* is called *yersiniosis*. *Y. pseudotuberculosis* sometimes but rarely causes disease. *Yersinia* may be associated with Crohn's disease, an inflammatory autoimmune condition of the gut. Iranian sufferers of Crohn's disease were more likely to have had earlier exposure to refrigerators at home, consistent with the unusual ability of *Yersinia* to thrive at low temperatures. *Yersinia* is implicated as one of the causes of reactive arthritis worldwide and is also associated with pseudoappendicitis, which is an incorrect diagnosis of appendicitis due to a similar presentation. Some species are *Y. aldovae, Y. aleksiciae, Y. bercovieri, Y. enterocolitica, Y. frederiksenii, Y. intermedia, Y. kristensenii, Y. mollaretii, Y. pestis, Y. pseudotuberculosis, Y. rohdei*, and *Y. ruckeri*.

Y. enterocolitica infections are sometimes followed by chronic inflammatory diseases such as arthritis. *Y. enterocolitica* seems to be associated with autoimmune Graves–Basedow thyroiditis. Although indirect evidence exists, direct causative evidence is limited, and *Y. enterocolitica* is probably not a major cause of this disease but may contribute to the development of thyroid autoimmunity arising for other reasons in genetically susceptible individuals. It has also been suggested that *Y. enterocolitica* infection is not the cause of autoimmune thyroid disease, but rather is only an associated condition, both having a shared inherited susceptibility. More recently, the role for *Y. enterocolitica* has been disputed.

VIBRIO. *Vibrio* is a genus of gram-negative bacteria possessing a curved rod shape (Fig. 5.50), several species of which can cause foodborne infection, usually associated with eating undercooked seafood. Typically found in saltwater, *Vibrio* are facultative anaerobes that test positive for oxidase and do not form spores. All members of the genus are motile and have polar flagella with sheaths. Recent phylogenies have been constructed based on a suite of genes (multilocus sequence analysis). The name *Vibrio* was derived from Filippo Pacini who isolated the microorganism from cholera patients and called them *vibrions* in 1854. Several species of *Vibrio* include clinically important human pathogens. Most disease-causing strains are associated with gastroenteritis but can also infect open wounds and cause septicemia. It can be carried by numerous sea-living animals, such as crabs or prawns, and has been known to cause fatal infections in humans during exposure. Pathogenic *Vibrio* include *V. cholerae* (the causative agent of cholera), *V. parahaemolyticus*, and *V. vulnificus*. *V. cholerae* is generally transmitted via contaminated water. Pathogenic *Vibrio* can cause foodborne infection, usually associated with eating undercooked seafood. *V. vulnificus* outbreaks commonly occur in warm climates, and small, generally lethal, outbreaks occur regularly. An outbreak occurred

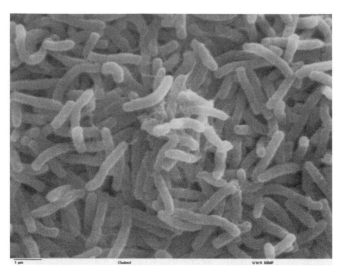

Figure 5.50. *Vibrio.*

in New Orleans after Hurricane Katrina, and several lethal cases occur most years in Florida.

V. parahaemolyticus is also associated with the Kanagawa phenomenon, in which strains isolated from human hosts (clinical isolates) are hemolytic on blood agar plates, while those isolated from nonhuman sources are nonhemolytic. Some common species are *V. adaptatus, V. aerogenes, V. aestuarianus, V. agarivorans, V. albensis, V. alginolyticus, V. anguillarum, V. brasiliensis, V. bubulus, V. calviensis, V. campbellii, V. chagasii, V. cholerae, V. cincinnatiensis, V. coralliilyticus, V. crassostreae, V. cyclitrophicus, V. diabolicus, V. diazotrophicus, V. ezurae, V. fischeri, V. fluvialis, V. fortis, V. furnissii, V.gallicus, V. gazogenes, V. gigantis, V. halioticoli, V. harveyi, V. hepatarius, V. hispanicus, V. hollisae, V. ichthyoenteri, V. indicus, V. kanaloae, V. lentus, V. litoralis, V. logei, V. mediterranei, V. metschnikovii, V. mimicus, V. mytili, V. natriegens, V. navarrensis, V. neonatus, V. neptunius, V. nereis, V. nigripulchritudo, V. ordalii, V. orientalis, V. pacinii, V. parahaemolyticus, V. pectenicida, V. penaeicida, V. pomeroyi, V. ponticus, V. proteolyticus, V. rotiferianus, V. ruber, V. rumoiensis, V. salmonicida, V. scophthalmi, V. splendidus, V. superstes, V. tapetis, V. tasmaniensis, V. tubiashii, V. vulnificus, V. wodanis,* and *V. xuii.*

Many *Vibrio* are also zoonotic. They cause disease in fish and shellfish and are common causes of mortality among domestic marine life. *V. fischeri, Photobacterium phosphoreum,* and *V. harveyi* are notable for their ability to communicate. Both *V. fischeri and P. phosphoreum* are symbiotes of other marine organisms (typically jellyfish, fish, or squid) and produce light via bioluminescence through the mechanism of quorum sensing. *V. harveyi* is a pathogen of several aquatic animals and is notable as the cause of luminous vibriosis in shrimps (prawns). The "typical," early-discovered *Vibrio* such as *V. cholerae* have a single polar flagellum (monotrichous) with sheath. Some species such as *V. parahaemolyticus* and *V. alginolyticus* have both a single polar flagellum with sheath and thin flagella projecting in all directions (peritrichous), and the other species such as *V. fischeri* have tufts of polar flagella with sheath (lophotrichous).

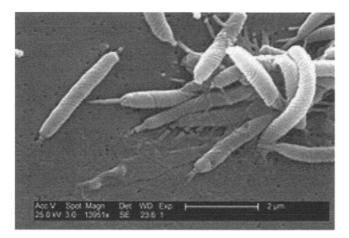

Figure 5.51. *Helicobacter*.

HELICOBACTER. *Helicobacter* is a genus of gram-negative bacteria possessing a characteristic helix shape (Fig. 5.51). They were initially considered to be members of the *Campylobacter* genus, but since 1989, they have been grouped in their own genus. Some species have been found living in the lining of the upper gastrointestinal tract, as well as the liver of mammals and some birds. The most widely known species of the genus is *H. pylori* that infects up to 50% of the human population. Some strains of this bacterium are pathogenic to humans, as it is strongly associated with peptic ulcers, chronic gastritis, duodenitis, and stomach cancer. It also serves as the type species of the genus.

Helicobacter spp. are able to thrive in the very acidic mammalian stomach by producing large quantities of the enzyme urease, which locally raises the pH from approximately 2 to a more biocompatible range of 6–7. Bacteria belonging to this genus are usually susceptible to antibiotics such as penicillin, microaerophilic (require small amounts of oxygen), and fast moving with their flagella.

The bacteria *H. pylori* usually do not cause problems in childhood. However, if left untreated, the bacteria can lead to digestive illnesses, including gastritis (the irritation and inflammation of the lining of the stomach), peptic ulcer disease (characterized by sores that form in the stomach or the upper part of the small intestine, called the *duodenum*), and even stomach cancer later in life. These bacteria are found worldwide, especially in developing countries, where up to 10% of children and 80% of adults can have laboratory evidence of an *H. pylori* infection, usually without having symptoms. In kids, symptoms of gastritis may include nausea, vomiting, and frequent complaints about pain in the abdomen. However, these symptoms are seen in many childhood illnesses.

H. pylori, which used to be called *Campylobacter pylori*, can also cause peptic ulcers (commonly known as *stomach ulcers*). In older kids and adults, the most common symptom of peptic ulcer disease is a gnawing or burning pain in the abdomen, usually in the area below the ribs and above the navel. This pain often gets worse on an empty stomach and improves as soon as the person eats food, drinks milk, or takes antacid medicine. Kids who have peptic ulcer disease can have ulcers that bleed, causing hematemesis (bloody vomit or vomit that looks like coffee grounds), or melena (stool that is black, bloody, or looks like tar). Younger children with peptic ulcer disease may not have symptoms as clear-cut, so their illness may be harder to diagnose.

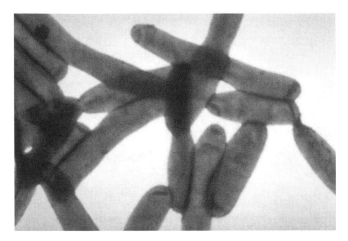

Figure 5.52. *Legionella.*

LEGIONELLA. *Legionella* is a gram-negative bacterium, including species that cause legionellosis or Legionnaires' disease, most notably *L. pneumophila* (Fig. 5.52). It may be readily visualized with a silver stain. *Legionella* is common in many environments, with at least 50 species and 70 serogroups identified. The side chains of the cell wall carry the bases responsible for the somatic antigen specificity of these organisms. The chemical composition of these side chains both with respect to components and arrangement of the different sugars determines the nature of the somatic or O antigen determinants, which are the essential means of serologically classifying many gram-negative bacteria.

Legionella acquired its name after the July 1976 outbreak of a then-unknown "mystery disease" sickened 221 persons, causing 34 deaths. The outbreak was first noticed among people attending a convention of the American Legion, a congressionally chartered association of US military veterans. The convention in question occurred in Philadelphia during the US Bicentennial year. This epidemic among US war veterans, occurring in the same city as, and within days of the two hundredth anniversary of, the signing of the Declaration of Independence, was widely publicized and caused great concern in the United States. On January 18, 1977, the causative agent was identified as a previously unknown bacterium. *Legionella* is traditionally detected by culture on buffered charcoal yeast extract (BCYE) agar. *Legionella* requires the presence of cysteine to grow and therefore does not grow on common blood agar media used for laboratory-based total viable counts or on-site displides. Common laboratory procedures for the detection of *Legionella* in water concentrate the bacteria (by centrifugation and/or filtration through 0.2-μm filters) before inoculation onto a charcoal yeast extract agar containing antibiotics (e.g., glycine vancomycim polymixin cyclohexamide, GVPC) to suppress other flora in the sample. Heat or acid treatment is also used to reduce interference from other microbes in the sample.

After incubation for up to 10 days, suspect colonies are confirmed as *Legionella* if they grow on BCYE containing cysteine, but not on agar without cysteine. Immunological techniques are then commonly used to establish the species and/or serogroups of bacteria present in the sample. Many hospitals use the Legionella urinary antigen test for initial detection when *Legionella pneumonia* is suspected. Some of the advantages offered by this test are that the results can be obtained in a matter of hours rather than the 5 days required for culture and that a urine specimen is generally more easily obtained than a sputum specimen. One disadvantage is that the urine antigen test only detects antibodies toward *L. pneumophila*; only a culture will detect infection by the other *Legionella* species. New techniques for the rapid detection of *Legionella* in water samples are emerging, including the use of PCR and rapid immunological assays. These technologies can typically provide much faster results. *Legionella* live within ameba in the natural environment. *Legionella* species are the causative agent of the human Legionnaires' disease and the lesser form, Pontiac fever. *Legionella* transmission is via aerosols—the inhalation of mist droplets containing the bacteria. Common sources include cooling towers, swimming pools (especially in Scandinavian countries and other countries such as Northern Ireland), domestic hot-water systems, fountains, and similar disseminators that tap into a public water supply. Natural sources of *Legionella* include freshwater ponds and creeks. Person-to-person transmission of *Legionella* has not been demonstrated.

Once inside the host, incubation may take up to 2 weeks. Initial symptoms are flu-like, including fever, chills, and dry cough. Advanced stages of the disease cause problems with the gastrointestinal tract and the nervous system and lead to diarrhea and nausea, respectively. Other advanced symptoms of pneumonia may also present.

However, the disease is generally not a threat to most healthy individuals and tends to lead to harmful symptoms only in those with a compromised immune system and in the elderly. Consequently, it should be actively checked for in the water systems of hospitals and nursing homes. The Texas Department of State Health Services provides recommendations for hospitals to detect and prevent the spread of nosocomial infections due to *Legionella*. According to the journal *Infection Control and Hospital Epidemiology*, hospital-acquired legionella pneumonia has a fatality rate of 28% and the source is the water distribution system.

OPPORTUNISTIC PATHOGENS. They are infectious microorganisms that are normally a commensal or do not harm their host but can cause disease when the host's resistance is low. The following are some common examples.

Candida albicans. It is an yeastlike fungus that abounds in Nature and is part of the gut flora (Fig. 5.53). However, it is an opportunistic pathogen that can cause infection to its host when the host immune system is compromised or when the balance in gut flora is disrupted leading to less competition and (eventually) to its overgrowth. *Candida albicans* is the species responsible for many *Candida*-related infections generally called *candidosis* (or *candidiasis*) of the mucous parts of the mouth, vagina, skin, esophagus, and other organs.

Staphylococcus aureus. S. aureus (Fig. 5.54) can cause a range of illnesses from minor skin infections, such as pimples, impetigo, boils (furuncles), cellulitis, folliculitis, carbuncles, scalded skin syndrome, and abscesses, to life-threatening diseases such as pneumonia, meningitis, osteomyelitis, endocarditis, toxic shock syndrome (TSS), chest pain, bacteremia, and sepsis. Its incidence is from skin, soft tissue, respiratory, bone, joint, endovascular to wound infections. It is still one of the five most common causes of nosocomial infections, often causing postsurgical wound infections. Abbreviated to *S. aureus* or *Staph aureus* in medical literature, *S. aureus* should not be confused with the similarly named and similarly dangerous (and also medically relevant) species of the genus *Streptococcus. S. aureus* may occur as a commensal on skin; it also occurs

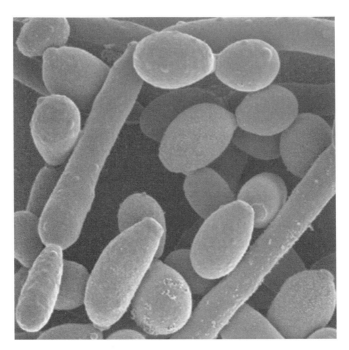

Figure 5.53. *Candida albicans.*

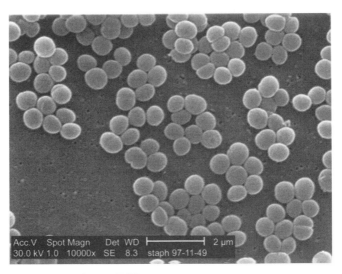

Figure 5.54. *Staphylococcus aureus.*

in the nose frequently (in about one-third of the population) and throat less commonly. The occurrence of *S. aureus* under these circumstances does not always indicate infection and, therefore, does not always require treatment (indeed, treatment may be ineffective and recolonization may occur). It can survive on domesticated animals, such as dogs, cats, and horses, and can cause bumblefoot in chickens. It can survive for hours to days, weeks, or even months on dry environmental surfaces depending on the strain. It can host phages, such as Panton–Valentine leukocidin, that increase its virulence.

S. aureus can infect other tissues when barriers have been breached (e.g., skin or mucosal lining). This leads to furuncles (boils) and carbuncles (a collection of furuncles). In infants, *S. aureus* infection can cause a severe disease called *staphylococcal scalded skin syndrome (SSSS)*. *S. aureus* alv avian leukosis virus infections can be spread through contact with pus from an infected wound, skin-to-skin contact with an infected person by producing hyaluronidase that destroys tissues, and contact with objects such as towels, sheets, clothing, or athletic equipment used by an infected person. Deeply penetrating *S. aureus* infections can be severe. Prosthetic joints put a person at particular risk for septic arthritis, staphylococcal endocarditis (infection of the heart valves), and pneumonia, which may spread rapidly.

Pseudomonas aeruginosa. An opportunistic, nosocomial pathogen of immunocompromised individuals, *P. aeruginosa* typically infects the pulmonary tract, urinary tract, burns, and wounds, and also causes other blood infections. *P. aeruginosa* secretes a variety of pigments, including pyocyanin (blue-green), fluorescein (yellow-green and fluorescent, now also known as *pyoverdin*), and pyorubin (red-brown). King, Ward, and Raney developed Pseudomonas Agar P (also known as *King A media*) for enhancing pyocyanin and pyorubin production and Pseudomonas Agar F (also known as *King B media*) for enhancing fluorescein production.

P. aeruginosa is often preliminarily identified by its pearlescent appearance and grape-like or tortillalike odor *in vitro*. Definitive clinical identification of *P. aeruginosa* often includes identifying the production of both pyocyanin and fluorescein, as well as its ability to grow at 42 °C. *P. aeruginosa* is capable of growth in diesel and jet fuel, hence it is known as *hydrocarbon-utilizing microorganism* (or HUM bug), causing microbial corrosion. It creates dark gellish mats sometimes improperly called *algae* because of their appearance.

5.11.1.2. Parasites

Parasitology is the study of parasites, their hosts, and the relationship between them. As a biological discipline, the scope of parasitology is not determined by the organism or environment in question, but by their way of life. This means it forms a synthesis of other disciplines and draws on techniques from fields such as cell biology, bioinformatics, biochemistry, molecular biology, immunology, genetics, evolution, and ecology. Parasites can provide information about host population ecology.

In fisheries biology, for example, parasite communities can be used to distinguish distinct populations of the same fish species coinhabiting a region. In addition, parasites possess a variety of specialized traits and life history strategies that enable them to colonize hosts. Understanding these aspects of parasite ecology, of interest in their own right, can illuminate parasite-avoidance strategies practiced by hosts.

PROTOZOA. The protozoa are single-celled animals and the smallest of all animals. Most of them can only be seen under a microscope (Fig. 5.55). They do breathe, move, and reproduce like multicelled animals. They live in water or at least where it is damp. Animals in this group include the paramecium, euglena, and ameba. Protozoa

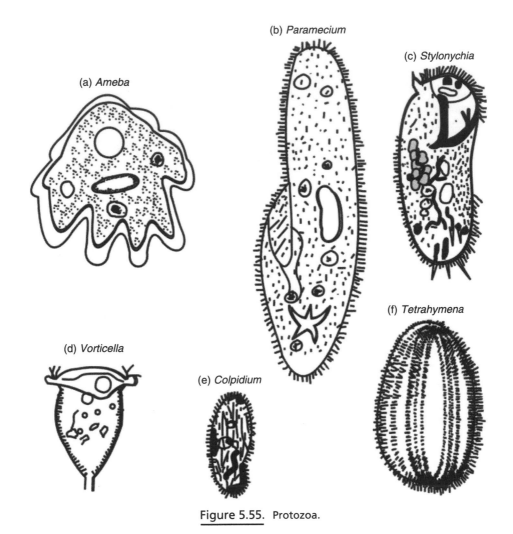

(a) *Ameba*

(b) *Paramecium*

(c) *Stylonychia*

(d) *Vorticella*

(e) *Colpidium*

(f) *Tetrahymena*

Figure 5.55. Protozoa.

is a subkingdom of microorganisms that are classified generally as unicellular nonfungal eukaryotes. Protozoans are a major component of the ecosystem.

The word protozoan is originally an adjective and is also used as a noun. While there is not any exact definition for the term *protozoan*, most scientists use the word to refer to a unicellular heterotrophic protist, such as the ameba and ciliate. The term *algae* is used for microorganisms that photosynthesize. However, the distinction between protozoa and algae is often vague. For example, the alga *Dinobryon* has chloroplasts for photosynthesis, but it can also feed on organic matter and is motile. Protozoans are referred to generally as *animal-like protists*. The protozoa are a paraphyletic group; some protozoa have life stages alternating between proliferative stages (e.g., trophozoites) and dormant cysts. As cysts, protozoa can survive harsh conditions, such as exposure to extreme temperatures or harmful chemicals, or long periods without access to nutrients, water, or oxygen. Being a cyst enables parasitic species to survive outside a host and allows their transmission from one host to another. When protozoa are in the form of trophozoites (in Greek, *tropho* means to nourish), they actively feed. The conversion of a trophozoite to cyst form is known as *encystation*, while the process of transforming back into a trophozoite is known as *excystation*.

Protozoa can reproduce by binary or multiple fission. Some protozoa reproduce sexually, some asexually, while some use a combination (e.g., Coccidia). An individual protozoan is hermaphroditic does not include all genetic relatives of the group). They constitute their own "kingdom" by the Integrated Taxonomic Information System 2009 classification. Protozoa were previously often grouped in the kingdom of Protista, together with the plantlike algae, funguslike slime molds, and animal-like protozoa. As a result of the twenty-first century systematics, protozoa, along with ciliates, mastigophorans, and apicomplexans, are arranged as animal-like protists. With the possible exception of myxozoa, protozoa are not categorized as metazoa. Protozoans are unicellular organisms and are often called the *animal-like protists* because they subsist entirely on other organisms for food. Most protozoans can move about on their own. Amebas, paramecia, and trypanosomes are all examples of animal-like protists.

NEMATODES. The nematodes are the most diverse phylum of pseudocoelomates and one of the most diverse of all animals. Nematode species are very difficult to distinguish; over 28,000 have been described, of which over 16,000 are parasitic. It has been estimated that the total number of nematode species might be approximately 1,000,000. Unlike cnidarians or flatworms, roundworms have a digestive system that is like a tube with openings at both ends. Nematodes commonly parasitic on humans include ascarids (*Ascaris*), filarias, hookworms, pinworms (*Enterobius*), and whipworms (*Trichuris trichiura*). The species *Trichinella spiralis*, commonly known as the *trichina worm*, occurs in rats, pigs, and humans and is responsible for the disease trichinosis. *Baylisascaris* usually infests wild animals but can be deadly to humans as well. *Dirofilaria immitus* are heartworms known for causing heartworm disease by inhabiting the heart, arteries, and lungs of dogs and some cats. *Haemonchus contortus* is one of the most abundant infectious agents in sheep around the world, causing great economic damage to sheep farms. In contrast, entomopathogenic nematodes parasitize insects and are considered by humans to be beneficial.

One form of nematode is entirely dependent on fig wasps, which are the sole source of fig fertilization. They prey on the wasps, riding them from the ripe fig of the wasp's birth to the fig flower of its death, where they kill the wasp, and their offspring await the birth of the next generation of wasps as the fig ripens.

Nematodes are structurally simple organisms. Adult nematodes are composed of approximately 1000 somatic cells and potentially hundreds of cells associated with the reproductive system.

Nematodes have been characterized as a tube within a tube (Fig. 5.56), referring to the alimentary canal, which extends from the mouth on the anterior end to the anus located near the tail. Nematodes possess digestive, nervous, excretory, and reproductive systems, but lack a discrete circulatory or respiratory system. In size they range from 0.3 mm to over 8 m.

CESTODES. Cestodes are tapeworms that belong to the class Cestoidea, phylum Platyhelminthes, and subclass Cestoda. They are specialized flatworms, looking very much like a narrow piece of adhesive tape. Tapeworms are the largest, and among the oldest, of the intestinal parasites that have plagued humans and other animals since time began. Found all over the world, tapeworms exist in many different forms, but they have no

Figure 5.56. Nematodes are characterized by the tubelike structure.

close relatives living outside animal hosts. Tapeworms have neither a mouth like the fluke nor a head or a digestive tract with digestive enzymes. The ends differ but neither have any organs or sensors that could be associated with what is commonly thought of being a "head" (Fig. 5.57). However, through a segment called *scolex*, they are able to absorb predigested food. The scolex attaches to the intestinal wall by hooks or suckers. The body contains hundreds of segments (proglottids), and each is a sexually complete unit that can reproduce, if necessary. Some tapeworms have reached lengths of more than 10 m (30 ft) with a life span, inside a host, of 30 years or more. Cestodaria is the unsegmented subclass of tapeworm, affecting various fishes and some reptiles.

Tapeworms are dependent on two hosts for their development, one human and the other animal. Larvae are found in animal hosts, while the adult worm is found in humans. However, there are two species where this development is reversed: *Echinococcus granulosis* and *Echinococcus multilocularis* differ from other tapeworms in that it is the adult worm that infects an animal host, while the larval form produces slow-growing cysts in humans. This condition is known as *echinococcosis* or *hydatid disease*, which requires surgical intervention to remove the cysts. Human infection comes as a result of eating insufficiently cooked meat (especially beef, pork, and fish), where the larvae are buried in the tissues of the animal involved. The fleas of both dogs and cats can also transmit tapeworm larvae. Rice-shaped particles that resemble pumpkin seeds in the stool can be a sign of a dog tapeworm, which can easily be mistaken for pinworms. It is important to deflea household animals frequently.

Some of the conditions and symptoms that tapeworms can cause include the following: mineral imbalances, abnormal thyroid function, intestinal gas, blood sugar imbalances, bloating, jaundice, fluid buildup, dizziness, fuzzy thinking, hunger pains, poor digestion, allergies, sensitivity to touch, weight changes, and symptoms of pernicious anemia. Treatment can take months before the entire worm is expelled. It is suggested that long periods of fasting not be undertaken since tapeworms cannot be starved but will only leave the person feeling weak and nauseated. It is better to eat foods that tapeworms do

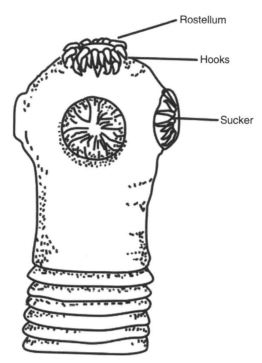

Figure 5.57. Cestodes: the head and the sucker are characteristics of the tapeworms.

not like, such as onions and garlic. This weakens the worm so that it loses its grip. Then it can be easily dislodged and expelled.

Tapeworms known as *Spirometra* and a roundworm called *Gnathostoma* can develop in humans after eating raw snake. *Spirometra* causes a bizarre eye disease called *sparganosis*. In the Far East, sometimes poultices of raw frogs or snake muscle are put directly on the eye to cure various ailments. Tapeworm larvae then migrate into the tissue around the eye. This is not solely a condition possible only in underdeveloped countries. For example, raw meat is often used in the western world for bruises and black eyes. The other unusual tapeworm is the gnathostomes worm, which attaches to the stomach wall of animals and humans. It eventually passes the eggs of a tiny one-eyed bug called *cyclops*. Fish, frogs, or snakes later eat these infected bugs, which ultimately infect that animal.

TREMATODES. Trematodes are flukes of the class Trematoda and phylum Platyhelminthes. Important ones affecting man belong to the genera *Schistosoma* (blood fluke), *Echinostoma* (intestinal fluke), *Fasciolopsis* (liver fluke), *Gastrodiscoides* (intestinal fluke), *Heterophyes* (intestinal fluke), *Metagonimus* (intestinal fluke), *Clonorchis* (Asiatic liver fluke), *Fasciola* (liver fluke), *Dicrocoelium* (liver fluke), *Opisthorchis* (liver fluke), and *Paragonimus* (lung fluke). Man usually becomes infected after ingesting insufficiently cooked fish, crustaceans, or vegetables that contain their larvae. The cycle begins when larvae are released into freshwater by infected snails. The free-swimming larvae can then directly penetrate the skin of humans while swimming or can be ingested after encysting in or on various edible vegetation, fish, or crustaceans.

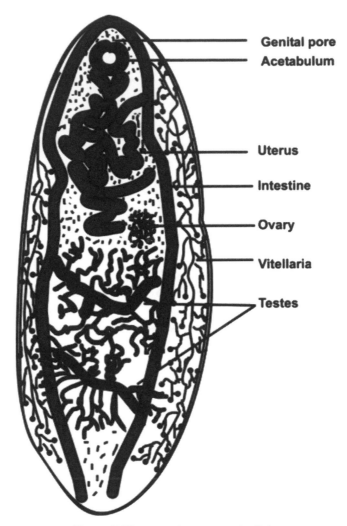

Genital pore
Acetabulum

Uterus

Intestine

Ovary

Vitellaria

Testes

Figure 5.58. Typical structure of a fluke.

The oval-shaped fluke (sometimes called a *flatworm*) has a tough outer body layer called a *tegument* that covers layers of circular, longitudinal, and diagonal muscles that protect it from the human digestive tract (Fig. 5.58). Some can inhabit the liver, bile duct, or lymph vessels. They can be several inches long, an inch or so wide, and only thick enough to hold themselves together. Below are just a very few examples of the thousands known.

Blood flukes. Schistosoma japonicum, S. mansoni, S. haematobium are the three species of blood flukes (schistosomes) that cause the disease schistosomiasis, which infects about 200 million people worldwide. *S. japonicum* is found in Asia; *S. mansoni* occurs in Africa, the Eastern Mediterranean, the Caribbean, and South America; and *S. haematobium* is found in Egypt. Freshwater snails act as intermediate hosts in the life cycle of these blood flukes. Snails release the larvae into the water, where they can penetrate the skin of swimmers or bathers. The parasites burrow into the skin and are then carried into the bloodstream to be taken to the liver, intestines, or bladder. There are two forms of schistosomiasis. First, in which inflammation begins when the worms lodge in the lining of the intestine or liver. Second, in which the bladder and urinary tract can become fatally infected by worms as they lodge in the walls. Travelers to Africa, especially, are warned not to bath, wade, or swim in freshwater because of possible infestations by blood flukes. Infection causes fever and chills and elevates the number of white blood cells (eosinophils), as well as produces abdominal pain resulting from enlargement of the liver and spleen. Often, these symptoms do not show up for 4–8 weeks after exposure and, therefore, may not be associated with the possibility of parasite infestation while on vacation.

Liver fluke. *Clonorchis sinensis* is common in the Orient and Hawaii and is transmitted through the ingestion of raw, dried, salted, pickled, or undercooked fish. Snails, carp, and 40 additional fish species have been known to be the intermediate host to this fluke. In humans, it inhabits the bile ducts of the liver, causing it to enlarge and become tender, as well as producing chills, fever, jaundice, and a type of hepatitis.

Oriental lung fluke. *Paragonimus westermani* is found mainly in the Far East, producing the disease called *paragonimiasis*. Humans acquire the fluke by ingesting infected crabs and crayfish that have not been sufficiently cooked or are served raw. The adult worms go to the lungs, and, sometimes, the brain, where seizures similar to epilepsy can occur. Symptoms include an occasional mild cough, producing peculiar rusty brown sputum. The lung fluke can perforate lung tissue and deplete oxygen supplies to the entire bloodstream. Symptoms often resemble those of pulmonary tuberculosis.

Sheep liver fluke. *Fasciola hepatica* is more common in Central and South America, parts of Africa, Asia, and Australia. Infection is usually acquired from eating the larva worms encysted on such aquatic vegetation as watercress. Worms migrate to the liver and bile ducts, where they produce upper right quadrant abdominal pain, liver abscesses, and fibrosis.

Intestinal fluke. *Fasciolopsis buski* is more common in Southeast Asia, Australia, and Latin America. Transmission occurs when individuals bite into the unpeeled outer skin of plants that harbor encysted larvae. Such plants can be water chestnuts, bamboo shoots, and lotus plant roots because they are often cultivated in ponds and streams infected by animal waste. Adult flukes live in the duodenum (the shortest and widest part of the small intestine) and jejunum (connects the duodenum and the ileum, which opens into the large intestine), where they cause ulceration. Symptoms include diarrhea, nausea, vomiting, abdominal pain, and facial and abdominal edema.

Emerging Pathogens. Human development and population growth exert many and diverse pressures on the quality and quantity of water resources and on the access to them. Nowhere are the pressures felt so strongly as at the interface of water and human health. Infectious, water-related diseases are a major cause of morbidity and mortality worldwide. Although a significant proportion of this immense burden of disease is caused by "classical" water-related pathogens, such as typhoid and cholera, newly recognized pathogens and new strains of established pathogens are being discovered that present important additional challenges to both the water and public health sectors. Between 1972 and 1999, 35 new agents of disease were discovered and many more have reemerged after long periods of inactivity or are expanding into areas where they have not previously been reported. Among this group are pathogens that may be transmitted by water. Understanding why pathogens emerge or reemerge is fundamental to effective water resource management and drinking-water treatment and delivery and has become a priority for many national and international organizations. It is also important to be able to gauge the risk from any emerging disease.

The perceived severity of risk and significance of an emerging infectious disease may be so far removed from reality that there is potential for inappropriate allocation of resources. This can have repercussions for countries at all stages of development. Emerging pathogens are those that have appeared in a human population for the first time or have occurred previously but are increasing in incidence or expanding into areas where they have not previously been reported, usually over the last 20 years (WHO, 1997). Reemerging pathogens are those whose incidence is increasing as a result of long-term changes in their underlying epidemiology. By these criteria, 175 species of infectious agents from 96 different genera are classified as emerging pathogens. Of this group, 75% are zoonotic species.

5.11.1.3. Viruses

A virus is a small infectious agent that can replicate only inside the living cells of organisms. Most viruses are too small to be seen directly with a light microscope (Fig. 5.59). Viruses infect all types of organisms, from animals and plants to bacteria and archaea. About 5000 viruses have been described in detail, although there are millions of different types. Viruses are found in almost every ecosystem on Earth and are the most abundant type of biological entity. The study of viruses is known as *virology*, a subspeciality of microbiology.

Virus particles (known as *virions*) consist of two or three parts: the genetic material made from either DNA or RNA, long molecules that carry genetic information; a protein coat that protects these genes; and in some cases, an envelope of lipids that surrounds the protein coat when they are outside a cell. The shapes of viruses range from simple helical and icosahedral forms to more complex structures (Fig. 5.60). The average virus is about one-hundredth the size of the average bacterium.

Viruses are found wherever there is life and have probably existed since living cells first evolved. The origin of viruses is unclear because they do not form fossils, so molecular techniques have been the most useful means of investigating how they arose. These techniques rely on the availability of ancient viral DNA or RNA, but, unfortunately, most of the viruses that have been preserved and stored in laboratories are less than 90 years old. Viruses display a wide diversity of shapes and sizes, called *morphologies*. Generally, viruses are much smaller than bacteria. Most viruses that have been studied have a diameter between 10 and 300 nm. Some filoviruses have a total length of up to 1400 nm; their diameters are only about 80 nm. Most viruses cannot be seen with a light microscope, so scanning and transmission electron microscopes are used to visualize virions. To increase the contrast between viruses and the background, electron-dense "stains" are used. These are solutions of salts of heavy metals, such as tungsten, that scatter the electrons from regions covered with the stain. When virions are coated with

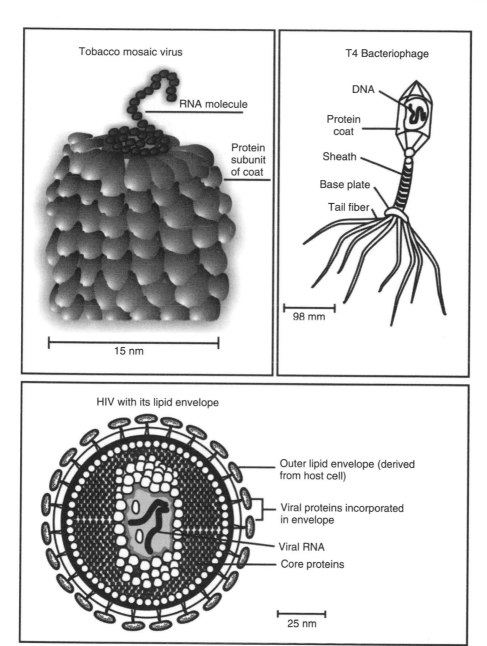

Figure 5.59. Different types of viruses.

stain (positive staining), fine detail is obscured. Negative staining overcomes this problem by staining the background only.

A complete virus particle, known as a *virion*, consists of nucleic acid surrounded by a protective coat of protein called a *capsid*. These are formed from identical protein subunits called *capsomers*. Viruses can have a lipid "envelope" derived from the host cell membrane. The capsid is made from proteins encoded by the viral genome, and its shape serves as the basis for morphological distinction. Virally coded protein subunits

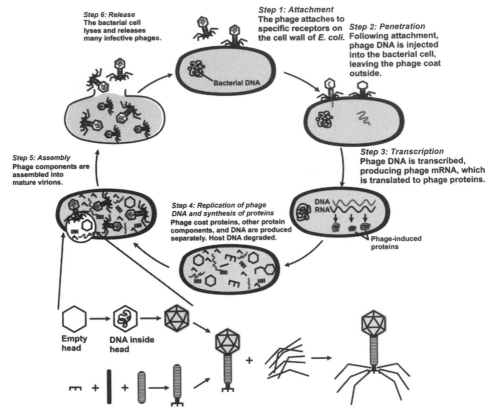

Figure 5.60. Life cycle of bacteriophage.

will self-assemble to form a capsid, generally requiring the presence of the virus genome. Complex viruses code for proteins that assist in the construction of their capsid. Proteins associated with nucleic acid are known as *nucleoproteins*, and the association of viral capsid proteins with viral nucleic acid is called a *nucleocapsid*. The capsid and entire virus structure can be mechanically (physically) probed through atomic force microscopy. The life cycle of a bacteriophage is shown in Fig. 5.60.

ENTERIC VIRUSES. Enteric viruses are extremely small particles ranging from 20 to 85 nm in diameter. In comparison, a human red blood cell averages 7600 nm in diameter. Each virus contains a single type of nucleic acid, either RNA or DNA, which is enclosed by a protein "shell" called a *capsid*. Replication of viruses can only take place in a living host cell. The virus is capable of using its genetic information to commandeer the host cell's machinery to produce more virus particles. In the case of enteric viruses, these particles are released into the host's feces and subsequently find their way into the external environment.

Transmission. Various modes of transmission are observed with the enteric viruses (Fig. 5.61). The most common mode of enteric virus transmission is by person-to-person contact. Small children are the most susceptible because of their close contact with other

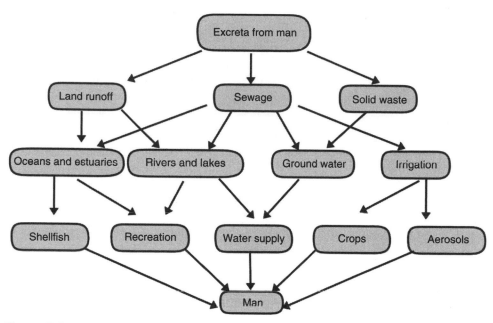

Figure 5.61. Routes of enteric virus transmission. *Source:* Adapted from Rao VC, Melnick JL. 1986.

TABLE 5.7. Diseases Caused by Human Enteric Viruses Transmitted by Water

Genus	Popular Name	Illnesses
Enterovirus	Poliovirus, coxsackieviruses A and B, echovirus	Paralysis, meningitis, fever, herpangina, meningitis, fever, respiratory disease, hand, foot and mouth disease, myocarditis, heart anomalies, rash, pleurodynia, diabetes, gastroenteritis
Hepatovirus	Hepatitis A	Hepatitis
Reovirus	Human reovirus	Unknown
Rotavirus	Human rotavirus	Gastroenteritis
Mastadenovirus	Human adenovirus	Gastroenteritis, respiratory disease, conjunctivitis
Calicivirus	Human calicivirus, Norwalk virus, hepatitis E	Gastroenteritis, fever, small round structured virus (SRSV), hepatitis
Coronavirus	Human coronavirus	Gastroenteritis, respiratory disease

children and their less than optimal hygienic habits. Adults are generally less subject to infection because of the immunity acquired by previous exposure to the virus. Enteric viruses can be transmitted by water (Table 5.7).

Diagnosis. Clinical diagnosis of a viral infection requires identification of the virus in feces or body fluids or the viral antibodies present in serum of the patient. Antibodies detected in the serum indicate present or past infection. Active infections are determined somewhat after the fact by comparing the concentration of antibodies in samples taken a few weeks apart (i.e., acute- and convalescent-phase serum samples).

Environmental detection. In treated wastewater, the secondary effluent (activated sludge) will usually have culturable virus concentrations of less than 10 per 1000 ml. The number of viruses may in actuality be greater, but based on the virus assay methods presently available, these are the orders of magnitude observed. Owing to the low numbers of viruses expected in water and wastewater and restrictions in the viral assay procedure, virus concentration estimation must be attempted. Large sample volumes, as much as 1500 l, are passed through filters in such a manner that viruses present are adsorbed to the filter medium. Most often these are 10-in spun-glass filters that are electropositive in charge, through which the sample water is passed at a rate of approximately 1 gal/min.

Astroviruses. Astroviruses infect animals and humans and cause gastroenteritis. Some strains replicate in cell culture. They are generally associated with infection in children younger than 1 year but may also cause a mild infection in adults. Epidemiological evidence of transmission by food is limited, but infections via contaminated shellfish and water have been reported. Their infectious dose is less than 100 virus particles. Astroviruses survive heating for 30 min at 50 °C.

Hepatitis E virus. Hepatitis E virus belongs to the calicivirus group and is nonculturable. It occurs widely in Asia, Africa, and Latin America, where waterborne outbreaks are common. It has rarely been identified elsewhere. The virus infects the liver, and symptoms of hepatitis are produced following a 22- to 60-day incubation period. The disease is self-limiting and does not progress to a carrier or chronic state. Transmission is generally via fecally contaminated water, and evidence for foodborne transmission has not been documented.

Picornaviruses. This group includes poliovirus, coxsackievirus B, and echoviruses, many of which are culturable. They do not cause gastroenteritis but are transmitted by the fecal-oral route and excreted in feces. Polioviruses were the first to be recognized as foodborne. Wild strains are now rare, and New Zealand is a registered WHO polio-free zone. Outbreaks of foodborne illness associated with coxsackievirus and echovirus have beenreported.

Adenovirus. Of the many types of adenovirus, only two types, 40 and 41, are generally associated with fecal-oral spread and gastroenteritis (especially in children). Most infections are subclinical or mild. The enteric adenoviruses, types 40 and 41, are difficult to grow in cell culture, whereas most other nonfecal types are culturable. Transmission is generally via fecally contaminated water, and evidence for foodborne transmission has not been documented.

RESPIRATORY VIRUSES. Viral infections commonly affect the upper or lower respiratory tract. Although these infections can be classified by the causative virus (e.g., influenza), they are generally classified clinically according to syndrome (e.g., the common cold, bronchiolitis, croup). Although specific pathogens commonly cause characteristic clinical manifestations (e.g., rhinovirus typically causes the common cold and respiratory syncytial virus (RSV) typically causes bronchiolitis), each can cause many of the viral respiratory syndromes (Table 5.8).

TABLE 5.8. Causes of Common Respiratory Syndromes

Syndrome	Common Causes	Less Common Causes
Bronchiolitis	RSV	Influenza, parainfluenza, adenoviruses, rhinoviruses
Common cold	Rhinoviruses, coronaviruses	Influenza, parainfluenza, enteroviruses, adenoviruses, human metapneumonviruses, RSV
Pneumonia	Influenza viruses, RSV, adenoviruses	Parainfluenza viruses, enteroviruses, rhinoviruses, human metapneumoviruses
Influenza-like illness	Influenza viruses	Parainfluenza viruses, adenoviruses

Abbreviation: RSV, respiratory syncytial virus.

Respiratory syncytial virus is spread by coughing and sneezing, close contact with sick patients, or hand contamination. Infection develops in caregivers who touch their eyes or nose with contaminated fingers.

Ulloa-Gutierrez reported that *metapneumovirus* was identified as a cause of acute upper and lower respiratory tract infections in children and adults worldwide, with most episodes occurring during winter. Most children have been infected by 5 years of age. The illness in young children may be life-threatening bronchiolitis or pneumonia. Patterns of adult infection are not well understood.

Influenza viruses cause epidemic respiratory illness every winter in most countries. Influenza often begins with cold symptoms and progresses to involve the lungs. Most patients develop a chronic cough that can last for weeks. Pneumonia can develop and is a common cause of death among more susceptible people.

Much publicity has been given to the possibility of an emerging especially virulent strain that will increase the death toll from thousands per year to millions in the United States and Canada. Some virologists were concerned that influenza virus epidemics in birds would produce a newly virulent human virus. WHO warned that the world is not prepared for the next pandemic. As of January 2006, the strain of avian influenza A (H5N1), has been identified in only 148 humans (of which, 79 were fatal), who acquired the infection via direct contact with infected birds. The strain was first detected in Hong Kong in 1997 and has spread through Southeast Asia and then in Russia and Turkey. In 2009, an H1N1 variant (swine flu) emerged and caused another media frenzy; the WHO declared a "pandemic," and despite reports of a relatively mild illness with a low mortality rate, news anchors began to refer to a *deadly virus*. The positive aspect of the scare tactics was increased international cooperation in monitoring the spread of the virus and increased funding of vaccine development. Some of the fear was generated by comparison with the 1917 flu pandemic caused by another H1A1 virus.

5.11.2. Indicator Organisms in Polluted Water

The validity of any indicator system is also affected by the relative rates of removal and destruction of the indicator versus the target hazard. So differences due to environmental resistance or even ability to multiply in the environment all influence their usefulness. Hence, viral, bacterial, parasitic protozoan, and helminth pathogens are unlikely to all behave in the same way as a single indicator group, and certainly not in all situations.

Furthermore, viruses and other pathogens are not part of the normal fecal microbiota but are only excreted by infected individuals. The higher the number of people, therefore, contributing to sewage or fecal contamination, the more likely the presence of a range of pathogens.

Coliforms. Gram-negative, non-spore-forming, oxidase-negative, rod-shaped facultative anaerobic bacteria that ferment lactose (with β-galactosidase) to acid and gas
within 24–48 h at $36 \pm 2\,°C$. Not specific indicators of fecal pollution.

Thermotolerant Coliforms. Coliforms that produce acid and gas from lactose at $44.5 \pm$
$0.2\,°C$ within 24 ± 2 h and are also known as *fecal coliforms* (FCs) due to their
role as fecal indicators.

Escherichia coli. Thermophilic coliforms that produce indole from tryptophan, but also
defined now as coliforms able to produce β-glucuronidase (although taxonomically
up to 10% of environmental *E. coli* may not). Most appropriate group of coliforms
to indicate fecal pollution from warm-blooded animals.

Fecal Streptococci (FS). Gram-positive, CAT-negative cocci from selective media
(e.g., azide dextrose broth or Endo agar) that grow on bile aesculin agar and at
$45\,°C$, belonging to the genera *Enterococcus* and *Streptococcus* possessing the
Lancefield group D antigen.

Enterococci. All fecal streptococci (FS) that grow at pH 9.6, at 10 and $45\,°C$, and
in 6.5% NaCl. They also fulfill the following criteria: resistance to $60\,°C$ for
30 min and ability to reduce 0.1% methylene blue. The enterococci are a subset of
FS that grow under the aforementioned conditions. Alternatively, enterococci can
be directly identified as microorganisms (Fig. 5.62) capable of aerobic growth at
$44 \pm 0.5\,°C$ and of hydrolyzing 4-methlumbelliferyl-β-D-glucoside (MUD, detecting β-glucosidase activity by blue fluorescence at 366 nm) in the presence of thal-
lium acetate, nalidixic acid, and 2,3,5-triphenyltetrazolium chloride (TTC, which
is reduced to the red formazan) in the specified medium (ISO/FDIS 7899-1, 1998).

Sulfite-reducing clostridia (SRC). Gram-positive, spore-forming, nonmotile, strictly
anaerobic rods that reduce sulfite to H_2S.

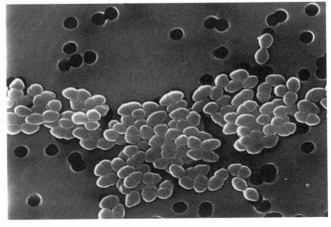

Figure 5.62. Electron microscopic image of enterococci.

Clostridium perfringens. Similar to the sulfite-reducing clostridia (SRC), but also ferment lactose, sucrose, and inositol with the production of gas, produce stormy clot fermentation with milk, reduce nitrate, hydrolyze gelatin, and produce lecithinase and acid phosphatase. Bonde (1963) suggested that not all SRC in receiving waters are indicators of fecal pollution, hence *C. perfringens* is the appropriate indicator.

5.11.2.1. Microbial Component of Water

Microbes constitute the foundation of the aquatic abundance that people see and enjoy (Fig. 5.63). Microbes include bacteria, bacterialike organisms called *archaea*, viruses, protozoa, helminths, and protists. Microbes are natural and vital members of all aquatic communities and are the foundation of lake and stream ecology—without them the natural water worlds would not be possible. Certain microbes, however, when present in excessive numbers, pose a threat to human health. Like all ecosystems, freshwater ecosystems require energy inputs to sustain the organisms within. In lakes and streams, plants and also certain microbes perform photosynthesis to harvest the sun's energy. Microbial photosynthesizers include protists (known as *algae*) and cyanobacteria. Other protists and animals feed on these organisms, forming the next link in the food chain. Plant material from the land also enters lakes and streams at their edges, providing an important nutrient source for many waterbodies.

Decomposers form an especially important part of freshwater ecosystems because they consume dead bodies of plants, animals, and other microbes. These microbial agents of decay are an important part of the ecosystem because they convert detritus (dead and

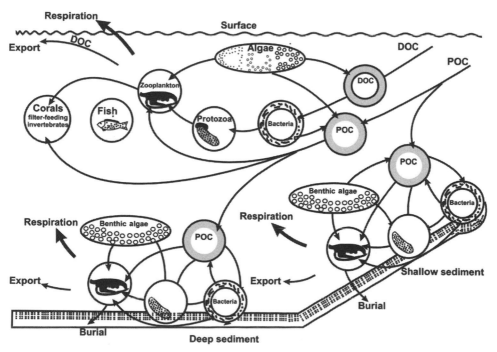

Figure 5.63. Microbial components of water and their metabolism; POC–particulate organic carbon; DOC–dissolved organic carbon.

decaying matter) and organic materials into needed nutrients, such as nitrate, phosphate, and sulfate. Decomposers and other microbes are thus essential to the major biogeochemical cycles by which nutrients are exchanged between the various parts of the ecosystem, both living and nonliving. Without microbial decomposers, minerals and nutrients critical to plant and animal growth would not be made available to support other levels of the freshwater food chain.

AEROBES AND ANAEROBES. Aerobic decomposers in water need oxygen to survive and do their work. The lapping waves and babbling brook help increase the level of dissolved oxygen that is crucial to so many creatures in lake and stream ecosystems, none more so than the bacteria. If there is not enough oxygen in the water, many parts of the system suffer: the aerobic decomposers cannot digest plant matter, insects cannot develop and mature, and the fish cannot play their part, whether browsing for small food particles or eating other fish. Eventually, the stream or pond will be changed, starting at the microbial level.

Human interaction can jeopardize parts of this system in a variety of ways. One principal way is through the runoff of fertilizers or sewage into a waterbody. Both contain nutrients that plants, algae, and cyanobacteria can use to grow, and excessive nutrient amounts can lead to very rapid growth. Interconnected sequences of physical, biological, and chemical events may eventually deplete the water's dissolved oxygen supply, leading to changes in the aquatic ecosystem. If the conditions become severe enough, only a few species (known as *anaerobes*) tolerant to low oxygen conditions will survive. This process, called *cultural eutrophication*, can have profound and lasting consequences on the waterbody.

MICROBES AND HUMAN HEALTH. Freshwater is the host to numerous microorganisms that affect human health directly. Polluted drinking water is a major source of illness and death throughout the world, particularly in developing countries, and in almost all cases, the responsible organisms cycle between the waterbody and the digestive tract of humans or other animals. Released in fecal waste of the infected host, they enter the water again to complete their life cycle. Most infections derived this way cause diarrhea, abdominal cramping, and potentially more serious symptoms, including fever, vomiting, and intestinal bleeding. Following are some common microbes in lakes and streams that are responsible for causing disease.

- The protist, *Giardia lamblia*, is found in freshwater bodies throughout the world. *Giardia* infection is a common waterborne illness in the United States.
- The bacterium *V. cholerae*, while rare in the United States, remains a significant source of disease and death in countries without advanced sewage treatment and with no potable water supplies. For example, a cholera epidemic in 1991 killed more than a thousand people in Peru (South America), where more than 150,000 cases of the illness were confirmed.
- The bacterium, *E. coli*, is a very common waterborne pollutant. Humans have a large and harmless population of *E. coli* in their lower, large intestines, and bacteria make up a large fraction of the volume of human feces. When released into drinking water or recreational water sources, *E. coli* can be ingested and enter the upper small intestine, causing diarrhea.
- Other bacteria known as *coliform* bacteria cause similar symptoms.

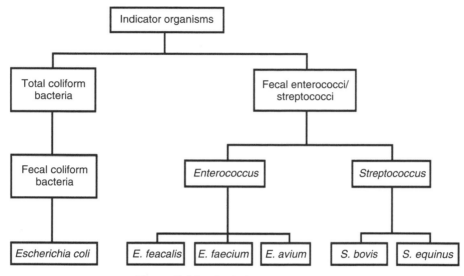

Figure 5.64. The indicator organisms.

The level of FC bacteria in pools, ponds, and other waterbodies is frequently measured during the summer months to assess the safety of recreation in these waters.

5.11.2.2. Concept of Indicator Organisms

Indicator organisms are used to measure potential fecal contamination of environmental samples (Fig. 5.64). The presence of coliform bacteria, such as *E. coli*, in surface water is a common indicator of fecal contamination. Coliform bacteria in water samples may be quantified using the MPN method, a probabilistic test that assumes cultivable bacteria meet certain growth and biochemical criteria. If preliminary tests suggest that coliform bacteria are present at numbers in excess of an established cut-off (the coliform index; Table 5.9), fecal contamination is suspected and confirmatory assays such as the Eijkman test are conducted.

Coliform bacteria selected as indicators of fecal contamination must not persist in the environment for long periods following efflux from the intestine, and their presence must be closely correlated with contamination by other fecal organisms. Indicator organisms need not be pathogenic.

Certain criteria should exist before an indicator organism can be considered reliable in predicting a health risk:

1. The organism must be exclusively of fecal origin and consistently present in fresh fecal waste.
2. It must occur in greater numbers than the associated pathogen.
3. It must be more resistant to environmental stresses and persist for a greater length of time than the pathogen.
4. It must not proliferate to any great extent in the environment
5. Simple, reliable, and inexpensive methods should exist for the detection, enumeration, and identification of the indicator organism.

TABLE 5.9. Coliform Index (MPN Determination from Multiple-Tube Test)

Number of Tubes Giving Positive Reaction Out of			MPN Index Per 100 ml	95% Confidence Limits	
3 of 10 ml Each	3 of 1 ml Each	3 of 0.1 ml Each		Lower	Upper
0	0	1	3	<0.5	9
0	1	0	3	<0.5	13
1	0	0	4	<0.5	20
1	0	1	7	1	21
1	1	0	7	1	23
1	1	1	11	3	36
1	2	0	11	3	36
2	0	0	9	1	36
2	0	1	14	3	37
2	1	0	15	3	44
2	1	1	20	7	89
2	2	0	21	4	47
2	2	1	28	10	150
3	0	0	23	4	120
3	0	1	39	7	130
3	0	0	64	15	380
3	1	0	43	7	210
3	1	1	75	14	230
3	1	2	120	30	380
3	2	0	93	15	380
3	2	1	150	30	440
3	2	2	210	35	470
3	3	0	240	36	1300
3	3	1	460	71	2400
3	3	2	100	150	4800

Source: From *Standard Methods for the Examination of Water and Wastewater*, 12th ed., p. 608. The American Public Health Association, Inc, New York.

Organisms that fit these criteria include the coliform bacteria, FS (enterococci), and the SRC (i.e., *C. perfringens*) (Fig. 5.64). Gastrointestinal pathogens known to have caused outbreaks of enteric disease are largely from the systematically defined family Enterobacteriaceae and include *Salmonella, Shigella, Y. enterocolitica, Klebsiella pneumoniae, Enterobacter*, and enterotoxigenic *E. coli. V. cholerae*, and *C. jejuni* are two other enteric pathogens often found in contaminated water. These organisms are spread by water contaminated with fecal material from humans and other warm-blooded animals.

5.11.2.3. Total Coliform Test

Coliform is not a taxonomic classification but rather a working definition used to describe a group of gram-negative, facultatively anaerobic, rod-shaped bacteria that ferments lactose to produce acid and gas within 48 h at 35 °C. In 1914, the U.S. Public Health Service adopted the enumeration of coliforms as a more convenient standard of sanitary significance.

Although coliforms were easy to detect, their association with fecal contamination was questionable because some coliforms are found naturally in environmental samples. This led to the introduction of the FCs as an indicator of contamination. FC, first defined based on the works of Eijkman, is a subset of total coliforms that grows and ferments lactose at an elevated incubation temperature, hence also referred to as *thermotolerant coliforms*. FC analyses are done at 45.5 °C for food testing, except for water, shellfish, and shellfish harvest water analyses, which use 44.5 °C. The FC group consists mostly of *E. coli*, but some other enterics such as *Klebsiella* can also ferment lactose at these temperatures and can, therefore, be considered as FCs. The inclusion of *Klebsiella* spp. in the working definition of coliforms diminished the correlation of this group with fecal contamination. As a result, *E. coli* has reemerged as an indicator, partly facilitated by the introduction of newer methods that can rapidly identify *E. coli*.

Currently, all three groups are used as indicators but in different applications. Detection of coliforms is used as an indicator of sanitary quality of water or as a general indicator of sanitary condition in the food-processing environment. FCs remain the standard indicator of choice for shellfish and shellfish harvest waters, and *E. coli* is used to indicate recent fecal contamination or unsanitary processing. Almost all the methods used to detect *E. coli*, total coliforms, or FCs are enumeration methods that are based on lactose fermentation. The MPN method is a statistical, multistep assay consisting of presumptive, confirmed, and completed phases. In the assay, serial dilutions of a sample are inoculated into the broth media. Analysts score the number of gas-positive (fermentation of lactose) tubes, from which the other two phases of the assay are performed and then use the combinations of positive results to consult a statistical table to estimate the number of organisms present. Typically, only the first two phases are performed in coliform and FC analyses, while all three phases are done for *E. coli*. The three-tube MPN test is used for testing most food products. The five-tube MPN is used for water, shellfish, and shellfish harvest water testing, and there is also a 10-tube MPN method that is used to test bottled water or samples that are not expected to be highly contaminated.

There is also a solid-medium plating method for coliforms that use Violet Red Bile Agar, which contains neutral red pH indicator, so that lactose fermentation results in formation of pink colonies. There are also membrane filtration tests for coliforms and FCs that measure aldehyde formation due to fermentation of lactose.

MPN. It is a presumptive test for coliforms, FCs, and *E. coli*.

PRESUMPTIVE TEST. In the presumptive test, a series of 9 or 12 tubes of lactose broth are inoculated with measured amounts of water to see if the water contains any lactose-fermenting bacteria that produce gas. If, after incubation, gas is seen in any of the lactose broths, it is presumed that coliforms are present in the water sample. This test is also used to determine the MPN of coliforms present per 100 ml of water (Figs. 5.65–5.67).

CONFIRMED TEST. In this test, plates of Levine EMB agar or Endo agar are inoculated from positive (gas-producing) tubes to see if the organisms that are producing the gas are gram negative (another coliform characteristic). Both these media inhibit the growth of gram-positive bacteria and thus help distinguish coliforms from noncoliforms. On EMB agar, coliforms produce small colonies with dark centers (nucleated colonies). On Endo agar, coliforms produce reddish colonies. The presence of coliformlike colonies confirms the presence of a lactose-fermenting gram-negative bacterium.

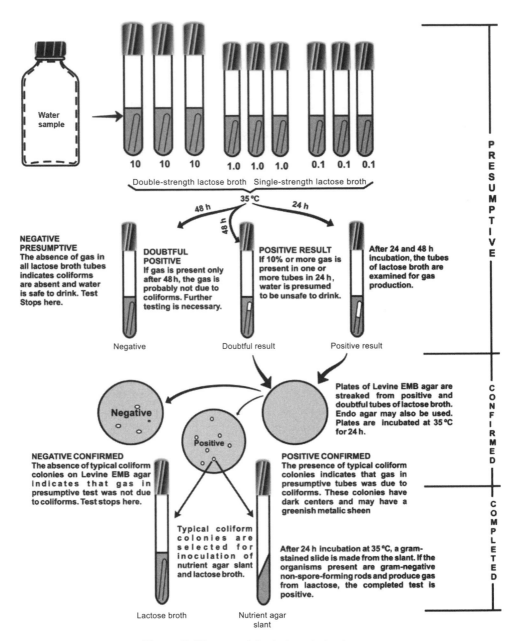

Figure 5.65. Bacteriological analysis of water.

COMPLETED TEST. In the completed test, our concern is to determine if the isolate from the agar plates truly matches the definition of a coliform. Media for this test include a nutrient agar slant and a Durham tube of lactose broth. If gas is produced in the lactose tube and a smear from the agar slant reveals that we have a gram-negative non-spore-forming rod, we can be certain that we have a coliform.

The completion of these three tests with positive results establishes that coliforms are present; however, there is no certainty that *E. coli* is the coliform present. The organism

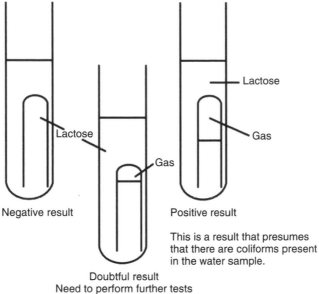

Figure 5.66. The presumptive test.

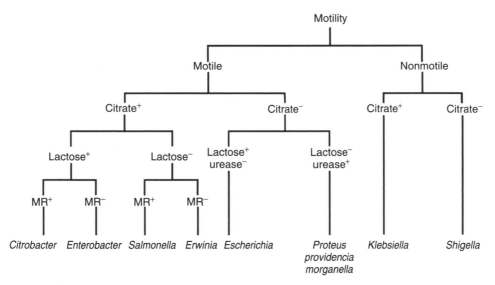

Figure 5.67. Separation outline for different gram-negative rods and cocci.

might be *Enterobacter aerogenesis*. Of the two, *E. coli* is the better sewage indicator since *E. aerogenesis* can be of nonsewage origin. To differentiate these two species, one must perform the IMViC test (Figs. 5.65–5.67).

IMViC Tests. In the differentiation of *E. aerogenes* and *E. coli*, as well as some other related species, four physiological tests have been grouped together into what are called the *IMViC test*. The i stands for indole, with the lowercase simply facilitating pronunciation; the M and V stand for methyl red and Voges–Proskauer tests, respectively; and the C signifies citrate utilization. In the differentiation of the two coliforms *E. coli* and *E. aerogenes*, the test results appear as charted below, revealing completely opposite reactions for the two mechanisms on all tests.

	I	M	V	C
E. coli	+	+	−	−
E.aerogenes	−	−	+	+

The significance of these tests is that when testing drinking water for the presence of the sewage indicator *E. coli*, one must be able to rule out *E. aerogenes*, which has many of the morphological and physiological characteristics of *E. coli*. Since *E. aerogenes* is not always associated with sewage, its presence in water would not necessarily indicate sewage contamination.

PRESENCE-ABSENCE TEST. The presence-absence (P-A) test is a presumptive detection test for coliforms in water. The test is a simple modification of the multiple-tube procedure. One test sample, 100 ml, is inoculated into a single culture bottle to obtain qualitative information on the presence or absence of coliforms based on the presence or absence of lactose fermentation. This test is based on the principle that coliforms and other pollution indicator organisms should not be present in a 100-ml water sample. Comparative studies with the MF procedure indicate that the P-A test may maximize coliform detection in samples containing many organisms that could overgrow coliform colonies and cause problems in detection. The P-A test is described in standard methods for water testing and by the US Environmental Protection Agency. Beef extract and peptones provide nitrogen, vitamins, and amino acids in the P-A Broth. Lactose is the carbon source in the formula.

The potassium phosphates provide buffering capacity; sodium chloride provides essential ions. Sodium lauryl sulfate is the selective agent, inhibiting many organisms except coliforms. Bromcresol purple is used as an indicator dye; lactose-fermenting organisms turn the medium from purple to yellow with or without gas production.

$$CH_2SH \quad\quad CH_2 \quad\quad CH_2 \quad\quad CH_2$$

Cysteine → H_2S + α-Amino acrylic acid → Imino acid → H_2O → Pyruvic acid + NH_3

MEMBRANE FILTER. The MF technique is highly reproducible, can be used to test relatively large volumes of sample (Fig. 5.68), and yields numerical results more rapidly than the multiple-tube procedure. The MF technique is extremely useful in monitoring drinking water and a variety of natural waters. However, the MF technique has limitations, particularly when testing waters with high turbidity or noncoliform (background) bacteria. For such waters or when the MF technique has not been used previously, it is desirable to conduct parallel tests with the multiple-tube fermentation technique to demonstrate applicability and comparability.

As related to the MF technique, the coliform group may be defined as comprising all aerobic and many facultatively anaerobic, gram-negative, non-spore-forming, rod-shaped bacteria that develop a red colony with a metallic sheen within 24 h at 35 °C on an Endo-type medium containing lactose. Some members of the total coliform group may produce a dark red or nucleated colony without a metallic sheen. When verified, these are classified as atypical coliform colonies. When purified cultures of coliform bacteria are tested they produce a negative cytochrome oxidase (CO) and positive β-galactosidase

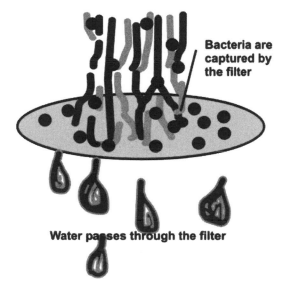

Bacteria are captured by the filter

Water passes through the filter

Figure 5.68. Membrane filtration method.

(ONPG) reaction. Generally, all red, pink, blue, white, or colorless colonies lacking sheen are considered noncoliforms by this technique.

FC bacterial densities may be determined by either the multiple-tube procedure or the MF technique. If the MF procedure is used for chlorinated effluents, it gives comparable information to that obtained by the multiple-tube test before accepting it as an alternative. The MF procedure uses an enriched lactose medium and incubation temperature of $44.5 \pm 0.2\,°C$ for selectivity and gives 93% accuracy in differentiating between coliforms found in the feces of warm-blooded animals and those from other environmental sources. Because incubation temperature is critical, the MF cultures are made waterproof (plastic bag enclosures) and submerged in a water bath for incubation at the elevated temperature or an appropriate, accurate solid heat sink incubator is used. Alternatively, an equivalent incubator that will hold the $44.5\,°C$ temperature within $0.2\,°C$ (throughout the chamber), over a 24-h period, while located in an environment of ambient air temperatures ranging from 5 to $35\,°C$ can be used.

Materials and Culture Medium

1. *M-FC Medium.* The need for uniformity dictates the use of dehydrated media. Never prepare media from basic ingredients when suitable dehydrated media are available. Follow manufacturer's directions for rehydration. Commercially prepared media in liquid form (sterile ampule or other) also may be used if known to give equivalent results.

2. *Culture Dishes.* Use tight-fitting plastic dishes because the MF cultures are submerged in water bath during incubation. Enclose groups of FC cultures in plastic bags or seal individual dishes with waterproof (freezer) tape to prevent leakage during submersion.

3. *Incubator.* The specificity of the FC test is related directly to the incubation temperature. Static air incubation may be a problem in some types of incubators because of potential heat layering within the chamber and the slow recovery of temperature each time the incubator is opened during daily operations. To meet the need for greater temperature control use a water bath, a heat sink incubator, or a properly designed and constructed incubator giving equivalent results. A temperature tolerance of $44.5 \pm 0.2°\,C$ can be obtained with most types of water baths that also are equipped with a gable top for the reduction of heat and water losses. A circulating water bath is excellent but may not be essential to this test if the maximum permissible variation of $0.2\,°C$ in temperature can be maintained with other equipment.

Procedure

1. *Selection of Sample Size.* Use sample volumes that will yield counts between 20 and 60 FC colonies per membrane. When the bacterial density of the sample is unknown, filter several decimal volumes to establish FC density. Estimate the volume expected to yield a countable membrane, and select two additional quantities representing one-tenth and 10 times this volume (Fig. 5.69).

2. *Filtration of Sample.* Follow the same procedure and precautions as described above.

3. *Preparation of Culture Dish.* Place a sterile absorbent pad in each culture dish and pipet approximately 2 ml M-FC medium, prepared as directed above, to saturate the

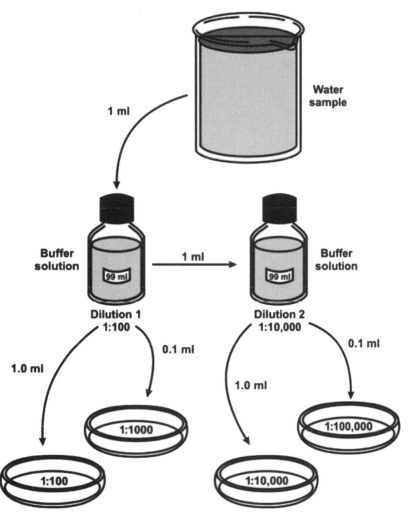

Figure 5.69. Method of preparing proper dilution samples.

pad. Carefully remove any excess liquid from the culture dish. Place the prepared filter on medium-impregnated pad. As a substrate substitution for the nutrient-saturated absorbent pad, add 1.5% agar to M-FC broth.

4. *Incubation*. Place prepared cultures in waterproof plastic bags or sealed petri dishes, submerge in water bath, and incubate for 24 ± 2 h at $44.5 \pm 0.2\,^\circ$C. Anchor dishes below the water surface to maintain critical temperature requirements. Place all prepared cultures in the water bath within 30 min after filtration. Alternatively, use an appropriate, accurate solid heat sink or equivalent incubator.

5. *Counting*. Colonies produced by FC bacteria on M-FC medium are various shades of blue. Pale yellow colonies may be atypical *E. coli*, which can be verified for gas production in mannitol at $44.5\,^\circ$C. Non-fecal-coliform colonies are gray to cream colored. Normally, few non-fecal-coliform colonies will be observed on M-FC medium because of selective action of the elevated temperature and addition

of rosolic acid salt reagent. Elevating the temperature to $45.0 \pm 0.2\,°C$ may be useful in eliminating environmental *Klebsiella* from the FC population. Count colonies with a low power (10–15 magnifications) binocular wide-field dissecting microscope or other optical device.

5.11.2.4. Fecal Coliform

Compute the density from the sample quantities that produced MF counts within the desired range of 20–60 FC colonies. This colony density range is more restrictive than the 20–80 total coliform range because of larger colony size on M-FC medium. Record densities as FCs per 100 ml. Compute the count by the following equation:

$$(total)\ coliform\ colonies/100\ ml = (coliform\ colonies\ counted \times 100)/(ml\ sample\ filtered).$$

5.11.2.5. Fecal Streptococci

The streptococci have been under consideration as indicators of fecal pollution for many years. Their poor acceptance as a measure of fecal pollution from human and warm-blooded animal excreta has been due in part to the relatively low recovery rates in comparison to coliform densities in polluted waters; the multiplicity of detection procedures; poor agreement between the various methods for their quantitative enumeration; and the lack of detailed and systematic studies on the sources, survival, and interpretation of streptococci in various types of waters. Furthermore, undue emphasis has been placed on the *Streptococcus faecalis* group (enterococci) with little or no regard for other streptococcal strains present in the gut of humans and warm-blooded animals or birds (Fig. 5.70). The predominating species of streptococci may vary markedly in various animal excreta. Some detection methods yield excellent results with human fecal samples but give poor quantitative recovery of the streptococci present in pig or cow feces.

The fecal streptococcus group consists of a number of species of the genus *Streptococcus*, such as *S. faecalis, S. faecium, S. avium, S. bovis, S. equinus*, and *S. gallinarum*.

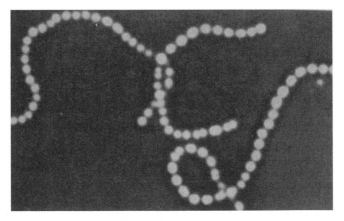

Figure 5.70. *Streptococcus faecalis*.

They all give a positive reaction with Lancefield group D antisera and have been iso-lated from the feces of warm-blooded animals. In addition, *S. avium* sometimes reacts with Lancefield group Q antisera. *S. faecalis* subsp. *liquefaciens* and *S. faecalis* subsp. *zymogenes* are differentiated based on the ability of these strains to liquefy gelatin and hemolyze red cells. However, the validity of these subspecies is questionable.

The normal habitat of FS is the gastrointestinal tract of warm-blooded animals. *S. faecalis* and *S. faecium* were once thought to be more human specific than other *Strep-tococcus* species. Other species have been observed in human feces but less frequently. Similarly, *S. bovis, S. equinus*, and *S. avium* are not exclusive to animals, although they usually occur at higher densities in animal feces. Certain streptococcal species predomi-nate in some animal species and not in others, but it is not possible to differentiate the source of fecal contamination based on the speciation of FS.

The FS have been used with FCs to differentiate human fecal contamination from that of other warm-blooded animals. Editions of *Standard Methods* previous to the seven-teenth edition suggested that the ratio of FCs to FS could provide information about the source of contamination. A ratio greater than four was considered indicative of human fecal contamination, whereas a ratio of less than 0.7 was suggestive of contamination by nonhuman sources. The value of this ratio has been questioned because of vari-able survival rates of fecal *Streptococcus* group species. *S. bovis* and *S. equinus* die off rapidly, once exposed to aquatic environments, whereas *S. faecalis* and *S. faecium* tend to survive longer. Furthermore, disinfection of wastewater appears to have a signif-icant effect on the ratio of these indicators, which may result in misleading conclusions regarding the source of contaminants. The ratio is affected also by the methods for enu-merating FS. The KF MF procedure has a false-positive rate ranging from 10% to 90% in marine water and freshwaters. For these reasons, the FC/FS ratio cannot be recom-mended and should not be used as a means of differentiating human and animal sources of pollution.

5.11.2.6. *Clostridium* perfringens

C. perfringens (formerly known as *C. welchii*) is a gram-positive, rod-shaped, anaerobic, spore-forming bacterium (Fig. 5.71). *C. perfringens* is ubiquitous in Nature and can be found as a normal component of decaying vegetation, marine sediment, the intestinal tract of humans and other vertebrates, insects, and soil.

The *C. perfringens* enterotoxin (CPE) mediating the disease is heat labile (dies at 74 °C) and can be detected in contaminated food, if not heated properly, and feces.

Incubation time is between 6 and 24 h (commonly 10–12 h) after ingestion of con-taminated food. Often, meat is well prepared but too far in advance of consumption. Since *C. perfringens* forms spores that can withstand cooking temperatures, if let stand long enough, germination ensues and infective bacterial colonies develop. Symptoms typically include abdominal cramping and diarrhea; vomiting and fever are unusual. The whole course usually resolves within 24 h. Very rare, fatal cases of clostridial necrotizing enteritis (also known as *Pig-Bel*) have been known to involve "type C" strains of the organism, which produce a potently ulcerative β-toxin. This strain is most frequently encountered in Papua, New Guinea.

It is likely that many cases of *C. perfringens* food poisoning remain subclinical, as antibodies to the toxin are common among the population. This has led to the conclu-sion that most of the population has experienced food poisoning due to *C. perfringens*.

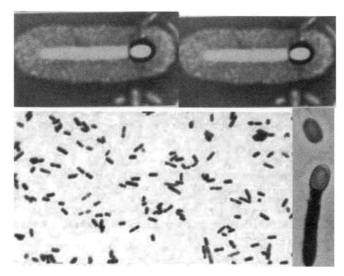

Figure 5.71. *Clostridium perfringens.*

On blood agar plates, *C. perfringens* grown anaerobically produces β-hemolytic, flat, spreading, rough, translucent colonies with irregular margins. Nagler agar, containing 5–10% egg yolk, is used to identify strains that produce α-toxin, a diffusible lecithinase that interacts with the lipids in egg yolk to produce a characteristic precipitate around the colonies. One half of the plate is inoculated with antitoxin to act as the control in the identification.

5.11.2.7. Heterotrophic Plate Count

Heterotrophic plate count (HPC) is a procedure used to estimate the number of live heterotrophic bacteria that are present in a water sample. A sample of water is put on a plate that contains nutrients that bacteria need to survive and grow. The nutrient media that is most often used for this test is called *R2A agar*, which is a gelatinelike substance that is best suited to the needs of water bacteria. After 5–7 days, the number of small spots on the plate, called *colonies*, is counted and a measure of how many bacteria are present in each milliliter of water can be determined (Fig. 5.72). The HPC results are generally reported as CFU/ml or colony forming units per milliliter. Each colony forming unit represents an initial single, live bacterium that was capable of multiplying until it could be observed on the plate. This test can provide useful information about water quality and supporting data on the significance of coliform test results. High concentrations of the general bacterial population may hinder the recovery of coliforms.

It is important to understand that the colony count, alone, does not allow one to draw conclusions about the risks to public health. However, it currently serves as a relatively easy way to measure filtration and disinfection efficiency, as well as the estimated numbers of bacteria in areas that have the potential for increased contamination.

Growth of bacteria following drinking-water treatment is normally referred to as *regrowth*. This type of growth is typically reflected in higher HPC values measured

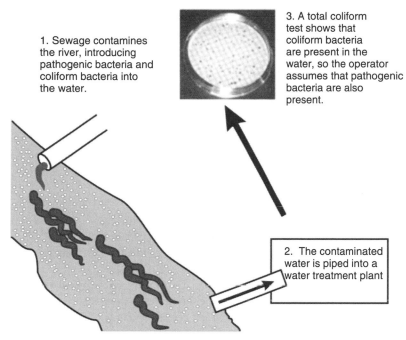

1. Sewage contamines the river, introducing pathogenic bacteria and coliform bacteria into the water.

3. A total coliform test shows that coliform bacteria are present in the water, so the operator assumes that pathogenic bacteria are also present.

2. The contaminated water is piped into a water treatment plant

Figure 5.72. Coliform test.

in water samples. Regrowth generally occurs in areas of distribution or plumbing systems where the water may remain stationary for a longer amount of time, bottled water, water softeners, or carbon filters. In order to ensure that regrowth of bacteria is kept to a minimum, general water safety practices such as maintenance protocols, regular cleaning, temperature management, and maintenance of a disinfectant residual (e.g., chlorine) should be in place. If the HPC points exceed recommendations, one should consider whether the system has been adequately cleaned, whether the disinfectant residual is effective, and the efficiency of temperature management. Failure in any of these areas could lead to elevated HPC levels.

Unlike other indicators, such as *E. coli* or total coliforms, low concentrations of HPC organisms will still be present after drinking-water treatment. In general, water utilities can achieve heterotrophic bacteria concentrations of 10 CFUs/ml or less in finished water. Within a distribution system, increases in the density of HPC bacteria are usually the result of bacterial regrowth. The density reached can be influenced by the bacterial quality of the finished water entering the system, temperature, residence time, presence or absence of a disinfectant residual, construction materials, surface-to-volume ratio, flow conditions, the availability of nutrients for growth and in chlorinated systems, the chlorine/ammonia ratio, and the activity of nitrifying bacteria.

5.11.2.8. Bacteriophage

Bacteriophage is any one of a number of viruses that infect bacteria. Bacteriophages are among the most common biological entities on Earth. The term is commonly used in its shortened form, phage. Typically, bacteriophages consist of an outer protein capsid enclosing the genetic material. The genetic material can be ssRNA, dsRNA, ssDNA, or dsDNA ("ss" and "ds" prefixes denote single strand and double strand, respectively) that is long with either circular or linear arrangement. Bacteriophages are much smaller than the bacteria they destroy.

Phages are estimated to be the most widely distributed and diverse entities in the biosphere. Phages are ubiquitous and can be found in all reservoirs populated by bacterial hosts, such as soil or the intestines of animals. One of the densest natural sources for phages and other viruses is seawater, where up to 9×108 virions/ml have been found in microbial mats at the surface and up to 70% of marine bacteria may be infected by phages. They have been used for over 60 years as an alternative to antibiotics in the former Soviet Union and Eastern Europe. They are seen as a possible therapy against multidrug-resistant strains of many bacteria.

Bacteriophages may have a lytic or lysogenic life cycle, and a few viruses are capable of carrying out both. With lytic phages such as the T4 phage, bacterial cells are broken open (lysed) and destroyed after immediate replication of the virion (Fig. 5.73). As soon as the cell is destroyed, the new phages can find new hosts. Lytic phages are the kind suitable for phage therapy.

In contrast, the lysogenic cycle does not result in immediate lysing of the host cell. Those phages able to undergo lysogeny are known as *temperate phages*. Their viral genome will integrate with the host DNA and replicate along with it fairly harmlessly, or may even become established as a plasmid. The virus remains dormant until host conditions deteriorate, perhaps due to depletion of nutrients, and then the endogenous phages (known as *prophages*) become active. At this point, they initiate the reproductive cycle, resulting in lysis of the host cell. As the lysogenic cycle allows the host cell to continue to survive and reproduce, the virus is reproduced in all copies of the cell. Sometimes prophages may provide benefits to the host bacterium while they are dormant

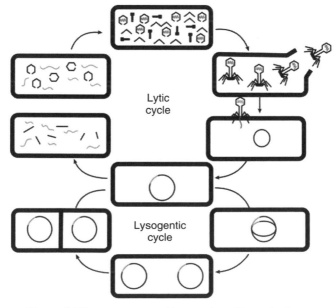

Figure 5.73. Lysogenic and lytic cycles of bacteriophage.

by adding new functions to the bacterial genome in a phenomenon called *lysogenic conversion* (Fig. 5.73). A famous example is the conversion of a harmless strain of *V. cholerae* by a phage into a highly virulent one, which causes cholera. This is why temperate phages are not suitable for phage therapy.

Bacteriophages attach to specific receptors on the surface of bacteria, including lipopolysaccharides, teichoic acids, proteins, or even flagella. This specificity means that a bacteriophage can only infect certain bacteria bearing receptors that they can bind to, which in turn determines the phage's host range. Host growth conditions also influence the ability of the phage to attach and invade bacteria. As phage virions do not move independently, they must rely on random encounters with the right receptors when in solution (blood, lymphatic circulation, irrigation, soil water, etc.). Complex bacteriophages use a hypodermic-syringe-like motion to inject their genetic material into the cell. After making contact with the appropriate receptor, the tail fibers bring the base plate closer to the surface of the cell. Once attached completely, the tail contracts, possibly with the help of ATP present in the tail (Prescott et al., 1993), injecting genetic material through the bacterial membrane.

5.11.2.9. Other Indicator Organisms

All organisms need certain conditions to survive and multiply. Since each aquatic organism has specific tolerances of chemical and physical conditions, the presence or absence of particular organisms can tell us a lot about the body of water we are studying. Compared to using complex chemical analysis, using microorganisms to determine water quality is relatively simple and effective. A species that is normally present in an aquatic ecosystem under specific conditions is called an *indicator organism*.

Bacteria are an example of a pollution indicator organism. Pathogens are of great importance in the study of water quality. Many diseases around the world are linked to water as a carrier for bacteria, viruses, or parasites. All water samples being inspected or tested for human consumption should be tested for coliform bacteria. An intestinal bacteria, *E. coli*, is often associated with fecal material.

It is important to note, however, that some aquatic communities are naturally low in oxygen or will go through low oxygen phases. For example, it would be unreasonable to expect a pond community to have the same habitat conditions as a fast flowing stream.

Ponds and still waters absorb more of the sun's heat energy and retain it for longer periods, so they tend to be warmer than fast flowing streams. Because of this heat retention and lack of motion that would incorporate oxygen, ponds are usually much lower in dissolved oxygen content than streams. This critical difference affects every aspect of the pond food web. Therefore, biological indexes designed for fast flowing streams cannot be used to accurately assess the water quality of ponds.

5.11.3. Microbiology of Sewage Treatment

Human activities generate a tremendous volume of sewage and wastewater that require treatment before discharge into waterways. Often this wastewater contains excessive amounts of nitrogen, phosphorus, and metal compounds, as well as organic pollutants, that would overwhelm waterways with an unreasonable burden. Wastewater also contains chemical wastes that are not biodegradable, as well as pathogenic microorganisms that can cause infectious disease.

5.11.3.1. Wastewater Treatment and Filling

The chemical and biological waste in sewage and water must be broken down before it is deposited to the soil and environment. This breakdown can effectively be controlled by managing the microbial population in waters and encouraging microorganisms to digest the organic matter. The water must then be purified before it is considered fit to drink. Water taken from ground sources must also be treated before consumption.

To purify water for drinking, a number of processes are conducted to reduce the microbial population and maintain that population at a safe level. First, the solid matter is allowed to settle out in a sedimentation tank. Flocculating materials such as alum are used to drag microorganisms to the bottom of the tank.

Then the filtration process is begun. Water is filtered through either a slow sand filter or a rapid sand filter. This process removes 99% of the microorganisms. The slow sand filter is composed of finer grains of sand, and the filtration process takes longer than the rapid sand filter, where larger grains are used.

Many communities then purify the water by chlorination. When added to water, chlorine maintains the low microbial count and ensures that the water remains safe for drinking purposes. Chlorine gas or hypochlorite (NaOCl) is used for chlorination purposes. The water is chlorinated until a slight residue of chlorine remains. Sewage treatment involves a more complex set of procedures that are needed for water purification because the volume of organic matter and the variety of microorganisms are much greater.

The first treatment, or primary treatment, of sewage and wastewater involves the removal in settling tanks of particulate matter such as plant waste. The solids that sediment are strained off, and the sludge is collected to be burned or buried in landfills. Alternatively, it can be treated in an anaerobic sludge-digesting tank, as follows.

During the secondary treatment of wastewater and sewage, the microbial population of liquid and sludge waste is reduced. In the anaerobic sludge digester, microorganisms break down the organic matter of proteins, lipids, and cellulose into smaller substances for metabolism by other organisms. Products of these breakdowns include organic acids, alcohols, and simple compounds. Methane gas is produced in the sludge tank, and it can be burned as a fuel to operate the waste treatment facility. The remaining sludge is incinerated or buried in a landfill, and its fluid is recycled and purified (Fig. 5.74).

Primary treatment is represented by the steps preceding secondary treatment, and tertiary treatment is performed in the chlorination tank at the conclusion of the process.

In aerobic secondary sewage treatment, the fluid waste is aerated and then passed through a trickling filter. In this process, the liquid waste is sprayed over a bed of crushed rocks, tree bark, or other filtering material. Colonies of bacteria, fungi, and protozoa grow in the bed and act as secondary filters to remove organic materials. The microorganisms metabolize organic compounds and convert them to carbon dioxide, sulfate, phosphates, nitrates, and other ions. The material that comes through the filter has been 99% cleansed of microorganisms.

Liquid waste can also be treated in an activated digester after it has been vigorously aerated. Slime-forming bacteria form masses that trap other microorganisms to remove them from water. Treatment for several hours reduces the microbial population significantly, and the clear fluid is removed for purification. The sludge is discarded in a landfill or into the sea.

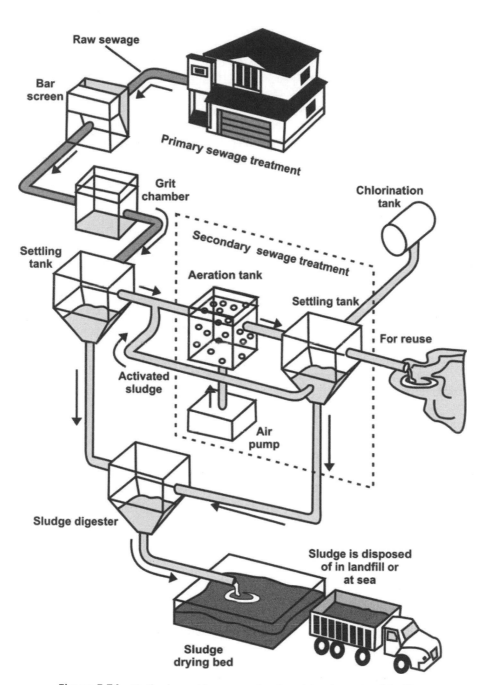

Figure 5.74. Methods used in sewage treatment in a large municipality.

In the tertiary treatment of sewage, the fluid from the secondary treatment process is cleansed of phosphate and nitrate ions that might cause pollution. The ions are precipitated as solids, often by combining them with calcium or iron, and the ammonia is released by oxidizing it to nitrate in the nitrification process. Adsorption to activated charcoal removes many organic compounds such as PCBs, a chemical pollutant.

The home septic system is a waste treatment facility on a small scale. In a septic tank, household sewage is digested by anaerobic bacteria, and solids settle to the bottom of the tank. Solid waste is carried out of the outflow apparatus into the septic field beneath the ground. The water seeps out through holes in tiles and enters the soil, where bacteria complete the breakdown processes. A similar process occurs in cesspools, except that sludge enters the ground at the bottom of the pool and liquids flow out through the sides of the pool.

5.11.3.2. Microorganisms in Solid Waste Rreatment

Microorganisms are the agents that bring about the conversion of wastes into useful products such as fuel gases, fuel alcohol, and also compost, which can be used as manure. The gases produced in landfills due to decomposition by anaerobic organisms also can be used as a source of energy. The major problem today in producing energy from waste is the cost factor. Efforts are being made to produce genetically modified organisms that will produce energy from wastes more efficiently and at a least cost. Microorganisms are omnipresent and are responsible for many good as well as bad things in the biosphere. They are present even in waste materials, where they carry out various biochemical processes to degrade waste materials. This process may be aerobic or anaerobic. Solid waste decomposition is carried out by bacteria, which decompose complex organic materials to simple water-soluble organic compounds. These are then converted to CO_2 and H_2O aerobically or to CH_4 anaerobically. Fungi are mostly aerobic and feed on decaying organic matter. Soil fungi play a vital role in stabilizing solid wastes in composting and landfilling processes by decomposing plant tissues such as cellulose and lignin. Protozoa are predators on bacteria. They are found wherever bacteria are prevalent. Thus, they help to maintain the equilibrium of microbial flora in solid waste disposal systems.

The rapid urbanization and change in life style has increased the waste load and thereby pollution loads on the urban environment to unmanageable and alarming proportions. The existing waste dumping sites are full beyond capacity and are under unsanitary conditions leading to pollution of water sources, proliferation of vectors of communicable diseases, foul smell and odors, release of toxic metabolites that cause anesthetic ambience and eye sore. It is difficult to get new dumping yards, and open dumping is prohibited by law. This is particularly true for countries with severe constraints of land availability, dense population, and environmental fragility, and where expectation for management of solid wastes relies on an overly centralized approach. In earlier days, municipal wastes, composed mainly of biodegradable matter, did not create much problem to the community, as the quantity of wastes generated was either recycled/reused directly as manure or was within the assimilative capacity of the local environment. The biodegradable waste of the urban centers was accepted by the suburban rural areas for composting in the agricultural fields. With increasing content of plastics and nonbiodegradable packaging materials, municipal wastes became increasingly unacceptable to cultivators. As a result, the excessive accumulation of solid wastes in the urban environment poses serious threat.

The solid waste is placed on slow moving conveyor belts. Materials such as corrugated paper are handpicked, and then ferrous materials are removed by magnetic separation. Thus, the materials that are not easily biodegradable are separated. The waste is then ground in hammer mills and converted to pulp. Then it is mixed with nutrient source, water, and fillers. Nutrient sources such as sewage sludge, night soil, or animal manure are used. Wood chips or ground corncobs may be used as fillers. Water is added to maintain 50% moisture. In many Indian compost cycles, fillers and other nutrient sources are not included.

5.12. MICROORGANISMS AS FRIENDS OF MAN

Human health is something that cannot be compromised for any other thing in the world. All the renowned health organizations suggest that cleanliness and hygiene are major factors that might affect a person's health. Therefore, it is essential to ensure clean and hygienic food, environment, and living. That is why most people prefer janitorial services to ensure a consistently clean environment in their buildings. Plus, cleaning supplies help in getting rid of germs and harmful bacteria. These microorganisms are a major cause behind many infections and diseases. However, not every microorganism is bad and harmful.

There are countless bacteria present in the human intestine. It is assumed that the amount of bacteria present in an intestine is more than the amount of cells lining the organ. These bacteria play an important role in the digestion of food and also destroy harmful organisms that might enter along with food. Some bacteria are really helpful for lactose intolerant people as they help in breaking down lactose in the digestive tract.

Lactobacillus and *Saccharomyces boulardii* are also helpful species of bacteria for children. They help fight diarrhea and also destroy many harmful organisms that might make a child sick. Bacteria that help in breaking lactose into lactic acid also play a significant role in reducing DNA damage.

Bacteria present on our skin help in fighting allergies and fungi that can cause irritation and infections. Bacteria present in our stomach help in maintaining the pH level inside the stomach. They are really health-friendly organisms that release useful hormones and vitamins such as vitamin B and K into the body.

Microorganisms are used in the production of antibiotics, wine, vinegar, butanol, and organic acids. Other industrial uses of bacteria include tanning of leather and curing of coffee and tea. Therefore, bacteria with a bad reputation for being harmful maintain a healthy balance not just in our body but also in the overall ecosystem. Experts have debated how to define probiotics. One widely used definition, developed by the WHO and the Food and Agriculture Organization of the United Nations, is that probiotics are "live microorganisms, which, when administered in adequate amounts, confer a health benefit on the host."

Probiotics are not the same thing as prebiotics, nondigestible food ingredients that selectively stimulate the growth and/or activity of beneficial microorganisms already in people's colons. When probiotics and prebiotics are mixed together, they form a synbiotic.

Probiotics are available in foods and dietary supplements (e.g., capsules, tablets, and powders) and in some other forms as well. Examples of foods containing probiotics are yogurt, fermented and unfermented milk, miso, tempeh, and some juices and soy beverages. In probiotic foods and supplements, the bacteria may have been present originally or added during preparation.

Most probiotics are bacteria similar to those naturally found in people's guts, especially in those of breastfed infants (who have natural protection against many diseases). Most often, the bacteria come from two groups, *Lactobacillus* and *Bifidobacterium*. Within each group, there are different species (e.g., *Lactobacillus acidophilus* and *Bifidobacterium bifidus*), and within each species, there are different strains (or varieties). A few common probiotics, such as *S. boulardii*, are yeasts, which are different from bacteria.

5.12.1. Microbes to Join the Oil Industry

Bacteria are among Nature's chief recyclers. Because of their ability to break down a variety of compounds into their basic elements, bacteria are used extensively in environmental biotechnology. One of the applications where bacteria are gaining greater use is in the oil industry.

There are more than 27,000 species of bacteria. Several different types are capable of breaking down both simple and complex hydrocarbons (organic compounds that contain only hydrogen and carbon), the components of crude oil. Two of the major genera in which such microbes are found are *Pseudomonas* and *Bacillus*. These bacteria are described as *oleophilic* for being attracted to oil. They use hydrocarbons as a food source, or simply break down hydrocarbons, with no obvious use for them. The bacteria convert the hydrocarbons into methanol (a type of alcohol), water, and carbon dioxide. *Pseudomonas* and *Bacillus* are commonly found in areas that naturally contain oil, such as near underground oil deposits. However, these bacteria can also be found where there is no oil. This is because the ability to break down hydrocarbons is not always the bacteria's main function, so these traits may lie dormant.

Also, these bacteria can live in a range of habitats. Although most live under conditions considered normal for microbes, there are also some new specimens that have been dubbed extremophiles. Bacteria and other microorganisms have been found in various types of rock several hundred meters (and in some cases, kilometers) below the Earth's surface. The oil industry is using hardy oil-eating bacteria in a number of ways. The most common applications are in bioremediation (using microbes to clean up pollutants) of oil spills or reduction of the environmental impact of waste products caused by oil production.

Water is used to help remove oil from the ground or refine it into petroleum products. As a result, wastewater containing dissolved hydrocarbons makes up the largest volume of waste material generated by the oil and gas industry. For example, a typical refinery uses about 18 barrels of water for every barrel of crude oil it processes. Some of the components of dissolved hydrocarbons, such as phenols, are highly toxic and do not break down in the environment. If the wastewater is to be recycled or released as surface water, it must be treated to remove virtually all the dissolved oil by use of bioremediation lagoons, where water is placed in a large pondlike enclosure and treated with oil-eating bacteria. However, the process tends to be slow, expensive, and inefficient.

All microorganisms synthesize certain amount of lipids for membranes and other functional and structural units and only a small number are able to accumulate lipids in an amount exceeding 20% of cell mass as reserve material. Among more than 600 species of yeasts and 70,000 molds, only about 125 are capable of accumulating lipid concentrations above this percentage. Oleaginous microorganisms are considered as attractive alternative source of lipids. They help to reduce the possibility of the cost of production processes using inexpensive substrates. They are also able to synthesize a range of different products. Moreover, better strains developed by genetic techniques with more efficacy are now available. It has been demonstrated that such microbial oils, also called *single cell oils* (SCO), can be used as feedstock for biodiesel production. Compared to other vegetable oils and animal fats, the production of SCO has many advantages, such as short life cycle; less labor requirement; less affected by venue, season, and climate; and easier to scale-up.

Neuron BioIndustrial, a division of Neuron BPh based in Granada, Spain, has developed a new process to metabolize raw glycerin to turn it into microbial oil that can be used as raw material for biodiesel production. Using microorganisms as biocatalysts, Neuron BioIndustrial makes possible the conversion of this by-product into a value-added compound. Currently, one of the most innovative bioprocesses of MicroBioTools is the development of MicroBioOil, a new process to metabolize raw glycerin from biodiesel production to turn it into microbial oil to produce biodiesel (Fig. 5.75). Neuron BioIndustrial has selected a specific type of microorganism that metabolizes large quantities of raw glycerin as the sole carbon source. The resulting microbial mass contains more than 50% of its dry weight as lipids. These microorganisms are nonpathogenic and are not genetically modified.

Microbial oil can be extracted by common extraction methods due to its similar composition to oils conventionally used for biodiesel production (e.g., soybean, sunflower oils). This oil can be used as new raw material to produce biodiesel, as its properties and fatty acid composition are in accordance with the European (EN14214) and American (ASTM D6751) standards. It is also possible to carry out methanolysis with the microbial biomass to directly obtain the mixture of methyl ester.

Both the process and the strains used for this technology have been fully developed and patented by Neuron BioIndustrial. MicroBioOil is marketed through licensing agreements and technical consultancy contracts.

5.12.2. Microbes and Biodiesel: Production and Feedstocks

Biodiesel refers to a vegetable oil or animal-fat-based diesel fuel consisting of long-chain alkyl (methyl, propyl, or ethyl) esters. Biodiesel is typically made by chemically reacting

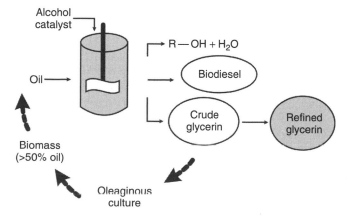

Figure 5.75. The process of formation of microbial oil.

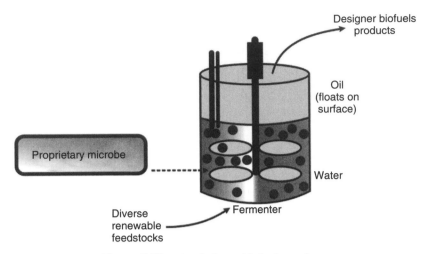

Figure 5.76. The designer biofuels products.

lipids (e.g., vegetable oil, animal fat (tallow)) with an alcohol. Biodiesel is meant to be used in standard diesel engines and is thus distinct from the vegetable and waste oils used in fuel-converted diesel engines. Biodiesel can be used alone, or blended with petrodiesel. A company, LS9, claims to have engineered their own microbes, lifting genes from other microbes and recombining them into an organism that does just what they want. In this way they can precisely tweak the characteristics of the resulting fuel. LS9 claims that by tweaking its microbes it can produce "designer biofuels" that are, in the lingo, "fit for purpose" (Fig. 5.76). That is to say, they can be matched precisely to the required use. One product is "bio-crude," which can substitute directly for crude oil. It can be refined into gas or used to make all the many petroleum products we know and love, such as plastics, fertilizers. Other products can go directly into tanks, including bioequivalents to gasoline, diesel, and even jet fuel (http://www.biodieselnow.com/general_biodiesel_21/f/5/t/17976.aspx). Chemically speaking, hydrocarbons are hydrocarbons; the products of LS9 are essentially identical to their fossil-based counterparts. They can do whatever oil products can do, without the need for special equipment.

Biodiesel is a fuel composed of monoalkyl esters traditionally derived from vegetable oils or animal fats. There is currently an unprecedented increase in interest and demand for biodiesel and other fuels derived from renewable biomass. However, pure vegetable or seed oils are expensive and constitute between 70% and 85% of the overall biodiesel production cost. Municipal sewage sludge is gaining traction in the United States and around the world as a lipid feedstock for biodiesel production. It is plentiful and consists of significant concentrations of lipids that can make production of biodiesel from sludge profitable. However, there are challenges to be faced by biodiesel production from waste sludge. Determining how best to collect the different fractions and treat them for maximum lipids extraction is a major challenge. To accelerate biodiesel production, cosolvents and high shear mixing have been proposed. Nevertheless, there is very little information on the cost-effective means of increasing lipid solubility. Alkali-catalyzed transesterification is much faster than acid-catalyzed transesterification and is most often used commercially. However, for lipid feedstocks with greater than 1% free fatty acids (FFAs), such as in sludge, acid catalysis followed by base catalysis is recommended because of soap formation with alkali-catalyzed transesterification and high FFA content. To boost biodiesel production, it is suggested that wastewater operators utilize microorganisms that are selected for their oil-producing capabilities. This could increase biodiesel production to the 10 billion gal mark. The presence of pharmaceutical chemicals in sludge poses a great challenge. This requires a careful selection of treatment technologies and microbes that are selective for pharmaceutical chemicals. Finally, biodiesel production from sludge could be very profitable in the long run. Currently, the estimated cost of production is $3.11 per gallon of biodiesel. To be competitive, this cost should be reduced to levels that are at or below the current petro diesel cost of $3.00 per gallon.

Algae fuel, also called *algal fuel, algaeoleum*, or *third-generation biofuel*, is a biofuel derived from algae. During photosynthesis, algae and other photosynthetic organisms capture carbon dioxide and sunlight and convert it into oxygen and biomass. Up to 99% of the carbon dioxide in solution can be converted, which was shown by Weissman and Tillett (1992) in large-scale open pond systems. Several companies and government agencies are funding efforts to reduce capital and operating costs and make algae fuel production commercially viable. The production of biofuels from algae does not reduce atmospheric carbon dioxide (CO_2) levels because any CO_2 taken out of the atmosphere by the algae is returned when the biofuels are burned. They do, however, eliminate the introduction of new CO_2 by displacing fossil hydrocarbon fuels.

High oil prices, competing demands between foods and other biofuel sources, and the world food crisis have ignited interest in algaculture (farming algae) for making vegetable oil, biodiesel, bioethanol, biogasoline, biomethanol, biobutanol, and other biofuels, using land that is not suitable for agriculture. Algal fuels' attractive characteristics include the following: they do not affect freshwater resources, they can be produced using ocean water and wastewater, and they are biodegradable and relatively harmless to the environment if spilled. Algae cost more per unit mass (as of 2010, food grade algae costs ~$5000/tonne) because of the high capital and operating costs, yet can theoretically yield between 10 and 100 times more energy per unit area than other second-generation biofuel crops. One biofuel company has claimed that algae can produce more oil in an area the size of a two-car garage than a football field of soybeans because almost the entire algal organism can use sunlight to produce lipids or oil. The US Department of Energy estimates that if algae fuel replaced all the petroleum fuel in the United States, it would require 15,000 mi^2 (40,000 km^2). This is less than one-seventh the area of corn harvested in the United States in 2000. However, these claims remain commercially unrealized.

5.12.3. Genetic Modification of Lignin Biosynthesis for Improved Biofuel Production

Finite petroleum reserves and the increasing demands for energy in industrial countries have created international unease. For example, the dependence of the United States on foreign petroleum both undermines its economic strength and threatens its national security. As highly populated countries such as China and India become more industrialized, they too might face similar problems. It is also clear that no country in the world is untouched by the negative environmental effects of petroleum extraction, refining, transportation, and use. For these reasons, governments around the world are increasingly turning their attention to biofuels as an alternative source of energy. Serious efforts to produce cellulosic ethanol on an industrial scale are already underway. Notably, in 2006, the US President George W. Bush announced the goal of reducing 30% of foreign oil requirements by 2030 using crop biomass for biofuel production.

As a result, the Department of Energy announced the funding of three major biofuel centers and the establishment of six cellulosic ethanol refineries, which, when fully operational, are expected to produce more than 130 million gal of cellulosic ethanol per year. Other than the Canadian Iorgen plant, no commercial cellulosic ethanol plant is yet in operation or under construction. However, research in this area is underway and funding is becoming available around the world for this purpose, from both governmental and commercial sources. For example, the British Petroleum has donated half a billion dollars to US institutions to develop new sources of energy, primarily biofuel crops. Presently, several problems face the potential commercial production of cellulosic ethanol. First, the high costs of production of cellulases in microbial bioreactors. Second, and most important, the costs of pretreating lignocellulosic matter to break it down into intermediates and remove the lignin to allow the access of cellulases to biomass cellulose. These two costs together make the price of cellulosic ethanol about two- to three-fold higher than the price of corn grain ethanol.

Plant genetic engineering technology offers great potential to reduce the costs of producing cellulosic ethanol. First, all necessary cell-wall-degrading enzymes such as cellulases and hemicellulases could be produced within the crop biomass so there would be no need, or only minimal need, for producing these enzymes in bioreactors. Second, plant genetic engineering technology could be used to modify lignin amount and/or configuration in order to reduce the needs for expensive pretreatment processes. Finally, future research on the upregulation of cellulose and hemicellulose biosynthesis pathway enzymes for increased polysaccharides will also have the potential to increase cellulosic biofuel production.

The factors that affect the suitability of potential new feedstock crops around the globe for bioethanol production are complex and relate to country- and region-specific agricultural practices, market forces, and political as well as biological issues. These factors include land availability, locally accepted cropping systems, and types and forms of transportation fuel. In addition, the current status of a particular species in terms of its development as a crop (e.g., the development of breeding strategies) is another important issue; in terms of biology, the feedstock crops that have so far been recommended for conversion to cellulosic ethanol have a high amount of cellulosic biomass. These crops include corn, rice, sugarcane, fast-growing perennial grasses such as switchgrass and giant miscanthus, and woody crops such as fast-growing poplar and shrub willow. Depending on where they are planted, the ideal characteristics of nonfood cellulosic crops are use of the C4 photosynthetic pathway, long canopy duration, perennial growth, rapid growth in spring (to outcompete weeds), high water usage efficiency, and possibly, partitioning of nutrients to subterranean storage organs in the autumn.

The source of lignocellulosic biomass is the plant cell wall, which has important roles in determining the structural integrity of the plant and in defense against pathogens and insects. The structure, configuration, and composition of cell wall vary depending on plant taxa, tissue, age, and cell type, and also within each cell wall layer. The basic structure of the primary cell wall is a scaffold of cellulose with cross-linking glycans, and there are two types of primary cell wall, which are classified according to the type of cross-links.

5.12.4. Microbes and Biogas Production

Presently, four different microbial groups are considered to be involved in the process of biogas production. The hydrolytic bacteria that catabolize carbohydrates, proteins, lipids, and other components of biomass to fatty acids, H_2, and CO_2. The hydrogen-producing acetogenic bacteria that catabolize certain fatty acids and neutral end products of the group "first" to acetic acid, CO_2, and H_2. The homoacetogenic bacteria synthesize acetate using H_2, CO_2, and formate or hydrolyzed multicarbon compounds to acetic acid. The methanogenic archaebacteria utilize acetic acid, CO_2, and H_2 to produce methane gas.

The first three microbial groups are represented by facultative as well as strict anaerobic bacteria such as *Bacillus*, *Cellulomonas*, *Clostridium*, *Eubacterium*, and *Ruminococcus*, while the fourth group, the methanogenic archaebacteria, is represented by *Methanosarcina*, *Methanothrix*, *Methanobacterium*, and *Methanospirillum*. Biogas typically refers to the gas produced by the biological breakdown of organic matter in the absence of oxygen. Biogas originates from biogenic material and is a type of biofuel.

Biogas is produced by anaerobic digestion or fermentation of biodegradable materials such as biomass, manure, sewage, municipal waste, green waste, plant material, and energy crops (Fig. 5.77). This type of biogas comprises primarily of methane and carbon dioxide. Another type of gas generated using biomass is wood gas, which is produced by gasification of wood or other biomass. This type of gas consists primarily of nitrogen, hydrogen, and carbon monoxide, with trace amounts of methane.

The gases methane, hydrogen, and carbon monoxide can be combusted or oxidized with oxygen. Air contains 21% oxygen. This energy release allows biogas to be used as a fuel. Biogas can be used as a low cost fuel in any country for any heating purpose, such as cooking. It can also be used in modern waste management facilities where it can be used to run any type of heat engine to generate either mechanical or electrical power.

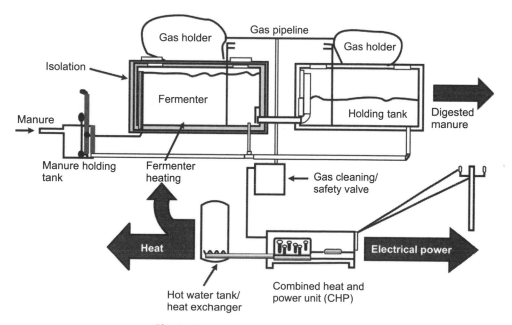

Figure 5.77. Biogas production plant.

Biogas can be compressed, much like natural gas, and used to power motor vehicles, and in the United Kingdom, for example, it is estimated to have the potential to replace around 17% of vehicle fuel. Biogas is a renewable fuel, so it qualifies for renewable energy subsidies in some parts of the world.

5.12.5. Biosensors for Environmental Monitoring

A biosensor is an analytical device composed of a biological sensing element (enzyme, receptor, antibody, or DNA) in intimate contact with a physical transducer (optical, mass, or electrochemical) that together relates the concentration of an analyte to a measurable electrical signal. In theory, and as verified to a certain extent in the literature, any biological sensing element may be paired with any physical transducer. The majority of reported biosensor research has been directed toward the development of devices for clinical markets; however, driven by the need for better methods for environmental surveillance, research into this technology is also expanding to encompass environmental applications.

Biosensors are being developed for different applications, including environmental and bioprocess control, quality control of food, agriculture, military, and, particularly, medical applications. A biosensor is a device that detects, records, and transmits information regarding a physiological change or the presence of various chemical or biological materials in the environment (Fig. 5.78). More technically, a biosensor is a probe that integrates a biological component, such as a whole bacterium or a biological product (an enzyme or antibody), with an electronic component to yield a measurable signal. Biosensors, which come in a large variety of sizes and shapes, are used to monitor changes in environmental conditions. They can detect and measure concentrations of specific bacteria or hazardous chemicals; they can measure acidity levels (pH). In short, biosensors can use bacteria and detect them too. For environmental control and monitoring, biosensors can provide fast and specific data of contaminated sites. They offer other advantages over current analytical methods, such as the possibility of portability and working on site and the ability of measuring pollutants in complex matrices with

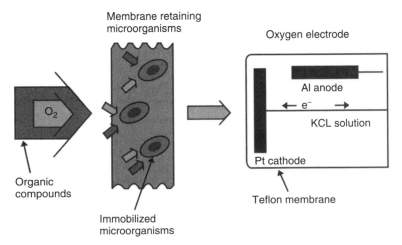

Figure 5.78. The principle of microbial electrode.

TABLE 5.10. Various Types of Biosensors Used

Analyte	Type of Interaction	Recognition Biocatalyzer	Transduction System
Pesticides			
Parathion	Biocatalytic	Parathion hydrolase	Amperometric
Propoxur and carbaryl	Biocatalytic	Acetyl cholinesterase	Fiber-optic
Diazinon and dichlorvos	Biocatalytic	Tyrosinase	Amperometric
Fertilizers			
Nitrate	Biocatalytic	Nitrate reductase	Amperometric
Nitrite	Biocatalytic	Nitrite reductase	Optical
Phosphate	Biocatalytic	Polyphenol oxidase and alkaline phosphatase, phosphorylase A, phosphoglucomutase, and glucose-6-phosphate dehydrogenase	Amperometric
Heavy Metals			
Copper and mercury	Biocatalytic	*Spirulina subsalsa*	Amperometric
Copper	Biocatalytic	Recombinant *Saccharomyces cerevisiae*	Amperometric
Cadmium and lead	Biocatalytic	*Staphylococcus aureus* or recombinant *Bacillus subtilis*	Optical
Arsenic, cadmium, and bismuth	Biocatalytic	Cholinesterase	Electrochemical
Cadmium, copper, chrome, nickel, zinc	Biocatalytic	Urease	Optical
Copper and mercury	Biocatalytic	Glucose oxidase	Amperometric

minimal sample preparation. On the other hand, biosensors offer the possibility of determining not only specific chemicals but also their biological effects, such as toxicity or endocrine-disrupting effects, which are sample information of great interest (Table 5.10).

Biosensors are usually classified into various basic groups, according to the signal transduction and the biorecognition principles. On the basis of the transducing element, biosensors can be categorized as electrochemical, optical, piezoelectric, and thermal. The electrochemical biosensors, and among them the amperometric and the potentiometric ones, are the best described in the literature; those based on optical principles are the next most commonly used transducers. In fact, most catalytic biosensors are based on electrochemical methods, whereas affinity biosensors have generally proved more amenable to optical detection methods. The various types of optical transducers exploit properties such as simple light absorption, fluorescence/phosphorescence, bio/chemiluminescence, reflectance, Raman scattering, and refractive index. Surface plasmon resonance (SPR) is another common transduction mechanism whose main advantage over most optical biosensors is that the analyte presence can be determined directly, without the use of labeled molecules. Finally, cantilever biosensors are an emerging group of biosensors, which are based on the bending of silicon cantilevers caused by the adsorption of target molecules onto the cantilever surface, where receptor molecules are immobilized. According to the biorecognition principle, biosensors are classified into immunochemical, enzymatic, nonenzymatic receptor, whole-cell, and DNA biosensors. Immunosensors present the advantages of sensitivity and selectivity inherent to the use of immuno-chemical interactions. Limitations are the troubles derived from the regeneration of the immunosurface and cross-reactivity, although a certain degree of cross-reactivity is often desirable in order to determine different congeners of the same family. Enzymes are suitable to act as recognition elements because of their specificity and the availability of a wide range of enzymes. In general, enzymatic biosensors are based on the selective inhibition of specific enzymes by different classes of compounds. The decrease of activity of the immobilized enzyme in the presence of the target analyte is frequently used for its quantification. Biosensors based on natural receptors can be built by integrating the specific receptor within a membrane and by coupling it to a transducing device. These natural receptors are proteins of noncatalytic or nonimmunogenic origin, which span cell membranes and can specifically bind certain compounds. The binding signal is detected as a structural change or an associated enzyme activity. Whole cells of living organisms, such as bacteria, yeast, fungi, plant and animal cells, or even tissue slices have been used as the recognition component by interrogating their general metabolic status. These biosensors are useful to determine the toxicity of the substrate to certain cells. Another application of whole-cell biosensors is the determination of the "biological oxygen demand" (BOD). In the case of DNA biosensors, two strategies are applied to detect pollutants: one is the hybridization detection of nucleic acid sequences from infectious microorganisms and the other one is the monitoring of small pollutants interacting with the immobilized DNA layer (drugs, mutagenic pollutants, etc.). One key step in the development of biosensors is the immobilization of the biological component at the transducer surface. The immobilization procures both the stabilization of the biomaterial and the proximity between the biomaterial and the transducer. The immobilization methods most generally used are physical adsorption at a solid surface, cross-linking between molecules, covalent binding to a surface, and entrapment within a membrane, surfactant matrix, polymer, or microcapsule. In addition to these conventional methods, sol–gel entrapment, Langmuir–Blodgett (LB) deposition, electropolymerization, self-assembled biomembranes, and bulk modification have also been used. Biosensors developed for environmental monitoring measure an effect such as toxicity and endocrine effect and detect a compound or a group of compounds based on the specific recognition of a biomolecule Bogue (2003). A wide range of compounds of environmental concern or under suspicion are being considered for biosensor development (Table 5.10). Genetically engineered bacteria can give off a detectable signal, such as light, in the presence of a specific pollutant they like to feed on. They may glow in the presence of toluene, a hazardous compound found in gasoline and other petroleum products. They can indicate whether an underground fuel tank is leaking or whether the site of an oil spill has been cleaned up effectively. These informer bacteria are called *bioreporters*.

5.12.6. Microbes as Mining Agents

Many microbes are central geological agents that influence elemental cycles and hence recycle key nutrients and degrade organic as well as inorganic matter (Ehrlich, 2002). "Microbial biogeochemistry is the study of microbially influenced geochemical reactions, enzymatically catalyzed or not, and their kinetics." In biohydrometallurgy, microbes are used as geological agents by mining and water treatment industries in order to find new approaches and technologies to process ores and concentrates, remediate wastewater, and recover and recycle metals. The type of microbe used in biohydrometallurgy to transform

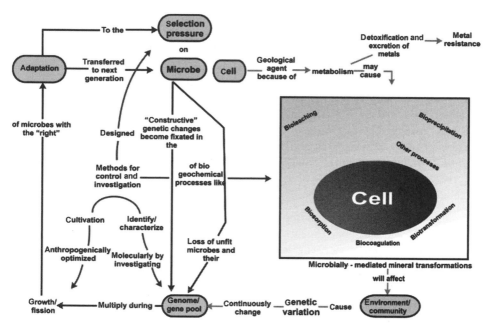

Figure 5.79. Microbes as mining agents (Ehrlich, 2002).

minerals is almost exclusively prokaryotes. The microbes used, in order to process and extract valuable metals from metal-bearing materials (bioleaching microbes), are different from those that are used for remediating and treating metal-containing solutions (SRB). The invention of this application generally relates to the recovery of mineral values from subterranean formations or strata and particularly to the recovery of mineral values that are soluble in sulfuric acid (Fig. 5.79).

Numerous methods of recovering mineral values from strata are known. These mineral values are recovered from the ore by a variety of methods that is generally referred to herein as *solubilizing*. These methods generally solubilize or dissolve the desired mineral value by solvent action, chemical conversion to a more soluble form, chemical reaction with an agent in a carrier fluid, or changing the conditions in the strata or source materials so that the desired mineral value is mobilized and can be suspended in a carrier fluid.

The process of this invention can be generally applied to any strata or source material containing mineral values soluble in sulfuric acid. Known processes of solubilizing such mineral values require high volumes of solubilizing fluid and displacement fluids for an even modest recovery.

The invention of this application provides a simple process that can be used to recover substantially all the desired minerals using *in situ* solubilizing methods and *in situ* generation of the solubilizing agent. The process of this invention avoids the loss of high mineral concentration liquor in stagnant areas of the stratum.

It has been discovered that a highly efficient process for leaching mineral values from a stratum by *in situ* leaching can be accomplished by generating the leaching reagent or solubilizing agent *in situ*, that is, generating the leaching agent in the leaching fluid after the fluid has contacted or saturated the stratum to be leached. As initially injected, the leaching liquid does not have a leaching reagent concentration sufficient to leach a significant quantity of the mineral value from the stratum. The leaching reagent is generated *in situ* in the leaching fluid after the fluid has contacted the stratum. In previous methods, a high concentration of leaching reagent was required in the leaching fluid to leach and keep the mineral value dissolved in the fluid that swept through the stratum. This formed a region of high mineral concentration in the leading portion of the fluid, which was eventually trapped in the stagnant areas of the flow pattern. Thus, much of the mineral values were lost, wasting both leaching agent and mineral values.

By this invention, a simple process has been developed for recovering mineral values from a source material such as a subterranean stratum in high yield using a relatively low intial volume of sulfuric acid and low volumes of sweep or displacement fluid. This is a process for recovering mineral values from a stratum leachable by a fluid containing a leaching or solubilizing reagent. The process comprises the following steps.

1. Injecting into the said stratum, to contact said mineral value, a leaching fluid that contains said leaching reagent at a concentration insufficient to substantially leach the said mineral value. The leaching fluid contains a means for generating the leaching reagent *in situ* in the said stratum to produce a concentration of leaching reagent sufficient to leach the desired mineral value contacted in the said stratum.

2. Leaching the said contacted mineral value.

3. Recovering from the said stratum the said leaching fluid containing said leached mineral value such that a major portion of the said leached mineral value is recovered, with the concentration of the said leached mineral value in the recovered leaching fluid being relatively uniform throughout the fluid recovery cycle.

5.12.7. Microbial Fuel Cells

Microbial fuel cells (MFCs), or biological fuel cells, use bacteria to convert chemical energy in biodegradable materials, such as wastewater pollutants into electricity. The bacteria consume the pollutants, releasing electrons that flow through a circuit and generate electricity. In this process of power generation, pollutants are broken down and clean water is produced.

An MFC or biological fuel cell is a bioelectrochemical system that drives a current by mimicking bacterial interactions found in Nature. Mediator-less MFCs are a much more recent development, and due to this the factors that affect optimum operation, such as the bacteria used in the system, the type of ion membrane, and the system conditions such as temperature, are not particularly well understood. Bacteria in mediator-less MFCs typically have electrochemically active redox enzymes, such as cytochromes, on their outer membrane, which can transfer electrons to external materials.

An MFC is a device that converts chemical energy to electrical energy by the catalytic reaction of microorganisms (Fig. 5.80). A typical MFC consists of anode and cathode compartments separated by a cation (positively charged ion)-specific membrane. In the anode compartment, fuel is oxidized by microorganisms, generating electrons and protons. Electrons are transferred to the cathode compartment through an external electric circuit, and the protons are transferred to the cathode compartment through the membrane. Electrons and protons are consumed in the cathode compartment, combining with oxygen to form water. In general, there are two types of MFCs: mediator and mediator-less MFCs.

An MFC follows the basic principle of a fuel cell. It consists of batterylike terminals: an anode and a cathode electrode. An external circuit connects the two electrodes, and

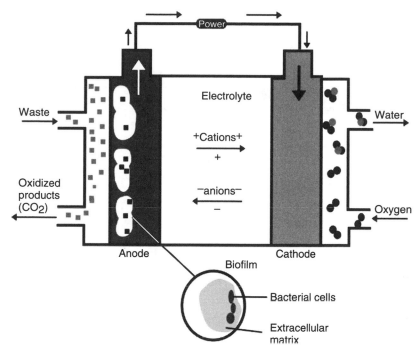

Figure 5.80. The microbial fuel cell.

an electrolyte solution helps conduct electricity. The anode and cathode are separated by a proton or cation exchange membrane. There is a difference in voltage between the two electrodes, and when electrons flow through the circuit, electric power is generated.

In an MFC, bacteria live at the anode and convert organic substrates in wastewater into CO_2, protons, and electrons. Under aerobic conditions, oxygen or nitrate acts as terminal electron receptors to produce water. However, oxygen is not available in an MFC, so the bacteria need an insoluble receptor such as the anode. The electrons are transferred by membrane-associated components, electron shuttles, or nanowires and then flow through the external circuit to the cathode, generating electricity. The protons flow across the proton exchange membrane to reach the cathode. At the cathode, oxygen is reduced as the electron receptor to produce water (Fig. 5.81).

Although MFCs did not receive much attention because the amount of energy they generate is too low to be of use, improvements over the years make this potential energy source a valuable research topic in the field of biotechnology.

MFCs have a number of potential uses. The first and most obvious is harvesting the electricity produced for a power source. Virtually any organic material could be used to "feed" the fuel cell. MFCs could be installed to wastewater treatment plants. The bacteria would consume waste material from the water and produce supplementary power for the plant. The advantages of using MFCs are that it is a very clean and efficient method of energy production. Chemical processing wastewater and designed synthetic wastewater have been used to produce bioelectricity in dual- and single-chambered mediator-less MFCs (noncoated graphite electrodes), apart from wastewater treatment. Higher power production was observed with biofilm-covered anode (graphite). A fuel cell's emissions are well below regulations. MFCs also use energy much more efficiently than standard

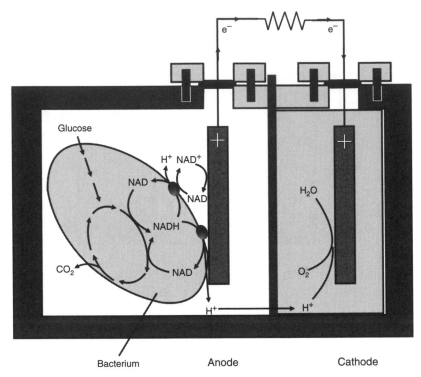

Figure 5.81. Microbial fuel cells: novel biotechnology for energy generation. (Rabaey and Verstraete, 2005).

combustion engines, which are limited by the Carnot cycle. In theory, an MFC is capable of energy efficiency far beyond 50%. According to the new research conducted by René Rozendal, using the new MFCs, conversion of the energy to hydrogen is eight times as high as conventional hydrogen production technologies.

However, MFCs do not have to be used on a large scale, as the electrodes in some cases need only be 7 μm thick by 2 cm long. The advantages to using an MFC in this situation as opposed to a normal battery is that it uses a renewable form of energy and would not need to be recharged like a standard battery. In addition to this, they could operate well in mild conditions, at 20–40 °C, and also at pH of around 7. They are more powerful than metal catalysts but are currently too unstable for long-term medical applications such as in pacemakers (Biotech/Life Sciences Portal).

Besides wastewater power plants, as mentioned before, energy can also be derived directly from crops. This allows the setup of power stations based on algae platforms or other plants, incorporating a large field of aquatic plants. The fields are best set up in synergy with existing renewable plants (e.g., offshore windturbines). This reduces costs, as the MFC plant can then make use of the same electricity lines as the wind turbines.

5.12.8. Microbes as a Source of Bioenergy

Considering, the prevalence, durability, and efficiency of microbes, the notion of utilizing the microscopic organisms as a means of generating sustainable energy sounds too good

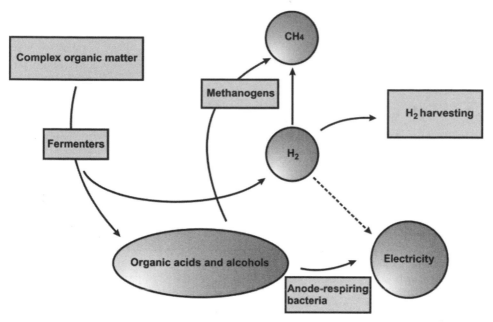

Figure 5.82. Microbes can be used as a source of bioenergy.

to be true. Rittmann et al. (2008) outlined paths where bacteria are the best hope in producing renewable energy in large quantities without damaging the environment or competing with our food supply.

Fermentation produces H_2, simple organic acids, and alcohols. This H_2 can be harvested as an energy source, but methanogens and, perhaps, anode-respiring bacteria consume H_2. Because the accumulation of too much H_2 inhibits fermentation, H_2 must be removed by harvesting, methanogens or anode-respiring bacteria to enable the conversion process to produce more H_2 and organic products (Fig. 5.82; Rittmann et al., 2008).

They can do it in two ways that are complementary. In the first way, communities of anaerobic bacteria convert the energy value (contained in electrons) of biomass into socially useful bioenergy forms: methane (natural gas), hydrogen gas, or electricity. The biomass can be wastes from agriculture, animals, industry (such as food industry), and humans (sewage). In this case, the capturing of the bioenergy is also a means to remove the pollution in the waste material. In addition, the biomass could be made especially for being a bioenergy source, which leads to the second way. In the second way, photosynthetic microorganisms capture sunlight energy through photosynthesis to make more of self. The microorganisms can be algae or cyanobacteria. Some of the photosynthetic microbes contain a high proportion of lipids, or oils, that can be used as a feedstock for liquid fuels, such as biodiesel. The nonlipid portions constitute biomass for feeding into the first way for energy conversion.

For the photosynthetic systems, we think that certain microbe types are more efficient for capturing sunlight energy and converting it to high value energy material, such as lipids, at a high rate. Cyanobacteria are reported to be very good at forming lipids and growing very fast relatively. For the other systems, one cannot rely on one type of microorganism but on communities of microorganisms that work synergistically to convert the biomass to, methane, hydrogen, or electricity. In some cases (mainly for making methane), one knows the key members of the community pretty well.

5.13. MICROBES AS A DISASTROUS ENEMY

Bacteria are an integral part of our environment and play many beneficial, but sometimes harmful, roles. They are found on all raw agricultural products. Harmful bacteria can be transferred from food to people, people to food, or from one food to another. Bacteria can grow rapidly at room temperature. Growth of harmful bacteria in food may be slowed or stopped by refrigerating or freezing. Foodborne illness can produce symptoms from mild to very serious. Illness can occur 30 min to 2 weeks after eating food containing harmful bacteria. People who are most likely to become sick from food-related illness are infants and young children, senior citizens, and people with weakened immune systems.

Humans evolved in the presence of numerous microbial communities that preceded the appearance of mammals on this planet. The role of these microbial communities in our own evolution is a matter of considerable interest. Indeed, comparative studies in germ-free and conventional animals have established that the intestinal microflora is essential for the development and function of the mucosal immune system during early life, a process that is now known to be important to overall immunity in adults. An absence of intestinal bacteria is associated with reduction in mucosal cell turnover, vascularity, muscle wall thickness, motility, baseline cytokine production, and digestive enzyme activity, as well as in defective cell-mediated immunity. Furthermore, the intestinal microflora makes an important metabolic contribution of vitamin K, folate, and short-chain fatty acids such as butyrate, a major energy source for enterocytes, and also mediates the breakdown of dietary carcinogens.

Marine algae represent an unexploited biomass source that could provide a cheap, clean, and renewable fuel source that is ideally suited for production in Florida. Algae are arguably the most promising nonfood source of biofuels, producing an yield that is 10 times higher than land crops, and using the ocean would mean farmland that could grow food would not be a part of the biofuel equation. The US Department of Energy estimates that in the United States, to replace all the petroleum fuel with algae-derived biofuel, under intensive cultivation, requires only 15,000 mi^2, roughly the size of the state of Maryland. With one of the largest coastlines in the United States and all population centers near the coast, Florida is positioned to develop a nearly unlimited supply of biomass to support biofuel production from its surrounding oceans. Marine algae use simple nutrients and energy from the sun to support the base of the food chain in the world's oceans, and much less research is available to demonstrate their fuel potential in comparison to land-based biomass sources (Debnath et al., 2007).

Microorganisms naturally decompose dead/dying algae in the oceans at temperatures ranging from -2 to more than $100\,°C$, and these microbes could be exploited to produce biomass-degrading enzymes specific to the processing of algal carbon to form soluble sugars that can be fermented to ethanol. Ethanol refining requires harsh (hot, acidic) conditions for microbes. Marine microbes are need to be exploited, as they grow in hostile or extreme environments, for discovering new enzymes that have evolved naturally to deal with harsh conditions (Debnath et al., 2007).

A state-of-the-art off-the-grid algal cultivation facility will allow us to produce algae with a minimal ecological footprint. Once cultivated, the algae yield lipids, carbohydrates,

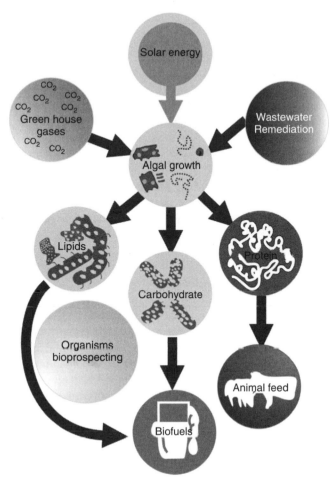

Figure 5.83. Algal cultivation facility allowing a minimal ecological footprint.

and protein, which can be processed into biofuels using microorganisms or used as biomass in animal feed (Fig. 5.83). Wastewater from the process (and other sources) can then be reintroduced into the system during cultivation, making this a highly efficient, ecologically friendly alternative for producing fuel from the sun.

The bacterial flora of the gastrointestinal tract varies on a longitudinal basis, with the oral cavity containing about 200 different species, the stomach being almost sterile, and the bacterial content increasing distally, with approximately 10^8 bacteria per gram (dry weight) of ileal content and up to 10^{12} bacteria per gram (dry weight) of colonic content. Studies carried out some decades ago on the bacterial flora are still thought to hold true. The large intestine contains organisms from over 30 identified genera and as many as 500 separate species or phenotypes. The main type of bacteria in the colon are obligate anaerobes, the most abundant being members of the genus *Bacteroides* and anaerobic gram-positive cocci such as *Peptostreptococcus* sp., *Eubacterium* sp., *Lactobacillus* sp., and *Clostridium* sp. More recent 45 studies of large bowel biopsies have confirmed that *Bacteroides* is a dominant isolate from these specimens. Thus, intestinal bacteria represent a complex and incompletely understood microbiome, and certain organisms are thought to play a role in the onset of inflammatory diseases of the bowel, whereas others are considered protective.

5.13.1. Microbes as Biowarfare Agents

Biological warfare (BW), also known as *germ warfare*, is the use of pathogens such as viruses, bacteria, other disease-causing biological agents, or the toxins produced by them as biological weapons (or bioweapons). A biological weapon may be intended to kill, incapacitate, or seriously impair a person, group of people, or even an entire population. It may also be defined as the material or defense against such actions. BW is a military technique that can be used by nation-states or nonnational groups. In the latter case, or if a nation-state uses it clandestinely, it may also be considered bioterrorism.

Analysis of the properties of microbes that are currently considered biological weapons against humans revealed no obvious relationship to virulence, except that all are pathogenic to humans. Until recently, the use of biological weapons was considered more from an academic than practical point of view. The list of agents and/or toxins that can be used as biological weapons is long (Tables 5.11 and 5.12). Bacteria, rickettsia, viruses, fungi, protozoa, and toxins can all be used as biological weapons (Burrows and Renner, 1999). Potential biological agents include organisms causing smallpox, anthrax, plague, tularemia, brucellosis, Q fever, viral encephalitis, hemorrhagic fever, and botulinum toxin and staphylococcal enterotoxin B (Leggiadro, 2000). The infection may be acquired by inhalation of aerosols, ingestion of contaminated water or food, or direct contact with infectious agents. Early recognition, diagnosis, and treatment of infected patients are of utmost importance. Special attention must be given to the use of genetically modified microorganisms. Medical protection from biological weapons and continuous education is very important (Bojić et al., 2007). Some of them are highly lethal, while others cause morbidity and disability. Notably, the weapon potential of a microbe rather than its pathogenic properties or virulence appeared to be the major consideration when categorizing certain agents as biological weapons. In an effort to standardize the assessment of the risk that is posed by microbes as BW agents using the basic principles of microbial communicability (defined here as a parameter of transmission) and virulence, a simple formula is proposed for estimating the weapon potential of a microbe (Casadevall and Pirofski, 2004). Fungi cause disease directly by infection or indirectly through mycotoxins. Fungi that are used as weapons might be targeted against humans, livestock, or crops. Humans and animals encounter fungi and mycotoxins

TABLE 5.11. Agents of Biological Warfare

Bacterial Agents	Viruses	Biological Toxins
Anthrax	Smallpox	Botulinum
Brucellosis	VEE	Staphylococcal enterotoxin B
Cholera	VHF	Ricin
Plague (pneumonic)		T-2 mycotoxins
Tularemia		
Q fever		

Abbreviations: VEE, Venezuelan equine encephalitis; VHF, viral hemorrhagic fever.

TABLE 5.12. Potential Airborne Biological Weapons

Airborne Microbe or Toxin	Type	Disease or Infection
Bacillus anthrax spores	Bacteria	Anthrax
Brucella		Brucellosis
Chlamydia pneumoniae		—
Clostridium perfringens		Toxicosis
Corynebacterium diphtheriae		Diphtheria
Coxiella burnetti		Q fever
Francisella tularensis		Tularemia
Klebsiella pneumoniae		—
Legionella spp.		—
Mycobacterium tuberculosis		TB
Neisseria meningitidis		—
Pseudomonas mallei		Glanders
Pseudomonas pseudomallei		—
Streptococcus pneumoniae		—
Yersinia pestis		Pneumonic plague
Coccidioides immitis	Fungi	Coccidiodomycosis
Histoplasma capsulatum		Histoplasmosis
Stachybotrys atra (*S. chartarum*)		—
Camelpox (GE)	Virus	—
Crimean hemorrhagic fever		Hemorrhagic fever
Ebola (GE)		Hemorrhagic fever
Hantaan		—
Influenza		Influenza
Junin		Hemorrhagic fever
Lassa		Lassa fever
Marburg		Hemorrhagic fever
Variola		Smallpox
EEE (VEE, WEE)		Encephalitis
Rickettsiae (various)	—	—
Aflatoxin	Toxin	Toxicosis
Botulinum		Toxicosis
Ricin		Toxicosis
Trichothecene mycotoxins		Toxicosis
Abrin		Toxicosis
Staphylococcus a		Toxicosis
Staphylococcus enterotoxin B		Toxicosis
Batrachotoxin A		Toxicosis
Diphtheria toxin		Toxicosis
Palytoxin		Toxicosis
Saxitoxin		Toxicosis
T-2 mycotoxin		Toxicosis

Abbreviations: EEE, eastern equine encephalitis; VEE, Venezuelan equine encephalitis; WEE, western equine encephalitis.

through inhalation, ingestion, and contact with skin and mucous membranes. Effective fungal bioweapons would require the ability to cause significant destruction and a means of delivery to target populations or farms. Effective countermeasures against fungal bioweapons would be able to prevent or treat this damage (Klassen-Fischer, 2006). None of the fungi are as toxic as the botulinum toxin from *Clostridium botulinum* and as dangerous as nuclear weapons (Paterson, 2006). With the exception of *Coccidioides* spp., human pathogenic fungi are not found among lists of microbes with potential for BW and bioterrorism against humans. However, many human pathogenic fungi are easily obtainable from the environment, are highly dispersible, and can cause significant disease after inhalation with relatively low inocula (Casadevall and Pirofski, 2006).

As a tactical weapon, the main military problem with a BW attack is that it would take days to be effective and therefore, unlike a nuclear or chemical attack, would not immediately stop an opposing force. Some biological agents (especially organisms causing smallpox, plague, and tularemia) have the capability of person-to-person transmission via aerosolized respiratory droplets, which can be undesirable, especially if they are transmitted to unintended target populations, including neutral or even friendly forces. Containment of transmission is less of a concern for terrorists, but it was very much a concern for post-WWII BW development by major powers. These biological weapons are also very unstable.

Although biological weapons may serve limited purpose as a battlefield weapon, they may act as an effective strategic deterrence on par with nuclear weapons. Any nation that can pose a credible threat of mass casualty has the ability to alter the terms in which other nations interact with it. Biological weapons allow for the potential to create a staggering level of destruction and loss of life (Table 5.13). As such, they can act as a strategic deterrent.

5.14. FUTURE CHALLENGES: MICROBES IN THE SPACE

On April 20, 1967, the unmanned lunar lander Surveyor 3 landed near Oceanus Procellarum on the surface of the moon. One of the things aboard was a television camera. Two and a half years later, on November 20, 1969, Apollo 12 astronauts Pete Conrad and Alan L. Bean recovered the camera. When NASA scientists examined it back on Earth, they were surprised to find specimens of *Streptococcus mitis* that were still alive. Because of the precautions the astronauts had taken, NASA could be sure that the germs were inside the camera when it was retrieved, so they must have been there before the Surveyor 3 was launched. These bacteria had survived for 31 months in the vacuum of the moon's atmosphere. Perhaps NASA should not have been surprised because there are other bacteria that thrive under near-vacuum pressure on the Earth today. Anyway, we now know that the vacuum of space is not a fatal problem for bacteria (http://www.panspermia.org/bacteria.htm).

Many microorganisms, including bacteria, can survive in the hostile space environment. This is the most extreme environment for microorganisms. Experiments to study the response of microorganisms to the space environment after exposure in Earth orbit have constantly been reported. Spores of *Bacillus subtilis* were exposed to selected factors of space (vacuum, solar UV radiation, heavy ions of cosmic radiation), and their response was studied by Horneck (1981) after recovery. Horneck (1993) studied spores of *B. subtilis* onboard several spacecraft (Apollo 16, Spacelab 1, and LDEF). Microbes were exposed to selected parameters of space, such as space vacuum, different spectral ranges of solar UV radiation, and cosmic rays and their survival and genetic changes were studied after retrieval. The spores survive in space up to several years, if protected against the high influx of solar UV radiation. Water desorption caused by the space vacuum leads to structural changes in their DNA; the consequences are an increased

TABLE 5.13. Potential Biological Weapons[a]

Agent/Disease	Weaponized	Water Threat	Infective Dose	Period Stable in Water	Chlorine Tolerance
Anthrax	Yes	Yes	6000 spores	2 yr (spores)	Spores are resistant
Brucellosis	Yes	Probable	10,000 organisms	20–72 d	Unknown
Cholera	Unknown	Yes	1000 organisms	Survives well	Easily killed
Clostridium perfringens	Probable	Probable	10^8 organisms	Common in sewage	Resistant
Glanders	Probable	Unlikely	3.2×10^6 organisms	Up to 30 d	Unknown
Melioidosis	Possible	Unlikely	Unknown	Unknown	Unknown
Plague	Probable	Yes	500 organisms	16 d	Unknown
Psittacosis	Possible	Possible	Unknown	18–24 h in seawater	Unknown
Q fever	Yes	Possible	25 organisms	Unknown	Unknown
Salmonella	Unknown	Yes	10^4 organisms	8 d in freshwater	Inactivated
Shigellosis	Unknown	Yes	10^4 organisms	2–3 d	Inactivated, 0.05 ppm, 10 min
Tularemia	Yes	Yes	10^8 organisms	Up to 90 d	Inactivated, 1 ppm, 5 min
Typhus	Probable	Unlikely	10 organisms	Unknown	Unknown
Encephalomyelitis	Probable	Unlikely	25 particles	Unknown	Unknown
Enteric viruses	Unknown	Yes	6 particles	8–32 d	Readily inactivated (rotavirus)
Hemorrhagic fever	Probable	Unlikely	106 particles	Unknown	Unknown
Smallpox	Possible	Possible	10 particles	Unknown	Unknown
Cryptosporidiosis	Unknown	Yes	132 oocysts	Stable for days or more	Resistant

[a]Burrows and Renner (1999).

mutation frequency and altered photobiological properties of the spores. UV effects, such as killing and mutagenesis, are augmented, if the spores are in space vacuum during irradiation.

In 1998, US astronauts participating in the NASA 6 and NASA 7 visits to Mir collected environmental samples from air and surfaces in Mir's control center, dining area, sleeping quarters, hygiene facilities, exercise equipment, and scientific equipment. Imagine their surprise when they opened a rarely accessed service panel in Mir's Kvant-2 Module and discovered a large free-floating mass of water. "According to the astronauts' eye witness reports, the globule was nearly the size of a basketball," says Ott.

Moreover, the mass of water was only one of several hiding behind different panels. Scientists later concluded that the water had condensed from humidity that accumulated over time as water droplets coalesced in microgravity. The pattern of air currents in Mir carried air moisture preferentially behind the panel, where it could not readily escape or evaporate.

The water was also not clean: two samples were brownish and a third was cloudy white. Behind the panels the temperature was a toasty 28 °C just right for growing all kinds of microbes. Indeed, samples extracted from the globules by syringes and returned to Earth for analysis contained several dozen species of bacteria and fungi, plus some protozoa and dust mites.

Aboard Mir, colonies of organisms were also found growing on "the rubber gaskets around windows, on the components of space suits, cable insulations and tubing, on the insulation of copper wires, and on communications devices," says Andrew Steele, a scientist at the Carnegie Institution of Washington working with other investigators at the Marshall Space Flight Center in Alabama. Aside from being unattractive or an issue for human health, microorganisms can attack the structure of a spacecraft itself.

"Microorganisms can degrade carbon steel and even stainless steel," continued Steele. "In corners where two different materials meet, they can set up a galvanic [electrical] circuit and cause corrosion. They can produce acids that pit metal, etch glass, and make rubber brittle. They can also foul air and water filters." In short, germs can be as bad for a spacecraft's health as for crew health (Bell, 2007).

Horneck et al. (2010) reported the responses of microorganisms (viruses, bacterial cells, bacterial and fungal spores, and lichens) to selected factors of space (microgravity, galactic cosmic radiation, solar UV radiation, and space vacuum) in space and laboratory simulation experiments. Space microbiology study may involve the analysis of the samples collected outside the Earth's biosphere or pure cultures of terristerial microorganisms that were evaluated either during or after exposed to components of the space flight environment or studies of autoflora of crew members or microflora of recoverable spacecraft, which were performed to evaluate changes in populations of microorganisms (Taylor, 1974). Three problem areas have been formulated to be concentrated on by space microbiology: (i) study of bacterial growth in weightlessness, (ii) study of chromosome–episome interaction in bacteria during development in weightlessness, and (iii) elucidation of the selective role of weightlessness in populations of microorganisms (Parfenov and Lukin, 1973).

A comprehensive analysis of both the molecular genetic and phenotypic responses of any organism to the spaceflight environment has never been accomplished because of significant technological and logistical hurdles. Moreover, the effects of spaceflight on microbial pathogenicity and associated infectious disease risks have not been studied.

The bacterial pathogen *Salmonella typhimurium* was grown aboard space shuttle mission STS-115 and compared with identical ground control cultures. Global microarray and proteomic analyses revealed that 167 transcripts and 73 proteins changed expression with the conserved RNA-binding protein *Hfq* identified as a likely global regulator involved in response to this environment. *Hfq* involvement was confirmed with a ground-based microgravity culture model. Spaceflight samples exhibited enhanced virulence in a murine infection model and extracellular matrix accumulation consistent with a biofilm. Strategies to target *Hfq* and related regulators could potentially decrease infectious disease risks during spaceflight missions and provide novel therapeutic options on Earth (Wilson et al., 2007).

REFERENCES

Alanis AJ. 2005. Resistance to antibiotics: are we in the post-antibiotic era? *Arch Med Res* **36**, 697–705.

Amann R, Springer N, Schönhuber W, Ludwig W, Schmid EN, et al. 1997. Obligate intracellular bacterial parasites of acanthamoebae related to Chlamydia spp. *Appl Environ Microbiol* **63**, 115–121.

Andreoni V, Gianfreda L. 2007. Bioremediation and monitoring of aromatic-polluted habitats. *Applied Microbiology & Biotechnology* **76**, 287–308.

Astaraei AR. 2008. Microbial count and succession, soil chemical properties as affected by organic debrises decomposition American-Eurasian. *J Agric Environ Sci* **4**, 178–188.

Badger JH, Hoover TR, Brun YV, Weiner RM, Laub MT, et al. 2006. Comparative genomic evidence for a close relationship between the dimorphic prosthecate bacteria *Hyphomonas neptunium* and *Caulobacter crescentus*. *J Bacteriol* **188**, 6841–6850.

Bagchi SN, Karamchandani A, Bisen PS. 1985. Isolation and preliminary characterization of cadmium and lead tolerant strains of *Lyngbya* IU 487. *Microbios Lett* **29**, 65–68.

Beachey EH. 1981. Bacterial adherence: adhesin-receptor interactions mediating the attachment of bacteria to mucosal surface. *J Infect Dis* **143**, 325–345.

Bell TE. 2007. Attack of the space. 16 May, 2007 Science@NASA microbes http://www.cosmosmagazine.com/features/online/1325/attack-space-microbes.

Bisen PS, Dev A, Gour RK, Jain RK, Sengupta LK. 1995. Study of vesicular-arbuscular mycorrhizal fungus *Glomus mosseae* in soil samples of Bhopal. In: Adholeya A, Singh S, eds., *Mycorrhizal Biofertilizers for the Future*, pp. 73–76. TERI, New Delhi.

Bisen PS, Gour RK, Jain RK, Dev A, Sengupta LK. 1996. VAM colonization in tree species planted in Cu, Al, and coal mines of Madhya Pradesh with special reference to *Glomus mosseae*. *Mycorrhiza News*, **8**, 9–11 (TERI, New Delhi).

Bisen PS, Khare P. 1991. Mitigating effect of physicochemical factors on Ni^{2+}, Hg^{2+} and Cu^{2+} toxicity in *Cylindrospermum* IU 942. *Environ Technol Lett* **12**, 297–301.

Bisen PS, Mathur S. 1993. Adaptive response of wild and mutant type *Synechococcus cedrorum* to a polychlorinated pesticide Endosulfan. *Biomed Environ Sci* **6**, 265–272.

Bisen PS, Shukla HD, Gupta A, Bagchi SN. 1987. Preliminary characterization of a novel *Synechococcus* isolate showing mercury, cadmium and lead tolerance. *Environ Technol Lett* **8**, 427–432.

Bogue RW. 2003. Biosensors for monitoring the environment. *Sens Rev* **23**, 302–310.

Bojić I, Vukadinov J, Minić S. 2007. Diseases caused by bacteria and rickettsia in biological warfare and bioterrorism. *Med Pregl* **60**, 195–197.

Bonde GJ. 1963. *Bacterial Indicators of Water Pollution: A Study of Quantitative Estimation*, 2nd ed., Teknish Forlag, Copenhagen.

Borgne SL, Paniagua D, Vazquez-Duhalt R. 2008. Biodegradation of organic pollutants by halophilic bacteria and archaea. *J Mol Microbiol Biotechnol* **15**, 74–92.

Boudou A, Ribeyre F. 1997. Aquatic ecotoxicology: from the ecosystem to the cellular and molecular levels. *Environ Health Perspect* **105** (Suppl 1), 21–35.

Burrows WD, Renner SE. 1999. Biological warfare agents as threats to potable water. *Environ Health Perspect* **107**, 975–984.

Campbell NA. 1993. *Biology*, 3rd ed., pp. 520. The Benjamin/Cummings Publishing Company, Inc, New York.

Carter MR, Gregorich EG. 2008. *Soil Sampling and Methods of Analysis*, 2nd ed., pp. 341. CRC Press Taylor & Francis, Boca Raton, FL.

Casadevall A, Pirofski LA. 2004. The weapon potential of a microbe. *Trends Microbiol* **12**, 259–263.

Casadevall A, Pirofski LA. 2006. The weapon potential of human pathogenic fungi. *Med Mycol* **44**, 689–696.

Ceylan O, Okmen G, Ugur A. 2008. Isolation of soil*Streptomyces* as source antibiotics active against antibiotic-resistant bacteria. *Eurasia J Biosci* **2**, 73–82.

Colberg PJS, Young LY. 1995. Anaerobic degradation of nonhalogenated homocyclic aromatic compounds coupled with nitrate, iron, or sulfate reduction. *Microbial Transformation and Degradation of Toxic Organic Chemicals*, pp. 307–330. Wiley-Liss, New York.

Colwell RR, Grimes DJ. 2000. *Nonculturable Microorganisms in the Environment*, pp. 343. American Society for Microbiology, ASM Press, Washington, DC.

Couch HB. 1995. *Diseases of Turfgrass*, 3rd ed., Krieger Publishing, Malabar, FL.

Danhorn T, Fuqua C. 2007. Biofilm formation by plant-associated bacteria. *Annu Rev Microbiol* **61**, 401–422.

Debnath M, Paul AK, Bisen PS. 2007. Natural bioactive compounds and biotechnological applications of marine bacteria. *Curr Pharm Biotechnol* **8**, 253–260.

Delong EF, Pace NR. 2001. Environmental diversity of bacteria and archaea. *Syst Biol* **50**, 470–478.

Dev A, Gour RK, Jain RK, Bisen PS, Sengupta LK. 1997. Effect of vesicular arbuscular mycorrhiza-*Rhizobium* (VAM-*Rhizobium*) interaction on heavy metal (Cu, Zn, Fe) uptake in soyabean (*Glycine max*, var. JS-335) under variable P doses. *Int J Trop Agric* **15**, 75–79.

Dopson M, Baker-Austin C, Koppineedi PR, Bond PL. 2003. Growth in sulfidic mineral environment: metal resistance mechanisms in acidophilic microorganisms. *Microbiology* **149**, 1959–1970.

Ehrlich HL. 2002. How microbes mobilize metals in ores: a review of current understandings and proposals for further research. *Miner Metall Process* **19**, 220–224.

Emerson D, Weiss JV, Megonigal JP. 1999. Iron-oxidizing bacteria are associated with ferric hydroxide precipitates (Fe-Plaque) on the roots of wetland plants. *Appl Environ Microbiol* **65**, 2758–2761.

Fenchel T, Finlay BJ. 2006. The diversity of microbes: resurgence of the phenotype. *Philos Trans R Soc Lond B Biol Sci* **361**, 1965–1973.

Folkow S. 2006. The prokaryotes: a handbook on the biology of bacteria. *Proteobacteria*, pp. 14.

Friedrich B, Schwartz E. 1993. Molecular biology of hydrogen utilization in aerobic chemolithotrophs. *Annu Rev Microbiol* **47**, 351–383.

Gadd GM, Griffiths AJ. 1978. Microorganisms and heavy metal toxicity. *Microb Ecol* **4**, 303–317.

Gothalwal R, Bisen PS. 1993. Isolation and physiological characterization of *Synechococcus cedrorum* 1191 strain tolerant to heavy metals and pesticides. *Biomed Environ Sci* **6**, 187–194.

Gothalwal R, Bisen PS. 1998. Degradation of chlorinated hydrocarbon insecticides by *Pseudomonas* spp. *Int J Trop Agric* **16**, 189–193.

Gothalwal R, Bisen PS. 2001. Bioremediation. In: Pimentel D, ed., *Encyclopedia of Pest Management*, pp. 89–93. Marcel Dekker, Inc, New York.

Granatstein D, Bezdicek D. 2003. *Monitoring Soil Microbes: Invisible Helpers*, Tilth Producers Quarterly, Spring, New York, p. 436.

Hodges CF. 1990. The microbiology of non-pathogens and minor root pathogens in high sand content greens. *Golf Course Management*, **58**, 60–75.

Horneck G. 1981. Survival of microorganisms in space: a review. *Adv Space Res* **1**, 39–48.

Horneck G. 1993. Responses of Bacillus subtilis spores to space environment: results from experiments in space. *Orig Life Evol Biosph* **23**, 37–52.

Horneck G, Klaus DM, Mancinelli RL. 2010. Space microbiology. *Microbiol Mol Biol Rev* **74**, 121–156.

http://en.wikipedia.org/wiki/Anoxygenic_photosynthesis.

http://www.tutorvista.com/content/biology/biology-iii/origin-life/origin-life-summary.php.

ISO/FDIS 7899-1. 1998. Water Quality–Detection and enumeration of intestinal enterococci in surface and waste water–Part 1. Miniaturised method (Most Probable Number) by inoculation in liquid medium. International Standards Organization, Geneva.

Jannasch HW. 1989. Chemosynthetically sustained ecosystems in the deep sea. In: Schlegel HG, Bowien B, eds., *Autotrophic Bacteria*, pp. 147–166. Springer, New York.

Jeffery WW, Simon S. 1984. Bacterial resistance and detoxification of heavy metals. *Enzyme Microb Technol* **49**, 1–112.

Klassen-Fischer MK. 2006. Fungi as bioweapons. *Clin Lab Med* **26**, 387–395.

Korn-Wendisch F, Kutzner HJ. 1992. The family streptomycetaceae. In: Balows A, Truper HG, Dworkin M, Harder W, Schleifer KH, eds., *The Prokaryotes*, pp. 921–995. Springer-Verlag, New York.

Kulkarni M, Chaudhari A. 2007. Microbial remediation of nitro-aromatic compounds: an overview. *J Environ Manage* **85**, 496–512.

Leggiadro RJ. 2000. The threat of biological terrorism: a public health and infection control reality. *Infect Control Hosp Epidemiol* **21**, 53–56.

Leyval C, Turnau K, Haselwandter K. 1997. Effect of heavy metal pollution on mycorrhizal colonization and function: physiological, ecological and applied aspects. *Mycorrhiza* **7**, 139–153.

Lindow SE, Hecht-Poinar EI, Elliot VJ. 2002. *Phyllosphere Microbiology*, p. 395. American Phytopathological Society, St. Paul, MN.

Lynch JM. 1990. Microbial metabolites. In: Lynch JM, ed., *The Rhizosphere*, pp. 177–206. John Wiley and Sons, Inc, Chichester, NY.

Madigan MT, Martinko JM, Parker J. 1997. Antibiotics: isolation and characterization. *Brock Biology of Microorganisms*, 8th ed., pp. 440–442. Prentice-Hall International Inc, New Jersey.

Mancino CF, Barakat M, Maricic A. 1993. Soil and thatch microbial populations in and 80% sand: 20% peat creeping bentgrass putting green. *HortScience* **28**, 189–191.

McBride MJ. 2001. Bacterial gliding motility: multiple mechanisms for cell movement over surfaces. *Annu Rev Microbiol* **55**, 49–75.

Mehta R, Shukla HD, Bisen PS. 1992. Physiological and genetical characterization of a novel *Synechococcus* sp. isolate showing resistance against cyanide, heavy metals and drugs. *Environ Tech Lett* **13**, 253–258.

Michael G. 1998. *Life on the Edge: Amazing Creatures Thriving in Extreme Environments*, Plenum, New York.

Montesinos E. 2003. Plant associated microorganisms: a view from the scope of microbiology. *Int Microbiol* **6**, 221–223.

Mueller JG, Cerniglia CE, Pritchard PH. 1996. Bioremediation of environments contaminated by polycyclic aromatic hydrocarbons. *Bioremediation: Principles and Applications*, pp. 125–194. Cambridge University Press, Cambridge.

Munn CB. 2004. *Marine Microbiology Ecology and Applications*, pp. 85–114. Garland Science/BIOS Scientific Publishers, UK.

Nelson EJ, Jason B, Harris J, Morris G Jr, Calderwood SB, Camilli A. 2009. Cholera transmission: the host, pathogen and bacteriophage dynamic. *Nat Rev Microbiol* **7**, 693–702.

Neu MP. 2000. Actinide interactions with microbial chelators: how desferrioxamine siderophores affect plutonium chemistry and facilitate microbial uptake, Los Alamos. *Science* **26** (2), 416–417.

Oren M. 2002. Bioterrorism (Article in Hebrew) *Harefuah*, **141** (Spec. No. 13–5), 124.

Parfenov GP, Lukin AA. 1973. Origins of life and evolution of biospheres. *Space Life Sci* **4**, 160–179.

Paterson RR. 2006. Fungi and fungal toxins as weapons. *Mycol Res* **110**, 1003–1010.

Peng JF, Song YH, Yuan P, Cui CY, Qiu GL. 2009. The emediation of heavy metals contaminated sediment. *J Hazard Mater* **161**, 633–640.

Pilon-Smits E, Pilon M. 2002. Phytoremediation of metals using transgenic plants. *Crit Rev Plant Sci* **21**, 439–456.

Prauser H. 1964. Aptness and application of colour for exact description of colours of *Streptomyces*. *Z Allg Mikrobiol* **4**, 95–98.

Prescott LM, Harley JP, Klein DA. 1993. *Microbiology*, 2nd ed., William C Brown Pub, New York.

Rabaey K, Verstraete W. 2005. Microbial fuel cells: novel biotechnology for energy generation. *Trends Biotechnol* **23**, 291–298.

Rittmann BE, Krajmalnik-Brown R, Halden RU. 2008. Pre-genomic, genomic and post-genomic study of microbial communities involved in bioenergy. *Nat Rev Microbiol* **6**, 604–612.

Ruggiero CE, Boukhalfa H, Forsythe JH, Lacl JG, Hersman LE, Neu MP. 2004. Actinides and metal toxicity to prospective bioremediation bacteria. *Environ Microbiol* **7**, 88–97.

Ruggiero CE, Neu MP, Matonic JH, Reilly SD. 2000. Dissolution of plutonium (IV) hydroxide by desferrioxamine siderophores and simple organic chelators. *Inorg Chem* **41**, 3593–3595.

Sharma SK, Bisen PS. 1992. Hg^{2+} and Cd^{2+} induced inhibition of light induced proton efflux in the cyanobacterium *Anabaena flos-aquae*. *Biometals* **5**, 163–167.

Sharma SK, Singh DP, Shukla HD, Ahmad A, Bisen PS. 2001. Influence of sodium ion on heavy metal-induced inhibition of light-regulated proton efflux and active carbon uptake in the cyanobacterium *Anabaena flos-aquae*. *World J Microbiol Biotech* **17**, 707–711.

Silver S. 1992. Plasmid-determined metal resistance mechanisms: Range and overview. *Plasmid*, **27**, 1–3.

Singh DP, Gothalwal R, Bisen PS. 1989a. Toxicity of sodium diethyl-dithiocarbamate (NaDDC) on photoautotrophic growth of *Anacystis nidulans* IU 625. *J Basic Microbiol* **29**, 685–694.

Singh DP, Khare P, Bisen PS. 1989b. Effect of Ni^{2+}, Hg^{2+} and Cu^{2+} on growth, oxygen evolution and photosynthetic electron transport in *Cylindrospermum* IU 942. *J Plant Physiol* **134**, 406–412.

Singh DP, Sharma SK, Bisen PS. 1993. Differential action of Hg^{2+} and Cd^{2+} on the photosynthetic apparatus of *Anabaena flos-aquae*. *Biometals* **6**, 125–132.

Shirling EB, Gottlieb D. 1966. *Methods, Classification, Identification and Description of Genera and Species*, **vol. 2**, pp. 61–292. The Williams and Wilkins Company, Baltimore.

Spormann AM. 1987. Unusual swimming behavior of a magnetotactic bacterium. *FEMS Microbiol Lett* **45**, 37–45.

Suzuki I, Lee D, Mackay B, Harahuc L, Oh JK. 1999. Effect of various Ions, pH, and osmotic pressure on oxidation of elemental sulfur by *Thiobacillus thiooxidans*. *Appl Environ Microbiol* **65**, 5163–5168.

Sylvia DM, Fuhrman JJ, Hartel PG, Zuberer DA. 1997. *Principles and Applications of Soil Microbiology*, Prentice-Hall, Inc, Englewood Cliffs, NJ.

Taylor GR. 1974. Space microbiology. *Ann Rev Microbiol* **28**, 121–137.

VGD, Inc. Balanced sensitivity medium (Difco–1863).

Waksman SA. 1954. In: Watham MASS, ed., *The Actinomycetes*, 1st ed., pp. 185–191. Academic Press, New York.

Weissman JC, Tillett DT. 1992. A Design and operation of an outdoor microalgae test facility: large-scale system results. 1989–1990 Aquatic Species Project Report, National Renewable Energy Laboratory, Golden, Colorado, NREL/MP-232-4174, pp 32–56.

WHO. 1997. World Health Report—Conquering suffering enriching humanity. World Health Organization, Geneva, pp. 168.

Williams ST, Goodfellow M, Alderson G, Wellington EMH, Sneath PHA, Sackin MJ. 1983a. Numerical classification of *Streptomyces* and related genera. *J Gen Microbiol* **129**, 1743–1813.

Williams ST, Goodfellow M, Wellington EMH, Vickers JC, Alderson G, Sneath PHA, Sackin MJ, Mortimer AM. 1983b. A probability matrix for identification of *Streptomyces*. *J Gen Microbiol* **129**, 1815–1830.

Wilson JW, Ott CM, Höner zu Bentrup K, Ramamurthy R, Quick L, et al. 2007. Space flight alters bacterial gene expression and virulence and reveals a role for global regulator Hfq. *Proc Natl Acad Sci U S A* **104**, 16299–16304.

6

MICROBES IN AGRICULTURE

6.1. PROLOGUE

Microbes are the tiny living organisms not visible to naked eye. However, they play a very important role in Nature and contribute a lot to plants, animals, and human beings. In general, they represent the groups, prokaryotes and eukaryotes. Bacteria, actinomycetes, and blue-green algae are the sole representatives of prokaryotes. Among eukaryotes, algae, protozoa, and fungi are placed as microbes. Farmers and ranchers often think of microbes as pests that are destructive to their crops or animals (as well as themselves), but many microbes are beneficial. Soil microbes (bacteria and fungi) are essential for decomposing organic matter and recycling old plant material. Some bacteria and fungi form relationships with plant roots that provide important nutrients such as nitrogen and phosphorus. Fungi can colonize upper parts of plants and provide many benefits, including drought and heat tolerance and resistance to insects plant diseases.

In agriculture, fungi, bacteria, algae, and viruses are important with respect to their contribution in the form of either loss or gain in the production of grains, fruits, vegetables, oil, milk, poultry, fodder, and livestock. Most of the nutrients, both minor and major, present in the soil are managed by microorganisms through integrated nutrient management. It has been observed that the efficiency of utilization of nutrients of high grade complex fertilizers in terms of biological chemicals and agronomic/economic efficiency has dropped from 20 to below 9 in terms of kg grain/kg of NPK. It is therefore important to manage the nutrients in a proper way.

Microbes: Concepts and Applications, First Edition. Prakash S. Bisen, Mousumi Debnath, Godavarthi B. K. S. Prasad
© 2012 Wiley-Blackwell. Published 2012 by John Wiley & Sons, Inc.

6.2. THE SOIL PLANT MICROORGANISMS

Microorganisms are known to be very important for plant growth. They multiply and actively help in making essential nutrients available for the plant through a symbiotic process by releasing the "locked-up" nutrients to be ready for uptake and utilization. Microorganisms have an active role in protecting plants against soilborne diseases (Fig. 6.1). It seems obvious that microorganisms are in the soil because there is food. Soil is an excellent culture media for the growth of many types of organisms, including bacteria, fungi, algae, protozoa, and viruses. In addition, various nematodes, insects, and so on are also present. A spoonful of soil contains billions of microorganisms. In general, the majority of microbial population is found in the upper 6–12 in of soil and the number decreases with depth. Higher number occurs in the organically rich surface layers than in the underlying mineral soils. Particularly, high numbers of microbes occur in association with plant roots. Fungal populations are favored in soils of low pH, and bacteria tend to occur in higher numbers in those of higher pH. There is about two and a half times more carbon in the soil than there is in the atmosphere.

Unseen soil microbes respond to and influence global climate change. It has been noted that the respiration of soil microbes returns to normal after a number of years under heated conditions. It was further argued that the microbes consumed so much of the available food under heated conditions that future levels of decomposition were reduced because of food scarcity. Some soil microbes are also adapted to the changed environment and reduce their respiration accordingly. The abundance of soil microbes decreased under warm conditions.

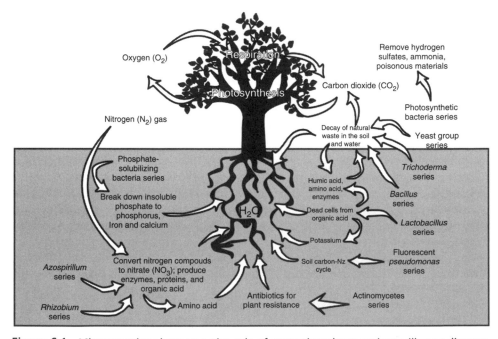

Figure 6.1. Microorganisms have an active role of protecting plants against soilborne diseases.

Various types of organisms and microorganisms live in soil. Some microorganisms also burrow and channel through soil, which improves soil structure and aggregation, while other microorganisms have the ability to break down resistant organic matter such as lignin, toxins, and pesticides. Microorganisms also have the ability to protect plants from antagonistic pathogens, and some can dissolve minerals, making nutrients available to plants. Fungi are able to break down resistant materials such as cellulose, gums, and lignin. They dominate in acidic, sandy soils and in fresh organic matter.

Actionomycetes also are able to decompose resistant substances in soil. One type, *Frankia*, helps plants to get nutrients needed from the air by breaking triple bond nitrogen to ammonium that plants can use. Antibiotics are made from soil actinomycetes.

Bacteria decompose a wider range of earth material than any other microbe group. Heterotrophs gain their energy and carbon from other organisms, while autotrophs synthesize their own energy from light or by chemical oxidation. Some bacteria can fix nitrogen into forms that plants can use. How quickly decomposition of dead organic matter occurs depends on soil temperature and soil moisture. Without the microorganisms, dead organic matter would pile high on Earth's surface.

6.2.1. Soil Fertility

Soil fertility is the capacity to supply proper amounts of different nutrients in the appropriate proportion for the growth of crops. The availability of both inorganic and organic matter determines the soil fertility. The inorganic matter of soil comprises the entire essential and trace minerals present in the soil in the form of salts (acidic and basic). The inorganic element either gets adsorbed onto the clay particles or gets dissolved in the soil water. Soil fertility refers to the amount of nutrients in the soil, which is sufficient to support plant life. It must contain organic matter and a relatively low pH value. The soil must also contain micro- and macroorganisms, and it must be well drained. The organic matter in the soil exists mainly as humus or as partially decomposed (plant and animal tissues). It is also prepared artificially as farmyard manure, green manure/greenleaf manure, compost, vermicompost, biofertilizers, and so on. The four strategies used by farmers for soil management (Fig. 6.2) show that they had a good empirical understanding of soil fertility and degradation but that this understanding was incomplete: they were entirely unaware of crucial soil processes occurring underneath the soil surface.

On the whole, the balanced availability of both inorganic and organic matter in the soil determines the soil fertility. Indirectly, these organic and inorganic matters help in the proliferation of various qualitative microflora that play a very vital role in maintaining the nutritional balance of the soil. Thus, microorganisms have a great role to play in determining soil fertility, for without a proper distribution of microflora, no soil can support plant growth which speaks of its fertility (Fig. 6.3). Microorganisms in soil affect the fertility of soil by means of physical or chemical changes.

Soil fertility is critical in sustainable farming and needs to be considered not only for crop productivity but also for the protection of aquatic environments. Fertile soil has an abundance of plant nutrients including nitrogen, phosphorus, and potassium; an abundance of minerals; as well as an abundance of organic matter. Mycorrhizae is being used to enhance the uptake of nutrients and water for establishment of seedlings on degraded lands. This will, however, not lead to improvement of soil fertility by mycorrhizae as such, although the success of plant growth will eventually lead to reclamation of degraded lands. However, nitrogen-fixing bacteria such as *Rhizobium* or the actinomycete genus

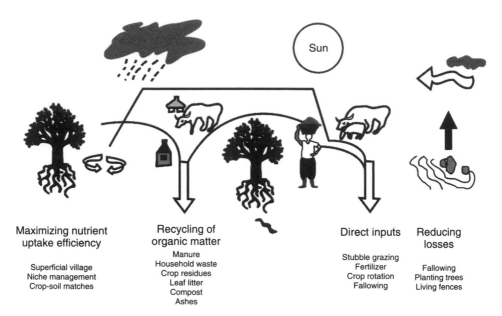

Figure 6.2. Aggregated and stylized farmers' mental model of soil fertility management.

Frankia can be used to induce nodule formation in a variety of plant species, so that they can be used for improving soil fertility of degraded lands. This nodule formation can be induced both in leguminous and nonleguminous plant species comprising annuals (cereal and legume crops) and perennials (trees). Efforts are underway to manipulate the genes of both host and rhizobia to obtain maximum efficiency of nodule formation. Strains are also being tailored for unusual soil environments representing degraded lands.

The key role of fertilizers and their judicious use in crop husbandry is well understood when one is familiar with the general facts about plant nutrition. It is now known that at least 16 plant food elements are necessary for the growth of green plants. These plant nutrients are called *essential elements*. In the absence of any one of these essential elements, a plant fails to complete its life cycle, although the disorder caused can be corrected by the addition of that element. The 16 elements are carbon (C), hydrogen (H), oxygen (O), nitrogen (N), phosphorus (P), sulfur (S), potassium (K), calcium (Ca), magnesium (Mg), iron (Fe), manganese (Mn), zinc (Zn), copper (Cu), molybdenum (Mb), boron (B), and chlorine (Cl). Green plants obtain carbon from carbon dioxide in the air and oxygen and hydrogen from water, whereas the remaining elements are taken from the soil. On the basis of their relative amounts normally found in plants, the plant nutrients are termed as *macronutrients*, if large amounts are involved, and micronutrients, if only traces are involved. The micronutrients essential for plant growth are iron, manganese, copper, zinc, boron, molybdenum, and chlorine. All other essential elements listed above are macronutrients.

As mentioned above, most of the plant nutrients, besides carbon, hydrogen, and oxygen, originate from the soil. The soil system is viewed by the soil scientists as a triple-phased system of solid, liquid, and gaseous phases. These phases are physically separable. The plant nutrients are based in the solid phase, and their usual pathway to the plant system is through the surrounding liquid phase, the soil solution, and then to

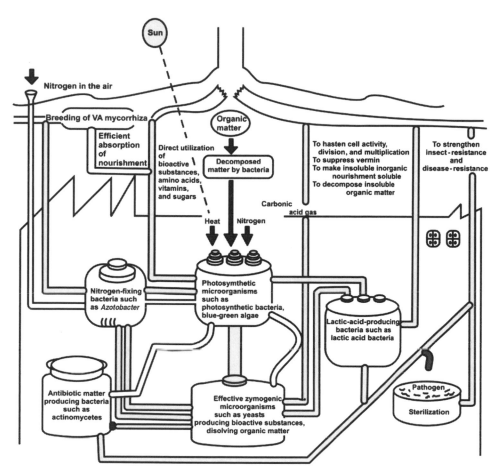

Figure 6.3. The function of effective microorganisms in the soil (http://www.agriton.nl/apnanman.html).

the plant root and plant cells. This pathway may be written in the form of an equation as M (solid) → M (solution) → M (plant root) → (plant top), where "M" is the plant nutrient element in continual movement through the soil-plant system. The operation of the above system is dependent on the solar energy through photosynthesis and metabolic activities. Soil is a mosaic of dynamic microenvironment that differs in physical, biological, and chemical properties. Hence, the microbial communities that govern ecosystem C and N cycling are spatially and temporally variable (Fig. 6.4). It has been observed that both fungi and earthworms, which are known to enhance the formation of soil aggregates, directly control the formation of microaggregates (within macroaggregates) (Kong et al., 2007). In addition, several studies have indicated that the microaggregate structure creates an operationally definable microenvironment for microorganisms; that is, the differences in microbial community are greater between macroaggregates and microaggregates within a soil type than among different soil types. This microenvironment is characterized by low predation pressure, relatively stable water potential, low oxygen availability, and low accessibility for exogenous toxic elements. As a result, the spatial compartmentalization associated with these microenvironments protects microbes

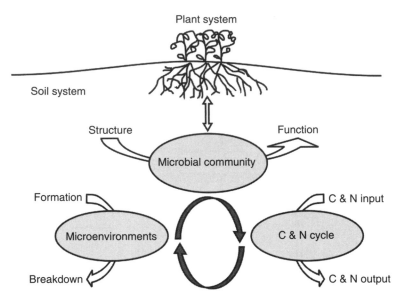

Figure 6.4. Interactions between soil microstructures and biota control on ecosystem functioning.

from contaminants, fosters a unique microbial community structure, and also reduces the activity level of the microflora. The last function of microaggregates directly induces the stabilization and storage of soil C and N. C, and especially C derived from fungal and bacterial cell wall components, is preferentially sequestered within microaggregates occluded within macroaggregates.

This is, however, an oversimplified statement for gaining a physical concept of the natural phenomenon, but one should bear in mind that there are many physical and physicochemical processes influencing the reactions in the pathway. The actual transfer in nature takes place through the charged ions, the usual form in which plant food elements occur in solutions (liquid phase of the system). Plant roots take up plant food elements from the soil in these ionic forms. The positively charged ions are called *cations*, which include potassium (K^+), calcium (Ca^{++}), magnesium (Mg^{++}), iron (Fe^{+++}), zinc (Zn^{++}). The negatively charged ions are called *anions*, and the important plant nutrients taken in this form include nitrogen (NO_3), phosphorus (H_2PO_4), sulfur (SO_4), and chlorine (Cl).

The process of nutrient uptake by plants refers to the transfer of the nutrient ions across the soil−root interfaces into the plant cell. The energy for this process is provided by the metabolic activity of the plant, and in its absence no absorption of nutrients take place. Nutrient absorption involves the phenomenon of ion exchange. The root surface, like soil, carries a negative charge and exhibits cation exchange property. The most efficient absorption of the plant nutrients takes place on the younger tissues of the roots, which are capable of growth and elongation.

In this respect, root systems are known to vary from crop to crop. Hence, their feeding power differs. The extent and the spread of the effective root system determine the soil volume trapped in the feeding zone of the crop plant. This is indeed an important information in a given soil-plant system that helps us to choose fertilizers and fertilizer use practices. The absorption mechanisms of the crop plants are fairly known now. There are three mechanisms in operation in the soil-water-plant systems: (i) the contact exchange and root interception, (ii) the mass flow or convection, and (iii) diffusion. In the case of contact exchange and root interception, the exchangeable nutrient ions from the clay-humus colloids migrate directly to the root surface through contact exchange when plant roots come into contact with the soil solids. Nutrient absorption through this mechanism is, however, insignificant, as most of the plant nutrients occur in the soil solutions. Scientists have found that plant roots actually grow to come into contact with only 3% of the soil volume exploited by the root mass, and the nutrient uptake through root interception is even still less. The second mechanism is mass flow or convection, which is considered to be the important mode of nutrient uptake. This mechanism relates to nutrient mobility with the movement of soil water toward the root surface where absorption through the roots takes place along with water. Some are called *mobile nutrients*. Others that move only a few millimeters are called *immobile nutrients*. Nutrient ions, such as nitrate, chloride, and sulfate, are not absorbed by the soil colloids and are mainly in solution. Such nutrient ions are absorbed by the roots along with soil water. The nutrient uptake through this mechanism is directly related to the amount of water used by plants (transpiration). It may, however, be mentioned that the exchangeable nutrient cations and anions other than nitrate, chloride, and sulfate, which are absorbed on soil colloids, are in equilibrium with the soil solution and do not move freely with water when they are absorbed by the plant roots. These considerations, therefore, bring out that there are large differences in the transport and root absorption of various ions through the mechanism of mass flow. Mass flow is, however, responsible for supplying the root with much of the plant needs for nitrogen, calcium, and magnesium, when present in high concentrations in the soil solution, but does not do so in the case of phosphorus or potassium. Nutrient uptake through mass flow is largely dependent on the moisture status of the soil and is highly influenced by the soil physical properties controlling the movement of soil water.

The third mechanism is diffusion. It is an important phenomenon by which ions in the soil medium move from a point of higher concentration to a point of lower concentration. In other words, the mechanism enables the movement of the nutrient ions without the movement of water. The amount of nutrient ion movement in this case is dependent on the ion concentration gradient and transport pathways, which, in turn, are highly influenced by the content of soil water. This mechanism is predominant in supplying most of the phosphorus and potassium to plant roots. It is important that the rhizosphere volume of soil in the immediate neighborhood of the effective plant root receives plant nutrients continuously to be delivered to the roots by diffusion.

The organic-phosphorus-containing compounds are derived from plants and microorganisms and are composed of nucleic acids, phospholipids, and phytin. Organic matter derived from dead and decaying plant debris is rich in organic sources of phosphorus. The deficiency of phosphorus may occur in crop plants growing in soils containing adequate phosphates. This may be partly due to the fact that plants are able to absorb phosphorus only in an available form. Soil phosphates are rendered available by either plant roots or soil microorganisms through secretion of organic acids. Therefore, sulfate- and phosphate-dissolving soil microorganisms play some part in correcting phosphorus deficiency of crop plants. They may also release soluble inorganic phosphate (H_2PO_4) and sulfide into soil through decomposition of sulfate- and phosphate-rich organic compounds. On the other hand, certain microorganisms, through assimilation, may immobilize available sulfate and phosphates in their cellular material (Figs. 6.5 and 6.6). Such immobilization processes in soil may also contribute to sulfur and phosphorus deficiency in

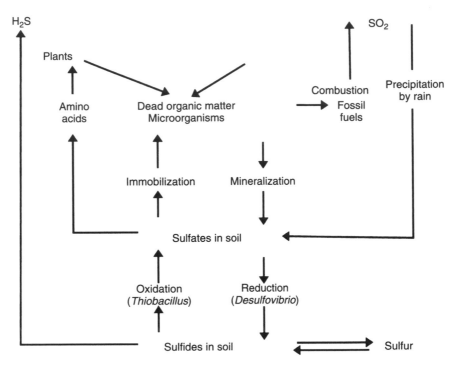

Figure 6.5. Sulfate-solubilizing property of soil microorganisms.

crop plants. Microorganisms and plant roots readily dissolve such sulfide and phosphates and render them easily available to plants. One of the ways to correct deficiency of phosphorus in plants is to inoculate seed or soil with phosphate-dissolving microorganisms along with phosphate fertilizers.

Many fungi and bacteria (e.g., *Aspergillus, Penicillium, Bacillus*, and *Pseudomonas*) are potential solubilizers of bound phosphates as revealed by experiments in pure culture. Although bacteria have been used in the commercial preparation of phosphate-dissolving cultures to improve the growth of plants, fungi seem to be better agents in the dissolution of phosphates. Phosphate-dissolving bacteria are known to reduce the pH of the substrate by secretion of a number of organic acids such as formic, acetic, propionic, lactic, glycolic, fumaric, and succinic acids. Some of these acids (hydroxy acids) may form chelates with cations such as Ca and Fe, and such chelation results in effective solubilization of phosphates. Organic acid production is invariably associated with phosphate solubilization in pure cultures of microorganisms. The correlation has not been established between the change in pH of the medium after microbial growth and the amount of phosphates solubilized.

6.2.2. Rhizosphere Environment

The rhizosphere is the zone of the soil surrounding a plant root where the biology and chemistry of the soil are influenced by the root. This zone is about 1 mm wide but has no distinct edge. Rather, it is an area of intense biological and chemical activity influenced

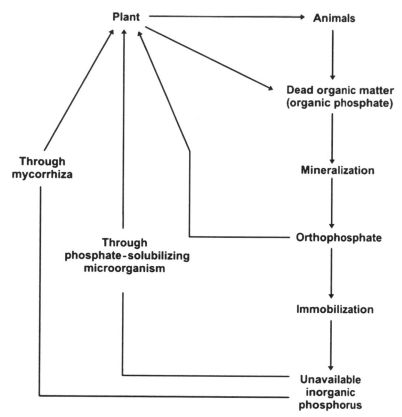

Figure 6.6. Availability of phosphate by the soil microorganisms.

by compounds exuded by the root and by microorganisms feeding on the compounds. As plant roots grow through soil, they release water-soluble compounds such as amino acids, sugars, and organic acids that supply food to the microorganisms. The food supply means microbiological activity in the rhizosphere, which is much greater in soil away from plant roots. In return, the microorganisms provide nutrients to the plants. All this activity makes the rhizosphere the most dynamic environment in the soil. Because roots are underground, rhizosphere activity has been largely overlooked, and it is only now that we are starting to unravel the complex interactions that occur. For this reason, the rhizosphere has been called the *last frontier* in agricultural science. Practically all ecological interactions, such as symbiosis, syntrophism, synergism, commensalism, and antagonism, between plants and microorganisms, and among different microorganisms, are found in this region (Fig. 6.7).

The roots exude water and compounds broadly known as *exudates*. Root exudates include amino acids, organic acids, carbohydrates, sugars, vitamins, mucilage, and proteins. The exudates act as messengers that stimulate biological and physical interactions between roots and soil organisms. They modify the biochemical and physical properties of the rhizosphere and contribute to root growth and plant survival. However, the fate of the exudates in the rhizosphere and the nature of their reactions in the soil remain poorly understood.

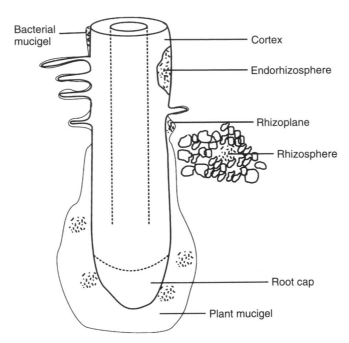

Figure 6.7. Distribution of microorganisms in the rhizosphere.

The exudates have several functions. They

• Defend the rhizosphere and root against pathogenic microorganisms
• Attract and repel particular microbe species and populations
• Keep the soil around the roots moist
• Obtain nutrients
• Change the chemical properties of the soil around the roots
• Stabilize soil aggregates around the roots
• Inhibit the growth of competing plant species.

The plant roots, which the rhizosphere is associated with, can affect the physical environment of the rhizosphere. As plants transpire water with more force during the day than during the night, they change the soil water potential immediately near their roots and so the rhizosphere undergoes fluctuations that the bulk soil avoids. Movement of organic matter away from the root as well as bacteria colonizing new locations occurs more readily in sandy soils than clayey soils. Sand has larger pores between each granule allowing microorganisms and exudates to travel. Therefore, the larger the granule size, the further the rhizosphere and microorganisms associated with it will extend into the surrounding soil.

Plant roots compact the soil on the short term as they grow, but once they die and decay, they can actually leave the soil more porous. Several factors can lower the pH in the rhizosphere. Respiration leads to carbon dioxide (and eventually to bicarbonate/carbonic acid) generation. In addition to respiration of the roots themselves, the rhizosphere is very rich in carbon resulting in other organisms, from prokaryotes to fungi to small animals, living and respiring in the rhizosphere more than in the bulk soil. Fungi are found in more acidic soils than alkaline, and bacteria have a very broad pH spectrum where they can survive. The influencing effect of pH on the rhizosphere is critical in supporting a biologically diverse microbial community. Plant-derived compounds are responsible for providing the additional carbon that allows the rhizosphere to host a large variety of organisms. These compounds fall into five categories: exudates, secretions, mucilages, mucigel, and lysates. Exudates include surplus sugars, amino acids, and aromatics that diffuse out of cells to the intercellular space and the surrounding soil. Owing to their diffusive nature, exudates are limited to compounds of low molecular weights. Secretions are by-products of metabolic activity. Because they are actively released from the cell, secretions can be of both low and high molecular weights. When an epidermal root cell dies and is broken open, lysates from within the cell become available to the surrounding microbial community. Mucilages are cells sloughed off the root cap as the root grows. Abrasive forces of the root against the soil's particulate matter are responsible for the removal of cells. These cells consist of cellulose, pectin, starch, and lignin. Mucigel is a slime coating the surface of a root, which increases the connectivity between plant roots and the surrounding soil. It is more common on the main body of the root and root hairs than the tip. During dry spells, mucigels are responsible for allowing plants to continue to uptake water and nutrients.

6.2.3. Ammonification in Soil

The nitrogen in most plants and animals exists in the form of protein. Most of the nitrogen in soil exists in the form of organic molecules, mostly proteins derived from the decomposition of dead plant and animal tissue. When an organism dies, its proteins are attacked by the proteases of soil bacteria to produce polypeptides (peptones) and amino acids ($C_2H_4NO_2$-R). This process is called *peptonization*. Then, the amino groups on the amino acids are removed by a process called *deamination*, producing ammonia (NH_3). In most soils, the ammonia dissolves in water to form ammonium ions (NH^{4+}). The process of the production of ammonia from organic compounds is called *ammonification*. In addition to the ammonification of amino acids, other compounds such as nucleic acids, urea, and uric acid go through the ammonification process. The bacteria that accomplish the process (*Bacillus, Clostridium, Proteus, Pseudomonas*, and *Streptomyces*) are called *ammonifying bacteria*. Ammonification of organic compounds is a very important step in the cycling of nitrogen in soil, since most autotrophs are unable to assimilate amino acids, nucleic acids, urea, and uric acid and use them for their own enzyme and protoplasm construction. Ammonification is an important stage in the nitrogen cycle, a natural cycle that makes the Earth's supply of nitrogen available to organisms that need it, such as plants. Like many other natural cycles, the nitrogen cycle can be disrupted by human activities, which can lead to imbalances at various stages, sometimes causing environmental problems. Ammonification can be a major problematic area in the nitrogen cycle when human intervention occurs, as buildups of ammonia can cause health problems and environmental issues. One of the most elementary of the ammonification reactions is the oxidation of the simple organic compound urea ($CO(NH_2)_2$), to ammonia through the action of a microbial enzyme known as *urease* (two units of ammonia are produced for every unit of urea that is oxidized). Urea is a commonly utilized agricultural fertilizer, used to supply ammonia or ammonium for direct uptake by plants, or as a substrate for the microbial production of nitrate through nitrification (Fig. 6.8).

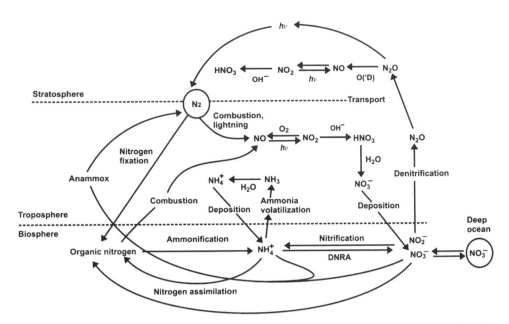

Figure 6.8. Nitrogen assimilation and ammonification in the soil; dissimilatory nitrate reduction to ammonium (DNRA).

Ammonium is a suitable source of nitrogen uptake for many species of plants, particularly those that live in acidic soils and waters. However, most plants that occur in nonacidic soils cannot utilize ammonium very efficiently. Nitrogen is one of the most abundant elements in the tissues of all organisms and is a component of many biochemicals, particularly amino acids, proteins, and nucleic acids. Consequently, nitrogen is one of the critically important nutrients and is required in relatively large quantities by all organisms. Animals receive their supply of nitrogen through the foods they eat, but plants must assimilate inorganic forms of this nutrient from their environment.

However, the rate at which the environment can supply inorganic nitrogen is limited and usually small in relation to the metabolic demands of plants. Therefore, the availability of inorganic forms of nitrogen is frequently a limiting factor for the productivity of plants. This is a particularly common occurrence for plants growing in terrestrial and marine environments and to a lesser degree, in freshwaters (where phosphate supply is usually the primary limiting nutrient, followed by nitrate). The dead biomass of plants, animals, and microorganisms contains large concentrations of organically bound nitrogen in various forms, such as proteins and amino acids. The process of decomposition is responsible for recycling the inorganic constituents of the dead biomass and preventing it from accumulating in large unusable quantities. Decomposition is, of course, mostly carried out through the metabolic functions of a diverse array of bacteria, fungi, actinomycetes, other microorganisms, and some animals. Ammonification is a particular aspect of the more complex process of organic decay, specifically referring to the microbial conversion of organic nitrogen into ammonia (NH_3) or ammonium (NH^{4+}). Ammonification occurs under oxidizing conditions in virtually all ecosystems and is carried out by virtually all microorganisms that are involved in the decay of dead organic matter. In situations where oxygen is not present, a condition referred to as *anaerobic*, different microbial decay reactions occur, producing nitrogen compounds known as *amines*. The microbes derive some metabolically useful energy from the oxidation of organic nitrogen to ammonium. In addition, much of the ammonium is assimilated and used as a nutrient for the metabolic purposes of the microbes. However, if the microbes produce ammonium in quantities that exceed their own requirements, as is usually the case, the surplus is excreted into the ambient environment (such as the soil) and is available for use as a nutrient by plants, or as a substrate for another microbial process known as *nitrification* (Fig. 6.9, Fig 6.10). Animals, in contrast, mostly excrete urea or uric acid in their nitrogen-containing liquid wastes (such as urine), along with diverse organic nitrogen compounds in their feces. The urea, uric acid, and organic nitrogen of feces are all substrates for microbial ammonification.

6.2.4. Nitrification in Soil

The biological conversion of ammonium to nitrate nitrogen is called *nitrification*. Nitrification is a two-step process (Fig. 6.11). *Nitrosomonas* convert ammonia and ammonium to nitrite. Bacteria called *Nitrobacter* finish the conversion of nitrite to nitrate. The reactions are generally coupled and proceed rapidly to the nitrate form; therefore, nitrite levels at any given time are usually low. These bacteria known as *nitrifiers* are strict "aerobes," meaning they must have free dissolved oxygen (DO) to perform their work. Nitrification occurs only under aerobic conditions at DO levels of 1 mg/l or more. At DO concentrations less than 0.5 mg/l, the growth rate is minimal. Nitrification requires a long retention time, a low food-to-microorganism ratio (F/M), a high mean cell residence time (measured as MCRT or sludge age), and adequate buffering (alkalinity). A plug-flow extended aeration tank is ideal. Temperature, as discussed below, is also important, but not really. The nitrification process produces acid. This acid formation lowers the pH of the biological population in the aeration tank and can cause a reduction in the growth rate of nitrifying bacteria. The optimum pH for *Nitrosomonas* and *Nitrobacter* is between 7.5 and 8.5; most treatment plants are able to effectively nitrify with a pH of 6.5–7.0.

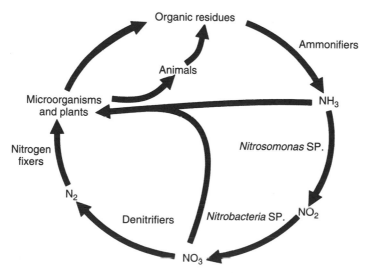

Figure 6.9. Role of microorganisms in the nitrogen cycle.

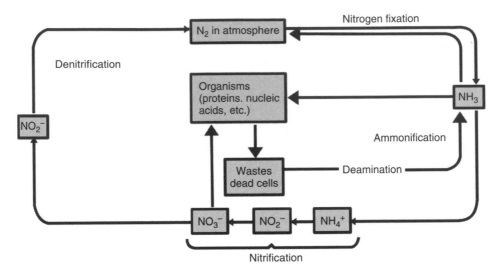

Figure 6.10. Nitrogen cycle.

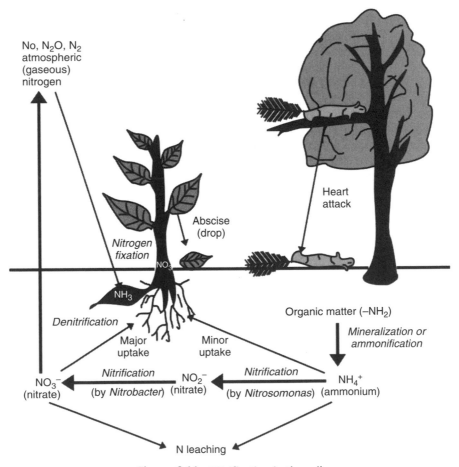

Figure 6.11. Nitrification in the soil.

Nitrification stops at a pH below 6.0. The nitrification reaction (i.e., the conversion of ammonia to nitrate) consumes 7.1 mg/l of alkalinity as $CaCO_3$ for each milligram per liter of ammonia nitrogen oxidized. An alkalinity of no less than 50–100 mg/l is required to ensure adequate buffering.

Water temperature also affects the rate of nitrification. Nitrification reaches a maximum rate at temperatures between 30 and 35 °C (86 and 95 °F). At temperatures of 40 °C (104 °F) and higher, nitrification rates fall to near zero. At temperatures below 20 °C, nitrification proceeds at a slower rate, but will continue at temperatures of 10 °C and less. However, if nitrification is lost, it will not resume until the temperature increases to well over 10 °C. Some of the most toxic compounds to nitrifiers include cyanide, thiourea, phenol, and heavy metals such as silver, mercury, nickel, chromium, copper, and zinc. Nitrifying bacteria can also be inhibited by nitrous acid and free ammonia.

6.2.5. Organic Compounds Released by Plants

Photosynthesis is a process that converts carbon dioxide into organic compounds, especially sugars, using the energy from sunlight. Photosynthesis is also the source of carbon in all the organic compounds within organisms' bodies, the process always begins when energy from light is absorbed by proteins called *photosynthetic reaction centers* that contain chlorophylls. In plants, these proteins are held inside organelles called *chloroplasts*, while in bacteria they are embedded in the plasma membrane. Some of the light energy gathered by chlorophylls is stored in the form of adenosine triphosphate (ATP). The rest of the energy is used to remove electrons from a substance, such as water. These electrons are then used in the reactions that turn carbon dioxide into organic compounds (Fig. 6.12).

One of the most important factors to shape plant communities is competition between plants, which affects the availability of environmental factors such as light, nutrients, and water. In response to these environmental parameters, plants adjust the emission of many different biogenic volatile organic compounds (BVOCs). BVOCs can also elicit responses in neighboring plants, thus constituting a platform for plant–plant interactions. Organohalogens are released to the groundwater from buried geological sediments. The organohalogens found in even very old sediments can be explained partly by the content of organic halogens in both terrestrial and marine plants and partly by microbial halogenation occurring during the partial degradation (humification) of dead plant material. Incorporation of inorganic iodine into humic substances has been demonstrated to produce organically bound iodine, and this process is probably causing the high concentrations of organic iodine found in an aquifer in old marine sediments. Chloroform found in the shallow groundwater below a spruce forest is probably the result of natural chlorination processes of the microorganisms in the soil. The chloroperoxidases can be excreted by microorganisms, in particular, by fungi, and can produce reactive halogen, thus causing the formation of chloroorganic compounds such as chloroform from inorganic chloride and soil organic matter.

6.2.6. Nitrogen-Fixing Bacteria

Nitrogen is one of the most important chemical elements for plants. If there is not enough nitrogen available in the soil, plants look pale and their growth is stunted. Nitrogen-fixing

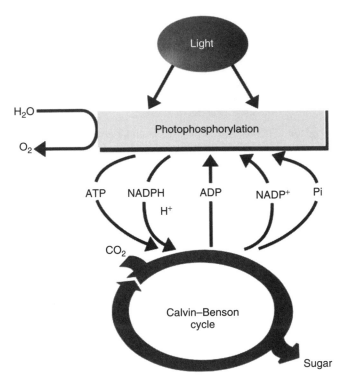

Figure 6.12. Process of photophosphorylation.

plants are called *legumes*. All peas and beans are legumes. The plants work together with nitrogen-fixing bacteria called *rhizobia* to "fix" nitrogen. The rhizobia chemically convert the nitrogen from the air to make it available to the plant. Leguminous plants live in a symbiotic relationship with the nitrogen-fixing bacteria. Rhizobia live in nodules in the plant's root. This way the plant can look after its own nitrogen needs and fertilizer is not required. In addition, when the crop is harvested and the plant cut back to ground level, the root nodules should release all the valuable fixed nitrogen.

Nitrogen-fixing leguminous plants have the unique ability to fix atmospheric nitrogen in the ground and make their own fertilizers. Actually, these plants do not pull off this feat on their own. They owe partial credit for this effort to their symbiotic relationship with nitrogen-fixing bacteria. The leguminous plants provide nutrients to the bacteria in return for which the bacteria fix atmospheric nitrogen through anaerobic processes (processes that work without oxygen). The primary function of nitrogen-fixing bacteria is "survival," and in their efforts to survive, they enter into a symbiotic relationship with leguminous plants or some survive on their own. As a part of their metabolic cycle, they fix nitrogen. The enzyme that nitrogen-fixing bacteria use is called *nitrogenase* (Fig. 6.13). It is a chemical responsible for nitrogen fixation and without which, this process is impossible. The process at chemical level that enables nitrogen fixation can be summarized in the following way:

$$N_2 + 6H^+ + 6e^- \rightarrow 2NH_3$$

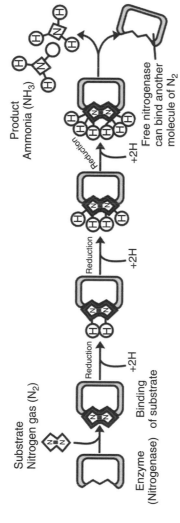

Figure 6.13. Nitrogenase is an enzyme complex found in nitrogen-fixing bacteria that convert atmospheric nitrogen to ammonia.

The end products are ammonia (NH_3) and water. Nitrogenase, the vital ingredient that makes nitrogen fixation possible, is destroyed when it comes in contact with oxygen. So the process of fixation in nitrogen-fixing bacteria occurs only in anaerobic (oxygen-deprived) conditions or oxygen is neutralized by its combination with chemicals such as leghemoglobin. Nitrogen fixation is one stage in the nitrogen cycle that maintains the balance of this element in nature (Fig. 6.14). The nitrogen-fixing bacteria and other microorganisms that fix nitrogen are collectively called *diazotrophs*. There are many strains of nitrogen-fixing bacteria in soil that carry out this process. They are important agents in the "nitrogen cycle." All the different types of diazotrophs have a nitrogen-fixing system based on iron-molybdenum nitrogenase. The following sections briefly discuss some of the symbiotic nitrogen-fixing bacteria.

6.2.6.1. Nitrogen-Fixing Bacteria Type 1: Rhizobia

Rhizobia are soil bacteria that fix nitrogen (diazotrophy) after becoming established inside root nodules of legumes (Fabaceae). Rhizobia require a plant host, as they cannot independently fix nitrogen. Morphologically, they are generally gram-negative, motile, nonsporulating rods (Fig. 6.15a,b). Their nitrogen fixation process cannot be executed without the help of their symbiotic partners, namely, the legume plants.

6.2.6.2. Nitrogen-Fixing Bacteria Type 2: Frankia

The bacteria belonging to the genus *Frankia* survive through their symbiotic relationship with actinorhizal plants, which are similar to leguminous plants (Fig. 6.16a,b).

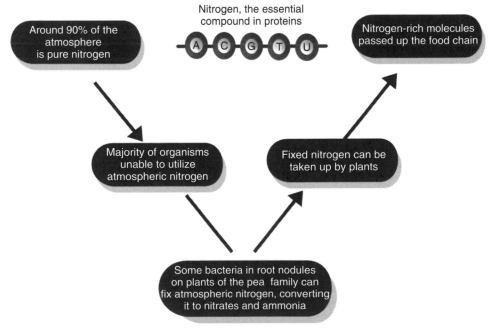

Figure 6.14. Nitrogen fixation.

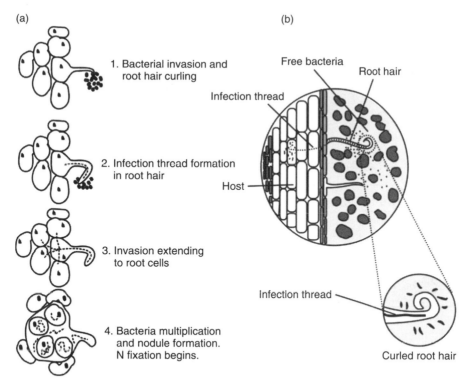

Figure 6.15. (a) The infection process of legume roots by rhizobia bacteria. (b) Closer view of the infection of root hair to form nodules.

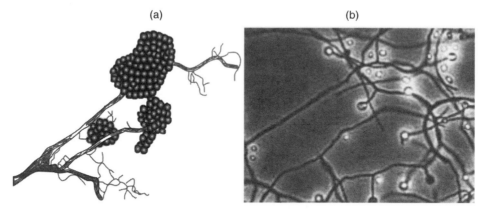

Figure 6.16. (a) Symbiotic relationship of *Frankia* with actinorhizal plants. (b) *Frankia* under SEM.

These bacteria form nodules in the roots of these plants. They wholly satisfy the nitrogen needs of these plants and indirectly enrich the soil with nitrogen compounds.

6.2.6.3. Nitrogen-Fixing Bacteria Type 3: Cyanobacteria

Some cyanobacteria show symbiotic behavior by their association with lichens; liverworts, a type of fern plant; and cycad plant. One example of this type of symbiotic

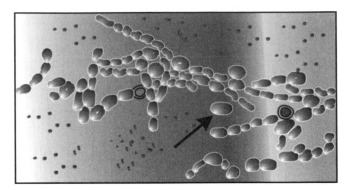

Figure 6.17. Thick-walled heterocysts of *Anabaena* in the leaves of *Azolla*.

nitrogen-fixing bacteria is *Anabaena*. Besides these, some of the nitrogen-fixing bacteria, which are free-living (nonsymbiotic) organisms, are *Clostridium, Desulfovibrio, Klebsiella pneumoniae, Bacillus polymyxa, Bacillus macerans, Escherichia intermedia, Azotobacter vinelandii, Anabaena cylindrica, Nostoc commune, Rhodobacter sphaeroides, Rhodopseudomonas palustris, Rhodobacter capsulatus*, among others.

Examination of an *Azolla* leaf reveals that it consists of a thick, greenish (or reddish) dorsal (upper) lobe and a thinner, translucent ventral (lower) lobe emerged in the water. It is the upper lobe that has an ovoid central cavity, the "living quarters" for filaments of *Anabaena*. The filaments of *Anabaena* with larger, oval heterocysts are normally visible around the crushed fern leaf (Fig. 6.17). The thick-walled heterocysts often appear more transparent and have distinctive "polar nodules" at each end of the cell.

Around 5–10% of the vegetative cells of a filament can transform into heterocysts when cyanobacteria are deprived of both nitrate and ammonia. Heterocysts are rounded, seemingly empty cells, usually distributed regularly along a filament or at one end of a filament. They develop as a result of the differentiation of the vegetative cells and are the sole sites of nitrogen fixation in heterocystous cyanobacteria. At the time of transformation of a vegetative cell into heterocyst, the former synthesizes a very thick new wall, reorganizes its photosynthetic membranes, discards its photosystem II and phycobiliproteins, and synthesizes the nitrogen-fixing enzyme nitrogenase. Photosystem I still operates and generates ATP (Fig. 6.18).

Although much of the nitrogen is removed when protein-rich grain or hay is harvested, significant amounts can remain in the soil for future crops. This is especially important when nitrogen fertilizer is not used, as in organic rotation schemes or some less-industrialized countries. Nitrogen is the most commonly deficient nutrient in many soils around the world, and it is the most commonly supplied plant nutrient. Supply of nitrogen through fertilizers has severe environmental concerns.

6.3. ROOT MICROBIAL INTERACTION

Plant roots exude an enormous range of potentially valuable small molecular weight compounds into the rhizosphere. Some of the most complex chemical, physical, and biological interactions experienced by terrestrial plants are those that occur between roots

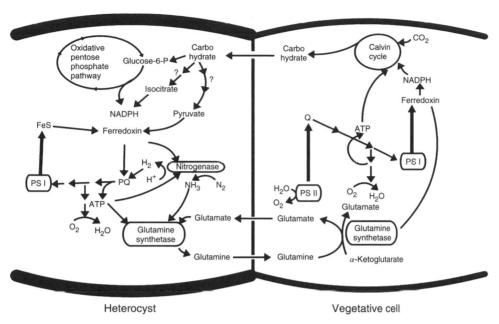

Figure 6.18. Metabolic interaction of heterocyst and vegetative cells.

and their surrounding environment of soil (i.e., the rhizosphere). Interactions involving plant roots in the rhizosphere include root–root, root–insect, and root–microbe interactions (Fig. 6.19; Bais et al., 2006). The most intense interactions between microbes and plants take place at the rhizosphere, which is the interface between plant roots and soil. The influence of plants on microbial population structure and function in the rhizosphere has important ecological implications for soil function, including biogeochemical cycles. Similarly, soil microbes have a tremendous influence on plant health and productivity. One straightforward and visible benefit for the plant is a better supply of and access to nutrients. The role of mutualistic nitrogen-fixing rhizobia has been well documented for decades, but recent data detail the intimate exchange of nutrients during the symbiosis of plant roots and bacteria. Plants attract nitrogen-fixing bacteria to invade the cells in the root and provide them with carbohydrates as a food source while the bacteria reduce nitrous compounds in the soil that are then used by plants.

Similarly, interactions between plants and fungi can also provide nutrients for the plant. Arbuscular mycorrhizal fungi (AMF), which form an intricate internal symbiosis with the roots of most flowering plants, are associated with the provision of phosphorus to plants in exchange for organic carbohydrates. Positive interactions include symbiotic associations with epiphytes and mycorrhizal fungi and root colonization by bacterial biocontrol agents and plant-growth-promoting bacteria (PGPB). Negative interactions include competition or parasitism among plants, pathogenesis by bacteria or fungi, and invertebrate herbivory. Elucidation is still required for the factors that determine whether the chemical signature of a plant's root exudates will be perceived as a negative or a positive signal. However, accumulated evidence suggests that root exudates have a major role in determining outcomes of interactions in the rhizosphere and, ultimately, plant and soil community dynamics. Microbes also indirectly help the nutrient uptake bacteria of the *Azospirillum* genus and promote increased root mass and more efficient nitrogen

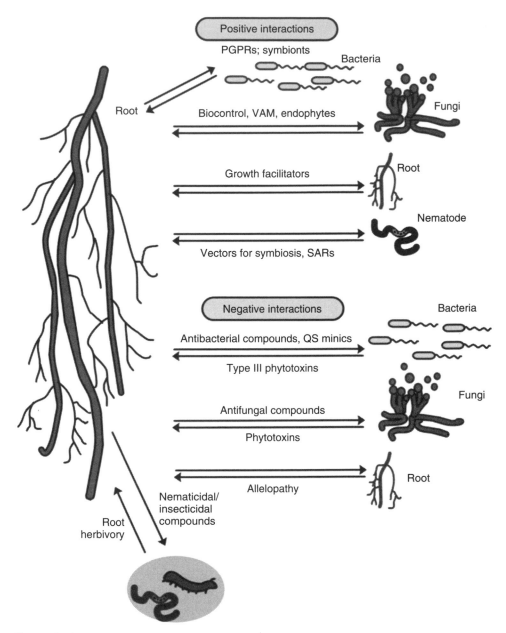

Figure 6.19. Root–root, root–microbe, and root–insect interactions; plant growth-promoting rhizobacteria (PGPR).

uptake from the soil in response to the plant hormone indole-3-acetic acid. Using these bacteria and fungi could provide significant environmental benefits, as they would allow a reduction in the application of nitrogen and phosphorus fertilizers.

The overuse of such fertilizers has become a major concern because they cause nitrate contamination of soil and groundwater by leachates and because microbial denitrification converts residual nitrogen into the greenhouse gas nitrous oxide. Equally, excess phosphorous compounds leach into groundwater, rivers, and streams, where they promote algal growth and cause other environmental problems. Although commercial fertilizer products based on rhizobia and *Azospirillum* are already available, their wider application is restricted by inconsistent performance and limitations of host range. A better understanding of plant–microbe symbiosis could help to overcome at least some of these problems.

In addition to enhancing the nutrient supply of plants, microbes also confer a degree of protection against plant diseases. In particular, various bacteria and fungi, especially of the genera *Pseudomonas, Bacillus*, and *Trichoderma*, produce a range of metabolites against other phytopathogenic fungi. Such biocontrol agents are already being used efficiently in the field, but they have not yet attained the degree of efficacy and consistency that is needed for large-scale commercialization. However, there is potential for improvement, and with development, such microbes could become a realistic alternative to the heavy fungicide regimens used at present in agriculture.

A reduction in the use of these chemicals would lead to obvious environmental benefits and it would also appeal to consumers who seek more natural produce. In addition to these direct effects on plant growth, rhizobacteria exert another health-promoting effect on the plants with which they interact. This phenomenon is known as *induced systemic resistance* (ISR) and arises when interactions with nonpathogenic bacteria confer better disease resistance on plants. ISR differs from systemic acquired resistance (SAR) in plants; the latter occurs in response to localized microbial attack, whereas the former is triggered by microbial interactions in the rhizosphere, and also enhances resistance of remote aerial plant parts against pathogens.

This systemic change in plant physiology bears conceptual similarities to other phenotypes, such as improved stress, drought, or disease resistance, that have been linked to plant–fungal associations. Little is known about why bacteria induce this condition, which is clearly beneficial to the plant. Although ISR is generally ascribed to interactions with bacteria, there is emerging evidence that some fungi, particularly endophytic fungi, may also be able to induce a similar response in plants.

The interactions of rhizosphere microbes with plants depend on the establishment of intimate associations between the two partners. This intimate cooperation between plant and bacteria displays a high level of host specificity. There is also a growing body of evidence suggesting that many other associations between plants and microbes show similar degrees of specificity; different plant species, and even different cultivars of the same plant species, establish distinct microbial populations in their rhizospheres when grown in the same soil.

The formation of these communities depends, at least in part, on the activation of specific programs of gene expression in the microbe in response to chemical signals secreted from the plant. A pertinent example is the induction of nodulation genes in receptive rhizobia (Fig. 6.20), which are triggered by the production and secretion of particular flavonoids by the plant. In the case of the rhizobia–legume interaction, the plant also responds to bacterial signals, and it is likely that this type of chemical cross talk is typical of other microbe–plant interactions. Other examples of plant-derived signals that influence microbial gene expression include phenolics exuded from plant wounds, which induce expression of virulence genes in pathogenic *Agrobacterium* spp., and compounds that mimic the quorum sensing signals used by bacteria to regulate gene expression.

In general, however, there is only scanty knowledge of signaling interactions between beneficial microbes and plants. Understanding how microbes respond to plant signals in terms of growth and gene expression and the role that plant signaling has in determining

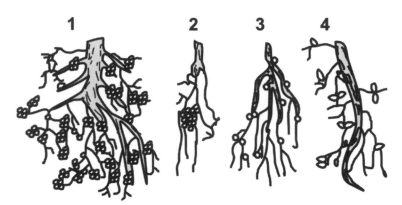

Figure 6.20. The series of events during the nodulation process.

interaction specificity or driving population selection is central to reaping the benefits of plant–microbe interactions. Before the advent of genomic technologies, there were limited options to investigate these interactions in detail, particularly with a view to their commercial exploitation.

6.3.1. Biological Dinitrogen Fixation

Biological fixation of dinitrogen is typically more important. A few species of microorganisms can synthesize an enzyme, called *nitrogenase*, which is capable of catalytically breaking the triple bond of N_2, generating two molecules of ammonia for each molecule of dinitrogen that reacted. Because the activation energy for this chemical reaction is rather high, it takes quite a lot of biological energy to fix atmospheric dinitrogen in this way, that is, about 12–15 mol of ATP per mole of dinitrogen that is converted. Despite the high energy costs, the dinitrogen fixation reaction is still very favorable ecologically because access is gained to ammonia and ammonium, the chemical forms of nitrogen that organisms can utilize for their nutrition. Because the nitrogenase enzyme is denatured by oxygen (O_2), the fixation reaction can only occur under anaerobic conditions, where oxygen is not present.

There are numerous species of free-living microorganisms that can fix atmospheric dinitrogen, including species of true bacteria, blue-green algae or bacteria, and actinomycetes. These microorganisms are most abundant in wet or moist environments, especially in situations where nutrients other than nitrate or ammonium are relatively abundant, for example, in rotting logs or other dead biomass, or in lakes that are well fertilized with phosphate from sewage. Such conditions are typically relatively deficient in available forms of nitrogen and are commonly anoxic, creating obviously favorable opportunities for species of microorganisms that can utilize dinitrogen.

Some species of plants live in an intimate and mutually beneficial symbiotic relationship with microorganisms that have the capability of fixing dinitrogen. The plants benefit from the symbiosis by having access to a dependable source of fixed nitrogen, while the microorganisms benefit from the energy and habitat provided by the plant. *Rhizobium japonicum* is the best known example of symbiosis involved with many species in the legume family (Fabaceae). Some plants in other families also have dinitrogen-fixing symbioses, for example, red alder (*Alnus rubra*) and certain actinomycetes, a type

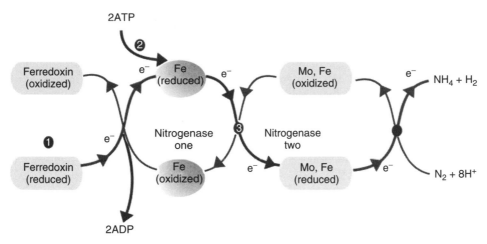

Figure 6.21. Mechanism of biological nitrogen fixation.

of microorganism. Many species of lichens, which constitutes a symbiotic relationship between a fungus and a blue-green bacterium (cyanobacteria), can also fix dinitrogen. Biological dinitrogen fixation (Fig. 6.21) is an ecologically important process, being ultimately responsible for most of the fixed nitrogen that occurs in the biomass of organisms and ecosystems. The only other significant sources of fixed nitrogen to ecosystems are atmospheric depositions of nitrate and ammonium with precipitation and dustfall and the direct uptake of NO_2 gas by plants.

6.3.2. Free-Living Dinitrogen Fixation

When soils are amended with large amounts of nitrogen-poor organic substrates, nitrogen and eventually other mineral nutrients become limiting for the microbial degraders. Thus, the addition of nitrogen fertilizers usually stimulates oil degradation in contaminated soils. Nevertheless, the nitrogen imbalance created by accidental oil spills or land farming oily sludges selectively favors the development of the soil asymbiotic dinitrogen-fixing microbes (Beauchamp et al., 2006)

Knowledge of the microbiology of dinitrogen (N_2)-fixing bacteria in compost rich in deinking paper sludge (DPS) is limited. Dinitrogen (N_2)-fixing bacteria from DPS composts were isolated and studied for their N_2-fixing activity *in vitro* and *in vivo*. Two gram-negative N_2-fixing isolates were identified as *Pseudomonas*. At 20 °C, both isolates revealed that N_2-fixing activity was higher than that of three arctic *Pseudomonas* strains. Their N_2-fixing activity was found to occur between 18 and 25 °C, a pattern that was similar to the reference isolate *Azotobacter* ATCC 7486. Composts successfully showed N_2-fixing activity after carbohydrate amendments both with and without inoculation of a N_2-fixing isolate. In *Pisum sativum* cultivated under standard growth conditions, the extent of N_2 fixation with time estimated by the acetylene reduction assay (ARA, PN_2F) and rates of the actual nitrogen accumulation (ANA) of plant biomass were calculated from six independent growth experiments. In the plants inoculated with indigenous soil *Rhizobium* populations and cultivated on 0.63 mmol/l nitrate level, the percentage of PN_2F/ANA ratios ranged from 25.7% to 61.5%. In peas inoculated with the inoculant strain, the PN_2F/ANA ratios were markedly higher, ranging from 59.8% to 65.1% (Škrdleta et al., 1993).

The effect of herbicides on legume nitrogen (N_2) fixation is of great importance. The effect depends on the herbicide, its concentration, and different weather conditions. There is a differentiated effect of herbicides on nodulation and dinitrogen fixation. In most cases, it was found that applied herbicide doses are safe for both nodulation and maintaining full dinitrogen fixation activity. Metribuzin is reported to reduce soybean nodulation and N_2 fixation rate (by ARA) by 50%, whereas linuron did not decrease the N_2 fixation rate (Rennie and Dubetz, 1984). It was also reported that post-emergence herbicides affect nodulation and N_2 fixation (Ozair and Moshier, 1988; Schnelle and Hensley, 1990; Wache, 1987).

6.3.3. Associative Nitrogen Fixation

Nitrogen, the most limiting nutrient, is the input required in the largest quantity for lowland rice production. The concerns on nitrogen economy and efficiency and its impact on environment have renewed interest in exploring alternative or supplementary nitrogen source for sustainable agriculture. Several studies have indicated the existence of significant rice genotypic differences in N_2-fixation-stimulating (NFS) traits. Rice genotypes with high NFS are desirable because they add N to the soil-water-plant system without additional farm inputs and reduce dependence on fertilizer. Large genotypic differences in percentage nitrogen derived from air (% Ndfa), such as 1.5% in Abang Basur, medium maturing genotype, to 21% in Oking Seroni, late maturing genotype, indicate the potential of isolating genotypes with high NFS for sustainable agriculture. The exogenous supply of nitrogenous fertilizer to lowland rice significantly inhibited nitrogen fixation but improved plant growth. Although phosphorus fertilizer did not affect atom% 15N excess and % Ndfa significantly, slight decrease in atom% 15N excess and increase in N_2 fixation was observed. Inhibitory effect of exogenous supply of nitrogen fertilizer indicates limited potential of associative N_2 fixation to significantly benefit agriculture. Farmers would have to withhold nitrogen fertilizer from their rice crop in order to increase biological N_2 fixation associated with rice. If they do such practice, the plants will be nitrogen deficient and might have a lower yield. However, the development of nitrogen fixation in response to a deficiency of available nitrogen may well be an integral part of the nitrogen cycle of the natural ecosystem and low input farming system, thereby maintaining a nitrogen balance in the environment (Shrestha and Maskey, 2005). Associative nitrogen fixation is carried out by a number of bacterial species living on the roots of nonleguminous plants (Pedersen et al., 1978, Rennie, 1980). Nitrogen-fixing enteric bacteria of the genera *Klebsiella* and *Enterobacter* have frequently been isolated from the roots of various plants (Haahtela et al., 1981) and are known to be able to adhere to the roots of grasses and cereals *in vitro*.

Fimbriae are involved in enterobacterial adhesion to grass roots (Haahtela et al., 1981, Hrabak et al., 1981). Two types of fimbriae occur on *Klebsiella* associated with plants: the so-called type 3 and type 1 fimbriae (Duguid and Old, 1980). The two fimbrial types differ morphologically and serologically and in their binding specificities: the type 1 fimbriae bind to mannosides (Old, 1972), but the receptor structure for the type 3 fimbria is not known. In associative nitrogen fixation, certain nitrogen-fixing bacteria establish themselves close to the root or can be found on the leaf surface. They use root exudates, secretions, and sloughed cells as an energy source. Some bacteria, such as *Acetobacter*, can occupy internal root tissues of sugarcane. The amount of nitrogen fixed by these microbes is indirect, as 90% of the nitrogen becomes available only when the bacteria die.

An example is *Azotobacter paspali* which grows as a sheath around the root of the sand grass *Paspalum notatum*. No nodules are formed, but the plant is benefited greatly. Survival of *A. paspali* away from the plant is poor. The acetylene test shows that the *A. paspali* and *P. notatum* association fixes as much nitrogen as a moderate legume crop. Whether this also takes place under natural conditions is yet to be established. *A. paspali* associated with only the five tetraploid ecotypes of *P. notatum* out of 33 stimulated. *A. paspali* lives outside the roots in the rhizosphere. This environment is ideal for heterotrophic nitrogen-fixing bacteria. It is almost anaerobic, a factor of importance to the association, which, although aerobic, is very sensitive to oxygen. The energy and carbon requirements of the heterotrophic free-living nitrogen fixers are met with by the root cell and soluble root exudates. These organisms occur in the rhizospheres of a variety of plants. Other examples of associative symbiosis are rice with *Beijerinckia* and the grass *Digitaria decumbens* with *Spirillum lipoferum*.

Mycorrhizal fungi and the symbiotic nitrogen-fixing actinomycete *Frankia* may occupy the inter- and intracellular spaces inside roots. They may establish a highly integrated mutualistic association with their plant hosts. Helically lobed bacteria have been reported to be found in the collapsed epidermal cells of *Ammophila arenaria* and within roots of *Zea mays*. They appear to enter by intercellular penetration of the epidermal cells.

6.3.4. Legume–Rhizobia Symbioses

In case of legume–*Rhizobium* symbiosis, a legume provides bacteria with energy-rich carbohydrates and some other compounds, while *Rhizobium* supplies the host legume with nitrogen in the form of ammonia. Unlike any plant, rhizobia (and some other microorganisms) can fix inert N_2 gas from the atmosphere and supply it to the plant as NH^{4+}, which can be utilized by the plant. The diversity seen in the legumes and their interacting partners is as wide ranging as the difference between the brains of canaries and humans. For the past 20 years, rhizobial and legume biologists have pursued a scientific investigation based on this biodiversity for the purposes of understanding the complexities of the agriculturally and environmentally important nitrogen-fixing symbiosis epitomized by nodulation (Young et al., 2001).

Legumes are unique in their response to *Nod* factors in that they actively promote entry of bacteria into the root (Fig. 6.22). However, as yet, we do not know how legumes evolved the ability to recognize such signals or how entry is actually accomplished. Genome projects and scientific pursuits that include a diversity of legumes and rhizobial species will better inform us as to which genes/proteins are conserved among all hosts and symbionts and help us to determine whether the ability to fix N_2 into ammonia can be transferred to crops other than legumes.

6.3.5. Mycorrhizal Association

In a mycorrhizal association, the fungus colonizes the host plants' roots, either intracellularly as in AMF or extracellularly as in ectomycorrhizal fungi (Fig. 6.23).

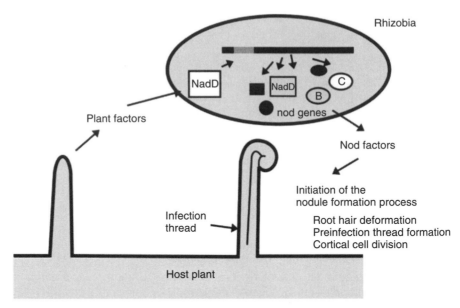

Figure 6.22. Response of Nod factors in legumes promote entry of bacteria into their root.

They are an important component of soil life and soil chemistry. Mycorrhizae are beneficial associations between the growing root system of a plant, which is the host, and a mycorrhizal fungus as a symbiont. In such a mycorrhizal association, the host and the symbiont belong not only to different species but also to different kingdoms. Mycorrhizal fungi develop and replicate by such manners typical to this kingdom. They may belong to the Basidiomycetes, Ascomycetes, or Zygomycetes. They all have characteristically filamentous structures and may produce visible fruiting bodies both below and on the soil surface. *Amanita muscaria* (the fly amanite), *Scleroderma citrinum, Laccaria laccata, Boletus edulis*, and *Cantharellus cibarius* are just a few of the widely known fruiting bodies of mycorrhizal fungi. This mutualistic association provides the fungus with relatively constant and direct access to carbohydrates, such as glucose and sucrose, supplied by the plant. The carbohydrates are translocated from their source (usually leaves) to root tissue and onto the fungal partners. In return, the plant gains the benefits of the mycelium's higher absorptive capacity for water and mineral nutrients (due to comparatively large surface area of mycelium/root ratio), thus improving the plant's mineral absorption capabilities.

Plant roots alone may be incapable of taking up phosphate ions that are demineralized, for example, in soils with a basic pH. The mycelium of the mycorrhizal fungus can, however, access these phosphorus sources and make them available to the plants they colonize. Mycorrhizal plants are often more resistant to diseases, such as those caused by microbial soilborne pathogens, and are also more resistant to the effects of drought. These effects are perhaps due to the improved water and mineral uptake in mycorrhizal plants.

6.4. PATHOGENIC MICROBES IN AGRICULTURE

Plant pathology is the scientific study of plant diseases caused by pathogens (infectious diseases) and environmental conditions (physiological factors). Organisms that

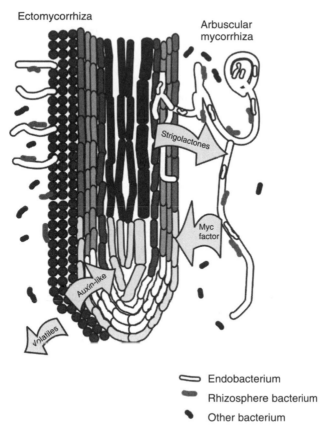

Figure 6.23. Mycorrihzal association of plant roots.

cause infectious disease include fungi, oomycetes, bacteria, viruses, viroids, viruslike organisms, phytoplasmas, protozoa, and nematodes. Bacteria pathogenic for plants are responsible for devastating losses in agriculture.

The great majority of microorganisms are harmless and many are beneficial. Some of the properties of nonpathogenic microorganisms have been used, for instance, in biotechnology to produce metabolites or enzymes. Nonpathogenic microorganisms have also been used as inoculum to protect from diseases (biocontrol and biofertilizers in agriculture, probiotics), to restore contaminated sites (bioremediation), or for fermentation in food processes. Pathogenic microorganisms, although they represent only a small part in the total microbial world, receive much attention because they represent a threat for the human or animal health or to agriculture. They can cause diseases of plague dimensions with serious economic and environmental consequences. From a practical perspective, pathogenicity or virulence is the capacity of some microorganisms to cause disease. However, microbiologists recognize that pathogenicity represents a form of versatility and specialization that enables certain microorganisms to replicate within a specific host (infectivity) and damage host cells. Although cellular damage is not clinically apparent in many cases, a significant proportion of infected hosts show signs of disease or eventually die. The outcome of the infection is dependent on not only the properties of the pathogen (virulence, invasiveness, toxic or allergenic effects) but also the host immunity status.

From this point of view, pathogens fall into two basic types: primary pathogens that cause disease among at least a portion of normal individuals and opportunistic pathogens that cause disease only in individuals who are compromised in either their innate or humoral immune defenses.

6.4.1. Microbes and Agriculture

In agriculture, fungi, bacteria, algae, and viruses are important with respect to their contribution in the form of either loss or gain in the production of grains, fruits, vegetables, oil, milk, poultry, fodder, and livestock. Microbes include fungi, bacteria, and viruses. Farmers and ranchers often think of microbes as pests that are destructive to their crops or animals (as well as themselves), but many microbes are beneficial. Soil microbes (bacteria and fungi) are essential for decomposing organic matter and recycling old plant material. Some soil bacteria and fungi form relationships with plant roots that provide important nutrients such as nitrogen and phosphorus. Fungi can colonize the upper parts of plants and provide many benefits, including drought and heat tolerance and resistance to insects and plant diseases.

Most bacteria that are associated with plants are actually saprotrophic, and do no harm the plant. However, a small number, around 100 species, are able to cause disease. Bacterial diseases are much more prevalent in subtropical and tropical regions of the world. Most plant pathogenic bacteria are rod shaped (bacilli). In order to be able to colonize the plant they have specific pathogenicity factors.

Phytoplasma and *Spiroplasma* are the genera of bacteria that lack cell walls and are related to the mycoplasmas, which are human pathogens. Together they are referred to as the *mollicutes*. They also tend to have smaller genomes than true bacteria. They are normally transmitted by sap-sucking insects, being transferred into the plants phloem where it reproduces.

Viruses are almost always thought of as agents of disease. They seem to be living in the plants without doing any harm. There are many types of plant viruses, and some are even asymptomatic. Normally, plant viruses only cause a loss of crop yield. Therefore, it is not economically viable to try to control them, the exception being when they infect perennial species, such as fruit trees. Most plant viruses have small, single-stranded RNA genomes (Fig. 6.24). These genomes may only encode three or four proteins: a replicase, a coat protein, a movement protein to allow cell to cell movement though plasmodesmata, and sometimes a protein that allows transmission by a vector. There are a few examples of plant diseases caused by protozoa. They are transmitted as zoospores, which are very durable and may be able to survive in a resting state in the soil for many years. They have also been shown to transmit plant viruses.

New viruses are also emerging each year to threaten the lives of both humans and animals.

Activities of microorganisms are very important to almost every sector of concern to mankind, such as agriculture, forestry, food, industry, medicine, and environment.

The scope and significance of microbiology has enlarged manifold, particularly when importance of environment was realized globally and the word environment was used in a much wider sense in terms of totality to include almost everything, every bit of Nature.

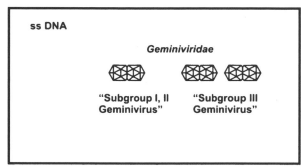

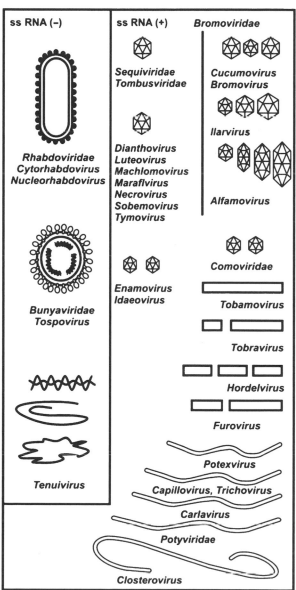

Figure 6.24. Plant viruses.

6.4.2. Biopesticides

Today, biological control is widely regarded as a desirable technique for controlling insects due to its minimal environmental impact and avoidance of problems of resistance in the vectors and agricultural pests. Biopesticides are naturally occurring substances (biochemical pesticides, biological agent) possessing the desirable properties of a chemical pesticide. It makes them highly toxic to the target organism, which can be mass produced on an industrial scale; they have a long shelf life; and they can be safely transported. Conventional pesticides, by contrast, are generally synthetic materials that directly kill or inactivate the pest. For example, a plant in the presence of chitosan will naturally promote ISR to allow the plant to defend itself against disease, pathogens, and pests. Biopesticides are considered ecofriendly and easy to use.

Biopesticides are certain natural plant products that belong to the so-called secondary metabolites that include thousands of alkaloids, terpenoids, phenolics, and minor secondary chemicals. Biopesticides are derived from natural materials such as animals, plants, bacteria, and certain minerals. For example, canola oil and baking soda have pesticidal applications and are considered biopesticides. At the end of 2001, there were approximately 195 registered biopesticide active ingredients and 780 products. Biopesticides have usually no known function in photosynthesis, growth, or other basic aspects of plant physiology; however, their biological activity against insect pests, nematodes, fungi, and other organisms is well documented. These biodegradable, economical, and renewable alternatives are used especially in organic farming systems.

The interest in biopesticides is based on the disadvantages associated with chemical pesticides, some of which are as follows: extensive pollution of the environment; serious health hazard due to the presence of their residues in food, fiber, and fodder; and increasing cases of insects developing resistance. Their chief disadvantages are very high specificity, which will require an exact identification of the pest/pathogen and may require multiple pesticides to be used, and often variable efficacy due to the influences of various biotic and abiotic factors. Since biopesticides are usually living organisms, they bring about pest/pathogen control by multiplying within the target insect pest/pathogen. The total world production of biopesticides is over 3000 tons per year, which is increasing at a rapid rate.

Bacillus anthracis the cause of anthrax poisoning is currently a great concern because of its use as a terror weapon. The virulence of *B. anthracis* depends on the presence of two large plasmids; strains lacking one or the other plasmid are not virulent. New strains of *B. anthracis* with unpredictable properties could arise. The δ-endotoxin was found to have some effects on nontarget organisms such as the monarch butterfly, nematode, earthworms, and *Chironomus riparius* larvae (Charbonneau et al., 1994; Kondo et al., 1992; Losey et al., 1999). Continuous use of this biopesticide, in the form of either Bt (*Bacillus thuringiensis*) suspensions or Bt corn, could contaminate the soil and aquatic environments and cause adverse effects on nontarget aquatic invertebrates. Thus, contamination of the environment by the δ-endotoxin gene and protein is certainly possible and warrants investigation.

B. thuringiensis is both a major pesticide and the source of the genes used to produce insect toxins in genetically modified (GM) crops. The endotoxins of *B. thuringiensis* (Bt toxins) are stored as inactive crystals in bacterial spores, which are activated in the insect gut to create pores on the cells of the insect gut, causing an inrush of water that bursts the cell (BenDov et al., 1999). These crystal proteins (Cry proteins) are insect

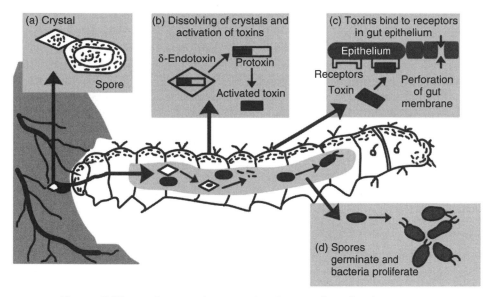

Figure 6.25. Mechanism of toxicity of endotoxin of *Bacillus thuringensis*.

stomach poisons that must be eaten to kill the insect (Fig. 6.25). Once eaten, an insect's own digestive enzymes activate the toxic form of the protein. The Cry proteins bind to specific "receptors" on the intestinal lining and rupture the cells. Insects stop feeding within 2 hours of the first bite and, if enough toxin is eaten, die within 2 or 3 days. There are several strains of Bt, each with differing Cry proteins. Scientists have identified more than 60 Cry proteins. Proteins have been found with insecticidal activity against the Colorado potato beetle (e.g., Cry3A, Cry3C), corn earworm (Cry1Ac, Cry1Ab), tobacco budworm (Cry1Ab), and European corn borer (Cry1Ab, Cry1Ac, Cry9C). Most of the Bt corn hybrids, targeted against the European corn borer, produce only the Cry1Ab protein; a few produce the Cry1Ac or the Cry9C protein.

A third bacterium, *Bacillus cereus*, is a common soil bacterium and a common cause of food poisoning. The three species of bacteria are closely related, differing mainly in their plasmids (plasmids are circular DNA molecules that contain genetic origins of replication that allow them to replicate independent of the chromosome). The plasmids of the three species may readily be transferred from one species to another (Helgason et al., 2000). The toxin genes from the three species are located on the plasmids and the genes tend to cluster in "islands" that sometimes are mobilized (caused to move) by lysogenic bacterial viruses (bacteriophages that integrate themselves into the bacterial genome or bacterial plasmid as prophage). The ready exchange of plasmids bearing toxin genes between the three species has raised some concern.

When *B. anthracis* was mated with *B. thuringiensis* to transfer plasmids, recombination could create plasmids bearing toxins both for anthrax and for killing insects. The insect killing ability of *B. thuringiensis* is based on the presence of an island of toxin genes (BenDov et al., 1999) on one of many (up to 17) plasmids in the bacterium (Andrup et al., 1994). The strain, *B. thuringiensis* sv. *israelensis*, has a plasmid borne prophage that is induced to multiply when the strain mates with phage-insensitive strains of *B. thuringiensis* or *B. cereus* (Kanda et al., 2000).

B. thuringensis nowadays serve as the basis for large-scale production of microbial insecticides. For more than 30 years, various liquid and granular formulations of Bt have been used successfully against European corn borer and other insect pests of a variety of crops. The widespread use of Bt is often challenged by production as well as formulation costs. *B. thuringiensis* var. *aizawai* strain GM-7 can be propagated (Ramos et al., 2000) in a bioreactor with molasses as base medium (Wong, 1993). The optimum fermentation parameters are 30 °C, pH 7, 1% of inoculum, 1 vvm aeration, and 500 rpm agitation. The resulting extract was obtained using the lactose-acetone method (Fig. 6.26) (Dulmage 1970). The extract was formulated into granules using various biopolymers as encapsulating matrices. Solid materials can be premixed with dried *B. thuringiensis* extract with a hand rotary flour sifter and then mixed with the liquid ingredients. The mixture was allowed to dry for 1–2 h at room temperature (26 °C), after which the matrices were chopped, sieved through a 20-mesh screen for the pectin and cornstarch and through a 10-mesh screen for the gelatin matrix, and allowed to dry overnight in a laminar flow hood.

Various alternative local media have successfully replaced costly synthetic media, but the actual constraint is the harvesting process and formulation costs. Harvesting efficacy governs the marketability of a product by affecting potency and aiding in further processing during formulation development. Formulation is a crucial link between production and application and dictates economy, longer shelf life, ease of application, and enhanced field efficacy. There are various environmental factors, such as ultraviolet radiation, rain, pH, temperature, and foliage physiology, that impede the efficacy of Bt formulations. There have been developments of various formulations depending on application target and feasibility, such as solid and liquid, to overcome the adverse environmental effects. Conventional formulations have been substituted by advanced versions such as microencapsulations and microgranules to enhance residual entomotoxicity.

The Bt toxin genes are used in crop genetic engineering. Transgenic maize having agronomically desirable traits have been achieved by incorporation of a gene that codes for the Bt toxin (Fig. 6.27). The *cry1Ah* gene was a candidate gene for insect-resistant transgenic maize research. To confirm the *cry1Ah* gene's function in transgenic maize, a plant expression vector pHUAh harboring *cry1Ah* was constructed and transferred into immature embryonic calli of maize Q31 × Z31 by microprojectile bombardment. The *hyg* gene was used as the selective marker gene. A total of 66 regenerated plants were obtained in T0 generation, 13 of which were PCR positive. Bioassay results of T1 transgenic plants showed that events B1 and B7 were highly resistant to the Asian corn borer. Currently, there has been little or no effort to evaluate the possible recombination between *B. anthracis* in the field and the endotoxin genes of crop plants. Such gene exchange could occur in the soil between GM plant debris and bacteria. Also, it is not unlikely that GM crops carrying anthrax genes could be produced for either vaccines or bioweapons.

6.4.3. Commercial Microbial Pesticides

There is a long history of hunting for new natural products for pharmaceutical purposes. For example, penicillin, cyclosporin, and streptomycin are natural products from microorganisms. However, the efforts dedicated to screening of microbial natural products for pesticides are small compared to pharmaceutical natural product programs. Approximately 400 insecticidal/miticidal, 30 herbicidal, and less than 20 nematicidal microbial

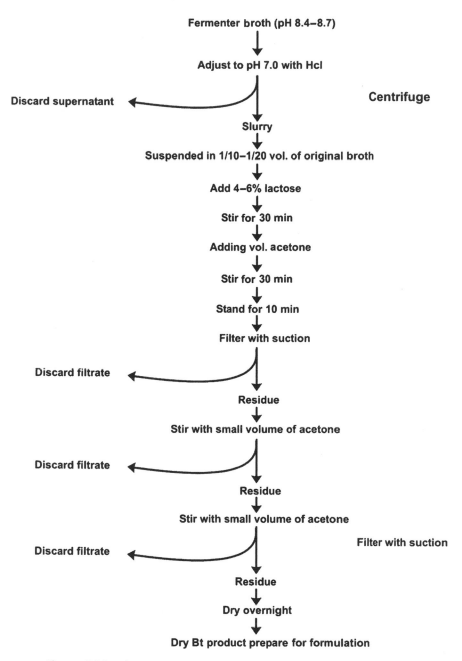

Fermenter broth (pH 8.4–8.7)

Adjust to pH 7.0 with Hcl

Discard supernatant ← **Centrifuge**

Slurry

Suspended in 1/10–1/20 vol. of original broth

Add 4–6% lactose

Stir for 30 min

Adding vol. acetone

Stir for 30 min

Stand for 10 min

Filter with suction

Discard filtrate ←

Residue

Stir with small volume of acetone

Discard filtrate ←

Residue

Stir with small volume of acetone

Discard filtrate ← **Filter with suction**

Residue

Dry overnight

Dry Bt product prepare for formulation

Figure 6.26. Small-scale processing for formulation of Bt bioinsecticide.

natural products are known compared to the tens of thousands of known pharmaceutically active natural products from microorganisms. The technology for discovering and commercializing microbial pesticides and other biopesticides, however, is proven and in some cases quite powerful. Many plant and microbial natural products have become commercial pesticides (e.g., pyrethrins and synthetic pyrethroids from *Pyrethrum daisy*,

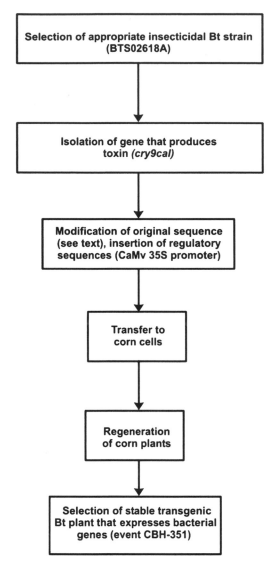

Figure 6.27. Transgenic corn (Bucchini and Goldman, 2002).

azadirachtin from the neem tree, avermectin from the microbe *Streptomyces avermitilis*, and spinosyn from the microbe *Saccharopolyspora spinosa*). Some of the bacteria formulated as products of biocontrol agents are shown in Table 6.1. The natural world can provide alternatives to existing chemicals in both the short and long term. Out of 1.5 million species of fungi, only 73,000 species have been described. A small percentage of these have been tested in pesticidal screens. In addition, other than *B. thuringiensis*, only a small number of strains of known genera have been tested as pesticides.

Plant protection against pathogens, pests, and weeds has been progressively reoriented from a therapeutic approach to a rational use of pesticide chemicals in which consumer health and environmental preservation prevail over any other productive or economic considerations. Microbial pesticides are being introduced in this new scenario of crop

TABLE 6.1. Bacteria Formulated as Products of Biocontrol Agents

Product (Brand Name)	Biocontrol Agent	Target Pathogen/Disease
AtEze	*Pseudomonas chlororaphis* 63-28	Wilt disease as well as stem and root rots
Ballad	*Bacillus pumilus* QST 2808	Asian soybean rust
Bio-Save 10LP, 110	*Pseudomonas syringae*	*Botrytis cinerea, Penicillium* spp., *Mucor pyroformis, Geotrichum candida*
BlightBan A506	*Pseudomonas fluorescens* A506	Frost damage, *Erwinia amylovora*, and russet-including bacteria
Companion	*Bacillus subtilis* GB03 Others include *B. subtilis, Bacillus lichenformis, Bacillus megaterium*	*Rhizoctonia, Pythium, Fusarium,* and *Phytophthora*
Deny	*Burkholderia cepacia* type Wisconsin	*Rhizoctonia, Pythium, Fusarium,* and disease caused by lesion, spiral, lance, and sting nematodes
EcoGuard	*Bacillus licheniformis* SB3086	Dollar spot, low and moderate disease pressure
Frostban A	*Pseudomonas fluorescens* strains A506 and 1629 RS, *Pseudomonas syringae* 742 RS	Frost-forming bacteria
Frostban B	*Pseudomonas fluorescens* A506	Frost-forming bacteria
Frostban C	*Pseudomonas syringae* 742 RS	Frost-forming bacteria
Frostban D	*Pseudomonas fluorescens* 1629RS	Frost-forming bacteria
Galltrol	*Agrobacterium radobacter* strain 84	Crown gall disease caused by *Agrobacterium tumefaciens*
GB34 biological fungicide	*Bacillus pumilus* GB34	Fungal pests *Rhizoctonia* and *Fusarium*
Green-Releaf	*Bacillus licheniformis* SB3086	Many fungal species, especially those causing leaf spot and blight diseases

protection, and currently, several beneficial microorganisms are the active ingredients of a new generation of microbial pesticides or the basis for many natural products of microbial origin (Table 6.2).

The development of a microbial pesticide requires several steps addressed to its isolation in pure culture and screening by means of efficacy bioassays performed *in vitro*, *ex vivo*, *in vivo*, or in pilot trials under real conditions of application (field, greenhouse, postharvest). For the commercial delivery of a microbial pesticide, the biocontrol agent must be produced at an industrial scale (fermentation), preserved for storage, and formulated by means of biocompatible additives to increase survival and to improve the application and stability of the final product. Despite the relatively high number of patents for biopesticides, only a few of them have materialized in a register for agricultural use.

The excessive specificity in most cases and biosafety or environmental concerns in others are major limiting factors. Nontarget effects may be possible in particular cases,

TABLE 6.2. Microbes Used as Pesticide

Microbe	Registered Product	Target Disease or Pest Being Controlled
NMV	ILCAR, GYPCHEK, VIRTUSS	Tobacco budworm, gypsy moth, Douglas fir
CPV	MATSUKEMIN	Pine caterpillar
Bacillus thuringiensis	THURICIDE and many more	Wide range of insects
Agrobacterium radiobacter K84	GALLTROL-A	Crown gall of stone fruits
Pseudomonas sp. and *Enterobacter* sp.	DAGGER-G	Damping off of above-ground parts of cotton seedlings
Pseudomonas sp. and *Bacillus* spp.	QUANTUM-4000	Wilt of cereals and vegetables
Beauveria bassiana	BOVERIN	Range of insects

such as displacement of beneficial microorganisms, allergenicity, toxigenicity (production of secondary metabolites toxic to plants, animals, or humans), pathogenicity (to plants or animals) by the agent itself or due to contaminants, or horizontal gene transfer of these characteristics to nontarget microorganisms. However, these nontarget effects should not be evaluated in an absolute manner, but relative to chemical control or the absence of any control of the target disease (e.g., toxins derived from the pathogen). Consumer concerns about live microbes due to emerging foodborne diseases and bioterrorism prevent from creating a socially receptive environment to microbial pesticides.

6.4.4. Bioweedicides

In irrigated agriculture, weed control through chemical herbicides creates spray drift hazards and adversely affects the environment. Besides, pesticide residues (herbicides) in food commodities directly or indirectly affect human health. These lead to the search for an alternate method of weed management that is ecofriendly. In this regard, the biological approach (a deliberate use of natural enemies to suppress the growth or reduce the population of the weed species) is gaining momentum. This approach involves two strategies: the classical or inoculative strategy and the inundative or bioherbicide strategy.

In the inoculative approach, an exotic biocontrol agent is introduced in an infested area. This method is slow and is dependent on favorable ecological conditions, which limits its success in intensive agriculture. While in the inundative approach, bioherbicides are used to control indigenous weed species with native pathogen, applying them in massive doses in the area infested with target weed flora. Bioherbicides offer many advantages including a high degree of specificity of target weed, no effect on nontarget and beneficial plants or man, absence of weed resistance development, and absence of residue buildup in the environment. Commercial bioweedicides first appeared in the market in the United States in the early 1980s with the release of the products Devine, Collego, and Biomal. Success stories of these products and the expectation of obtaining perfect analogs of chemical herbicides have opened a new vista for weed management. Plant pathologists and weed scientists have identified over 100 microorganisms that are candidates for development as commercial bioherbicides. Devine, developed by Abbott Laboratories,

USA, and the first mycoherbicide derived from fungi (*Phytophthora palmivora* Butl.), is a facultative parasite that produces lethal root and collar rot of its host plant *Morrenia odorata* (strangler wine) and persists in soil saprophytically for extended periods of residual control. It was the first product to be fully registered as a mycoherbicide. It infects and kills strangler wine (control 95–100%), a problematic weed in citrus plantation in Florida.

Commercially Collego, a formulation of the endemic anthracnose fungus *Colletotrichum gloeosporioides* f.sp. *aeschynomene* (cga), was developed to control northern joint vetch (*Aeschynomene virginica*) in rice and soybean fields. Dry powder formulation containing 15% spores (conidia) of cga as an active ingredient was registered in 1982 under the trade name Collego, having a shelf life of 18 months. It is the first commercially available mycoherbicide for use in annual weed in annual crops with more than 90% control efficiency.

Besides the many advantages of bioherbicides, certain factors have been reported to limit the development of bioherbicides into commercial products. These factors include biological constraints (host variability, host range resistance mechanisms, and interaction with other microorganisms that affect efficacy), environmental constraints (epidemiology of bioherbicides dependent on optimum environmental conditions), technical constraints (mass production and formulations development of reliable and efficacious bioherbicide), and commercial limitations (market size, patent protection, secrecy, and regulations). The bioherbicides approach is gaining momentum. New bioherbicides will find place in irrigated lands, wastelands, as well as in mimic parasitic weeds or resistant weed control. Research on synergy test of pathogens and pesticides for inclusion in IPM, developmental technology, fungal toxins, and application of biotechnology, especially genetic engineering, is required. However, bioherbicides should not be viewed as a total replacement to chemicals, but rather as a complement in integrated weed management systems.

6.4.5. Diseases Caused by Bacteria

Bacteria like many other types of microorganisms are highly parasitic and are found to interact with many host of varying characteristics and physiologies including and not limited to plants. The major families of bacteria that are pathogenic to plants include *Agrobacterium, Clavibacter, Erwinia, Pseudomonas, Xanthomonas, Streptomyces*, and *Xylella*. With the exception of the species of *Streptomyces*, all other species of the genera are small, rod-shaped, gram-negative bacteria with length approximately 0.5–1.0 μm. A few bacterial diseases are described in Table 6.3.

The four main categories of bacterial plant disease symptoms based on the extent of damage caused to plant tissue include vascular wilt, rotting, necrosis, and tumor formation. Tikka disease of groundnut is caused by *Cercospora personata*. The disease is characterized by the development of circular black spots on the leaf and the yield of groundnuts is affected (Fig. 6.28a). Blast disease of rice is caused by *Pyricularia oryzae*, in which gray spots appear on the leaves and husk (Fig. 6.28b).

Some of the diseases are briefly discussed in the following sections.

6.4.5.1. Bacterial Leaf Spot

Bacterial leaf spot is easily identified by the symptoms on the infected plant's foliage. Plants that are infected with bacterial leaf spot will develop dark, water-soaked spots that

TABLE 6.3. Bacterial Diseases of Plants

Disease	Causative Agent	Hosts	Symptoms and Signs	Additional Features
Granville wilt	*Pseudomonas solanacearum*	Tobacco, tomato, potato, eggplant, pepper, and other plants	Stunting, yellowing, and wilting of parts above ground; roots decay and become black or brown	Occurs in most countries in temperate and semitropical zones, causes crop losses of hundreds of millions of dollars
Fire blight	*Erwinia amylovora*	Apple and pear	Blossoms appear water soaked and shrivel; spreads to leaves and stems causing rapid dieback	First plant disease proved to be caused by a bacterium
Wildfire of tobacco	*Pseudomonas syringae*	Tobacco	Yellowish green spots on leaves	Wildfire of tobacco occurs worldwide; causes losses in seedlings and field plants
Blight of beans	*Xanthomonas campestris* *Pseudomonas syringae*	Beans (common blight) Beans (brown spot)	Yellowish green spots on leaves Lower side of leaves enlarge, coalesce, and become necrotic	Most phytopathogenic xanthomonads and pseudomonads cause necrotic spots on green parts of susceptible hosts may be localized or systemic

Disease	Pathogen	Hosts	Symptoms	Notes
Soft rot	*Erwinia carotovora*	Many fleshy tissues, fro example, cabbage, carrot, celery, onion	Soft decay of fleshy tissues that become mushy and soft	Occurs worldwide; causes major economic losses
Crown gall	*Agrobacterium tumefaciens*	More than 100 genera of woody and herbaceous plants	Initially a small enlargement of stems or roots usually at or near the soil line, increasing in size, becoming wrinkled, and turning brown to black	The conversion of a normal cell to one that undergoes excessive multiplication is caused by a plasmid (a small circular piece of DNA) carried by the pathogenic bacterium
Aster yellows	MLO	Many vegetables, ornamentals, and weeds	Chlorosis, dwarfing, malformation	Greatest losses suffered by carrots; transmission by leafhoppers
Citrus stubborn disease	*Spiroplasma citri* (MLO)	Citrus and stone fruits and vegetables	Chlorosis, yellowing of leaves, shortened internodes, wilting	First MLO pathogen of plant disease cultured

Abbreviation: MLO, mycoplasma-like organism.

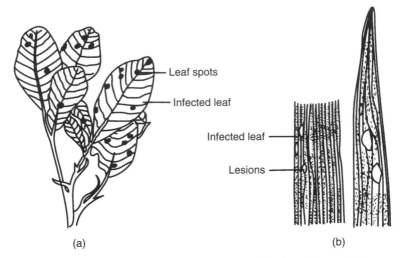

Leaf spots

Infected leaf

Infected leaf

Lesions

(a)   (b)

Figure 6.28. (a) Tikka disease of groundnut. (b) Blast disease of rice.

Figure 6.29. Bacterial leaf spot (*Pseudomonas cichorii*) on *Epipremnum aureum*.

are accompanied by encasing yellow halos (Fig. 6.29). Continuous rain and moisture will cause the coalescence of the spots. Severely infected leaves will defoliate prematurely. Bacterial leaf spot is a common nuisance of citrus and stone fruit trees and vegetables, as well as other indoor and outdoor foliage plants. There is no cure for bacterial leaf spot. However, the potential for bacterial leaf spot can be reduced by keeping the area free of decomposing debris and watering the plants at soil level. The spread of bacterial leaf spot can be controlled with a copper-based fungicidal spray if applied at first signs of infection.

Susceptible plants include *E. aureum* (*Pothos*), *Philodendron panduriforme* (fiddle-leaf philodendron), *Aglaonema* spp. (Chinese evergreen), and *Monstera* spp. (split-leaf philodendron). Symptoms are varied and may include brownish-black lesions, light and dark zones on *E. aureum* leaves, and a yellow halo around affected areas on *Monstera deliciosa* leaves.

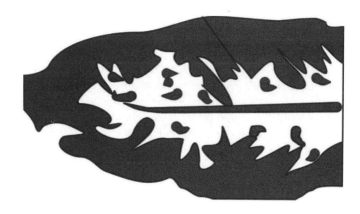

Figure 6.30. Bacterial leaf spot.

6.4.5.2. Bacterial Leaf Blight (Xantomonas spp.)

Susceptible plants include *Syngonium* spp., *Aglaonema roebelenii* (often called *Schismatoglottis*), and possibly other *Aglaonema* spp. (Chinese evergreens). *Syngonium* spp. is most often attacked by this bacterium. Symptoms include translucent lesions at the leaf tip and along the leaf margin. The lesions may elongate and extend into the middle of the leaf. Lesions are dark green at first, then turn yellow, and eventually turn brown when dead. The diseased area is often surrounded by a bright yellow halo that separates it from the healthy portion of the leaf. White flakes of dried bacterial exudate are often seen generally on older lesions on the undersides of leaves.

6.4.5.3. Bacterial Blight (Erwinia chrysanthemi)

This disease affects many plants, including *Aglaonema* spp. (Chinese evergreen), *Dieffenbachia* spp., *Philodendron* spp., and *Syngonium* spp. The bacteria attack some plants systemically (internally), especially *Dieffenbachia* spp. Symptoms of systemic infection are the yellowing of new leaves, wilting, and a mushy, foul-smelling stem rot. Aerial spread of this bacterium can cause foliar infection. Symptoms may appear as rapid, mushy leaf collapse on *Philodendron* spp., definite leaf spots on *Syngonium* spp., or all of these symptoms on *Philodendron selloum*. *E. chrysanthemi* grows best in warm-to-hot, wet, and humid environments. Attack by these bacteria often results in the death of foliage plants.

6.4.5.4. Bacterial Leaf Spot and Stem Canker
(Xanthomonas campestris pv. hederae)

These bacteria attack the English ivy, *Hedera helix*. Leaf spots are light green and translucent with a reddish margin; older spots turn brown or black (Fig. 6.30). Leaf stalks become black and shriveled. This decay may extend down to twigs and woody stems, and definite cankers may be seen.

6.4.5.5. Crown Gall

Crown gall is a root and stem disease that is most commonly found on woody plants (Fig. 6.31). Roses and flowering fruit trees are common victims of crown gall. Crown gall disease can infect more than 140 genera of plants and trees. Infected plants will develop smooth, light galls on their roots and stems. As the galls age, they develop into hardened, discolored galls that eventually slough off to make room for new, secondary galls. These formations inhibit the plant's ability to transport nutrients and water throughout the plant. This lack of transport results in the plant's loss of vigor, which is also accompanied by growth stunt and branch and twig dieback. Crown gall disease is a soilborne bacterial disease for which there is no cure, and the causative agent can thrive in the soil for several years without a host. Planting resistant plants is the best protection against crown gall disease.

6.4.5.6. Fire Blight

Fire blight is a destructive bacterial disease that is especially threatening to bushes and pome fruit trees. The causative agent lies dormant in the plant and in decomposing matter that lies around the plant. The bacterium begins its growing season as the plant enters its growing season. It enters the natural openings of the plant through its twigs and branches and is often transported by insect and honeybee bites. Trees and plants that are infected with fire blight will display tan bacterial ooze near the points of infection (Fig. 6.32). The infected areas become necrotic, turn black, wilt, and become deformed.

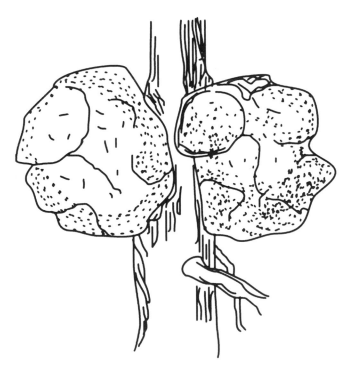

Figure 6.31. Crown gall.

Figure 6.32. Fire blight.

Unlike many diseases, vigorously growing trees are more susceptible to fatal infections than slow-growing ones.

Fire blight can be controlled by pruning away the diseased and infected areas of the plant. Pruning should be completed with sharp, sterile pruning shears that are sterilized between each cut. The potential for fire blight infections and repeat infections can be reduced by applying a copper-based fungicidal spread several times throughout the year.

6.4.5.7. Phytopathogenic Bacteria

Bacteria gain entry into the plants through mechanical openings such as wounds or pruning cuts or through natural openings such as the stomata on leaves or the hydathode openings of the vascular system. After entry, they proliferate rapidly in the apoplast, the intercellular space associated with movement of water and nutrients, thereby competing with neighboring host cells. Plants lack specialized adaptive immune system and cells found in most vertebrates, and hence, they solely rely on their innate immune system. Primary immune response is induced by the perception of microbe-associated molecular patterns (MAMPs) or pathogen-associated molecular patterns (PAMPs) using distinct membrane-bound pattern recognition receptors (PRRs) that could activate an array of different intracellular defensive signaling pathways.

6.4.6. Diseases Caused by Virus

Viruses have been in existence since time immemorial, as early as the 1800s; however, virologists could not identify them positively during the 1800s, and proper identification took place in the 1900s with advancement in technology. The word virus originates from a Latin word meaning a poison or toxin. Majority of these viruses are well known for their virulence and are disgracefully hard to treat because they mutate faster and quite effectively.

A virus cannot survive or reproduce on its own; it would need a host so that it can pass on its genes. Typically, the particle of a virus will consist of a protein cover that contains the genetic material of a virus. This explains the reason why virologists today are reluctant to categorize viruses as living organisms. Still on point, another reason why it is so difficult to classify viruses is because of their unique behavior that is different from any other form of life known today.

Viruses also cause many important plant diseases and are responsible for huge losses in crop production and quality in all parts of the world. Infected plants may show a range of symptoms depending on the disease, but often there is leaf yellowing (of the whole leaf or in a pattern of stripes or blotches), leaf distortion (e.g., curling), and/or other growth distortions (e.g., stunting of the whole plant, abnormalities in flower or fruit formation). Sometimes the virus is restricted to certain parts of the plant (e.g., the vascular system; produces discrete spots on the leaf), but in other cases, it spreads throughout the plant causing a systemic infection. Infection does not always result in visible symptoms (as witnessed by names such as carnation latent virus and lily symptomless virus, both members of the genus Carlavirus). Occasionally, virus infection can result in symptoms of ornamental value, such as "breaking" of tulips or variegation of *Abutilon*. Some of the viruses are described in the following sections.

6.4.6.1. Tobacco Mosaic Virus

The first plant virus discovered, tobacco mosaic virus (TMV), attacks members of the nightshade, or Solanaceae, family (Fig. 6.33). These include tobacco, pepper, potato, tomato, eggplant, cucumber, and *Petunia*. The virus gains entry through the breaks in cell walls caused by insects or other physical damage. It has another claim to fame as the first virus imaged with an electron microscope.

6.4.6.2. Cucumber Mosaic Virus

The cucumber mosaic virus infects cucumber, tomato, peppers, melons, squash, spinach, celery, beet, and other plants. Aphids spread the virus, and they cause physical damage to the plant, which allows entry of the virus via wind, splashing or dripping sap. The

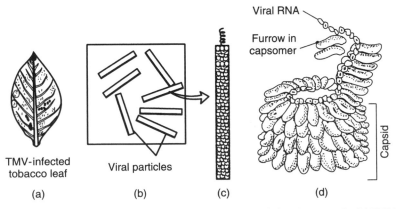

Figure 6.33. Tobacco mosaic disease. (a) Infected tobacco leaf showing mosaic. (b) TMV particles as seen under electron microscope. (c) One TMV particle enlarged. (d) Model of TMV to show capsid and RNA.

virus causes twisting in young leaves, stunts growth of the entire plant, and causes poor fruit or leaf production.

6.4.6.3. Barley Yellow Dwarf

The barley yellow dwarf virus infects several grains and staple crops, including wheat. Aphids primarily spread the virus. The virus causes discoloration of leaves and the tips of the plants, which reduces photosynthesis, stunts growth, and decreases production of seed grains.

6.4.6.4. Bud Blight

The bud blight virus infects soybeans, a staple crop. It causes the stem to bend at the top and the buds to turn brown and drop off the plant. Nematodes spread this virus.

6.4.6.5. Sugarcane Mosaic Virus

The sugarcane mosaic virus discolors leaves of the sugarcane plant, restricting its ability to feed itself through photosynthesis and grow. It stunts the growth of young plants. Aphids and infected seed spread the virus.

6.4.6.6. Cauliflower Mosaic Virus

The cauliflower mosaic virus (Fig. 6.34) infects members of the *Brassica*, or mustard, family, which includes cabbage, brussels sprouts, cauliflower, broccoli, and rapeseed. It causes a mosaic or mottle on the leaves, which stunts growth. Aphids and mechanical exposure spread the virus.

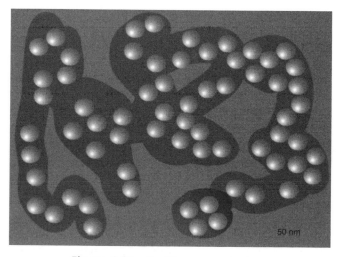

Figure 6.34. Cauliflower mosaic virus.

6.4.6.7. Lettuce Mosaic Virus

The lettuce mosaic virus mottles the leaves of almost all types of lettuce, stunting its growth and eliminating its market appeal. Aphids and infected seeds spread the virus.

6.4.6.8. Maize Mosaic Virus

The maize mosaic virus causes yellow spots and stripes on the leaves of corn, stunting growth. Leafhoppers spread the virus.

6.4.6.9. Peanut Stunt Virus

The peanut stunt virus causes discoloration and distortion of the leaves of peanuts and some other rhizomes, stunting their growth. Aphids and sap spread the virus

6.4.7. Soil Biological Control and Plant Diseases

Plant diseases result when a susceptible host and a disease-causing pathogen meet in a favorable environment. If any one of these three were not present, there would be no disease. Many intervention practices (fungicides, methyl bromide fumigants, etc.) focus on taking out the pathogen after its effects become apparent. Integrated pest and disease management is used these days (Fig. 6.35). Organic farming uses various practices to manage the pest population at a safe level (one that does not cause economic injury) rather than completely destroying the pests using synthetic chemicals.

These practices include the following.

Invert the soil after harvesting a crop to expose pests.

Clean bunds and channels of grasses that harbor pests.

Grow pest-tolerant varieties.

Sow the crops at the right time.

Sow healthy seeds.

Increase the seed rate so that uprooting insect and disease-infected plants later does not affect optimum plant populations.

Handpick and destroy egg masses, gregarious larvae, caterpillars, and adult beetles.

Use light traps.

Apply sticky grease bands on fruit trees to stop insects from crawling up the trunk.

Release insect parasites and predators, and apply biological control agents such as *B. thuringiensis*.

When pesticides are used, restrict these to a few mainly plant-based pesticides: neem, karanj products, derris (also known as *rotenone*), and pyrethrum.

Plant diseases may occur in natural environments, but they rarely run rampant and cause major problems. In contrast, the threat of disease epidemics in crop production is constant. The reasons for this are becoming increasingly evident.

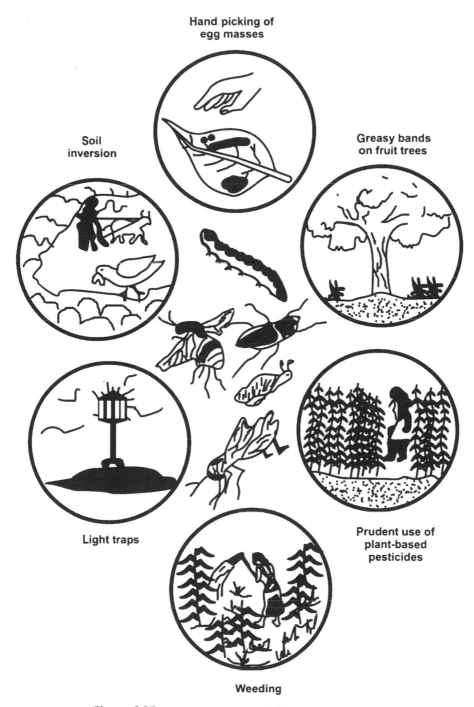

Figure 6.35. Integrated pests and disease management.

Until we improve the soil life, we will have to continue on this pesticide treadmill. The general principle is to add the beneficial soil organisms and the food they need—the ultimate goal being the highest number and diversity of soil organisms. The higher the diversity, the more stable the soil biological system. These beneficial organisms will suppress disease through competition, antagonism, and direct feeding on pathogenic fungi, bacteria, and nematodes. We cannot restore the balance of organisms that was present under native, undisturbed circumstances, but we can build a new, stable balance of soil organisms that will be adapted to the altered soil conditions. This is a proactive plan that moves us toward the desired outcome of disease prevention.

There are two types of disease suppression: specific and general. Specific suppression results from one organism directly suppressing a known pathogen. These are cases where a biological control agent is introduced into the soil for the specific purpose of reducing disease incidence. General suppression is the result of a high biodiversity of microbial populations that creates conditions unfavorable for plant disease development. A good example of specific suppression is the strategy used to control one of the organisms that cause damping off, that is, *Rhizoctonia solani*. *R. solani* kills young seedlings under cool temperature and wet soil conditions. The beneficial fungus *Trichoderma viride* locates and then attacks *R. solani* through a chemical released by the pathogen. Beneficial fungal strands (hyphae) entangle the pathogen and release enzymes that dehydrate *Rhizoctonia* cells, eventually killing them. Currently, *Trichoderma* cultures are sold as biological seed treatments for damping off disease of several crops.

Introducing a single organism to soils seldo achieves disease suppression for very long. If not already present, the new organism may not be competitive with existing microorganisms. If food sources are not abundant enough, the new organism will not have enough to eat. If soil conditions are inadequate, the introduced beneficial organism will not survive. This practice is not sufficient to render the soil "disease suppressive"; it is like planting flowers in the desert and expecting them to survive without water. With adequate soil conditions, inoculation with certain beneficial organisms should only be needed once. A soil is considered suppressive when, in spite of favorable conditions for disease to occur, a pathogen cannot become established, establishes but produces no disease, or establishes and produces disease for a short time and then declines (Schneider, 1982).

Suppressiveness is linked to the types and number of soil organisms, fertility level, and nature of the soil itself (drainage and texture). The mechanisms by which disease organisms are suppressed in soil include induced resistance, direct parasitism (one organism consuming another), nutrient competition, and direct inhibition through antibiotics secreted by beneficial organisms.

In addition, the response of plants growing in the soil contributes to suppressiveness. This is known as *induced resistance* and occurs when the rhizosphere (soil around plant roots) is inoculated with a weakly virulent pathogen. After being challenged by the weak pathogen, the plant develops the capacity for future effective response to a more virulent pathogen. In most cases, adding mature compost to a soil induces disease resistance in many plants. The level of disease suppressiveness is typically related to the level of total microbiological activity in a soil. The larger the active microbial biomass, the greater the soil's capacity to use carbon, nutrients, and energy, thus lowering their availability to pathogens. In other words, competition for mineral nutrients is high, as most soil nutrients are tied up in microbial bodies. Nutrient release is a consequence of grazing

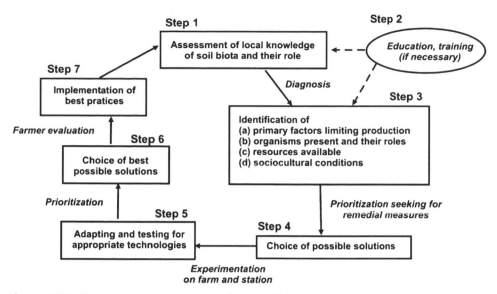

Figure 6.36. The seven-step process to optimum soil biological management and conservation.

by protozoa and other microbial predators: once bacteria are digested by the predators, nutrients are released in their waste.

A seven-step process is proposed for the assessment, management, and conservation of soil biodiversity (Fig. 6.36), in which all the different stakeholders, especially farmers, are involved in each step (Swift, 1997).

6.4.7.1. First Step

Recognizing that soil biota play a key role in sustaining agricultural production is the first step toward proper management and conservation. The level of knowledge in the farmer/farming community should be ascertained and appropriate awareness raising, training, and education provided to promote due recognition of soil biota and its functions, highlighting the importance of different soil biota roles at various levels of the ecosystem/landscape in the step.

6.4.7.2. Second Step

This step requires field observation and demonstration of the effect of different components of soil biota and an introduction to ecological principles to illustrate the interactions among different components of the system, for instance, soil-water-plant relationships and among soil, chemical, physical, and biological properties of different soil types.

6.4.7.3. Third Step

This step requires a rapid assessment of soil quality and function. In the past, soil quality assessments focused on chemical and physical properties, and complex classifications and soil survey/testing procedures were developed, requiring high level soil science expertise. These also include a characterization of the current farming system and practices of different farmer groups, including human resources (labor, time, sociocultural habits), available organic resources (manures, household wastes, leaves, wood and other residues, composts), and biological indicators of soil quality and function. The most useful bioindicators associated with soil function, and the ones that are easiest to measure, are those consisting of plants, soil organisms, and their processes.

6.4.7.4. Fourth Step

This step relates to choosing possible remedial solutions. It is essential that the stakeholders involved determine how the different limitations to agricultural production and sustainability can be overcome using local or imported resources, knowledge, and capacity. The limitations or constraints may include social, cultural, economic, political, agronomic, biological, environmental, edaphic, and/or genetic factors. They will vary from one group of farmers to another. In addition, in determining best soil biological management options, and other solutions to overcome constraints, adequate knowledge is required on how agricultural practices (e.g., tillage, pesticides, fertilizers, liming, rotation, residue disposal, irrigation, and drainage) affect soil biota and their activity.

Soil biological management has been proposed (Table 6.4). This would guide farmers and land use decision makers to the potential practices that can be adapted or changed. These include both direct interventions such as inoculation for disease and pest control and soil fertility improvement (such as rhizobia, actinomycetes, mycorrhizae, diazotrophs) and indirect interventions through, for example, cropping system design and management, organic matter management, and genetic control of soil function (manipulating resistance to disease, organic matter, and root exudates).

6.4.7.5. Fifth Step

It involves testing potential solutions. After choosing a number of different possible solutions, these must be tested on farm using an iterative and participatory screening process through adaptive experimentation. In this adaptive process, different treatments and techniques are tested simultaneously and repeated over several cropping cycles to identify the most resistant, economic, practical, and socially acceptable practices. Their approach consists of

- Extensive farm site characterizations to identify various ecological, social, economic, and cultural constraints, as well as to define opportunities for intervention (steps 3 and 4)
- Participatory design and implementation of on-farm and off-farm experiments
- Evaluation of results and sharing of experiences by participants
- Feedback of results and experiences to design further experiments; to refine databases, models, and predictions; and to develop new technologies
- Introduction of new technologies into on-farm testing after a satisfactory number of iterations of the above process.

TABLE 6.4. Effects of Different Management Practices on Soil Biota and
Soil Function and Constraints to Using Them

Management Practice	Effect on Biota and Function	Constraints to Use
Tillage	More rapid decomposition, higher ratio of bacteria/fungi, lower populations of macro- and mesofauna, short-term increase in nutrient availability but increase in long-term losses, better root growth in tilled layer, higher erosion risks	Labor, tools and machinery, cost, soilborne diseases, sloping lands
No tillage	Higher populations of macro-meso- and microfauna, greater ratio of fungi/bacteria, organic matter accumulation on soil surface, nutrient conservation, lower runoff and erosion, increase in presence and incidence of pests and diseases	Machinery, cost, soil compaction and heavy textures, pest management
Organic matter input	Changes in decomposition rates and organism populations (some increase and others decrease, depending on the type of material), increased nutrient availability, storage and exchanges, improved soil physical structure and water relations, reduction in acidity and Al toxicity, greater microbial and fauna activity, especially detritivores	Availability, labor, livestock presence, cost
Fertilization	Usually reduction in mycorrhization and N_2 fixation (with P and N, respectively), mineralization-immobilization balance changes, increased plant production and organic matter inputs, increases in populations of some organisms through greater food supply	Availability, cost
Pesticides	Reduced incidence of diseases and pest; parasites and pathogenic organisms, but negative effects on nontarget biota such as beneficial insects and earthworms; improved plant production but often creation of dependence; destabilization of nutrient cycles; loss of soil structure; long-term increased resistance of target biota	Cost, environmental and health impacts

(*continued*)

TABLE 6.4. (*Continued*)

Management Practice	Effect on Biota and Function	Constraints to Use
Irrigation/flooding	Increased water availability, pH neutralization, changes in nutrient cycling (often higher anaerobic processes) and availability, higher asymbiotic N_2 fixation, increased populations of drought-stressed biota, lower numbers of sensitive biota, lower organic matter decomposition rates, depression of soilborne diseases and weeds	Cost, available water, labor, tools
Crop rotations	The "rotation effect"; improved pest and disease management; more efficient soil nutrient utilization; greater diversity above and below ground; higher populations, biomass, and activity of most organisms (especially with legumes); improved soil aggregation and infiltration; reduced bulk density; higher organic matter	Social acceptability, opportunity costs, agroecosystem compatibility, climate, soil conditions
Inoculation of selected soil biota (e.g., rhizobia, mycorrhizae, earthworms, rhizobacteria, antagonists, biocontrol agents)	Increased N fixation; nutrient availability in soil; water uptake and efficiency of nutrient acquisition by plants; higher yields; increased heavy metal tolerance; better resistance to plant diseases, pests, and parasites; increased soil porosity, aeration, aggregate stability, water infiltration and holding capacity; faster decomposition rates and nutrient cycling	Cost, availability, environmental adaptability, competition with and/or replacement of native biota, adequate soil conditions

6.4.7.6. Sixth Step

It involves choosing the best possible solutions. The feasibility of each management option should be evaluated and adapted according to local agricultural, edaphic, climatic, socioeconomic, and cultural conditions. The success of this process depends on the ability of the farmers and other stakeholders to discern the best management strategies and to eliminate any potential loopholes that may lead to future failure in the adopted practices. These include assuring the appropriate incentives for, and compensation of, any opportunity costs or other losses associated with the practices to be adopted. A major point to recognize is that the final decision of the stakeholder groups as to which practices are to be implemented will be substantially different for small-scale versus large-scale farmers and for resource-poor versus resource-rich farmers. This is because of the inherent differences in the problems and priorities of each group. Once the best solutions have been identified by each group of stakeholders, these should be taken to the final step of implementation and further long-term testing.

6.4.7.7. Seventh Step

The final step involves the implementation of integrated soil biological management. The final adoption of the best practices for integrated soil biological management is a long-term process, resulting from the assessment, learning, identification, prioritization, choice, testing, adaptation, participation, discussion, agreement, and decision of the best management options. A final selection process occurs as the farmers evaluate the best choices in the field and decide whether to implement these practices on larger- and long-term scales or to revert to their traditional management strategies. This is a critical step, where the hard work of carefully following the previous steps is often at risk of being permanently lost. To ensure wider adoption of the best adapted soil biological and associated management techniques, appropriate incentive measures and long-term monitoring are required, so that the improvements in agricultural production and human well-being can be effectively demonstrated and sustained.

6.5. MICROBES AS A TOOL OF GENETIC ENGINEERING

Bacteria are useful tools in genetic modification. Using plasmids, circular portions of interchangeable bacterial DNA, organisms, such as other bacteria, can be transformed genetically. The main use of microbes (e.g., *Escherichia coli*) in the field of genetic engineering is as a "vector." "Vectors" are microbes that carry the "gene of interest" to the "target" genome. The recombinant DNA of the vector is called *plasmid DNA*.

Munching bacteria is the genetically engineered bacteria that are used for cleaning oil spills. These bacteria are produced by modifying the DNA structure of the bacterium called *B. thuringiensis*. Microorganisms have high reproduction rate and less reproduction time. Also, they need cheaper nutrients and less space for growth. The physical conditions for their cultivation are also not that costly. Genetic engineering or genetic manipulation, as it should properly be called, relies essentially on the ability to manipulate molecules *in vitro*. Most biomolecules exist in low concentrations and as complex, mixed populations, which is not possible to work with effectively.

This problem was solved in 1970 using the molecular biologist's favorite bug, *E. coli*, and a normally innocuous commensal occupant of the human gut. By inserting a piece of DNA of interest into a vector molecule, that is, a molecule with a bacterial origin of replication, when the whole recombinant construction is introduced into a bacterial host cell, a large number of identical copies is produced. Together with the rapid growth of bacterial colonies, all derived from a single original cell bearing the recombinant vector, in a short time (e.g., a few hours) a large amount of the DNA of interest is produced. This can be purified from contaminating bacterial DNA easily and the resulting product is said to have been "cloned."

6.5.1. *Agrobacterium*: The Friendly Bacteria

A. tumefaciens (Young et al., 2001) is the causal agent of crown gall disease (the formation of tumors) in over 140 species of dicot. It is a rod-shaped, gram-negative soil bacterium. Symptoms are caused by the insertion of a small segment of DNA (known as the *T-DNA*, for "transfer DNA") into the plant cell (Chilton et al., 1977), which is incorporated at a semirandom location into the plant genome.

   A. tumefaciens is an alphaproteobacterium of the family Rhizobiaceae, which includes the nitrogen-fixing legume symbionts. Unlike the nitrogen-fixing symbionts, tumor-producing *Agrobacterium* are pathogenic and do not benefit the plant. The wide variety of plants affected by *Agrobacterium* makes it of great concern to the agriculture industry (Moore et al., 1997). In order to be virulent, the bacterium must contain a tumor-inducing plasmid (Ti plasmid or pTi), of 200 kb, which contains the T-DNA and all the genes necessary to transfer it to the plant cell. Many strains of *A. tumefaciens* do not contain a pTi.

   A. tumefaciens infects the plant through its Ti plasmid. The Ti plasmid integrates a segment of its DNA, known as *T-DNA*, into the chromosomal DNA of its host plant cells (Fig. 6.37). *A. tumefaciens* has flagella that allow them to swim through the soil toward photoassimilates that accumulate in the rhizosphere around roots. Some strains may chemotactically move toward chemical exudates from plants, such as acetosyringone and sugars. The former is recognized by the VirA protein, a transmembrane protein encoded in the VirA gene on the Ti plasmid. Sugars are recognized by the chvE protein, a chromosomal gene encoded protein located in the periplasmic space. At least 25 vir genes on the Ti plasmid are necessary for tumor induction. In addition to their perception role, VirA and chvE induce other vir genes. The VirA protein has kinase activity: it phosphorylates itself on a histidine residue. Then the VirA protein phosphorylates the virG protein on its aspartate residue. The virG protein is a cytoplasmic protein produced from the virG Ti plasmid gene. It is a transcription factor, inducing the transcription of the vir operons. The chvE protein regulates the second mechanism of the vir gene activation. It increases VirA protein sensibility to phenolic compounds.

   The T-DNA can be transferred to a plant cell. The T-DNA must be cut out of the circular plasmid. A VirD1/D2 complex nicks the DNA at the left- and right-border sequences. The VirD2 protein is covalently attached to the 5'-end. VirD2 contains a motif that leads to the nucleoprotein complex being targeted to the type IV secretion system (T4SS).

   In the cytoplasm of the recipient cell, the T-DNA complex becomes coated with VirE2 proteins, which are exported through the T4SS independent of the T-DNA complex. Nuclear localization signals (NLSs) located on the VirE2 and VirD2 are recognized by the importin alpha protein, which then associates with importin beta and the nuclear pore complex to transfer the T-DNA into the nucleus. VIP1 also appears to be an important protein in the process, possibly acting as an adapter to bring the VirE2 to the importin. Once inside the nucleus, VIP2 may target the T-DNA to areas of chromatin that are being actively transcribed, so that the T-DNA can integrate into the host genome.

   Since the Ti plasmid is essential to cause disease, prepenetration events in the rhizosphere occur to promote bacterial conjugation, which is exchange of plasmids between bacteria. In the presence of opines, *A. tumefaciens* produces a diffusible conjugation signal, a quorum sensing signal called *N-3-oxo-octanoyl homoserine lactone* (3OC8HSL) or the *Agrobacterium autoinducer*. This activates the transcription factor TraR, positively

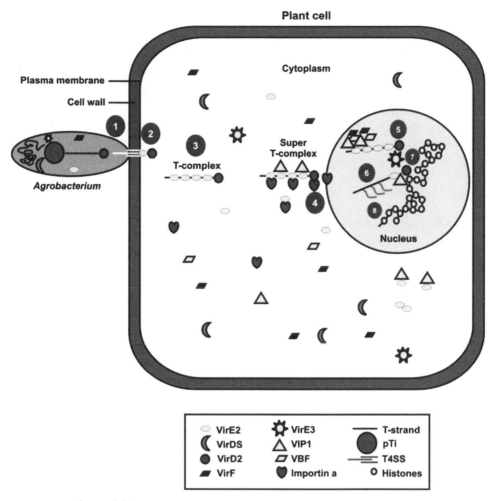

Plant cell

○	VirE2	✿	VirE3	—	T-strand
☾	VirDS	△	VIP1	═	pTi
●	VirD2	▱	VBF	═	T4SS
▰	VirF	♥	Importin a	O	Histones

Figure 6.37. Mode of action of Ti plasmid present in *Agrobacterium*.

regulating the transcription of genes required for conjugation. Economically, *A. tumefaciens* is a serious pathogen of walnuts, grapevines, stone fruits, nut trees, sugar beets, horse radish, and rhubarb.

6.5.2. Transformation Technology: Refined Tools for Genetic Transformation

In molecular biology, transformation is the genetic alteration of a cell resulting from the uptake, incorporation, and expression of the exogenous genetic material (DNA) that is taken up through the cell wall(s). Transformation occurs most commonly in bacteria and occurs naturally in some species. Transformation can also be effected by artificial means.

Bacteria that are capable of being transformed, whether naturally or artificially, are called *competent*. Genetic material can also be transferred between cells by conjugation or transduction. Conjugation involves the direct contact of two different bacterial cells with the DNA being transferred from one cell to the other. In transduction, viruses called *bacteriophages* inject the foreign DNA into their host. Introduction of foreign DNA into eukaryotic cells is usually called *transfection*. Transformation is also used to describe the insertion of new genetic material into nonbacterial cells including animal and plant cells. Bacterial transformation may be referred to as a stable genetic change brought about by the uptake of naked DNA (DNA without associated cells or proteins), and competence refers to the state of being able to take up exogenous DNA from the environment.

Agrobacterium-mediated transformation (Fig. 6.38) is the easiest and most simple plant transformation. Plant tissue (often leaves) is cut into small pieces, for example, 10×10 mm, and soaked for 10 minutes in a fluid containing suspended *Agrobacterium*. Some cells along the cut will be transformed by the bacterium that inserts its DNA into the cell. Placed on selectable rooting and shooting media, the plants will regrow (Fig. 6.39). Some plants species can be transformed just by dipping the flowers in a suspension of *Agrobacterium* and then planting the seeds in a selective medium.

Unfortunately, many plants are not transformable by this method. Viral transformation (transduction) and packaging the desired genetic material into a suitable plant virus allows this modified virus to infect the plant. If the genetic material is DNA, it can recombine with the chromosomes to produce transformant cells. However, genomes of most plant viruses consist of single-stranded RNA that replicates in the cytoplasm of infected cell. For such genomes, this method is a form of transfection and not a real transformation since the inserted genes neither reach the nucleus of the cell nor integrate into the host genome. The progeny of the infected plants is virus free and also free of the inserted gene. To achieve genetic transformation in plants, we need the construction of a vector (genetic vehicle), which transports the genes of interest, flanked by the necessary controlling sequences, that is, promoter and terminator, and delivers the genes into the host plant. The two kinds of gene transfer methods in plants are discussed below.

6.5.2.1. Vector-Mediated or Indirect Gene Transfer

Among the various vectors used in plant transformation, the Ti plasmid of *A. tumefaciens* has been widely used. This bacterium is known as *natural genetic engineer* of plants

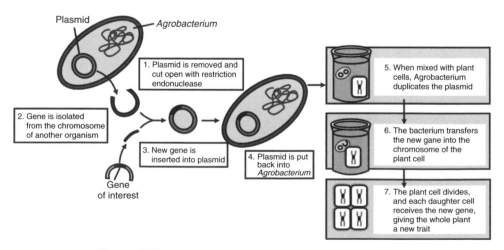

Figure 6.38. Bacterial transformation using *Agrobacterium*.

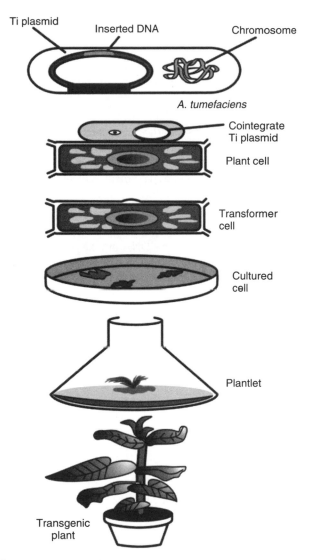

Ti plasmid

Inserted DNA

Chromosome

A. tumefaciens

Cointegrate Ti plasmid

Plant cell

Transformer cell

Cultured cell

Plantlet

Transgenic plant

Figure 6.39. Method of genetic transformation using *Agrobacterium tumefaciens*.

because of its natural ability to transfer T-DNA of its plasmids into plant genome on infection of cells at the wound site and cause an unorganized growth of a cell mass known as *crown gall*. Ti plasmids are used as gene vectors for delivering useful foreign genes into target plant cells and tissues. The foreign gene is cloned in the T-DNA region of Ti plasmid in place of unwanted sequences. To transform plants, leaf disks (in case of dicots) or embryogenic callus (in case of monocots) are collected and infected with *Agrobacterium* carrying recombinant disarmed Ti plasmid vector. The infected tissue is then cultured (cocultivation) in a shoot regeneration medium for 2–3 days during which time the transfer of T-DNA along with foreign genes takes place. After this, the transformed tissues (leaf disks/calli) are transferred onto selection cum plant regeneration medium supplemented usually with a lethal concentration of an antibiotic to selectively eliminate nontransformed tissues. After 3–5 weeks, the regenerated shoots (from leaf disks) are transferred to root-inducing medium, and after another 3–4 weeks, complete plants are transferred to soil following the hardening (acclimatization) of regenerated plants. The molecular techniques such as PCR and Southern hybridization are used to detect the presence of foreign genes in the transgenic plants.

6.5.2.2. Vectorless or Direct Gene Transfer

In the direct gene transfer methods, the foreign gene of interest is delivered into the host plant cell without the help of a vector. The methods used for direct gene transfer in plants are discussed below.

CHEMICAL-MEDIATED GENE TRANSFER. In this method, chemicals such as polyethylene glycol (PEG) and dextran sulfate induce DNA uptake into plant protoplasts. Calcium phosphate is also used to transfer DNA into cultured cells.

MICROINJECTION. Here, the DNA is directly injected into plant protoplasts or cells (specifically into the nucleus or cytoplasm) using fine-tipped (0.5–1.0 μm diameter) glass needle or micropipette. This method of gene transfer is used to introduce DNA into large cells such as oocytes, eggs, and the cells of early embryo.

ELECTROPORATION. This method involves a pulse of high voltage applied to protoplasts, cells, or tissues to make transient (temporary) pores in the plasma membrane, which facilitate the uptake of foreign DNA. The cells are placed in a solution containing DNA and subjected to electrical shock to create holes in the membrane. The foreign DNA fragments enter through the holes into the cytoplasm and then into nucleus.

PARTICLE GUN/PARTICLE BOMBARDMENT. In this method, the foreign DNA containing the genes to be transferred is coated onto the surface of minute gold or tungsten particles (1–3 μm) and bombarded onto the target tissue or cells using a particle gun (also called as *gene gun, shot gun*, or *microprojectile gun*). The *microprojectile bombardment method* was initially named as *biolistics* (Fig. 6.40). Two types of plant tissue are commonly used for particle bombardment, primary explants and the proliferating embryonic tissues.

TRANSFORMATION. This method is used for introducing foreign DNA into bacterial cells, for example, *E. coli*. The transformation frequency (the fraction of cell population that can be transferred) is very good in this method. The uptake of plasmid DNA by *E. coli* is carried out in ice-cold $CaCl_2$ (0–5 °C) followed by heat shock treatment at 37–45 °C for about 90 s. The transformation efficiency refers to the number of transformants per microgram of added DNA. The $CaCl_2$ breaks the cell wall at certain regions and binds the DNA to the cell surface.

CONJUGATION. It is a natural microbial recombination process and is used as a method for gene transfer. In conjugation, two live bacteria come together and the single-stranded DNA is transferred via cytoplasmic bridges from the donor bacteria to the recipient bacteria.

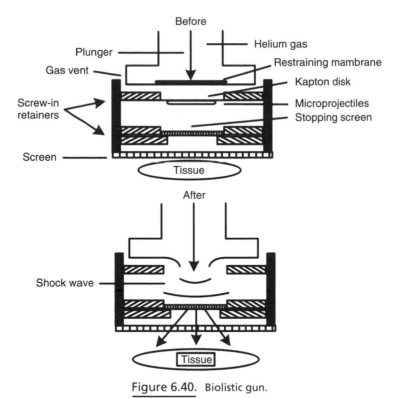

Figure 6.40. Biolistic gun.

LIPOSOME-MEDIATED GENE TRANSFER OR LIPOFECTION. Liposomes are circular lipid molecules with an aqueous interior that can carry nucleic acids. They encapsulate the DNA fragments and then adhere to the cell membranes and fuse with them to transfer DNA fragments. Thus, the DNA enters the cell and then the nucleus. This is a very efficient technique used to transfer genes in bacterial, animal, and plant cells.

6.5.2.3. Selection of Transformed Cells from Untransformed Cells

This is an important step in the plant genetic engineering. For this process, a marker gene (e.g., for antibiotic resistance) is introduced into the plant along with the transgene followed by the selection of an appropriate selection medium (containing the antibiotic). The segregation and stability of the transgene integration and expression in the subsequent generations can be studied by genetic and molecular analyses (Northern, Southern, and Western blots, and PCR).

6.5.3. Transgenic Cellulosic Biofuel Crops

Currently, most of the ethanol produced in the United States is derived from maize kernel, at levels in excess of 4 billion gal per year. Plant lignocellulosic biomass is renewable, cheap, and globally available at 10–50 billion tonnes per year. At present, plant biomass is converted to fermentable sugars for the production of biofuels using pretreatment processes that disrupt the lignocellulose and remove the lignin, thus allowing the access of microbial enzymes for cellulose deconstruction. Both the pretreatments and the production of enzymes in microbial tanks are expensive. Recent advances in plant genetic engineering could reduce biomass conversion costs by developing crop varieties with less lignin, crops that self-produce cellulase enzymes for degradation and ligninase enzymes for lignin degradation, or plants that have increased cellulose or an overall biomass yield (Updegraff, 1969; Sticklen, 2006).

Lignocellulosic plant matter shows tremendous potential for the production of chemicals and renewable fuels. However, the cost of making products, such as alcohols, from lignocellulosic material is much higher than that of traditional production processes due in part to the difficulty in degrading complex lignocellulosic biomass. Using biotechnology, improved varieties of the plant species are proposed to be used as feedstocks for production of ethanol or other biofuels. As the time lines for plant biotechnology tend to be longer than those in microbial or even algal biotechnology, due to the longer life cycle of the organism, there is the ordinary need for years of field testing and (in commercial agriculture) the frequent need for repeated cycles of crossbreeding to ensure that introduced traits are expressed in the desired genetic background. There are also several companies planning to use GM plants as the production platform to manufacture industrial enzymes for use in preprocessing feedstocks for cellulosic fuel manufacture. The following are the companies that are using biotechnology to create new plant varieties as improved feedstocks for biofuel production.

Conventional Feedstocks

Agrivida: corn
ArborGen: purpose-grown trees
CanaVialis S.A.: sugarcane
Edenspace Systems: corn
FuturaGene: hybrid poplar and willow
Targeted Growth: corn

Newer Feedstocks

Agragen: *Camelina*
Agrisoma: *Brassica* and *Jatropha*
Agrivida: switchgrass, sugarcane, sorghum, and others
Ceres: nonfood grasses
Chromatin: switchgrass, *Miscanthus*, sorghum, and sugarcane
Edenspace Systems: switchgrass
Evogene: canola, soybean, and others
Farmacule BioIndustries: sugarcane, tobacco
FuturaGene: switchgrass, *Miscanthus*
Kaiima: castor beans (non-GMO (genetically modified organism))
Mendel Biotechnology: grasses, and others
Metabolix: switchgrass, oil crops, and others
Rahan Meristem: *Jatropha*, castor beans
SG Biofuels: *Jatropha*
Targeted Growth: *Camelina*, canola, and others

Enzyme Manufacture in Plants

Infinite Enzymes (transgenic plants)

Medicago (transient expression)

Syngenta (transgenic plants).

Many of these companies have already begun field-testing their new varieties, and in one case, that of Syngenta, approvals are in hand to commercialize a transgenic variety in some countries of the world. But the products of most of these companies are still several years away, creating the challenge for the smaller companies to obtain and conserve sufficient capital to stay in business long enough to survive until products can be introduced. When it comes to selecting the right plant source for future cellulosic biofuel production, the solution will not be "one size fits all," and it certainly does not have to involve food and feed crops. Even corn, the largest global source of grain and feed and a feedstock for ethanol, is given consideration to have potential as cellulosic fuel sources.

First-generation biofuel feedstock sources such as sugarcane and cereal grains can be used to produce bioethanol and biobutanol and oilseeds to produce biodiesel and can compete directly with needs for world food security. The heavy use of oilseed rape releases quantities of methyl bromide to the atmosphere, which can be prevented by gene suppression.

Second-generation bioethanolic/biobutanolic biofuels will come from cultivated lignocellulosic crops or straw wastes. The technology requires heat and acid to remove lignin, which could be partially replaced by transgenically reducing or modifying lignin content and upregulating cellulose biosynthesis. Nonprecipitable silicon emissions from burning could be reduced by transgenically modulating silicon content. Before extensive cultivation, the shrubby *Jatropha* and castor beans should have highly toxic protein components transgenically removed from their meal, cancer-potentiating diterpenes removed from the oils, and allergens removed from the pollen.

Third-generation processes include algae and cyanobacteria for biodiesel production and transgenic manipulation to deal with "weeds," light penetration, photoinhibition, carbon assimilation, and so on.

The possibilities of producing *fourth-generation* biohydrogen and bioelectricity using photosynthetic mechanisms are being explored. There seem to be no health or environmental impact study requirements when the undomesticated biofuel crops are grown, yet there are illogically stringent requirements should they transgenically be rendered less toxic and more efficient as biofuel crops (Gressel, 2008). Only second-generation and beyond biofuels will make a real dent in the amount of fossil of petroleum used. The biofuel crops will only be cost-effective in the long run if they are further domesticated transgenically to remove toxins and environmental contaminants, and to be more productive and have the right properties as fuels, as well as have residues that have value. The "third-generation fuel" would consist of engineering crops in such a way that their very properties are tailored to particular conversion processes to yield fuels and bioproducts. Examples of this are engineered trees with low lignin content.

Large-scale planting of many of these species before they are domesticated is of questionable value, unless the presently planted material can serve as rootstocks on which newer and better transgenic varieties can be grafted. Unfortunately, it is unlikely that it will be possible to graft euphorbs such as castor bean and *Jatropha*, as they all have mature stems that are hollow. Crops that cannot be wholly used will lose even more value after transgenic algae and cyanobacteria come into use, as every bit of value must be derived from such crops. The plantations of perennial lignocellulosics and straws will then only be of value if they have been engineered to have lower/modified lignin, such that they can be used as fodder for ruminants. Whether or not the efforts to develop transgenic or other modified plant varieties as energy crops proves successful, it is undoubtedly a good thing that research programs pursued by companies like these are shifting attention away from "traditional" biofuel species such as corn and sugarcane toward nonconventional nonfood species such as switchgrass, *Camelina*, *Jatropha*, and woody plants. Most observers would agree with the need for the biofuel industry to increasingly look toward fast-growing, nonfood species as the source of biomass for biofuel production, and significant germplasm improvements in these species can be made even without advanced biotechnology.

Until now, no attempts to create transgenic plants that express cellulases, enzymes that degrade lignocellulosic material, have been successful. Genetically engineered transgenic corn plants have been developed that produce enzymes that can turn their leaves and stems into sugar by breaking down cellulose. The enzymes would not break down cellulose while the plants were still alive. Part of the solution was to use an enzyme found in bacteria that live in hot springs. The enzyme is only active at high temperatures higher than those that a plant's cells would reach while it is alive. As a result, the enzyme remains dormant until it is heated to about $50\,^{\circ}C$. The plants could lower the cost of creating ethanol from these sources, making such biofuel more competitive than that produced from corn kernels, the primary source of ethanol in the United States today. Cellulosic sources of ethanol, such as waste biomass and switchgrass, are attractive because they are cheap and abundant. But converting cellulose, a complex carbohydrate, into sugars that can be fermented to make ethanol is more expensive than converting the starch in corn grain into sugar. Breaking down the cellulose typically requires expensive enzymes extracted from genetically engineered microbes. Thus, this is one of the several promising approaches to address the central obstacle impeding establishment of a cellulosic biofuel industry, namely, the absence of low cost technology to overcome the recalcitrance of cellulosic biomass.

Transgenic plants do not have negative environmental effects. Hence, this provided the transgenic plants to express cellulose-degrading enzymes (Fig. 6.41). The cellulases are expressed in recoverable quantities and can be isolated from the plants. Alternatively, plant matter containing the expressed cellulose-degrading enzymes can be used directly as a feedstock for conversion to fuel. The plants may also be fed to livestock to aid in the digestion of lignocellulosic substrates. Isolated and purified cellulases from the transgenic plants can be used in place of conventionally produced cellulases, for example, in fermentation processes such as brewing and wine making.

Current biomass-based production of ethanol or other fuels typically relies on plant species already grown and harvested for food or other purposes or in some cases, uses cellulosic waste products. The business model that companies can develop and sell plant varieties (or seeds) to be grown in a dedicated manner solely for use in fuel production is one that is just beginning to be utilized by the small number of companies that have already introduced classically bred, nontransgenic plant lines tailored for biofuel use (e.g., high yielding forage sorghum varieties developed by Edenspace, new *Camelina* varieties introduced by Great Plains Oil and Exploration, and *Jatropha* lines introduced by SG Biofuels). Some of these companies are developing plant species that have never

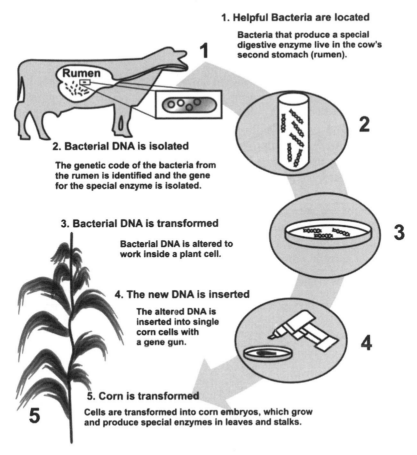

1. Helpful Bacteria are located

Bacteria that produce a special digestive enzyme live in the cow's second stomach (rumen).

2. Bacterial DNA is isolated

The genetic code of the bacteria from the rumen is identified and the gene for the special enzyme is isolated.

3. Bacterial DNA is transformed

Bacterial DNA is altered to work inside a plant cell.

4. The new DNA is inserted

The altered DNA is inserted into single corn cells with a gene gun.

5. Corn is transformed

Cells are transformed into corn embryos, which grow and produce special enzymes in leaves and stalks.

Figure 6.41. Production of anaerobic ruminal bacterial (*Butyrivibrio fibrisolvens* H17c) cellobiose in corn biomass.

been used and sold in commercial agriculture to any significant extent (e.g., switchgrass, *Miscanthus*).

6.5.4. Expression of Microbial Enzymes in Crops

In the past decades, genetic engineering has offered an alternative to chemical synthesis, using bacteria, yeasts, and animal cells as factories for the production of therapeutic proteins (Liénard et al., 2007). More recently, molecular farming has shifted its focus to plants among the major players in recombinant protein production systems. Plants provide an inexpensive, contamination-free, and convenient system for the large-scale production of valuable recombinant proteins. More than 100 recombinant proteins have now been produced in a range of different species. Indeed, therapeutic protein production is safe and extremely cost-effective in plants. Unlike microbial fermentation, plants are capable of carrying out posttranslational modifications, and unlike production systems based on mammalian cell cultures, plants are devoid of human infective viruses and prions. Furthermore, a large panel of strategies and new plant expression systems are currently developed to improve the plant-made pharmaceuticals' yields and quality.

Recent advances in the control of posttranslational maturations in transgenic plants will allow them, in the near future, to perform humanlike maturations on recombinant proteins and, hence, make plant expression systems suitable alternatives to animal cell factories.

Molecular farming in plants offers efficient plant-based expression systems, and now more than 100 biopharmaceutical recombinant proteins have been produced in a range of different species. With an increasing number of products in development, molecular farming in plants is finally coming of age (Sengupta et al., 2001; Twyman et al., 2003). These products include antibodies, vaccines, human blood products, hormones, and growth regulators. For such products, plants offer practical and safety advantages, as well as lower production costs, compared with traditional systems based on microbial or animal cells or transgenic animals (Commandeur et al., 2003). However, plant-made pharmaceuticals differ from their mammalian homologs by the structure of their N-linked glycans (Paccalet et al., 2007). For instance, most mammalian glycoproteins harbor terminal sialic acids that control their half-life in the bloodstream. The absence of the whole sialylation machinery in plants is of major concern, as nonsialylated plant-made pharmaceuticals may not perform at their full potential in humans because of their removal from the circulation through the involvement of hepatic cell receptors. They have investigated the synthesis of N-acetylneuraminic acid (Neu5Ac) in the cytosol of plants by either the rerouting of the endogenous 3-deoxy-D-manno-2-octulosonic acid (Kdo) biosynthetic pathway or the expression of microbial Neu5Ac-synthesizing enzymes. They have also expressed genes encoding Neu5Ac lyase from *E. coli* and Neu5Ac synthase (neuB2) from *Campylobacter jejuni* in plants in the production of functional enzymes in the cytosol, which in turn can catalyze the synthesis of Neu5Ac *in vitro*. Experiments were carried out on two models, Bright Yellow 2 (BY2) tobacco cells and Medicago sativa (alfalfa), the perennial legume crop.

Plants are advantageous because several types of recombinant protein can be used in unprocessed or partially processed material, therefore removing many of the downstream costs. For example, recombinant subunit vaccines produced in plants can be administered by the consumption of raw or part-processed fruits and vegetables (Mason et al., 2002) and antibodies for passive immunization can be administered as topical pastes following minimal purification. Similarly, industrial enzymes such as glucanase (which is used to break down cellulose in animal feed) and phytase (which breaks down phytic acid and releases bioavailable phosphorus) can be introduced into the industrial process either as part-processed plant material or by expressing directly in the plant that needs to be processed (Denbow et al., 1998; Dai et al., 2000; Ziegler et al., 2000; Ziegelhoffer et al., 2001). One of the advantages of recombinant protein expression in the seeds of transgenic cereal plants is that high levels of the product can accumulate in a small volume, which minimizes the costs associated with processing (Perrin et al., 2000; Stoger et al., 2000). An example is the oleosin fusion platform developed by SemBioSys Genetics (http://www.sembiosys.com/), in which the target recombinant protein is expressed in oilseed crops as a fusion with oleosin. The fusion protein can be recovered from oil bodies using a simple extraction procedure, and the recombinant protein is separated from its fusion partner by endoprotease digestion (Moloney et al., 2003). Similarly, a strategy has been devised in which recombinant proteins are expressed as fusion constructs containing an integral membrane-spanning domain derived from the human T-cell receptor (Schillberg et al., 2000). The recombinant protein accumulates at the plasma membrane and can be extracted in a small volume using appropriate buffers and detergents. Maize was chosen by ProdiGene (http://www.prodigene.com/) as the first plant species for commercial molecular farming. The major factors in this decision were high biomass yield, ease of transformation and *in vitro* manipulation, and convenience of scale-up. Maize has been used for the commercial production of recombinant avidin and β-glucuronidase (Hood et al.,1997, 1999; Witcher et al., 1998), and its use for the production of recombinant antibodies (Hood et al., 2002) and further technical enzymes, such as laccase, trypsin, and aprotinin, is being explored (Hood, 2002). Tobacco has a long history as a successful crop system for molecular farming and is therefore one of the strongest candidates for the commercial production of recombinant proteins (Schillberg et al., 2002; Stoger et al., 2002). The major advantages of tobacco include the well-established technology for gene transfer and expression, high biomass yield, prolific seed production, and the existence of a large-scale processing infrastructure. Because tobacco is neither a food nor a feed crop, there is little risk that tobacco material will contaminate either the food or feed chains. Although many tobacco cultivars produce high levels of toxic alkaloids, there are low alkaloid varieties that can be used for the production of pharmaceutical proteins, and these metabolites are absent from tobacco cell suspensions, which can also be used to produce recombinant proteins. Several different crops have been investigated for seed-based production, including cereals (rice, wheat, and maize) and legumes (pea and soybean) (Stoger et al., 2002; Hood, 2002; Vierling and Wilcox, 1996). Potatoes are the major system for vaccine production (Tacket et al., 1998, 2000; Richter et al., 2000), and transgenic potato tubers have been administered to humans in at least three clinical trials carried out to date. A recent report describes the production of rotavirus VP6 capsid protein in transgenic potatoes for vaccination against acute viral gastroenteritis (Yu and Langridge, 2003). Artsaenko and colleagues showed how potatoes could be used for antibody production (Artsaenko et al., 1998), and this system has been investigated as a possible bulk antibody production platform (De Wilde et al., 2002). Potatoes have also been used for the production of glucanases (Dai et al., 2000), diagnostic antibody fusion proteins (Schunmann et al., 2002), and proteins from human milk (Chong and Langridge, 2000; Chong et al., 1997). Tomatoes are more palatable than potatoes and have other advantages, including high biomass yield and use of greenhouses for increased containment. Tomatoes were first used for the production of a rabies vaccine candidate, and have also been used to produce antibodies, although yields thus far are less than 3 mg/g fresh weight (Mason et al., 1996; McGarvey et al., 1995; Stoger et al., 2002). The progress of plant-derived pharmaceutical proteins through preclinical development and clinical trials is a significant bottleneck. These crucial issues will probably have more of an impact on the commercial success of molecular farming than the technological hurdles (Sengupta et al., 2001; Twyman et al., 2003).

6.5.5. Genetic Transformation for Crop Improvement: Biotic Stress Tolerance

Advancement in biotechnological techniques has opened many possibilities for breeding crops. Thus, Mendelian genetics allowed plant breeders to perform some genetic transformation in few crops. *Stress* is defined as an influence that is outside the normal range of homeostatic control in a given genotype (Lerner, 1999a). Where stress tolerance is exceeded, response mechanisms are activated (Lerner, 1999b). Both nonspecific (activated by reactive oxygen species) and specific (e.g., osmotic stress) responses depend on perception of the stress, signal transduction, activation of transcription factors, and gene expression (Krauss, 2001). The production of the stress response include the production and/or upregulation of metabolic pathways resulting in changes in the metabolome, for example, the formation of compatible compounds (antioxidants, phytoalexins, protein protectants, cryoprotectants) (Bohnert and Shen, 1999), and in the proteome, for example, increased expression of constitutive defense proteins and production of novel defense proteins and protein chaperones (Cushman and Bohnert, 2000; Grover et al., 1999).

Modern agriculture strives to achieve the highest possible crop yields in order to overcome the continuously growing land limitation. Uniformity, as well as growth density, renders modern crops susceptible to quickly spreading damage of many pathogens such as nematodes, bacteria, fungi, viruses, viroids, and phytoplasms. A growing resistance exists to the use of chemical pesticides due to many disadvantages brought forth by chemical abuse including negative environmental effects, and diminishing affectivity. For example, the magnitude of fungicidal treatments has provoked the appearance of resistant strains, necessitating the development of new treatments (Leroux et al., 2002). On the other hand, fighting pathogens by utilizing the biological inert plant mechanisms is environmentally safer and less prone to become ineffective by the emergence of resistant pathogens.

An example of a damaging plant pathogen is *Botrytis cinerea*. This phytopathogenic fungus has a broad host range, more than 200 plant species, including tomato (Elad et al., 2004). *B. cinerea* causes rapid destruction of the host plant tissues as it proceeds to colonize it (a pathology called *necrotrophy*). Together with other filamentous fungi, it is considered to be the principal pathogenic agent of plants. Estimated losses for vineyards in France amount to 15–40% of the harvest, depending on climatic conditions. Other losses are estimated at 20–25% of the strawberry crops in Spain and cut flowers in Holland. Fungicidal treatments against *B. cinerea* cost about €540 million in 2001, which represents 10% of the world fungicide market (Annual Report, UIPP, 2002). Other plant pathogens are viruses, for example, TMV, the potato virus Y (PVY), and tomato yellow leaf curl virus (TYLCV). Plant virus diseases pose severe constraints to the production of a wide range of economically important crops worldwide (Agrios, 1997). Some estimates put total worldwide damage due to plant viruses as high as US$ 60 billion per year. Diseases caused by plant viruses are difficult to manage and their control mainly involves the use of insecticides to kill insect vectors, the use of virus-free propagating materials, and the selection of plants with appropriate resistance genes. Virus-free stocks are obtained by virus elimination through heat therapy and/or meristem tissue culture, but this approach is ineffective for viral diseases transmitted by vectors. While vectors can be controlled by insecticides, often the virus has already been transmitted to the plant before the insect vector is killed. The use of resistant cultivars has been the most effective means of control; however, plant virus resistance genes are frequently unavailable and their introgression into some crops is not straightforward.

Specific responses to biological stresses involve the induction of antimicrobial proteins and phytoalexins (Slusarenko et al., 2000; Boller and Keen, 2000). Following stress perception, stress signal transduction takes place (Bolwell, 1999; Ellis et al., 2000; Heath, 2000; Nurnberger and Scheel, 2001). In the case of necrotizing pathogens, this may lead to a local hypersensitive response and an SAR with the production of AMPs (Broekaert et al., 2000) and phytoalexins (Mansfield, 2000). Nonnecrotizing pathogens, biocontrol organisms and insects may induce systemic resistance whose basis is uncertain (Van Loon et al., 2000). SAR involves ethylene and salicylic acid as signaling molecules, whereas ISR involves jasmonic acid. There is known to be cross talk between the respective signaling systems (Fig. 6.42), with suppression of one by the other leading to cross-susceptibility between pathogens and pests in some cases (Pieterse et al., 2001).

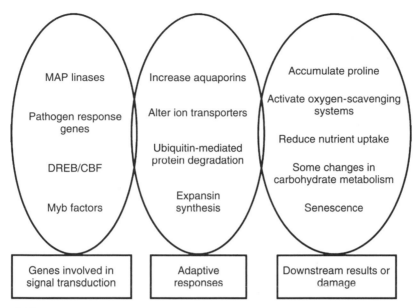

Figure 6.42. Interrelationship between genes involved in signal transduction, adaptive responses, and downstream results or damage.

On average, plants contain approximately 80,000 genes, which recombine during sexual hybridization. The offspring may therefore inherit around 1000 new genes as a result of this recombination. This is equivalent to a 0.0125% change in the genome. By contrast, only one or two new genes are transferred during plant transformation. As this represents a 0.0025% change in the genetic information of the plant, it is argued that plant transformation provides a more precise approach to crop improvement than sexual hybridization.

At present, legislation prevents any gene of unknown function and sequence from being introduced into plants. This means that before carrying out the transformation, the novel gene must be thoroughly characterized. It is therefore possible to predict and monitor the behavior of the gene following its incorporation into the plant DNA. This is critical to enable the long-term effects of plant transformation to be monitored. Currently, there is no evidence that "transgenes" behave any differently from the "resident" genes present naturally in the plant. The position and behavior of a transgene can be identified and analyzed more easily than any of the 80,000 resident genes.

Genetic engineering is often presented as a one-step rapid solution to the improvement of stress tolerance in plants. Foreign genes have been introduced into cells of several woody crops by using the *A. tumefaciens* Ti plasmid (Horsch et al., 1985). Biotic and abiotic stress-resistant mutants are not obtainable in tea, making conventional breeding inadequate and time consuming for crop improvement (Mondal et al., 2001). While it may benefit from, but not necessitate, the requirement for backcrossing for gene introgression, it does not reduce the requirement for field trials. The introduction of herbicide and pest resistance into plants has been an applied success, and these characteristics both separately and combined still dominate the applications for trial permits. However, they do not represent the complexity of the challenge for engineering for durable biotic and abiotic stress resistance. The dissection of stress responses in plants is showing high levels of complexity and redundancy at the perception, signaling, and expression levels, with cross-regulation (cross talk) between stress pathways and overlapping functions between stress metabolites and stress proteins in different stresses (Fig. 6.42).

To date, plant transformation has been commercialized for three major traits, namely, insect, herbicide, and virus tolerance (James, 2004). Directly or indirectly, the goal of these transformations has been to reduce pesticide use, enhance soil conservation, and increase yields and farm income. With a few exceptions such as transgenic papaya and rice in Hawaii and China, respectively, seeds have largely been produced by the private sector in the United States and have found their way into markets in some developed countries such as Argentina and South Africa. As a result, farmers in over 17 countries are now growing transgenic crops, and the number is growing in both developed and, especially, in developing countries (James, 2004). At least five European countries (Spain, Germany, France, Portugal, and the Czech Republic) have also started to grow transgenic crops on a commercial scale. Furthermore, many public research sector institutions in over 15, poor, developing countries, including Africa, have also started developing a diverse array of crops and phenotypes using various genes that confer both biotic and abiotic stress tolerances (Cohen, 2005). The overall goal of most of these plant transformation activities is to increase the yield potential of crops by increasing the gross physiological capacity of plants to produce harvestable yield and ameliorating the negative consequences of biotic and abiotic stresses (Sinclair et al., 2004). From the standpoint of private sector technology and product developers, the profit motive is part of the equation since the resources used for R&D, regulatory approvals, IPR, marketing, and other costs must be recovered at the end of the commercial product development chain.

Many plant genomes have already been sequenced, including that of *Arabidopsis* (a model dicot) and rice (a model monocot). As a result, based on homology, many genes have been identified in important crops, some of which have great potential for use in plant transformation with respect to goals highlighted above.

Between 1996 and 2000, there was an increase from 4.2 to 104.7 million acres of transgenic plants grown globally (Cockburn, 2002). In terms of the selection of "viable" transgenic plants for commercial production, the foreign gene has to function in the desired way and the chosen elite variety (the transformant) should be free from pleiotropic effects. For most, if not all, of the current transformation protocols, a tissue culture stage is a requirement with plants being derived through either somatic embryogenesis or organogenesis. With respect to the engineering of insect resistance in plants, high expression of the Bt toxin can be achieved via chloroplast engineering as the number of chloroplast genomes per cell is between 5000 and 10,000 (Kota et al., 1999). The two Bt genes that are found in most of the commercial transgenic crops are Cry1Ab and Cry1Ac.

As a consequence of their amino acid sequence similarity (90% homology), if a resistance allele appears in the insect population to one of these proteins, the chances of it conferring resistance to the other Bt protein is quite high (Kota et al., 1999). The overexpression of the *B. thuringiensis* Cry2Aa2 protein in chloroplasts demonstrated resistance to plants against both Bt-susceptible and Bt-resistant insects (Kota et al., 1999). Therefore, it may be necessary in some cases to increase the number of Bt proteins in use in the production of transgenic crops to preempt problems such as the development of resistance alleles in the insect population.

Fusarium wilt is economically the most destructive disease of lentil (*Lens culinaris* Medik) and can cause up to 100% yield loss. Cold tolerance is an important trait for winter lentil cultivation at high altitudes. Different DNA marker systems including restriction fragment length polymorphism (RFLP), RAPD, and amplified fragment length polymorphism (AFLP) were used to construct a genetic linkage map of *Lens* sp. (Eujayl et al., 1998). F6-derived F8 recombinant inbred lines were genotyped with 257 morphological, RFLP, RAPD, and AFLP markers. The linkage map was exploited to identify markers linked to Fusarium wilt resistance and radiation frost tolerance (Eujayl et al., 1998, 1999). The population was evaluated for two seasons for radiation frost injury and three seasons for Fusarium wilt. Both traits were monogenically inherited. Four RAPD markers linked to the Fusarium wilt resistance locus were identified and located in the present map. Likewise, one RAPD marker was linked with the radiation frost tolerance locus. Fine mapping was done to develop more closely linked markers. Traits such as Fusarium resistance and radiation frost tolerance were transferred to other adapted lines in backcross programs. Transformations were confirmed by PCR analysis and Southern hybridization (Krishnamurthy et al., 2000).

A major transcription system that controls abscisic acid independent gene expression in response to dehydration and low temperature has been described (Qiang et al., 1998). The system includes the DRE/CRT (dehydration-responsive element/C-repeat) cis-acting elements and its DNA binding protein, DREB/CBF (DRE-binding protein/C-repeat binding factor), which has an AP2 domain. Overexpression of the cDNA encoding DREB1A in transgenic *Arabidopsis* plants activates the expression of many genes and results in improved tolerance to drought, salt loading, and freezing. The DREB constructs are being used to test the effect of the *Arabidopsis* transcription factors on ICARDA (International Center for Agricultural Research in the Dry Areas)-mandated crop plants.

Up until the beginning of 1998, transgenic herbicide-tolerant crops accounted for about 35% of all GM crops released (Gray and Raybould, 1998). Various problems associated with gene escape have been identified, particularly with outbreeding crops. The literature up to 1998 suggests that gene flow had occurred between the following crops and their wild relatives: sugar beet, maize, sunflower, carrot, sorghum, strawberries, quinoa, and squash (Gray and Raybould, 1998). Biotic defense compounds are divided into phytoprecipitins, which are constitutive, and phytoalexins, which are induced on pathogen stress perception. These compounds are products of many metabolic pathways and have been extensively reviewed by Mansfield (2000). The transfer of stilbene synthase from grapevine to tobacco, resulting in resveratrol synthesis was reported to confer resistance against *B. cinerea*, but predictable variability in expression of the transgene was reported (Hain et al., 1993).

Stress metabolite engineering is complicated by a lack of knowledge of pathways and their regulation and poses the question of how metabolite fluxes between shared pathways can be controlled, indeed redundant homeostatic mechanisms may be discovered. In the case of stress proteins, there are limits on genes of known function that are available, but perhaps, more important is the issue of whether single or multiple gene transformations will confer stable resistance. There are technical limitations in multigene engineering but more important is the global character of stress responses. Some have argued that the solution lies in engineering for constitutive expression of stress pathways, but this may confer a yield penalty and plants have evolved to rely on inducible responses. There is also the complication that at least some plant stress pathways are subject to reciprocal regulation; for example, the salicylic acid pathway for pathogen resistance may suppress the jasmonic acid pathway for pest resistance (Fig. 6.42). Furthermore, there is evidence that different pathogens may induce different stress responses in the same host, implying a higher level of stress interpretation and customization of stress responses. Some stress metabolites and stress proteins are antinutritional and allergenic, respectively. This poses a potential risk to consumers where these are used as the basis of transgenic resistance or where their expression is increased due to the presence of transgenes (Cassells and Doyle, 2003).

6.5.6. Genetic Transformation for Crop Improvement: Altering Plant Development

Plant genetic engineering uses the same germplasm that plant breeders use to improve crops. In the process of generating improved seeds, the practice has been to transform a few useful plant lines using the most easily transformable plant genotype and later introgress that trait into better adaptable genotypes, usually elite inbred lines. In this way, genetic engineering gets readily integrated into existing plant-breeding programs instead of overriding breeding. This is possible because most traits introduced by plant transformation are dominant and can be moved from one plant to the other as a single locus. Since only a few well-defined genetic elements are involved, undesirable genes, for example, expressing possible allergens and toxins, are avoided in the breeding program. As we move into the future, we can expect a diversity of genetically enhanced crops becoming integrated into breeding for sustainable agricultural production. The integration of conventional and molecular breeding also provides opportunities for conventional breeding to achieve new objectives delineated by the power of biotechnology. Integration of systems biology (the many interactions of genes, binding sites, metabolic pathways, and so on) and breeding and physiological processes (phenomics) under diverse environmental and management conditions will also aid in narrowing the gap between genotype and phenotype (Shasha, 2003; Thethewey, 2004; Yin et al., 2004).

Plant genetic improvement, most of it deliberate, has been taking place for over 10,000 years through selection of desirable phenotypes. Molecular and cellular biologists now have the opportunity to work closely and sensibly with plant breeders and agronomists to improve the yields and sustainability of agricultural production systems. In the twenty-first century, we find ourselves in the uniquely privileged position of being able to analyze more quantitatively the nature of changes derived by "conventional" plant breeding and selection, as well as plant transformation (Gutterson and Zhang, 2004). The use of plant transformation as a tool to modify and enhance plant traits represents a significant advance in plant science since it allows for the transfer into plants of specific, characterized genes under known regulatory control. However, when transformation procedures were developed, allowing the introduction of DNA into an organism, almost any gene became available.

The use of A. tumefaciens as a versatile vector for transformation was an important breakthrough, even more so in the 1990s when it was convincingly shown that it could also be used on cereals, which, like the seed legumes, had been among the most recalcitrant to transformation. Other technological breakthroughs were the use of cell- and plant-selectable markers and the development of novel transformation techniques. The latter techniques include the use of biolistics and simpler techniques for A. tumefaciens mediated transformation, such as explant transformation, and for Arabidopsis, namely, the extremely efficient and simple floral dip or vacuum infiltration procedure.

The precision of this technology and knowledge of the specific nature of the manipulated genetic information make the effect of this type of gene transfer more predictable than the random mixing of genes that occur during classical breeding (Machuka, 2005). For this reason, the next generation of plant-biotechnology-derived crops is likely to include traits that could not be generated solely by the tools of conventional plant breeding. Such biotech crops include crops that will produce nutraceuticals, biopharmaceuticals, plastics and protein polymers, modified plant lipids and starches, phytoremedial agents, and biofuels (Maliga and Graham, 2004).

Useful genes for plant breeding are already abundant and could be used to solve previously impossible or very difficult-to-solve problems. The use of transgenes can further simplify the genetic architecture for desirable traits in ways that may be superior to or not possible even when perfect markers are available for robust quantitative trait loci (QTLs) of large effect. Transgenes typically condition strong genetic effects at operationally single loci, which also exhibit dominant gene action where only one copy of the event is needed for maximal trait expression in a hybrid cultivar. These features of transgenes can reduce complex quantitative improvement to a straightforward, often dramatic, solution. Excellent examples are provided by the expression in transgenic corn hybrids of insecticidal toxin proteins from Bt to reduce feeding damage by larvae of the European corn borer (*Ostrinia nubilalis*) or the corn rootworm beetle (*Diabrotica* spp.). Partial resistance in maize germplasm to these insect pests had been previously characterized as quantitatively inherited traits with low heritability. In addition to improving the tolerance and/or resistance to biotic and abiotic stress, genetic modification offers many possibilities for modifying the plant development and their chemical composition. Examples of developmental changes include the engineering of male sterility, the modification of fruit ripening, and alterations in flowering behavior and plant architecture. The chemical compositions of fruits and seeds can even be modified so that they can produce nonplant compounds such as antibodies and biodegradable plastics. The possibility to obtain rice varieties with a high level of vitamin A and better iron uptake, which could alleviate the nutritional problems of many people, is very appealing. Among the novel traits introduced with this technology was herbicide resistance, which raised an emotional aversion against transgenic plants among consumers because they did not want (more) herbicides to be used. Although, it is obvious that this technology has tremendous possibilities for plant breeding and human well-being, it appears that this is the first time that the introduction of a novel biological technique became the subject of such public scrutiny. It is remarkable that similar concerns about transgenics are virtually absent in the area of medical applications.

Apart from genes that confer specific traits, selectable marker genes, promoters for cell and tissue specificity, and other regulatory sequences such as terminators and intracellular targeting signal sequences are the key in plant transformation. The criteria for choosing selectable markers and other elements will depend on regulatory familiarity, patent status, likelihood of approval, suitability for use in transformation, and market acceptance (König, 2003). The genes encoding functional traits can be deployed singly for monogenic traits or they can be stacked or pyramided to specify more than two traits or products in the same crop plant or other plant production platforms (Halpin, 2005). However, transformation of specific sequences can only deal with relatively small number of genes. Population of mutants, in which a series of gene knockouts occur, are being developed to deal with the identification of a large number of genes. These are then screened for changes in phenotype (changes in function in a defined environment) and any changes identified traced back to the specific DNA sequence that has been mutated.

Transformation allows genes from all organisms to be considered for crop improvement, and the early products on the market, conferring resistance to insects or tolerance to herbicides, are based on genes derived from microorganisms. The genes used have not all been direct copies of those in the microorganism, in one case the complete gene has been synthesized in the laboratory to produce the exact protein sequence of a cleaved *B. thuringiensis* insect toxin (Koziel et al., 1993). So far, the major targets for transformation to improve crops have been related to crop protection to either enhance the possibilities of using broad-spectrum herbicides or provide resistance against insects. Genes have also been introduced that affect quality traits such as fruit ripening, fatty acid composition of oils, and starch properties. Traits that are more complex, for example, broad-spectrum disease resistance, or which may require several genes to bring about the desired changes, are difficult to achieve and require further research. While an almost limitless number of structural genes to express different proteins are available, there are currently considerable limitations on the availability of promoters and regulators to ensure the reliable expression of the genes at the desired times during development and in the desired organs and tissues. There is also a need for much more understanding as to how complex regulatory and signaling controls work and can be modified. At present, determining what might be achieved in crop improvement through transformation is only at the very early stage.

6.5.7. Genomics: Gene Mining and Gene Expression Profiling

The advent of DNA microarray technology has offered the promise of obtaining new insights into the secrets of life by monitoring the activities of thousands of genes simultaneously. These measurements or gene profiles provide a "snapshot" of life that maps to a cross section of genetic activities in a four-dimensional space of time and the biological entity. Although microarray experiments (DeRisi et al., 1997; Golub et al., 1999) hold the promise of the innovative technology being used to classify biological types, development of powerful and efficient analysis strategies to perform more complex biological tasks, such as mining disease-relevant genes and building genetic networks, remains a significant demand. According to the modes of learning algorithms, current analysis strategies can be classified into two groups. An unsupervised learning method, such as a typical cluster analysis, by ignoring the biological attribute (label) of a DNA example (instance), directly works on genes themselves and is a useful tool to study functional genomics, but it is unable to efficiently relate differential gene expression profiles to phenotypes. On the other hand, the second strategy of supervised learning is a target-driven process, in that a suitable induction algorithm is used to identify the most contributed genes for a specific target, for example, classification of biological types, gene mining, or data-driven gene networking (Xia et al., 2004).

Gene mining is the process of exploiting the DNA sequence of one genotype to isolate useful genes from related genotypes. The principle reason for gene mining is to identify and isolate genes that are characterized for conferring essential traits. The widespread use and availability of molecular biological techniques have allowed for the rapid development and identification of nucleic-acid-derived sequences. With the availability of integration of laboratory equipment with advanced computer software, we are able to conduct advanced quantitative analyses, database comparisons, and computational algorithms to seek and identify gene sequences. Genetic databases for organisms such as *E. coli, Haemophilus influenzae, Mycoplasma genitalium*, and *Mycoplasma pneumoniae* are available for the public. These biological databases store information that is searchable and from which biological information may be retrieved. Publicly available sequence databases on the Internet have been exploited for the identification of useful genes. The resources mainly used are Gene Bank at the National Centre for Biotechnology Information (http://www.ncbi.nlm.nih.gov.) and the *Arabidopsis thaliana* TC database at TIGR (The Institute for Genomic Research, http://www.tigr.org/tdt/tgi/agi). The work was carried out using Internet connection and a DOS-based sequence analysis software package and the facilities of the basic molecular laboratory for PCR verification. It is recommended to carry out this type of analysis in Windows or apply Macintech Environment since they are most readily available in the public domain of the sequence analysis.

Various methods for gene mining are DNA extraction and PCR-based gene mining, data mining, using genetic algorithm, peptide mass fingerprinting, from biomedical literature, studying interactions and connectivity with the help of GENEWATCH and ORIEL, and DNA chip analysis by microarray (Fig. 6.43). The applications of gene mining includes, allele mining for stress tolerance, sequence diversity, drug discovery, mining colon tumor relevant genes, mining molecular signatures for leukemia subtypes, classification of biological types, uncultivable microbes, and novel genes.

Allele mining is the process of identification and access to allelic variation that affects the plant phenotype and is of the utmost importance for the utilization of genetic resources, such as in plant variety development. Gene-specific primers amplify the DNA

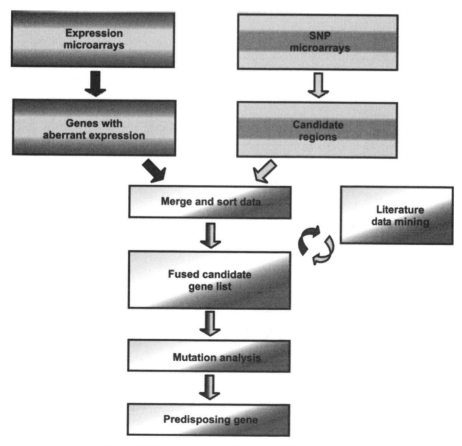

Figure 6.43. Analysis and gene profiling of a gene.

of each accession, and the amplified product represents either the entire allele or some functional component of the allele, such as the promoter or the coding sequences. Considering the huge number of accessions that are held collectively by gene banks, genetic resource collections are deemed to harbor a wealth of undisclosed allelic variants. The challenge is how to unlock this variation. Allele mining is a research field aimed at identifying allelic variation of relevant traits within genetic resource collections. For identified genes of known function and basic DNA sequence, genetic resource collections may be screened for allelic variation by the "tiling strategy" using DNA chip technology. In that approach, the basic DNA sequence of a gene is spotted on a chip in the form of a large series of sequence-overlapping probes consisting of 15–20 bases. Each base position in a fluorescently labeled sample is then interrogated for the presence of point mutations by monitoring hybridization signals with the spotted probes. Because the sequence of samples is determined in comparison with the primary composition of a gene, this method is also known as *resequencing*. With this method, new point mutations, in relatively large DNA fragments, can be detected. Once allelic variants of interest have been identified, the approach can be optimized by focusing on target sets of polymorphisms, for example, using SNP detection methods.

Some of the examples of gene mining are discussed below.

1. *Allele Mining for Stress-Tolerance Genes in Oryza Species and Related Germplasm*. The international project to sequence the genome of *Oryza sativa* L. cv. Nipponbare has made allele mining possible for all genes of rice. Scientists used a rice calmodulin gene, a rice gene encoding a late-embryogenesis-associated protein, and salt-inducible rice gene to optimize the PCR for allele mining of stress-tolerance genes on identified accessions of rice and related germplasm. Two sets of PCR primers were designed for each gene. Primers based on the 5′ and 3′ untranslated region of genes were found to be sufficiently conserved so as to be effective over the entire range of germplasm in rice for which the concept of allelism is applicable. However, the primers based on the adjacent amino (N) and carboxyl (C) termini amplify additional loci. Field-based phenotyping of germplasm identifies tolerant accessions and biochemical and physiological analysis groups. The existing and emerging tools of genomics and proteomics help to identify key genes or key members of a gene family involved in each mechanism. The technique of choice for allele mining is PCR. Gene-specific primers amplify the DNA of each accession, and the amplified product represents either the entire allele or some functional component of the allele, such as the promoter or the coding sequences (Latha et al., 2004).

2. *Allele Mining and Sequence Diversity at the Wheat Powdery Mildew Resistance Locus Pm3*. The production of wheat is threatened by a constantly changing population of pathogen races. Considering the capability of many pathogens to overcome genetic resistance, the identification and implementation of new sources of resistance is essential. Landraces and wild relatives of wheat have played an important role as genetic resources for the improvement of disease resistance. The allele mining approach was studied to characterize and utilize the naturally occurring resistance diversity in wheat using 1320 hexaploid wheat landraces selected on the basis of ecogeographical parameters favoring growth of powdery mildew. The landraces were infected with a set of differential powdery mildew isolates, which

allowed the selection of resistant lines. The molecular tools derived from Pm3 haplotype studies were applied to study the genetic diversity at this locus. From the known Pm3 R alleles, Pm3b was the only one frequently identified. The new interesting and functional alleles can be transferred to susceptible but economically important wheat varieties as single genes or R gene cassettes to achieve efficient control of mildew (Kaur et al., 2008).

3. *Gene-Mining the A. thaliana Genome.* The landraces were infected with a set of differential powdery mildew isolates, which allowed the selection of resistant lines. The molecular tools derived from Pm3 haplotype studies were applied to study the genetic diversity at this locus. From the known Pm3 R alleles, Pm3b was the only one frequently identified. In the same set, a high frequency of landraces carrying a susceptible haplotype were found. This analysis allowed the identification of candidate-resistant lines that were further tested for the presence of new potentially functional alleles. On the basis of transient expression assays as well as virus-induced gene silencing (VIGS), at least two new functional Pm3 alleles were identified. The new interesting and functional alleles can be transferred to susceptible but economically important wheat varieties as single genes or R gene cassettes to achieve efficient control of mildew (Berger, 2004).

The future scenario of gene mining is

- Advancement in gene mining companies
- Metagenomics for mining new genetic resources of microbial communities
- Gene mining with the hierarchical clustering algorithm
- "Gene mining" strategies of drug discovery
- Global gene mining and the pharmaceutical industry
- Synthetic life and gene mining.

Gene mining is not only a boon for plant biotechnology but also equally good for animal sciences and the work is progressing in exponential scale worldwide. Gene mining provided molecular biologists with a powerful and usable tool for extracting disease-relevant genes, a major theme in the postgenomic era. This technique leaves the target-driven gene functioning uncertain.

6.6. FUTURE CHALLENGES: FUNCTIONAL GENOMICS APPROACH FOR IMPROVEMENT OF CROPS

In recent years, advances in genetics and genomics have greatly increased our understanding of structural and functional aspects of plant genomes but at the same time have challenged us with many compelling avenues of investigation. The complete genome sequences of *Arabidopsis*, rice, sorghum, and poplar, as well as an enormous number of plant expressed sequence tags (ESTs), have become available. *Arabidopsis* genome sequence will expedite map-based cloning in tomato on the basis of chromosomal synteny between the two species and will facilitate the functional analysis of tomato genes (Mysore et al., 2001).

DNA sequencing technology is undergoing a revolution with the commercialization of second-generation technologies capable of sequencing thousands of millions of nucleotide bases in each run. The data explosion resulting from this technology is likely to continue to increase with the further development of second-generation sequencing and the introduction of third-generation single-molecule sequencing methods over the coming years. The question is no longer whether we can sequence crop genomes that are often large and complex, but how soon can we sequence them? The increasing availability of DNA sequence information enables the discovery of genes and molecular markers associated with diverse agronomic traits, creating new opportunities for crop improvement. However, the challenge remains to convert this mass of data into knowledge that can be applied in crop breeding programs (Edwards and Batley, 2010). Plant genomics has allowed us to discover and isolate important genes and to analyze functions that regulate yields and tolerance to environmental stress. A summary of methods used for assigning functions to genes of higher plants is shown in Table 6.5. Metabolomics represents an important addition to the tools currently used in genomics-assisted selection for crop improvement. These tools have recently been turned to evaluation of the natural variance apparent in metabolite composition (Fernie and Schauer, 2009).

Functional genomics involves determination of the function of genes, and several innovative techniques are available to decipher the functions of these genes. Some of the approaches are specific to plant genomes and cannot be used in animal or microbial systems. The different approaches of functional genomics that can be used for the study of the functions of genes in plants are shown in Figure 6.44.

Transgenic methods have been successfully applied to trait improvement in a number of crops. A nontransgenic method for reverse genetics called *targeting induced local lesions in genomes* (TILLING) has been developed as a method for inducing and identifying novel genetic variation and has been demonstrated in the model plant, *A. thaliana*. TILLING has been extended to the improvement of crop plants and

TABLE 6.5. Summary of Methods Used for Assigning Functions to Genes of Higher Plants

Method	Strategy	Criterion for Assigning Function
BLAST search	Computer scanning of databases	Similarity with known gene
Synteny	Compare genetic maps from different species	Similarity in sequence and genetic map position
Chimeraplasty	Insert DNA/RNA hybrids into cells	Altered phenotype due to specific gene mutation
TUSC*	Insert DNA that jumps into genes	Generation of wholesale mutations in maize
Activation tagging	Insert DNA enhancers via bacteria	Generation of wholesale mutations in plants
RNA silencing	Infect tobacco with genetically altered TMV	Switching on or switching off individual genes
Microarray	DNA chips	Tracking activity using mRNA profile
Proteomics	2D gels for protein expression	Tracking activity using protein profile
Metabolomics	Study of all metabolites	Tracking activity using metabolite expression

Abbreviation: TUSC, trait utility system in corn.

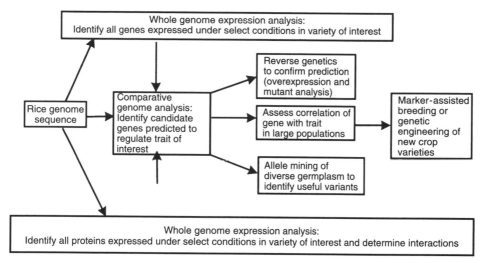

Figure 6.44. Approaches of functional genomics to study the functions of genes.

shows great promise as a general method for both functional genomics and modulation of key traits in diverse crops (Slade and Knauf, 2005). However, improved varieties, not sequences per se, contribute to improve economic return to the farmer. Functional genomics and systems biology research are facilitating the identification of gene networks that are involved in controlling genetic variation for agronomically valuable traits in elite breeding populations. Furthermore, combining the new knowledge from genomic research with conventional breeding methods is essential for enhancing response to selection, hence crop improvement. Superior varieties can result from the discovery of novel genetic variation, improved selection techniques, and/or the identification of genotypes with improved attributes due to superior combinations of alleles at multiple loci assembled through marker-assisted selection. Although it is clear that genomics research has great potential to revolutionize the discipline of plant breeding, high costs invested in or associated with genomics research currently limit the implementation of genomics-assisted crop improvement, especially for inbreeding and/or minor crops.

In the next few years, the entire genome or at least gene space will likely be sequenced for most major crops. Genome sequencing and the data generated by the "omics" tools will deliver candidate genes that need to be analyzed functionally, either using natural mutants or by applying a transgene approach in which the gene is (ectopically) overexpressed or downregulated. Systems biology research should be established for all basic biological processes relevant to crop productivity and quality, such as metabolic pathways and signaling systems, that is, sensors and networks. Furthermore, the plant evaluation platforms should allow analysis of the performance of plants under controlled or highly monitored conditions in order to select genotypes or alleles and/or genes that maximize agricultural output. Application of phenotyping platforms to breeding lines and other large plant populations will allow the identification of major genetic determinants for important agricultural traits, such as grain yield, harvesting index, flowering time, nitrogen- and water-use efficiency, drought tolerance, pathogen resistance. New tilling-based collections with mutations in all genes, in conjunction with phenotyping platforms, will help to identify the alleles in genes affecting major crop traits.

Scientists and breeders should be able to explore different plant architectures *in silico* so as to grow diverse forms according to known relationships in gene, protein, and metabolite networks (Alisdair and Schauer, 2008). Ultimately, computer models should allow us to predict which genes need to be altered in order to obtain a desired trait in a specific environmental scenario. The applications of such prediction could be tremendous and are likely to revolutionize current plant breeding and crop improvement. Ultimately, it will be possible to improve plants by design.

New technologies must be developed to make the functional analysis of transgenes more effective by eliminating variability due to position effects. These technologies should allow site-specific integration of transgenes in predetermined chromosomal locations or even homologous recombination. Improved transgene analysis will become important as scientists grow to realize that the successful engineering of pathways will require the modification of multiple genes or their simultaneous introduction. To this end, new gene delivery systems in which multiple genes can be tested in parallel will need to be developed. Finally, new technologies, such as resequencing, promise to accelerate dramatically the process of mutation detection.

REFERENCES

Agrios GN. 1997. General overview of plant pathogenic organisms. In: Ruberson JR, ed., *Handbook of Pest Management*, CRC Press, Boca Raton, FL.

Alisdair FR, Schauer N. 2008. Metabolomics-assisted breeding: a viable option for crop improvement. *Trends Genet* **25**, 39–48.

Andrup L, Damgaard J, Wasserman K, Boe L, Madsen SM, Hansen F. 1994. Complete nucleotide sequence of the *Bacillus thuringiensis* sub sp isrealensis plasmid pTX14-3 and its correlation with biological properties. *Plasmid* **31**, 72–88.

Artsaenko O, Kettig B, Fiedler U, Conrad U, During K. 1998. Potato tubers as a biofactory for recombinant antibodies. *Mol Breed* **4**, 313–319.

Bais HP, Tiffany L, Weir LG, Perry SG, Vivanco JM. 2006. The role of root exudates in rhizosphere interactions with plants and other organisms. *Annu Rev Plant Biol* **57**, 233–266.

Beauchamp CJ, Lévesque G, Prévost D, Chalifour FP. 2006. Isolation of free-living dinitrogen-fixing bacteria and their activity in compost containing de-inking paper sludge. *Bioresour Technol* **97**, 1002–1011.

BenDov E, Nissan G, Pelleg N, Manasherab R, Bousiba S, Zaritsky A. 1999. Refined circular restriction map of the *Bacillus thuingiensis* subspisrealensis plasmid carrying the mosquito larvicidal genes. *Plasmid* **42**, 186–193.

Berger DK. 2004. Gene-mining the Arabidopsis thaliana genome: applications for biotechnology in Africa. *S Afr J Bot* **70**, 173–180.

Bohnert HJ, Shen B. 1999. Transformation and compatible solutes. *Sci Hortic* **78**, 237–260.

Boller T, Keen NT. 2000. Resistance genes and the perception and transduction of elicitor signals in host-pathogen interactions. In: Slusarenko A, Fraser RSS, Van Loon LC, eds., *The Mechanisms of Resistance to Plant Disease*, pp. 189–229, Kluwer, Dordrecht.

Bolwell GP. 1999. Role of active oxygen species and NO in plant defence response. *Curr Opin Plant Biol* **2**, 287–294.

Broekaert WF, Terras FRG, Cammue BPA. 2000. Induced and preformed antimicrobial proteins. In: Slusarenko A, Fraser RSS, Van Loon LC, eds., *The Mechanisms of Resistance to Plant Disease*, pp. 371–477, Kluwer, Dordrecht.

Bucchini L, Goldman LR. 2002. Starlink corn: a risk analysis. *Environ Health Perspect* **110**, 5–13.

Cassells C, Doyle BM. 2003. Genetic engineering and mutation breeding for tolrance to abiotic and biotic stresses science and technology safety. *Bulg J Plant Physiol* 52–82. Special Issue.

Charbonneau CS, Drobney RD, Rabeni CF. 1994. Effects of *Bacillus thuringiensis* var. *israelensis* on nontarget benthic organisms in a lentic habitat and factors affecting the efficacy of the larvicide. *Environ Toxicol Chem* **13**, 267–279.

Chilton MD, Drummond MH, Merio DJ, Sciaky D, Montoya AL, Gordon MP, Nester EW. 1977. Stable incorporation of plasmid DNA into higher plant cells: the molecular basis of crown gall tumorigenesis. *Cell* **2**, 263–271.

Chong DKX, Langridge WHR. 2000. Expression of full-length bioactive antimicrobial human lactoferrin in potato plants. *Transgenic Res* **9**, 71–78.

Chong DK, Roberts W, Arakawa T, Illes K, Bagi G, Slattery CW, Langridge WH. 1997. Expression of the human milk protein beta-casein in transgenic potato plants. *Transgenic Res* **6**, 289–296.

Cockburn A. 2002. Assuring the safety of genetically modified, GM. foods, the importance on an holistic, integrative approach. *J Biotechnol* **98**, 79–106.

Cohen JI. 2005. Poorer nations turn to publicly developed GM crops. *Nat Biotechnol* **23**, 27–33.

Commandeur U, Twyman RM, Fischer R. 2003. The biosafety of molecular farming in plants. *AgBiotechNet* **5**, 110.

Cushman JC, Bohnert HJ. 2000. Genomic approaches to plant stress tolerance. *Curr Opin Plant Biol* **3**, 117–124.

Dai ZY, Hooker BZ, Anderson DB, Thomas SR. 2000. Improved plant-based production of E1 endoglucanase using potato: expression optimization and tissue targeting. *Mol Breed* **6**, 277–285.

De Wilde C, Peeters K, Jacobs A, Peck I, Depicker A. 2002. Expression of antibodies and Fab fragments in transgenic potato plants: a case study for bulk production in crop plants. *Mol Breed* **9**, 2871–2882.

Denbow DM, Grabau EA, Lacy GH, Kornegay ET, Russell DR, Umbeck PF. 1998. Soybeans transformed with fungal phytase gene improve phosphorus availability for broilers. *Poult Sci* **77**, 878–881.

DeRisi JL, Iyer VR, Brown PO. 1997. Exploring the metabolic and genetic control of gene expression on a genomic scale. *Science* **278**, 680–686.

Duguid JP, Old DC. 1980. Adhesive properties of Enterobacteriaceae. In: Beachey EH, ed., *Bacterial Adherence. Receptors and Recognition*, pp. 185–217, Chapman and Hall, London.

Dulmage HT. 1970. Insecticidal activity of HD-1, a new isolate of *Bacillus thuringiensis* var *alesti*. *J Invert Pathol* **15**, 232–239.

Edwards D, Batley J. 2010. Plant genome sequencing: applications for crop improvement. *Plant Biotechnol J* **8**, 2–9.

Elad Y. 2004. In: Elad Y, Williamson B, Tudzynski P, Delen N, eds., *Botrytis: Biology, Pathology and Control*, pp. 1–8, Kluwer Academic Publishers, Dordrecht, The Netherlands.

Ellis J, Dodds P, Pryor T. 2000. The generation of plant disease resistance gene specificities. *Trends Plant Sci* **5**, 373–379.

Eujayl I, Baum M, Powell W, Erskine W, Pehu E. 1998. A genetic linkage map of lentil (*Lens sp*) based on RAPD and AFLP markers using recombinant inbred lines. *Theor Appl Genet* **97**, 83–89.

Eujayl I, Erskine W, Baum M, Pehu E. 1999. Inheritance and linkage analysis of frost injury in lentil. *Crop Sci* **39**, 639–642.

Fernie AR, Schauer N. 2009. Metabolomics-assisted breeding: a viable option for crop improvement? *Trends Genet* **25**, 39–48.

Golub TR, Slonim DK, Tamayo P, Huard C, Gaasenbeek M, et al. 1999. Molecular classification of cancer: class discovery and class prediction by gene expression monitoring. *Science* **286**, 531–537.

Gray AJ, Raybould AF. 1998. Reducing transgene escape routes. *Nature* **392**, 653–654.

Gressel J. 2008. Transgenics are imperative for biofuel crops. *Plant Sci* **174**, 246–263.

Grover A, Sahi C, Sanan N, Grover A. 1999. Taming abiotic stress in plants through genetic engineering, current strategies and perspective. *Plant Sci* **143**, 101–111.

Gutterson N, Zhang JZ. 2004. Genomics applications to biotech traits: a revolution in progress? *Cur Opin Plant Biol* **7**, 226–230.

Haahtela K, Wartiovaara T, Sundman V, Skujins J. 1981. Root-associated N2 fixation (acetylene reduction) by enterobacteriaceae and azospirillum strains in cold-climate spodosols. *Appl Environ Microbiol* **41**, 203–206.

Hain R, Reif HJ, Langebartels R, Kindl H, Vornam B, et al. 1993. Disease resistance results from foreign phytoalexin formation in a novel plant. *Nature* **361**, 153–156.

Halpin C. 2005. Gene stacking in transgenic plants–the challenge for 21st century plant biotechnology. *Plant Biotechnol J* **3**, 141–155.

Heath MC. 2000. Nonhost resistance and non-specific plant defenses. *Curr Opin Plant Biol* **3**, 315–319.

Helgason E, Okstad OA, Caugant DA, Johansen HA, Fouet A, Mock M, Hegna I, Kolsto A. 2000. *Bacillus anthracis, Bacillus cereus*, and *Bacillus thuringiensis*-one species on the basis of genetic evidence. *Appl Environ Microbiol* **66**, 2627–2630.

Hood EE. 2002. From green plants to industrial enzymes. *Enzyme Microb Technol* **30**, 279–283.

Hood EE, Kusnadi A, Nikolov Z, Howard JA. 1999. Molecular farming of industrial proteins from transgenic maize. *Adv Exp Med Biol* **464**, 127–147.

Hood EE, Witcher D, Maddock S, Meyer T, Baszczynski C, et al. 1997. Commercial production of Avidin from transgenic characterization of transformant, production, processing extraction and purification. *Mol Breed* **3**, 291–306.

Hood EE, Woodardet SL, Horn ME. 2002. Monoclonal antibody manufacturing in transgenic plants–myths and realities. *Curr Opin Biotechnol* **13**, 630–635.

Horsch RB, Fry JF, Hoffmann NL, Eichholtz D, Rogers SG, Fraley RT. 1985. A simple and general method for transferring genes into plants. *Science* **227**, 1229–1231.

Hrabak EM, Urbano MA, Dazzo FB. 1981. Growthphase-dependent immunodeterminants of Rhizobium trifolii lipopolysaccharide which bind trifoliin A, a white clover lectin. *J Bacteriol* **148**, 697–711.

James C. 2004. Preview: global status of commercialized biotech/GM crops: 2004. ISAA Brief No. 32, *International Service for the Acquisition of Agri-Biotech Applications*, Ithaca (NY).

Kanda K, Takada Y, Kawasaki F, Kato F, Murata A. 2000. Mating in *Bacillus thuringiensis* can induce plasmid integrative prophage J7W-1. *Acta Virol* **44**, 189–192.

Kaur N, Street K, Mackay M, Yahiaoui N, Keller B. 2008. Allele mining and sequence diversity at the wheat powdery mildew resistance locusPm3. *Plant Mol Biol* **65**, 93–106.

Kondo S, Ohba M, Ishii T. 1992. Larvicidal activity of *Bacillus thuringiensis* serovar *israelensis* against nuisance chironomid midges (Diptera: Chironomidae) of Japan. *Lett Appl Microbiol* **15**, 207–209.

Kong AYY, Fonte SJ, van Kessel C, Six J. 2007. Soil aggregates control N cycling efficiency in long- term conventional and alternative cropping systems. *Nutr Cycl Agroecosys* **79**, 45–58.

König A. 2003. A framework for designing transgenic crops–science, safety and citizen's concerns. *Nat Biotechnol* **21**, 1274–1279.

Kota M, Daniell H, Varma S, Garczynski SF, Gould F, Moar WJ. 1999. Overexpression of the Bacillus thuringiensis, Bt. Cry2Aa2 protein in chloroplasts confers resistance to plants against susceptible and Bt-resistant insects. *Proc Natl Acad Sci U S A* **96**, 1840–1845.

Koziel MG, Beland GL, Bowman C, Carozzi NB, et al. 1993. Field performance of elite transgenic maize plants expressing an insecticidal protein derived from *Bacillus thuringiensis*. *Biotechnology* **11**, 194–200.

Krauss G. 2001. Signal transmission via ras proteins. *Biochemistry of Signal Transduction and Regulation*, John Wiley & Sons, Inc., New York. doi: 10.1002/3527600051.ch9.

Krishnamurthy KV, Suhasini K, Sagare AP, Meixner M, de Kathen A, Pickard T, Schieder O. 2000. Agrobacterium mediated transformation of chickpea (*Cicer arietinum* L) embryo axes. *Plant Cell Rep* **19**, 235–240.

Latha R, Rubia L, Bennett J, Swaminathan MS. 2004. Allele mining for stress tolerance genes in Oryza species and related germplasm. *Mol Biotechnol* **27**, 101–108.

Lerner HR. 1999a. Introduction to the response of plants to environmental stresses. In: Lerner HR, ed., *Plant Responses to Environmental Stresses*, pp. 1–26, Marcel Dekker, New York.

Lerner HR, ed. 1999b. *Plant Responses to Environmental Stresses*, Marcel Dekker, New York.

Leroux P, Fritz R, Debieu D, Albertini C, Lanen C, Bach J, Gredt M, Chapeland F. 2002. Mechanisms of resitance of fungicides in field strains of *Botrytis cinerea*. *Pest Manag Sci* **58**, 876–888.

Liénard D, Sourrouille C, Gomord V, Faye L. 2007. Pharming and transgenic plants. *Biotechnol Annu Rev* **13**, 115–147.

Loscy JE, Ravor LS, Carter ME. 1999. Transgenic pollen harms monarch larvae. *Nature* **399**, 214.

Machuka J. 2005. Plant transformation: Goals, strategies and relevance for African crop improvement and production of high value biologics. *Afr Crop Sci Conf Proc* **7**, 1297–1303.

Maliga P, Graham I. 2004. Molecular farming and metabolic engineering promise a new generation of high-tech crops. *Curr Opin Plant Biol* **7**, 149–151.

Mansfield JW. 2000. Antimicrobial compounds and resistance. In: Slusarenko A, Fraser RSS, Van Loon LC, eds., *The Mechanisms of Resistance to Plant Disease*, pp. 325–370, Kluwer, The Netherlands.

Mason HS, Ball JM, Shi JJ, Jiang X, Estes MK, Arntzen CJ. 1996. Expression of Norwalk virus capsid protein in transgenic tobacco and potato and its oral immunogenicity in mice. *Proc Natl Acad Sci U S A* **93**, 5335–5340.

Mason HS, Warzecha H, Tsafir MS, Arntzen CJ. 2002. Edible plant vaccines: applications for prophylactic and therapeutic molecular medicine. *Trends Mol Med* **8**, 324–329.

McGarvey PB, Hammond J, Dienelt MM, Hooper DC, Fu ZF, Dietzschold B, Kaprowski H, Michaels FH. 1995. Expression of the rabies virus glycoprotein in transgenic tomatoes. *Biotechnology* **13**, 1484–1487.

Moloney M, Boothe J, Van Rooijen G. 2003. Oil bodies and associated proteins as affinity matrices. US Patent 6509453.

Mondal KC, Banerjee D, Jana M, Pati BR. 2001. Colorimetric assay method for determination of the tannin acyl hydrolase (EC 3.1.1.20) activity. *Anal Biochem* **295**, 168–171.

Moore LW, Chilton WS, Canfield ML. 1997. Diversity of opines and opine-catabolizing bacteria isolated from naturally occurring crown gall tumors. *Appl Environ Microbiol* **63**, 201–207.

Mysore KS, Tuori RP, Martin GB. 2001. Arabidopsis genome sequence as a tool for functional genomics in tomato. *Genome Biol* **2**, 1003.

Nurnberger T, Scheel D. 2001. Signal transmission in the plant immune response. *Trends Plant Sci* **6**, 372–379.

Old DC. 1972. Inhibition of the interaction between fimbrial hemagglutinins and erythrocytes by D-mannose and other carbohydrates. *J Gen Microbiol* **71**, 149–157.

Ozair CA, Moshier LJ. 1988. Effect of postemergence her bicides on nodulation and nitrogen fixation in soybeans *(Glycine max)*. *Appl Agric Res* **3**, 214.

Paccalet T, Bardor M, Rihouey C, Delmas F, Chevalier C, et al. 2007. Engineering of a sialic acid synthesis pathway in transgenic plants by expression of bacterial Neu5Ac-synthesizing enzymes. *Plant Biotechnol J* **5**, 16–25.

Pedersen WL, Chakrabarty K, Klucas RV, Vivader AK. 1978. Nitrogen fixation (acetylene reduction) associated with roots of winter wheat and sorghum in Nebraska. *Appl Environ Microbiol* **35**, 129–135.

Perrin Y, Vaquero C, Gerrard I, Sack M, Drossard J, Stoger E, Christou P, Fischer R. 2000. Transgenic pea seeds as bioreactors for the production of a single-chain Fv fragment (scFV) antibody used in cancer diagnosis and therapy. *Mol Breed* **6**, 345–352.

Pieterse CMJ, Van Pelt JA, Van Wees SCM, Ton J, Leon-Kloosterziel KM, et al. 2001. Rhizobacteria–mediated induced systemic resistance: triggering, signaling and expression. *Eur J Plant Pathol* **107**, 51–61.

Qiang L, Kasuga M, Sakuma Y, Abe H, Miura S, Yamaguchi-Shinozaki K, Shinozaki K. 1998. Two transcriptional factors DREB1 and DREB2 with an EREBP/AP2 DNA binding domain separate two cellular signal transduction pathways in drought- and low-temperature-responsive gene expression respectively in *Arabidopsis*. *Plant Cell* **10**, 1391–1406.

Ramos LHM, McGuire MR, Wong LJG, Castro-France R. 2000. Evaluation of pectin, gelatin and starch granular formation of *Bacillus thuringiensis*. *South West Entomol* **25**, 59–67.

Rennie RJ. 1980. Dinitrogen-fixing bacteria: computer-assisted identification of soil isolates. *Can J Microbiol* **26**, 1275–1283.

Rennie RJ, Dubetz S. 1984. Effect of fungicides and herbici des on nodulation and N2 fixation in soybean fields lacking indigenous *Rhizobium japonicum*. *Agron J* **76**, 451.

Richter LJ, Thanavala Y, Arntzen CJ, Mason HS. 2000. Production of hepatitis B surface antigen in transgenic plants for oral immunization. *Nat Biotechnol* **l18**, 1167–1171.

Schillberg S, Emans N, Fischer R. 2002. Antibody molecular farming in plants and plant cells. *Phytochem Rev* **1**, 45–54.

Schillberg S, Zimmermann S, Findlay K, Fischer R. 2000. Plasma membrane display of anti-viral single chain Fv fragments confers resistance to tobacco mosaic virus. *Mol Breed* **6**, 317–326.

Schneider RW. 1982. In: Schneider RW, ed., *Suppressive Soils and Plant Disease*, p. 88, The American Phytopathological Society, St. Paul, MN.

Schnelle MA, Hensley DL. 1990. Effect of pesticides upon nitrogen fixation and nodulation by dry bean. *Pestic Sci* **28**, 83.

Schunmann PHD, Coia G, Waterhouse PM. 2002. Biopharming the SimpliRED (TM) HIV diagnostic reagent in barley, potato and tobacco. *Mol Breed* **9**, 113–121.

Sengupta S, Sengupta LK, Bisen PS. 2001. Bioengineered crops: the commercial and ethical considerations. *Curr Genom* **2**, 181–197.

Shasha DE. 2003. Plant systems biology: lessons from a fruitful collaboration. *Plant Physiol* **132**, 415–416.

Shrestha RK, Maskey SL. 2005. Associative nitrogen fixation in lowland rice. *Nepal Agric Res J* **6**, 112.

Sinclair TR, Purcell LC, Sneller CH. 2004. Crop transformation and the challenge to increase yield potential. *Trends Plant Sci* **9**, 70–75.

Škrdleta V, Němcová M, Lisá L, Novák K. 1993. Dinitrogen fixation (Acetylene reduction) and nitrogen accumulation in peas cultivated under the standard growth conditions. *Folia Microbiol* **38**, 229–234.

Slade AJ, Knauf VC. 2005. TILLING moves beyond functional genomics into crop improvement. *Transgenic Res* **14**, 109–115.

Slusarenko A, Fraser RSS, Van Loon LC. 2000. In: Slusarenko A, Fraser RSS, Van Loon LC, eds., *The Mechanisms of Resistance to Plant Disease*, Kluwer, Dordrecht, Kluwer, Dordrecht.

Sticklen M. 2006. Plant genetic engineering to improve biomass characteristics for biofuels. *Curr Opin Biotechnol* **17**, 315–319.

Stoger E, Sack M, Perrin Y, Vaquero C, Torres E, Twyman RM, Christou P, Fischer R. 2002. Practical considerations for pharmaceutical antibody production in different crop systems. *Mol Breed* **9**, 149–158.

Stoger E, Vaquero C, Torres E, Sack M, Nicholson L, 2000. Cereal crops as viable production and storage systems for pharmaceutical scFv antibodies. *Plant Mol Biol* **42**, 583–590.

Swift MJ. 1997. Biological management of soil fertility as a component of sustainable agriculture: perspectives and prospects with particular reference to tropical regions. In: Brussard L, Cerrato RF, eds., *Soil Ecology in Sustainable Agriculture Systems*, pp. 137–159, Lewis Publishers, Boca Raton (FL).

Tacket CO, Mason HS, Losonsky G, Clements JD, Levine MM, Arntzen CJ. 1998. Immunogenicity in humans of a recombinant bacterial-antigen delivered in transgenic potato. *Nat Med* **4**, 607–609.

Tacket CO, Mason HS, Losonsky G, Estes MK, Levine MM, Arntzen CJ. 2000. Human immune responses to a novel Norwalk virus vaccine delivered in transgenic potatoes. *J Infect Dis* **182**, 302–305.

Thethewey RN. 2004. Metabolite profiling as an aid to metabolic engineering in plants. *Curr Opin Plant Biol* **7**, 196–201.

Twyman RM, Stoger E, Schillberg S, Christou P, Fischer R. 2003. Molecular farming in plants: host systems and expression technology. *Trends Biotechnol* **21**, 570–578.

Updegraff DM. 1969. Semimicro determination of cellulose in biological materials. *Anal Biochem* **32**, 420–424.

Van Loon LC, Fraser RSS, Van Loon LC. 2000. Systemic induced resistance. In: Slusarenko A. ed., *The Mechanisms of Resistance to Plant Disease*, pp. 521–574. Kluwer, Dordrecht, The Netherlands.

Vierling RA, Wilcox JR. 1996. Microplate assay for soybean seed coat peroxidase activity. *Seed Sci Technol* **24**, 485–494.

Wache H. 1987. Effect of herbicides on symbiotic nitrogen fixation of lucerne *(Medicago sativa* L.). *Zentralbl Microbiol* **142**, 349.

Witcher DR, Hood E, Peterson D, Bailey M, Marchall L, et al. 1998. Commercial production of b-glucuronidase (GUS): a model system for the production of proteins in plants. *Mol Breed* **4**, 301–312.

Wong LJG. 1993. Seleccion de cepas nativas y de ex 'tractos de fermentaci6n de *Bacillus thuringiensis* contra *Trichoplusia ni* (Hubne), *Heliothis virescens* (Fabricius) (Lepidoptera: Noctuidae) Tesis de Doctorado en Ciencias especialidad en Microbiologia. Fac. de Ciencias Biolgicas. Divisin de Estudios de Posgrado. U.A.N.L. Monterrey. N.L. Mex.

Xia L, Shaoqi R, Yadong W, Binsheng G. 2004. Gene mining: a novel and powerful ensemble decision approach to hunting for disease genes using microarray expression profiling. *Nucleic Acids Res* **32**, 2685–2694.

Yin X, Struik PC, Kropff MJ. 2004. Role of crop in predicting gene-to-phenotype relationships;. *Trends Plant Sci* **9**, 426–432.

Young JM, Kuykendall LD, Martinez-Romero E, Kerr A, Sawada H. 2001. A revision of *Rhizobium* Frank 1889, with and emended description of the genus and the inclusion of all species of *Agrobacterium* Conn 1942 and *Allorhizobium undicola* de Lajudie et al. 1989 as new combinations: *Rhizobium radiobacter, R. rhizogenes, R. rubi, R. undicola* and *R. vitis*. *J Syst Evol Microbiol* **51**, 89–103.

Yu J, Langridge W. 2003. Expression of rotavirus capsid protein VP6 in transgenic potato and its oral immunogenicity in mice. *Transgenic Res* **12**, 163–169.

Ziegelhoffer T, Raasch JA, Austin-Phillips S. 2001. Dramatic effects of truncation and sub-cellular targeting on the accumulation of recombinant microbial cellulase in tobacco. *Mol Breed* **8**, 147–158.

Ziegler MT, Thomas SR, Danna KJ. 2000. Accumulation of a thermostable endo-1, 4-b- D-glucanase in the apoplast of Arabidopsis thaliana leaves. *Mol Breed* **6**, 37–46.

7

MICROBES AS A TOOL FOR INDUSTRY AND RESEARCH

7.1. PROLOGUE

The use of microorganisms for large-scale industrial processes is not new, although it has assumed renewed emphasis in recent years. Centuries ago, people in Asia and Africa learned to make wine, beer, vinegar, and saki with bacteria and yeast, without knowing the scientific basis of such productions. Bacteria have many properties that are industrially useful. The diversity of the bacterial kingdom is reflected by the diverse applications of bacteria as a cheap labor force. *Thiobacillus ferrooxidans* can concentrate gold trapped in rock minerals. Biomining may be the way of mining in the future, and researchers are now trying to modify the bacteria to collect the ores of interest.

Certain bacteria are used to clean our waste, be it pollution, compost heaps, or sewage, bacteria can get rid of things. That industrial waste can be cleaned with bacteria has been known for over 30 years. Bacteria have a taste for mining wastewaters no matter how toxic the contaminants are for animals and humans. Specialized bacteria metabolize these toxic chemicals into nontoxic, or less toxic compounds. Some bacteria can clean up mining pollutants.

7.2. HISTORICAL DEVELOPMENT

The study of microorganisms, or microbiology, began when the first microscopes were developed in 1665 by the English scientist, Robert Hooke, who viewed many

Microbes: Concepts and Applications, First Edition. Prakash S. Bisen, Mousumi Debnath, Godavarthi B. K. S. Prasad
© 2012 Wiley-Blackwell. Published 2012 by John Wiley & Sons, Inc.

Figure 7.1.   Robert Hooke and his compound microscope.

small objects and structures using a simple lens that magnified approximately 30 times (Fig. 7.1). His specimens included the eye of a fly, a bee stinger, and the shell of a protozoan. Hooke also examined thin slices of cork, which was the bark of a particular type of oak tree. He found that cork was made of tiny boxes that Hooke referred to as *cells*. He published his work in a book titled "Micrographie," which contained a miscellany of his thoughts on chemistry as well as a description of the microscope and its uses. Microorganisms are relevant to all our lives in a multitude of ways. Sometimes, the influence of microorganisms on human life is beneficial, whereas at other times, it is detrimental. For example, microorganisms are required for the production of bread, cheese, yogurt, alcohol, wine, beer, antibiotics (e.g., penicillin, streptomycin, chloramphenicol), vaccines, vitamins, enzymes, and many more important products. Many products of microbes contribute to public health as aids to nutrition, other products are used to interrupt the spread of disease, and still others hold promise for improving the quality of life in the years ahead. Microbes are also an important and essential component of an ecosystem. Molds and bacteria play key roles in the cycling of important nutrients in plant nutrition, particularly those of carbon, nitrogen, and sulfur. Bacteria referred to as *nitrogen fixers* live in the soil where they convert vast quantities of nitrogen in air into a form that plants can use. Microorganisms also play major roles in energy production. Natural gas (methane) is a product of bacterial activity, arising from the metabolism of methanogenic bacteria. Microorganisms are also being used to clean up pollution caused by human activities, a process called *bioremediation* (the introduction of microbes to restore stability to disturbed or polluted environments). Bacteria and fungi have been used to consume spilled oil, solvents, pesticides, and other environmentally toxic substances.

7.2.1. Era of Pasteur and Fermentation

Fermentation is the chemical transformation of organic substances into simpler compounds by the action of enzymes, which are complex organic catalysts produced by microorganisms such as molds, yeasts, or bacteria. Enzymes act by hydrolysis, the process of breaking down or predigesting complex organic molecules to form smaller (and in the case of foods, more easily digestible) compounds and nutrients. For example, the enzyme protease (all enzyme names have the suffix "ase") breaks down huge protein molecules first into polypeptides and peptides, then into numerous amino acids, which are readily assimilated by the body. The enzyme amylase works on carbohydrates, reducing starch and complex sugars to simple sugars. The enzyme lipase hydrolyzes complex fat molecules into simpler free fatty acids. These are but three of the more important enzymes. There are thousands more, both inside and outside of our bodies. In some fermentation, important by-products such as alcohol or various gases are also produced. The word "fermentation" is derived from Latin, meaning "to boil" since the bubbling and foaming of early fermenting beverages seemed closely akin to boiling.

Fermented foods often have numerous advantages over the raw materials from which they are made. As applied to soy foods, fermentation not only makes the end product more digestible but also creates improved (in many cases meatlike) flavor and texture, improves appearance and aroma, synthesizes vitamins (including vitamin B_{12}, which is difficult to get in vegetarian diets), destroys or masks undesirable or beany flavors, reduces or eliminates carbohydrates believed to cause flatulence, decreases the required cooking time, increases storage life, transforms what might otherwise be agricultural wastes (such as okara) into tasty and nutritious human foods (such as okara tempeh), and replenishes intestinal microflora (as with miso or acidophilus soy milk).

Fermentation is usually activated by molds, yeasts, or bacteria, working singularly or together. The great majority of these microorganisms come from a relatively small number of genera, roughly eight genera of molds, five of yeasts, and six of bacteria. An even smaller number are used to make fermented soy foods: the molds used are *Aspergillus, Rhizopus, Mucor*, and *Neurospora* species; the yeasts used are *Saccharomyces* species; and the bacteria used are *Bacillus* and *Pediococcus* species plus any or all of the species used to make fermented milk products. Molds and yeasts belong to the fungus kingdom, the study of which is called *mycology*. Fungi are as distinct from true plants as they are from animals. While microorganisms are the most intimate friends of the food industry, they are also its ceaseless adversaries. They have long been used to make foods and beverages, yet they can also cause them to spoil. However, when used wisely and creatively, microorganisms are an unexploitable working class, whose very nature is to labor tirelessly day and night, never striking or complaining, ceaselessly providing human beings with new foods. Like human beings, but unlike plants, microorganisms cannot make carbohydrates from carbon dioxide, water, and sunlight. They need a substrate to feed and grow on. The fermented foods they make are created incidentally as they live and grow.

Human beings are known to have made fermented foods since neolithic times. The earliest types were beer, wine, and leavened bread (made primarily by yeasts) and cheeses (made by bacteria and molds). These were soon followed by East Asian fermented foods, namely, yogurt and other fermented milk products, pickles, sauerkraut, vinegar (soured wine), butter, and a host of traditional alcoholic beverages. More recently, molds have been used in industrial fermentation to make vitamins B_2 (riboflavin) and B_{12}, textured protein products (from *Fusarium* and *Rhizopus* in Europe), antibiotics (such as penicillin), citric acid, and gluconic acid. Bacteria are now used to make the amino acids lysine and glutamic acid. Single cell protein (SCP) foods such as nutritional yeast and microalgae (*Spirulina, Chlorella*) are also made in modern industrial fermentations.

For early societies, the transformation of basic food materials into fermented foods was a mystery and a miracle, for they had no idea what caused the sudden, dramatic transformation. Some societies attributed this to divine intervention; the Egyptians praised Osiris for the brewing of beer and the Greeks established Bacchus as the god of wine. Likewise, at many early Japanese miso and shoyu breweries, a small shrine occupied a central place. In ancient times, fermentation included smoking, drying, and freezing as basic and widely practiced food preservation techniques. Wang and Hesseltine (1979) note "probably the first fermentation were discovered accidentally when salt was incorporated with the food material, and the salt selected certain harmless microorganisms that fermented the product to give a nutritious and acceptable food." The process was taken a step further by the early Chinese who first inoculated the basic foods with molds, which created enzymes; in salt-fermented soy foods such as miso, soy sauce, soy nuggets, and fermented tofu, these aided salt-tolerant yeasts and bacteria.

The early 1800s saw a great increase of interest in microbiology in Europe. The scientific period began with great advances in botany, increased interest in microscopy, and willingness to investigate individual organisms. The two major problems challenged the researchers in the new field of microbiology were the basic nature of the fermentation process and the basic nature of enzymes. The scientific breakthroughs that led to the unraveling of the mysteries of fermentation starting in the 1830s were made primarily by French and German chemists.

Although showing that fermentation was generally the result of the action of live microorganisms was an epic breakthrough, it neither explained the basic nature of the fermentation process nor proved that it was caused by the microorganisms that were apparently always present. As early as the late 1700s, it had been recognized that there was another type of chemical change that resembled yeast fermentation in some respects. This was the sort of changes that occur, for example, in the digestion of food. In 1752, Reamur, in studying the digestive processes of a falcon, showed that its digestive juices were able to dissolve meat. In 1785, William Irvine discovered that aqueous extracts of sprouted barley caused liquefaction of starch. The first clear recognition of what were later called *enzymes* came in 1833 when two French chemists, A. Payen and J. F. Persoz, made a more detailed investigation of the process of solubilizing starch with a malt extract to form a sugar that they called *maltose*. They called the agent responsible for this transformation *diastase*. They also showed that it was destroyed or inactivated by boiling; that without undergoing a permanent change itself, a small amount of diastase could convert a large amount of starch to sugar; and that it could be concentrated and purified by precipitation with alcohol. In 1835, the German naturalist Swann, mentioned above for his early work with fermentation, isolated a substance from gastric juice that could bring about the dissolution of meat but was not an acid. He called it *pepsin* from a Greek word meaning "digestion." It soon became fashionable to call organic catalysts such as diastase and pepsin "ferments" because digestion and fermentation, both allied with life, seemed to be somewhat similar processes. Under the influence of the vitalists, ferments were grouped into two types: those involved with life processes were called *organized ferments* and those that were not (such as pepsin) were merely *unorganized ferments*. A relation between the two types of ferments was suspected by many, and in 1858, M. Traube put forward the theory that all fermentations were due to ferments, definite chemical substances he regarded as related to the proteins and produced in the cells by the organism. In 1876, to reduce confusion that existed concerning the two types of ferments, the German physiologist Wilhelm Kuehne suggested that an unorganized

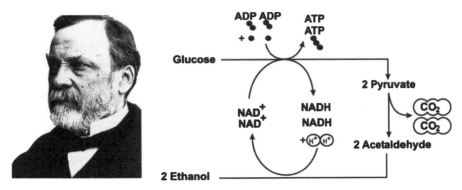

Figure 7.2. Louis Pasteur and fermentation reaction.

ferment, acting in the absence of life, be called an *enzyme*, after the Greek words meaning "in yeast"; in 1881, this term was anglicized to "enzyme" by William Roberts, and it had begun to catch on by the 1890s.

Many scientists, including Pasteur, had attempted unsuccessfully to extract the fermentation enzyme from yeast (Fig. 7.2). Success came finally in 1897 when the German chemist Eduard Buechner ground up yeasts, extracted a juice from them, then found to his amazement that this "dead" liquid would ferment a sugar solution, forming carbon dioxide and alcohol just like living yeasts. Clearly, the so-called unorganized ferments behaved just like the organized ones. From that time on, the term *enzyme* came to be applied to all ferments. The term *ferment* dropped out of the scientific vocabulary altogether and the vitalist position collapsed, never to recover. Thereafter, it was agreed that only one set of laws applied to all things, both animate and inanimate, and that there was no special vital force that characterized living things and acted under different laws. And it was finally understood that fermentation is caused by enzymes produced by microorganisms. In 1907, Buechner won the Nobel Prize in chemistry for his work, which opened a new era in enzyme and fermentation studies.

The fields of microbiology, biochemistry, fermentation technology, mycology, and bacteriology all shared a deep interest in the nature and working of enzymes. Yet still by the early 1900s, no one knew exactly what enzymes were or how they acted. As the agricultural microbiologist, Conn asked in 1901, "How can they produce chemical actions without being acted upon or entering into the reactions? Are enzymes fully lifeless or semi-living? We still do not know the fundamental mystery of fermentation." Gradually, an understanding of enzymes and catalysts developed. In 1905, Harden and Young discovered coenzymes, agents necessary for the action of enzymes. In 1926, the American biochemist J. B. Sumner first purified and crystallized an enzyme (urease) and showed that it was a protein, more precisely a protein catalyst. Eventually enzymes came to be seen as the key catalysts in all the life processes, each highly specialized in its catalytic action and generally responsible for only one small step in complex, multistep biochemical reactions. Enzymes are still produced only by living organisms, both animals and plants.

Advances in microbiology and fermentation technology have continued steadily up until the present. For example, in the late 1930s, it was discovered that microorganisms could be mutated with physical and chemical treatments to be higher yielding, faster growing, tolerant of less oxygen, and able to use a more concentrated medium. Strain selection and hybridization developed as well, affecting most modern food fermentations.

Pasteur solved the mysteries of rabies, anthrax, chicken cholera, and silkworm diseases and contributed to the development of the first vaccines. Pasteur was responsible for some of the most important theoretical concepts and practical applications of modern science. Although not a physician, Pasteur was undoubtedly the most important medical scientist working in the nineteenth century. He gave a new meaning to medicine. He was one of the forerunners in the study of microorganisms. He not only explained the causes for contagious diseases but also recommended ways of avoiding them. Pasteur was a founder of the germ theory. He laid the foundations for three distinct sciences: immunology, microbiology, and stereochemistry. It was Pasteur who brought to an end the debate on spontaneous generations, which had continued for centuries. He clearly demonstrated that spontaneous generation was not possible, and by doing so, Pasteur set the stage for modern biology and biochemistry. Pasteur described the scientific basis for fermentation—the process of production of wine, beer, and vinegar (Fig. 7.2). He clearly demonstrated that the nature of fermentation was organic (a product of a certain type of living organism) and not inorganic, as proposed and defended by Justus von Liebig. Pasteur developed a vaccine against anthrax, a particularly deadly, highly communicable disease of domestic animals.

7.2.2. Discovery of Antibiotics

Infections are very common and responsible for a large number diseases adversely affecting human health. Most of the infectious diseases are caused by bacteria. Infections caused by bacteria can be prevented, managed, and treated through antibacterial group of compounds known as *antibiotics*. Antibiotics can be loosely defined as the variety of substances derived from bacterial sources (microorganisms) that control the growth of or kill other bacteria. However, synthetic antibiotics, usually chemically related to natural antibiotics, have since been produced, which accomplish comparable task. A common scheme of classifications for antibiotics is shown in Figure 7.3:

Antibiotics can also be classified based on their chemical structure. A similar level of effectiveness, toxicity, and side effects is rendered by the antibiotics of the same structural group. Broad-spectrum antibiotics are effective against a broad range of microorganisms

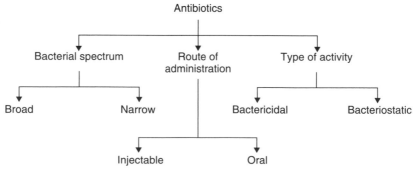

Figure 7.3. Classification of antibiotics.

Figure 7.4. Sir Alexander Fleming and *Penicillium*.

in comparison to narrow-spectrum antibiotics. Bactericidal antibiotics kill the bacteria, whereas bacteriostatic antibiotics halt the growth of bacteria.

Sir Alexander Fleming, a Scottish biologist, defined new horizons for modern antibiotics with his discoveries of the enzyme "lysozyme" and the antibiotic substance "penicillin" (Fig. 7.4). The discovery of penicillin from the fungus *P. notatum* perfected the treatment of bacterial infections such as syphilis, gangrene, and tuberculosis. He also contributed immensely toward medical sciences with his writings on the subjects of bacteriology, immunology, and chemotherapy.

Various types of antibiotics work in either of the following two ways:

- A bactericidal antibiotic kills the bacteria generally by interfering with the formation of either the bacterium's cell wall or its cell contents. Penicillin, daptomycin, fluoroquinolones, metronidazole, nitrofurantoin, and cotrimoxazole are some examples of bactericidal antibiotics.
- A bacteriostatic antibiotic stops bacteria from multiplying by interfering with bacterial protein production, DNA replication, or other aspects of bacterial cellular metabolism. Some bacteriostatic antibiotics are tetracyclines, sulfonamides, spectinomycin, trimethoprim, chloramphenicol, macrolides, and lincosamides.

A brief time line of the origin of antibiotics is shown in Table 7.1.

7.2.3. Growth of Industrial Fermentation

Industrial fermentation is the intentional use of fermentation by microorganisms such as bacteria and fungi to make products useful to humans. Fermented products have applications as food and in the general industry. Ancient fermented food processes, such

TABLE 7.1. History of Origin of Antibiotics

Year	Origin	Description
1640	England	John Parkington recommended using mold for treatment in his book on pharmacology
1870	England	Sir John Scott Burdon-Sanderson observed that culture fluid covered with mold did not produce bacteria
1871	England	Joseph Lister experimented with the antibacterial action on human tissue on what he called *Penicillium glaucium*
1875	England	John Tyndall explained antibacterial action of the *Penicillium* fungus to the Royal Society
1877	France	Louis Pasteur postulated that bacteria could kill other bacteria (anthrax bacilli)
1897	France	Ernest Duchesne healed infected guinea pigs from typhoid using mold (*P. glaucium*)
1928	England	Sir Alexander Fleming discovered the enzyme lysozyme and the antibiotic substance penicillin from the fungus *Penicillium notatum*
1932	Germany	Gerhard Domagk discovered sulfonamidochrysoidine (Prontosil)

During the 1940s and 1950s, streptomycin, chloramphenicol, and tetracycline were discovered and Selman Waksman (in 1942) used the term *antibiotics* to describe them

as making bread, wine, cheese, curd, idli, dosa, can be dated back to more than 6000 years. They were developed long before man had any knowledge of the existence of the microorganisms involved.

There are four major groups of commercially important fermentation.

- *Microbial Cells or Biomass*. SCP, baker's yeast, lactobacillus, *Escherichia coli*, etc.
- *Microbial Enzymes*. Catalase, amylase, protease, pectinase, glucoseisomerase, cellulase, hemicellulase, lipase, lactase, streptokinase, etc.
- *Recombinant Products*. Insulin, HBV, interferon, GCSF, streptokinase.
- *Biotransformations*: Phenylacetylcarbinol, steroid biotransformation, etc.

When a particular organism is introduced into a selected growth medium, the medium is inoculated with the particular organism. Growth of the inoculum does not occur immediately, but takes a little while. This is the period of adaptation, called the *lag phase*. Following the lag phase, the rate of growth of the organism steadily increases for a certain period called the *log* or *exponential phase* (Fig. 7.5).

After a certain time of exponential phase, the rate of growth slows down due to the continuously falling concentrations of nutrients and/or a continuously increasing (accumulating) concentration of toxic substances. This phase, where the increase of the rate of growth is checked, is called the *deceleration phase*. After the deceleration phase, growth ceases and the culture enters a stationary phase or a steady state. The biomass remains constant, except when certain accumulated chemicals in the culture lyse the cells (chemolysis). Unless other microorganisms contaminate the culture, the chemical constitution remains unchanged. Mutation of the organism in the culture can also be a source of contamination, called *internal contamination*.

The acetone-butanol (AB) fermentation process (Fig. 7.6) was widely used in the earlier part of this century for the industrial production of solvents (Jones and Woods,

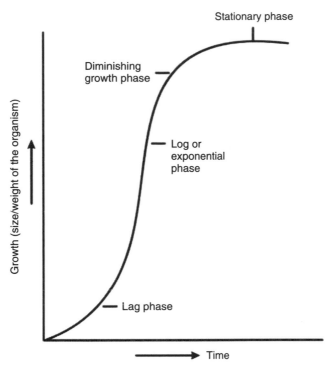

Figure 7.5. Growth of inoculum.

1986). The fermentation process, developed during the First World War for ammunition production, utilized starch as the fermentation substrate. The commercial AB fermentation process originally established by the Commercial Solvents Corporation (CSC) in the United States continued to use maize as the fermentation substrate until the early 1930s when the widespread availability of cheap molasses from the sugar industry provided a strong incentive to switch substrates. Although considerable effort was invested in attempting to utilize the existing starch fermenting strains for use on molasses, this was never successfully achieved (McCutchan and Hickey, 1954). The poor performance obtained with the maize strains prompted a quest for new isolates that would provide economically viable solvent yields.

In 1935, the patent protecting the CSC process monopoly lapsed, and a number of new companies entered the market worldwide. During the 1930s and 1940s, a substantial number of new saccharolytic clostridial strains were isolated and patented by different companies (Beesch, 1952; McCutchan and Hickey, 1954; Ross, 1961; Walton and Martin, 1979; Jones and Keis, 1995). Among the first strains to be patented were a number that, under optimal conditions, were able to utilize between 4% and 6% fermentable sugars, producing solvent concentrations of 14–18 g/l with solvent yields from 25% to 30%. Early studies revealed that molasses was nutritionally deficient as a fermentation substrate, and supplementation with an additional source of nitrogen and a buffering agent became common practice (Beesch, 1952; McCutchan and Hickey, 1954). It is apparent from both the scientific and patent literature that many of the newly isolated strains had serious limitations. Although some were used commercially, in a number of instances, sucrose utilization was poor and it was necessary to invert the sugars. With some strains,

Figure 7.6. Acetone-butanol fermentation process.

long fermentation times (up to 72 h) resulted in lower productivity, and sugar concentrations were limited to around 6% set sugars (McCutchan and Hickey, 1954). The quest for more efficient strains that could use higher concentrations of fermentable sugars to produce higher solvent yields with shorter fermentation times continued.

7.2.4. Production of Strains

Metabolic engineering can be defined as the purposeful modification of cellular activities with the aim of strain improvement (Bailey, 1991). Development of microbial strains for the production of metabolites is based primarily on the application of rDNA technology to alter the properties of the metabolic network by modifying the level of activity or the properties of specific enzymes (Fig. 7.7).

These principles have been applied to the generation of a large number of *E. coli* strains designed for the production of commercially important compounds (Cameron and Tong, 1993). Cultures with these engineered strains usually use media containing glucose. This sugar is nowadays the most utilized raw material in industrial fermentations with *E. coli*, mostly because it is relatively inexpensive and is the preferred carbon and energy source for this bacterium.

As a component of culture media, glucose provides carbon atoms for biomass and product generation. The cell's capacity to uptake and metabolize this carbohydrate has a profound impact on its growth rate and productivity. Thus, it is expected that modifications to glucose transport systems should have an important impact on the cell's physiology, and this, in turn, can either improve or become detrimental in an industrial production context (Fig. 7.8).

At present, almost all the pectinolytic enzymes used for industrial applications are produced by the fungi, namely, *Aspergillus* sp., *Aspergillus japonicus, Rhizopus stolonifer, Alternaria mali, Fusarium oxysporum, Neurospora crassa, Penicillium italicum* ACIM F-152. There are a few reports of pectinase production by bacterial strains.

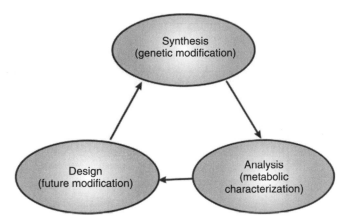

Figure 7.7. Principles of metabolic engineering.

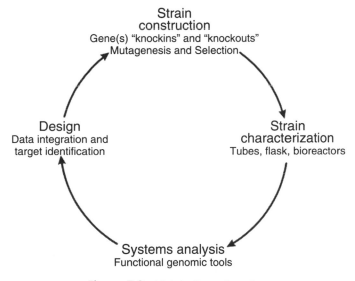

Figure 7.8. Metabolic engineering.

Some of the bacterial species producing pectinases are *Agrobacterium tumefaciens, Bacteroides thetaiotaomicron, Ralstonia solanacearum*, and *Bacillus* sp. (Jayani et al., 2005).

Microbial strains can be manipulated to improve their technological properties. The classical examples are improved product yield and/or improved growth characteristics. Furthermore, probiotic and other properties can be improved using screening technology.

7.2.5. Strain Development

The major motivation for industrial strain development is economic, since the metabolite concentrations produced by wild strains are too low for economical processes. Depending on the system, it may be desirable to isolate strains that require shorter fermentation times,

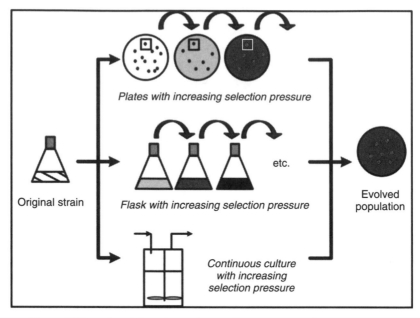

Figure 7.9. Industrial strain production by increasing selection pressure.

do not produce undesirable pigments, have reduced oxygen needs, exhibit decreased foaming during fermentation, or are able to metabolize inexpensive substrates. Wild strains frequently produce a mixture of chemically closely related substances. Mutants synthesizing one component as the main product are preferable since they make possible a simplified process for product recovery. Changes in the genotype of microorganisms can lead to the biosynthesis of new metabolites. Thus, mutants synthesizing modified antibiotics are normally selected. After an organism producing a valuable product is identified, it becomes necessary to increase the product yield from fermentation to minimize production costs (Fig. 7.9).

Product yields can be increased by developing a suitable medium for fermentation, refining the fermentation process, and improving the productivity of the strain.

Generally, major improvements arise from the last approach. Hence, all fermentation enterprises place a considerable emphasis on this activity. The techniques and approaches used to genetically modify strains to increase the production of the desired product are called *strain improvement* or *strain development*.

Strain improvement is based on the following three approaches:

1. Mutant selection
2. Recombination
3. Recombinant DNA (rDNA) technology.

7.3. CLINICAL DIAGNOSTICS IN A NEW ERA

Molecular diagnostics has become a growing part of the clinical laboratory. It includes all tests and methods to identify a disease and understand the predisposition for a disease,

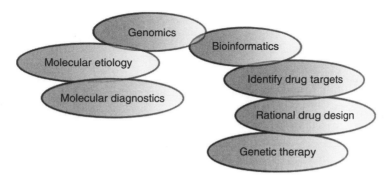

Figure 7.10. The clinical impact of genomics and bioinformatics on diagnosis and therapy.

analyzing the DNA or RNA of an organism. Rapid advances in molecular diagnostics enable basic research and results in practical diagnostic tests. The basic application is to determine changes in sequence or expression levels in crucial genes involved in disease. Genomics and bioinformatics play a major role in molecular diagnostics (Fig. 7.10). The use of molecular diagnostics, such as preimplantation diagnostics or predictive genetic testing, still has technical problems as well as novel, and to date unclear, social, ethical, and legal implications. The scope of molecular diagnostics in molecular medicine could be expanded well beyond current nucleic acid testing. It plays an important role in practice of medicine, public health, pharmaceutical industry, forensics, biological warfare, and drug discovery. The molecular diagnostic marketplace offers a growth opportunity, given the interest in utilizing molecular tools to precisely target therapeutics. Molecular studies allow the laboratory to become predictive. Now statements can be made about events that occur to a patient in the future. This new technology provides results that give an indication that the patient may be at risk for a disease long before the patient becomes symptomatic.

Molecular pathology is in a state of rapid evolution, featuring continuous technology development and new clinical opportunities for drug selection, predicting efficiency, toxicity, and monitoring disease outcome. Thus, major advances are being made in genetics, resulting in increased use of molecular technology in clinical laboratory.

7.3.1. Genes and Diseases

The human genome is composed of three billion base pairs of DNA, which are divided among the 46 chromosomes that are present in every human cell. We inherit 22 autosomal (nonsex) chromosomes and 1 sex chromosome (X or Y) from each parent. These chromosomes vary in length from 50 million to 250 million base pairs. These 46 chromosomes are found in the nucleus of every one of the body's trillions of cells (Fig. 7.11).

Each chromosome contains many genes, the basic physical and functional units of heredity. Each gene contains the instructions of how to make a protein or proteins. The exact protein that is made is determined by the order of the DNA bases within the gene. It is believed that there are between 30,000 and 40,000 genes within the human genome. This is far fewer than the number that was predicted originally and only about twice the amount of genes that are present in a fly. Less than 2% of the human genome contains genes (coding regions). The rest of the genome is believed to be involved in providing

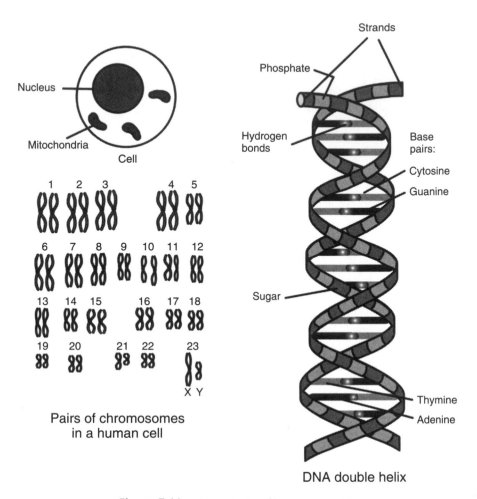

Figure 7.11. The anatomy of human genome.

chromosomal structural support and in regulating the amount of proteins produced in different cells of the body and at different times. Although each cell in the body contains a full complement of the 3 billion base pairs of DNA, the genes are used selectively. The genes that make proteins that are needed for basic functions (called *housekeeping genes*) are active in most cell types. Genes that code for proteins that have more specific functions are only activated in cell types where that protein is needed. It is by expressing different genes and hence producing different combinations of proteins that the different cell types can fulfill their various roles, for example, the genes expressed in a brain cell differ from those of a skin cell that differ from those of a liver cell. So at any one time, a normal cell will activate only those genes that code for the proteins it needs and it will actively suppress all the other genes.

Gene-based molecular diagnostics is changing the practice of medicine and will continue to do so for the foreseeable future. The primary molecular diagnostics approaches are based on two major database sciences (genomics and proteomics) and their relationship to disease and metabolic processes (functional genomics and functional proteomics).

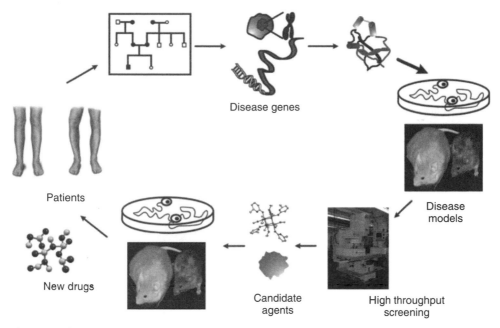

Figure 7.12. The schematic representation of the use of molecular diagnostics in the present era.

The genome or proteome that forms the basis for these tests is very often of human origins, but may also be from a pathogen. As a result, the breadth of the application in molecular diagnostics is virtually unlimited. With all this opportunity, however, progress has been slow in the development, introduction, and commercialization of molecular diagnostics by identifying new candidate genes and high throughput screening (Fig. 7.12).

Proteins are in fact what perform all of life's essential functions. Changes in the DNA sequence of our genome, both coding and noncoding, can have disastrous consequences. These changes can lead to the production of faulty malformed proteins that are incapable of performing their correct function or over- or underproduction of a protein resulting in the complete disruption of certain cellular processes. These disruptions to protein structure and regulation result in the expression of many of the common diseases that we observe today (cancers, heart disease, and diabetes). Many, if not most, diseases have their roots in our genes. Genes through the proteins they encode determine how efficiently we process foods, how effectively we detoxify poisons, and how vigorously we respond to infections. More than 4000 diseases are thought to stem from mutated genes inherited from one's mother and/or father. Common disorders such as heart disease and most cancers arise from a complex interplay between multiple genes and between genes and factors in the environment (Fig. 7.13).

A complex relationship exists between human genetics and various disease states. On the basis of their genetic contribution, human diseases can be classified as monogenic, chromosomal, and multifactorial. Monogenic diseases are caused by alterations in a single gene, and they segregate in families according to the traditional Mendelian principles of inheritance. Chromosomal diseases, as their name implies, are caused by alterations in chromosomes; for instance, within an individual's genome, some chromosomes may be missing, extra chromosome copies may be present, or certain portions of chromosomes

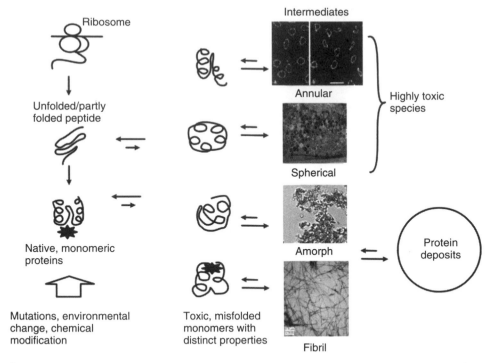

Figure 7.13. A disease/other protein may not correctly fold because it looses or is unable to attain its native, closely packed three-dimensional structure.

may be deleted or duplicated. In monogenic diseases, mutations in a single gene are both necessary and sufficient to produce the clinical phenotype and to cause the disease. The impact of the gene on genetic risk for the disease is the same in all families. In complex disorders with multiple causes, variations in a number of genes encoding different proteins result in a genetic predisposition to a clinical phenotype. Pedigrees reveal no Mendelian inheritance pattern, and gene mutations are often neither sufficient nor necessary to explain the disease phenotype. Environment and lifestyle are major contributors to the pathogenesis of complex diseases. In a given population, epidemiological studies expose the relative impact of individual genes on the disease phenotype. However, between families, the impact of these genes might be totally different. In one family, a rare gene C (family 3) might have a large impact on genetic predisposition to a disease. However, because of its rarity in the general population, the overall population effect of this gene would be small. Some genes that predispose individuals to a disease might have minuscule effects in some families (gene D, family 3; Peltonen and McKusick (2001); Fig. 7.14).

7.3.2. Biomedical Research in the Postgenomic Era

In a very short time, within a decade, we have advanced from having very little information about the genetic details of biology to possessing an immense amount of structural information about individual genes. Currently, the complete genome sequences of more than 60 species are available in databases, and the prediction that there will be a total

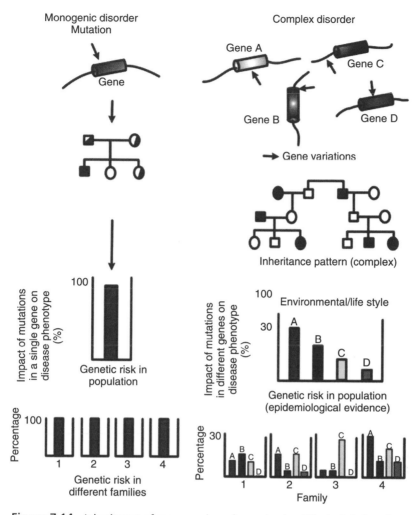

Figure 7.14. Inheritance of monogenic and complex (multifactorial) disorders.

of 100 sequenced genomes in databases within the next few months seems realistic. This dramatic increase in the amount of genomic information will have a tremendous impact on biomedical research and on the way that medicine is practiced. When all the human genes are truly known, scientists will have produced a Periodic Table of Life, containing the complete list and structure of all genes and providing us with a collection of high precision tools with which to study the details of human development and disease (Fig. 7.15). New technologies will facilitate analyses of individual variations in the whole genome and the expression profiles of all genes in all cell types and tissues. The way will thus be paved for systems biology and for deciphering the genetic repertoires of many organisms. The complete genome sequence of humans and of many other species provides a new starting point for understanding our basic genetic makeup and how variations in our genetic instructions result in disease.

The number of disease genes discovered so far is 1112 (Fig. 7.16). This number does not include at least 94 disease-related genes identified as translocation gene fusion

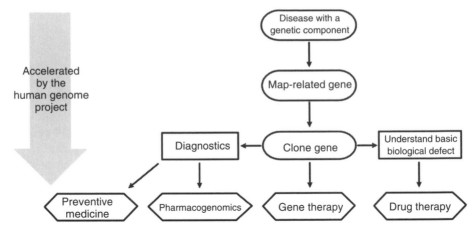

Figure 7.15. Mapping and cloning a gene can lead to strategies that reduce the risk of disease.

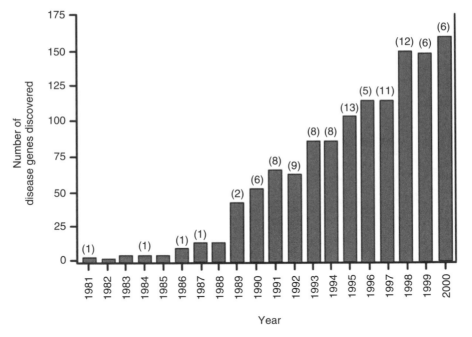

Figure 7.16. Pace of disease gene discovery.

partners in neoplastic disorders. Numbers in parentheses indicate disease-related genes that are polymorphisms.

As molecular biology developed and grew in influence through the 1980s, the field of molecular pharmacology also took center stage. With the invention of cloning of genes encoding multiple receptors, ion channels, regulatory proteins, drug-metabolizing enzymes (DMEs), and proteins involved in other aspects of drug disposition, the field of molecular pharmacology offered powerful tools for understanding mechanisms of drug action and disposition at the molecular level. However, with the drive and progress toward

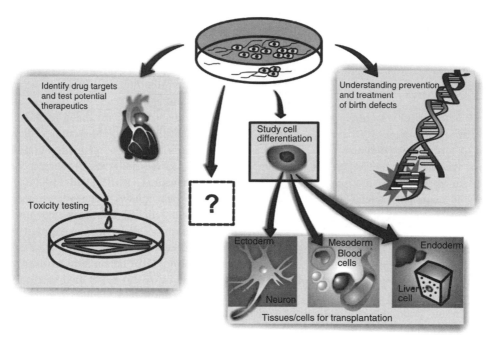

Figure 7.17. The promises of stem cell research.

understanding the basic molecular mechanisms of drug action, many of the most pressing questions of molecular pharmacology remained somewhat distinct from the broader questions of how these molecular phenomena functioned in a context of organ systems and whole animal physiology. The tools of molecular biology have now permeated every aspect of biomedical science. Molecular pharmacology can no longer be viewed as a niche discipline in the broader pharmacology arena. Molecular tools are now integral to virtually all major aspects of modern pharmacology and drive fundamental breakthroughs that span molecular to complex cellular, organ, and it *in vivo* systems (Fig. 7.17). Many of the most pressing challenges for today's molecular pharmacologists involve placing advances we have made at a molecular level in the context of complex systems. Probably all disease states except trauma are influenced by genetic factors. In the last decade of the twentieth century, all avenues of biomedical research were directed to the gene. The human genome contains all the information necessary from conception until death. The completion of the draft sequence of the human genome in the Human Genome Project has resulted in sequencing the entire human genome. This remarkable achievement will reveal the genetic instructions that specify the molecular components, the design, and the operating software of the human body (Fig. 7.18). This knowledge will transform medicine, giving us the means to see and to understand human anatomy, physiology, and pathophysiology in molecular detail.

This development will dramatically accelerate the development of new strategies for the diagnosis, presentation, and treatment of diseases. Especially for common complex diseases such as cancer, genetic differences contribute to the risk of contracting the disease, clinical course of disease, and responsiveness to different treatments. Endeavors in the Human Genome Project and technical advances in molecular biology are expected to have a revolutionary effect on oncology practice, including how anticancer drugs are

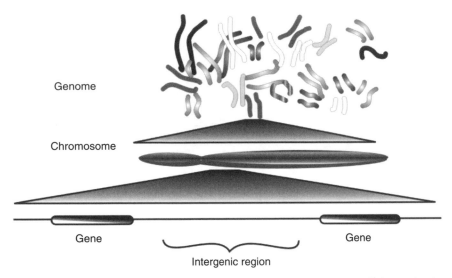

Figure 7.18. The Human Genome Project has resulted in sequencing of the entire human genome.

discovered and developed, how cancers are detected and classified, and finally, how patients are treated and monitored. Since the discovery of oncogenes, tumor suppressor genes, and, more recently, genes of apoptosis, cancer has become one of the most important diseases in the design of approaches based on genetics and genomic research (Sah et al., 2006; Khan et al., 2009, 2010a, 2010b). The explosion of information generated by large-scale genomic technologies has resulted in an exponential increase in the number of genes and proteins available for pharmaceutical and diagnostic research development. The increasing understanding of complex molecular pathways involved in cancer will shift clinical practice from empirical treatment to treatment based on molecular taxonomy of disease. The two important emerging fields in cancer in the postgenomic era include the development of new diagnostic tests based on microarray technology and the development of new drugs in the emerging field of pharmacogenomics (Fig. 7.19).

7.3.3. Molecular Diagnostics Technology and Health Care Industries

Molecular diagnostics is being combined with therapeutics and forms an important component of integrated health care. Molecular diagnostic technologies are also involved in the development of personalized medicine based on pharmacogenetics and pharmacogenomics. Currently, there has been a considerable interest in developing rapid diagnostic methods for point of care and biowarfare agents such as anthrax.

The number of companies involved in molecular diagnostics has increased remarkably during the past few years. More than 500 companies have been identified to be involved in developing molecular diagnostics, and 250 of these are profiled in the report along with tabulation of 573 collaborations. Despite the strict regulation, most of the development in molecular diagnostics has taken place in the United States, which has the largest number of companies. Some of the companies are enlisted in Table 7.2.

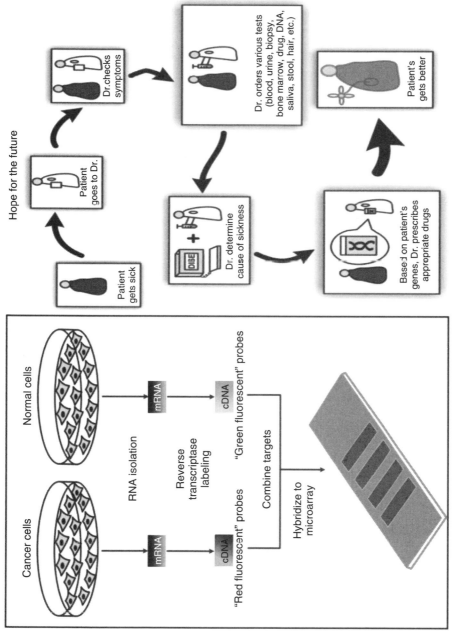

Figure 7.19. Microarray and pharmacogenomics: a solution to cancer diagnostics.

TABLE 7.2. Companies Dealing with Molecular Diagnostic Instruments and Reagents

Vendors	Description
Abbott Diagnostics	Abbott is a global provider of *in vitro* diagnostic instruments and tests, which include solutions for molecular diagnostics.
Beckman Coulter	Beckman Coulter is a Vidiera NsP nucleic acid sample preparation instrument and a Vidiera NSD nucleic sample detection instrument for research in genetic analysis and clinical diagnostics.
Becton Dickinson	In its range of molecular diagnostic products, BD offers various comprehensive solutions including BD Viper, BD Probe Tec, and BD Affirm VPiii.
bioMerieux	bioMerieux offers NucliSens range of solutions for various molecular diagnostic applications.
Bio-Rad	Bio-Rad offers solutions for nucleic acid sample preparation, electrophoresis, PCR, having applications in molecular diagnostics
Biotron	Biotron's unique solutions for molecular diagnostics include POC molecular diagnostic system, comprehensive assay for oncology applications, assays for hospital-acquired infections, assays for molecular typing, sequence-based high resolution typing, multiplex assays, and automated nucleic acid extraction.
CPC Diagnostics	CPC offers a wide range of products for molecular diagnostics, including FISH assays, PCR of infectious diseases, specialized media for prenatal, postnatal, and tumor cytogenetics, especially microbial culture media and automated systems. CPC has recently carved its molecular diagnostics and research division in Hyderabad. It represents the following companies: Cytocell, UK; Argene, France; Cytogen, Germany; Biomed, USA; and Giles Scientific, USA.
DSS Imagetech	DSS recently added a division for molecular diagnostic applications, offering a wide array of advanced products.
Eppendorf	Eppendorf offers Mastercycler thermal cyclers for various molecular diagnostic applications.
HiMedia Laboratories	HiMedia specializes in manufacturing high quality raw materials and culture media for the diagnostic segment. It offers the widest range of molecular diagnostics, cell biology reagents, and kits.
Professional Biotech	Professional Biotech offers Rotor-Gene real-time PCR amplification system. It offers a comprehensive range of indigenously manufactured reagents for the molecular diagnostic and cell biology segments.
Roche Diagnostics	Roche's products for molecular diagnostics include *in vitro* diagnostic kits and kits for screening blood. Roche also offers automated platforms that enable full automation of the PCR process.
Suyog Diagnostics	Suyog Diagnostics, a part of the Noble group, offers the model FASTPREP-6002-500 DNA/RNA/protein homogenizer, DNA lysing matrix, and a range of molecular diagnostic and cell biology chemicals.
Siemens Healthcare Diagnostics	Siemens offers Versant 440 Molecular System and OpenGene DNA Sequencing System for molecular diagnostic applications.
Trivitron	Trivitron offers complete solutions for molecular diagnostics with nucleic acid extraction system with kits, real-time PCR system and kits, capillary electrophoresis systems for post-PCR analysis, and so on.
Other players	Arun & Co., BioScreen, Genetix Biotech, and Imperial Life Sciences.

Abbreviation: POC, point of care.

Through collaborations, mergers, or acquisitions, pharmaceutical firms seem to be focusing on deals right now that involve biotechnology. That's not surprising considering that sales in the biotech segment grew 16.8% to approximately $70 billion in the 12 months through June of 2007, compared with only 6–7% growth in the pharma market as a whole, according to the IMS Health Company.

In fact, pharmaceutical companies are aggressively competing for biotechnology deals in an attempt to snare potential drug candidates that may enrich their pipelines. In the process, they have been willing to pay premium prices for biotech firms even earlier in the development cycle.

The markets for molecular diagnostics technologies are difficult to estimate. Molecular diagnostics markets overlap with markets for nonmolecular diagnostic technologies in the *in vitro* diagnostic market and are less well defined than those for pharmaceuticals.

Undoubtedly, the United States with 45% of sales contributes to dominate the pharmaceutical market. But countries such as Japan, France, Germany, United Kingdom, Italy, Spain, Canada, China, and Brazil also show sales in billions (Table 7.3).

Molecular diagnostic markets are analyzed for 2007 according to technologies, applications, and geographical regions. Among the top therapies, oncologics exhibits highest stellar growth and tops the list. The other therapy classes as suggested by the World Health Organization are shown in Table 7.4.

A major portion of the molecular diagnostic market can be attributed to advances in genomics and proteomics. Biochip and nanobiotechnology are expected to make a significant contribution to the growth of molecular diagnostics (Fig. 7.20). Advancing technologies are pushing the diagnostics industry to the fore, and the molecular diagnostics industry, in particular, is emerging as a powerful health care player with tremendous potential. Advances in PCR (polymerase chain reaction), multiplexing, sequencing, and other technologies are propelling both new and old companies forward with novel capabilities (Fig. 7.21). The molecular diagnostic industry is characterized by a very diverse, constantly changing technology base that continuously produces new opportunities and applications. Similarly, a growing understanding of the molecular basis of cancer and

TABLE 7.3. Global Sales Data of the Pharmaceutical Market

Country	Sales Till June 2007, billion dollars	Share of Global Sales, %	12-mo Change in Sales, %
United States	283.3	44.9	7.7
Japan	56.4	9.0	0.6
France	36.2	5.7	4.5
Germany	33.7	5.4	0.0
United Kingdom	22.2	3.5	5.2
Italy	21.3	3.4	−0.2
Spain	17.8	2.8	9.3
Canada	16.0	2.5	6.8
China	12.0	1.9	15.9
Brazil	9.1	1.4	10.9
Top 10 markets	508.1	80.5	5.9

Note: Sales are in US dollars for the 12 mo ending June 2007.
Source: IMS Health, MIDAS.

TABLE 7.4. Molecular Diagnostic Market Showing the Various Therapies and Global Growth in Sales

Therapies	Sales Till June 2007, billion dollars	Share of Global Sales, %	12-mo Change in Sales, %
Oncologics	37.5	6.0	17.0
Lipid regulators	34.2	5.4	−1.7
Respiratory agents	26.5	4.2	12.2
Acid pump inhibitors	24.6	3.9	1.8
Antidiabetics	22.6	3.6	12.6
Antidepressants	20.2	3.2	−1.5
Antipsychotics	19.3	3.1	10.9
Angiotensin II antagonists	17.8	2.8	13.8
Antiepileptics	14.0	2.2	13.1
Erythropoietins	13.9	2.2	5.7
Top 10 therapy classes	230.6	36.6	7.9

Note: All therapy classes are World Health Organization code groups. Sales are in US dollars for the 12 mo ending June 2007.
Source: IMS Health, MIDAS.

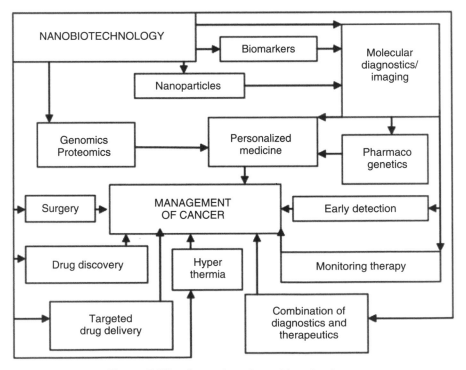

Figure 7.20. The market of nanobiotechnology.

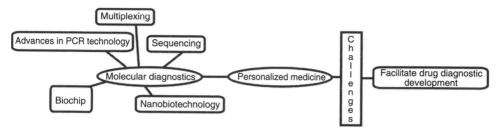

Figure 7.21. Challenges of molecular diagnostics.

other chronic diseases is opening up new realms of medicine to the possibilities of molecular diagnostic testing. While the infectious disease sector, particularly viral load testing, continues to occupy the largest sector of the molecular diagnostics market, other areas, particularly oncology, will see greater rates of growth in the near future. An evaluation of the emergence and growth of personalized, or pharmacogenomic, medicine, which is providing tremendous revenue opportunities for molecular diagnostics. The challenge for molecular diagnostics companies will be to generate profits while forming cooperative relationships with pharma, regulatory agencies, consumers, and other stakeholders in ways that facilitate drug-diagnostic codevelopment and end in a win-win situation for all parties.

7.3.3.1. Molecular Techniques

Technological revolutions in microbiology and molecular biology have significantly expanded and improved the capabilities of diagnostic microbiology (Tang, 2003). Molecular methods, replacing biological amplification by enzymatic amplification of specific nucleic acid sequences, have dramatically changed the way we detect and characterize infectious agents. These methods not only have enhanced diagnostic validity and decreased the turnaround time for patient results but also have increased clinical relevance of the information provided by the laboratory. As one technological milestone in biotechnology, PCR has simplified and accelerated the *in vitro* process of nucleic acid amplification and significantly broadened the microbiologists' diagnostic arsenal. Commercial kits and "home-brewed" procedures have been developed and applied to the detection of microbial pathology, the identification of clinical isolate, and strain subtyping. They are also useful for physicians who take care of patients with atypical pneumonia. The detection and identification of amplification products, or amplicons, has become a routine procedure in the molecular diagnostic laboratory, which not only "visualize" the amplified DNA molecules but also enhance test sensitivity and specificity. Such visualization techniques included classical agarose gel electrophoresis with or without a Southern blot hybridization, colorimetric microtiter-plate system, direct sequencing, matrix hybridization, and recently developed "real-time" system in which amplification and identification happen simultaneously.

The rapid, *in vitro* enzymatic amplification characteristic of PCR indicates its primary application in the detection of organisms causing atypical pneumonia, which are usually unculturable, slow growing, or fastidious. Microbial nucleic acids extracted from a respiratory specimen may be analyzed for the presence of various organism-specific nucleic acid sequences regardless of the physiologic requirements or viability of the organism.

For example, a sequence homology between the animal coronavirus and the newly identified SARS virus formed the basis to rapidly detect and identify the latter pathogen. A colorimetric microtiter-plate RT-PCR system was successfully used to detect and subtype respiratory syncytial virus (RSV) in nasal wash specimens. It is an advantage for molecular techniques to have one universal multiplex procedure to detect human adenoviruses, which contain at least 51 different serotypes. A real-time RT-PCR test kit is available commercially for the rapid diagnosis of SARS virus-caused atypical pneumonia.

7.3.3.2. Laboratory Monitoring of Infections

Many bacteria can exist in both a pathogenic and a nonpathogenic state. Merely finding the organism, especially in the normal-flora-colonized upper respiratory tract environment, does not imply that it is causing the disease. In this scenario, molecular methods can be used to detect virulence determinants. Not all virulence determinants are chromosomally mediated, but molecular methods can be used to detect and identify these virulence factors carried by plasmids. The RT-PCR procedure was successfully applied to the differentiation of, for example, viable from nonviable *Legionella pneumophila*, which is especially useful for chemotherapy efficacy monitoring. A PCR-based test targeting *Pneumocystis carinii* in sputum samples from AIDS patients has been used to monitor treatment with pentamidine. There has been a growing demand for the quantitation of nucleic acid targets, which has been used to monitor therapeutic response and provide prognostic information. Quantitative detection of respiratory *Chlamydia pneumoniae* infection was performed by a real-time PCR for the purpose of monitoring atypical pneumonia therapy. Similarly, real-time RT-PCR assays were used to quantitate RSV and SARS virus RNA in nasal aspirate specimens.

RAPID IDENTIFICATION OF EMERGING PATHOGENS. Molecular methods have won superfluous credits regarding the discovery and characterization of novel pathogens causing atypical pneumonia. Within the past decade, PCR followed by a sequencing method successfully identified and characterized hantavirus, human metapneumovirus, and SARS viruses. In addition to the detection of bacterial pathogens directly from respiratory specimen, nucleotide sequence analysis of the small-subunit (16S) bacterial rRNA gene allows characterization of previously unrecognized bacterial species causing atypical pneumonia. Since viruses lack ribosomal genes, several subtractive technologies allied to amplification methods have been used to identify novel viruses. Probably due to the "nonsterile" characteristic of respiratory tract specimens, these techniques have not been widely used to hunt for novel viruses causing atypical pneumonia.

GENOTYPIC DETERMINATION OF ANTIMICROBIAL RESISTANCE. Antimicrobial susceptibility testing is one of the most important tasks in a clinical microbiology laboratory, which provides an *in vitro* estimate of the probability that an infection will respond to chemotherapy it *in vivo*. Molecular techniques are starting to play a role in the rapid detection of resistance. In some cases, such techniques offer the opportunity to reduce the time required for the institution of definitive therapy, thus reducing the use of inappropriate antibiotics. Rapid detection may also allow early recognition of carriers infected by resistant organisms and the appropriate implementation of isolation, epidemiological investigation, and integrated infection control practices. An RT-PCR-based method has been reported for antimicrobial susceptibility testing of *Chlamydia trachomatis*.

The detection of a tetM gene by molecular methods has been used to determine tetracycline resistance in *Mycoplasma* species. Molecular approaches have been used to detect influenza gene mutations related to reduced susceptibility to neuraminidase inhibitors and resistance to amantadine. The emergence of erythromycin-resistant *Bordetella pertussis* has been traced to one mutation in the 23S rRNA gene, which can be detected by a PCR-based assay.

EPIDEMIOLOGIC INVESTIGATION ENHANCEMENT. Microorganism typing using molecular methods has important implications for the epidemiologic investigation of atypical pneumonia. A bacterial restriction endonuclease analysis of bacterial chromosomal DNA was used to incriminate a water system as the source of a 32-case Legionnaires' disease outbreak. Gene sequence analysis is the ultimate discriminatory tool, and PCR followed by direct sequencing analysis was used to determine the possible epidemiologic relatedness between the SARS viruses recovered from humans and other wild animals. A genetic analysis was used to type *Bacillus anthracis* isolates and traces the possible resource that resulted in the 2001 bioterrorism-associated anthrax outbreak in the United States. Realtime PCR and microarray assays have been applied for the typing and subtyping of influenza viruses directly in respiratory samples.

7.3.4. Therapeutic Applications of rDNA Pharmaceuticals

rDNA technology has made a revolutionary impact in the area of human health care by enabling mass production of safe, pure, and effective rDNA expression products (Bhopale and Nanda, 2005). Currently, several categories of rDNA products, namely, hormones of therapeutic interest, hemopoietic growth factors, blood coagulation products, thrombolytic agents, anticoagulants, interferons, interleukins, and therapeutic enzymes are being produced for human use. Prokaryotic (bacteria) or eukaryotic (yeast, mammalian cell culture) systems are generally used as a host for the production of usable quantities of the desired rDNA product (Fig. 7.22). Most of the rDNA products approved by the FDA (Food and Drug Administration) are being produced using these systems. Bacteria such as *E. coli* are widely used for the expression of rDNA products. They offer several advantages due to high level of recombinant protein expression, rapid growth of cell, and simple media requirement. However, there are some limitations such as intracellular accumulation of heterologous proteins, the potential for product degradation due to traces of protease impurities, and production of endotoxin. Yeast such as *Saccharomyces cerevisiae*, *Hansenula polymorpha*, and *Pichia pastoris* are among the simplest eukaryotic organisms. They grow relatively quickly and are highly adaptable to large-scale production (Hahn-Hägerdal et al., 2005). These organisms do not produce endotoxin. They are capable of glycosylating proteins up to a certain extent like mammalian cells. Mammalian systems such as Chinese hamster ovary (CHO) cell and baby hamster kidney (BHK) cell systems are often the choice for production of human therapeutic proteins. The CHO and BHK cell systems are an ideal choice, as these are capable of glycosylating the protein at the correct sites. However, cost of production of the products using these systems is high because of slow growth and expensive nutrient media. The choice of expression system can influence the character, quantity, and cost of a final product.

The use of genetically engineered plants to produce valuable proteins is increasing slowly (Fischer et al., 2004, Hellwing et al., 2004). The system has potential advantages of economy and scalability. However, variation in product yield, contamination with

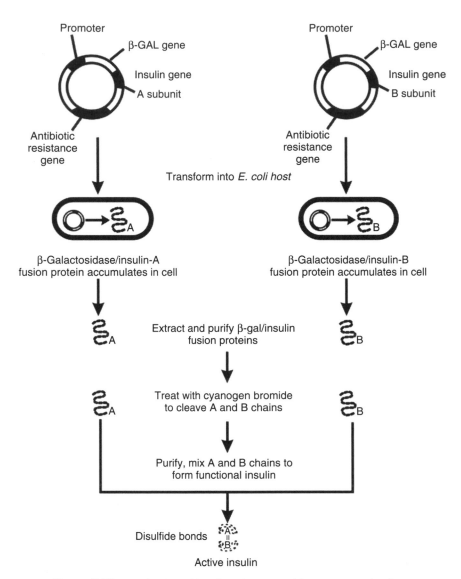

Figure 7.22. Production of insulin using recombinant DNA technology.

agrochemicals and fertilizers, impact of pest and disease, and variable cultivation conditions should also be considered. Plant cell culture system combines the advantages of whole plant system as well as animal cell culture (Magnuson et al., 1998). Although no recombinant products have yet been commercially produced using plant cell culture, several companies are investigating the commercial feasibility of such a production system. rDNA technology or genetic engineering is an umbrella term for a set of experimental techniques that enable individual genes and DNA sequences to be manipulated resulting in genetically modified organisms (GMOs) and products. There have been many potential applications of rDNA in medicine, agriculture, and industry. Conventionally, proteins and other biological products, processed from human or animal serum or tissues, often are of low purity. Production of therapeutic products by the rDNA technology (Fig. 7.22) has several advantages such as provision of drugs that could not be produced by conventional methods, manufacture of sufficient quantities of drugs, and provision for manufacture of safe drugs (Ma et al., 2003).

The future of rDNA products as a human therapeutic is looking good. More than 110 companies are involved in the discovery, development, and marketing of rDNA products. These companies have more than 80 therapeutics in clinical development with a combined portfolio of 73 marketed products (Pavlou and Reichert, 2004; Reichert, 2004). During 2004, two rDNA products, insulin glulisine (Apidra, Sanofi Aventis, Strasbourg, France) and the fertility drug, lutropin alfa (Luveris, Serono, Geneva, Switzerland) have been approved by the US FDA. Three products, insulin detemir (Lantus, Novo Nordisk, Begsvard, Denmark) for diabetes; calcitonin, for treating osteoporosis (Unigene Laboratories, Fairfield, NJ, USA); and palifermin, a keratinocyte growth factor used for treating mucosites (Amgenwoodland Hills, CA, USA), are undergoing FDA review. Analysis of therapeutic market segmentation from 2001 to 2008 reveals that there are five therapy areas—hematology, diabetes and endocrinology, oncology, central nervous system (CNS) disorders, and infectious diseases—that are the key market shareholders. Products of hematology will continue to lead sales over the next coming years. Products of oncology will become the second most important revenue generator. As for the future market, erythropoietin will continue to lead sales, interferon will follow, and insulin will maintain its third place (Fig. 7.23).

7.3.5. Aptamers as Therapeutics

Aptamers are single-stranded nucleic acids that directly inhibit a protein's function by folding into a specific three-dimensional structure that dictates high affinity binding to the targeted protein. They are oligonucleic acids or peptides that bind to a specific target molecule. Aptamers are usually created by selecting from a large random sequence pool, but natural aptamers also exist in riboswitches. Aptamers can be used for both basic research and clinical purposes as macromolecular drugs. Aptamers can be combined with ribozymes to self-cleave in the presence of their target molecule. These compound molecules have additional research, industrial, and clinical applications.

Nucleic acid aptamers can be selected from pools of random sequence oligonucleotides to bind a wide range of biomedically relevant proteins with affinities and specificities that are comparable to antibodies. Aptamers exhibit significant advantages relative to protein therapeutics in terms of size, synthetic accessibility, and modification by medicinal chemistry. Despite these properties, aptamers have been slow to reach the marketplace, with only one aptamer-based drug receiving approval so far. A series of aptamers currently in development may change how nucleic acid therapeutics are perceived. It is likely that in the future, aptamers will increasingly find use in concert with other therapeutic molecules and modalities (Keefe et al., 2010).

Antisense compounds are single-stranded nucleic acids that, in principle, disrupt the synthesis of a targeted protein by hybridizing in a sequence-dependent manner to the mRNAs that encode it. The mechanism of inhibition by nucleic acid aptamers is fundamentally different.

Through *in vitro* selection techniques, aptamers can be generated that bind essentially any protein (or small molecule) target (Fig. 7.24). A high affinity, specific inhibitor can theoretically be made to order, provided that a small quantity of pure target is available.

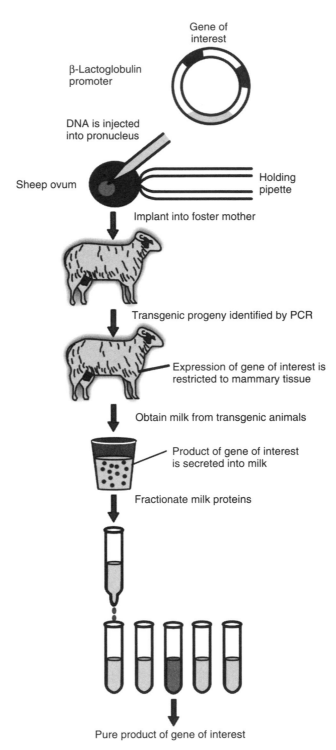

Figure 7.23. Production of gene of interest in transgenic progeny.

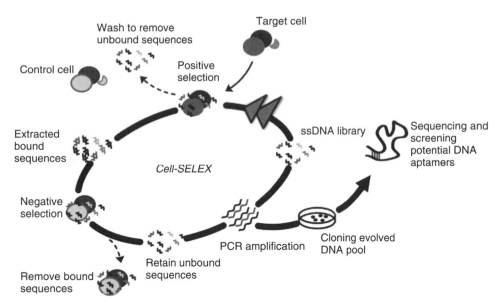

Figure 7.24. Selection of DNA sequences by an *in vitro* selection process called *SELEX*.

Because they inhibit the activity of existing proteins directly, aptamers are more similar to monoclonal antibody or small-molecule drugs than to antisense compounds, and this property greatly increases the number of clinical indications that are potentially treatable by nucleic acid-based compounds.

Aptamers are oligonucleotides derived from an *in vitro* selection process called *SELEX*. Aptamers have been evolved to bind proteins which are associated with a number of disease states. Using this method, many powerful antagonists of such proteins have been found. In order for these antagonists to work in animal models of disease and in humans, it is necessary to modify the aptamers. First of all, sugar modifications of nucleoside triphosphates are necessary to render the resulting aptamers resistant to nucleases found in serum. Changing the 2′OH groups of ribose to 2′F or 2′NH$_2$ group yields aptamers that are long lived in blood. The relatively low molecular weight (8000–12,000) of aptamers leads to rapid clearance from the blood.

Aptamers can be kept in the circulation from hours to days by conjugating them to higher molecular weight vehicles. When modified, conjugated aptamers are injected into animals, and they inhibit physiological functions known to be associated with their target proteins. A new approach to diagnostics is also described. Aptamer arrays on solid surfaces will become available rapidly because the SELEX protocol has been successfully automated. The use of photo-cross-linkable aptamers will allow the covalent attachment of aptamers to their cognate proteins, with very low backgrounds from other proteins in body fluids. Finally, protein staining with any reagent that distinguishes functional groups of amino acids from those of nucleic acids (and the solid support) will give a direct readout of proteins on the solid support (Brody and Gold, 2000). Following is a discussion of how aptamers in various cases can be used to combat diseases.

Case 1: β-Lactamase can confer bacterial resistance to many antibiotics. Introduction of a tightly binding RNA aptamer along with the normal antibiotic would then be

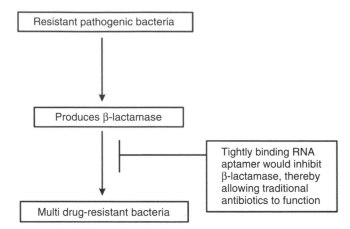

Figure 7.25. Multidrug-resistant bacteria can be produced from resistant pathogenic bacteria.

sufficient to inhibit the growth of the bacteria and eventually kill antibiotic-resistant bacteria. Without this aptamer, the antibiotic alone would be broken down by the bacteria. This tightly binding nucleic acid can subsequently interfere with protein function. Consequently, by finding a therapeutic aptamer that binds tightly to β-lactamase, traditional antibiotics can be used instead of constantly developing new antibiotics. Development of new antibiotics is not only a lengthy process but also very costly. Many new bacterial strains evolve every year with increased resistance toward many of the antibiotics on the market today. This fact emphasizes the need for a method that allows the use of antibiotics already in circulation. SELEX method of aptamer selection against β-lactamase cannot solve the problem of multidrug-resistant bacteria but can help to find new ways of fighting these pathogens (Fig. 7.25).

There are many types of β-lactamase that can be used for potential targets. Selection has been performed on metallo β-lactamase (Kim, 2004), which can break down penicillin, cephalosporin, and carbapenem. However, different findings may come of this particular proposal, as a different RNA sequence could be found. This sequence could work on different strains of bacteria than originally thought.

Case 2: Selection of RNA aptamers against EphA2. Lack of binding between the ephrin A1 ligand and the EphA2 receptor, 130-kDa tyrosine kinase receptor found in adult human epithelial cells, causes unstoppable cell growth and subsequently, development of tumors associated with epithelial cancers (Kinch, 2005; Ansuini et al., 2009). In a normal cell, EphA2 receptor kinase activity is inhibited when the cell membrane EphA2 receptor can bind to ephrin A ligands. Conversely, in a cancerous cell, EphA2 receptor is damaged and cannot bind correctly to ligands, causing EphA2 to continually phosphorylate and thus increasing number of malignant tumor cells (Walker-Daniels et al., 2002). Inhibiting the kinase activity and phosphorylation of EphA2 protein receptor has been associated with a decrease in the growth of malignant cells (Ansuini et al., 2009).

Increased levels of EphA2 phosphorylation due to decreased amount of EphA2/ephrin A1 binding has been linked to increased tumor growth. Thus, selection of RNA aptamers conjugated with a bacterial toxin can be used as a therapeutic tool, degrading the malfunctioning EphA2 protein and thus decreasing kinase activity and tumor proliferation.

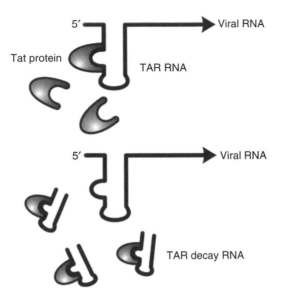

Figure 7.26. Activity of decoy RNA in inhibiting HIV RNA transcription and reducing viral replication.

Such an aptamer would be much more useful than its antibody equivalent due to modifications that can make the aptamer less resistant to enzyme degradation, which will increase therapeutic delivery time and potentially decrease number of treatments.

Case 3: The idea of using nucleic acid molecules as therapeutic agents was conceived in the 1970s with the development of antisense strategies. Numerous nucleic acid ligands, also termed *decoys* or *aptamers*, have been developed during the past 15 years that can inhibit the activity of many pathogenic proteins (Fig. 7.26). Decoy RNA corresponding to the TAR RNA sequence competes with the virus-encoded sequence for binding of tat protein. By sequestering the tat protein from the real TAR RNA sequence present in the HIV RNA, the decoy RNA can inhibit HIV RNA transcription and thus reduce viral replication (White et al., 2000).

Two of them, Macugen and E2F decoy, are in phase III clinical trials. Several properties of aptamers make them an attractive class of therapeutic compounds. Their affinity and specificity for a given protein make it possible to isolate a ligand to virtually any target, and adjusting their bioavailability expands their clinical utility. The ability to develop aptamers that retain activity in multiple organisms facilitates preclinical development. Antidote control of aptamer activity enables safe, tightly controlled therapeutics. Aptamers may prove useful in the treatment of a wide variety of human maladies, including infectious diseases, cancer, and cardiovascular disease (Nimjee et al., 2005).

7.3.6. Immunodiagnostics

Immunodiagnostics is a diagnostic methodology that uses antigen–antibody reaction as their primary means of detection (Fig. 7.27). The concept of using immunology as a diagnostic tool was introduced in 1960 as a test for serum insulin.

It is difficult to cover all areas of clinical applications using antibody detection. However, clinical application of immunodiagnostics can be best demonstrated in available

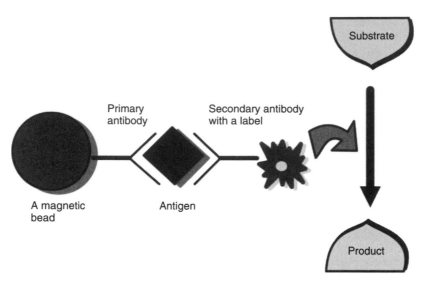

Figure 7.27. Immunodiagnostics that uses an antigen–antibody reaction.

immunoassays for HIV and hepatitis. Immunoassays have been developed to detect anti-HIV antibodies or viral antigens present in serum, plasma, dried blood spots, urine, and saliva.

Assay formats range from enzyme immunoassays (EIAs), enzyme-linked immunosorbent assay (ELISA)-based Western blot assays, and even rapid handheld immunoassays. In general, however, the EIA remains the most widely used serologic test for detecting antibodies to HIV-1. Thus, HIV-1 immunoassays represent the advances in antibody detection technologies to detect and identify infectious agents.

Typically, an EIA is used to screen the blood samples for the presence of HIV-specific antibodies and involves using a whole-virus lysate, synthetic peptides, or recombinant viral proteins as the antigen bound to plastic in a microtiter-plate format. Each blood sample is tested as a singlet with this system; reactive samples are referred to as *initially reactive* (IR) and must be verified by repeating the testing in duplicate using the same EIA test kit. If the IR sample is found to be reactive on the second test run, the sample is labeled as *repeatedly reactive* (RR), or positive. In each RR specimen, the presence of HIV-1-specific antibody is confirmed or ruled out either by an indirect immunofluorescence assay (IFA) using virus-infected target cells or by the Western blot assay, which can identify antibody to specific virus antigens. More sophisticated and expensive procedures not yet approved by the US FDA, such as *in vitro* culture techniques for isolating virus or detecting HIV-specific DNA or RNA by PCR, are also available through a few specialized laboratories (Fig. 7.28).

Another study comparing ELISA methods with Western blotting, microagglutination, IFA, and flow cytometry (FC) for detection of antibodies to *Francisella tularensis* and diagnosis of tularemia is another source to demonstrate the use of antibody detection techniques. ELISA, also known as an *EIA*, is a biochemical technique used mainly in immunology to detect the presence of an antibody or an antigen in a sample. ELISA has been used as a diagnostic tool in medicine and plant pathology, as well as a quality control (QC) check in various industries. In simple terms, in ELISA, an unknown amount of antigen is affixed to a surface, and then a specific antibody is applied over the surface

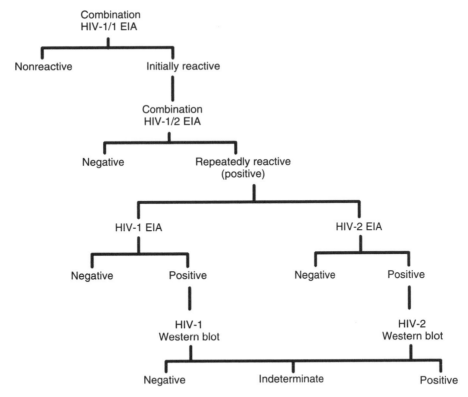

Figure 7.28. Pattern of *in vitro* culture techniques for isolating virus or detecting HIV-specific DNA or RNA by the polymerase chain reaction (PCR).

so that it can bind to the antigen (Fig. 7.29). This antibody is linked to an enzyme, and in the final step, a substance is added that the enzyme can convert to some detectable signal, most commonly a color change in a chemical substrate.

Performing ELISA involves at least one antibody with specificity for a particular antigen. The sample with an unknown amount of antigen is immobilized on a solid support (usually a polystyrene microtiter plate) either nonspecifically (via adsorption to the surface) or specifically (via capture by another antibody specific to the same antigen, such as in a "sandwich" ELISA). After the antigen is immobilized, the detection antibody is added, forming a complex with the antigen. The detection antibody can be covalently linked to an enzyme or can itself be detected by a secondary antibody that is linked to an enzyme through bioconjugation. Between each step, the plate is typically washed with a mild detergent solution to remove any proteins or antibodies that are not specifically bound. After the final wash step, the plate is developed by adding an enzymatic substrate to produce a visible signal, which indicates the quantity of antigen in the sample.

Traditional ELISA typically involves chromogenic reporters and substrates that produce some kind of observable color change to indicate the presence of antigen or analyte. Newer ELISA-like techniques utilize fluorogenic, electrochemiluminescent, and real-time PCR reporters to create quantifiable signals. These new reporters can have various advantages including higher sensitivities and multiplexing. Technically, newer assays of this type are not strictly ELISAs as they are not "enzyme linked" but are instead linked to

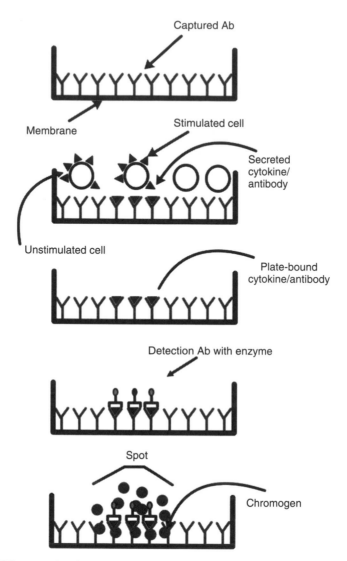

Figure 7.29. ELISA has been used as a diagnostic tool in medicine and plant pathology.

some nonenzymatic reporter. However, given that the general principles in these assays are largely similar, they are often grouped in the same category as ELISAs.

7.3.7. Reporter Genes in Molecular Diagnostics

Reporter genes have become an invaluable tool in studies of gene expression. Reporter gene technology is widely used to monitor the cellular events associated with signal transduction and gene expression. On the basis of the splicing of transcriptional control elements to a variety of reporter genes (with easily measurable phenotypes), it "reports" the effects of a cascade of signaling events on gene expression inside cells. The principal

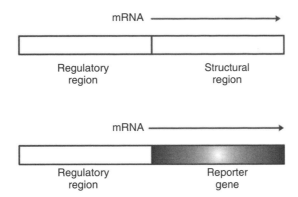

Figure 7.30. Regulatory region controlling the structural region of the gene.

advantage of these assays is their high sensitivity, reliability, convenience, and adaptability to large-scale measurements. They are widely used in biomedical and pharmaceutical research and also in molecular biology and biochemistry. Genetic reporter systems represent an extensive toolbox for the study of regulatory promoter and enhancer sequences as well as for the study of transcription factors (Alam and Cook, 1990; Kain and Ganguly, 2001). A gene consists of two functional parts: one is a DNA sequence that gives the information about the protein that is produced (coding region) and the other is a specific DNA sequence linked to the coding region, which regulates the transcription of the gene (promoter). The promoter either activates or suppresses the expression of the gene. A reporter gene is one that codes for a product that is easy to measure. These genes are often used in situations in which the product of some gene of interest is difficult to assay quantitatively.

Let us consider a typical gene as being composed of a regulatory region, lying upstream of the transcription start point, and a structural region (Fig. 7.30), including the open reading frame (ORF) and any 5′ or 3′ untranslated regions (UTRs). They are nucleic acid sequences encoding easily assayed proteins. They are used to replace other coding regions whose protein products are difficult to assay. The most typical use of a reporter gene is to analyze how a gene is regulated.

Reporter genes have become an invaluable tool in studies of gene expression (Naylor, 1999). They are widely used in biomedical and pharmaceutical research and also in molecular biology and biochemistry. A reporter gene enables researchers to track and study another gene in cell cultures, animals, and plants. Because most gene therapy techniques only work on a small number of individuals, researchers need to use a reporter gene to identify which cells have taken up the gene currently under study and which cells have incorporated it into their chromosomes. Regulatory sequences of interest are combined with a reporter construct of choice and are subsequently assayed in conjunction with the relevant transcription factors (Phippard and Manning, 2003). Reporter genes can be attached to other sequences so that only the reporter protein is made or so that the reporter protein is fused to another protein (fusion protein). Reporter genes can "report" many different properties and events such as

- Strength of promoters, whether native or modified for reverse genetics studies
- Efficiency of gene delivery systems

- Intracellular fate of a gene product, a result of protein traffic
- Interaction of two proteins in the two-hybrid system or of a protein and a nucleic acid in the one-hybrid system
- Efficiency of translation initiation signals
- Success of molecular cloning efforts.

If the reporter system is well chosen then the level of reporter gene expression will correlate with the transcriptional activity of the introduced transgenic factors. In order to assure such a correlation, it is important that the reporter gene does not disturb the metabolism of the transformed cells and that the gene is not endogenously expressed by the target cells creating background signals.

Expression of reporter genes can be measured by enzyme activity assay of the expressed enzyme encoded by the reporter gene using chromo-, fluoro-, or luminogenic substrates; immunological assay of the expressed protein encoded by the reporter gene (reporter gene ELISA); or by histochemical staining of cells or tissues typically to localize enzymatic activity ectopically expressed from reporter gene constructs in transformed cells.

Reporter genes are widely used to study gene expression and regulation mechanisms in living cells (Fig. 7.31). Not all expressed enzymes are easily detectable, so reporter genes were introduced into cellular DNA to investigate gene function by means of a measurable property, the luminescence. Examples of these reporter genes are firefly luciferase, β-glycosidase, alkaline phosphate, β-glucuronidase, and β-glucosidase. By far, the most popular reporter is the firefly luciferase from the American firefly (*Photinus pyralis*). The high sensitivity, easy handling, short process time, and high quantum yield of the bioluminescence reaction make this the "method of choice" to understand gene regulation. Typical commercial assay kits are optimized for extended half-life of more than 5 min. Genetically, most reporter genes are placed downstream of the promoter region, but close to the gene under study.

7.3.8. Human Genome Project and Health Services

A genome is an organism's complete set of DNA, including all its genes. Each genome contains all the information needed to build and maintain that organism. In humans, a copy of the entire genome of more than 3 billion DNA base pairs is contained in all cells that have a nucleus. The Human Genome Project was an international research effort to determine the sequence of the human genome and identify the genes that it contains. The project was coordinated by the National Institutes of Health (NIH) and the US Department of Energy. Additional contributors included universities across the United States and international partners in the United Kingdom, France, Germany, Japan, and China. The Human Genome Project formally began in 1990 and was completed in 2003, 2 years ahead of its original schedule.

The work of the Human Genome Project has allowed researchers to begin to understand the blueprint for building a person. As researchers learn more about the functions of genes and proteins, this knowledge will have a major impact in the fields of medicine, biotechnology, and the life sciences (Fig. 7.32). The major scientific thrust of the Human Genome Project begins with the isolation of human genomic and cDNA clones (by cell-based cloning or PCR-based cloning). These are then used to construct high resolution genetic and physical maps before obtaining the ultimate physical map, the complete

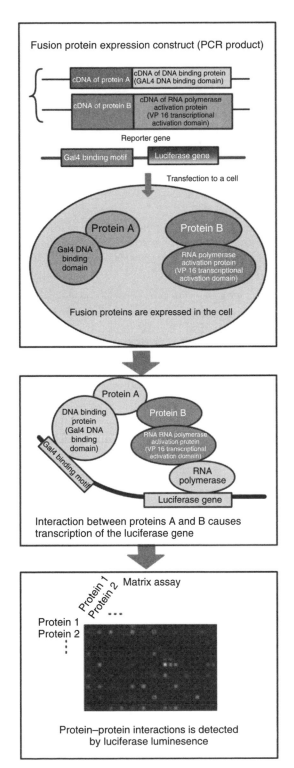

Figure 7.31. Fusion protein expression construct can detect the gene expression and correlate the transcriptional activity.

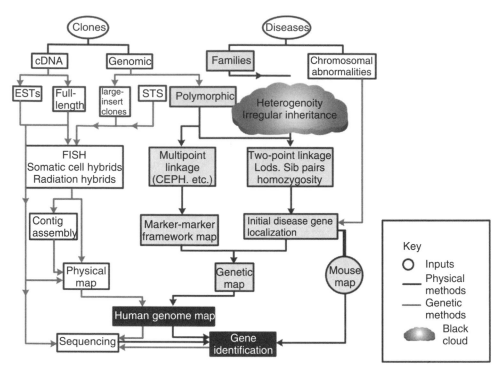

Figure 7.32. Major scientific strategies and approaches being used in the Human Genome Project.

nucleotide sequence of the 3300 Mb nuclear genome. Inevitably, the project interacts with research on mapping and identifying human disease genes. In addition, ancillary projects include studying genetic variation, genome projects for model organisms, and research on ethical, legal, and social implications. The data produced are being channeled into mapping and sequence databases, permitting rapid electronic access and data analysis.

The Human Genome Project has already fueled the discovery of more than 1800 disease genes. As a result of the Human Genome Project, today's researchers can find a gene suspected of causing an inherited disease in a matter of days, rather than the years it took before the genome sequence was in hand. There are now more than 1000 genetic tests for human conditions. These tests enable patients to learn their genetic risks for disease and also help health care professionals diagnose disease.

At least 350 biotechnology-based products resulting from the Human Genome Project are currently in clinical trials. Having the complete sequence of the human genome is similar to having all the pages of a manual needed to make the human body. The challenge now is to determine how to read the contents of these pages and understand how all of these many, complex parts work together in human health and disease. One major step toward such comprehensive understanding was the development in 2005 of the HapMap, which is a catalog of common genetic variation, or haplotypes, in the human genome. HapMap data have accelerated the search for genes involved in common human diseases and have already yielded impressive results in finding genetic factors involved in conditions ranging from age-related blindness to obesity. The tools created through the Human Genome Project continue to underlie efforts to characterize the genomes of important organisms used extensively in biomedical research, including fruit flies, roundworms, and mice. Much work still remains to be done. While the cost of DNA sequencing has dropped by three orders of magnitude in the past 15 years, sequencing an individual's genome for medical purposes is still prohibitively expensive and the NIH is supporting innovative research to lower the cost.

Since its outset, the project has been especially justified by the expected medical benefits of knowing the structure of each human gene. Inevitably, this information will provide more comprehensive prenatal and presymptomatic diagnoses of disorders in individuals judged to be at risk of carrying a disease gene. The information on gene structure will also be used to explore how individual genes function and how they are regulated. Such information will provide sorely needed explanations for biological processes in humans. It would also be expected to provide a framework for developing new therapies for diseases, in addition to simple gene therapy approaches. More importantly, as mutation screening techniques develop, an expected benefit would be to alter radically the current approach to medical care, from one of treating advanced disease to preventing disease based on the identification of individual risk (Cantor, 1998).

Exciting though such possibilities are, there may be unexpected difficulties in understanding precisely and comprehensively how some genes function and are regulated (cautionary precedents are the lack of progress in predicting protein structure from the amino acid sequence and the imperfect understanding of the precise ways in which the regulation of globin gene expression is coordinated, decades after the relevant sequences have been obtained). In addition, the single gene disorders that should be the easiest targets for developing novel therapies are very rare; the most common disorders are multifactorial. Hence, although the data collected in the Human Genome Project will inevitably be of medical value, some of the most important medical applications may take some time to be developed.

The UK Wellcome Trust has been a major supporter of genome projects for microbial pathogens and its Beowulf Genomics. Various other organizations have supported similar programs. In some cases, prokaryotes have been selected for genome sequencing because of their known associations with chronic diseases viz. *Helicobacter pylori* (associated with peptic ulcers) and *C. pneumoniae* (associated with respiratory disease as well as coronary heart disease) (Danesh et al., 1997). Other completed projects (Table 7.5) have yielded the genomes of prokaryotes known to be causative agents of disease, such as *Mycobacterium tuberculosis, Treponema pallidum* (the causative agent of syphilis), and *Rickettsia prowazekii* (Cole et al., 1998). In addition to having a more complete understanding of these organisms, the new information can be expected to lead to more sensitive diagnostic tools and new targets for establishing drugs and vaccines.

7.3.9. Prognostics, Diagnostics, and Therapeutics of Disease-Causing Pathogens

Molecular methods used directly on clinical material play an important role in the diagnosis. Molecular diagnostic tests that allow timely and accurate detection of pathogens are already implemented in many laboratories (Fig. 7.33). The early molecular diagnosis of gray mold pathogen should significantly aid in the development of therapies tailored to this clinically and pathogenetically distinctive subgroup of bioprevention. It is therefore appropriate to analyze these issues in detail, especially the extent to which molecular biology has revised traditional diagnostic criteria. The emergence of speedy, DNA-based

TABLE 7.5. Germ Wars: Examples of Pathogenic Microorganisms for
Which Genome Projects have been Developed

Organism	Type	Genome Size, Mb	Associated Disease
Bordetella pertussis	Bacterium	3.88	Whooping cough
Borrelia burgdorferi	Spirochete bacterium	0.95	Lyme disease
Chlamydia pneumoniae	Intracellular bacterium	1.0	Respiratory disease, coronary heart disease
Chlamydia trachomatis	Intracellular bacterium	1.7	Trachoma is a major cause of blindness
Clostridium difficile	Bacterium	4.4	Antibiotic-associated diarrhea, pseudomembranous colitis
Helicobacter pylori	Bacterium	1.67	Peptic ulcers
Leishmania major	Parasitic protozoan	34	Leishmaniasis
Mycobacterium leprae	Bacterium	2.8	Leprosy
Mycobacterium tuberculosis	Bacterium	4.4	Tuberculosis
Plasmodium falciparum	Parasitic protozoan	30	Malaria
Rickettsia prowazekii	Bacterium	1.1	Typhus
Salmonella typhi	Bacterium	4.5	Typhoid fever
Treponema pallidum	Spirochete bacterium	1.1	Syphilis
Trypanosoma brucei	Parasitic protozoan	30	African trypanosomiasis (sleeping sickness)
Trypanosoma cruzi	Parasitic protozoan	30	American trypanosomiasis
Vibrio cholerae	Bacterium	2.5	Cholera
Yersinia pestis	Bacterium	4.38	Plague, schistosomiasis, filariasis

Source: Strachan T, Read AP. 1999. *Human Molecular Genetics*. 2nd ed. Copyright © 1999, Bios Scientific Publishers, an imprint of Taylor & Francis Group. Chapter 13, Genome Projects.

diagnostic tests adds another tool that doctors can use to diagnose patients. Some of the sampling tests used in hospitals today is based on findings of infectious disease experts (Table 7.6).

Over the last two decades, the pathogenic basis for the most common heritable cardiovascular disease, hypertrophic cardiomyopathy (HCM), has been investigated extensively. Affecting approximately 1 in 500 individuals, HCM is the most common cause of sudden death in young athletes. In recent years, genomic medicine has been moving from the bench to the bedside throughout all medical disciplines including cardiology. Now, genomic medicine has entered clinical practice as it pertains to the evaluation and management of patients with HCM. The continuous research and discoveries of new HCM-susceptibility genes, the growing amount of data from genotype–phenotype correlation studies, and the introduction of commercially available genetic tests for HCM make it essential that the modern day cardiologist understand the diagnostic, prognostic, and therapeutic implications of HCM genetic testing.

Infectious diseases are a major cause of mortality both in human and animal kingdoms (Table 7.7). The emergence of infectious diseases has become more serious, as represented by new pathogens such as the viruses causing AIDS, Nipah virus encephalitis,

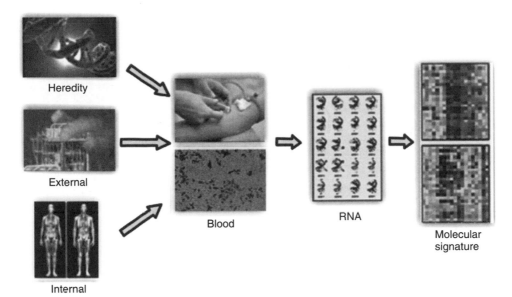

Heredity

External

Internal

Blood

RNA

Molecular signature

Figure 7.33. Strategy used for molecular diagnostic tests for timely and accurate detection of pathogen.

TABLE 7.6. Sampling of Tests Used in Hospitals Based on Infectious Disease Experts

Test Type	How It Works	Comment
Cell culture	Organisms grown in a dish and examined under the microscope	Most commonly used type of test. Results may take days. Not all organisms can be cultured. But method is proven after years of use. The least expensive bacterial testing method.
Molecular rapid test	Organisms are identified by their genetic makeup. A snippet of DNA from an infectious agent is compared with a culture taken from the patient. If there is a match, the doctor can make a diagnosis.	Results are available within a few hours. Able to detect specific strains of organisms, so treatment options are more effective. Technology is costly and requires expertise, which limits its use to hospitals and major clinics. Tests can be expensive, relative to other diagnostic tools.
Immunoassay	Tests the sample for antibodies, which are the body's response to an infection, or proteins from the bug itself.	Commonly used in tests that consumers can use at home (e.g., pregnancy test). Results can be available within minutes. Generally not as sensitive as DNA-based tests.

TABLE 7.7. Major Diseases Responsible for Mortality of Humans and Animals

Category	Organism	Disease
Bacteria	*Bacillus anthracis*	Anthrax
	Bordetella species, including *B. pertussis,B. parapertussis*, and *B. holmesii*	Whooping cough and others
	Brucella species	Brucellosis
	Burkholderia pseudomallei	Melioidosis
	Francisella tularensis	Tularemia
	Gram-negative bacilli	Atypical (nosocomial) pneumonia
	Legionella pneumophila	Legionnaires' disease
	Leptospira interrogans	Leptospirosis
	Pasteurella multocida	Atypical pneumonia and other
	Yersinia pestis	Plague
Bacteria-like	*Chlamydia pneumoniae* (TWAR)	Atypical pneumonia
	Chlamydia trachomatis	Atypical pneumonia and others
	Chlamydia psittaci	Psittacosis
	Mycoplasma pneumoniae	Atypical (walking) pneumonia
Rickettsia	*Coxiella burnetii*	Acute Q fever
	Rickettsia rickettsiae	Rickettsiosis
Viruses	Adenoviruses	Atypical pneumonia and others
	Enterovirus	Atypical pneumonia and others
	Hantavirus	Hantavirus pulmonary syndrome
	Herpes virus, including HSV, CMV, VZV, EBV, HHV-6, HHV-7, and HHV-8	Atypical pneumonia and others
	Influenza viruses A, B, and C	Influenza
	Measles virus	Atypical pneumonia and others
	Metapeumoviruses	Atypical pneumonia and others
	Non-SARS coronavirus	Atypical pneumonia and others
	Parainfluenza viruses 1, 2, and 3	Parainfluenza
	Respiratory syncytial virus	Atypical pneumonia and others
	Rhinoviruses	Atypical pneumonia and others
	SARS virus	SARS (atypical pneumonia)
Fungi	*Aspergillus* species	Aspergillosis and others
	Candida species	Candidiasis and others
	Coccidioides immitis	Coccidioidomycosis
	Cryptococcus neoformans	Atypical pneumonia and others
	Histoplasma capsulatum	Histoplasmosis
Protozoa	*Pneumocystis carinii*	Atypical pneumonia
	Toxoplasma gondii	Toxoplasmosis

avian influenza, dengue fever and West Nile encephalitis; reemerging pathogens such as those causing malaria, measles; foodborne pathogens; methicillin-resistant *Staphylococcus aureus* (MRSA); and multidrug-resistant *M. tuberculosis*. Microarray profiling offers many potential advances in diagnostic and therapeutic intervention in human disease because of its unparalleled ability to conduct high throughput analysis of gene expression. However, limitations of this technique relate in part to issues regarding the various methodologies and experimental designs as well as difficulties in the interpretation of results (Berzsenyi et al., 2006). Despite these limitations, microarray has been used efficiently in disease diagnosis. Lee et al. (2003) developed a DNA microarray called

PathoChip™ for the detection of 44 highly prevalent and fastidious pathogenic bacteria. They used a variety of clinical isolates collected from blood, sputum, stool, cerebral spinal fluid, pus, and urine to evaluate the technique. Another patented array composing of DNAs amplified with 35 kinds of primers used to amplify 16S–18S sequences, specific antigen, toxin, and virulent genes of pathogens causing respiratory infectious disease was developed by Ezaki (2003).

DNA microarray studies of complex diseases and gene-interaction networks may contain modules of coregulated or interacting genes that have distinct biological functions. DNA microarray helps in understanding the pathogenesis of these complex diseases, which often implicates hundreds of genes (Yoo et al., 2009). Garaizar et al. (2006) summarized the past usage for microbial strain characterizations and the future utilization of DNA microarray technology in epidemiological studies and molecular typing of bacterial pathogens. A more focused assay concentrating on genomic regions of variability previously detected by genome-wide microarrays will find broad applications in routine bacterial epidemiology.

7.3.10. Promise of Pharmacogenomics

Personalized medicine and pharmacogenomics are inextricably linked. Pharmacogenomics is the use of genetic variations (such as SNP, gene expression variability, or other molecular signatures) to understand and correlate with differential response to pharmaceutical agents (drugs). Pharmacogenomics can be deployed clinically to stratify patients into responders and nonresponders, and this practice is termed *personalized medicine*. To frame the context of current approaches, pharmacogenomics seeks to identify and validate the signature(s) of molecular analytes and these are converted to assays using the tools of molecular diagnostics. The deployment of molecular diagnostic assays on defining and targeting patient populations is the domain of personalized medicine (Fig. 7.34). The generally accepted imaging activities that are currently used in personalized medicine are pharmacogenomics, genomics, and theranostics. Personalized medicine has taken a prominent role in cancer treatment and cardiology and neurology segments. Medicine, as we move into the third millennium, still targets therapy to the broadest patient population that might possibly benefit from it, and it relies on statistical analysis of this population's response for predicting therapeutic outcome in individual patients. Therapists, of necessity, make decisions about the choice of drug and appropriate dosage based on information derived from population averages. This "one drug fits all" approach could, with the fruits of pharmacogenomic research, evolve into an individualized approach to therapy where optimally effective drugs are matched to a patient's unique genetic profile. This involves classifying patients with the same phenotypic disease profile into smaller subpopulations, defined by genetic variations associated with disease, drug response, or both. The assumption underlying this approach is that drug therapy in genetically defined subpopulations can be more efficacious and less toxic than in a broad population.

Individualizing drug therapy raises a number of issues with enormous practical consequences. Currently, the pharmaceutical industry is in a consolidation and merger phase, with ever larger corporations emerging at a steady pace. This consolidation is done in the expectation that many novel drugs can be brought to market with high efficacy against major diseases, driven by genomics-based drug discovery. Indeed, large corporations

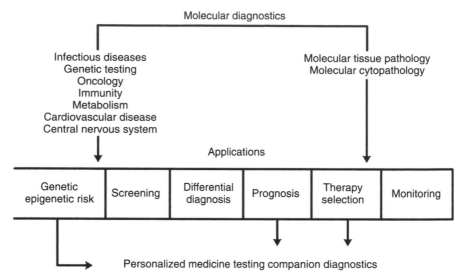

Figure 7.34. Various applications of molecular diagnostics showing that personalized medicine is the major outcome of this diagnostics.

depend on generating "blockbuster" drugs—drugs that raise in excess of a billion dollars in revenue each year by targeting large patient populations. However, it remains to be seen whether betting on a "one drug fits all" approach is realistic. Certainly, a few blockbuster drugs continue to emerge, for example, the Cox-2 selective inhibitors in the therapy of inflammatory joint diseases. Efficacy does not appear to exceed substantially that of traditional nonsteroidal anti-inflammatory drugs (NSAIDs), which inhibit both Cox-1 and Cox-2 to varying degrees; however, the incidence of gastrointestinal lesions is reduced. Yet, only a portion of patients receiving conventional NSAIDs develop these lesions, and the traditional drugs are much less expensive. Moreover, it remains to be seen what long-term sequelae arise from treatment with Cox-2 selective inhibitors. These sequelae might be beneficial (e.g., the possible prevention of colon cancer or neurodegenerative disorders associated with inflammation in the CNS), but the physiological functions of Cox-2 remain poorly understood. Trials over longer time periods will be needed to address these questions fully. As three-quarters of all health care costs are used for the treatment of chronic illness, mostly of the aged, long-term issues will be the battleground where optimal therapies will be decided (Khan et al., 2011).

Whether a single drug emerges superior to others in a broad patient population or best clinical response requires differential therapy of small subpopulations is the subject of fierce debates. Bringing a new drug to the market currently costs approximately $500 million, making it economically impossible to target small patient populations. If smaller patient populations are to be served, we need to change the entire process, up to the final regulatory agency approval for clinical use. Conceivably, targeting well-defined patient populations will sharpen our analysis of risk/benefit ratios and permit clinical trials to be substantially reduced in size. Laws and FDA regulations may have to be changed to accommodate the need for targeting patients with rare diseases or with subtypes of otherwise common diseases. This approach will set the stage for testing whether targeting small patient populations with select drugs is superior to treating many patients with the best drug available for a given disease. The outcome may vary from one case to another.

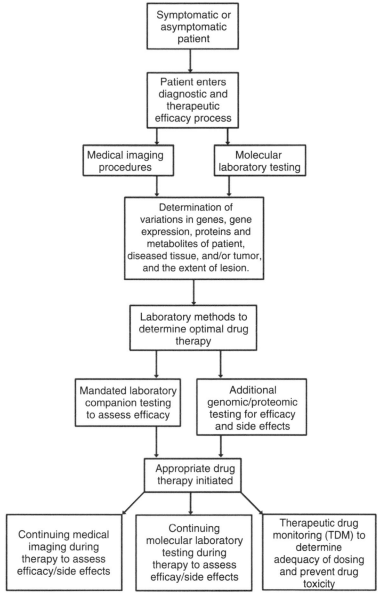

Figure 7.35. Flow diagram of personalized medicine.

Thus, individualizing drug therapy with the use of pharmacogenomics holds the potential to revolutionize medical therapeutics by challenging the "one drug fits all" approach (Fig. 7.35). Furthermore, pharmacogenomics could also enhance the value of currently approved drugs with limited market share because of significant toxicity or limited efficacy, enabling prescribers to identify patients for whom they will be both effective and safe.

The emergence of personalized medicine raises issues for those who pay for treatment. The cost of new diagnostic tests and individualized medications may be more expensive, but it is hoped that the predictive potential of personalized medicine could avert more costly treatments required after the onset of a disease. Currently, less than 5% of all US private companies reimburse for genetic tests, indicating that the current health care delivery system may not be able to deliver effective "personalized medicine."

Pharmacogenetics applies not only to traditional drugs but also to bioengineered proteins and gene therapy. Human genetic variability can be expected to affect all treatment modalities. For example, breast cancer treatment with trastuzumab (Herceptin), a humanized monoclonal antibody against the HER2 receptor developed by Genentech, Inc., is linked to HER2 overexpression. This reaction correlates with poor clinical prognosis and serves as a marker for responsiveness to trastuzumab therapy, either alone or in combination with chemotherapy.

The cytochrome $P450$ monooxygenase system of enzymes is responsible for a major portion of drug metabolism in humans. Although commonly serving to detoxify xenobiotics, these enzymes are also principally responsible for the activation of procarcinogens and promutagens in the human body. This scenario is particularly important for lipophilic drugs such as CNS-active drugs, which generally must be lipophilic to penetrate the blood–brain barrier. Because renal excretion is minimal for these compounds, $P450$ metabolism provides the primary means of drug elimination. This large family of genes has been intensely studied, and among the numerous $P450$ subtypes, CYP2D6, 3A4/3A5, 1A2, 2E1, 2C9, and 2C19 play particularly critical roles in genetically determined responses to a broad spectrum of drugs. Determination of a patient's CYP2D6 phenotype/genotype may prove useful in treatment with antipsychotic drugs, while comprehensive genotyping assays for all relevant $P450$ isotypes and their main sequence variants are being developed.

Cytochrome $P450$s inactivate or in some cases activate xenobiotics. Therefore, $P450$ polymorphisms affect an individual's susceptibility to environmental toxins. As a result, sequence variation of $P450$ isotypes attracts special attention in toxicogenetics. The US National Institute of Environmental Health Sciences launched the Environmental Genome Project with the stated goal of understanding the genetic factors governing an individual's response to the environment on a genome-wide scale. This effort parallels the study of genetic variability in drug response.

Genetic variability is seen in the area of both pharmacokinetics (absorption, distribution, metabolism, and excretion (ADME)) and pharmacodynamics (drug effects). These differences in drug responses are largely caused by genetic polymorphisms. For example, genetic variations can influence the activity or have an effect on the expression of the following proteins:

- DMEs
- Drug transporters
- Drug receptors
- G proteins.

Figure 7.36 shows the relevant pathways including the positions of genes that are important for drug response and drug effects. The aim is to discriminate responders and nonresponders to certain drugs and to identify individuals at increased risk for adverse drug reactions each based on variations in relevant genes. Hence, pharmacogenetics and pharmacogenomics are contributing to improve drug treatment and to enable and support the development of drugs, which are safer, more targeted, and individualized.

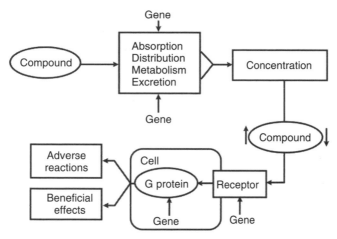

Figure 7.36. Genes determine drug effects.

7.4. INDUSTRIAL MICROORGANISMS AND PRODUCT FORMATION

During the past decades, the industrial application of biotechnologies was characterized by a rapid development. Microbiologists, biotechnologists, ecologists, engineers, and specialists of different areas have experimented and developed methods for efficient usage of the products from the activity of a lot of microorganisms from the environment. The availability of microorganisms with high specific activity, that is, strains possessing useful characteristics, is the basic element for the development of the applied and fundamental biotechnology. Observing different fermentation processes in nature, man has chosen a range of microorganisms for production of bakery products, cheese, wine, beer, and so on. The modern microbiology operates with thousands of strains of actinomycetes, bacteria, yeasts, microscopic fungi, viruses, and other microorganisms whose morphological, taxonomic, and biochemical characteristics are well studied. A huge number of these microorganisms are used in industrial microbiology. Genetic engineers have performed a lot of experiments for selection of high productive strains and construction of microorganisms with new characteristics. The process of selection of just a single industrial strain is a result of the efforts of tens of researchers or sometimes several generations of scientists.

The development of the modern biotechnology in a given country can be estimated through the quality and quantity of the applied strains, as the economy, the industrial development, the ecological culture, the human health, and the conditions of life are to a great extent determined by them. As a result of the purposeful research, a number of immunological and diagnostic, prophylactic and curative, and hormonal and stimulating bioproducts are prepared and applied in the human and veterinary medicine. Nowadays by means of bacterial and viral vaccines, mankind successfully fights against a lot of infectious diseases. Microbiology is important for agriculture due to its application in forage and silage preparation. The soil fertility is improved and the processes of biosynthesis and plant bioconservation are regulated and directed by the introduction of microbiological products (Fig. 7.37).

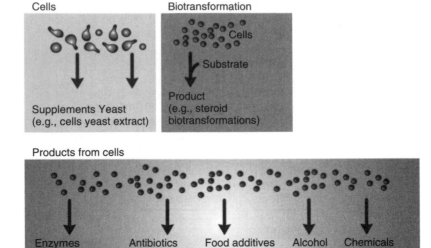

Figure 7.37. Modern biotechnology helps in the synthesis of enzymes, antibiotics, food additives, alcohol, and chemicals by biotransformation.

7.4.1. Microorganisms and Their Products

In a broadest sense, industrial microbiology is concerned with all aspects of business that relate to microbiology. In a more restricted sense, industrial microbiology is concerned with employing microorganisms to produce a desired product and preventing microbes from diminishing the economic value of various products (Table 7.8).

Various commercial products of economic value made by microbes are

- Pharmaceuticals, including antibiotics, steroids, human protein, vaccines, and vitamins
- Organic acids
- Amino acids
- Enzymes
- Alcohols
- Organic solvents
- Synthetic fuels.

Microbes and/or microbial products are being used in drain lines and grease traps; septic tanks; odor neutralizers; environment friendly biologic cleaning and maintenance products; the quick start-up wastewater treatment plants; prevention of active sludge loss; treatment of toxic industrial effluents; plant growth and increase of fertilizer efficiency; degradation of ammonia in breeding; fish farming; treatment of polluted lakes, lagoons, and running waters; and recovery and bioremediation of water and soil after oil spills. These characteristics of microbes could be exploited further in obtaining some valuable products of daily use. The cheap raw materials available in nature as a waste may be

TABLE 7.8. Industrial Use of Microbial Products

Microbial Product	Role in Enhanced Oil Recovery	Some of the Effects
Gases (H_2, N_2, CH_4, CO_2)	Reduce oil viscosity and improve flow characteristics Displace immobile oil in place	Improved oil recovery by gases Miscible CO_2 flooding
Acids (low molecular weight acids, primarily low molecular weight fatty acids)	Improve effective permeability by dissolving carbonate precipitates from pore throat. Significant improvement of permeability and porosity. CO_2 produced from chemical reactions between acids and carbonate reduce oil viscosity and causes oil droplet to swell	Enhanced oil flooding
Solvents (alcohols and ketones that are typical cosurfactants)	Dissolve in oil and reduce viscosity Dissolve and remove heavy, long-chain hydrocarbons from pore throat (increase effective permeability) Involved in stabilizing and lowering the tension that promotes emulsification Reduce interfacial tension	Emulsification promotion for increased miscibility
Biosurfactants	Reduce interfacial tension between oil and rock/water surface, which causes emulsification, improving pore scale displacement Alter wet ability	Microbial surfactant Flooding
Biopolymers	Improve the viscosity of water in water flooding and direct reservoir fluids to previously upswept areas of the reservoir Improve the sweep efficiency of water flood by plugging high permeability zones or water-invaded zones Control of water mobility	Microbial permeability modification (selective plugging)
Biomass (microbial cells)	Physically displace oil by growing between oil and rock/water surface Reverse wet ability by microbial growth Can plug high permeability zones Perform selective partial degradation of whole crude oil Act as selective and nonselective plugging agents in wetting, alteration of oil viscosity, oil power point, desulfuration	Same biopolymers

converted into useful commercial products by the activity of microbes. Microbes thus serve a dual purpose. First, they are good agents of disposal of these wastes, and second, the resultant end products of their breakdown are useful commercial products.

7.4.2. Primary and Secondary Metabolites

Metabolism is defined as the sum of all the biochemical reactions carried out by an organism. Primary metabolic pathways converge too few end products, while secondary metabolic pathways diverge too many products. Primary requires the cell to use nutrients in its surroundings such as low molecular weight compounds for cellular activity. There are three potential pathways for primary metabolism: the Embden-Meyerhof-Parnas (EMP) pathway, the Entner–Doudoroff pathway, and the hexose monophosphate (HMP) pathway. The EMP pathway produces two molecules of pyruvate via triose phosphate intermediates. This pathway occurs most widely in animal, plant, fungal, yeast, and bacterial cells. Many microorganisms, however, use this pathway solely for glucose utilization. During primary metabolism, hexoses such as glucose are converted to SCP by yeasts and fungi. This is generally done by using a combination of EMP and HMP pathways, followed by the TCA cycle and respiration. Yeasts from the *Saccharomyces* species produce alcohol as cells grow during the log phase (during trophophase) using an anaerobic primary metabolic pathway. These account for most of the alcohol found in Nature and are widely used in the fermentation industry to produce beer, wine, and spirits.

In the citric acid fermentation process involving *Aspergillus niger*, hexoses are converted via the EMP pathway to pyruvate and acetyl CoA that condenses with oxaloacetate to form citrate in the first step of the TCA cycle. Ethanol, lactic acid, and acetic acid were the first commercial products of the fermentation industry. Several of these products have applications as alternative energy sources; for example, alcohol has been used to produce a cheaper alternative to petrol in developing countries such as Brazil and in Europe between World Wars I and II.

Secondary metabolism synthesizes new compounds. Secondary metabolites are not vital to the cell's survival itself but are more so for the entire organism. Relatively few microbial types produce the majority of secondary metabolites. Secondary metabolites are produced when the cell is not operating under optimum conditions, for example, when the primary nutrient source is depleted. Secondary metabolites are synthesized for a finite period by cells that are no longer undergoing balanced growth. A single microbial type can produce very different metabolites. *Streptomyces griseus* and *Bacillus subtilis* each produce more than 50 different antibiotics. Most secondary metabolites are produced by families as closely related compounds. The chemical structure and their activities cover a wide range of possibilities, including antibiotics, ergot alkaloids, naphthalenes, nucleosides, peptides, phenazines, quinolines, terpenoids, and some complex growth factors. The production of economically important metabolites such as antibiotics by microbial fermentation is one of the major activities of the bioprocess industry. Secondary metabolites such as penicillin are produced during the stationary phase (idiophase) of cell growth. Most of the knowledge concerning secondary metabolism comes from the study of commercially important microorganisms. There are some similarities between the pathways that produce primary and secondary metabolites, namely, that the product of one reaction is the substrate for the next and the first reaction in each case is the rate-limiting step. Also the regulation of secondary metabolic pathways is interrelated in complex ways to primary metabolic regulation.

7.4.2.1. Primary Metabolites

Fermentation products of primary metabolism such as ethanol, acetic acid, and lactic acid were the first commercial products of the fermentation industry. These industrial revelations were soon followed by citric acid production along with other products of fungal origin. Owing to the high product yield and the low reproducibility costs, major interest has been shown in the respective markets. Production of cell constituents, that is, lipids, vitamins, polysaccharides, as well as intermediates in the synthesis of cell constituents such as amino acids and nucleotides is also of great economic importance in the present day industry. The effectiveness of yeasts along with other microorganisms as sources of the B group vitamins has been recognized for more than 50 years and like the products of catabolic primary metabolism, for example, ethanol, citric acid, are of great commercial importance.

7.4.2.2. Secondary Metabolites

Antibiotics were first defined as a chemical compound produced by a microorganism, which has the capacity to inhibit the growth of and even destroy bacteria and microorganisms in dilute solutions. Antibiotic production includes 160 different products and has a total annual worldwide market of $23 billion. Sir Alexander Fleming first discovered the antibiotic properties of the mold *P. notatum* in 1929 at St. Mary's hospital in London, when he noticed that *P. notatum* destroyed *Staphylococcus* bacteria in culture. Penicillin is bactericidal to a number of gram-positive bacteria and acts by inhibiting transpeptidation thus preventing new cells from forming walls. It belongs to the β-lactam family of antibiotics. A team at Oxford University further proved penicillin's value as a drug by developing methods of growth, extraction, and purification. During World War II, research was moved to the United States where large-scale growth of the mold began. First, penicillin molds were grown in small shallow containers containing nutrient broth. Methods of growth were improved by using deep fermentation tanks with continuous sterile air supply and corn steep liquor as a nutrient source. In 1943, a cantaloupe mold, *Penicillium chrysogenum* was found to produce twice the amount of penicillin than *P. notatum*. Since then, researchers continued to find higher yielding penicillin molds and have also improved yields further by exposing molds to X-rays and UV light. The first type of penicillin produced was penicillin G, which had to be administered to patients parenterally because it is broken down by stomach acid. Penicillin V was later formulated so that it could be taken orally; unfortunately, it was less active than penicillin G.

7.4.3. Characteristics of Large-Scale Fermentation

Fermentation is the process of deriving energy from the oxidation of organic compounds, such as carbohydrates, and using an endogenous electron acceptor, which is usually an organic compound. In contrast, respiration is where electrons are donated to an exogenous electron acceptor, such as oxygen, via an electron transport chain. Fermentation is important in anaerobic conditions where there is no oxidative phosphorylation to maintain the production of ATP (adenosine triphosphate) by glycolysis. During fermentation, pyruvate is metabolized to various compounds. Homolactic fermentation is the production of lactic acid from pyruvate, alcoholic fermentation is the conversion of pyruvate into ethanol and carbon dioxide, and heterolactic fermentation is the production of lactic acid as well as other acids and alcohols. Fermentation does not necessarily have to be carried out in an anaerobic environment. For example, even in the presence of abundant oxygen, yeasts greatly prefer fermentation to oxidative phosphorylation, as long as sugars are readily available for consumption (Fig. 7.38).

Parameters that must be precisely regulated are temperature, pH, rate and nature of mixing, oxygenation, and sterility and containment. Stepwise scale-up of culture requires

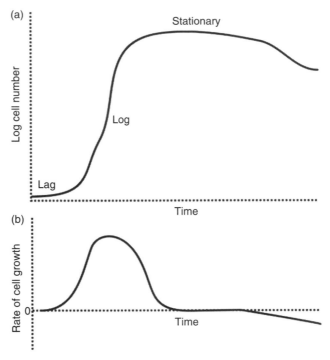

Figure 7.38. The growth of cells in a fermentor showing fermentation carried out (a) under anaerobic conditions and also (b) in the presence of oxygen.

5–10 ml stock, 200- to 1000-ml shake flask, 10- to 100-l seed fermentor, and 1000- to 100,000-l production fermentor (e.g., a Pichia fermentor). The types of cultures are batch, fed batch, and continuous (for indigo production).

Fed-batch fermentation is described (Bungay, 1993) as the type of system where "nutrient is added when its concentration falls below some set point." Usually, the addition of the nutrient is controlled by a computer for precision. The best way to control the addition of the feed is to monitor the concentration of the nutrient itself in the fermentor or reactor vessel, so that exactly how much more of it needs to be added is known. The nutrient is added in several doses to ensure that there is not too much of the nutrient present in the fermentor at any time. If too much of a nutrient is present, it may inhibit the growth of the cells. By adding the nutrient a little bit at the time, the reaction can proceed at a high rate of production without getting overloaded.

Advantages of Continuous Fermentation
    Smaller bioreactors
    Smaller scale of equipment for cell harvesting
    Avoids "downtime" between batches
    Uniform physiological state of cells.

Disadvantages
    500–1000 h duration

Overgrowth of variants lacking plasmid

Maintenance of sterility over long term

Batch to batch variation of medium is problematic.

The exponential growth of biomass and the frequent changes of environmental conditions (fluid characteristics, addition of antifoam, etc.) create problems in dissolved oxygen (DO) control of fermentations in laboratory and pilot scale. Traditionally DO is controlled by application of conventional control algorithms, such as proportional integral derivative (PID), connected to a control cascade. The performance of the control loop depends on the oxygen consumption. For DO loop optimization of *E. coli* fermentation, empirical adaption of tuning parameters and fuzzy tuning was applied.

- Optimal agitation
  - adequate mixing of nutrients
  - prevention of accumulation of toxic metabolites
  - uniformity of pH and temperature
  - rate of transfer of oxygen from bubbles to liquid medium.
- Excessive agitation
  - hydromechanical stress (shear)
  - temperature increase in culture.

7.4.4. Fermentation Scale-Up

Two of the most common phrases in fermentation technology research is "scaling up" and "scaling down" studies. However, the phrase "scaling up" is more commonly understood and practiced during the designing of industrial-scale fermentors, whereas "scaling down" studies are rarely heard. In reality, many fermentation technologists are not aware that during most times of their work, they are doing "scale-down studies," may be the phrase "scale-up" has more impact factor than "scale-down" studies.

Scale-up studies are carried out at the laboratory or even pilot-plant-scale fermentors to yield data that could be used to extrapolate and build the large-scale industrial fermentors with the sufficient confidence that it will function properly with all its behaviors anticipated (Fig. 7.39). More important during scale-up exercises is to try to build industrial-size fermentors capable of or close to producing the fermentation products as efficient as those produced in small-scale fermentors.

Most scale-up studies are usually carried at different phases involving different scales of fermentors. Preliminary work are carried out at the level of petri dishes and small-scale laboratory fermentors to establish whether the process is technically viable, meaning it is possible to produce such fermentation process and the products on the small scale. Additional parameters not provided by petri dish studies and far more confidence are obtained by carrying further studies on submerged liquid fermentation using various-size laboratory-scale fermentors and even a pilot plant fermentor.

There are a few rules of the thumb followed when doing scale-up studies:

1. Similarity in the geometry and configuration of fermentors used in scaling up.

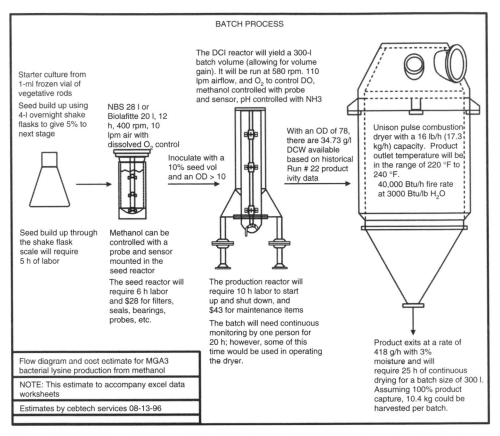

Figure 7.39. Scale-up studies carried out for large-scale industrial fermentors. DCW, dry cell weight.

2. A minimum of three or four stages of increment in the scaling up of the volume of fermentation studies. Each jump in scale should be by a magnitude or power increase and not an increase of a few liters capacity.

Slight increase in the working volume would not yield significant data for scale-up operation. The size of fermentation increases during scale-up; various measured parameters might not show predictable linear corelationships. Some parameters need to be modified and adjusted during scale-up studies. The objective is to try to get the same fermentation efficiency as obtained in small-scale fermentors at the most economical values.

7.5. MAJOR INDUSTRIAL PRODUCTS FOR HEALTH AND INDUSTRY

Use of microorganisms, usually grown on a large scale, to obtain valuable commercial products by way of significant chemical transformations is called *industrial microbiology*. This discipline of microbiology dates back and originated with beer and wine making fermentation processes (alcoholic fermentation) and subsequently expanded in the area of production of pharmaceuticals (e.g., antibiotics), food additives (e.g., amino acids), organic acids (e.g., butyric acid and citric acid), enzymes (e.g., amylases, proteases), and vitamins. All these products are obtained by enhancing the metabolic reactions that microorganisms were already capable of carrying out in natural conditions. But, at present, in addition to this traditional industrial microbiology, a new era of microbial biotechnology is expanding in which the genes of the microorganisms responsible for such and other metabolic reactions are being manipulated to produce many new products at commercial level.

7.5.1. Immobilized Enzymes

Immobilization of biocatalysts helps in their economic reuse and in the development of continuous bioprocesses. Biocatalysts can be immobilized using either the isolated enzymes or the whole cells. Immobilization often stabilizes structure of the enzymes, thereby allowing their applications even under harsh environmental conditions of pH, temperature, and organic solvents and thus enables their uses at high temperatures in nonaqueous enzymology and in the fabrication of biosensor probes. In the future, development of techniques for the immobilization of multienzymes along with cofactor regeneration and retention system can be gainfully exploited in developing biochemical processes involving complex chemical conversions.

Immobilization means associating the biocatalysts with an insoluble matrix, so that it can be retained in proper reactor geometry for its economic reuse under stabilized conditions thus allowing, by essence, to decouple the enzyme location from the flow of the liquid carrying the reagents and products. This technique helps in the development of continuous processes, allowing more economic organization of the operations, automation, decrease of labor, and investment/capacity ratio. Immobilized biocatalysts offer several other advantages, notable among them is the availability of the product in greater purity. Purity of the product is very crucial in food processing and pharmaceutical industry since contamination could cause serious toxicological, sensory, or immunological problems. The other major advantages include greater control over enzymatic reaction and high volumetric productivity with lower residence time, which are of great significance in the food industry, especially in the treatment of perishable commodities as well as in other applications involving labile substrates, intermediates, or products. The various methods of immobilization of whole cells by different methods are given in Table 7.9.

Immobilization of whole cells has been shown to be a better alternative to immobilization of isolated enzymes. Doing so avoids the lengthy and expensive operations of enzyme purification and preserves the enzyme in its natural environment, thus protecting it from inactivation either during immobilization or its subsequent use in continuous system. It may also provide a multipurpose catalyst, especially when the process requires the participation of a number of enzymes in sequence. The major limitations that may need to be addressed while using such cells are the diffusion of substrate and products through the cell wall and unwanted side reactions due to the presence of other enzymes.

The cells can be immobilized in either a viable or a nonviable form. Immobilized nonviable cell preparations, which are normally obtained by permeabilizing the intact cells, for the expression of intracellular activity are useful for simple processes that require a single enzyme with no requirement for cofactor regeneration, such as hydrolysis of sucrose or lactose. On the other hand, immobilized viable cells, which serve as

TABLE 7.9. Immobilization of Whole Cells by Different Methods

Support Material	Cells	Reaction
Adsorption		
Gelatin	Lactobacilli	Lactose/lactic acid
Porous glass	*Saccharomyces carlsbergensis*	Glucose/ethanol
Cotton fibers	*Zymomonas mobilis*	Glucose/ethanol
Vermiculite	*Z. mobilis*	Glucose/ethanol
DEAE-cellulose	*Nocardia erythropolis*	Steroid conversion
Covalent Bonding		
Cellulose + cyanuric chloride	*S. cerevisiae*	Glucose/ethanol
Ti(IV) oxide, etc.	*Acetobacter* sp.	Wort/vinegar
Carboxymethyl cellulose + carbodiimide	*Bacillus subtilis*	L-histidine/uronic acid
Cross-linking of cell to cell		
Diazotized diamines	*Streptomyces*	Glucose/fructose
Glutaraldehyde	*E. coli*	Fumaric acid/L-aspartic acid
Flocculation by chitosan	*Lactobacillus brevis*	Glucose/fructose
Entrapment		
Al alginate	*Candida tropicalis*	Phenol degradation
Ca alginate	*S. cerevisiae*	Glucose/ethanol
Mg pectinate	Fungi	Glucose/fructose
K-carrageenan	*E. coli*	Fumaric acid/L-aspartic acid
Chitosan alginate	*S. cerevisiae*	Glucose/ethanol
Encapsulation		
Cellulose acetate	*Comamonas* sp.	7-Aminocephalosporinic acid production
Ethyl cellulose	*Streptomyces* sp.	Glucose/fructose
Polyester	*Streptomyces* sp.	Glucose/fructose
Alginate-polylysine	Pancreas cells	—

"controlled catalytic biomass," have opened new avenues for continuous fermentation on heterogeneous catalysis basis by serving as self-proliferating biocatalysts.

Most of the enzymes used at industrial scale are normally the extracellular enzymes produced by the microbes. This has been mainly due to their ease of isolation as crude enzymes from the fermentation broth. Moreover, the extracellular enzymes are more stable to external environmental perturbations compared to the intracellular enzymes. However, over 90% of the enzymes produced by a cell are intracellular. The economic exploitation of these, having a variety of biochemical potentials, has been limited in view of the high cost involved in their isolation. Also, compared to extracellular enzymes, the intracellular enzymes are more labile. Delicate and expensive separation methods are required to release the enzymes undamaged from the cell and to isolate them. This increases the labor and the cost of the enzyme. These problems could now be obviated by the use of permeabilized cells as a source of enzyme. Permeabilization of the cell removes the barrier for the free diffusion of the substrate/product across the cell membrane and also empties the cell of most of the small molecular weight cofactors, thus minimizing the unwanted side reactions. The permeabilized *Kluyveromyces fragilis* cells, which are nonviable, convert lactose in milk to glucose and galactose, whereas the viable (nonpermeabilized) cells convert the lactose to ethanol and are useful in the complete desugaration of milk.

A variety of physical, chemical, and enzymatic techniques have been developed for the permeabilization of cells. Some techniques, such as entrapment in polyacrylamide, have been shown to result in permeabilization of the cells. It is evident that most of the major industrially important intracellular enzymes can now be immobilized using whole cells.

7.5.1.1. Techniques and Supports for Immobilization

A large number of techniques and supports are now available for the immobilization of enzymes or cells on a variety of natural and synthetic supports (Fig. 7.40). The choice of the support and the technique depends on the nature of the enzyme, nature of the substrate, and the enzyme's ultimate application. Techniques for immobilization have been broadly classified into four categories: entrapment, covalent binding, cross-linking, and adsorption. A combination of one or more of these techniques has also been investigated. Both the activity and the operational stability of the biocatalysts are important in terms of economy of the process.

A novel biotinylated and enzyme-immobilized nano-bioelement was first prepared by using heterobifunctional poly(glycidyl methacrylate-*co*-divinylbenzene)/polystyrene

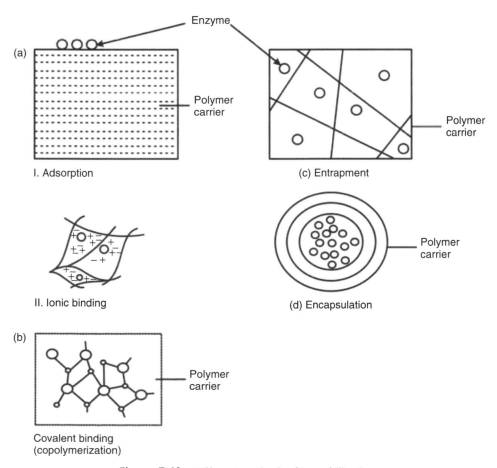

Figure 7.40. Different methods of immobilization.

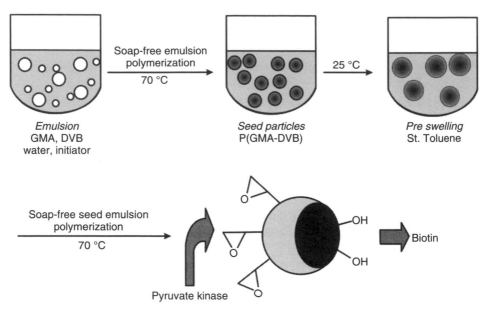

Figure 7.41. Preparation procedure of biotinylated and enzyme-immobilized latex beads.

(P(GMA-DVB)/PSt) composite latex beads (Fig. 7.41). 5-(*n*-succinimidyloxycarbonyl) pentyl D-biotinamide (Biotin-X-NHS) was first reacted with the hydroxyl group on the polystyrene domain of latex beads. Pyruvate kinase was then directly immobilized to the biotinylated latex beads through the epoxy groups on the latex bead surface. The concentration of pyruvate was monitored spectrophotometrically to obtain the maximum velocity (V_{max}) and the Michaelis constant (K_m) of covalently immobilized pyruvate kinase. The enzyme activity was roughly half of the free one when the concentration of the substrate was 100 μM, which remained almost unchanged even after storing at 4 $^\circ$C for 48 days (Yong-Zhong, 2004).

7.5.2. Antibiotics

Antibiotic is a substance produced by certain bacteria or fungi that kills other cells or interferes with their growth. In nature, these substances help some microbes survive by limiting the multiplication of other microbes that share the same environment. Antibiotics that attack pathogenic (disease-causing) microbes without severely harming normal body cells are useful as drugs.

Antibiotics are especially useful for treating infections caused by bacteria. Antibiotics came into widespread use during the 1940s. At that time, they were often called *wonder drugs* because they cured many bacterial diseases that were once fatal. The number of deaths caused by meningitis, pneumonia, tuberculosis, and scarlet fever declined drastically after antibiotics became available. Today, physicians prescribe antibiotics to treat many bacterial diseases.

In addition, some antibiotics are effective against infections caused by fungi and protozoa, and a few are useful in treating cancer. Antibiotics are also used to treat infectious diseases in animals. Farmers sometimes add small amounts of antibiotics to livestock feed. The antibiotics support the animals' growth for reasons that are not entirely understood.

Antibiotics are not effective against cold, influenza, or other viral diseases. In addition, the effectiveness of antibiotics is limited because both pathogenic microbes and cancer cells can become resistant to them.

7.5.2.1. The Manufacturing Process

Although most antibiotics occur in Nature, they are not normally available in the quantities necessary for large-scale production (Fig. 7.42).

For this reason, a fermentation process was developed. It involves isolating a desired microorganism, fueling growth of the culture, and refining and isolating the final antibiotic product. It is important that sterile conditions be maintained throughout the manufacturing process, because contamination by foreign microbes will ruin the fermentation.

7.5.2.2. Starting the Culture

Before fermentation can begin, the desired antibiotic-producing organism must be isolated and its number must be increased by many times. To do this, a starter culture from a sample of previously isolated, cold-stored organisms is created in the laboratory. In order to grow the initial culture, a sample of the organism is transferred to an agar-containing plate. The initial culture is then put into shake flasks along with food and other nutrients necessary for growth. This creates a suspension, which can be transferred to seed tanks for further growth.

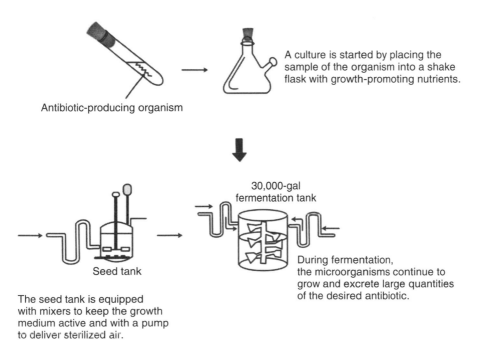

Figure 7.42. Schematic diagram of the manufacturing process of antibiotics.

The seed tanks are steel tanks designed to provide an ideal environment for growing microorganisms. They are filled with all the things the specific microorganism would need to survive and thrive, including warm water and carbohydrates such as lactose or glucose sugars. In addition, they contain other necessary carbon sources, such as acetic acid, alcohols, or hydrocarbons, and nitrogen sources such as ammonia salts. Growth factors such as vitamins, amino acids, and minor nutrients round out the composition of the seed tank contents. The seed tanks are equipped with mixers, which keep the growth medium moving, and a pump to deliver sterilized, filtered air. After about 24–28 h, the material in the seed tanks is transferred to the primary fermentation tanks.

7.5.2.3. Fermentation

The fermentation tank is essentially a larger version of the steel, seed tank, which is able to hold about 30,000 gal. It is filled with the same growth media found in the seed tank and also provides an environment inductive to growth. Here, the microorganisms are allowed to grow and multiply. During this process, they secrete large quantities of the desired antibiotic. The tanks are cooled to keep the temperature between 73 and 81 °F (23–27 °C). It is constantly agitated, and a continuous stream of sterilized air is pumped into it. For this reason, antifoaming agents are periodically added. Since pH control is vital for optimal growth, acids or bases are added to the tank as necessary.

7.5.2.4. Isolation and Purification

After 3–5 days, the maximum amount of antibiotic will have been produced and the isolation process can begin. Depending on the specific antibiotic produced, the fermentation broth is processed by various purification methods. For example, for antibiotic compounds that are water soluble, an ion-exchange method may be used for purification. In this method, the compound is first separated from the waste organic materials in the broth and then sent through an equipment that separates the other water-soluble compounds from the desired one. To isolate an oil-soluble antibiotic, such as penicillin, a solvent extraction method is used. In this method, the broth is treated with organic solvents such as butyl acetate or methyl isobutyl ketone, which can specifically dissolve the antibiotic. The dissolved antibiotic is then recovered using various organic chemical means. At the end of this step, the manufacturer is typically left with a purified powdered form of the antibiotic, which can be further refined into different product types.

7.5.2.5. Refining

Antibiotic products can take on many different forms. They can be sold in solutions for intravenous bags or syringes, in pill or gel capsule form, or as powders, which are incorporated into topical ointments. Depending on the final form of the antibiotic, various refining steps may be taken after the initial isolation. For intravenous bags, the crystalline antibiotic can be dissolved in a solution and put in the bag, which is then hermetically sealed. For gel capsules, the powdered antibiotic is physically filled into the bottom half of a capsule then the top half is mechanically put in place. When used in topical ointments, the antibiotic is mixed into the ointment.

From this point, the antibiotic product is transported to the final packaging stations. Here, the products are stacked and put in boxes. They are loaded up on trucks and transported to various distributors, hospitals, and pharmacies. The entire process of fermentation, recovery, and processing can take anywhere from 5 to 8 days.

7.5.2.6. Quality Control

QC is of utmost importance in the production of antibiotics. Since it involves a fermentation process, steps must be taken to ensure that absolutely no contamination is introduced at any point during production. To this end, the medium and all the processing equipment are thoroughly steam sterilized. During manufacturing, the quality of all the compounds is checked on a regular basis. Of particular importance are frequent checks of the condition of the microorganism culture during fermentation. These are accomplished by using various chromatography techniques. Also, various physical and chemical properties of the finished product are checked, such as pH, melting point, and moisture content.

In the United States, antibiotic production is highly regulated by the FDA. Depending on the application and type of antibiotic, more or less testing must be completed. For example, the FDA requires that for certain antibiotics, each batch must be checked by it for effectiveness and purity. Only after the FDA has certified the batch, it can be sold for general consumption.

7.5.2.7. Kinds of Antibiotics

Antibiotics are selectively toxic, that is, they damage some types of cells without harming others. Medically useful antibiotics attack infectious microbes or cancer cells without excessively hurting human cells. Antibiotics fight different types of illnesses in a variety of ways.

ANTIBACTERIAL ANTIBIOTICS. Antibiotics are selectively toxic against bacteria because bacterial cells differ greatly from human cells. One of the chief differences is that bacteria, unlike animal cells, have a cell wall (Fig. 7.43). This wall is a rigid structure that forms the cell's outer boundary.

The type of cell wall a bacterium has is one factor that determines which antibiotics can kill it. Scientists use a process called *Gram staining* to classify cell walls of bacteria. Hans C. J. Gram, a Danish bacteriologist of the late 1800s, developed this process. This method classifies bacteria as gram positive (G+) or gram negative (G−). Some antibiotics selectively kill either gram-positive or gram-negative bacteria. These substances are called *narrow-spectrum antibiotics*. The antibiotic vancomycin selectively kills gram-positive bacteria such as *Staphylococcus, Streptococcus*, and *Enterococcus*. Aztreonam is a narrow-spectrum antibiotic that kills only gram-negative bacteria, such as *E. coli* and *Pseudomonas aeruginosa*. Other antibiotics can kill both gram-positive and gram-negative bacteria. These drugs are called *broad-spectrum antibiotics*. Ceftriaxone is one example of a broad-spectrum antibiotic. No broad-spectrum antibiotic can kill all bacteria, and no narrow-spectrum antibiotic can kill all gram-positive or all gram-negative bacteria.

OTHER KINDS. Some antibiotics are effective against infections caused by fungi and protozoans, whose cells differ from human cells. Antibiotics that fight fungi include

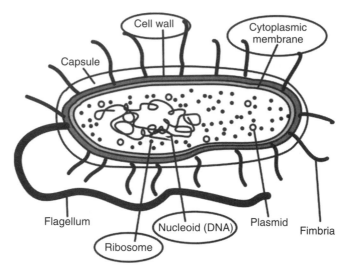

Figure 7.43. Antibiotics are targeted to kill the bacterial cells.

miconazole and amphotericin. Paromomycin is used to treat amebiasis, an intestinal disease caused by a protozoan.

Anticancer antibiotics attack cells while they are dividing. These drugs are somewhat selectively toxic because cancer cells generally divide much more frequently than normal cells. But some normal cells such as blood-forming cells divide rapidly. Anticancer antibiotics also affect these cells. The antibiotic doxorubicin is used to treat certain types of leukemia, breast cancer, and other tumors.

Some Widely Used Antibiotics

Ampicillin is effective against G+ and G− respiratory and urinary tract infections.

Azithromycin is effective against different kinds of pneumonia and certain other G+ infections.

Ceftriaxone is effective against G+ and G− infections, including gonorrhea.

Chloramphenicol is used against Rocky Mountain spotted fever and G+ and G− infections, including meningitis.

Ciprofloxacin is used in urinary tract infections and acute diarrheal diseases caused by certain G− bacteria.

Dicloxacillin is used against penicillin G-resistant and methicillin-sensitive staphylococcal infections.

Doxycycline is used against pneumonia, G+, and G− infections, bite wounds, acne.

Fluconazole is used in fungal infections of skin, mucous membranes, blood, and brain.

Gentamicin: is used in serious infections, especially G- infections.

Neomycin is used against G+ and G− infections, especially skin infections and those resulting from burns.

Penicillin G is used to treat syphilis, strep throat, and other G+ infections.

Rifampin is used to treat tuberculosis.

Streptomycin is used in the treatment of tuberculosis and bubonic plague.

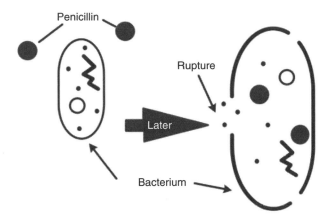

Figure 7.44. Mode of action of penicillin.

Vancomycin is used in case of serious staphylococcal, enterococcal, and streptococcal infections that resist other drugs.

7.5.2.8. Mode of Action of Antibiotics

Antibiotics fight microbes and cancer cells by interfering with normal cell functions (Fig. 7.44). In most cases, this interference occurs in one of three ways: (i) prevention of cell wall formation, (ii) disruption of the cell membrane (covering), and (iii) disruption of chemical processes.

PREVENTION OF CELL WALL FORMATION. Penicillins and some other antibiotics destroy microbes by interfering with their cell wall formation. Animal cells do not form walls. As a result, these antibiotics do not damage them.

DISRUPTION OF THE CELL MEMBRANE. All cells have a membrane that controls the movement of substances in and out of the cell. Some antibiotics, including amphotericin and nystatin, disrupt the cell membrane of certain microbes. A damaged membrane might allow vital nutrients to escape or poisonous substances to enter and kill the cell. These antibiotics do not harm human cells because the drugs affect membrane components found only in microbial cells.

DISRUPTION OF CHEMICAL PROCESSES. All cells produce proteins and nucleic acids, which are vital to life. Human cells produce these substances in much the same way as microbial cells do. But in some cases, these processes differ enough so that antibiotics interfere with the chemical activities in microbial cells, but not in human cells. For example, streptomycin and tetracycline prevent certain kinds of microbes from producing proteins and rifampin interferes with the formation of nucleic acids.

7.5.2.9. Dangers of Antibiotics

Many antibiotics are regarded among the safest drugs when properly used. But antibiotics can sometimes cause unpleasant or dangerous side effects. The three main dangers are

(i) allergic reactions, (ii) destruction of helpful microbes, and (iii) damage to organs and tissues.

ALLERGIC REACTIONS. In most cases, these reactions are mild and produce only a rash or fever. But severe reactions can occur and can even cause death. All antibiotics are able to produce allergic reactions, but such reactions occur most often with penicillins. A physician usually asks if a patient has ever had an allergic reaction to an antibiotic before prescribing that drug. Most people who are allergic to one antibiotic can take other antibiotics that have significantly different chemical compositions.

DESTRUCTION OF HELPFUL MICROBES. Certain areas of the body commonly harbor both harmless and pathogenic microbes. These two types of microbes compete for food, and so, the harmless microorganisms help restrain the growth of those that cause disease. Many antibiotics, especially broad-spectrum drugs, do not always distinguish between harmless and dangerous microbes. If a drug destroys too many harmless microorganisms, the pathogenic ones will have a greater chance to multiply. This situation can lead to a new infection called a *superinfection*. Physicians usually prescribe a second drug to combat a superinfection.

DAMAGE TO ORGANS AND TISSUES. This is rare in people using antibiotics that act only against the cells of pathogenic microbes. Extensive use of some antibiotics, however, may damage tissues and organs. For example, streptomycin is known to cause kidney damage and deafness. Physicians prescribe drugs with such known risks only if no other drug is effective (Fig. 7.45).

Anticancer antibiotics act against all cells that divide rapidly and so can affect both normal and cancer cells. For example, cells in the bone marrow divide constantly to produce fresh blood cells. Anticancer antibiotics can damage the bone marrow. Such damage increases the risk of infection by reducing the number of white blood cells, which help the body to fight against disease.

7.5.2.10. Resistance to Antibiotics

Some pathogenic microbes develop an ability to resist the effects of certain antibiotics. The most widespread and worrisome resistance in pathogenic microbes occurs in bacteria.

Bacteria can become resistant to antibiotics through a type of evolution. In bacteria, as in other living things, genes carry instructions controlling life processes. Occasionally, a gene in a bacterium naturally changes in a way that enables the microbe to resist the effects of an antibiotic. Such a change is called a *mutation*. The change may provide resistance to one specific antibiotic or to a group of chemically similar antibiotics, for example, the penicillins. Bacteria can also acquire resistance from other bacteria by the transfer of genetic material. The antibiotics also show differences in their mode of action (Table 7.10).

7.5.3. Organic Acids

Microbial production of organic acids is a promising approach for obtaining building-block chemicals from renewable carbon sources. Although some acids have been produced for some time and in-depth knowledge of these microbial production processes

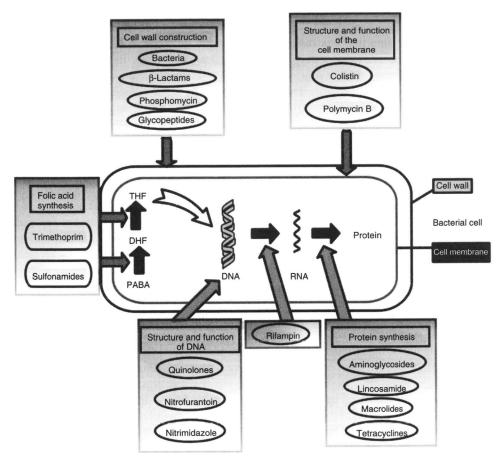

Figure 7.45. Action of antibiotics. PABA, *para*-aminobenzoic acid; DHF, dihydrofolate; THF, tetrahydrofolate.

TABLE 7.10. Various Antibiotics have Different Modes of Action

Antibiotic Type	Mode of Action	Bacterial Resistance Mechanism
Aminoglycosides: gentamicin	Block protein synthesis	Inactivation
β-Lactams: penicillins, cephalosporins	Block cell wall formation	Inactivation, mutation
Glycopeptides: vancomycin	Block cell wall formation	Mutation of binding molecules
Macrolides	Block protein synthesis	Ribosome protection
Quinolones	Inhibit DNA replication	Mutation of binding molecules
Rifampin	Inhibits bacterial RNA polymerase	Mutation of binding molecules
Tetracyclines	Block protein synthesis	Inactivation
Trimethoprim	Block formation of nucleic acids and f-met	Mutation of binding molecules

has been gained, further microbial production processes seem to be feasible, but large-scale production has not yet been possible. Citric, lactic, and succinic acid production exemplify three processes in different stages of industrial development. Although the questions being addressed by current research on these processes are diverging, a comparison is helpful for understanding microbial organic acid production in general (Saue et al., 2008). The oldest microbial process for production of a high volume, low cost organic acid is the production of citric acid by the filamentous fungus *A. niger*. Currently, the yearly production of citric acid is approximately 1.6 million tons. Unlike most of the other bioderived acids that are considered industrial products, citric acid was produced industrially before the development of a microbial process.

The industrial production relied on extraction from Italian lemons until it was discovered that *Aspergillus* accumulates this acid in high amounts under certain conditions. The crucial parameters resulting in efficient production of citric acid by *A. niger* have been determined empirically and include high substrate concentration, low and finite content of nitrogen and certain trace metals, thorough maintenance of high DO, and low pH. The exact definition of these parameters enabled the development of highly efficient biotechnological processes. However, many of the biochemical and physiological mechanisms underlying the process remain unknown. These mechanisms are currently undergoing investigation to enable improvement of the citric acid production process, for which significant improvement is no longer possible through traditional means, such as mutagenesis or cultivation optimization. In addition to the well-established filamentous fungal species, the yeast *Yarrowia lipolytica* has been developed as a microbial cell factory for citric acid. The starting point for this line of research was to gain access to *n*-paraffins and fatty acids (as animal fats) that are not converted by *A. niger* as carbon sources. However, *Y. lipolytica* also proved efficient in the production of citric acid from other carbon sources, such as glucose and sucrose. Citric acid concentrations of 140 g/l are now easily reached, and *Y. lipolytica* is probably used on an industrial scale, although few details are known of the actual production methods.

Lactic acid and its production by lactic acid bacteria (LAB) have a long history in the food industry, and microbial processes for lactic acid production were established early in the past century. However, the large-scale commercial production of the purified acid by microorganisms is relatively new. The production of the biodegradable plastic polylactide (e.g., used in food containers) led to increased interest in optically pure lactic acid. This accounts for the recent shift from chemical to microbial production processes. Approximately 150,000 tons of lactic acid were produced in 2002, 90% of which was by fermentation with LAB. LAB have complex nutrient requirements, and they ferment sugars via different pathways, resulting in homo-acid, hetero-acid, or mixed acid fermentation (Fig. 7.46). However, it is not only bacteria that accumulate lactate. The filamentous fungus *Rhizopus oryzae* is another natural producer that has the advantage of growing on mineral medium and carbon sources such as starch or xylose.

The market for succinic (amber) acid is currently small and 16,000 tons per year. However, if the price becomes competitive, succinic acid could replace petroleum-derived maleic anhydride, which has a market volume of 213,000 tons per year. An even higher market volume is conceivable for succinic acid, as it is a versatile building-block chemical suitable for many uses. Replacing petroleum-derived chemicals and taking into account that succinic acid formation consumes CO_2 (theoretically 1 mol CO_2 per mole of succinic acid produced), the introduction of succinic acid as a commodity building block has the potential to lead to large reductions in environmental pollution. To date, no industrial

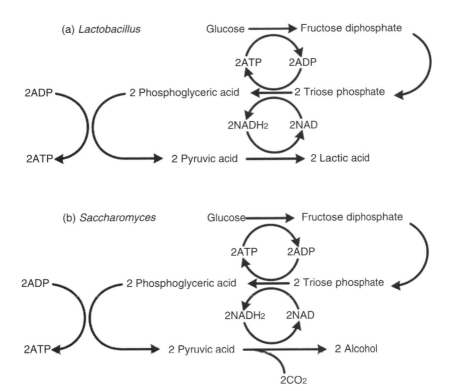

Figure 7.46. Lactic acid and alcohol fermentation.

process for microbial succinic acid production has been established; however, calculations show that such a process can be competitive provided that some of the outlined issues are resolved. The first approach for microbial production of succinic acid was the engineering of the mixed acid fermentation of *E. coli. Apfl ldhA* double mutant *E. coli* strain NZN111 was used to produce succinic acid by overexpressing the *E. coli* malic enzyme gene (*sfcA*). This strain, however, produced a large amount of both malic acid and succinic acid (Hong and Lee, 2004).

Later it was discovered that several anaerobic rumen bacteria naturally produce large amounts of succinic acid. However, cultivation of such bacteria is dependent on expensive and complex nutrient sources, and by-product formation is a general problem that remains to be solved.

7.5.4. Enzymes

Microbial enzymes are most widely used in the food and beverage industries and to a lesser extent in clinical and analytical laboratories and as protease detergents in washing powders. The most economical and convenient method of producing these enzymes is by microbial fermentation. *Bacillus stearothermophilus* produces amylases as secondary metabolites, but most other microbes produce enzymes as primary metabolites, during exponential growth. The major commercial utilization of microbial enzymes is in the food and beverage industries (Jeenes et al., 1991), pharmaceutical industry, textile and leather industries, analytical processes, detergents, and dairy industry. Most enzymes are synthesized in the logarithmic phase of the batch culture and may, therefore, be considered as primary metabolites. However, some, for example, amylases of *B. stearothermophilus* are produced by idiophase-type cultures and may be considered as equivalent to secondary metabolites (Manning and Campbell, 1961). Enzymes may be produced from animals and plants, as well as microbial sources, but the production by microbial fermentation is the most economical and convenient method. For example, microbial rennin is being used instead of calf rennin and proteases used in detergents are produced by microbes. It is now possible to engineer microbial cells to produce animal or plant enzymes. Most of the enzymes in industrial use are extracellular proteins produced by *Aspergillus* sp. or *Bacillus* sp. and include α-amylase, β-glucanase, cellulase, dextranase, lactase, lipase, pectinase, proteases. The extent of purification required for most of these enzymes is minimal, and they can be produced in tons without serious problems.

7.5.4.1. Commercial Production

Most of the commercial enzymes are obtained from aerobic microorganisms. The following two processes are generally employed.

1. *Surface Culture Method.* This method was largely used before the World War II. In this method, shallow trays were used to grow microorganisms. Hence, large-scale productions of enzymes were difficult and costly.
2. *Semisolid or Solid Culture Method.* In this method (Fig. 7.47), the enzyme production and fermentations are carried out in closed, deep tank fermentors varying between 1000- and 30,000-gal capacity. A substrate (e.g., wheat bran, sugar beet, cellulose are used depending on the type of microorganisms to be taken to excrete the enzyme) moistened with a suitable nutrient solution is sterilized and then inoculated with mold spores (most of the enzymes are of fungal origin). The substrate allows air to penetrate and the mold mycelia to grow throughout. When the enzyme level reaches its maximum, the fermented mass is extracted with water and the enzyme is separated out with a suitable solvent. It is now filtered, concentrated, and dried.

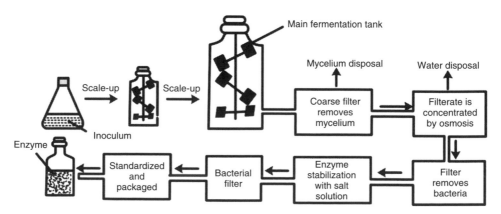

Figure 7.47. Schematic representation of enzyme production.

TABLE 7.11. Some Important Microbial Enzymes and Their Applications

Source (Genus)	Enzymes	Reaction	Application
Bacillus	α-Amylase	Starch hydrolysis	Converts starch to glucose or dextran in food industry
	Proteases	Protein digestion	Help laundering
Escherichia	Penicillin acylase	Benzoyl cleavage	6-Aminopenicillanic acid production
	L-asparaginase	Removal of L-asparagine involved in tumor growth	Leukemia/cancer treatment
Aspergillus	Amyloglucosidase	Dextrin hydrolysis	Glucose production
	β-Galactosidase	Lactose hydrolysis	Lactose hydrolysis in milk or whey
	Aminoacylase	Hydrolysis of acylated L-amino acids	Resolution of racemic mixtures
	Glucose oxidase	Oxidation of glucose	Glucose detection in blood
Streptomyces	Glucose isomerase	Conversion of glucose to fructose	Production of high fructose syrups
Several marine bacteria	Luciferase	Bioluminescence	Assay for ATP
Several bacteria and cyanobacteria	Nucleases (restriction endonucleases)	Hydrolysis of phosphodiester bonds in nucleic acids	Genetic engineering

Microbial enzymes are also used for the production of synthetic polymers. Plastic industry mostly uses chemical methods for producing alkene oxides used in the production of plastics. It is now possible to synthesize alkene oxides using microbial enzymes, and genetically engineered strains would make commercial production feasible. Some microbial enzymes and their applications are shown in Table 7.11.

The synthesis of alkene oxides from alkenes requires the sequential action of three enzymes: pyranose-2-oxidase from the fungus *Oudemansiella mucida*, a haloperoxidase from the fungus *Caldariomyces*, and an epoxidase from a *Flavobacterium* sp.

7.5.5. Solvents

Although the production of acetone and butanol by *Clostridium* strains was a thriving industrial fermentation process, it is no longer competitive with the chemical synthesis of solvents and has been discontinued. However, studies on the molecular biology of *Clostridium* strains suggest that genetic engineering for improved solvent production is feasible and could result in the revival of the industrial fermentation process. Acetone and butanol were historically produced through fermentation of carbohydrate raw materials. Conventional feedstocks such as grain and molasses, and the energy required to recover products by distillation, are too costly for traditional batch fermentation to compete with petrochemical synthesis. Common uses of organic solvents are in dry cleaning (e.g., tetrachloroethylene), as a paint thinner (e.g., toluene, turpentine), as nail polish removers and glue solvents (acetone, methyl acetate, ethyl acetate), in spot removers (e.g., hexane, petrol ether), in detergents (citrus terpenes), in perfumes (ethanol), and in chemical synthesis. The use of inorganic solvents (other than water) is typically limited to research chemistry and some technological processes. An important property of solvents is boiling point. This also determines the speed of evaporation. Small amount of low boiling point solvents such as diethyl ether, dichloromethane, or acetone will evaporate in seconds at room temperature, while high boiling point solvents such as water or dimethyl sulfoxide need higher temperatures, an airflow, or the application of vacuum for fast evaporation.

> Low boiling temperature: below 100 °C (boiling point of water).
>
> Medium boiling temperature: between 100 and 150 °C.
>
> High boiling temperature: above 150 °C.

Most organic solvents have a lower density than water, which means they are lighter and will form a separate layer on top of water. Most of the halogenated solvents such as dichloromethane or chloroform will sink to the bottom of a container, leaving water as the top layer.

7.5.6. Amino Acids

In the 1950s, *Corynebacterium glutamicum* was found to be a very efficient producer of L-glutamic acid. Since this time, biotechnological processes with bacteria of the species *Corynebacterium* developed to be among the most important in terms of tonnage and economical value. L-glutamic acid and L-lysine are bulk products nowadays. L-valine, L-isoleucine, L-threonine, L-aspartic acid, and L-alanine are among other amino acids produced by corynebacteria. Applications range from feed to food and pharmaceutical products. The growing market for amino acids produced with corynebacteria (Fig. 7.48) led to significant improvements in bioprocess and downstream technology, as well as in molecular biology. During the past decade, big efforts were made to increase the productivity and to decrease the production costs. At present, the amino acid industry has come to occupy an important role in world chemical industries. China is an agricultural country and has a very large population. Its annual demand for amino acids used in feed additives and pharmaceutical products is huge. Most of the amino acids are currently manufactured in China. But the industrial production process has not been set up for a few limited kinds of amino acids such as L-tryptophan, L-histidine, and L-arginine. The extraction method is still an industrial process for L-cysteine. However, the extraction method depends on the availability of natural protein-rich resources such as hair keratin and feather, and the production process is not environment friendly as unpleasant odors are produced and there are problems of waste treatment. So there is an urgent need for establishing the production process to meet the demand for these amino acids.

The fermentation method is being applied to industrial production of most L-amino acids. Coryneform bacteria have played a principal role in the progress of amino acid fermentation industry. However, the precise genetic and physiological changes resulting in increased overproduction of amino acids in various coryneform bacterial strains have remained unknown. Success in attempts to further increase the productivity and yields of already highly productive strains will depend on the availability of detailed information on the metabolic pathways, their regulation, and mutations.

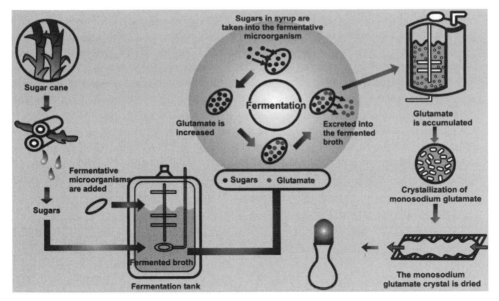

Figure 7.48. Production of monosodium glutamate by fermentation.

C. glutamicum, Corynebacterium crenatum, and *Corynebacterium pekingese* are used in amino acid production process. But the mechanism of amino acid accumulation of those mutants has not been extensively investigated. So far, the strain improvement has mainly been carried out by an iterative process of mutagenesis and screening. It is difficult to increase the production yield further by these methods.

7.5.6.1. Microbial Production of L-Tryptophan

A microorganism with amino acid racemase activity was screened and identified as *Pseudomonas putida*. The amino acid racemase has low substrate specificity. It has activity toward most aliphatic amino acids. A technique of L-tryptophan production was established by coupling the reactions catalyzed by amino acid racemase and tryptophan synthase. Optimization of cell cultivation and transformation conditions was carried out. In addition, two distinct amino acid racemase genes from *P. putida* were cloned. Only alanine racemase activity was found in recombinant *E. coli* TG1 containing DadX gene. Another amino acid racemase encoded by alr gene showed low substrate specificity.

7.5.6.2. Enzymatic Synthesis of L-Cysteine

DL-2-amino-A2-thiazoline-4-carboxylic acid (DL-ATC) is a precursor of L-cysteine synthesis. A microorganism having the ability to utilize DL-ATC as a sole carbon and nitrogen source and accumulating L-cysteine was screened and identified as *Pseudomonas* sp. A 6-kb DNA fragment isolated from the genome DNA of *Pseudomonas* sp. was shown to be involved in the conversion of DL-ATC. Sequence analysis showed that this DNA fragment contained genes encoding ATC hydrolase and *N*-carbamoyl-L-cysteine hydrolase.

Metabolic Engineering for *C. crenatum*

1. Aspartokinase (AK) gene, phosphenolpyruvate carboxylase (PPC) gene, and pyruvate carboxylase (PYC) gene from wild-type *C. crenatum* and aminoethyl cysteine (AEC)-resistant mutant strain were cloned and sequenced. Comparison of the two AK gene sequences showed that there happened a single point mutation L80P in the β-subunit of AK of the AEC-resistant mutant. This mutation resulted in antifeedback regulation of the enzyme activity. Overexpressions of those three genes in *C. crenatum* were investigated. The simultaneous amplification of the activities of both AKfbr and PYC resulted in growth further with about 50% increase of L-lysine production in the middle phase and 18% increase in the late phase.

2. *N*-acetyl glutamate kinase genes from wild-type strain of *C. crenatum* and L-arginine-producing mutant were cloned and sequenced. The accumulation of L-arginine in the mutant was due to the increase in enzyme activity. Overexpression of argB yielded about 25% increase of L-arginine production.

7.5.6.3. Studies on the Metabolic Pathway of Cyclic Imide in *Alcaligenes* eutrophus

A hydantoin-cleaving microorganism 112R4 was screened and identified as *Alcaligenes eutrophus*. The *A. eutrophus* can utilize succinimide as a sole carbon and nitrogen source, indicating that a complete transformation pathway of succinimide existed and imidase and half-amidase were suggested to be involved in the metabolic pathway. A 6-kb EcoRI-EcoRI fragment isolated from the genomic DNA of *A. eutrophus* 112R4 was shown to be involved in the transformation of succinimide. Sequence analysis showed the existence of five continuous ORFs in this fragment. The functions of the five ORFs were confirmed by deletion analysis, functional analysis, and homology search confirming encoded half-amidase, amide transport protein, imidase, and two dehydrogenases subunits. An extensive investigation of imidase and half-amidase as well as those two encoding genes was carried out. Another 6-kb DNA fragment also involved in succinimide hydrolysis was cloned. These two 6-kb fragments have short overlap sequences indicating that other metabolic pathways of imide existed in this bacterium.

7.5.7. Vitamins

Out of all vitamins now available commercially, vitamin B_{12} and vitamin B_2 (riboflavin) are the main vitamins produced by microbial fermentations.

7.5.7.1. Vitamin B_{12} (Cyanocobalamine; Cobalbumin)

This vitamin is recovered as a by-product of streptomycin and aureomycin antibiotic fermentations. A soluble cobalt salt is added to the fermentation reaction as a precursor to vitamin B_{12}. Relatively high amount of this vitamin accumulates in the fermentation medium at concentrations not toxic to *Streptomyces* species. Vitamin B_{12} (cyanocobalamine) is also produced on a large scale by direct fermentation. *Propionibacterium shermanii* or *Pseudomonas denitrificans* are the bacteria that are used nowadays for fermentation processes. *P. shermanii* is grown in anaerobic culture for 3 days at 30 °C and in aerobic culture for 4 days; the fermentation medium (growth medium) contains glucose, corn steep liquor (a waste product of starch manufacture), ammonia, and cobalt chloride.

Ammonia is used in the form of ammonium hydroxide that maintains the pH of the medium at 7. *P. denitrificans* is grown for 2 days in an aerobic culture medium containing sucrose, betaine, glutamic acid, cobalt chloride, 5,6-dimethylbenzimidazole, and salts. The vitamins produced by fermentation are retained within the cells. The cells are, therefore, collected by high speed centrifugation, and the vitamin B_{12} is recovered by releasing the vitamin from the cells by treatment with acid, heat, and so on. The vitamin thus released from the cell is absorbed on ion-exchange resin IRC-50 or charcoal. It is then purified further by phenol and water. The vitamin is finally crystallized from aqueous-acetone solutions. The normal yield is 23 mg/l when *Propionibacterium* is used.

7.5.7.2. Vitamin B₂ (Riboflavin)

Riboflavin (vitamin B_2) is recovered as a by-product of acetone-butanol fermentation and is produced by various *Clostridium* species. But, the yeasts *Eremothecium ashbyii* (also known as *Ashbya gossypii*) and *Ashnua gossypii* are used to commercially produce this vitamin by direct fermentation. The requirement nutrient medium (growth medium) for fermentation contains glucose, corn oil, soy oil, and glycine. When *E. ashbyii* is used, the fermentation is normally carried out for 4–6 days at 36 °C in aerobic conditions and at pH 6–7.5. The recovery of the vitamin is at the rate of 4.25 g/l. Later on, the yeast cells are recovered and used as a feed supplement for domestic animals to supply the riboflavin required by them.

7.5.8. Yeast

The yeast *S. cerevisiae* has been used in baking and in fermenting alcoholic beverages for thousands of years. It is also extremely important as a model organism in modern cell biology research and is one of the most thoroughly researched eukaryotic microorganisms. Researchers have used it to gather information about the biology of the eukaryotic cell and ultimately human biology. Other species of yeast, such as *Candida albicans*, are opportunistic pathogens and can cause infections in humans. The useful physiological properties of yeasts have led to their use in the field of biotechnology. Fermentation of sugars by yeast is the oldest and largest application of this technology. Many types of yeasts are used for making many foods: baker's yeast in bread production, brewer's yeast in beer fermentation, and in wine fermentation and xylitol production. Yeasts have recently been used to generate electricity in microbial fuel cells and produce ethanol for the biofuel industry (Fig. 7.49).

Yeasts are chemoorganotrophs, as they use organic compounds as a source of energy and do not require sunlight to grow. Carbon is obtained mostly from hexose sugars such as glucose and fructose or disaccharides such as sucrose and maltose. Some species can metabolize pentose sugars such as ribose, alcohols, and organic acids. Yeast species either require oxygen for aerobic cellular respiration (obligate aerobes) or are anaerobic but also have aerobic methods of energy production (facultative anaerobes). Unlike bacteria, there are no known yeast species that grow only anaerobically (obligate anaerobes). Yeasts grow best in a neutral or slightly acidic pH environment.

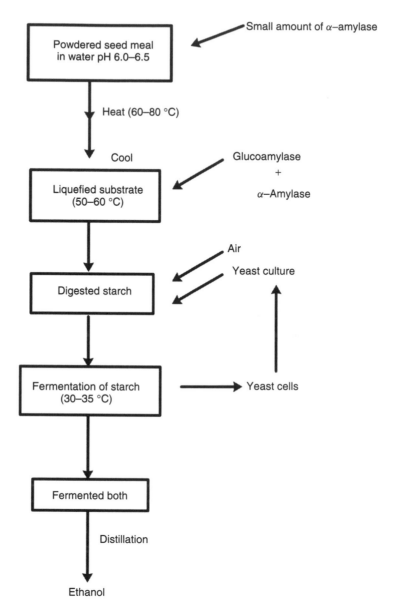

Figure 7.49. Production of ethanol for biofuel industry.

Yeasts vary in what temperature range they grow best. For example, *Leucosporidium frigidum* grows at $-2-20\,°C$ ($28-68\,°F$), *Saccharomyces telluris* at $5-35\,°C$ ($41-95\,°F$), and *Candida slooffi* at at $28-45\,°C$ ($82-113\,°F$). The cells can survive freezing under certain conditions, with viability decreasing over time.

Yeasts are generally grown in the laboratory on solid growth media or in liquid broths. Common media used for the cultivation of yeasts include potato dextrose agar (PDA) or potato dextrose broth, Wallerstein Laboratories nutrient (WLN) agar, yeast peptone dextrose (YPD) agar, and yeast mold (YM) agar or broth. Home brewers who cultivate yeast frequently use dried malt extract and agar as a solid growth medium. The antibiotic cycloheximide is sometimes added to yeast growth media to inhibit the growth of *Saccharomyces* yeasts and select for wild/indigenous yeast species. This will change the yeast process. Yeasts have asexual and sexual reproductive cycles. The most common mode of vegetative growth in yeasts is asexual reproduction by budding, where a small bud, or daughter cell, is formed on the parent cell. The nucleus of the parent cell splits into a daughter nucleus and migrates into the daughter cell. The bud continues to grow until it separates from the parent cell, forming a new cell. Some yeasts, including *Schizosaccharomyces pombe*, reproduce by binary fission instead of budding.

7.5.9. Single Cell Protein

SCP is the name given to a variety of microbial products that are produced by fermentation. When properly produced, these materials make satisfactory pertinacious ingredients for animal feed or human food (Fig. 7.50). The production of protein from hydrocarbon wastes of the petroleum industry is the most recent microbiological development in industry. Yeast, fungi, bacteria, and algae are grown on hydrocarbon wastes, and cells are harvested as sources of protein. As per reports, 100 lb of yeast will produce 250 tons of proteins in 24 h, whereas a 1000-lb steer will synthesize only 1 lb of protein in 24 h and this is after consuming 12–20 lb of plant proteins. Similar algae grown in ponds

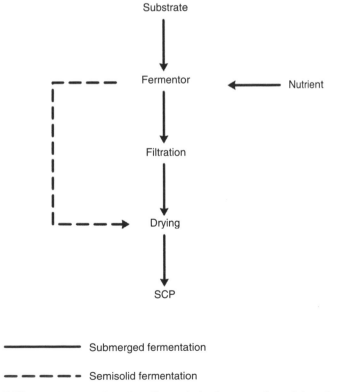

Figure 7.50. Production of single cell protein by fermentation of the substrate.

can produce 20 tons (dry weight) of protein per acre per year. This yield is 10–15 times higher than that of soybean and 25–50 times higher than that of corn. There are both advantages and disadvantages in using microorganisms for animal or human consumption. Advantages are that bacteria are usually high in protein (50–80%) and have a rapid growth rate. The principal disadvantages are as follows:

1. Bacterial cells have small size and low density, which make harvesting from the fermented medium difficult and costly.
2. Bacterial cells have high nucleic acid content relative to yeast and fungi. This can be detrimental to human beings, tending to increase the uric acid level in blood. This may cause uric acid poisoning or gout. To decrease the nucleic acid level, an additional processing step has to be introduced and this increases the cost.
3. The general public thinking is that all bacteria are harmful and produce disease. An extensive education program is required to remove this misconception and to make the public accept bacterial protein.

Since World War I, rapid development took place in biotechnological applications of *S. cerevisiae*, as far as culture development, process optimization, and scale-up of products are concerned. World production of yeast biomass is of the order of 0.4 million metric tons per annum including 0.2 million tons baker's yeast alone.

Yeasts synthesize amino acids from inorganic acids and sulfur supplemented in the form of salts. They get carbon and energy sources from the organic wastes, for example, molasses, starchy materials, milk whey, fruit pulp, wood pulp, and sulfite liquor.

Yeasts have the advantages of larger size (easier to harvest), lower nucleic acid content, high lysine content, and the ability to grow at acid pH. However, the most important advantage is familiarity and acceptability because of the long history of its use in traditional fermentations. Disadvantages include lower growth rates, lower protein content (45–65%), and lower methionine content than in bacteria. Yeast cells are recovered by decantation centrifugation (including washing) and drying treatment methods. After washing, undesirable traces of medium are removed, which are again recycled for economic reasons (Fig. 7.51). As a result of final harvesting by rotary vacuum filter, a cake containing 20–40% dry matter is obtained, which is then dried to get a product of 6–10% water content.

Filamentous fungi have advantages in ease of harvesting but have their limitations in lower growth rates, lower protein content, and acceptability. Algae have disadvantages of having cellulosic cell walls that are not digested by human beings and they also concentrate heavy metals. SCP basically is composed of proteins, fats, carbohydrates, ash ingredients, water, and other elements such as phosphorus and potassium. The composition depends on the organism and the substrate on which it grows. If SCP is to be used successfully, there are some criteria to be satisfied. The SCP must be safe to eat. The nutritional value of SCP depends on high composition of amino acids and should be acceptable to the general public. It must have the functionality, that is, characteristics, which are found in common staple foods. The economic viability of the SCP process is extremely complex and is yet to be demonstrated. The worldwide food protein deficiency is becoming alarming day to day. During World War II, when there was shortage of proteins and vitamins in the diet, the Germans produced yeasts and a mold (*Geotrichum candidum*) in some quantity for food; this led to the idea to produce edible proteins on a large scale by means of microorganisms during 1970s. Several industrial

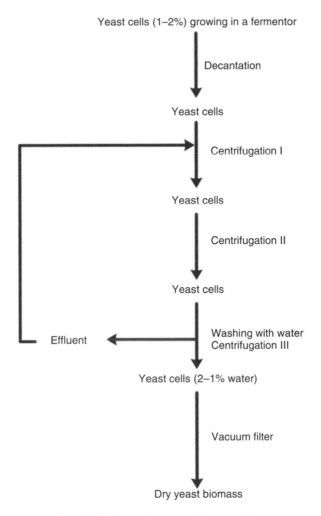

Figure 7.51. Production of dry yeast biomass.

giants investigated the possibility of converting cheap organic materials into proteins using microorganisms. The term *single cell protein* was coined at the Massachusetts Institute of Technology by Prof. C. L. Wilson in 1966 and represents microbial cells (primary) grown in mass culture and harvested for use as protein sources in foods or animal feeds. Many scientists believe that SCP production is a possible solution to meet out the shortage of protein.

7.5.10. Fermented Food

Ancient fermented food processes, such as making bread, wine, cheese, curds, idli, dosa, can be dated to more than 6000 years ago. They were developed long before man had any knowledge of the existence of the microorganisms involved. Fermentation is also a powerful economic incentive for semi-industrialized countries in their willingness to produce bioethanol. Fermented food is prepared by involving a step where microorganisms (or enzymes) alter the properties of the food, for example, yogurt or bread. The purpose of fermenting food is often to get a better taste or texture, but one important reason is that food is kept better when fermented. All the different types of cultured milk have evolved from the fact that fresh milk rapidly deteriorates, and a controlled fermentation with LAB gives food a longer shelf life.

There is a growing attention from the research community on the fermentation of cereal products. Examples are koko (ogi of Nigeria, akasa of Ghana), sour cereal porridge, and tuo zaafi, a thick sorghum or millet porridge that is a staple food in African savanna areas. New products are also being developed in this area.

Research has shown that fermentation can inhibit pathogenic bacteria that otherwise could cause, for example, diarrhea (1 out of 10 children in developing countries dies due to dehydration caused by diarrhea). Toxins and antinutritive factors can also be reduced, and the nutritive value can be ameliorated. The foods that are highest in nutrition are those that are eaten in their fresh, natural, and unprocessed state. As soon as a food is tampered with in any way, nutrient loss results. The longer a food is held in storage, the lower it becomes in nutrition.

Fermented foods are usually processed or destroyed in some manner. After that, they are often stored and used over a period of weeks or even months. Many times, foods are first heated to a high temperature before fermentation is allowed to occur. Milk is first heated or pasteurized to kill all bacteria. Then it is inoculated with a specific bacterial strain to ferment it into yogurt. The milk serves merely as a bacterial culture ground. Table 7.12 shows a list of few ethnic fermented foods from the Hindu Kush.

If heat is not used, then the food is often chopped, sliced, smashed, or blended. A whole head of cabbage does not readily "ferment," but if you bruise and chop it to pieces, then

TABLE 7.12. Ethnic Fermented Foods of Nepal

Product	Substrate	Nature and Use	Major Consumer
Gundruk	Leafy brassica vegetable	Fermented and dried sour; high in vitamin B complex	All Nepalese, Sikkim, and Bhutanese
Meseura	Black gram	Dry nuggets mixed with taro petioles; rich in protein and taste	Newar community
Sinki	Radish tap root	Fermented underground in the field; dried and use as soup with many vegetables in dry period	Eastern Nepal
Kinema	Soya bean	Sticky soya and used as curry	Non-Brahmin Nepalese
Chhurpi	Yak/cow milk	Head cheese; masticator	All Nepalese, Sikkim, and Bhutanese
Chang/tongba/ jannd	Finger millet	Mild alcoholic beverage	Non-Brahmins Nepalese, Bhutias, and Lepchas
Faapar kojard	Buckwheat	Mild alcoholic and acidic beverage	Non-Brahmin Nepalis, Bhutias, and Lepchas
Tho/jaanr	Rice	Semiliquid thick alcoholic drink	Newar Nepalis
Chaulani	Rice (annga, bayerni)	5- to 6-h soaked rice water and believed to be nutritious and to cure heat stroke	Brahmin-Chhetris

the bacteria will do their natural job of finishing the decomposition process. Whenever foods are cut, chopped, or sliced to start the fermentation process, rapid oxidation of the food and nutrient loss occur. Another reason for eating fermented foods is that they are high in B vitamins or that they may somehow encourage the body to produce more vitamin B_{12} in the intestine.

According to research, the levels of vitamin B_{12} may be reduced by fermented foods. A Bulgarian report indicates that the bacteria within yogurt use the B_{12} for their own growth. The B_{12} in kefir (a fermented milk drink) decreases in proportion to its fermentation.

7.5.11. Mushroom

There are at least 12,000 species of fungi that can be considered as mushrooms, with at least 2000 species showing various degrees of edibility. Furthermore, over 200 species of mushroom have been collected from the wild and utilized for various traditional medical purposes mostly in the Far East. To date, about 35 mushroom species have been cultivated commercially, and of these, about 20 are cultivated on an industrial scale. Small-scale mushroom production represents an opportunity for farmers interested in an additional enterprise and is a specialty option for farmers without much land (Sanodiya et al., 2009; Bisen et al., 2010).

Mushroom production can play an important role in managing farm organic wastes when agricultural and food processing by-products are used as the growing media for edible fungi (Fig. 7.52). The spent substrate can then be composted and applied directly back to the soil. Mushroom production is completely different from growing green plants. Mushrooms do not contain chlorophyll and therefore depend on other plant material (the "substrate") for their food. The part of the organism that we see and call a mushroom is really just the fruiting body. Unseen are the tiny mycelium threads that grow throughout the substrate and collect nutrients by breaking down the organic material. This is the main body of the mushroom. Generally, each mushroom species prefers a particular growing medium, although some species can grow on a wide range of materials. Many mushroom suppliers sell several kinds of spawn, and the beginning mushroom farmer should take advantage of this selection in early trials to determine which species grow best on available materials. Eventually, learning to produce spawn might reduce the cost of production. This possibility should be evaluated only after mastering the later stages of cultivation.

While the mycelium is growing and until it fully occupies the substrate, the mushroom farmer typically manipulates the growing environment to favor mycelial growth. The atmospheric conditions are then changed to initiate "pinheads," and then to complete fruiting. For example, in oyster mushroom production, under closely controlled conditions, the grower lowers the temperature and the CO_2 in the growth room to initiate fruiting. Each species has specific requirements for its stages of development.

7.5.12. Beverages

An alcoholic beverage is a drink that contains ethanol, commonly known as *alcohol* (although in chemistry, the definition of "alcohol" includes many other compounds).

Beer has been a part of human culture for 8000 years. In Germany, Ireland, the United Kingdom, and many other European countries, drinking beer (and other alcoholic

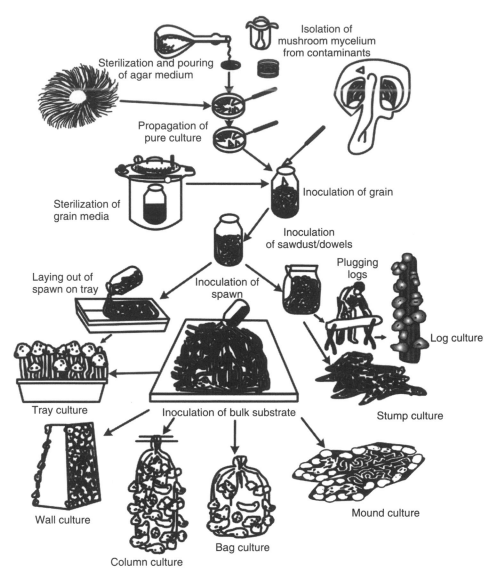

Figure 7.52. Commercial production of mushrooms.

beverages) in a local bar or pub is a cultural tradition. Nonalcoholic beverages are drinks that usually contain alcohol, such as beer and wine, but contain less than 0.5% alcohol by volume. This category includes low alcohol beer, nonalcoholic wine, and apple cider.

Wines are made from a variety of fruits, such as grapes, peaches, plums, or apricots. The most common wines are produced from grapes. The soil in which grapes are grown and the weather conditions in the growing season determine the quality and taste of the grapes, which in turn affect the taste and quality of wines. When ripe, the grapes are crushed and fermented in large vats to produce wine. Beer is also made by the process of fermentation. A liquid mix, called *wort*, is prepared by combining yeast and malted cereal, such as corn, rye, wheat, or barley. Fermentation of this liquid mix produces alcohol and

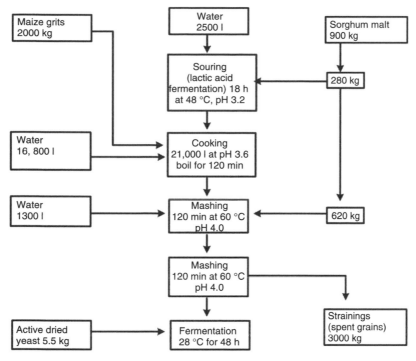

Figure 7.53. Production of beer.

carbon dioxide. The process of fermentation is stopped before it is completed in order to limit the alcohol content. The alcohol so produced is called *beer* (Fig. 7.53). It contains 4–8% of alcohol.

Whisky is made by distilling the fermented juice of cereal grains such as corn, rye, or barley (Fig. 7.54). Scotch whisky was originally made in Scotland. The word *Scotch* has become almost synonymous with whisky of good quality. Rum is a distilled beverage made from fermented molasses or sugarcane juice and is aged for at least 3 years. Caramel is sometimes used for coloring. Brandy is distilled from fermented fruit juices. Brandy is usually aged in oak casks. The color of brandy comes from either the casks or the added caramel. Gin is a distilled beverage. It is a combination of alcohol, water, and various flavors. Gin does not improve with age, so it is not stored in wooden casks. Liqueurs are made by adding sugar and flavoring such as fruits, herbs, or flowers to brandy or to a combination of alcohol and water. Most liqueurs contain 20–65$ alcohol. They are usually consumed in small quantities after dinner.

7.6. FOOD DIAGNOSTICS, FOOD PRESERVATION, AND FOODBORNE MICROBIAL DISEASES

Food diagnostics is an emerging field that applies "modern" methods of detection of bacteria, viruses, parasites, chemicals, biotoxins, heavy metals, and prions in all steps of the food chain, from raw materials to end products. Using molecular diagnostic techniques,

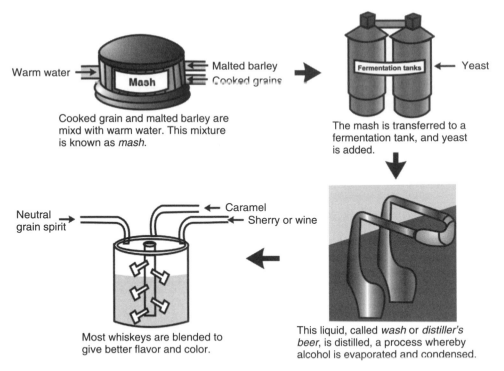

Warm water ➡ | Mash | ⬅ Malted barley
⬅ Cooked grains

Cooked grain and malted barley are mixd with warm water. This mixture is known as *mash*.

Fermentation tanks ⬅ Yeast

The mash is transferred to a fermentation tank, and yeast is added.

Neutral grain spirit ➡ | | ⬅ Caramel
⬅ Sherry or wine

Most whiskeys are blended to give better flavor and color.

This liquid, called *wash* or *distiller's beer*, is distilled, a process whereby alcohol is evaporated and condensed.

Figure 7.54. Production of whisky.

the detection of a fragment of genetic material (nucleic acids) that is unique to the target pathogenic organism can be successfully accomplished (Debnath et al., 2010).

One of the most practical and useful applications of molecular tools is their specificity, as they target genetic regions unique to the organism, and depending on the gene target, they can also yield valuable information about virulence properties of the organism. They are also invaluable in detecting and identifying infectious agents. DNA-based methods are also used for the analysis of foods. In the past decade, the need for methods to detect and to quantify DNA from GMOs has been a major driver for the development and optimization of PCR-based techniques.

Throughout the world, food production, preparation, and distribution have become increasingly complex, and raw materials are often sourced globally. Changes in food processing techniques and food distribution and the emergence of new food pathogens have changed the epidemiology of foodborne diseases. Foodborne microorganisms are continuously changing due to their inherent ability to evolve and their amazing capacity to adapt to different forms of stress. New primary production technologies and food manufacturing practices are introduced all the time; food consumption patterns and the demographic structure of many countries continue to change.

7.6.1. Food Preservation and Microbial Growth

The nutrients that give us energy and help us maintain good health also cause our food to spoil. There are innumerable microorganisms in the atmosphere that derive their nutrition from these nutrients by breaking them into simpler forms. As these minute life-forms start disintegrating the nutrients, they set off the process of food spoilage. With the knowledge of the role that microorganisms play in spoiling food, a number of methods of food preservation have been developed by man. All these methods work by altering conditions such as temperature, availability of water or oxygen in the food, or in the environment in which the food is stored. Changing or altering these factors hinders the growth of these minute organisms and hence prevents food spoilage.

7.6.1.1. Bacterial Counts of Food

Bacteria can be found in nearly any location on the Earth's surface that has a supply of moisture, such as lakes and ponds, soil, the surfaces of plants and skin, and the digestive tract of humans and animals. While the vast majority of bacteria pose no threat to us and in some cases offer considerable benefits, a few types of bacteria may produce serious diseases. In order to reduce the incidence of such diseases, various substances that routinely harbor potentially harmful organisms are tested to determine their bacterial content or titer.

Three methods are used to determine the bacterial titer:

1. When bacterial content is very low (no more than a few cells per liter), a known volume of sample is run through a small-pore (0.2 μm) filter that traps bacterial cells. The filter is then placed on an agar plate, and each cell grows up to form a visible colony. The number of colonies is divided by the sample volume to give bacterial concentration. For example, if a 10 l sample yielded 140 colonies, the titer would be 14 bacteria per liter.

2. For samples of intermediate bacterial content (drinking water, samples from public beaches), various volumes of the sample (10, 1, and 0.1 ml) are added to the growth medium and incubated for 24 h. Bacterial content is determined statistically on the basis of the smallest sample size that will produce growth in liquid culture. The current standard for drinking water is less than 1 coliform bacterium per 100 ml sample (1 coliform per 100 ml is the lowest titer that can be reliably detected using this method).

3. Food samples and liquid bacterial cultures have such a high bacterial content that even one drop spread on an agar plate will produce too many colonies to count. The serial dilution method is used to determine the bacterial titer of such cultures. A series of 1:10 dilutions (one part sample diluted with nine parts liquid) of such samples are plated and the resulting colonies are counted. The number of colonies on a plate is then multiplied by the dilution factor (the number of times that the 1:10 dilution was done) for that plate to obtain the bacterial count in the original sample.

The intestinal tract of warm-blooded animals contains a variety of bacterial species. When an animal is butchered, these bacteria may be spread from the gut to the edible portions of the carcass. While such bacteria generally are not harmful to humans (beef is most often contaminated with *E. coli*, a normal inhabitant of the human intestine), some types of bacteria (such as *Salmonella*, found frequently in chickens) can cause food poisoning. It is, therefore, essential that meat be thoroughly cooked before eating,

Figure 7.55. Various food that are susceptible or resistant to spoilage.

The nonpathogenic varieties of bacteria in foods may cause chemical changes that alter the flavor and texture of food (Fig. 7.55). Because of this, meat is refrigerated or frozen to retard the growth of bacteria. Since some bacteria can grow over a wide range of temperatures (from <10 °C to >45 °C for *E. coli*), refrigeration is adequate for only short-term storage of meat.

7.6.1.2. Microbial Spoilage of Canned Food

Canned foods may spoil due to either biological or chemical reasons. Biological spoilage of canned food occurs due to the action of various microorganisms. Spore-forming bacteria, for example, *Clostridium, Bacillus*, represent the most important group of canned-food-spoiling microorganisms because of their heat-resistant nature (thermophilic nature). In addition, there are other microorganisms that are not heat resistant (mesophilic) but enter through the leakage of the container during cooling and spoil the food. In this way, we can divide biological spoilage of canned food into the following two categories.

Biological Spoilage by Thermophilic Bacteria. Underprocessing of canned foods results in spoilage by thermophilic bacteria, which grow best at a temperature of 50 °C or higher. Five types of this spoilage can be recognized.

1. *Flat Sour Spoilage*. In canned foods, production of acid and no gas is referred to as *flat sour spoilage* because the food becomes sour, but the can shows no evidence of food spoilage because no gas is produced, that is, the can remains flat. Thus, the spoilage cannot be detected unless the can is opened. The spoilage is caused by *Bacillus* spp. such as *B. coagulans* and *B. stearothermophilus*, resulting in sour, abnormal odor; sometimes cloudy liquor is found in the food content of the can.

2. *Thermophilic Anaerobic (TA) Spoilage*. *Clostridium thermosaccharolyticum*, an obligate thermophile, causes spoilage. The can swells and may burst due to production of CO_2 and H_2. The food becomes fermented, sour, cheesy and develops butyric odor.

3. *Sulfide Spoilage*. *Clostridium nigricans* is involved in this spoilage. It produces H_2S gas, which is absorbed by the food product. The food becomes usually blackened and gives "rotten egg" odor.

4. *Putrefactive Anaerobic Spoilage*. *Clostridium sporogenes* causes spoilage through putrefaction. The can swells and may burst. Putrefaction may result from partial digestion of the food. The food develops typical "putrid" odor.

5. *Aerobic Spore Former's Spoilage*. *Bacillus* spp., the aerobic bacteria, causes spoilage. If the canned food is cured meat, swelling of the can is observed.

Biological Spoilage by Mesophilic Microorganisms. *Bacillus* spp., *Clostridium* spp., yeasts, and other fungi that are mesophilic (an organism growing best at a moderate temperature range of 25–40 °C) are mainly responsible for this type of canned food spoilage. As stated earlier, these organisms enter through the leakage of the container during cooling. *Clostridium butyricum* and *Clostridium pasteurianum* result in butyric acid type of fermentation in acidic (tomato juice, fruits, fruit juices, etc.) or medium acidic (corn, peas, spinach, etc.) food with swelling of the container due to the production of CO_2 and H_2. *B. subtilis* and *Bacillus mesenteroides* have been reported as spoiling canned seafood, meat, among others. Other mesophilic bacteria that have been reported in cans are *Bacillus polymyxa, Bacillus macerans, Streptococcus* sp., *Pseudomonas*, and *Proteus*. Yeasts and molds have also been found in canned foods. Yeasts result in CO_2 production and swelling of the cans.

7.6.1.3. Microbial Spoilage of Refrigerated Meat

Lipases are class of enzymes that catalyze the hydrolysis of long-chain triglycerides. These are glycerol ester hydrolases that also catalyzes esterification, interesterification, acidolysis, and aminolysis in addition to the hydrolytic activity on triglycerides. Beside these activities, there are two more classes of enzymes, mostly secreted by microorganisms, and these are the oxygenases and the reductases that also bring about changes in lipids.

The widespread use of refrigeration to store and preserve foodstuff provides great diversity of nutrient-rich habitat for psychrophillic and psychrotolerant food spoilage microorganisms. The lipases produced by certain organisms are significant to the food industry in that they improve the traditional chemical processes of food manufacture. However, certain microorganisms like *Arthobacter* sp., *Pseudomonas fragi, Pseudomonas fluorescens*, and *Serratia marcescens*, which produce cold active lipases, were isolated from refrigerated milk, meat products, and other spoiled food samples.

Dressed poultry is highly susceptible to spoilage by many microorganisms. The flesh and other parts, such as the liver, have hardly any microorganisms in any living animals; therefore, most of the contamination comes during the slaughtering processes such as bleeding, defeathering, removal of viscera, washing, and subsequent handling. In order to prevent spoilage by these microorganisms, meat is refrigerated usually at -10 to $-15\ ^\circ$C. However, during selling, these are stored at -2 to $-5\ ^\circ$C. Sometimes it has been observed that the temperature of the counter is maintained at 0–$2\ ^\circ$C and the material is held in this temperature for more than 8 h. Such sort of practice leads to increase in aerobic microbial count, especially the count of the psychrophiles. These microorganisms bring about biochemical changes, which can be termed as *spoilage*, as the material is not accepted for human consumption. Some of the spoilage-causing microorganisms get inactivated during refrigeration, whereas others survive the low temperature. The meat is also subjected to changes by its own enzymes. Such autolytic changes include proteolysis and lipolysis. The excessive autolysis is called as *souring*. The spoilage activities of microorganisms include the hydrolysis of fats and sometimes subsequent oxidation of the fatty acids liberated, leading to loss of flavor and production of off-odors due to certain aldehyde and acids. This may even change the normal color of the poultry meat to shades of green, brown, or gray as a result of the production of oxidized compounds, for example, peroxides. It is well known that the chicken meat flavor is primarily due to arachidonic acid, which, like all meat flavors, is fat soluble, is present as a glyceride, and gets liberated during processing. However, this arachidonic acid is β-oxidized by microbes leading to the formation of different N-hydroperoxyeicosatetraenoic acids with hydroperoxide substitution at C5, C8, C11, C12, and C15.

Another very nutritionally important part of chicken is the liver, and in many oriental countries, it is considered as a delicacy. It is a major source of vitamin A. One of the important fatty acids found in the liver is palmitic acid or hexadecanoic acid, $CH_3(CH_2)_{14}COOH$. Palmitic acid is the first fatty acid produced during lipogenesis and from which other longer fatty acids can be produced.

Reduction of palmitic acid yields cetyl alcohol. This is responsible for keeping vitamin A, which is present in the liver, intact. However, it has been reported that *Penicillium* sp. is capable of oxidizing a saturated fatty acid like palmitic acid to 2-undecanone and 2-pentadecanone. The large factors involved in food and meat spoilage are shown in Fig. 7.56.

7.6.1.4. Microbiology of Fermented Food

Approximately one-third of all food manufactured in the world is lost to spoilage. Microbial content of foods (microbial load) can be evaluated. On the basis of the shelf life, food can be classified as nonperishable foods (e.g., pasta), semiperishable foods (e.g., bread), and perishable foods (e.g., egg). The conditions for spoilage relate to water, pH, physical structure, oxygen, and temperature. Various intrinsic factors are also responsible, such as composition, pH, presence and availability of water, oxidation–reduction potential, alterations due to cooking, physical structure, and presence of antimicrobial substances.

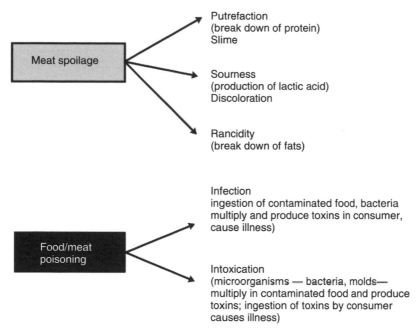

Figure 7.56. Factors involved in meat and food spoilage.

7.6.1.5. Microbiology of Fermented Milk Products

Milk and dairy products constitute an important item of our food. These products are very suitable for microbial growth. It thus becomes necessary to know the chemistry of milk, its spoilage, method of preservation, and different dairy products where microbes play a positive rather than negative role. Milk is considered as a complete food, and it contains proteins, fat, carbohydrates, minerals, vitamins, and water. It is also a good medium for the growth of microorganisms. It is, therefore, important to know the type of microorganisms present in milk, their control, and their use for beneficial purposes. Milk contains relatively few bacteria when it is secreted from the udder of a healthy animal. However, during milking operations, it gets contaminated from the exterior of the upper and the adjacent areas of the udder, dairy utensils, and milking machines. In this way, bacteria, yeasts, and molds get into the milk and constitute its normal flora. The number of contaminants added from various sources depends on the care taken to avoid contamination. The presence of these nonpathogenic organisms in milk is not serious, but if these organisms multiply quickly, they can cause spoilage of milk, such as souring or putrefaction, and develop undesirable odors. Control of their multiplication in milk is, therefore, very essential. The various approaches to remove microorganisms are enumerated in Table 7.13.

7.6.1.6. Direct Microscopic Count of Organisms in Milk

The microscope has been used to observe and count bacteria and somatic cells in milk since the early 1900s. It has proven to be a valuable tool to the dairy industry. Direct

TABLE 7.13. Approaches to Remove Microorganisms

Approach	Examples of Process
Removal of microorganisms	Avoidance of microbial contamination, physical filtration, centrifugation
Low temperature	Refrigeration, freezing
High temperature	Partial or complete heat inactivation of microorganisms (pasteurization and canning)
Reduced water availability	Water removal, as with lyophilization or freeze drying; use of spray dryers or heating drums; decreasing water availability by addition of solutes such as salt or sugar
Chemical-based preservation	Addition of specific inhibitory compounds (e.g., organic acids, nitrates, sulfur dioxide)
Radiation	Use of ionizing (gamma rays) and nonionizing (UV) radiation
Microbial-product-based inhibition	Addition of substances such as bacteriocins to foods to control foodborne pathogens

microscopic counts are possible using special slides known as *counting chambers*. Dead cells cannot be distinguished from living ones. Only dense suspensions can be counted (>107 cells/ml), but samples can be concentrated by centrifugation or filtration to increase sensitivity. For somatic cell counting, the direct microscopic somatic cell count (DMSCC) is considered an official reference method used for regulatory purposes for direct milk counts and/or for calibration of approved electron instruments. The regulatory procedure for somatic cells is outlined in detail in the most recent FDA 2400 form, and with the exception of the type of cells that are counted, this 2400 form procedure can be used for bacteria as well. While the direct microscopic clump count (DMCC) is not considered an official test for bacterial counts, it is used throughout the dairy industry to estimate bacteria colony forming units (i.e., "clumps") in raw milk samples taken from the farm, the tank truck, or the plant storage facility. DMCC is most widely used to screen incoming raw milk supplies (i.e., tank trucks) to determine whether the milk has an acceptable or legal bacterial load. The DMCC has become accepted in some states as a legal method for rejection of unacceptable milk.

In addition to providing estimated counts of bacteria and somatic cells, the direct microscopic method has also been used as a trouble-shooting guide in attempts to identify the general types of bacteria present in a milk sample. Narrowing down the predominant types of contaminants in a sample can sometimes provide a lead as to the potential source or cause of a microbial defect. It is to be noted that certain types of bacteria from very different sources can appear very similar under a direct microscopic smear resulting in a guess work at best.

When large numbers of pairs (diplococci) or short chains (streptococci) of spherical bacteria (cocci) are observed in raw milk, possible causes include poor cooling and/or dirty equipment. Environmental streptococci that cause mastitis may also appear as short chains or pairs, while very long chains are typical of *Streptococcus agalactiae*. When mastitis is the cause, bacterial cells may be observed in association with somatic cells (leukocytes). With mastitis and poor cooling, milk smears may appear as mostly one type of bacteria, while high counts from dirty equipment would be more likely to contain a mixture of bacterial types (rods).

TECHNICAL INFORMATION. Gram positive; cocci; 0.5–1.2μm in diameter; occur in pairs or chains of varying length; catalase negative; usually oxidase negative; colonies on SPC usually white, small, and subsurface; some strains survive pasteurization, most do not; and some strains grow slowly under refrigeration.

Streptococcus, Lactococcus, and *Enterococcus* species are the most common gram-positive organisms. They are easily recognized in milk smears, although distinguishing between specific types may be difficult.

Lactococcus (lactic streptococci) are involved in dairy fermentations (e.g., cheese) as well as in the spoilage of dairy products. They are common in Nature and in dairy environment and are often associated with plant materials including feeds and bedding materials. They may also thrive on milk soil of poorly cleaned equipment. These organisms do not grow or grow slowly under refrigeration, although they grow very well in milk at higher temperatures.

Poor cooling, especially when temperatures exceed 50–60 °F, often results in proliferation of these organisms, seen in milk smears as pairs and/or chains of cocci. These organisms may be responsible for "sour" (high TA) or "malty" defects in milk.

Enterococcus (fecal streptococci) are often associated with fecal matter, although they survive well in other environments. They appear similar to *Lactococcus*. They may be associated with poor cooling, dirty equipment, and in rare cases, mastitis.

Streptococcus strains considered as common causes of mastitis include contagious strains (*S. agalactiae*), spread from cow to cow, and environmental strains (*S. uberis, S. dysgalactiae*) contracted from the environment (e.g., bedding). *S. agalactiae* often appears in long chains, which may be seen associated with somatic cells. Other mastitis streptococci may be seen as pairs or chains of varying lengths resembling organisms above two figures.

The direct microscopic method offers the advantage of determining the extent of bacterial contamination of raw milk or cream samples taken directly from the dairy farm bulks or from pooled supplies in tankers or storage tanks. Individual samples can be examined in 10–15 min. The cause of high counts may be suggested by evaluating the morphology of the cells, the clumps, or for the presence of somatic cells. Such interpretations, however, need to be made by well-trained analysts.

7.6.1.7. Biosensors as Analytical Tools in Food and Drink Industries

The detection of pathogenic bacteria is key to the prevention and identification of problems related to health and safety. Biosensors offer advantages as alternatives to conventional methods due to their inherent specificity, simplicity, and quick response (Mello and Kubota, 2002).

Legislation is particularly tough in areas such as the food industry, where failure to detect an infection may have terrible consequences. Miniaturization of biosensors enables biosensor integration into various food production equipment and machinery. Potential uses of biosensors for food microbiology include on-line process microbial monitoring to provide real-time information in food production and analysis of microbial pathogens and their toxins in finished food. Biosensors can also be integrated into the Hazard Analysis and Critical Control Point (HACCP) programs, enabling critical microbial analysis of the entire food manufacturing process (Rasooly and Herold, 2006). Application of the

TABLE 7.14. Biosensors Used in Food Processing and Quality Control

Analyte	Matrix	Recognition Enzyme	Transduction System
Glucose	Grape juice, wine, juices, honey, milk, and yogurt	Glucose oxidase	Amperometric
Fructose	Juices, honey, milk, gelatin, and artificial edulcorants	Fructose dehydrogenase D-fructose 5-dehydrogenase	Amperometric
Lactose	Milk	β-Galactosidase	Amperometric
Lactate	Cider and wine	Transaminase and L-lactate dehydrogenase	Amperometric
Lactulose	Milk	Fructose dehydrogenase and β-galactosidase	Amperometric
L-amino acids	Milk and fruit juices	D-amino acid oxidase	Amperometric
L-glutamate	Soya sauce and condiments	L-glutamate oxidase	Amperometric
L-lysine	Milk, pasta, and fermentation samples	Lysine oxidase	Amperometric
L-malate	Wine, cider, and juices	Dehydrogenated malate others	Amperometric
Ethanol	Beer, wine, and other alcoholic drinks	Alcohol oxidase, alcohol dehydrogenase, NADH oxidase	Amperometric
Glycerol	Wine	Glycerophosphate oxidase and glycerol kinase	Amperometric
Catechol	Beer	Polyphenol oxidase	Amperometric
Cholesterol	Butter, lard, and egg	Cholesterol oxidase and peroxidase	Amperometric
Citric acid	Juice and athletic drinks	Citrate lyase	Amperometric
Lecithin	Egg yolk, flour, and soya sauce	Phospholipase D and choline oxidase	Electrochemical

biosensor technique in the field of food processing and QC is promising (Table 7.14; Cock et al., 2009).

A nanobiosensor device has been reported (Liu et al., 2007), which works as a molecular transistor triggered by the presence of specific pathogens on an immunosensor. This device can also measure the amount of pathogen contamination on a particular food or machine, giving processors more data to determine the extent of a problem. The transistor works by processing data through fundamental logic gates. The logic gates operate by converting binding events between an antigen and an antibody into a measurable electrical signal using polyaniline nanowires as the transducer. The logic gates are created by patterning antibodies at different spatial locations in an immunosensor assay. Immunosensors are biosensors that use antibodies to recognize the presence of a pathogen (Fig. 7.57).

Graphite electrodes fabricated by screen printing have also been used as amperometric detectors in biosensors based on NAD(+)-dependent dehydrogenases, tyrosinase, or genetically modified acetylcholinesterases. The monoenzyme sensors have been optimized as disposable or reusable devices for detection of a variety of substrates important in the food industry (D-lactic acid, L-lactic acid, and acetaldehyde) or in environmental pollution control (phenols and dithiocarbamate, carbamate, and organophosphorus pesticides). Tests on real samples have been performed with the biosensors; D-lactic acid and

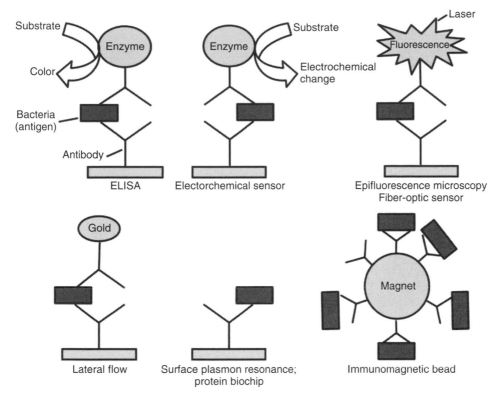

Figure 7.57. Various types of antigen–antibody reactions in foodborne pathogen detection.

acetaldehyde have been detected in wine and phenols in air. The affinity biosensors field has expanded significantly over the past decade, with a projected global biosensors market growth from $6.1 billion in 2004 to $8.2 billion in 2009, representing major industrial sectors such as pharma, Medicare, and food (Patel, 2006, Debnath et al., 2010). Potential markets include the medical, military, food, and environmental industries. Those industries combined have a market size of $563 million for pathogen-detecting biosensors and are expected to grow at a compounded annual growth rate of 4.5%. The food market is further segmented into different food product industries. The overall food pathogen testing market has grown to $192 million and 34 million tests in 2005. The trend in pathogen testing emphasizes the need to commercialize biosensors for the food safety industry as legislation creates new standards for microbial monitoring (Evangelyn et al., 2003; Debnath et al., 2010). With quicker detection time and reusable features, biosensors will be important to those interested in real-time diagnostics of disease-causing pathogens. As the world becomes more concerned with safe food and water supply, the demand for rapid detecting biosensors will only increase.

7.6.1.8. Food Preservation

The main aim of food preservation is to minimize the growth of microorganisms during the storage period, thus promoting longer shelf life and reduced hazard from eating the food. Fruits and vegetables are an important supplement to the human diet as they provide the essential minerals, vitamins, and fiber required for maintaining health. For various reasons, this abundance of production is not fully utilized and about 25–30% of it is wasted due to spoilage. Most of fruits and vegetables are seasonal crops and perishable in nature. In a good season, there may be a local glut, particularly of fruit, but because of insufficient transport facilities, lack of good roads, and poor availability of packing materials, the surplus cannot be taken quickly enough to the natural markets in urban areas. Moreover, the surplus often cannot be stored for sale in the off-season because of inadequate local cold storage facilities. Thus, the cultivators do not get a good price for their produce because of the glut and some of it is spoiled resulting in complete loss. Two approaches are possible for solving this problem. One is the creation/expansion of cold storage facilities in the fruit and vegetable producing regions themselves, as also in the major urban consumption centers, to ensure supply of fresh fruits and vegetables throughout the year. Another approach is to process the fruits and vegetables into various products that could be preserved for a long time and add to the value of the product. With increasing urbanization and rise in middle-class purchasing power, a change in food habits has been observed. There is a dying out of the practice of making and preserving dehydrated foods, pickles in individual homes.

Preservation processes include

1. Heating to kill or denature organisms (e.g., boiling)
2. Oxidation (e.g., use of sulfur dioxide)
3. Toxic inhibition (e.g., smoking, use of carbon dioxide, vinegar, alcohol, etc.)
4. Dehydration (drying)
5. Osmotic inhibition (e.g., use of syrups)
6. Low temperature inactivation (e.g., freezing)
7. Ultrahigh water pressure (e.g., fresherized, a kind of "cold" pasteurization, the pressure kills naturally occurring pathogens, which cause food deterioration and affect food safety)
8. Combinations of these methods
9. Chelation.

FOOD PRESERVATION METHODS.

Drying. One of the oldest method of food preservation is drying, which reduces water activity sufficiently to delay or prevent bacterial growth. Most types of meat can be dried. This is especially valuable in the case of pig meat, since it is difficult to keep without preservation. Many fruits can also be dried, for example, apples, pears, bananas, mangos, papaya, and coconut. Zante currants, sultanas, and raisins are all forms of dried grapes. Drying is also the normal means of preservation for cereal grains such as wheat, maize, oats, barley, rice, millet, and rye.

Smoking. Meat, fish, and some other foods may be both preserved and flavored through the use of smoke, typically in a smokehouse. The combination of heat to dry the food without cooking it and the addition of the aromatic hydrocarbons from the smoke preserve the food.

Freezing. Freezing is also one of the most common process used commercially and domestically for preserving a very wide range of foodstuff including prepared foodstuff, which would not have required freezing in their unprepared state. For example, potato waffles are stored in the freezer, but potatoes themselves require only a cool dark place to ensure many months' storage. Cold stores provide large volume long-term storage for strategic foodstocks held in case of national emergency in many countries.

Vacuum Packing. Vacuum packing stores food in a vacuum environment, usually in an airtight bag or bottle. The vacuum environment strips bacteria of oxygen needed for survival, hence preventing the food from spoiling. Vacuum packing is commonly used for storing nuts.

Salt. Salting or curing draws moisture from the meat through the process of osmosis. Meat is cured with salt or sugar, or a combination of the two. Nitrates and nitrites are also often used to cure meat.

Sugar. Sugar is used to preserve fruits, either in syrup with fruit such as apples, pears, peaches, apricots, plums or in crystallized form where the preserved material is cooked in sugar to the point of crystallization and the resultant product is then stored dry. This method is used for the skins of citrus fruit (candied peel), angelica, and ginger. A modification of this process produces glacé fruit such as glacé cherries where the fruit is preserved in sugar but is then extracted from the syrup and sold, the preservation being maintained by the sugar content of the fruit and the superficial coating of syrup. The use of sugar is often combined with alcohol for preservation of luxury products such as fruit in brandy or other spirits. These should not be confused with fruit-flavored spirits such as cherry brandy or sloe gin.

Pickling. Pickling method of preserving food is by placing it or cooking it in a substance that inhibits or kills bacteria and other microorganisms. This material must also be fit for human consumption. Typical pickling agents include brine (high in salt), vinegar, ethanol, and vegetable oil, especially olive oil but also many other oils. Most pickling processes also involve heating or boiling so that the food being preserved becomes saturated with the pickling agent. Frequently pickled items include vegetables such as cabbage (to make sauerkraut and curtido), peppers, and some animal products such as corned beef and eggs. EDTA may also be added to chelate calcium. Calcium is essential for bacterial growth.

Lye. Sodium hydroxide (lye) makes food too alkaline for bacterial growth. Lye will saponify fats in the food, which will change its flavor and texture. Lutefisk and hominy use lye in their preparation, as do some olive recipes.

Canning and Bottling. Canning involves cooking fruits or vegetables, sealing them in sterile cans or jars, and boiling the containers to kill or weaken any remaining bacteria as a form of pasteurization. Various foods have varying degrees of natural protection against spoilage and may require that the final step occur in a pressure cooker. High acid fruits such as strawberries require no preservatives to can and only a short boiling cycle, whereas marginal fruits such as tomatoes require longer boiling and addition of other acidic elements. Many vegetables require pressure canning. Food preserved by canning or bottling is at immediate risk of spoilage once the can or bottle has been opened.

7.6.2. Microbial Sampling and Food Poisoning

Food poisoning is any illness brought on by eating contaminated food. It is caused by pathogenic microbes or (rarely) chemicals. Symptoms vary according to the type of microbe involved, and some factors are characteristic to each disease.

Infective Dose. The disease can be triggered by a few cells or may require millions of cells.

Incubation Time. The time between infection and symptoms could be hours or weeks.

Duration of Illness. It can last for hours or months.

Cause of the Symptoms. Toxins in the food and/or infection.

7.6.2.1. Foodborne Diseases and Microbial Sampling

Myriad microbes and toxic substances can contaminate foods. There are more then 250 known foodborne diseases. The majority is infectious and is caused by bacteria, viruses, and parasites. Other foodborne diseases are essentially poisonings caused by toxins and chemicals contaminating the food. All foodborne microbes and toxins enter the body through the gastrointestinal tract and often cause the first symptoms there. Nausea, vomiting, abdominal cramps, and diarrhea are frequent in foodborne diseases. A large number of food poisoning cases have been reported worldwide. One study shown in Table 7.15 represents the severity of such cases in England from 1996 to 2002.

Many microbes can spread in more than one way, so it may not be immediately evident that a disease is foodborne. The distinction matters, because public health authorities need to know how a particular disease is spreading to take the appropriate steps to stop it. For example, infections with *E. coli* O157:H7 can be acquired through contaminated food, contaminated drinking water, contaminated swimming water, and from toddler to toddler at a day care center. Depending on the means of spread, the measures to stop other cases from occurring could range from removing contaminated food from stores, chlorinating a swimming pool, to closing a child day care center.

The most common foodborne infections in the European Union are caused by the bacteria such as *Campylobacter, Salmonella, Listeria* and viruses. They enter the body through the gastrointestinal tract and the first symptoms often occur there. Many reported foodborne illnesses are not part of recognized outbreaks but are registered as individual cases. In addition, foodborne diseases can be caused by bacterial toxins. Bacterial

TABLE 7.15. Food Poisoning Incidents Reported in England and Wales in 1996–2002

Year	Number of Cases
1996	94,923
1997	105,579
1998	105,060
1999	96,866
2000	98,076
2001	85,752
2002	81,562

toxins are toxins generated by bacteria and may be highly poisonous in many cases. These include toxins from *S. aureus, Clostridium botulinum*, and *Bacillus cereus*. The most commonly identified bacterial cause of diarrheal illness in the European Union is *Campylobacter*. Raw poultry meat is often contaminated with *Campylobacter* since these bacteria can live in the intestines of healthy birds. Eating undercooked chicken, or ready-to-eat food in contact with raw chicken, is the most common foodborne source of this infection. It causes fever, diarrhea, and abdominal cramps. *Salmonella* is also commonly found in the intestines of birds and mammals. It can spread to humans via foods, especially through meat and eggs. The illness caused by it is known as *salmonellosis*, which usually involves fever, diarrhea, and abdominal cramps. It can cause life-threatening infections if it invades the bloodstream. *Listeria* cases in humans, although less common than *Campylobacter* and *Salmonella*, have a high mortality rate, particularly among vulnerable groups such as the elderly. It is also very dangerous to pregnant women as it can cause fetal infections, miscarriages, and stillbirths. Ready-to-eat foodstuff, such as chees and fish or meat products, are often found to be at the origin of human infections. Foodborne diseases can also be caused by viruses, such as calicivirus (including norovirus), rotavirus, and hepatitis A virus. These are primarily transmitted by food or water contaminated with human waste. Calicivirus causes approximately 90% of epidemic nonbacterial outbreaks of gastroenteritis in the world. Parasites can also be present in food or water. Parasites that may be transmitted through consumption of contaminated food or drinking water include *Trichinella, Giardia, Sarcocystis*, and *Cryptosporidium*.

7.6.2.2. *Staphylococcal Food Poisoning*

Staphylococcal food poisoning is a gastrointestinal illness. It is caused by eating foods contaminated with toxins produced by *S. aureus*. It is one of the most common foodborne illnesses in the United States. The onset of symptoms in staphylococcal food poisoning is usually rapid and in many cases severe, depending on the individual's susceptibility to the toxin, the amount of contaminated food eaten, the amount of toxin in the food ingested, and the general health of the victim. The most common symptoms are nausea, vomiting, abdominal cramps, and prostration.

Some people may not always demonstrate all the symptoms associated with the illness. In more severe cases, headache, muscle cramping, and transient changes in blood pressure and pulse rate may occur. Recovery generally takes 2 days. However, it is not unusual for complete recovery to take 3 days and sometimes longer in severe cases. Death from staphylococcal food poisoning is very rare, although such cases have occurred among the elderly, infants, and severely debilitated persons. Staphylococcal food poisoning is caused by eating food contaminated with *S. aureus*. *S. aureus* is able to grow in a wide range of temperatures, pH (4.2–9.3), and sodium chloride (salt) concentrations. These characteristics enable *S. aureus* to grow in a wide variety of foods and conditions. Often this type of food poisoning occurs when cooked food is allowed to cool slowly and/or sit at room temperature for some time. The warm food allows *S. aureus* to grow.

S. aureus produces a toxin, an enterotoxin, that remains in the food even when reheated. The symptoms are caused by the toxin, not by the bacteria themselves; hence, staphylococcal food poisoning is sometimes called *food intoxication*.

Staphylococci are ubiquitous. They are present in air, dust, dirt, sewage, water, milk, and food, or on food equipment, environmental surfaces, humans, and animals. Humans and animals are the primary reservoirs. Staphylococci are present in the nasal passages

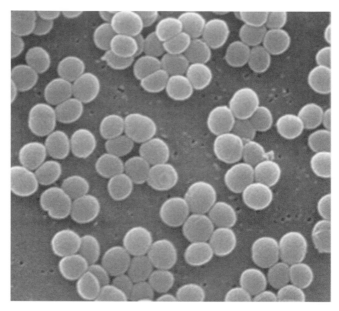

Figure 7.58. *S. aureus* colonies under SEM.

and throats and on the hair and skin of 50% or more of healthy individuals. The colonization rate is even higher for those who work with or who come in contact with sick individuals and hospital environments. The short incubation period, brevity of illness, and usual lack of fever help distinguish staphylococcal food poisoning from other types of food poisoning, except that caused by *B. cereus*. Toxin-producing *Staphylococcus aureus* can be identified in stool or vomit, and the toxin can be detected in food items (Fig. 7.58). Diagnosis of staphylococcal food poisoning in an individual is generally based only on the signs and symptoms of the patient. Testing for the toxin-producing bacteria or the toxin is not usually done in individual patients. Testing is usually reserved for outbreaks involving several people. It is important to prevent the contamination of food with *Staphylococcus* before the toxin can be produced. *S. aureus* is often present on skin, under fingernails, in the nose and throat, and in cuts, abrasions, boils, and abscesses. *S. aureus* can also be found on contaminated surfaces and food preparation utensils.

7.6.2.3. Clostridium Food Poisoning

Botulism is a rare but serious paralytic illness caused by a nerve toxin that is produced by the bacterium *C. botulinum*. *C. botulinum* is the name of a group of bacteria commonly found in soil (Fig. 7.59). The bacteria are anaerobic, gram-positive, spore-forming rods that produce a potent neurotoxin. These rod-shaped organisms grow best in low oxygen conditions. The bacteria form spores that allow them to survive in a dormant state until exposed to conditions that can support their growth. The organism and its spores are widely distributed in Nature. They occur in both cultivated and forest soils; bottom sediment of streams, lakes, and coastal waters; in the intestinal tracts of fish and mammals; and in the gills and viscera of crabs and other shellfish.

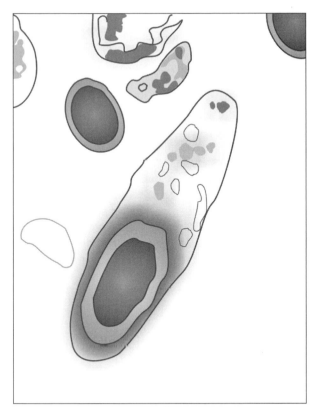

Figure 7.59. *Clostridium botulinum.*

Foodborne botulism is a severe type of food poisoning caused by the ingestion of foods containing the potent neurotoxin formed during the growth of the organism. The incidence of the disease is low, but the disease is of considerable concern because of its high mortality rate if not treated immediately and properly. Most of the 10–30 outbreaks that are reported annually in the United States are associated with inadequately processed, home-canned foods, but occasionally commercially produced foods are implicated as the source of outbreaks. Sausages, meat products, canned vegetables, and seafood products have been the most frequent vehicles for foodborne botulism.

SYMPTOMS. Classic symptoms of botulism include double vision, blurred vision, drooping eyelids, slurred speech, difficulty swallowing, dry mouth, and muscle weakness. Infants with botulism appear lethargic, feed poorly, are constipated, and have a weak cry and poor muscle tone. These are all symptoms of the muscle paralysis caused by the bacterial toxin. If untreated, these symptoms may progress to cause paralysis of the arms, legs, trunk, and respiratory muscles. In foodborne botulism, symptoms generally begin 18–36 h after consuming contaminated food, but they can occur as early as 6 h or as late as 10 days after consumption.

Botulinum toxin causes flaccid paralysis by blocking motor nerve terminals at the myoneural junction. The flaccid paralysis progresses symmetrically downward, usually starting with the eyes and face, then moving to the throat, chest, and extremities. When the diaphragm and chest muscles become fully involved, respiration is inhibited and unless the patient receives treatment in time, death from asphyxia results.

DETECTION AND TREATMENT. Although botulism can be diagnosed by clinical symptoms alone, differentiation from other diseases may be difficult. The most direct and effective way to confirm the clinical diagnosis of botulism in the laboratory is to demonstrate the presence of toxin in the serum or feces of the patient or in the food the patient consumed. Currently, the most sensitive and widely used method for detecting toxin is the mouse neutralization test, which involves injecting serum or stool into mice and looking for signs of botulism. This test typically takes 48 h. Culturing the specimens takes 5–7 days. Some cases of botulism may go undiagnosed because symptoms are transient or mild or are misdiagnosed as Guillain–Barré syndrome.

If diagnosed early, foodborne botulism can be treated with an antitoxin that blocks the action of toxin circulating in the blood. This can prevent the patient's condition from worsening, but recovery still takes many weeks. Physicians may try to remove contaminated food still in the gut by inducing vomiting or using enemas.

Although botulism has been known to cause death due to respiratory failure, in the past 50 years, the proportion of patients with botulism who die has fallen from about 50% to 8%. The respiratory failure and paralysis that occur with severe botulism may require a patient to be on a ventilator for weeks, with intensive medical and nursing care. After several weeks, the paralysis slowly improves.

PREVENTION. The types of foods implicated in botulism outbreaks vary according to food preservation and eating habits in different regions. Any food that is conducive to outgrowth and toxin production, that when processed allows spore survival, and that is not subsequently heated before consumption can be associated with botulism. Almost any type of food that is not very acidic (pH above 4.6) can support growth and toxin production by *C. botulinum* (Fig. 7.59). Botulinal toxin has been demonstrated in a considerable variety of foods, such as canned corn, peppers, green beans, soups, beets, asparagus, mushrooms, ripe olives, spinach, tuna fish, chicken and chicken livers and liver pate, and luncheon meats, ham, sausage, stuffed eggplant, lobster, and smoked and salted fish.

7.6.3. Food Diagnostics

Quality and risk assessments of food and water are the most important tasks for microbial diagnostic laboratories worldwide. The development of a global market for food industries and the intensive use of water resources make these tasks a major mission for researchers too. The discovery of new routes of transmission and the emergence of several foodborne and waterborne pathogenic bacteria revealed a serious health hazard for both developed and developing countries. The use of molecular based technologies in microbial diagnostic has greatly enhanced the ability to detect and quantify pathogenic bacteria in food and water (Fig. 7.60).

7.6.3.1. Tools and Methods to Detect Microbes in Food

Many of the molecular tools have been accepted and implemented in standard protocols for detection and quantification of the most important pathogenic bacteria, such as

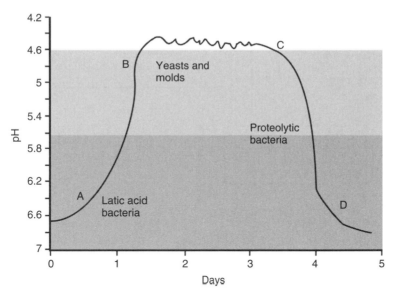

Figure 7.60. Effect of acidic food on the growth and development of various pathogens.

Salmonella sp. and *Listeria monocytogenes*. Despite the rapid diffusion of molecular tools in microbiology laboratories, there are still many drawbacks and obstacles concerning specificity, reproducibility, and reliability of nucleic acids and antibody-based technologies for microbial detection. This led, to some extent, to an underestimation of those methods as a permanent alternative to conventional culture-based detection techniques that sill represent the "golden standard" for microbial diagnostic. The complexity of food and water matrices and the cross-reaction of some molecular probes to target sites of innocuous bacteria closely related to pathogenic ones are the main enthralling challenges that researchers are still arguing with. In order to maintain food safety standards, conventional microbiological methods are still being used to detect bacteria and other organisms in food. However, these techniques are not ideal, as often it can be many days before results are known, which may be of particular economic importance for those foods with a short shelf life. The introduction of newer technology, such as nucleic acid probe and related amplification technology in other fields, has transformed the detection of many organisms. PCR allows nucleic acid probes, with their inherent specificity, to be used to detect organisms present in very low numbers within a short period of time. However, at present, in food microbiology, there are technical problems with using PCR, as certain components in food interfere with the reaction (Fig. 7.61).

7.6.3.2. Foodborne Disease Threats

Newly recognized microbes emerge as public health problems for several reasons: microbes can easily spread around the world, new microbes can evolve, the environment and ecology are changing, food production practices and consumption habits change, and better laboratory tests can now identify microbes that were previously unrecognized. Some of these organisms pose severe food threat. Some of the foodborne threats are discussed in the following sections.

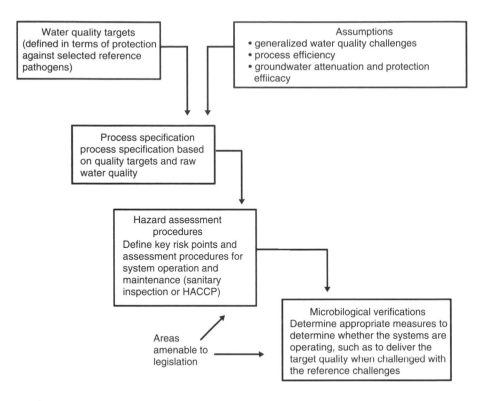

Figure 7.61. Process of the evaluation of water quality against various pathogens.

SALMONELLOSIS. This is a gastrointestinal disease due to foodborne *Salmonella* infection, the pathogen colonizes the intestinal epithelium (Fig. 7.62). *Salmonella* are gramnegative facultatively aerobic rods related to *E. coli, Shigella*, and other enteric bacteria. *Salmonella* normally inhabit the gut of animals and are thus found in sewage. *Salmonella* are pathogenic to humans. *Salmonella typhi*, causes typhoid fever, rare in the United States, with most of the 500 foodborne cases imported from other countries. In all, over 2000 serovars are pathogenic to humans. The American scientist, Daniel E. Salmon, is credited with the discovery of the *Salmonella* family of bacteria in the late 1800s. Since then, scientists have identified more than 2400 types of *Salmonella*. They have also figured out where *Salmonella* live, how they spread to humans, and how to reduce their spread among the general public. Even so, each year, the United States has about 40,000 cases of salmonellosis and many more cases go unreported.

Salmonella are often found in the feces (poop) of some animals, particularly reptiles. Iguanas, for example, carry *Salmonella marina*. People who have these animals as pets are at more risk of getting salmonellosis because the bacteria from a reptile's feces can get on its skin. Then, when people handle the reptiles, they get the bacteria on their hands (hand washing is a good way to reduce the risk of getting salmonellosis). Other strains of *Salmonella* can spread to people via foods that have come into contact with infected animal feces. These exposures happen when foods such as poultry, eggs, and beef are not cooked enough. Fruits and vegetables can also become contaminated from feces in the soil or water where they are grown.

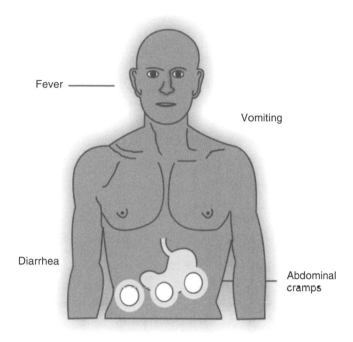

Fever

Vomiting

Diarrhea

Abdominal cramps

Figure 7.62. Symptomology of salmonellosis.

PATHOGENIC *E*. COLI. This is one of the most common bacteria found in the gut of animals and humans (Fig. 7.63). There are other animal species that contain this organism in the gut, including reptiles and fish. *E. coli* colonizes the gut within a few hours or days of birth depending on the species involved. In humans, the intestines can be colonized within 40 h of birth. *E. coli* adheres to the mucus overlaying the large intestine.

Other bacteria that are growing in the gut may be intestinal pathogens including *Salmonella, Shigella, Yersinia, Enterobacter*, and *Klebsiella*. All these pathogens are associated with human illness. Physiologically, *E. coli* can grow in the presence or absence of oxygen (O_2), and this is referred to as being *facultatively anaerobic. E. coli* can respond to environmental signals, such as chemicals, pH, temperature, osmolarity, and can swim toward or away from them. In response to changes in the environment, *E. coli* can change the size of the outer membrane pores to accommodate larger molecular nutrients or exclude inhibitory substances. Treatment of water with chlorine, bromine, or ozone interferes with the membrane's ability to transport nutrients. *E. coli* is the predominant facultative organism in the human intestinal tract; however, it makes up a very small proportion of the total bacterial count. Other bacteria, such as *Bacteroides* out number *E. coli* by 20:1. It is the regular presence of *E. coli* in the intestine that leads to its use as an indicator of fecal contamination of water and wastewater. Over 700 serotypes of pathogenic *E. coli* have been recognized based on the O (body antigen), H (flagellar antigen), and K (capsular antigen) antigens. Serotyping is important in distinguishing the small number of strains that cause disease. *E. coli* can cause infection in the urinary tract and brain stem (meningitis) as well as intestinal diseases referred to as *gastroenteritis*. There are five classes of *E. coli* that produce disease. They are classified by the method of pathogenesis: (i) toxins (enterotoxigenic), (ii) invasive (enteroinvasive), (iii) hemorrhagic

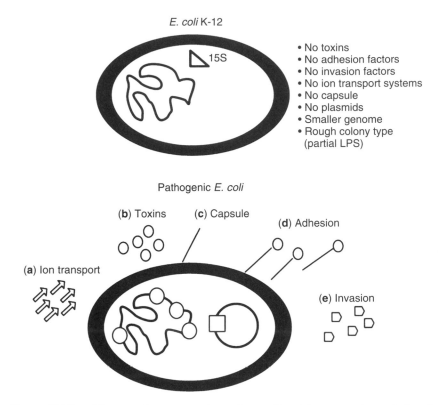

E. coli K-12

15S

- No toxins
- No adhesion factors
- No invasion factors
- No ion transport systems
- No capsule
- No plasmids
- Smaller genome
- Rough colony type (partial LPS)

Pathogenic *E. coli*

(**b**) Toxins

(**c**) Capsule

(**d**) Adhesion

(**a**) Ion transport

(**e**) Invasion

Figure 7.63. Differences between pathogenic *E. coli* and the nonpathogenic *E. coli.*

(enterohemorrhagic), (iv) pathogenic (enteropathogenic), and (v) aggregative (clumping or enteroaggregative).

CAMPYLOBACTER. It is usually transmitted through contaminated food or water and can infect the gastrointestinal tract and cause diarrhea, fever, and cramps. Hand washing and good food safety habits will help to prevent *Campylobacter* infections (or campylobacteriosis), which usually clear up on their own but sometimes are treated with antibiotics. *Campylobacter* infects over 2 million people each year and is a leading cause of diarrhea and foodborne illness (Fig. 7.64). Babies younger than 1 year, teens, and young adults are most commonly affected.

Causes. Campylobacter is found in the intestines of many wild and domestic animals (Fig. 7.65). The bacteria are passed in their feces, which can lead to infection in humans via contaminated food, contaminated meat (especially chicken), water taken from contaminated sources (streams or rivers near where animals graze), and milk products that have not been pasteurized.

Bacteria can be transmitted from person to person when someone comes into contact with fecal matter from an infected person, especially a child in diapers. Household pets can carry and transmit the bacteria to their owners. Once inside the human digestive system, *Campylobacter* infects and attacks the lining of both the small and large intestines. The bacteria also can affect other parts of the body. In some cases—particularly in very

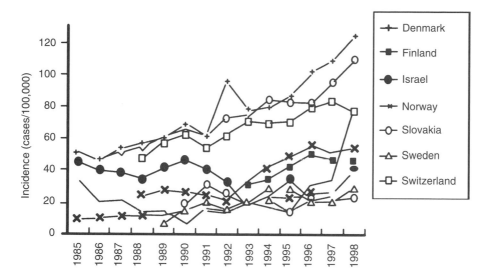

Figure 7.64. Incidence of campylobacteriosis in Europe in 1985–1998.

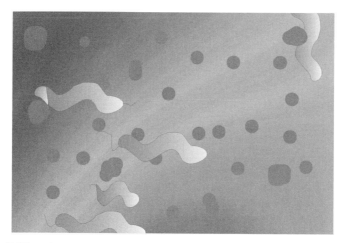

Figure 7.65. Diagrammatic view of electron microscopic view of *Campylobacter*.

young patients and those with chronic illnesses or a weak immune system—the bacteria can get into the bloodstream (called *bacteremia*). In rare cases, campylobacteriosis can lead to Guillain–Barré syndrome, a rare autoimmune disorder. Symptoms usually appear in 1–7 days after ingestion of the bacteria. The main symptoms of campylobacteriosis are fever, abdominal cramps, and mild to severe diarrhea. Diarrhea can lead to dehydration, which should be closely monitored. Signs of dehydration include thirst, irritability, restlessness, lethargy, sunken eyes, dry mouth and tongue, dry skin, fewer trips to the bathroom to urinate, and (in infants) a dry diaper for several hours.

In cases of campylobacteriosis, the diarrhea is initially watery, but later may contain blood and mucus. Sometimes, abdominal pain appears to be a more significant symptom

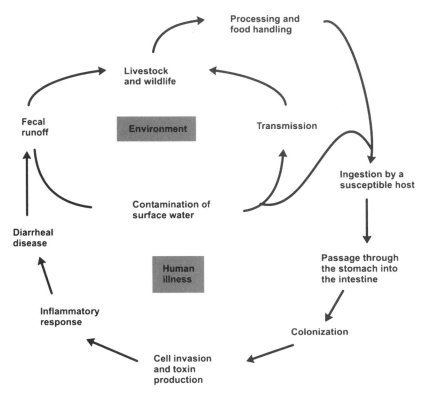

Figure 7.66. *Campylobacter jejuni* infections are commonly acquired by handling and consuming chicken and drinking unpasteurized milk and polluted water.

than the diarrhea. When this happens, the infection may be mistaken for appendicitis or a problem with the pancreas. We can prevent campylobacteriosis by using drinking water that has been tested and approved for purity, especially in developing countries, and by drinking milk that has been pasteurized. While hiking and camping, avoid drinking water from streams and from sources that pass through land where animals graze. Kill any bacteria in meat by cooking thoroughly and eating while still warm. While preparing food, wash the hands well before and after touching raw meat, especially poultry. Clean cutting boards, countertops, and utensils with soap and hot water after contact with raw meat (Fig. 7.66).

LISTERIOSIS. Listeriosis is an illness caused by bacteria found in food such as raw fish, unpasteurized milk, soft cheese, undercooked poultry, raw vegetables, and precooked chilled foods. It is one of the bacterium that causes food poisoning and is found in soil and water. Animals such as poultry may be carriers of the bacteria without showing any symptoms of poisoning. One might not even know that he or she has been infected until days or even weeks after exposure to the bacteria, which makes it hard to work out what it was that he or she ate that caused the problem. Symptoms of listeriosis could include a raised temperature, feeling hot and cold by turns, muscle aches, back pain, nausea, vomiting, and diarrhea.

Listeriosis is diagnosed in pregnant women as well. Experts think that changes in metabolic activity and a suppressed immune system may make pregnant women more

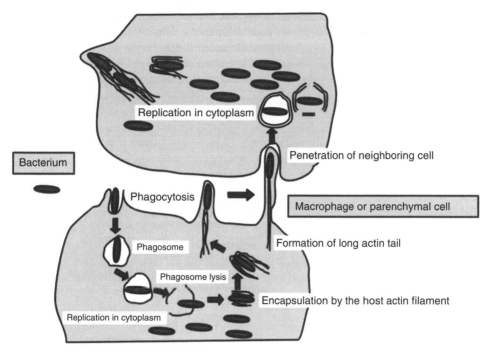

Figure 7.67. Steps in the invasion of cells and intracellular spread by *L. monocytogenes.*

susceptible to all types of food poisoning. Research indicates that listeriosis can affect the placenta, amniotic fluid, and the baby, and if left untreated, it can lead to miscarriage or stillbirth. Babies born to infected mothers can develop breathing problems, hypothermia, and meningitis (a serious brain infection). Babies who seem healthy at birth may develop symptoms later on, such as feeding problems and a high temperature. The bacterium apparently invades via the intestinal mucosa. It is thought to attach to intestinal cells by means of D-galactose residues on the bacterial surface, which adhere to D-galactose receptors on susceptible intestinal cells. The bacterium is taken up (including by nonphagocytic cells) by induced phagocytosis, which is thought to be mediated by a membrane-associated protein called *internalin*. Once ingested, the bacterium produces listeriolysin (LLO) to escape from the phagosome. The bacterium then multiplies rapidly in the cytoplasm and moves through the cytoplasm to invade adjacent cells by polymerizing actin to form long tails (Fig. 7.67).

OTHER FOODBORNE INFECTIOUS DISEASES. Calicivirus, or Norwalk-like virus, is an extremely common cause of foodborne illness, though it is rarely diagnosed because a laboratory test is not widely available (Fig. 7.68). It causes an acute gastrointestinal illness, usually with more vomiting than diarrhea, and resolves within 2 days.

Unlike many foodborne pathogens that have animal reservoirs, it is believed that Norwalk-like viruses spread primarily from one infected person to another. Infected kitchen workers can contaminate a salad or sandwich as they prepare it, if they have the virus on their hands. Infected fishermen have contaminated oysters as they harvest them.

Some of these organisms pose severe food threat and many microorganisms are responsible for foodborne Illness (Table 7.16). These include infections caused by *Shigella*,

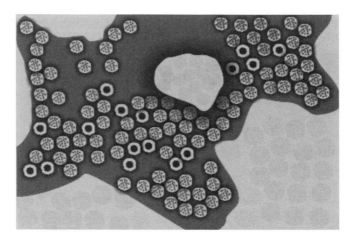

Figure 7.68. Microscopic view of calicivirus.

hepatitis A, and the parasites *Giardia lamblia* and cryptosporidia. Even strep throats have been transmitted occasionally through food. Other toxins and poisonous chemicals can cause foodborne illness. People can become ill if a pesticide is inadvertently added to a food or if naturally poisonous substances are used to prepare a meal. Every year, people become ill after mistaking poisonous mushrooms for safe species or after eating poisonous reef fishes. The spectrum of foodborne diseases is constantly changing. A century ago, typhoid fever, tuberculosis, and cholera were common foodborne diseases. Improvements in food safety, such as pasteurization of milk, safe canning, and disinfection of water supplies, have conquered those diseases. Today, other foodborne infections have taken their place, including some that have only recently been discovered. For example, in 1996, the parasite *Cyclospora* suddenly appeared as a cause of diarrheal illness related to Guatemalan raspberries. These berries had just started to be grown commercially in Guatemala and somehow became contaminated in the field with this unusual parasite. In 1998, a new strain of the bacterium *Vibrio parahemolyticus* contaminated oyster beds in Galveston Bay and caused an epidemic of diarrheal illness in people who ate the oysters raw. The affected oyster beds were near the shipping lanes, which suggested that the bacterium arrived in the ballast water of freighters and tankers coming into the harbor from distant ports (Table 7.16).

7.6.3.3. Commercial Food Safety Diagnostics

The implementation of food safety should be seen as an ongoing process, which is influenced by environmental, socioeconomical, political, and cultural factors. Food safety issues need to be addressed on a continuous basis, from a regional, national, and global point of view. New, flexible tools are required for evaluating and managing new food safety challenges (Fig. 7.69). To guarantee the safety of foodstuff, producers have therefore shifted their focus toward the use of food safety management tools, most importantly, HACCP, and the consequent application of hygienic measures, based on good manufacturing practice/good hygienic practice (GMP/GHP). Food safety management tools use the input of scientific information to identify critical contamination points in the food chain and the production process and design measures to control them. However, the

TABLE 7.16. Characteristics of Microorganisms Responsible for Foodborne Illness

Microorganism	Foodborne Illness	Symptoms	Common Food Sources	Incubation Period
Bacillus cereus	Intoxication	Watery diarrhea and cramps, or nausea and vomiting	Cooked product that is left uncovered, milk, meat, vegetables, fish, rice, and starchy foods	0.5–15 h
Campylobacter jejuni	Infection	Diarrhea, perhaps accompanied by fever, abdominal pain, nausea, headache, and muscle pain	Raw chicken, other foods contaminated by raw chicken, unpasteurized milk, untreated water	2–5 d
Clostridium botulinum	Intoxication	Lethargy, weakness, dizziness, double vision, difficulty speaking, swallowing, and/or breathing; paralysis; possible death	Inadequately processed, home-canned foods; sausages; seafood products; chopped bottled garlic; kapchunka; molona; honey	18–36 h
Clostridium perfringens	Infection	Intense abdominal cramps, diarrhea	Meat, meat products, gravy, Tex-Mex type foods, other protein-rich foods	8–24 h
Escherichia coli group	Infection	Watery diarrhea, abdominal cramps, low grade fever, nausea, malaise	Contaminated water, undercooked ground beef, unpasteurized apple juice and cider, raw milk, alfalfa sprouts, cut melons	12–72 h
Listeria mono-cytogenes	Infection	Nausea, vomiting, diarrhea; may progress to headache, confusion, loss of balance, and convulsions; may cause spontaneous abortion	Ready-to-eat foods contaminated with bacteria, including raw milk, cheese, ice cream, raw vegetables, fermented raw sausages, raw and cooked poultry, raw meat, and raw and smoked fish	Unknown, may range from a few days to 3 wk
Salmonella species	Infection	Abdominal cramps, diarrhea, fever, headache	Foods of animal origin; other foods contaminated through contact with feces, raw animal products, or infected food handlers. Poultry, eggs, raw milk, meat are frequently contaminated	12–72 h

(Continued)

TABLE 7.16. (*Continued*)

Microorganism	Foodborne Illness	Symptoms	Common Food Sources	Incubation Period
Shigella	Infection	Fever, abdominal pain and cramps, diarrhea	Feces-contaminated foods	12–48 h
Staphylococcus aureus	Intoxication	Nausea, vomiting, abdominal cramping	Foods contaminated by improper handling and holding temperatures—meats and meat products, poultry and egg products, protein-based salads, sandwich fillings, cream-based bakery products	1–12 h
Hepatitis A	Infection	Jaundice, fatigue, abdominal pain, anorexia, intermittent nausea, diarrhea	Raw or undercooked molluscan shellfish or foods prepared by infected handlers	15–50 d
Norwalk-type viruses	Infection	Nausea, vomiting, diarrhea, abdominal cramps	Shellfish grown in feces-contaminated water; water and foods that have come into contact with contaminated water	12–48 h
Giardia lamblia	Infection	Diarrhea, abdominal cramps, nausea	Water and foods that have come into contact with contaminated water	1–2 wk
Trichinella spiralis	Infection	Nausea, diarrhea, vomiting, fatigue, fever, abdominal cramps	Raw and undercooked pork and wild game products	1–2 d

lack of reliable data is often limiting the usefulness of this approach, and therefore, data collection is one of the priorities for future food safety strategies. In the absence of relevant data, other strategies might still have to be used to control food hazards. The FDA has been involved in the regulation of *in vitro* diagnostic devices (IVDs or laboratory tests) since the introduction of the Medical Device Amendments of 1976. IVDs developed as kits or systems intended for use in multiple laboratories require review by the FDA before being marketed to ensure appropriate performance and labeling.

So, what is there in existence to manage quality and safety, and how do they relate to each other? A list of the most well-known methods to manage quality and/or safety is provided, and these methods are briefly discussed individually and then how they integrate with each other (Fig. 7.70).

1. GHP/GMP or Sanitation Standard Operating Procedures (SSOPs) or prerequisite programs
2. HACCP

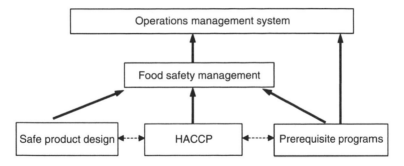

Figure 7.69. Procedure for food safety management tools.

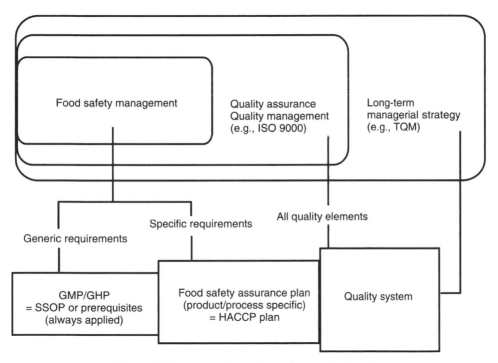

Figure 7.70. Food safety rules and their relationship.

3. QC
4. Quality assurance (QA)/quality management (QM)—ISO standards
5. Quality systems
6. Total quality management (TQM).

The food safety rules and their relationship are shown in Figure 7.70.

GOOD HYGIENIC PRACTICES/GOOD MANUFACTURING PRACTICES. The terms GHP and GMP refer to the measures and requirements that any establishment should meet to produce safe food. These requirements are prerequisites to other and more specific approaches, such as HACCP, and are often now called *prerequisite programs*. In recent years, the term *Standard Sanitary Operating Procedures (SSOPs)* has also been used in the United States to encompass basically the same issues, that is, best practices.

HAZARD ANALYSIS AND CRITICAL CONTROL POINT. HACCP is a systematic approach that identifies, evaluates, and controls hazards that are significant for food safety. HACCP ensures food safety through an approach that is built on foundations provided by GMP. It identifies the points in the food production process that require constant control and monitoring to make sure the process stays within identified limits. Statistical process control systems are relevant to this operation.

HACCP is legislated in many countries, including the United States and the European Union. The combination of GHP/GMP and HACCP is particularly beneficial in that the efficient application of GHP/GMP allows HACCP to focus on the true critical determinants of safety.

QUALITY CONTROL. It is an important subset of any QA system and is an active process that monitors and, if necessary, modifies the production system so as to consistently achieve the required quality. It can be argued that QC is used as part of the HACCP system, in terms of monitoring the critical control points in the HACCP plan. However, traditional QC is much broader than purely this focus on critical control points for safety systems.

QUALITY ASSURANCE/QUALITY MANAGEMENT. This can be defined as all the activities and functions concerned with the attainment of quality in a company. In a total system, this would include the technical, managerial, and environmental aspects. The best known of the QA standards is ISO 9000 and for environmental management, ISO 14000.

The term *quality management* is often used interchangeably with QA. In the seafood industry, the term *quality management* has been used to focus mostly on the management of the technical aspects of quality in a company, for instance, the Canadian Quality Management Programme, which is based on HACCP but covers other technical issues such as labeling.

ISO Standards. The International Organization for Standardization (ISO) in Geneva is a worldwide federation of national standard bodies from more than 140 countries. ISO's work results in international agreements that are published as International Standards. The vast majority of ISO standards are highly specific to a particular product, material, or process. However, two standards, ISO 9000 and ISO 14,000, mentioned earlier, are known as *generic management system standards*.

Over half a million ISO 9000 certificates have been awarded in 161 countries and economies around the world, and in 2001 alone, over 100,000 certificates were awarded, 43% of which were the new ISO 9001:2000 certificate.

Historically, the ISO 9000 series of standards of relevance to the seafood industry included the following:

ISO 9001 Quality Systems. Model for QA in design/development, production, installation, and servicing.

ISO 9002 Quality Systems. Model for QA in production and installation.

More recently, the new ISO 9001:2000 certificate is the only ISO 9000 standard against whose requirements a quality system can be certified by an external agency and replaces the old ISO 9001, 9002, and 9003 with one standard.

It is important to note that the ISO 9000 standards relate to QM with customer satisfaction as the endpoint and that they do not specifically refer to technical processes only. ISO 9000 gives an assurance to a customer that the company has developed procedures (and adheres to them) for all aspects of the company's business.

ISO 14,000 is primarily concerned with environmental management. Introduced much later than the ISO 9000 series, there are now over 35,000 ISO 14 000 certificates awarded in 112 countries or economies of the world. During 2001, nearly 14,000 certificates were awarded, around 40% of the total awarded since the introduction of the standard.

In most countries, implementation of ISO 9000 QM systems or ISO 14 000 environmental systems are voluntary.

QUALITY SYSTEMS. This term covers organizational structure, responsibilities, procedures, processes, and the resources needed to implement comprehensive QM (Jouve et al., 1998). They are intended to cover all quality elements. Within the framework of a quality system, the prerequisite program and HACCP provides the approach to food safety.

TOTAL QUALITY MANAGEMENT (TQM). TQM is an organization's management approach, centered on quality and based on the participation of all its members and aimed at long-term success through customer satisfaction and benefits to the members of the organization and to society (Jouve et al., 1998). Thus, TQM represents the organizations' "cultural" approach and together with the quality systems provides the philosophy, culture, and discipline necessary to commit everybody in the organization to achieve all the managerial objectives related to quality.

7.6.3.4. Business Outlook for Food Diagnostics

Consumer demands safe food today. The commercial food industry's ability to identify bacterial pathogens and unsafe residues has resulted in an almost five-fold increase in food recalls by major manufacturers since 1988. New technology allows government regulatory agencies to identify a bacterial pathogen and trace it back to its source more rapidly (Fig. 7.71). The key to this new technology is the availability of rapid food safety diagnostics.

These rapid food safety diagnostics provide a quick "positive or negative" answer before the food product enters the distribution system. Tests take up to 30 h to complete because of the requirement for a bacterial growth enrichment period. This growth enrichment period is necessary to increase the total number of bacteria so they can be detected using current technology. A negative answer means that the product does not contain that particular bacteria or toxin and no further testing is required. A positive answer means that further testing is needed at a reference laboratory using standard laboratory methods to confirm the actual presence of the bacteria or toxin. For example, rapid screening for *Salmonella typhimurium* requires approximately 24 h to complete.

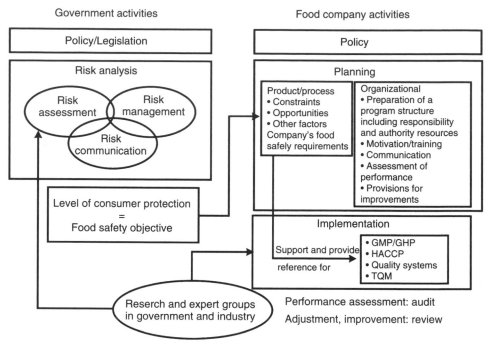

Figure 7.71. New policies and activities of the government and food company for food safety.

7.7. FUTURE CHALLENGES: NEXT GENERATION DIAGNOSTICS INDUSTRY

The most promising breakthroughs of the development of on-line or on-site, sensitive, low cost, rapid methods for routine use are expected to be made in the area of sensor technology. Many prototypes for food diagnostic application in the food and drink industry are currently being developed. They have high potential for automation and allow the construction of simple and portable equipment for fast analysis. These properties will open up many applications within quality and process control, control of fermentation processes, quality and safety control of raw materials, and for HACCP monitoring.

A new method must have high sensitivity, high specificity, high precision (repeatability), at the same time rapid, robust, and cheap. There is currently no method that will fulfill all requirements. Antibody-based methods as represented by the ELISA assay and DNA-based methods such as PCR are the most widely used technologies in food diagnostics today. Also immunomagnetic separation (IMS) is exploited in a number of commercially available kits and will become an even more important technology in the future. While some microarray-based systems are commercially available, most of the biosensors and microarray-based applications are still on the level of prototypes. For other promising methods with future potential—but with less momentum for commercial development, specifically in food diagnostics, such as FC, bacteriophage technology, or adenylate kinase—only a basic description of the method is given together with brief background information on the current stage of its development. Sensor technology covers a wide area of diverse techniques, including optochemical sensors and biosensors. Biosensors are a subgroup of chemical sensors where the analytical devices are composed of a biological recognition element such as enzymes, antibodies, receptors, proteins, oligonucleotides, or even a whole cell coupled to a chemical or physical transducer. A transducer measures the changes that occur when the sensor couples to its analyte. The sensitivity of the system is determined by the type of transducers used. Biosensors can be used for the detection of very different analytes such as pathogens, pesticides, and toxins. Biosensors can be grouped according to their 15 biological recognition elements into immunosensors using antibodies and hybrid sensors using DNA or RNA probes. There have been many sensors developed for the detection of foodborne pathogens with the goal to overcome problems associated with traditional microbiological detection techniques such as being time and labor intensive (Baeumner, 2003). In fact, biosensor advancements have greatly improved our ability to detect minute quantities of analytes, as research into biosensors has mainly focused on detection platforms with very low detection limits (Rider et al., 2003). It has been estimated that 38% of reported pathogen biosensors in the past 20 years were designed for the food industry. However, only a limited amount of methods are combined and currently exploited for their use in food diagnostics. As recognition elements, bioaffinity-based receptors that use the selective interaction between ligand and receptor, antibody, or nucleic acid are most widely used. As transducers, electrochemical and optical systems have gained practical importance. As nanobiotechnology progresses, sensors to detect pathogens or their constituents become smaller and more sensitive. Owing to the nature of these nanoscale sensors, the sample size from which the detection is being made is typically a microliter or smaller. Therefore, the challenge for scientists developing detection methods for pathogens in food is in the sample preparation. Although the sample preparation requirements will vary from one food product to another, research into this step is required to bridge the emerging field of nanosensors with the food industry. Thus, while the organism with the largest number of diagnosed cases may fluctuate from year to year, the food industry will always be looking for detection systems that will help identify all pathogens of concern in its food products (Rider et al., 2003).

Optical biosensors have been developed for rapid detection of contaminants in foods, including pathogens, and several have evolved into commercial prototype systems. The analyte in the food interacts with the bioactive molecule, usually an antibody. Antibodies can be immobilized directly on the fiber, either on the blunt end or along the sides of a fiber tip. The binding of antibody and analyte is detected as a change in an optical signal measured through the fiber-optic assembly. The light from a laser travels to the fiber tip and penetrates into the area outside the tip. A fluorescent-labeled complex binds to the antibodies on the tip. The fluorescent signal then radiates in all directions, and some of it travels back up the fiber tip to the detector. Detection of molecules in solution can be made either by direct binding to the biosensor coating the molecules or by competition binding with soluble capture molecules added together with the sample. Future applications might include protein quality and the detection of allergens, genetically modified proteins, BSE prions, pathogens, and biocide residues.

REFERENCES

Alam J, Cook JL. 1990. Reporter genes: application to the study of mammalian gene transcription. *Anal Biochem* **188**, 245–254.

Ansuini H, Meola A, Gunes Z, Paradisi V, Pezzanera M, et al. 2009. Anti-EphA2 antibodies with distinct in vitro properties have equal it *in vivo* efficacy in pancreatic cancer. *J Oncol* **2009**, 1–10.

Baeumner AJ. 2003. Biosensors for environmental pollutants and food contaminants. *Anal Bioanal Chem* **377**, 434–445.

Bailey JE. 1991. Toward a science of metabolic engineering. *Science* **252**, 1668–1675.

Beesch SC. 1952. Acetone-butanol fermentation of sugars. *Ind Eng Chem* **44**, 1677–1682.

Berzsenyi MD, Roberts SK, Beard MR. 2006. Genomics of hepatitis B and C infections: diagnostic and therapeutic applications of microarray profiling. *Antivir Ther* **11**, 541–552.

Bhopale GM, Nanda RK. 2005. Recombinant DNA expression products for human therapeutic use. *Curr Sci* **89**, 614–622.

Bisen PS, Baghel RK, Sanodiya BS, Thakur GS, Prasad GBKS. 2010. *Lentinus edodes*: a macro-fungus with pharmacological activities. *Curr Med Chem* **17**, 2419–2430.

Brody EN, Gold L. 2000. Aptamers as therapeutic and diagnostic agents. *Rev Mol Biotechnol* **74**, 5–13.

Bungay HR. 1993. *Basic Biochemical Engineering*, Biline Associates: Try, New York.

Cameron DC, Tong IT. 1993. Cellular and metabolic engineering. An overview. *Appl Biochem Biotechnol* **38**, 105–140.

Cantor CR. 1998. How will the human genome project improve our quality of life? *Nat Biotechnol* **16**, 212–213.

Cock SL, Zetty A, Ana M, Alaya AA. 2009. Use of enzymatic biosensors as quality indices: a synopsis of present and future trends in the food industry. *Chilean J Agric Res* **69**, 270–280.

Cole ST, Brosch R, Parkhill J. 1998. Deciphering the biology of *Mycobacterium tuberculosis* from the complete genome sequence. *Nature* **393**, 537–544.

Danesh J, Newton R, Beral V. 1997. A human germ project? *Nature* **389**, 21–24.

Debnath M, Prasad GBKS, Bisen PS. 2010. *Molecular Diagnostics: Promises and Possibilities*, Springer Science + Business Media, New York, pp. 467–481.

Evangelyn C, Alocilja S, Radke M. 2003. Market analysis of biosensor for food safety. *Biosens Bioelectron* **18**, 841–846.

Ezaki T. 2003. Detection of respiratory infectious diseases by using DNA microarray comprising specific target gene amplicons species specific primers. US20030091991A1.

Fischer R, Stoger E, Schillberg S, Christou P, Twyman RM. 2004. Plant based production of biopharmaceuticals. *Curr Opin Plant Biol* **7**, 152–158.

Garaizar J, Rementeria A, Porwollik S. 2006. DNA microarray technology: a new tool for the epidemiological typing of bacterial pathogens? *FEMS Immunol Med Microbiol* **47**, 178–189.

Hahn-Hägerdal B, Karhumaa K, Larsson CU, Gorwa-Grauslund MF, Görgens J, van Zyl WH. 2005. Role of cultivation media in the development of yeast strains for large scale industrial use. *Microb Cell Fact* **4**, 31.

Hellwing S, Drossard J, Twyman RM, Fischer R. 2004. Plant cell cultures for the production of recombinant proteins. *Nat Biotechnol* **22**, 1415–1422.

Hong SH, Lee SY. 2004. Enhanced production of succinic acid by metabolically engineered Escherichia coli with amplified activities of malic enzyme and fumarase. *Biotechnol Bioprocess Eng* **9**, 252–255.

Jayani RS, Saxena S, Gupta R. 2005. Microbial pectinolytic enzymes: a review. *Process Biochem* **40**, 2931–2944.

Jeenes DJ, MacKenzie DA, Roberts DA, Archer DB. 1991. Heterologous protein production by filamentous fungi. *Biotechnol Gen Eng Rev* **9**, 327–336.

Jones DT, Keis S. 1995. Origins and relationships of industrial solvent producing clostridial strains. *FEMS Microbiol Rev* **17**, 223–232.

Jones DT, Woods DR. 1986. Acetone-butanol fermentation revisited. *Microbiol Rev* **50**, 484–524.

Jouve JL, Stringer MF, Baird-Parker AC. 1998. *Food Safety Management Tools. ILSI Europe Risk Analysis in Microbiology*, Brussels, Belgium.

Kain SR, Ganguly S. 2001. Overview of genetic reporter systems. *Curr Protoc Mol Biol* Chapter 9, Unit9.6.

Keefe AD, Pai S, Ellington A. 2010. Aptamers as therapeutics. *Nat Rev Drug Discov* **9**, 537–550.

Khan Z, Khan N, Tiwari RP, Patro IK, Prasad GBKS, Bisen PS. 2010a. Down-regulation of survivin by oxaliplatin diminishes radioresistance of head and neck squamous carcinoma cells. *Radiother Oncol* **96**, 267–273.

Khan Z, Khan N, Varma AK, Tiwari RP, Mouhamad S, Prasad GBKS, Bisen PS. 2010b. Oxaliplatin-mediated inhibition of survivin increase sensitivity of head and neck squamous cell carcinoma cell lines to paclitaxel. *Curr Cancer Drug Target* **10**, 660–669.

Khan Z, Khan N, Tiwari RP, Sah NK, Prasad GBKS, Bisen PS. 2011. Biology of Cox-2: an application in cancer. *Curr Drug Targets* **12**, 1082–1093.

Khan Z, Tiwari RP, Mulherkar R, Sah NK, Prasad GBKS, Shrivastava BR, Bisen PS. 2009. Detection of survivin and p53 in human oral cancer: correlation with clinicopathological findings. *Head Neck* **31**, 1039–1048.

Kim KM. 2004. Inhibition of metallo-beta-lactamase by RNA. Master's thesis. Tech University, Texas.

Kinch MS. 2005. Targeted drug delivery using EphA2 or EphA4 binding moieties. Patent No. 20050153923. Laytonsville, MD

Lee SY, et al. 2003. Detection of 44 bacterial and 2 fungal pathogens using species specific probes. WO03095677A1.

Liu Y, Chakrabartty S, Alocilja EC. 2007. Fundamental building blocks for molecular bio-wire based forward-error correcting biosensors. *Nanotechnol J* **18**, 424017. 6pp.

Ma JKC, Drake PMW, Christou P. 2003. The production of recombinant pharmaceutical proteins in plants. *Nat Rev Genet* **4**, 794–805.

Magnuson NS, Linzmaier PM, Reeves R, An G, HayGlass K, Lee JM. 1998. Secretion of biologically active human interleukin-2 and interleukin-4 from genetically modified tobacco cells in suspension culture. *Protein Expr Purif* **13**, 45–52.

Manning GB, Campbell LL. 1961. Thermostable a-amylase of Bacillus stearothermophilus. I. Crystallization and some general properties. *J Biol Chem* **236**, 2952–2957.

McCutchan WN, Hickey RJ. 1954. The butanol-acetone fermentations. In: Underkofler A, Hickey RJ, eds., *Industrial Fermentations*, pp. 347–388, Chemical Publishing, New York.

Mello LD, Kubota LT. 2002. Review of the use of biosensors as analytical tools in the food and drink industries. *Food Chem* **77**, 237–256.

Naylor LH. 1999. Reporter gene technology: the future looks bright. *Biochem Pharmacol* **58**, 749–757.

Nimjee SM, Rusconi CP, Sullenger BA. 2005. Aptamers: an emerging class of therapeutics. *Annu Rev Med* **56**, 555–583.

Patel PD. 2006. Overview of affinity biosensors in food analysis. *J AOAC Int* **89**, 805–818.

Pavlou AK, Reichert JM. 2004. Recombinant protein therapeutics—Success rates, market trends and values to 2010. *Nat Biotechnol* **22**, 1513–1519.

Peltonen L, McKusick VA. 2001. Dissecting human disease in the postgenomic era. *Science* **291**, 1224–1229.

Phippard D, Manning AM. 2003. Screening for inhibitors of transcription factors using luciferase reporter gene expression in transfected cells. *Methods Mol Biol* **225**, 19–23.

Rasooly A, Herold KE. 2006. Biosensors for the analysis of food- and waterborne pathogens and their toxins. *J AOAC Int* **89**, 873–883.

Reichert JM. 2004. Biopharmaceutical approvals in the US. *Reg Aff J Pharm* **15**, 491–497.

Rider TH, Petrovick MS, Nargi FE, Harper JD, Schwoebel ED, et al. 2003. A B-cell based sensor for rapid identification of pathogens. *Science* **301**, 213–215.

Ross D. 1961. The acetone-butanol fermentation. *Prog Ind Microbiol* **3**, 73–89.

Sah NK, Khan Z, Khan GJ, Bisen PS. 2006. Structural, functional and therapeutic biology of survivin. *Cancer Lett* **244**, 164–171.

Sanodiya BS, Thakur GS, Baghel RK, Prasad GBKS, Bisen PS. 2009. *Ganoderma lucidum*: a potent pharmacological macrofungus. *Curr Pharm Biotechnol* **10**, 717–742.

Saue M, Porro D, Mattanovich D, Branduardi P. 2008. Microbial production of organic acids: expanding the markets. *Trends Biotechnol* **26**, 100–110.

Tang YW. 2003. Molecular diagnostics of atypical pneumonia. *Acta Pharmacol Sin* **24**, 1308–1313.

Walker-Daniels J, Riese DJ, Kinch MS. 2002. c-Cbl dependent EphA2 protein degradation is induced by ligand binding. *Mol Cancer Res* **1**, 79–87.

Walton MT, Martin JL. 1979. Production of butanol-acetone by fermentation. In: Peppler HJ, Perlman D, eds., *Microbial Technology*, 2nd ed., pp. 187–209. Academic Press Inc, New York.

Wang HL, Hesseltine CW. 1979. Mold-modified foods. In: Peppler HJ, Perlman D, eds., *Microbial Technology*, 2nd ed., pp. 95–129. Academic Press Inc, New York.

White RR, Sullenger BA, Rusconi CP. 2000. Developing aptamers into therapeutics. *J Clin Invest* **106**, 929–934.

Yong-Zhong DU. 2004. Biotinylated and enzyme immobilized hetero-bifunctional latex beads. *AIST Today* **4** (4), 17.

Yoo SM, Jong HC, Lee SY, Yoo NC. 2009. Applications of DNA microarray in disease diagnostics. *J Microbiol Biotechnol* **19**, 51–54.

INDEX

Note: Page numbers in *italics* refer to figures; those in **bold** to tables

Microbes: Concepts and Applications, First Edition. Prakash S. Bisen, Mousumi Debnath, Godavarthi B. K. S. Prasad
© 2012 Wiley-Blackwell. Published 2012 by John Wiley & Sons, Inc.